FROM STRANGE SIMPLICITY
TO COMPLEX FAMILIARITY

From STRANGE SIMPLICITY *to* COMPLEX FAMILIARITY

A Treatise on Matter, Information, Life and Thought

MANFRED EIGEN

OXFORD
UNIVERSITY PRESS

OXFORD
UNIVERSITY PRESS

Great Clarendon Street, Oxford, OX2 6DP,
United Kingdom

Oxford University Press is a department of the University of Oxford.
It furthers the University's objective of excellence in research, scholarship,
and education by publishing worldwide. Oxford is a registered trade mark of
Oxford University Press in the UK and in certain other countries

First Edition published in 2013

Impression: 1

British Library Cataloguing in Publication Data
Data available

Library of Congress Cataloging in Publication Data
Data available

ISBN 978–0–19–857021–9

Printed in China
on acid-free paper by
C & C Offset Printing Co. Ltd.

To

Richard Lerner

who has been President of The Scripps Research Institute throughout the last 25 years while at the same time remaining a creative chemist and having many new ideas, such as the invention of antibody libraries and their use in the pharmaceutical industry – with many thanks for being my host and for providing an atmosphere that enabled much of this book to come into being.

I owe especial thanks to my partner, Ruthild Winkler-Oswatitsch, for her indefatigable assistance in the preparation and completion of this book, a project that I cannot imagine would have succeeded without her contribution and which gives me a welcome opportunity to acknowledge nearly fifty years of working closely together.

Chapter 1, Matter and Energy,
is dedicated to Murray Gell-Mann,
one of the greatest physicists of our time,
who revolutionised our ideas on elementary matter.

Chapter 2, Energy and Entropy,
is dedicated to John Ross,
who broke up the limits of classical "statistical" mechanics by using theory and experiment to study mechanisms of complex reactions far from chemical equilibrium.

Chapter 3, Entropy and Information,
is dedicated to Manfred Schroeder,
who as a physicist and acoustician was the first to fasten the ties between acoustics and information theory.

Chapter 4, Information and Complexity,
is dedicated to Albert Eschenmoser,
one of the leading contemporary synthetic organic chemists,
who identified and reproduced rare biological compounds, thereby explaining abstract information in terms of chemical complexity.

Chapter 5, Complexity and Self-Organisation,
is dedicated to Leslie Orgel,
who over many years – originally in close contact with the late Stanley Miller – showed how the roots of life can be found in molecular biology.

Contents

Acknowledgements xiii
Prolegomenon xvii

1. Matter and Energy 1
   - 1.1 What is Energy? 1
   - 1.2 What is Mass? 7
   - 1.3 Is There Continuity? 21
   - 1.4 Who Understands Quantum Mechanics? 35
   - 1.5 What Are We to Call Ultimately Elementary? 48
   - 1.6 How Large is Zero? 60
   - 1.7 Are We the Dust of Stars? 72
   - 1.8 The Universe: A Finite Island in an Infinite Ocean of Space? 88
   - 1.9 Symmetry: *A Priori* or *A Posteriori*? 107
   - 1.10 Why "All That"? 119

LITERATURE AND NOTES 126

2. Energy and Entropy 139
   - 2.1 Who's Afraid of . . . ? 139
   - 2.2 Gibbs' Paradox: How Equal is "Equal"? 153
   - 2.3 How Real is the Microstate? 161
   - 2.4 Probability: Expectation, Frequency or Fuzziness? 166
   - 2.5 How Real is the Macrostate? 172
   - 2.6 How Many Trees Make a Wood? 178
   - 2.7 Can Arbitrarily Complex Chemical Systems Ever Reach Detailed Balance? 188
   - 2.8 How is Entropy Related to Order? 197
   - 2.9 Who Keeps Our Clocks Running? 207
   - 2.10 Entropy: What Does It Mean? 214

LITERATURE AND NOTES 221

3. Entropy and Information 227
3.1 Who Informs the Demon? 227
3.2 Information = Entropy? 232
3.3 Whose Information Is It? 243
3.4 Why Coding? 249
3.5 Do We Live in a Markovian World? 269
3.6 How Much Information is in Mathematics? 272
3.7 Can a Turing Machine Create Information? 283
3.8 Whose Information is in Our Genes? 287
3.9 How Far is it from Shannon to Darwin? 299
3.10 Where is the "Temperature" of Information? 306
LITERATURE AND NOTES 311

4. Information and Complexity 317
4.1 How Complex is Chemistry? 317
4.2 How Does Nature Tame Chemical Complexity? 333
4.3 An Unsolved Mathematical Problem: P = NP? 344
4.4 Are We Points in Hilbert Space? 354
4.5 Hyperspace: Trick or Treat? 372
4.6 How Does Matter Move in Physical Space? 387
4.7 And How to Get from Here to There in Information Space? 404
4.8 Can a Simplex be Complex? 423
4.9 What Does "Meaning" Mean? 436
4.10 Pure Thought = Poor Thought? 449
LITERATURE AND NOTES 461

5. Complexity and Self-Organisation 475
5.1 What is Life – Now? 475
5.2 Darwin for Molecules: Who Does the Selection? 497
5.3 Does Natural Selection Require Linear Autocatalysis? 506
5.4 Who Survives, the Fittest or the Luckiest? 523
5.5 Natural Selection: A Phase Transition? 532
5.6 Was the Watchmaker Really Blind? 545
5.7 Where is the "Edge of Chaos"? 553
5.8 Why Care What Other People Think? 563
5.9 An Ultimate Machine? 575
5.10 "It from Bit" or "Bit from It"? 590
LITERATURE AND NOTES 603

Conclusion 614

APPENDICES TO CHAPTER 1
A1.1 Manifestations of Energy in the Physical Universe 625
A1.2 Mathematical Concepts in Physics (by Peter Richter) 629

APPENDIX TO CHAPTER 2
A2 The Nature of Physical Phase Transitions 639

APPENDIX TO CHAPTER 3
A3 On the Nature of Mathematical Proof 651

APPENDIX TO CHAPTER 4
A4 The Mathematics of Darwinian Systems (by Peter Schuster) 667

APPENDIX TO CHAPTER 5
A5 Kinetics of Multistep Replication 701

Author Index 717
Subject Index 724

Acknowledgements

This book originated some 15 years ago when I retired from my position at the Max Planck Institute for Biophysical Chemistry in Göttingen, Germany. Let me start by thanking the Max Planck Society and its presidents for supporting my work during some 50 years of active scientific research. With regard to this book, my thanks include in particular the Society's current president, Peter Gruss, and my colleagues at the Göttingen Institute for generously allowing me continued use of space and facilities at the Institute after my retirement. At the same time I should like to express my gratitude to the Scripps Research Institute and its president, Richard Lerner, for hosting my extended annual stays at La Jolla, California. There I have enjoyed a unique scientific atmosphere in the field of molecular biology, which exerted a catalytic influence upon large parts of the manuscript, which I often referred to as the "Scripps book".

By European standards, Göttingen's University is not particularly old. It was founded in 1737 by George II, King of England, who at the same time was Elector of Hanover (which explains his involvement in German academia). Göttingen, one of the first secular universities in Europe, became eminent as a centre of the Enlightenment in Germany. Its scientific tradition is based on mathematics; in physics it was the birthplace of quantum mechanics, and in chemistry it was associated with pioneers such as Wöhler and Windaus. From early on it housed one of the most important libraries in Europe, which caused Benjamin Franklin to come to Göttingen for an extended stay – not in order to study *in* its university library, but rather to study the library itself. Today Göttingen is also the home of numerous research institutions at which these disciplines are also represented. The Göttingen Academy of Science, today still a meeting place for scientific discussion in an almost family atmosphere, has left its mark on Göttingen as a town of science.

La Jolla can be seen in a similar way. Its strongest field is modern biology. This was started by institutes like the Salk and the Scripps and has been perpetuated by the university, which is one of the youngest, but also among the largest, in the USA. Scientific discussion often leads to surprising new ideas. However, La Jolla also offers seclusion and time to oneself. For the thinking and writing that went into this book, the two places – Göttingen and La Jolla – were ideal. Now let me turn from places to people.

This book owes much to my partner Ruthild Winkler-Oswatitsch, who has been closely associated with my work over forty years, and to Paul Woolley, whose contributions in editing the manuscript were especially valuable. Without their dedicated support this book would never have seen the light of day. Furthermore, I thank my friends and former colleagues Peter Richter and Peter Schuster for contributing the two articles that appear in Appendices A1.2 and A4. Peter Richter also devoted much time to thoroughly checking the equations in the page proof.

The preparation of the manuscript received invaluable aid from many persons. Claus-Peter Adam, occasionally supported by Hartmut Sebesse, transformed my hundred-odd manual sketches into well-designed computer-graphic art. Guido Böse, and in part Bernhard Reuse, were most helpful in unearthing the literature references quoted in the text. Most of the reproduction and photographic work was provided by Peter Goldmann, Irene Böttcher-Gajewsky and Heidemarie Wegener. I am particularly grateful to Ingeborg Lechten, who did most of the typing work, in which she showed special skill in setting up complicated mathematical formulae. Anja Zembrzycki, assistant to Peter Vogt in La Jolla, contributed valuable text-processing in later phases of the book's preparation. Josephine Stadler, assistant to Stefan Hell in Göttingen, managed the text transfer between various computer systems and organised numerous print-outs. Last-minute assistance with the computer files was kindly given by Heinz Winkler, Svea Steinhauer and Silvia Schirmer. Paula Foley, assistant to the president of the Scripps Institute, provided help in many aspects of the daily work. All this technical assistance was indispensable in the production of the final manuscript, and I owe a great debt of gratitude to all those involved, including many not mentioned here.

A contributing factor of great importance for this book has been the continuous discussion with colleagues and friends all over the scientific world – exchanges that have covered a great breadth of problems, spanning from fundamental physics to biology and the relationship between them. This is the skein of ideas underlying this book, which accordingly starts with a description of fundamental physical principles. The two cartoons in Section 1.2 related to Einstein's theory of relativity refer to a letter of Einstein's that is in the possession of the Einstein Archive at the Hebrew University of Jerusalem. The physics chapters (1 and 2) stimulated many discussions about biology with physicists, such as the one with Richard Feynman mentioned in the text, or later ones with Hermann Haken, John Wheeler, Leonard Susskind, David Gross and others about analogies between physics and biology. In this connection I should mention my studies of James William Rohlf's excellent textbook *Modern Physics from α to Z^0*, which provided many suggestions for dealing with the problems in Chapter 1. In this connection, I also cherished numerous discussions with my son Gerald, who is a practitioner in experimental particle physics. Reinhard Genzel informed me of his newest results about a black hole at the centre of our galaxy (see Section 1.7). Rudolf Kippenhahn supplied me with the newest data on spectral shifts observed in distant galaxies, and Theodor Hänsch provided me with new results in laser physics (see Figure 3.9.3). With Hans-Joachim Queisser I had a very stimulating exchange of ideas on possible mechanisms of an evolution of our universe. The mathematician Andreas Dress, founding director of a joint Institute of Mathematical Biology (in Shanghai) of the Chinese Academy and the Max Planck Society, was of great help in formulating some of the problems in mathematical terms, while the theoretical physicist Walther Thirring of Vienna University was a stimulating partner in the discussion of physical models.

Against the backdrop of physics, this book moves on to deal with chemical and biological topics. At my home Institute I received much help from Mary Osborne in obtaining material. Karl-Ernst Kaissling of the Max Planck Institute for Ornithology at Seewiesen supplied us with many examples of pheromones (Section 4.1). We also enjoyed discussions with Brigitte and Harald Jokusch.

In chemistry, it was the Swiss school that made the deepest impression on me. I was fortunate in being able to spend much time during my La Jolla stays in personal contact with Albert Eschenmoser, who like me regularly spent time at the Scripps Institute. Chapter 4 takes up many of the problems we discussed. Other members of the Zürich School – André Dreiding, Jack Dunitz, Giulio Arigoni and Martin Quack – joined this round. Kurt Wüthrich, who also spends much of his time at La Jolla, is another participant. And then there are the many scholars who are or were based at La Jolla: the late Francis Crick, Sydney Brenner, the late Leslie Orgel, Gerald Joyce, Gustav Arrhenius, Peter Vogt, Paul Schimmel, Kyriacos Nicolaou (who kindly provided Figure 4.1.6), Gerald Edelman and (in earlier times) the late Stanley Miller. However, that list should go on to include the names of many colleagues whom I met at various places, of whom I can mention only a few, such as David Hubel, the late Sol Spiegelman, Rudolf Rigler, Jean-Pierre Changeux, Larry Loeb, Alex Rich, Fritz Melchers, David Rumschitzki, the late Shneior Lifson and Hans Wolfgang Bellwinkel. I cannot name all of them, but they are all included in my thanks.

Let me add one final remark. The person who had the greatest influence upon my work was the late Leslie Orgel. He reviewed much of his own work about the complexity of evolution in a late paper that was prepared for posthumous publication by Gerald Joyce, who also wrote an obituary in *Nature*. Leslie's earlier pupil, the late Christof Biebricher, who joined my group at Göttingen, provided numerous sophisticated experimental verifications of ideas that originated in our common work.

Last but not least, let me include the editors of Oxford University Press in my acknowledgements. I have never experienced such a pleasant cooperation between publisher and author, and have been delighted with the patience shown to me – which I needed, as the content of the book touches many fields and gave rise to many a delay as I followed up, or read up, on some new idea. The working relationship with Sönke Adlung at OUP was invariably cordial and understanding – as it is among friends.

Thanks to all of those mentioned by name – and to many others!

Prolegomenon

The seemingly all-embracing title of this book might create the impression that it is going to deal with (almost) everything. This is by no means my intention. In order to avoid any misunderstanding concerning what this book is about, the reader is invited to start by reading these introductory remarks before proceeding to the main text. One can even start by reading the conclusion at the end of the book (depending on how one likes to read detective stories).

The focal point of this first volume is the concept of information, which is meant to include both the quantitative and the qualitative (or semantic) aspects of information, thereby providing a link between physics and biology. Why, then, do I start in the first chapter by talking about the elementary states of matter, ranging from particle physics to cosmology? And if I think that matter constitutes an important part of my subject, why do I not leave its description to the experts in the field? In fact, I strongly suggest that the reader consult other writings on modern physics for a better understanding of what I in many cases only can hint at. My reason for starting with the physics of matter is to emphasise the difference in the ways of thinking that we encounter when we try to gain an understanding of the physical nature of elementary, inanimate matter as compared with the complex states of animate matter.

Elementary structures are uniquely fixed *a priori* by physical principles. The structures of living entities, by contrast, are results *a posteriori* of a protracted evolutionary process. They are a means to an end. It is true that the realisation of animate matter requires certain (molecular) structures upon which the higher-order structures are built, but the guiding principle is function rather than structure. The periodic tables of quarks and leptons reflect fundamental physical symmetries that are given *a priori*, in a way similar to that in which the periodic table of the chemical elements reflects the immediate consequences of those symmetries. In contrast, the table of the genetic code does not offer any obvious *a priori* principle, although it, too, possesses a logical structure. Where symmetries or logical principles are found in biology, they usually turn out to be of an *a posteriori* nature, in that they arise from a functional advantage that they offer, rather than from a more general principle that requires them to be the way they are. In this way, they also reflect their own historical origin.

Given this fact, it may seem surprising how much physical regularity is still involved in Nature's design of life. Here I do not so much mean the material structures that we discover. Of course, these are made of molecules, and these must fulfil the physical and chemical criteria of existence and stability, which in turn are the result of physical laws. However, this is the realm of biochemistry and biophysics, and it answers questions such as that of how a protein should be designed for functional activity, or how a permeable membrane can best be constructed. The problem I wish to address is a different one: it is that of the design of the unimaginably complex blueprints of the living state that have now been continuously in existence for some 4000 million years. How did this information originate, and how could it eventually bring about structures as complex as the human brain? This question is a special version of an unsolved mathematical problem: how can problems of exponential complexity be solved within polynomial time? The solution in this case, of course, has to be found by physical means. This is what I call the "physics of biology". It differs from the physics of inanimate matter and also from what we call biophysics, i.e. the physics behind the structure and function of all the "gadgets" that operate in a living organism.

If we ask a biologist how the miracle of life was able to originate on our planet, he will most probably refer to Darwin. He will refer to the myriads of small steps in an evolutionary process. The great developmental biologist Theodosius Dobzhansky once stated: "Nothing in biology makes sense except in the light of evolution." The fact that life had to evolve was already widely accepted in Darwin's time, but before Darwin no mechanism had been suggested that could place such a process on a durable, rational basis. Darwin provided this mechanism: the principle of natural selection. I cannot get away from the impression that physicists have somehow never accepted natural selection as a true physical principle. Even though it has at last been exonerated from the (completely false) accusation of being tautological in character, its physical foundation has remained suspect up to the present day. Erwin Schrödinger spoke explicitly of his regret that biologists had to accept the validity of Darwin's principle while for him, as a physicist, Lamarck's ideas seemed so much more attractive. And in the six volumes of the *Lexikon der Physik*, a remarkably well-edited reference work that appeared in Germany a few years ago, one finds under "Darwin" only a biography of the physicist Sir Charles Galton Darwin, a grandson of the famous naturalist. Nevertheless, Charles Darwin's contemporary, the physicist Ludwig Boltzmann, praised him as the man of the century, calling the 1800s the "century of Darwin".

The biologist, on the other hand, has problems with accepting "natural selection" as a principle based on physics. Ernst Mayr, the doyen of 20th-century biology, called it a "biological theory". This would certainly have been correct if, at the same time, he had not explicitly contrasted it with "physical theory", which "can be written down in mathematical terms". I cannot agree with this distinction, and I am not even sure that Darwin himself would have agreed with it. If I understand Mayr correctly, he was referring to the incredible complexity of any biological situation, which cannot be accounted for other than by "biological experience"; however, theories always refer to an abstraction of reality.

Take, for example, quantum mechanics, the hallmark of theories in 20th-century physics. It makes general assertions, such as particle–wave dualism and fundamental "uncertainty" in the description of physical processes in space and time, and these have far-reaching consequences for the behaviour of matter at the elementary level. Owing to its mathematical exactitude we are able not only to obtain detailed insight into the structure of matter, but also to see the consequences of this theory when we perform experiments under strictly defined initial and boundary conditions. This has created a new understanding of the physical world, which includes a deep understanding of chemistry – despite the fact that certain chemical systems may be too complicated for carrying out precise calculations. Quantum mechanics as such provided a deep physical understanding of chemistry without putting experimental (and theoretical) chemists out of their job.

In the same way, I see the usefulness of a physical understanding of biology. It will allow us to apply the stringency of theory directly to systems of defined composition and structure in a manner testable by experiments. In this way it will provide us with precise knowledge about fundamental mechanisms of selection and evolution. Yes, selection and evolution are processes of self-organisation, based on non-linear dynamics in macromolecular systems that have certain reactive properties. In this book, we are going to see what properties are required in order to endow these systems with information-generating behaviour. We shall visit the new physical world of information space, where systems emerge out of randomness and stabilise themselves by reproductive feedback and – through a series of (first-order as well as critical) phase transitions – end up as highly adapted systems, developing all sorts of functions.

To what overall functions are they adapted? The answer is: existence – under any habitable environmental conditions! In this respect, a theory with workable solutions for the existence of life is just as fundamental for the existence of life as quantum mechanics is for the existence of matter. Such a theory can be formulated in definite terms and applied for definite initial and boundary conditions. The existence of life, then, is dependent on the existence of conditions for self-organisation of information-gathering systems. This can be, and has been, tested experimentally in chemically relevant environments.

Now let me be somewhat more concrete: The physical basis of natural selection is to be found in non-linear molecular dynamics. But how can selection for a certain performance – often some very "unphysical" property – be based on the physical mechanisms of dynamical behaviour? Selection must be an internally self-organising process; there is no external selecting agent other than selection pressure for existence, exerted by the environment. Hence, the internal feedback mechanism of selection must include some relationship that brings the idea of purpose onto the physical level of dynamics. The magic word that does this is "information", used here in the sense both of absolute quantity, representing "entropic complexity", and of semantic quality, representing "specified complexity". The latter turns out to be uniquely linked to reproduction. If it is to offer any advantage, specified information must be capable of (1) conservation, (2) proliferation, (3) variation and (4) selection. The common factor that links all these four requirements is *error-prone replication.*

Information – more precisely, semantic information – represents a particular choice from among a tremendous variety of alternative structures with finite lifetime. Replication, as an inherent autocatalytic property of the class of material information carriers, is the only way to conserve such semantic information, which otherwise would disintegrate without any hope of recovery. Being autocatalytic in nature, replication automatically results in the proliferation of this conserved information. Proneness to error introduces the requisite variation. Selection arises as a consequence of replication under conditions of growth limitation. Both the stability of information and the selection of advantageous alternatives require that the error be kept below a defined threshold value. Violation of the error threshold causes selection in the form of a phase transition in information space. All these properties can be described by a system of non-linear differential equations which, according to a theorem of Perron and Frobenius (Chapter 4), can be shown to be generally solvable.

Can such a straightforward physical process bring about anything as complex as the human brain? A positive answer rests on the fact that the evolutionary process as a whole is practically unlimited – even though any individual step may represent only very slight progress. Natural selection utilises *advantage*, regardless of whether the advantage is the result of a property at the molecular level or the higher level of a complex integrated network of molecules, cells or even organisms.

"Problem-solving without knowing the problem" sets almost no bounds to the complexity of the problem-solving system. Thus, systems can arise that appear to behave in a highly logical way. However, logic of this kind is an *a posteriori* property, that is, it has arisen in response to a selective requirement.

Here we see most clearly the difference between the "physics of biology" referred to above and the physics of matter. Consider an elementary structure, such as an atom. The atom is a direct and inevitable consequence of general natural laws that already embody the logic and symmetry of its structure. This structure reflects in an inevitable manner the cause of its existence. Consequently, it is not dependent upon "historical" events. Up to the level of atoms and small molecules, there is no alternative to unique solutions. However, their "simplicity" gets stranger and stranger the more we try to approach the origin of matter.

In this respect, biological structures are completely different. They result from protracted processes whose history is reflected in the variant chosen – one among a huge variety of complex structures – which ultimately becomes selected. The solving of each problem of adaptation can proceed along many alternative paths and may lead to many alternative outcomes. At the beginning of the path towards a solution there is only the challenge of interaction. Symmetry and logical behaviour may develop, when they are of advantage for existence, and they will then appear as *a posteriori* achievements; logic or symmetry is thus an outcome and not a cause.

But where is the connection between these two realms of physics? Biological systems are undeniably large-scale physical structures. A cell is macroscopic; even a virus is macroscopic, composed as it is of relatively few, but large, macromolecules. In Chapter 2 we shall therefore examine the physics of larger, composite systems and their statistical mechanics. I shall appeal to the concept of entropy, which is primarily a measure of complexity. It demonstrates how many alternative "microstates" (i.e. states exactly defined at the microscopic level) determine the "macrostate" (i.e. states observed at the macroscopic level) of a physical system. Statistical mechanics is basically a method for averaging microstates in order to predict thermodynamic functions; it yields highly reliable results, despite the fact that the averaging procedures cover only minute fractions of the microstates that actually define any given macrostate. In fact, even for relatively small systems, the probability of occurrence of a particular microstate is vanishingly small, even if we wait for a time equal to the age of the universe.

In spite of its success, the statistical-mechanical approach fails to deal with the problem of information, as addressed in this book. Why? Let me give a short answer here and postpone the details until Chapters 2 and 3. My short answer is: an "average telephone number" is no telephone number at all. If we ring Directory Enquiries to ask for a particular subscriber's telephone number, then an answer such as "21 bits" would not help us in the least.

Nevertheless, statistical physics provides a bridge to the subject that Léon Brillouin christened *information theory*, although Claude Shannon, the originator of this theory (he called it "communication theory"), developed it from a purely mathematical point of view, without reference to physics. Shannon's theory does not deal with information as such, but rather with digits whose order encodes a message that has to be conserved while being transmitted through a communication line or processed by a computer.

Yet even this highly successful theory does not meet our ends, as little as statistical thermodynamics does. For this reason, I shall explain in some detail in the first three chapters why we need a new physical discipline to deal with the problems of semantic information in molecular biology, cellular biology and neurobiology. To develop a "physics of information" we have to set out from new premises. As I shall show in Chapter 3, there is no logical path from Shannon to Darwin, just as there is no (land) route between the two towns that bear these names.

What we gain is a physics that deals with a non-material substrate, namely, information. We define a new kind of space, *information space*, and use it to describe the non-linear dynamic processes that take place within it. Although we touch upon equilibrium and chaos, the important point is that we are dealing with phenomena that must *avoid* equilibrium, randomness and chaos. Our investigation leads to astonishing results and includes the crossroads of C.P. Snow's "two cultures". But this crossroad appears in the realm of physics. Is it real? Well, if we had sensory organs that could experience that world then we might call it real, depending upon what we choose to call "really real". Restricted as we are to the sensory organs with which we are equipped, we experience our world as complex familiarity.

Yes, we are familiar with this complexity, but we do not understand it. We do not understand how a complex organ, such as the human brain, can reproducibly appear in our world. This is a question of physical nature. Life with all its miraculous outcomes is a "regularity among natural processes" and as such – according to a definition given by Eugene Wigner (Chapter 3) – an object of physical space and time; it is rather a problem of non-material nature, requiring a physics of complexity and (semantic) information.

This book is written for a wide circle of readers. It is not just a description of facts already known to the experts. It also tries to offer new insights and therefore in some places requires one to dig very deep. It tries to give answers to the 50 questions raised in the section titles of this volume, although in many cases the answer still is simply: "We don't know!" Moreover, since the subject is very broad, and since it is the aim of the book to find solutions to fundamental problems of Nature, some readers might get discouraged at certain points – but don't give up! The various sections can be read independently. One can even start anew at any section of the book. References to other sections, where necessary, are given explicitly. In fact, the book is not just one large detective story which one has to follow step by step. It is more like a collection of short stories about different problems that, in the end, may have a common solution.

1 Matter and Energy

1.1.	What is Energy?	1
1.2.	What is Mass?	7
1.3.	Is There Continuity?	21
1.4.	Who Understands Quantum Mechanics?	35
1.5.	What Are We to Call Ultimately Elementary?	48
1.6.	How Large is Zero?	60
1.7.	Are We the Dust of Stars?	72
1.8.	The Universe: A Finite Island in an Infinite Ocean of Space?	88
1.9.	Symmetry: *A Priori* or *A Posteriori*?	107
1.10.	Why "All That"?	119

1.1. What is Energy?

If in modern times the story of creation were to be written by a physicist, it would start with the build-up of energy creating a space-time filled with radiation and matter. Similarly, when matter advanced to animation it had to start with a convergence of entropy to specify information that carries meaning. Energy and information are the two irreducible prerequisites of our existence.

When I say "irreducible" I mean "not deducible from anything more elementary" – like the axioms in the logical framework of mathematics. And when I use the term "creation" I avoid spelling it with a capital C, because I wish to stress an internal consistency with the physical cause to which it is inherently linked. This is not intended to detract from the grandeur that we associate with this term in our religious beliefs, whatever these may be.

Matter in all its manifestations is closely related to energy. So let me start with the question in the title of this section. What is energy?

Energy is encountered in many variations, such as motional impact that produces work, or electric power, or heat, or light, or just plain mass. Yes, even a cold piece of matter at rest is the equivalent of a stupendous amount of energy. Providentially, under the existing natural conditions of our planet, this form of energy does not easily convert itself to its violent alternative of kinetic energy. The insight that mass is equivalent to energy (and *vice versa*) appeared at the beginning of the 20th century as the most surprising result of Albert Einstein's theory of special relativity.

The huge amount of kinetic and radiation energy released in a nuclear explosion corresponds to the difference between the masses of the nuclear reactants and their reaction products. By the same token, the enormous amount of radiation energy that reaches us from sun – a small fraction of which suffices to "feed" all life on earth – results from the decomposition of mass, produced mainly in the interior of the sun. If four hydrogen nuclei (each having 1.0080 mass units) condense into one helium nucleus (with 4.0026 mass units), then the difference of 0.0294 mass units (i.e. almost 1%) is converted into energy, much of which goes out into space in the form of electromagnetic radiation.

A mechanism for the condensation of hydrogen to helium in fixed stars was proposed independently by Hans Bethe[1] and by Carl Friedrich von Weizsäcker.[2] The catalytic cycle, shown in Figure 1.1.1, is only one of several possible reaction schemes. It includes isotopes of the elements carbon, nitrogen and oxygen. Although the single constituents change in individual reaction steps, the cycle as a whole is conserved, in congruence with the definition of a catalyst. It is, so to speak, the "engine", while the hydrogen atoms, in the form of four protons, represent both the "fuel" and the raw material for the production of a ^{4}He nucleus.

The fusion reaction, for which alternative mechanisms will be discussed in Section 1.7, works only at very elevated temperatures, such as are present in the interior of active fixed stars. The bulk temperature of our sun reaches values of about 15 million degrees Kelvin – a little low for the above mechanism, which dominates in fixed stars that are ten times larger than our sun and can attain temperatures above hundreds of millions of Kelvin. The requirement of large dimensions for such a nuclear "reactor", which is only realisable in cosmic dimensions, answers the question of why we cannot simply utilise fusion in order to solve our energy problems on earth.

Catalysis is a major ingredient in chemical kinetics and, as such, it is also the basis of chemical turnover in animated matter. I choose this picture in order to demonstrate the elementary nature of catalysis, which appears at the most fundamental levels of matter and of life, and which recurs at all stages of organisation up to the level of human culture.

Mass-energy is "everything" that constitutes the physical structure of our universe. However, when asked what "that stuff", which manifests itself in such diverse and

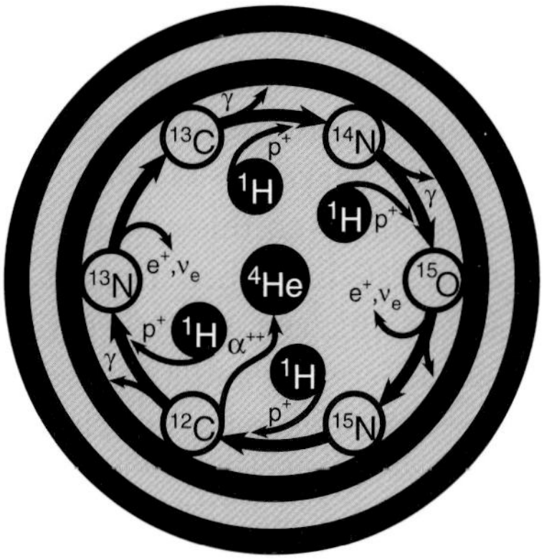

Figure 1.1.1

The carbon cycle

The carbon cycle represents the most efficient catalytic mechanism for the fusion of four hydrogen-1 nuclei (four protons p^+) to form one helium-4 nucleus (i.e. the twofold positively charged α particle). It is the major energy source of fixed stars that are more than ten times as large as our sun, requiring core temperatures of 100 million K (or more) for maximum efficiency. One cycle produces in addition three γ quanta, two positrons (e^+) and two electron neutrinos (ν_e), which contribute to the balance of charges and mass-energies. Most of the energy produced under stationary conditions is radiated away into space (see Section 1.3). Energy production in our sun, at core temperatures of some ten million Kelvin, is dominated by a fusion mechanism involving three protons that form a He-3 nucleus, with the deuteron (d^+) as an intermediate (i.e. $p^+ + p^+ \rightarrow d^+ + e^+ + \nu_e$; $d^+ + p^+ \rightarrow^3 He^{++} + \gamma$), followed by a combination of two $^3He^{++}$ to yield one $^4He^{++}$ nucleus plus two protons (see Section 1.7). The carbon cycle plays only a minor role at temperatures as "low" as in the core of our sun. I must therefore admit that the yellow colour, which refers to the surface temperature of the sun of ca. 5000 K, was chosen merely for aesthetic reasons. At temperatures to be expected for fixed stars with much higher temperatures, where the carbon cycle is dominant, the peak wavelength would lie at an "invisible colour" far in the ultraviolet.

mutually intertransformable appearances, really is, science replies with an embarrassed silence. We know how to measure it, but we simply do not know what it is. To quote Richard Feynman, from his legendary *The Feynman Lectures on Physics*:[3] "It is important to realise that in physics, today, we have no knowledge of what energy is."

However, we do have every reason to ask this question because all forms of energy are governed by a unifying principle. If we convert energy from one form into another, its amount – which can be determined experimentally and expressed in universal units (see Table A in Appendix A1.1) – remains strictly conserved. Not even the tiniest amount of the "stuff of the universe", called mass-energy, can be lost or gained. Nature balances her accounts meticulously when trading in mass-energy.

The law of conservation of energy has a curious history. Julius Robert Mayer formulated this principle on the basis of observations he made in the tropics while serving as a physician on a Dutch merchant ship. His interpretation, in which he tried to correlate chemical energy – taken in as food – with heat produced by the body, was anything but precise, and even this entailed a conceptual leap. Perhaps unsurprisingly, his paper was rejected. He then decided to study physics and soon came up with a second paper,[4] in which he correctly derived the mechanical equivalent of heat. Rudolf Clausius in Germany and John Tyndall in England made sure that credit was given to Robert Mayer when the principle was rediscovered by other prominent scientists.

Meanwhile the "energy principle" stands out as one of the primal verities that applies at all levels of physics. As a physical axiom it rests entirely upon experience and cannot be derived from anything more fundamental. The Göttingen mathematician Emmy Noether[5] demonstrated that it expresses the invariance of physical processes under translation along the time axis, just as conservation of momentum expresses a corresponding symmetry under translation along space coordinates. Her theorem states that the invariance of a physical process under symmetry transformation is necessary and sufficient for the existence of a conserved quantity.

Doubts about a universal validity of the law of energy conservation have arisen repeatedly, most distinctly in 1927, when the energy distribution in the so-called β-decay of atomic nuclei was measured exactly for the first time. Beta particles are electrons that are emitted by certain radioactive nuclei. Inside the nucleus a neutron is transformed into a proton, whereby the energy that separates the two discrete quantum states should be shared between the emitted electron and the recoiling nucleus. The electrons resulting from such a two-body decay should be essentially mono-energetic. What is observed, however, is a continuous distribution of the momenta and energies of the emitted electrons (see Figure 1.1.2).

Physicists were quite disturbed by this seeming contradiction and some, notably the doyen of quantum physics Niels Bohr, were almost willing to abandon the principle of energy conservation – at least for the realm of atomic physics. It was Wolfgang Pauli, with his down-to-earth attitude, who rejected categorically such ideas[6] and, instead, proposed in 1930 the existence of a hitherto unknown particle. Being uncharged (and therefore undetectable with the help of electric or magnetic fields) and massless (or nearly so), it was assumed to be ejected together with the electron (cf. Section 1.6). The principle of energy (and momentum) conservation could then be saved because it would now apply to the sum of both particles. Enrico Fermi developed a theory of β-decay and called the new "invisible" particle the neutrino.[7]

Nowadays the neutrino emitted together with the electron in β^--decay is more correctly called the electron antineutrino $\bar{\nu}_e$, while its antiparticle, the electron neutrino ν_e, is associated with the positron, which is emitted in what is called

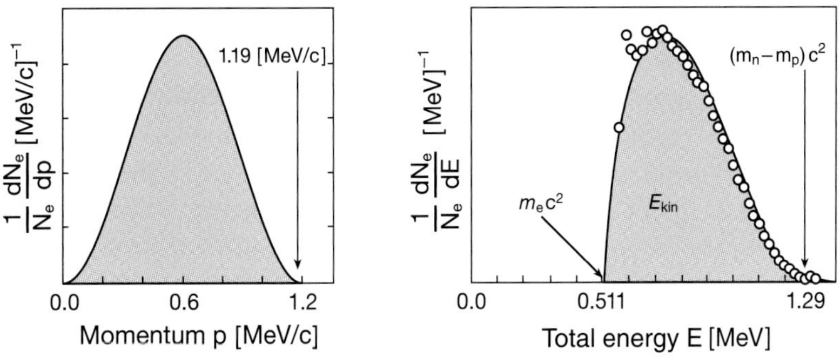

Figure 1.1.2

Momentum and energy conservation in β^--decay

β^--decay consists of the conversion of a neutron into a proton and an electron, usually occurring inside an atomic nucleus. The simplest example is that of the free neutron, which decays with a half-life of 618 s. The difference between the mass-energies of a neutron and a proton $(m_n - m_p)c^2$ amounts to 1.2933 MeV, while the mass-energy of the electron $m_e c^2$ is 0.5110 MeV, yielding $Q = (m_n - m_p - m_e)c^2 = 0.7823$ MeV. If the proton and the electron were the only decay products they should have opposite momenta and the recorded electrons should be (nearly) monoenergetic. The proton, with its almost 2000-fold larger mass, should change its velocity much less than the electron, the momentum of which has to be calculated according to special relativity, yielding a value of 1.188 MeV c^{-1}, while its kinetic energy should be 0.7819 MeV, a value close to Q as given above (see Section 1.2). What is observed instead is a continuous (nearly parabolic) distribution of both the momentum and the kinetic energy of the electron, ranging from zero up to the maximum value expected for a two-body decay. In the rest frame of the neutron, the proton and the electron do not have opposite momenta and the electrons are not monoenergetic. Their momenta (left-hand diagram) show the behaviour expected for three resulting particles, such as a proton, an electron and what was eventually called the "neutrino" (for which zero charge and nearly zero mass were assumed). The same is true for the energy distribution (right-hand diagram[7]). The experimental values behave according to Fermi's theory, which predicts a parabolic type of distribution. The square root of the transition rate, divided by the electron's momentum, according to Fermi's "Golden Rule", turns out to be a linear function of the electron's energy.[8]

β^+-decay. Because of its cryptic properties of zero charge and vanishingly small mass, the neutrino evaded observation for another 25 years, until in 1956 it was detected by Clyde L. Cowan, Frederick Reines and their collaborators[8] in work that was recognised by the 1995 Nobel prize for physics.

Conservation is not the only universal characteristic of energy. Fifteen years before Robert Mayer started to ponder about it, the French engineer Sadi Carnot, a son of Napoleon's minister of war, considered in detail the flow of heat that takes place between two bodies at different temperatures.[9] Being an engineer, he wanted to exploit this heat by converting it into useful work. Ever since James Watt had invented the steam engine (1765), thereby triggering the first industrial revolution, engineers had tried hard to maximise the amount of work that such machines could produce. It was Carnot who came up with the abstract concept of an ideal "generalised heat engine".[10] He defined its efficiency as the ratio of work done by the machine to the heat absorbed in the process. The amount of heat absorbed is clearly the amount of heat that flows from the high-temperature heat source (let us call its absolute temperature T_2) into the machine, minus the amount of heat that flows from the machine into a low-temperature sink (of absolute temperature T_1). By abstracting from reality, Carnot imagined a "reversible machine", that is, one in which all operations are carried out in a cycle where conversion of heat into work is as efficient as possible and no energy is wasted by losses or friction. He showed that such a machine – nowadays we refer to it as the "Carnot engine" – has a maximum efficiency of $(1-T_1/T_2)$, that is, the efficiency depends solely on the ratio of the cold to the hot temperature. This means that no work can be produced if the temperatures of the heat source and the heat sink are equal. The smaller the ratio T_1/T_2, the larger the amount of work that can be extracted, if one neglects the irreversible losses due to heat conduction and convection that increase with increasing temperature difference between source and sink.

The Carnot engine teaches us that only part of the energy is utilised for doing work and that some fraction of it is always rendered useless because it is dissipated. We encounter here – for the first time in physics – a principle of evaluation; i.e. *useful* versus *useless*. The abstraction of this principle led to the concept of entropy. We shall encounter the need for evaluation again when discussing complex self-organising systems.

For Carnot's contemporaries in the age of the industrial revolution, the important lesson was that not only is it impossible to construct a machine that produces more work than the equivalent energy that it absorbs (a "perpetual-motion machine of the first kind"), but, in addition, there can be no machine that is capable of converting precisely all the energy to work ("perpetual-motion machine of the second kind"). Some physicists nowadays refer to these two principles by saying: "You can't win, and neither can you break even." One may add to the latter statement: "Except at absolute zero temperature, which, however, can never be reached."[11] In the words of Alan Guth,[12] "You can't get a free lunch" (1st law). The 2nd law would then read: "You always have to pay for more than you get."

The fascination with the mysterious stuff called heat, which behaved as if it were an invisible fluid, was shared by many of the 19th-century physicists – among them James Prescott Joule, William Thomson (better known as Lord Kelvin) and Rudolf Clausius. Their efforts culminated in the construction of a logically consistent theoretical edifice called *thermodynamics* (more correctly *equilibrium thermodynamics*), which rests upon the two principles of energy conservation and dissipation. These principles are, in the words of Clausius:[13]

"The energy content of the world is constant."
"The entropy content of the world tends to a maximum."

This succinct formulation certainly goes well beyond the limits of 19th-century knowledge. As axioms, these two laws are based solely on experience. At best, they could have been tested in the very limited "universe" of a laboratory, where forces and fluxes can be kept under the strict control of the experimenter, who decides which parameters are to be varied and which are to be kept constant. Equilibrium thermodynamics is one of our best-tested and logically most consistent theories that applies both to energy and to matter. Later it was given a strict axiomatic formulation by the mathematician Constantin Carathéodory.[14] To extrapolate the laws of thermodynamics to the realms of time and space, which were largely uncharted at the time of Clausius, seems to me extremely brave as well as risky. What could Clausius have known about the universe, to make such a bold statement?

If we take a moment to consider the universe from the present-day point of view, we cannot escape the following questions: Where did all this mass-energy initially come from? Did the universe start in an ordered form? Who produced all the work that created the universe as we perceive it now? Will energy eventually end up in an entirely useless, orderless form? What will be the nature of this final form of energy? All these questions are involved if we desire to find even a provisional answer to our opening question: What *is* energy? If the universe began with a total amount of energy equal to zero, then conservation must mean a "zero-sum game", i.e. the compensation of the large amount of (positive) mass energy by a corresponding amount of (negative) potential energy.

Hence, what we can say so far is: We do not know what energy *is*. It must be a "something" of universal nature that appears in material and non-material forms and has the propensity to distribute itself among all states that are accessible without losing or gaining one iota; the amount of energy is perfectly accountable for. Some data about energies on the various scales in which physics operates are shown in Appendix A1.1.

Let me close this opening section by uttering a word of caution about asking straightforward-sounding questions such as "What is energy?" An answer may create the impression: "That is it!" But "it" – as Parmenides reminded us as long ago as 450 BC – is nothing but "the thought that recognises the 'It' ". However, the question provokes the thought.

1.2. What is Mass?

Mass is at least something that we can, in principle, touch. Consequently, Webster's dictionary calls it a "fundamental property of a body". Classical mechanics defines it as "the measure of a body's resistance to acceleration when a given force is applied" or simply as the ratio of force to acceleration. The "body" may be as large as our sun (some 10^{30} kg), or as small as a single electron (about 10^{-30} kg). However, characterising mass by the units of weight reminds us that, apart from resistance to acceleration, mass has another property, namely that of "being heavy" in a gravitational field. The (non-trivial) fact that there is an equivalence between "inertial" and "gravitational" mass will concern us in a later section.

The equivalence between mass and energy is today an experimentally well-established fact. In the previous section I mentioned the mass loss in a thermonuclear reaction, which is the source of the huge amount of energy released in an H-bomb explosion. Because this correspondence between mass and energy is reciprocal, i.e. each bit of energy is equivalent to a certain amount of mass, the conservation law applies to mass-energy, rather than to energy or mass alone. This is as important when we are dealing with the reaction of a single nucleus as it is when we are making assertions about the universe as a whole. However, it does not mean that mass can unrestrictedly be turned into other forms of energy. Let us look at an example.

A particle and an antiparticle, having the same mass and equal but opposite values of some other internal properties (such as charges or magnetic moments), annihilate one another if they collide. Conservation of mass-energy requires that an amount of energy, equivalent to their rest mass m (see below) plus their kinetic energy, is dissipated, either as photons, such as in the case of the electron and the positron, or in the form of mesons, as is the case for nucleons.

In gauge theories of modern particle physics (see Section 1.6) one even assumes the existence of a particular symmetry-breaking field, associated with a heavy and electrically neutral boson of spin zero as its quantum, the Higgs boson, which is supposed to equip a material particle with a finite mass. Excitingly, the Higgs boson was observed experimentally just as this book was in press. Even though it may not tell us what mass "is", it may tell us in more detail "how" mass is related to energy.

I am running a bit ahead of my agenda, but this book is not supposed to be a systematically organised graduate course.

The equivalence of mass and energy was a non-intended, though gratifying, result of a theory that appeared in 1905[15] and carries the name "special relativity". Its author was the then 26-year-old junior clerk at the Swiss patent office in Berne, Albert Einstein, who had to use his spare time in order to finish his doctoral thesis on the theory of Brownian motion. Einstein's theory of special relativity was a stroke of genius, formulated and written up within a few months. As its title *Zur Elektrodynamik bewegter Körper* (*On the Electrodynamics of Moving Bodies*) shows, the author did not originally set out to revolutionise Newtonian mechanics. He ultimately did so not by introducing any new theoretical concept or conjecture, but rather by giving full regard to an experimental fact that had been discovered 18 years earlier and was overdue for consideration by theoreticians. It turned out that Einstein was not the only scientist who thought about this problem. His competitor was the great French mathematician Jules Henri Poincaré, whose paper[16] on the dynamics of the electron, in which he independently developed similar ideas, appeared in 1906.

The basis of classical mechanics is provided by the so-called Galilean transformations: In a frame that is moving with a speed υ in the x direction, the x coordinate is to be replaced by $x' = x - \upsilon t$ while all other coordinates (t, y and z) remain unchanged. Galilean transformations assume an additivity of velocities. A head-wind, for instance, decreases the speed of an aeroplane relative to the ground.

However, let me now first discuss the results of the above-mentioned experiment. It is the Michelson–Morley experiment, perhaps the most famous experiment ever that had a negative result. In 1881 Albert Abraham Michelson, a German-born US physicist, had completed the construction of a sensitive interferometer.[17] This instrument was able to record fringe shifts, i.e. shifts in the black and white band interference pattern of two beams of light of identical frequency, and thereby to detect sensitively very small changes of light velocity, possibly associated with the motion of the light source relative to an "aether", the then supposed medium for the propagation of electromagnetic waves. For his work Albert Michelson was to become, in 1907, the first US citizen to receive the Nobel prize for physics.

Together with his American colleague Edward Williams Morley he conducted in 1887 the famous experiment, in which they measured the velocity of light in the direction of the motion of earth and looked for changes in this velocity when the direction of the light beam is varied up to 90°, i.e. perpendicular to the direction of the earth's movement. The length of the light beam, using a rotating drum faced with several mirrors, amounted to 10 m of light path. Given a light velocity of 300 million m/s and the speed of earth in its orbit of 30 000 m/s, the fringe shift expected according to the Galilean transformation was 0.4.[18]

The result of the experiment, confirmed later by other, even more precise, measurements, was: "no fringe shift at all!" (cf. Figure 1.2.1). The speed of light turned out to be independent of the motion of the source, and to be a frame-independent universal constant. Electromagnetic waves can travel in a vacuum; the assumption of an aether is superfluous. It is a property of physical space.

Let us now look somewhat more closely at Einstein's theory of special relativity, which raises the velocity of light *in vacuo* "c" to the rank of a fundamental natural constant. The theory rests on two postulates:

i) The laws of physics are the same in all inertial frames (i.e. frames of reference in which bodies move in straight lines with constant speeds unless forces act upon them).
ii) The speed of electromagnetic waves in a vacuum is independent of any motion of their source.

The first statement is identical with the one on the basis of Galilean transformation. The second postulate is required by the results of the Michelson–Morley

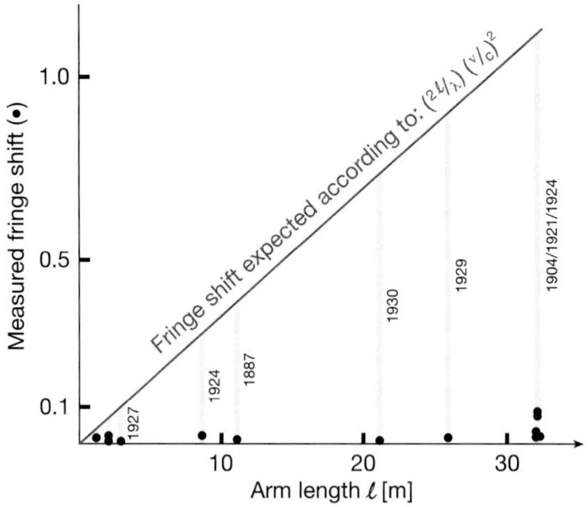

Figure 1.2.1
Michelson–Morley experiments

Experiments showing that the velocity of light is independent of the earth's motion. The arm length refers to the optical path, the fringe shift to the change of the interference pattern of light, which does not respond to the direction of the earth's motion (black dots = measured values). The year numbers refer to the dates when the experiments were performed.

experiment. It is responsible for all deviations from Newton's theory, which disappear when the source velocities υ are small compared with the speed of light c, but which become considerable when υ approaches c, as is the case for many processes in particle physics. The second statement requires a replacement of the Galilean transformation by transformations worked out first[19] (in 1890) by Hendrik Antoon Lorentz. His objective was to leave Maxwell's equations unchanged in all inertial frames of reference. Maybe Einstein, who seems to have been unacquainted with Lorentz' paper, started his work on relativity with similar intentions, as the title of his 1905 paper still suggests. The Lorentz transformations are expressed by a factor $\gamma \geq 1$, which equals the reciprocal square root of the term $(1-\beta^2)$, β being υ/c where υ is the velocity of the inertial frame and c is the velocity of light *in vacuo*. This factor, which appears explicitly in all relativistic expressions for time, length, momentum and energy, approaches unity if $\upsilon \ll c$, yielding the classical solution. Let us look first at some of its consequences according to Einstein's theory.

1) *Time dilation* is a "slowing down" of all clocks in an inertial frame that is moving with a speed υ relative to an observer in a system assumed to be at rest. For the observer in the (correspondingly defined) resting system, any time interval Δt he measures appears to be lengthened to $\gamma \Delta t$ in the system moving with respect to the clock he used in his own system. This is true regardless of which of the two systems houses the observer, and is hence defined as "the reference system at rest".

2) *Length contraction*, by the same token, occurs in the frame that is in relative motion with respect to the observer. Any length interval Δl in the direction of motion appears to be shortened in the moving system to $\Delta l/\gamma$. In each of the two systems an observer can measure time intervals with his clocks and hence

determine which events occur simultaneously. Another observer, however, who is in relative motion to him will not agree with his decisions. It is important to note that the disagreement is not based on individual illusions, but – according to special relativity – can be tested quantitatively by experiments (see below).

3) The *space–time diagrams* of observable processes are limited to the famous light cone, celebrated in popular representations of Einstein's theory. I refrain from reproducing it at this stage because the "cone" refers to four-dimensional space time (difficult to reproduce in a two-dimensional diagram), where the "face" of the hypercone would have an inclination of 45° with respect to a vertical central time axis. If the tip or vertex of the (upward-opening) cone is identified with the origin of the coordinates, all observable processes in future must lie inside the light cone, the trajectories of particles moving with constant velocities being straight lines. The range outside the light cone is not accessible to us, as the speed c, whose exact value amounts to 2.99792458×10^8 m/s (with the standard of the metre defined accordingly), cannot be exceeded in any physical process.

4) As a consequence of (1) to (3), the notion of *universal simultaneity* is to be abandoned, dooming Newton's concept[20] of an "absolute, true and mathematical time, of itself and from its own nature, which flows equably without regard to anything external" to a limited "relative" validity.

5) The *relativistic momentum*, given in classical physics as the product of mass and velocity, $p = m\upsilon$, also involves the relativistic correction. In an elastic collision of two particles the total momentum – like energy – is subject to a conservation law. It is the factor γ which takes care of the validity of this law under relativistic conditions, i.e. if υ is not negligible compared with c.

6) *Relativistic energy* follows from relativistic momentum when the laws of mechanics are applied. A force acting upon a freely movable particle along a certain distance in space causes an acceleration of the particle, and thereby provides it with a defined amount of kinetic energy E_{kin}. If F is the force and x the distance, then according to the laws of mechanics Fdx can be replaced by υdp, where υ is the velocity and p the momentum of the particle. After some rearrangement and integration when distance (and time) have advanced from 0 to x (or t), one obtains for E_{kin} the difference γmc^2, which yields for $\upsilon \ll c$ the classical value of the kinetic energy, with $E = E_0 + E_{kin} = mc^2 + m\upsilon^2/2$.

So far so good. As in the case of time, length and momentum (and force), the relativistic values approach the classical values in the limit $\upsilon \ll c$. However, in the above case of energy the classical value results from a difference of two values which themselves are not subject to this approximation. They are to be interpreted as the "total" relativistic energy $E = \gamma mc^2$ and a finite (i.e. non-zero) rest energy $E_0 = mc^2$, where the term "rest" refers to the "resting" state. In fact, under normal conditions the latter term is huge compared with the kinetic energy $E_k = m\upsilon^2/2$ because $\upsilon \ll c$.

Einstein immediately realised that this unprecedented result was of a fundamentally novel character. In the same year (1905) he augmented his relativity paper with a short note with the title: "Does the Inertia of a Body Depend upon its Energy Content?" where he concludes:[21]

> "The mass of a body is a measure of its energy content: If the energy changes by L, the mass changes in the same sense by L/c^2, if the energy is measured in ergs and the mass in grammes. Perhaps it will prove possible to test this theory by using bodies whose energy content is variable to a high degree (e.g. salts of radium). If the theory agrees with the facts, then radiation transmits inertia between emitting and absorbing bodies."

Einstein's concluding remarks, quoted above, were meant to refer to the energy content of a resting mass m. Nevertheless, the popular form $E = mc^2$ is usually interpreted as also applying to the total energy of a moving body, tacitly assuming m to include the factor γ (which is formally correct). Some years ago I encountered a lovely cartoon by Sidney Harris, which is reproduced in Figure 1.2.2. Through the internet I got in contact with the artist and, after I had explained the problem to him, he was kind enough to provide me with a new version of his former cartoon, which is shown in Figure 1.2.3. Einstein's view, taken from a letter he wrote in 1949 to Lincoln Barnett, an editor of the *New York Times*,[22] is reproduced in the legend of this figure. The equation on the blackboard is called the "master equation of special relativity".[23]

It is, in fact, more lucid than its (superficially simpler) popular form $E = mc^2$, where one has tacitly redefined m as the "relativistic mass" (i.e. γm). A particle at rest has a zero momentum p, so only the second term under the square root has any effect, yielding $E_0 = mc^2$. The other extreme is represented by a photon, which has no mass, but has a finite momentum $p = \hbar/\lambda$ (where λ is its wavelength and $\hbar$ is Planck's constant divided by 2π). Here only the first term is relevant, yielding $E = pc = h\nu$, where $\nu = c/\lambda$ is the frequency of the (monochromatic) light (see Section 1.3). This relationship is experimentally verified by Compton scattering (see Section 1.3), showing the transfer of momentum from photons to electrons.

The Pythagorean form of the "master equation" (Figure 1.2.3) triggers some afterthoughts with regard to the joke in the cartoon in Figure 1.2.2. In fact, there is some deeper sense in this, as a mere recollection of school mathematics may suggest. The master equation owes its form to some vectorial addition. And with special relativity we are still in compliance with the Euclidean metric, even though we are now applying it to a four-dimensional space.

Space–time as a four-dimensional continuum was introduced by the Göttingen mathematician Hermann Minkowsky soon after the advent of special relativity. In his famous lecture on "Space and time",[24] only a few months before his untimely

Figure 1.2.2

Cartoon by Sydney Harris © 1991.

death, he said: "From now on, space by itself and time by itself will sink into the shadows, while only a union of the two will retain independence." Einstein was at first not particularly impressed by the concept of space-time, but his view changed when he realised its usefulness for developing the theory of general relativity (see Section 1.8). It is true that for special relativity space-time just represents a conceptual tool that forms the basis of the four-vector representation of space-time and energy-momentum. I will now give a brief introduction to four-dimensional spaces, to which I shall return at various times throughout the book (see Section 1.8, Appendix A3, and also Sections 4.4 and 4.5).

Figure 1.2.4 introduces the space concept using a model of discrete numbers 1 to n. A glance at the picture tells us immediately that the concept involves a lot more than just adding another coordinate to three-dimensional space (which explains my reservation above with regard to drawing a "picture" of the celebrated "light

Figure 1.2.3

Cartoon by Sydney Harris © 2004.

Albert Einstein in a letter to Lincoln Barnett, science editor of the New York Times. "It is not good to introduce the concept of the mass $M = m/(1-v^2/c^2)^{1/2}$ of a moving body for which no clear definition can be given. It is better to introduce no other mass concept than the "rest mass" m. Instead of introducing M it is better to mention the expression for the momentum and energy of a body in motion." (Courtesy of Einstein Archive, Jerusalem)

cone"). The six drawings of the four-dimensional hypercube shown in Figure 1.2.4 are intended to demonstrate some of its peculiarities, which otherwise are hard to imagine.

Take a three-dimensional cube for comparison. It has eight vertices, twelve edges and six faces. Each of the twelve edges is shared by two faces and each of the six vertices is the common origin of three edges. If the two adjoining faces have an opposite sense of rotation, their common edges will cancel out, verifying a principle

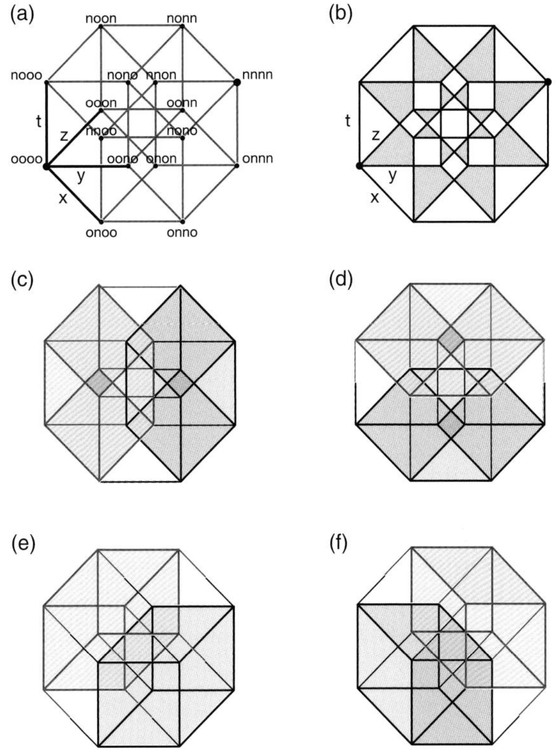

Figure 1.2.4

Four-dimensional space–time diagrams

According to John Archibald Wheeler[24]. Views of the four non-overlapping (shaded) subspaces (c, d, e and f). For details, see text.

of algebraic geometry which says: "The boundary of a boundary is zero." The idea goes back to the French mathematician Elie Joseph Cartan and has applications in Maxwell's electrodynamics as well as in Einstein's theory of general relativity, where it explains the conservation of mass-energy as the source of the curvature of space.[25]

The foregoing remarks may have sharpened our eyes for a closer inspection of the space–time diagrams depicted in Figure 1.2.4. Diagram (a) introduces the four coordinates whose discrete values run from 1 to n. I could not resist emphasising the two stars in diagram (b) in order to stress the complete symmetry of the whole figure. The boundary of a four-dimensional hypercube consists of eight different cubes which one can discern by focusing on diagrams (c) to (f). The projections chosen stretch the cubes to a slightly oblong shape, so that we may distinguish a pair of vertically "standing" cubes (c) from another pair of horizontally "lying" ones (d), and these from two pairs that extend from the fore- to the background (e and d). The four pairs of similarly oriented cubes, thus defined, do not have any face in common. All other pairwise combinations of cubes with differing orientation share one (and only one) face. Examples are shown in Figure 1.2.5.

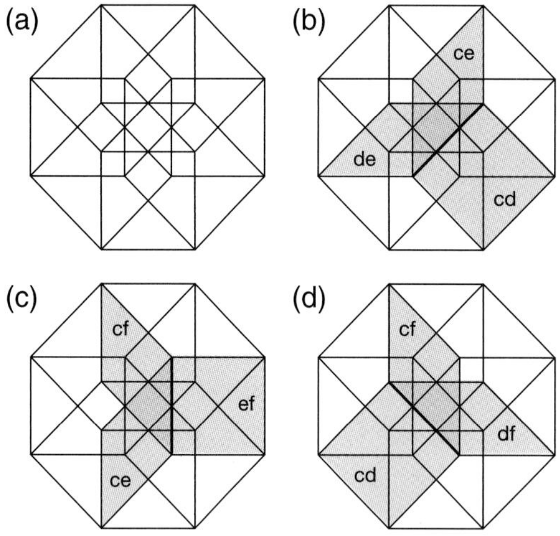

Figure 1.2.5
Other views of four-dimensional space-time

Some of the face-sharing cubes that make up most of the subspaces (see text).

Given eight different cubes, there are 28 possible pairwise combinations, of which only four do not show any overlap. Hence, there remain 24 different pairs, in which the two cubes share one common face. This matches the fact that eight isolated cubes having $8 \times 6 = 48$ faces, if joined to the hypercube, enclose 24 faces, each of which is common to two different cubes. The analogy to the three-dimensional case, where each of the 12 edges of the cube is shared by two faces, stresses the general nature of Cartan's principle of algebraic geometry, which in electrodynamics secures the conservation of charge and in geometrodynamics that of mass-energy (see Sections 1.3 and 1.8).

The true four-dimensional shape of the hypercube, of course, defies our imagination, which is biassed by the planar projection of three dimensions. What the diagrams reproduce correctly is the manner in which the substructures match one another in order to allow their folding into a closed four-dimensional structure.

In his life work, Minkowski laid the foundations of a geometry of numbers which constitutes an important aspect of number theory. The manuscript of his lecture on "Space and time", from which I quoted above, was sent to the editor of *Physikalische Zeitschrift* in late December 1908. Three weeks later, at the age of 44, Minkowski died from a ruptured appendix. His last paper entails a number of notable predictions. He recognised geometry as an emerging "quality" of physics. In his manuscript we find the sentence: "Three-dimensional geometry becomes a chapter of four-dimensional physics." That is what he meant when he declared, as quoted above, "Space in itself and time in itself will sink into the shadows."

Another point is the four-vector representation of space-time, one in which time (multiplied by c) and the three space coordinates x, y and z are the components of a four-dimensional space-time vector. The length of this vector – equalling the square root of the difference $(ct)^2-(x^2 + y^2 + z^2)$, which is always greater than zero because

$c > \upsilon$ – is to be an invariant, which is valid for every inertial frame. The same is true for the four-dimensional energy–momentum vector, the length of which is the square root of $E^2-(p_x^2 + p_y^2 + p_z^2)c^2 = E^2-(\mathbf{p}\,c)^2$ and as such is given by the invariant mc^2. (E is the total energy and p_x, p_y and p_z are the momentum coordinates, all given as Lorentz transforms.) The Lorentz transformation itself appears as a multiplication of the 4×1 column vector by a 4×4 matrix, yielding a new column vector, the components of which constitute the transformed space–time vector. This is the physical basis of the "master equation" in its "Pythagorean form" on the blackboard in Figure 1.2.3.

It is noteworthy that Minkowski, towards the end of his 1909 paper (on p. 109), mentions another, quite different, way of looking at the problem, namely by normalising the length/time ratio, i.e. setting $c \equiv 1$, and choosing an imaginary time axis: $t' = t\sqrt{-1}$. This would change the square term $(x^2 + y^2 + z^2)-(ct)^2$ into the entirely symmetrical form $x^2 + y^2 + z^2 + (ct')^2$ and recover Euclidean geometry. Minkowsky says explicitly: "Such a complete symmetry would transmit itself to the mathematical expression of any physical law, which is in accordance with the world postulate." The latter words in his terminology mean the manifold of all possible combinations of the variables x, y, z and t, which he called "world lines". Isn't this an anticipation of a world model proposed by Stephen Hawking and James Hartl in the 1980s, which also assumes an imaginary time axis leading to what Hawking calls the "no-boundary boundary condition" (see Section 1.9)? It symbolises a universe that is finite but has no "boundary", i.e. no beginning or end in imaginary time, the four-dimensional analogue of the two-dimensional surface of our (three-dimensional) sphere-like planet in which the north and south pole do not represent singularities or which (in Hawking's words) "is finite in extent but ...doesn't have a boundary or edge: if you sail off into the sunset, you don't fall off the edge or run into a singularity".[26] (Note that the above-quoted symmetric quadratic form could describe a four-dimensional "sphere" having the volume $\pi^2 r^4/2$, the surface of which does not include any edge or singularity; see Section 4.5.)

Special relativity reveals the rest mass m, appearing in the form of mc^2, as an equivalent of energy. Not knowing what energy is, can we prove this equivalence experimentally? Yes, we can do so, even in the way Einstein envisaged in his 1905 note quoted above. The mass defect associated with radioactive decay is well documented, and so is the mass annihilation of matter and antimatter mentioned earlier in this section, which is balanced by an equivalent amount of radiation energy. An excellent example proving its quantitative correlation with special relativity is presented by muon decay. The muons μ^- and μ^+ are the more massive siblings of the electron and the positron (see Sections 1.5 and 1.6). In nature they are produced by cosmic rays in the upper atmosphere, several thousand metres above sea level, and they travel with nearly light velocity, so that their γ factor is about 20. The mean lifetime of the muon in its rest frame would be $2.2\,\mu$s, and during that time it could travel on average only 660 m. However, a large fraction of the muons from the upper atmosphere is still observed at sea level. This is possible only if most of them stay "alive" during a period that is "time-dilated" by a factor of about 20. Correspondingly, their path is "length-contracted" by the same factor. In the frame of the resting muon (to which the 2.2-μs lifetime refers) the earth moves towards the muon with a relative speed near to light velocity. This "reduces" the true distance of several thousand metres to a "contracted length" of some hundred metres, as provided by the relativistic correction.

All these more or less direct inferences were made quite some time after the appearance of Einstein's theory. A direct proof of his results today, 50 years later, is almost routine in particle physics. It consists of a simultaneous determination of the velocity and the momentum of an electrically charged particle travelling at a speed close to that of light. All that was necessary to perform such an experiment was a focusable electron source, a variable electrical field (with potentials of several thousand volts) for deflecting the fast-moving electrons, an electromagnet producing a uniform magnetic flux density, perpendicular to the electric field, and a (photographic) detection screen, all to be placed in an evacuable apparatus. With such a device, utilising the principle of present-day mass spectrometers, the speed υ of the electrons can be varied and detected, while their momentum can be independently determined from the radius of curvature (r) of the circular path, enforced by the magnetic flux density (B) acting on the charge (e) perpendicular to its direction of motion ($p = reB$). If special relativity is valid, there should be a direct proportionality between the momentum p (or radius of curvature r) and the relativistic factor $\beta\gamma$, which is a unique function of the velocity υ. A plot using suitably normalised coordinates should yield a straight line with a slope of 45°.

In fact, precisely this experiment was carried out four years before the advent of Einstein's theory – almost unnoticed by the international community of physicists. In 1901, a young assistant professor by the name of Walter Kaufmann, working at the physics department of Göttingen University, performed this truly exciting experiment, with a conclusive result in favour of special relativity. But, of course, this theory was not yet in existence, and Kaufmann – although getting precisely the "right" results – did not really know what they meant.

However, Kaufmann knew very well what he was trying to achieve, namely, to show – as the title of his publication tells us – that for electrons moving under the influences of an electric and a magnetic field an "apparent electrodynamic mass" must exist which is different from the electron's inertial mass. His experimental device was similar to the one used by Joseph John Thomson in the discovery of the electron, the principle being routine in mass-spectrometric studies today. But we are in the year 1901, and Thomson's discovery, which won him the 1906 Nobel prize for physics, had been made only four years before. Kaufmann was just as up-to-date in recruiting an efficient electron source. No more than five years earlier, Antoine Henri Becquerel had discovered spontaneous radioactivity in uranium salts. In 1900, just in time for Kaufmann's experiment, Becquerel showed that the radiation

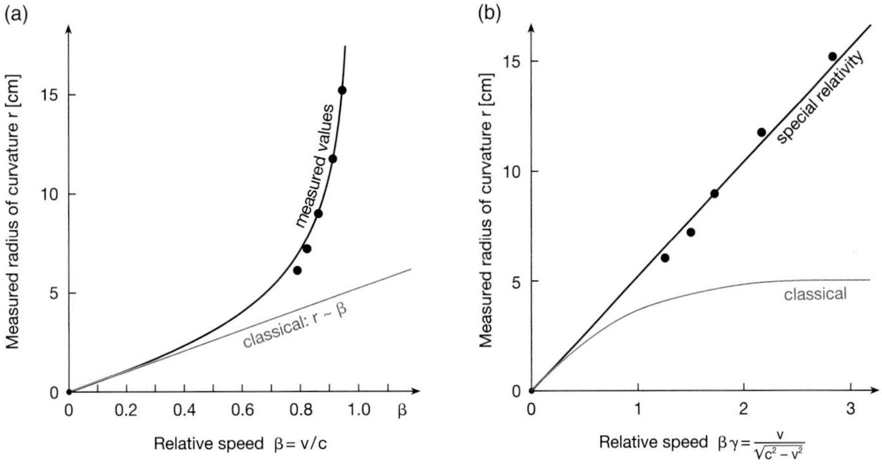

Figure 1.2.6

Experimental data for the speed dependence of the momentum of an electron

Obtained in 1901 by Walter Kaufmann. Measured values of the radius (r) of curvature of the electron in a fixed magnetic field as a function of its speed relative to that of light ($\beta = v/c$). (a) Data plotted according to classical theory. (b) Data plotted according to the expectations of Einstein's theory (J.W. Rohlf[23]).

emanating from radium consisted of electrons, later identified as products of β^--decay. Becquerel, together with Marie and Pierre Curie, who had discovered and studied other radioactive materials, were awarded the 1903 Nobel prize for physics. It was the Curies who kindly provided a grain of purified, highly active radium chloride for the experiments carried out at Göttingen. Kaufmann focused the beam of electrons using a kind of "pin hole" as a secondary source, the way it is done for producing a point-like light source, required for the confocal optics used in modern laser instruments. Kaufmann's design demonstrated a high level of sophistication for the time at which the experiment was done.

Let us now discuss his results from a modern point of view. We find his pertinent data in two consecutive papers in *Göttinger Nachrichten*, i.e. the proceedings of the Göttingen Academy of Sciences,[27] a source of many novel ideas in late 19th- and early 20th-century science.

In his first paper (1901) Kaufmann dealt more objectively with the data that he had directly measured, such as the velocity υ of the electrons and the radius of curvature of their cyclical path when they are exposed to an electrical and a magnetic field oriented perpendicularly to one another. Classical theory would have predicted a linear relationship between the velocity and the momentum ($p = m \times \upsilon$), represented by the radius of curvature ($r = \frac{p}{eB}$, e = electric charge, B = magnetic flux density) of the electron in the (constant) magnetic field. However, what Kaufmann found was a drastic increase in the momentum at velocities approaching the velocity of light (see Figure 1.2.6a); this he interpreted as an increase of the electron's mass.

The quantity $\beta\gamma$, characteristic of Einstein's theory, is defined uniquely by the measured velocity υ of the electrons (with $\beta = \upsilon/c$) as being $\upsilon/\sqrt{c^2 - \upsilon^2}$. According to special relativity, p is equal to $\beta\gamma$ mc, yielding $r = \beta\gamma\, mc/eB$. Hence, a plot of the measured r (or p) versus $\beta\gamma$ should yield a straight line, originating at $r = 0$ and having a slope of 45° if the ordinate is re-scaled by setting mc/eB equal to 1. This I have done in Figure 1.2.6b, where Kaufmann's data indeed demonstrates irrefutably the consequences of Einstein's theory.

However, Kaufmann could not know anything about a theory that was not yet in existence. For him – as he stated in his first paper – his data were just evidence for some "electromagnetic" mass that was not identical with the inertial mass of mechanics. In his second paper (1902) he tried to explore in more detail the nature of this new type of "pseudo-mass".

Meanwhile, his colleague in theoretical physics, Max Abraham,[28] had derived an equation on the basis of Maxwell's and Lorentz' ideas in which he postulated two types of masses: "longitudinal" and "transverse". The longitudinal mass represents inertia in the direction of motion, while the transverse mass opposes a force acting perpendicular to this direction (as the Lorentz force does in a magnetic field). Kaufmann in his second paper tried to prove Abraham's model by introducing corresponding parameters which carried him away from the clear-cut interpretation later offered by Einstein's theory. Abraham, throughout his short life, was opposed to relativity. His arguments were based on what he thought to be common sense, which – as modern physics has shown – may fail when we penetrate new dimensions. Kaufmann's close association with the ideas of his former colleague may be part of the reason why his early brilliant experimental work has fallen into almost complete oblivion.

What about the question in the title of this section? Is mass just some condensed form of energy? This certainly is an important insight, but not an answer to the question which would become identical with that asked in the first section. Obviously, the phrase "what is" makes little sense as long as we are dealing with first principles or elementary, i.e. non-derivable, properties. On the other hand, the question "What is entropy?" has a clear answer, since entropy is a quantity that is based on a definition, both in phenomenological and in statistical terms (see Chapter 2). Nevertheless, most people feel far more comfortable dealing with energy than with entropy.

Our mind does not inherit any *a priori* knowledge. It has to adapt to experience and observation, which are consciously reflected and embodied into a consistent scheme. Talking about what we observe and experience is helpful in creating "understanding". It reminds me of a critical remark made by Bertrand Russell[29] in his foreword to Ludwig Wittgenstein's *Tractatus Logico-Philosophicus*,[30] which concludes with the categorical statement: "Whereof one cannot speak, thereof one must be silent." Russell's plain reaction to this conclusion was: "What causes hesitation is the fact that, after all, Mr Wittgenstein manages to say a good deal about what cannot be said."

1.3. Is There Continuity?

Continuity is certainly a mathematical abstraction. In a historic talk entitled "Mathematical Problems", which David Hilbert[31] delivered in Paris at the International Congress of Mathematicians in 1900, continuity was the first of the 23 unsolved problems which he addressed.

All sets that can be put into a one-to-one correspondence with the set of natural numbers, as is true for the set of rationals, must be distinguished from the set of all real numbers, which include the irrationals and hence span a continuum, as all points of a continuous line (see Appendix A3). According to Cantor's concept of equivalence[32] there should exist (at least) two infinite sets, the denumerable one and the continuous one, where the latter is of superior potency (Hilbert used the word *Mächtigkeit*). Hence the essence of Cantor's idea was that infinities of different potency must exist. A proof of Cantor's hypothesis, as Hilbert pointed out, would provide a bridge between denumerable and continuous sets. However, the possibility of proof fell victim to Gödel's criteria of undecidability (cf. Section 3.6). Both the continuum hypothesis and its negation are compatible with the axioms of set theory.

Does continuity in the mathematical sense exist in physical space? The idea that matter cannot be divided indefinitely goes back to the Greek philosophers Leukippos and Demokritos in the 5th century BC, while the smallest material entities became accessible to experimental scrutiny only in the last century. But what about other forms of energy, such as radiation?

There was much in 19th-century thinking that intuitively suggested continuity at the finest level of organisation. Perhaps the greatest achievement in 19th-century physics was the understanding of the relationships between electric charges and electric and magnetic fields. The findings of Michael Faraday,[33] André Marie Ampère,[34] and many others were crowned in a unifying theory conceived by James Clerk Maxwell.[35] All knowledge derived from classical electrodynamics is contained in the four equations that bear Maxwell's name (see Vignette 1.3.1). These equations presuppose a universal speed of electromagnetic radiation in any inertial system, therefore they are consistent with special relativity. The notion of continuous electric and magnetic fields, or the assumption of an aether, a medium in which these fields could propagate, nourished the idea of continuity. Even classical mechanics, which dealt with discrete, massive objects, yielded continuous energies because of a continuous variation of velocities, a notion that was carried over to the theories of viscous and caloric "fluids" pioneered by James Clerk Maxwell,[35] Ludwig Boltzmann[36] and Josiah Willard Gibbs.[37]

Vignette 1.3.1 Maxwell's Equations

Maxwell's equations are among the most important contributions of nineteenth-century physics. They bring together electricity, magnetism and light – the first of the great unifications that characterise modern physics. Maxwell's equations are usually stated in the form:

$$\nabla \bullet \mathbf{E} = 4\pi\rho \qquad (1)$$

$$\nabla \bullet \mathbf{B} = 0(2) \qquad (2)$$

$$\nabla \times \mathbf{E} + \dot{B}/c = 0 \qquad (3)$$

$$\nabla \times \mathbf{B} - \dot{E}/c = 4\pi I/c \qquad (4)$$

E is the electric field density vector, **B** the magnetic flux density, $\dot{E}$ and $\dot{B}$ are the temporal derivatives $\partial\mathbf{E}/\partial t$ and $\partial\mathbf{B}/\partial t$, ρ is the (volume) electric charge density, c the velocity of light and t time. The symbol ∇ designates the gradient operator and, in combination with the dot symbol, the scalar or dot product with any vector **u** (e.g. **u** = **E** or **B**). The scalar product is called divergence: $\nabla \bullet \mathbf{u} = \text{div}\,\mathbf{u} = \partial u_x/\partial x + \partial u_y/\partial y + \partial u_z/\partial z$, where u_x, u_y and u_z are the x-, y- and z-components of the vector **u**. Likewise ∇ in combination with the cross symbol designates the vector or cross product of ∇ with the vector **u**, called curl (or rotation: rot):

$$\nabla \times \mathbf{u} = \text{curl}\,\mathbf{u} = (\partial u_z/\partial y - \partial u_y/\partial z)_x + (\partial u_x/\partial z - \partial u_z/\partial x)_y + (\partial u_y/\partial x - \partial u_x/\partial y)_z$$

The bracket terms are components of the curl vector that refer to directions perpendicular to the directions of both components involved in the bracket terms. This is indicated by the subscripts added to the brackets. Hence each of the three terms in Eqns. (3) and (4) yields a separate equation, e.g. for the x-component in Eqn. (3): $\partial Ez/\partial y - \partial Ey/\partial z + \partial Bx/\partial t = 0$. Thus the "true" number of Maxwell's equations is eight, rather than four, and Eqns. (1) to (4) are a condensed representation which may be further compressed in relativistic notation.

The physical content of Eqns. (1) to (4) is the following:

The first equation is a reflection of Coulomb's law, in which the electric charge is considered to be the source of the electric field. In (charge-)free space div **E** is zero. The scalar product of ∇ with the field vector **E** can be interpreted as the excess of "field lines" leaving the volume element over those entering it, the sources of the field lines being charges. The fact that isolated magnetic charges do not exist in any (classical) physical scenario explains Eqn. (2).

Eqns. (3) and (4) explain the inherent coupling between electric and magnetic fields. Eqn. (3) is a reformulation of Faraday's law, which was supposed to explain electric induction as a consequence of a rapid change of a magnetic field. Eqn. (4) is

Cont.

Cont.

an augmented form of Ampère's law, which as such did not contain the term $-\dot{E}/c$, because it was supposed to describe the "electromagnet" which operates using a D.C. voltage source. Maxwell first had simply reproduced Ampère's law in its original form, but he soon realised the lack of a symmetry, which turned out to be instrumental in deriving a wave equation.

The existence of electromagnetic waves, which may propagate uninhibitedly in "free" space, follows from Eqns. (1) to (4), in which all right-hand terms can be set to zero (and which therefore have been written in this form). Combination of Eqn. (3) and (4) yield $\nabla \times (\nabla \times E) = -c^{-2}(\partial^2\mathbf{E}/\partial t^2)$ which is already the form of a wave equation. With both $\nabla\mathbf{E}$ and $\nabla\mathbf{B}$ equalling zero one can reduce this equation, as can be easily shown for any of the x, y or z components, yielding for the wave equation the more familiar form (in **E** or **B**); $\nabla^2\mathbf{E} = c^{-2}(\partial^2\mathbf{E}/\partial t^2)$ where ∇^2 is the Laplace operator: $\nabla^2 = \partial^2/\partial x^2 + \partial^2/\partial y^2 + \partial^2/\partial z^2$. In "free" space $c = \sqrt{\varepsilon_0\mu_0}$, where ε_0 and μ_0 are the electric and magnetic permittivities of "free" space and hence characterise the velocity of light as a universal constant. Electromagnetic waves include radio waves as well as light and X-rays, up to the range of γ radiation. The fact that electromagnetic radiation remains unattenuated in free space provides the basis for radio-, light and X-ray astronomy.

Maxwell's equations (3) and (4) demonstrate a "moment of rotation" and thus yield another example of Elie Cartan's principle of algebraic geometry $\partial\partial = 0$ ("The boundary of a boundary is zero"; see also Section 1.8). In the case of Maxwell's equations it applies to 3-dimensional space and automatically takes care of a conservation of the field sources, which will turn out to be equally important in the 4-dimensional case of general relativity.

Remark: **B** is defined as having the dimension Tesla $=$ volt $\cdot$ s $\cdot$ m^{-2}. A more symmetric form of Maxwell's equation is obtained by multiplying B by the velocity of light in a vacuum, yielding the dimension of E, i.e. volt $\cdot$ m^{-1}.

However, towards the close of the 19th century there were certain observations that did not fit into the picture of a continuous world. I shall discuss two such findings that triggered a true revolution in the world picture painted by 19th-century physics.

A revolution is like an avalanche. Nobody realises precisely when it starts, but it is often an event caused by just a tiny, unpredictable fluctuation. Yet, once the avalanche has been recognised, it has already reached a point where it can no longer be stopped. The avalanche of quantum physics, unnoticed by most physicists of the time, started in the "year zero" of the 20th century.

In the years leading up to 1900, Max Planck studied the spectral distribution of the intensity of electromagnetic radiation emitted by a "black body". What is a black body? Black is the colour of a surface that absorbs all radiation that reaches it. However, this absorbing property is also possessed by a hole through which radiation passes without the possibility of being reflected out again. Think of a box with a tiny hole. The walls of the box are held at a fixed temperature. Such a box is a "black body". Most of the radiation that enters through the hole will be absorbed by the walls. The radiation inside the box reaches thermal equilibrium with the walls. A fraction of this radiation at thermal equilibrium can be emitted by the hole. The energy distribution of the radiation inside the box is the same as the energy distribution of radiation emitted through the hole.

The energy per unit area of black-body radiation was calculated, as a function of the radiation's wavelength, by Lord Rayleigh[38] and James H. Jeans,[39] but their radiation formula describes the experimental results correctly only at long wavelengths. With decreasing wavelengths, the radiation intensity predicted by the Rayleigh–Jeans formula, being inversely proportional to the fourth power of the wavelength, tends to exceedingly large values, a prediction called the "ultraviolet catastrophe". For human existence on earth it would be a true catastrophe indeed. Experimentally, however, one finds an exponential decrease, causing a hot "black body" to glow red in the dark and not blue as was suggested by the Rayleigh–Jeans formula.

Planck found a solution to the problem.[40] At first he regarded his solution as a mere mathematical artifice, and it took him years to realise that his derivation was not merely an interpolation. It turned out to be the correct solution, not only for the particular problem of black-body radiation, but also one that had universal validity for the distribution of energy. What was Planck's "revolutionary" idea?

Planck made the assumption – "just on paper" – that the energy of the atomic oscillators that produce the black-body radiation may not be continuously distributed, but may rather be delivered in discrete portions. He assumed the spacing between the energy levels to be proportional to the frequency of the oscillator that emits its energy in the form of radiation. This led him to the formula stating that the energy of the emitted bundle, or "quantum", of radiation is proportional to the frequency of the radiation (ν). The proportionality constant is called "Planck's constant" ($h = 6.63 \times 10^{-34}$ J $\cdot$s or 4.14×10^{-34} eV $\cdot$s). It has turned out to be one of the fundamental constants of physics. The difference between Planck's formula and the one of Rayleigh and Jeans lies in the replacement of an average thermal energy (proportional to the temperature) by the idea that an oscillator with discrete energy levels can, with a certain probability, be excited by thermal motion. The latter involves the Boltzmann factor, $\exp(-h\nu/kT)$, where T is the absolute temperature and k is Boltzmann's constant. This takes care of the exponential decay of the radiation intensity at shorter wavelengths (see Figures 1.3.1 and 1.3.2).

Planck's constant shares its fundamental nature with another non-derivable constant, the velocity of light in a vacuum ($c = 2.9979 \times 10^8$ m/s), which was promoted to this rank through Einstein's theory of special relativity. And there may be one more

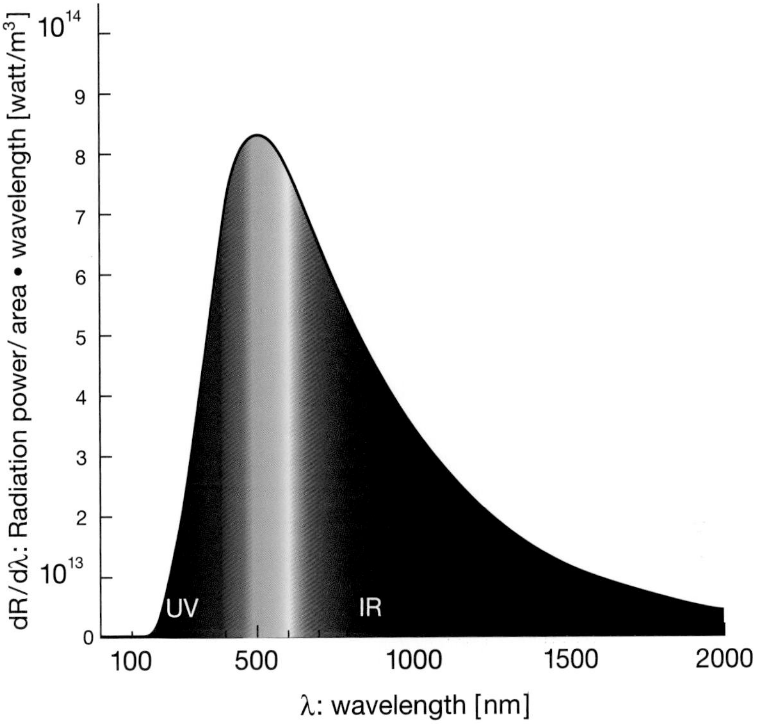

Figure 1.3.1

Planck's curve for sunlight ($T \approx 5000$ K)

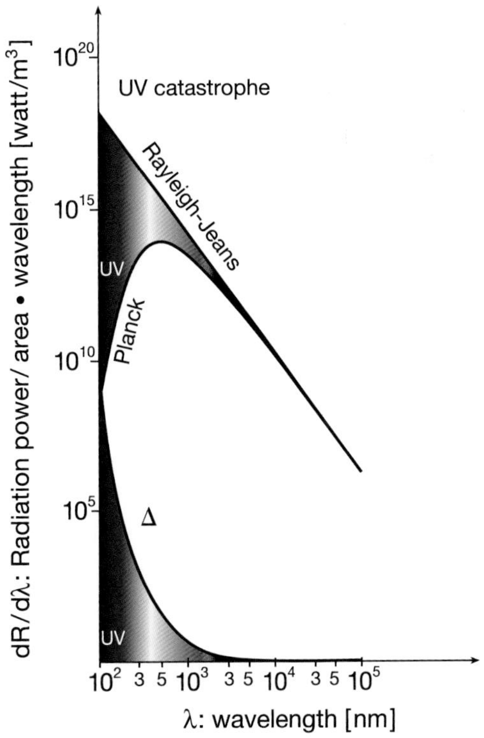

Figure 1.3.2

Deviation of Planck's curve from the classical radiation formula of Rayleigh and Jeans
Designed by Claus-Peter Adam.

fundamental constant of this nature: Newton's gravitational constant, which reappears in Einstein's theory of general relativity. All other physical constants, among them Boltzmann's constant (which will be referred to repeatedly in later chapters) are of a different nature. They just tell us about certain facts, such as elementary charges or masses, or the numbers of such particles in a given macroscopic amount of matter. Others connect different systems of physical (notably energy) units, such as the joule, the calorie and the electron volt. Boltzmann's constant, for instance, could have been set to a value of 1 (as Gibbs, not knowing of Boltzmann's work, in fact did) if the temperature had been measured in appropriately-scaled energy units.

Let us take a closer look at Planck's formula, which is one of the "master equations of physics" (J.W. Rohlf),[23] because it establishes the quantum as a fundamental property representing a discontinuous energy distribution. The position of the maximum of Planck's curve for $dR/d\lambda$ (where R is the radiation power per m^2 and λ the wavelength) is generally fixed to occur at a wavelength slightly larger than $hc/5kT$, the deviations being negligible for nearly all practical purposes. Likewise, the Planck curve for $dR/d\nu$ (where ν is the frequency) also shows a maximum fixed to a common value $\nu_{max} = 3kT/h$. The two universal distribution curves are shown in Figure 1.3.3.

They appear with different shapes, but their integrals yield the same value, having the form σT^4, which had been experimentally established in 1879 by the Austrian physicist Josef Stefan[41] and theoretically derived in 1889 by Ludwig Boltzmann[42] from thermodynamic considerations (see Section 1.10). The calculation of the precise value of the Stefan–Boltzmann constant σ had to await the formulation of Planck's law.

The universal curves in Figure 1.3.3 immediately allow a reproduction of the radiation curve of Figure 1.3.1 if the average surface temperature of the sun ($T = 5800$ K) is inserted for T. Likewise, the radiation curve of the Earth is reproduced by using a T value slightly lower than 300 K. A wonderful demonstration of the timeless fundamental nature of Planck's law is represented in Figure 1.3.4, which contains the results for the radiation background of space with a temperature of $T = 2.74$ K, as recorded quite recently by the space satellite COBE (Cosmic Background Explorer, Figure 1.3.5; see the reference to George Smoot in Section 1.8),[43] confirming (with greatly increased precision) the original data of the Nobel laureates Arno Penzias and Robert W. Wilson[44] of about 3 K, a cornerstone of the present model of the universe.

Max Planck was one of the most tragic personalities of 20th-century physics. Being of unquestionable moral integrity, he personally went to Hitler in 1933 in an attempt to persuade him to reverse his devastating racial policies – with the result that Hitler flew into a fit of rage, disgusting the 75-year-old scholar. Being the president of the Prussian Academy of Sciences (till 1938) Planck was forced to "persuade" Einstein to "resign". After Planck had lost his first wife and three of his children at an early age, his younger son, who was involved in active resistance against Hitler, was murdered by

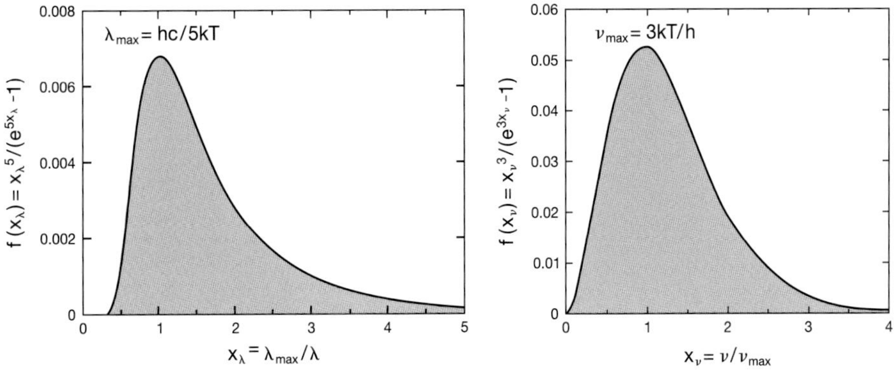

Figure 1.3.3
Planck curves

The Planck curve can be shown in two ways. One form refers to wavelength (λ), the other to frequency (ν). In each case the function $f(x_\lambda)$ or $f(x_\nu)$ is plotted for a given value of temperature. These functions are related to $dR/d\lambda$, the dependence of radiation power R on λ, and to $dR/d\nu$, the dependence of radiation power on ν, as follows:

$$\frac{dR}{d\lambda} = \frac{2\pi hc^2}{\lambda^5\left(e^{hc/\lambda kT}-1\right)} \approx 1.9 \cdot 10^{-3} T^5 f(x_\lambda)\ \mathrm{W \cdot m^{-3} \cdot K^{-5}}, \text{and}$$

$$\frac{dR}{d\nu} = \frac{2\pi h}{c^2}\frac{\nu^3}{e^{h\nu/hT}-1} \approx 1.13 \cdot 10^9 T^3 \cdot f(x_\nu)\ \mathrm{W \cdot s \cdot m^{-2} \cdot K^{-3}}.$$

For T = 300 K (room temperature) the values are:

> $\frac{dR}{d\lambda} \approx 3 \cdot 10^7\ \mathrm{Wm^{-3}}$ or $300\ \mathrm{Wm^{-2}}$ per $\lambda = 10\mu m$ at $\lambda_{max} \approx 9.6\mu$ m, and
$\frac{dR}{d\nu} = 1.6 \cdot 10^{15}\ \mathrm{W \cdot s \cdot m^{-2}}$ or $160\ \mathrm{W \cdot m^{-2}}$ per $\nu = 10^{13}\ s^{-1}$ at $\nu_{max} = 1.73 \cdot 10^{13}\ s^{-1}$

The Stefan–Boltzmann law yields $R = \sigma T^4$ with $\sigma = 5.67 \cdot 10^{-8} = \frac{2\pi^5 k^4}{15h^3c^2}\frac{W}{m^2 \cdot K^4}$ (according to Planck).

W = watt (measure of power = energy/time), m = metre, s = second, K = Kelvin, h = Planck′s constant, c = light velocity, k = Boltzmann′s constant, T = (absolute) temperature.

the Gestapo shortly before the end of the war in 1945. American officers brought Planck to Göttingen at the end of the war, where I had the pleasure, as a young student, of listening to two lectures he gave at the age of 88. His plain gravestone in the Göttingen cemetery carries the inscription:

MAX PLANCK

$$h = 6.64 \cdot 10^{-34}\ \mathrm{J \cdot s}$$

The picture in Figure 1.3.6 is by the Berlin artist Ellen Fuhr (née Hertz), a grandchild of Gustav Hertz, who in 1914, together with James Franck, had shown that in collisions of electrons with atoms energy is taken up by the atom only in quantised portions. For this supplement to Planck's work, they were awarded the 1925 Nobel prize for physics. I regard Ellen Fuhr's drawing as reflecting the memory of her grandfather's relation to Max Planck.

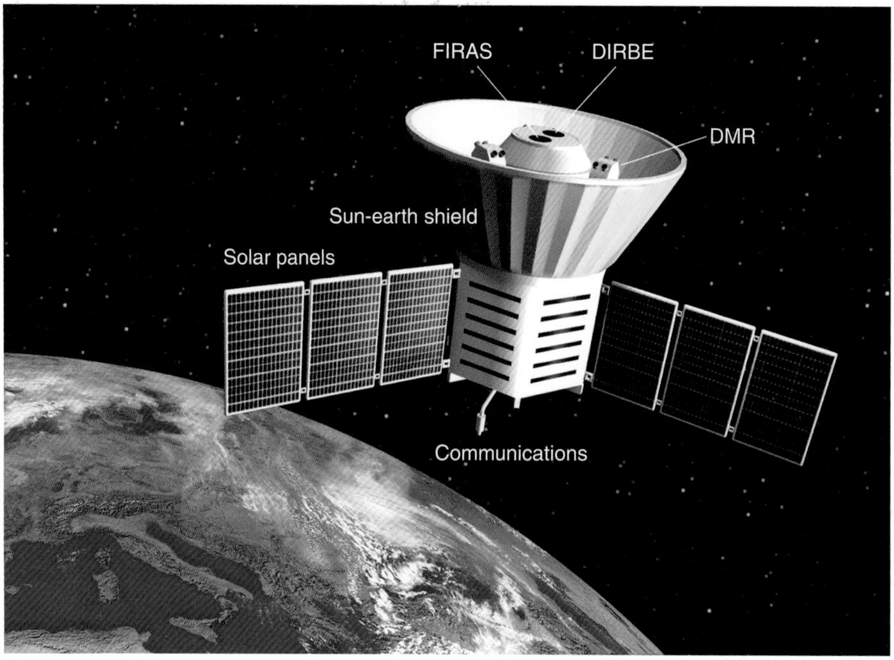

Figure 1.3.4

The COBE experiment

Artist's impression, constructed from the description given by George Smoot and Kean Davidson.[44]

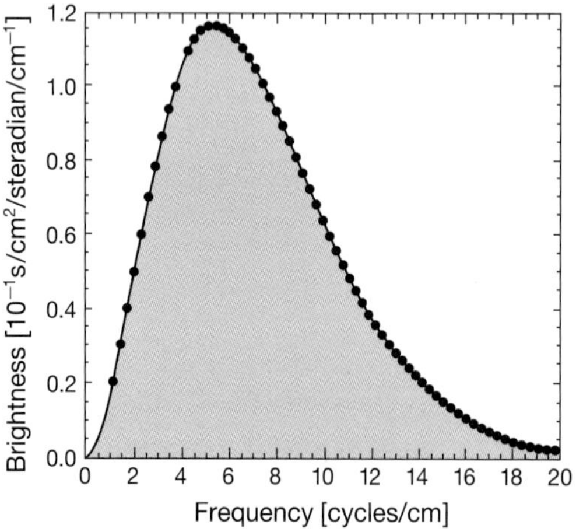

Figure 1.3.5

The Planck curve of "background radiation" ($T = 2.74\,K$)

Measured in the COBE experiment (after J.C. Mather *et al.*, *Astro. J.* **354** L 37 (1990)). The curve has the form of $f(x_\nu)$ as a function of ν/ν_{max} (see Figure 1.3.3).

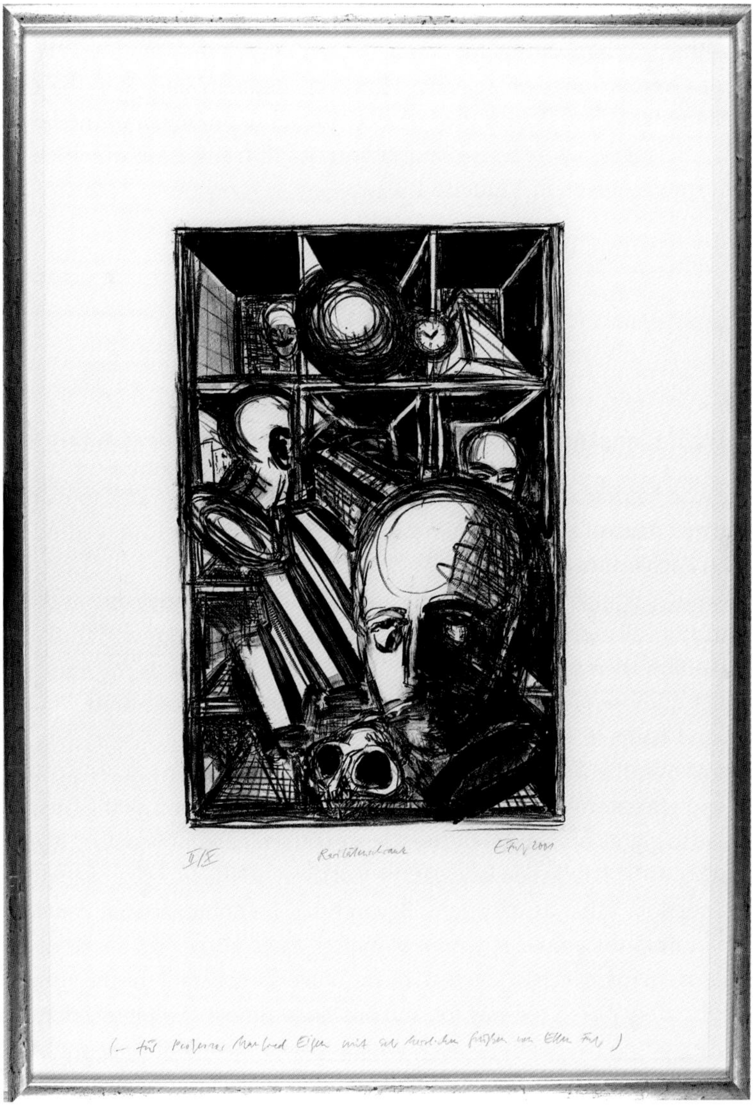

Figure 1.3.6

Raritätenschrank

A lithograph by Ellen Fuhr (2001).

Albert Einstein was at first somewhat indignant about Planck's decision, in spite of everything that was happening, to remain in Germany in order to preserve whatever he could of German physical science. However, Einstein later saw the tragedy of Planck's dilemma. When Planck died in 1947 Einstein, on behalf of the US National Academy of Sciences, wrote a touching obituary,[45] a translation of which from the original German is given in Vignette 1.3.2.

Vignette 1.3.2

To the Memory of Max Planck
The National Academy of Sciences of the United States of America

Anyone who has had the privilege of donating a great, creative idea to mankind is in no further need of accolades from posterity. Higher honours are granted to him by his very achievements.

And yet it is right and proper, even necessary, that on this day and at this place scientists from all parts of the world, who strive after knowledge and truth, should unite to affirm their undiminished allegiance to the ideal of discovery – even in times such as these, when political passions and brute force have brought so much sorrow and harm to mankind. This ideal, which throughout history has united thinkers all over the world, was personified in rare perfection by Max Planck.

Although the atomic nature of matter was already perceived by the Greeks, and was raised to higher probability by the science of the nineteenth century, it was Max Planck who, with his radiation law, was the first to reveal the true size of atoms, without the need for other assumptions. Beyond this, he demonstrated convincingly that, apart from the atomic structure of matter, there also exists a kind of atomic structure of energy, which is governed by the universal constant that he introduced.

This discovery paved the way for, and has since almost completely dominated, the physics of the twentieth century. Without it, no meaningful theory of atoms and molecules, and the energetic principles that govern their behaviour, would have been possible. Moreover, it broke out of the framework of classical mechanics and electrodynamics, confronting science with the task of finding a new conceptual basis for the whole of physics – a task that, despite important partial successes, has not yet been satisfactorily accomplished.

By making its obeisance before this great man, the National Academy of Sciences gives expression to the hope that the freedom of scientific research, conducted for the sake of pure knowledge, may remain with us, unfettered and unrestricted.

Albert Einstein

Translation by Paul Woolley

The establishment of the *quantum* as a new concept also owes much to Einstein, whose Nobel prize reflects this fact. (Planck received his Nobel prize for the radiation law in 1918, while Einstein was awarded the 1921 prize in 1922 with special reference to the "photoelectric effect". The laudatio explicitly excluded his two theories of relativity.)

Einstein was led to the quantum concept when solving a puzzle that concerned experimental physicists around the turn of the century.[46] If light is shone onto a metal surface, electrons are set free, and their energy distribution can be measured. This phenomenon is called the "photoelectric effect". Philipp Lenard[47] had previously discovered cathode rays, which could be observed by passing them through a specially constructed window of an X-ray tube. At the end of the 19th century this allowed him to start quantitative measurements of the photoelectric effect. The results he obtained puzzled him greatly, and his fellow experimental physicists, who had not yet accepted Planck's quantum "hypothesis", were equally at a loss to explain them. In 1905 Lenard received the Nobel prize for his discovery of cathode rays. In the same year Einstein found that he could explain Lenard's results on the basis of the photoelectric effect, by relating them to Planck's findings. We can only speculate that this may have been one reason why Lenard developed an ill-natured antagonism to all of Einstein's ideas, in particular to relativity, which resulted in personal hatred. He laid himself open to ridicule by proposing a specifically "German" (meaning "anti-Jewish") physics.

What did Lenard find that seemed so confusing? By increasing the intensity of light of a given frequency (and hence its total energy), the (maximum) kinetic energy of the emitted electrons remained the same, and only the number of electrons increased. On the other hand, if the frequency of the light was increased, then the energy of the emitted electrons rose as well. Moreover, below a certain threshold of light frequency, no electrons were emitted at all.

Einstein realised at once that this type of behaviour carried the signature of the new quantum concept. From his theory of relativity he knew that "particles" devoid of mass can carry a momentum that is proportional to their energy, which in the case of light quanta is proportional to their frequency. Hence, the energy transferred to a single electron must be independent of the light's intensity and only related to its frequency.

Electrons hit by a light quantum would bounce through the metallic phase and leave its surface. This explains three observations. First of all, the frequency (energy) of the light required to eject an electron from the metal surface has to exceed a certain value; otherwise the electrons remain "trapped" in the metal lattice. Secondly, if the light has a frequency higher than this threshold value, then for a given frequency each electron emitted has the same (excess) kinetic energy, and this corresponds to what Lenard had observed. Thirdly, shining more light (more quanta) onto the metal surface increases the number of electrons ejected without affecting their individual kinetic energy. Einstein was able to calculate both the minimum energy required for an electron to leave the metal (which turned out to be a property specific for each metal), and – what is more important – the maximum (frequency-dependent) kinetic energy of the electrons observed. The latter yielded the proportionality factor between the photon's energy (E) and its frequency (ν), and this factor turned out to be identical with Planck's constant (h).

Note that this approach to the quantum concept differs from Planck's in an essential point. Planck could not immediately conclude that light itself has a quantum structure and thus can be considered as a "particle". The black-body radiation curve rather shows that light is *emitted* and therefore *generated* through a quantised process. In Planck's theory it is the energy levels of the oscillators that emit the light which are quantised. Einstein, on the other hand, had to infer that light energy is delivered to matter in a discrete manner. By combining the two results, the inevitable conclusion is reached that light, by its very nature, *is* quantised.

It should be noted that in 1909 Einstein wrote a paper on energy fluctuations in black-body radiation.[48] The derivation of an expression for the mean square of energy fluctuations was based solely on Boltzmann's principle (see Section 2.3) and Planck's radiation formula. In this paper Einstein noted that to accept Planck's formula means to accept the corpuscular (particle-like) nature of light.

As a consequence of the work of Planck and Einstein, physics was facing a severe dilemma. Light can undoubtedly be diffracted and exhibits interference, which clearly indicates that it has a wave-like nature, in full agreement with Maxwell's equations. Yet Einstein had now shown that light is corpuscular as well. This dual nature of light was a mystery that had to be resolved. Moreover, it soon became known that material particles behave identically.[49] Electrons can be diffracted, like waves, and yet they are particles that possess a finite mass. It was Louis de Broglie[50] who, in his doctoral thesis, first proposed the idea of material waves. He derived an equation for a wavelength associated with a moving particle that is inversely proportional to the particle's momentum; the proportionality factor is Planck's constant, just as the wavelength of a photon is correlated with its momentum (which according to special relativity has a realistic physical meaning). Why does such behaviour appear strange to us? The answer is simply: We do not observe it in daily life, so it is outside the scope of our experience. Why do we not observe it? Is it that the laws applying to matter at its finest level of detail are *in principle* different from those that apply to macroscopic bodies? Take a tennis ball. A tennis ball travelling at high speed from a slam would – according to de Broglie's theory – have a wavelength of 10^{-32} to 10^{-33} m. In such cases the classical laws are sufficiently precise approximations to the quantum laws.

Perhaps the most important experiment demonstrating the dual nature of the quantum of electromagnetic radiation – i.e. that it is both particle-like and wave-like – was performed by the US physicist Arthur Holly Compton in 1922–23.[51] Moreover, it introduced the wave-like nature of material particles, almost coinciding with the development of quantum mechanics (see Section 1.4), the theory of the

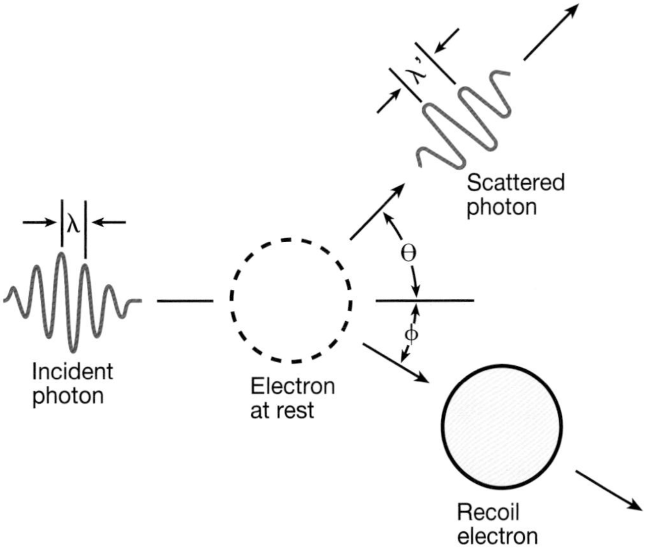

Figure 1.3.7

Schematic representation of Compton scattering

A photon of wavelength λ_1 and momentum $p_1 = h/\lambda_1$ is scattered by an electron (assumed to be initially at rest). After collision the momentum of the photon is $p_2 = h/\lambda_2$. The scattering angle of the photon is θ and that of the electron (which gains finite momentum in the collision) is ϕ.

20th century that was to revolutionise all physics. Compton's experiment is described schematically in Figure 1.3.7. It is based on the fact that electromagnetic radiation interacts with all particles that carry an electric charge. This scattering of photons that collide with a charged particle is termed *Compton scattering*. Compton allowed energy-rich photons from radioactive decay to be scattered by electrons: in the process $\gamma_1 + e \rightarrow \gamma_2 + e$ the incoming photon γ_1 and the electron e exchange energy and momentum, so that the outgoing photon γ_2 has a wavelength different from that of γ_1.

In order to describe the process we use Einstein's master equation on the blackboard in Figure 1.2.3 and we remember that a photon, although it has zero mass, exhibits a finite momentum $p_i = h/\lambda_i$, where h is Planck's constant and λ_i is the wavelength of the photon i. The rest of the calculation involves applying the laws of conservation of (total) energy and momentum for the electron and the photons, yielding altogether three equations. In this way, Compton showed that the change in the wavelength of the photon before and after its collision with an electron (assumed to be in the resting state up to the moment of collision) can be written as given in the legend of Figure 1.3.8. The constant hc/m_ec^2 is called the Compton wave-

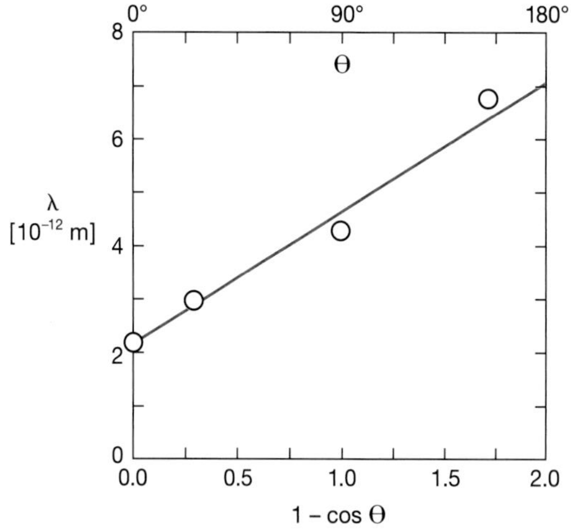

Figure 1.3.8
Compton's original data

The wavelength of the incident energetic photon (λ_1) was $2.2 \cdot 10^{-12}$m. The scattered radiation was measured at angles $\theta = 0$, 45, 90 and 135 degrees. Using the conservation laws for total energy and the momenta of photon and electron, Compton derived a relation expressing the change of the photon's wavelength as $(\lambda_2 - \lambda_1) = (hc/mc^2)(1 - \cos\theta)$. The plot of the wavelength as a function of $(1 - \cos\theta)$ yielded a straight line with a slope of $2.4 \cdot 10^{-2}$ m, representing the term hc/mc^2. This quantity was named the Compton wavelength, λ_c.

length λ_c. The mass-energy of the resting electron $m_e c^2$ is 0.511 MeV and hc equals 1240 eV · nm, yielding a value for λ_c of 2.4×10^{-12} m. In Figure 1.3.8 we see that Compton's original data indeed obey the relationship given in the legend.

What about a concise answer regarding the distribution of energy in our world? Is it continuous or is it discrete? The "classical" quantum hypothesis has answered this question for radiant energy in the relation

$$E = h\nu.$$

For a given frequency, energy can only appear in integral multiples of $h\nu$.

This insight provided the basis for one of the most remarkable conclusions of the quantum physics of the early 20th century: Bohr's model[52] for the structure of the hydrogen atom, which integrated two other important findings. Ernest Rutherford,[53] by his particle-scattering experiments (see Section 1.5), had identified the atomic nucleus, and Johann Jakob Balmer, a secondary-school teacher of mathematics in Basle, Switzerland, had found a simple formula that represented the wavelengths of the spectral lines of hydrogen, later called the Balmer series. In Balmer's time nobody knew why such a formula held true, and this mystery remained until 1913, when Bohr discovered its importance for building up the atomic model named after himself and Rutherford.

Can we now answer the question of whether there is continuity in physics? In the seventies I met Paul Adrien Maurice Dirac at the celebration of his 70th birthday in Trieste[54] and again shortly afterwards at one of the Lindau meetings of Nobel prize-winners. In his lecture the grand old man raised what he regarded as the three great questions of our time. One of these was: "Is there continuity?" In other words, he asked: "Is there anything in physics which is not quantised – or, is continuity a mathematical abstraction that is foreign to the physical world? Does it include space-time and its geometry itself, which according to Einstein is an ingredient of physics rather than the stage on which physics is performed? Are there smallest units of space and time?" Dirac would not have asked the question if it could have been answered, a situation that has prevailed until the present day. It is a question that will be revisited repeatedly throughout this part of the book.

To return to history, the quantum universe certainly presented a very uncomfortable situation to physicists in the early 20th century. Whenever an experiment dealt with the question "Is it a corpuscle?" The answer turned out to be "yes"! However, whenever an alternative experiment asked the question "Is it a wave?", the answer was "yes" as well! If both answers are correct at the same time, then neither of them can be correct on its own.

This situation reminds me of a story told to me by an Israeli friend. Two men who had a quarrel went to a rabbi. The rabbi, trying to calm down the excitement, asked them to state their positions independently, one after the other. After the first man had finished his story the rabbi said. "Yes, my friend, you must be right." The other person immediately jumped up: "How can you say this, not having heard my case?" The rabbi said: "Relax, my son, and tell me your story." After listening to the second story the rabbi conceded: "You are obviously right." At this moment the voice of the rabbi's wife came from the kitchen: "But rabbi, they can't both be right." The rabbi answered: "Yes dear, you are right too."

The situation in physics during the early 1920s cried out for a revolutionary change of concepts. The issue at stake was not just the new paradigm of a quantum universe. Rather, it was the clear recognition that energy and matter behave in a way fundamentally different from the way in which we intuitively expect them to.

1.4. Who Understands Quantum Mechanics?

The reader will have noticed that I still owe him an answer to the question of whether continuity is something that ultimately exists in the physical world, or whether it is a mathematical abstraction and as such just a useful assumption. A point in space can be considered as a sphere with a vanishing radius ($r \to 0$). A continuous space then requires that each point be surrounded by an infinite number of neighbouring points. However, as we shall see, there seems to be a smallest size below which physical routine does not make sense any more. Formerly this was regarded as a limit due to the finite resolution of any measurement. However, nowadays theory tells us that the fact that we cannot detect "zero size" is not merely a matter of experimental precision, but is rather a fundamental concept of physics. Our world at high resolution differs from what we perceive in our everyday experience.

The theory I refer to is quantum mechanics, which after more than 80 years of existence still defies our imagination, "even for those of us who have used it daily in our work for decades" (Murray Gell-Mann).[55]

Quantum mechanics is the answer to the classical quantum puzzle outlined in the previous section. The protagonists of classical physics never really accepted it. Planck somehow reconciled himself with it; Einstein acknowledged it as, at most, a preliminary answer. Quantum mechanics appeared in the mid-1920s, nearly simultaneously in two (seemingly) unrelated versions: wave mechanics and matrix mechanics.

Peter Debye, who himself made important contributions to the quantum concept, thereby preparing the ground for the new theories, told me the following story. One day in the mid-1920s a colleague at the Zürich physics department turned up in his office and reported very enthusiastically about de Broglie's idea of material waves. Debye remained cool and commented laconically: "Where is the wave equation? Without a wave equation, no wave theory!" The colleague went away, but soon came back and presented his wave equation (Figure 1.4.1).

The colleague was Erwin Schrödinger. The Schrödinger equation,[56] published in 1926, has become an indispensable tool in nearly all present-day applications of

Figure 1.4.1
Schrödinger's equation

This is one of the master equations of 20th century physics. The upper equation is the time-dependent, the lower one the stationary form.

quantum mechanics. Among its solutions are the eigenvalues that describe discrete energy levels – an elegant answer to the wave-particle dualism of matter and radiation.

At about the same time we find Werner Heisenberg at Göttingen University in the congenial atmosphere built up among mathematicians and physicists around David Hilbert, Max Born and Niels Bohr, who regularly visited from Copenhagen, for meetings that the students called the *Bohrfestspiele* (Bohr festivals).[57] The main actors in these "festivals", however, were young physicists in their 20s, such as Werner Heisenberg, Pascual Jordan, Friedrich Hund, Wolfgang Pauli, Paul Dirac and many other celebrities of 20th-century physics.

The Bohr festivals owed their existence to the generous donation of the Wolfskehl prize in 1905, consisting of 100,000 goldmarks, quite a respectable sum for pre-World-War-I times. It had been endowed by the mathematician Paul Wolfskehl with the proviso that it be used as prize money for the first complete proof of what is called "Fermat's Last Theorem". Fermat had stated that the equation $a^n + b^n = c^n$ has no non-trivial solution for integers a, b and c, if n is a positive integer greater than or equal to 3. The prize committee under the chairmanship of David Hilbert, knowing that a solution of the problem was not around the corner, very wisely used the interest on the money to finance meetings such as the Bohr festivals. Unfortunately, not much of the money survived the two world wars. In 1948 the sum had shrunk to DM 7,500, but it eventually managed to recover to DM 70,000. The prize was awarded by the Göttingen Academy of Sciences in 1996 to Andrew Wiles of Princeton University.[58]

As mentioned, the actors at the *Bohrfestspiele* in Göttingen – apart from the seniors Niels Bohr and Max Born – were young physicists in their 20s. Older colleagues referred to their work disparagingly as "Knabenphysik" (boy's physics). Formally, their approach differed quite distinctly from the wave picture, although later the two approaches were shown to be entirely equivalent.

Heisenberg's idea was to develop an algebraic scheme in which only "observables" appear. A key element of his theory is the "commutation relation" applied to the positional variable q and the momentum variable p, recorded in the upper line on the blackboard in Figure 1.4.2.

In his first paper in 1925, Heisenberg[59] had used it as a multiplication rule for two Fourier amplitudes. Max Born, to whom Heisenberg had shown his paper, was at first puzzled by this strange multiplication rule, but he then remembered that he had encountered it as a student in a lecture on linear algebra and identified it as a matrix multiplication. So Heisenberg's "reinvention" of matrix calculus (apparently previously unknown to him) revolutionised physics in a way similar to that in which Einstein's "reinvention" of Riemannian geometry had done 10 years earlier.

Max Born, together with his assistant Pascual Jordan, then reformulated Heisenberg's expressions in terms of matrix calculus. The new "matrix mechanics" was published in the famous paper of 1926, known as the *Dreimännerarbeit* (three-man paper) of Born, Heisenberg and Jordan.[60]

Figure 1.4.2

Heisenberg's master equations of "matrix mechanics"

The essence of Heisenberg's quantum mechanics is the commutation operator: [A, B] = AB–BA. If A and B are just numbers the commutator is zero, because AB = BA. In quantum mechanics A and B are matrices, the components of which are attributed to observables for which the equality sign does not necessarily hold. In Heisenberg's relation A and B are conjugate variables, such as momentum and position of a particle, where the operator AB – BA equals $\hbar/i$.

Heisenberg's uncertainty relation, for conjugate variables, such as E and t in the lower equation, is a consequence of such a non-commutability. Schrödinger's equation (Figure 1.4.1) instead uses differential operators which lead to the same general results as Heisenberg's method and in many cases are more practical to apply. Although Schrödinger's paper came out a little later than Heisenberg's, the two works originated independently. Both approaches are nowadays called quantum mechanics.

A matrix is a "table" of numbers a_{ik} where the indices i and k refer to the horizontal and vertical "coordinates" (rows and columns) of the table in which a_{ik} is not necessarily equal to a_{ki}. As a consequence, the product of two matrices A and B is not necessarily commutative, i.e. $AB \neq BA$ (see Appendix A1.1). This turned out to be the case for the so-called conjugate variables q (position) and p (momentum). In classical mechanics the Irish mathematician and astronomer William Rowan Hamilton[61] had derived an expression for the total energy in terms of these two conjugate canonical variables. The Hamiltonian is then to be written as an operator (as which it also applies to Schrödinger's equation) in order to obtain the corresponding matrix expressions, for which the commutative rule $PQ-QP = \hbar/i$ holds, in which i denotes the imaginary unit $i = \sqrt{-1}$.

The famous uncertainty relation of quantum mechanics, written for the conjugate variables E (energy) and t (time) in the second line on the blackboard in Figure 1.4.2, was published by Heisenberg in 1927,[62] and is a direct consequence of the commutation rule. Heisenberg first derived it in a rigorous way for the two variables p and q. This rule perhaps answers most clearly the question: Where does the duality of quantum and wave behaviour, which was most obvious in the eigenvalues of Schrödinger's wave equation, show up in Heisenberg's theory? Let me tell a little story which relates to it.

Being a student of physics at Göttingen (from 1945 to 1950) I attended Heisenberg's lectures on statistical thermodynamics and quantum mechanics. Heisenberg was at that time an honorary professor at Göttingen University. I still remember vividly how he introduced quantum-mechanical uncertainty early in his lectures. He said: "You only have to be moderately musical in order to be able to understand this relation. If you want to identify a tone you have to listen to it at least for a time span Δt which is equivalent to half of the period of the underlying oscillation of frequency ν, which in turn is the reciprocal of the time needed for a half period of oscillation, requiring $2\pi\Delta\nu \cdot \Delta t \geq 1/2$ if $\Delta\nu$ denotes the uncertainty of frequency. Now multiply both sides of the (weak) inequality by Planck's constant and you get $\Delta E \times \Delta t \geq \hbar/2$, which is almost the uncertainty relationship of quantum mechanics." And he added: "If you happen to play the violin and you have trouble with pitch, then just play as fast as you can".

Of course, the above derivation is pseudomathematics and Heisenberg had only been joking – but with a grain of physical truth. He himself was a pianist, and at that time I didn't know that years later I would play Mozart's piano sonatas for four hands together with him. We then agreed that the uncertainty relationship doesn't help very much for the piano because its tones are "quantised" already, the smallest $\Delta\nu$ being a semitone, corresponding to an "uncertainty" of almost 8% in frequency. I doubt whether any pianist can play that fast.

Heisenberg's lectures were very persuasive, like the melodies of the Pied Piper of Hamlin. The professor of theoretical physics of Göttingen University, Richard Becker – like Heisenberg, an early student of Sommerfeld in Munich – attended Heisenberg's 1946 lectures on quantum mechanics. I can still see him leaving the lecture hall, shaking his head, but when he returned the next time, he said: "Very well, he was right." I attended Becker's own lectures, too. They were much harder to understand, but it was a lot easier to do the problems afterwards than after Heisenberg's lectures. At the end of the semester Heisenberg made an excursion to the surroundings of Göttingen with all who survived the course. There he asked me what my private interests were. He was very satisfied when I said "music"; but when I added "chess", he stopped short and said, you must not do that, either physics or chess – to do both is a waste of time. I later found out that this was exactly what Sommerfeld had said to young Heisenberg before accepting him as a student.

The two approaches to quantum mechanics – which now is the common term for both of them – first looked quite disparate and were, I think, originally intended by their authors to do so. However, they turned out to be entirely equivalent. Do their results represent some "ultimate truth"?

Well, quantum mechanics – so far at least – has never failed. It is the theoretical basis of chemistry (see Section 1.5) and it provides the precise framework for quantum electrodynamics and its analogues for strong and weak nuclear forces, quantum chromodynamics and quantum flavour dynamics (see Section 1.6). In view of the fact that quantum mechanics is a linear theory – a point emphasised by Steven Weinberg – this is surprising. Of course, as a linear theory it automatically ensures the superposition of wave amplitudes. Weinberg[63] tried to construct a physically meaningful theory that in certain details would represent an alternative to quantum mechanics, but he failed to find one that was logically consistent. It should, however, be added that the linearity of quantum mechanics is traded off against an infinite-dimensional function space in which the non-linearities of mechanics may be hidden.

In response to the question in the title of this section, and before I come to different interpretations, I want to stress the experimental evidence. Observations of the diffraction of material waves include the early experiments with electrons by George Paget Thomson and Alexander Reid in England,[64] and those by Clinton Joseph Davisson and Lester Halbert Germer in the USA,[65] both performed in 1927 and recognised by the 1937 Nobel prize for physics.

The most obvious manifestation of the fundamental nature of the uncertainty principle is provided by the existence of atoms, with a positively charged nucleus of a size of about 10^{-15} m and a negatively charged cloud of electrons that extends over distances of about 10^{-10} m. As an example, take the hydrogen atom. According to the classical picture, the electron should radiate away its energy and plunge into the nucleus (and would do so within 10^{-11} s). Why does it not do this? The answer lies in the uncertainty principle. If the electron were absorbed by the nucleus, then it would be confined to a space of only 10^{-15} m diameter. Quantum-mechanical uncertainty would then require a huge momentum (the equivalent of a γ quantum with 10–50 million times the frequency of an ultraviolet quantum), which in turn would cause the hydrogen atom to break apart. As we shall see later (Section 1.7) gravitational forces can provide such stupendous pressures as to cause a collapse of all these scaffoldings, including those of the nucleus and its constituents.

Let me supplement these deliberations with some more quantitative data. The lifetime of any energy state is related to its natural spread of energy by the uncertainty principle. For instance, the lifetime of an excited state can be determined from the width of its corresponding emission line in the fluorescence spectrum. Applications will be encountered in different sections of this book. The same is true in particle physics. The lifetime of an unstable particle is correlated with its natural energy spread, where "natural" means that the line shape is determined solely by the uncertainty principle and not by superimposed processes.

The US physicist Gregory Breit and the Nobel laureate Eugene Wigner,[66] using S-matrix theory (the quantum mechanical theory of scattering processes, the early part of which goes back to Heisenberg's "Streumatrix"), have derived an expression which correlated the width of the cross-section of the particle reaction (Breit–Wigner

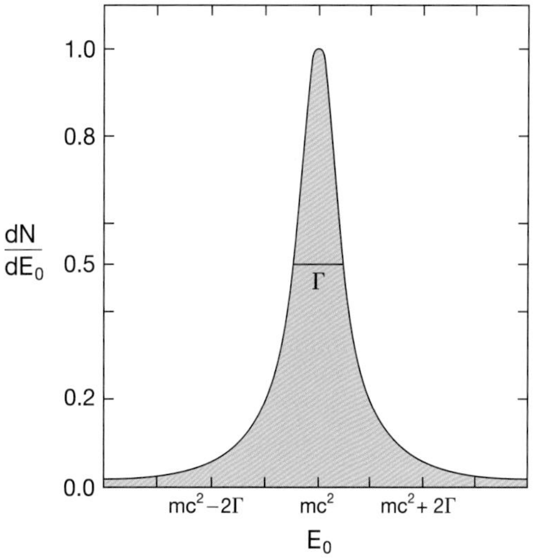

Figure 1.4.3

Breit–Wigner curve

For the distribution dN/dE_0 of mass-energy E_0 of an unstable particle. Γ is representative of the "natural line shape" of a particle.

resonance) with the lifetime of the (unstable) particle. Instead of reproducing the formula, I show in Figure 1.4.3 the energy distribution of the Breit–Wigner resonance line.

Near its maximum it is bell-shaped, like a Gaussian, but its tails approach the zero line more slowly, as made visible by the coloured shading. Γ is the width of the curve when it has decayed to half its maximum value. It is related to the uncertainty principle by $\Gamma\tau \approx \hbar$, τ being the natural lifetime of the unstable particle. Γ is called the "natural" width in order to distinguish it from line-broadening by superimposed processes.

Figure 1.4.4 shows a log–log plot of Γ (eV) versus τ (s). The experimental values included are a representative choice from many more data available.

Table 1.4.1 presents a glossary of particle names, which may assist the reader in identifying the particles in Figure 1.4.4.[24] What immediately strikes us is the fact that all the values fit a curve that extends over 50 orders of magnitude. However, the amazing-looking agreement between theory and experiment is only an apparent one. It is true that all points on the curve are based on experimental data. However, these refer to either Γ or τ, while the complementary value of either τ or Γ is calculated from the uncertainty relation $\Gamma\tau = \hbar$. Hence, the values included have no choice but to fit this relation. In particular, the α-decay of the radioactive ^{232}Th and the β-decay of the neutron, as well as the disintegration of the muon and of the pions, refer to measured decay times, while in the case of the mesons, ϕ and ρ, as well as that of the boson of weak interaction Z^0, it was the natural line width which offered itself to direct experimental detection (see below).

Despite this fact, the close fit to the curves for measuring Γ or τ in Figure 1.4.4 yields a consistent picture of particle physics and is an excellent demonstration of

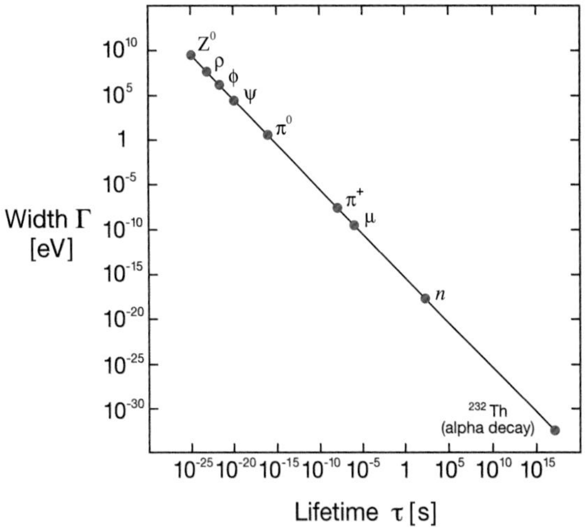

Figure 1.4.4
Particle width Γ as related to lifetime τ for decaying particles in a broad range

The relationship is governed by the uncertainty principle $\Gamma \times \tau = \hbar$.

Table 1.4.1

^{232}Th	Thorium^{232}	1.4×10^{10} years	Isotope of Thorium Decay by α-particle emission
n	Neutron	889 seconds	Baryon
μ	Muon	2.2×10^{-6} sec	Heavier sibling of electron
$\pi+$	Pion	2.6×10^{-8} sec	Meson*
π^0	Pion	9×10^{-17} sec	Meson*
ψ	Psi-particle	1×10^{-20} sec	Forming from $e^+ + e^-$
ϕ	Meson	1.6×10^{-22} sec	Meson (decay: $K^+ + K^-$)
ρ	Meson	4.3×10^{-24} sec	Meson (decay: $\pi^+ + \pi^-$)
Z^0	Z^0-Particle	2.6×10^{-25} sec	Neutral weak boson (weak interaction of quarks with leptons)

* Mesons involve quark-antiquark binding.

Only the two first examples show life times which could be determined by watching the decay process by our sensory organs. All other cases show existence at or below a millionth of a second. The isotope "232" of Thorium, on the other hand, with its lifetime exceeding ten thousand million years does not show any "visible" change.

the far-reaching consequences of the quantum-mechanical principle of uncertainty. First of all, the width Γ is not related in any simple way to the total mass-energy of the decaying particle. The neutron, for instance, has a mass-energy of nearly 1 GeV and, as was shown in Section 1.1, the products take up nearly all of this, so that only something like 10^{-18} eV, not accessible to direct detection, remains for the ΔE of the rate-limiting step. Similar disparities are seen for the muon and the pion.

On the other hand, for Z^0 the natural line width Γ reaches a noticeable fraction (i.e. 2.49 GeV) of the boson's mass energy of 91.17 GeV. In every case, the width Γ is determined by the energy of something like the "transition state" of the decay reaction and therefore depends strongly on the reaction path and the products of the reaction, including their energy distribution. This explains why in most cases only one of the two parameters is accessible to measurement, e.g. for the neutron the lifetime of 900 s and for the Z^0 boson only the line width of $\Gamma = 2.49$ GeV. Its lifetime of 2.6×10^{-25} s is not detectable by any present-day standard means, although – using tricks of all sorts – lifetimes as short as 9×10^{-17} s (π^0) have been determined.

For the latter example I show the experimental data for the total width Γ of Z^0-decay (Figure 1.4.5) and the contributions of different competing decay mechanisms to Γ (see the Breit–Wigner curve in Figure 1.4.3). These studies were done with the LEP (large electron–proton collider) machine at CERN in Geneva, Switzerland. The studies allowed a decision for the assignment to mechanisms in which three families of neutrinos (e, μ and τ), no more and no less, occur (see Figure 1.4.5 and consult Sections 1.5 and 1.6).

The amazing success of quantum mechanics in explaining the experimental facts of contrasts with our inability to visualise the underlying physical situation that is the cause of the phenomenon we observe. The original question was: How can something be spatially confined, like a particle, and at the same time non-confined, like a wave? How can one explain uncertainty or the intrinsically probabilistic nature of all quantum processes which as such are deterministic? Explanation is based on experience; therefore explaining something is to a large extent a matter of getting used to it. Our brain is an adaptive organ and has no *a priori* capability of recognising the truth. Here I return to the title of this section: Do we understand quantum mechanics?

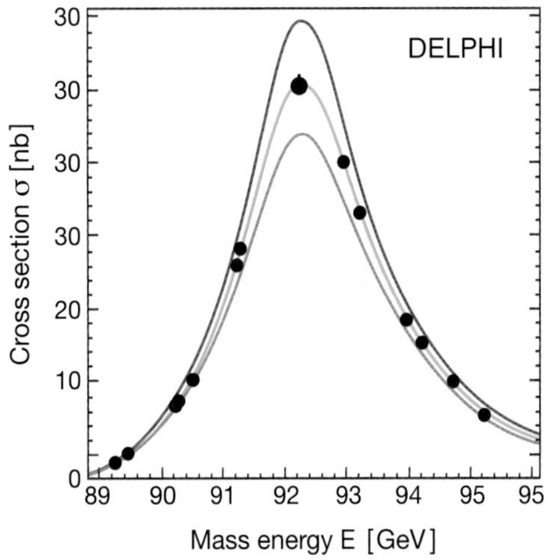

Figure 1.4.5

Measurement of the electroweak parameters for Z-decays into Fermion pairs.

The data were obtained by the DELPHI collaboration at the CERN laboratories in Geneva. The three curves represent the numbers of existing neutrino families: red = two neutrinos, green = three neutrinos, blue = four neutrinos. The data for the Z^0 line width show that there are three and only three families of neutrinos. (Courtesy of Gerald Eigen.)

One attempt to do this was the so-called Copenhagen interpretation, which saw the two non-congruent aspects of reality as being connected through their intrinsically dual character. This view was summed up in the term "complementarity", first applied to this problem by Niels Bohr. For a long time many physicists – and even today there are quite a number – were appalled by this notion that uncertainty is resolved (whereby probability amplitudes become "destroyed") through interference by an (intelligent?) observer. Schrödinger did not like the dominant role which the Copenhagen interpretation assigns to the observer. He expressed his dislike by a parody, the famous "Schrödinger cat". In a *Gedankenexperiment* he shut a cat into a box that contained a device capable of delivering a poisonous gas when triggered by an uncertain quantum event. Once the box is closed, the state of the cat (dead or alive) is therefore described by probability amplitudes – in other words, the poor animal is both alive and dead – until the box is opened, i.e. until the "as well as" is destroyed by interference through the observer. I think that this famous experiment was meant to ridicule the anthropomorphic Copenhagen view; yet for some physicists it became a subject for deep discussion, while in others it aroused equally deep feelings of consternation.

The true problem is how to understand the peculiar nature of the quantum-mechanical probability, in contrast to its classical analogue, to which the physicists had been accustomed since the end of the 19th century. Classical probability always refers to some lack of knowledge. The phenomenon *per se* is not uncertain; we merely lack sufficient information about a reality that is otherwise completely determinate. Einstein, together with his coworkers Boris Podolski and Nathan Rosen, proposed another *Gedankenexperiment*, usually referred to as the EPR experiment, which addresses this problem and which was meant to prove the incompleteness of quantum mechanics.[67] Einstein did not doubt the probabilistic nature of quantum theory, but he also refused to acknowledge any interpretation other than the classical one, involving a local and determinate process. A probabilistic interpretation would then require the existence of some further "hidden" variables.

The EPR experiment is very lucidly described by Abraham Pais in his biography of Einstein.[68] The gist is as follows: Two particles are initially in close interaction, and their total momentum and relative distance are known. They move apart and then observation is restricted to only one of the two particles. Knowledge of the momentum of one particle would imply a knowledge of the momentum of the other one, supposing that it has been kept away from interactions that destroy the coherence. Physicists nowadays call both undisturbed particles, or quanta, "entangled". Quantum mechanics emerged as an intrinsically non-local theory.

Einstein was not satisfied with any such interpretation. In his rebuttal, Niels Bohr was unable to convince him with his arguments that experimental procedures cannot provide any classical explanation of the phenomenon of complementarity as introduced by quantum mechanics. The (friendly) quarrel between Bohr and Einstein about "objective reality" would have remained a historical episode if it had not become possible to convert the *Gedankenexperiment* into an objective experimental test, using both light quanta and material particles. In the latter case, one may utilise the pair production of matter and antimatter particles and their mutual annihilation. This type of experiment was pioneered by (Madame) Chien-Shiung Wu and her coworkers[69] and has since been repeated in various forms in several laboratories around the world.

In Figure 1.4.6 I show such an experiment described in the thesis of my son Gerald Eigen. The experiment was carried out in the 1970s jointly with a fellow student,

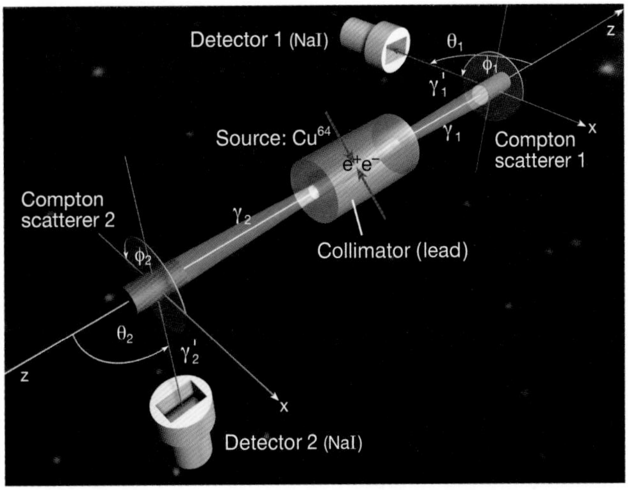

Figure 1.4.6

Experiment for testing the EPR hypothesis

The objective of the experiment is a measurement of the polarisation correlation of two photons (γ_1 and γ_2, each 511 keV) produced in the annihilation of an electron–positron pair at rest. The positron is produced by a $^{64}Cu\beta^+$-decay, located in the centre of a lead collimator (centre of the picture). The polarisation of each photon is measured by Compton scattering (see Section 1.3; Figures 1.3.7 and 1.3.8). What is being measured is the energy and azimuth angle of the scattered photons (γ_1 and γ_2) by sodium iodide detectors (NaI scintillating crystal + photomultipliers). The scattering occurs in a plastic scintillator (red). The energy of the recoiled electrons is measured by photomultipliers attached at the ends of the two (red) plastic scatterers (not shown in this schematic representation). Optimum detection of polarisation efficiency occurs at angles (Θ_1, Θ_2) of 82°. The connecting rate (four-fold coincidence of the signals at the two scatterers and the two NaI detectors) is measured as a function of the difference between angles ϕ_1 and ϕ_2 (see Figure 1.4.9).
(Courtesy of Gerald Eigen. Figure drawn by Claus-Peter Adam.)

Klaus Meisenheimer, in the laboratory of Professor Kai Runge at the University of Freiburg (Germany). Its purpose was to clarify inconsistencies of earlier experiments, in particular results which contradicted the non-locality suggested by the quantum mechanical interpretation.

The figure shows a schematic representation of the method used by the two students. It is based on measuring the polarisation of the 511-keV photons produced in the annihilation of a positron and an electron, using Compton scattering (see Section 1.3) as means of detection. What is of interest in the present context are the results, an example of which is presented in Figure 1.4.7. All measurements clearly turned out to be in favour of quantum mechanics describing the non-local behaviour of the entangled photons resulting from annihilation and detected by coincidence measurements. The peak of the experimental curve shown in Figure 1.4.7 according to classical theory could have reached half its value. The main conclusion, however,

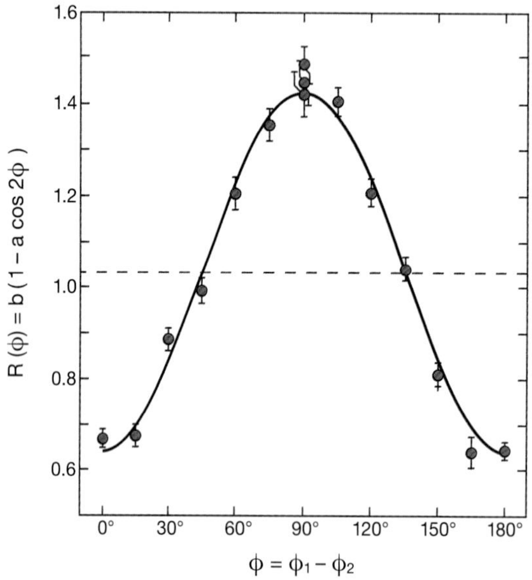

Figure 1.4.7

Normalised counting rate $R(\phi)$ as a function of the azimuthal angle difference $\phi = \phi_1 - \phi_2$.

(ϕ) is a combination of coincidences $(N_4 \cdot N_2)/(N_{31} \cdot N_{32})$, where the fourfold coincidence (N_4) is the decisive quantity to be detected. The appearance of two-fold (N_2) and three-fold $(N_{31}$ and $N_{32})$ coincidences in $R(\phi)$ is for the purpose of normalisation, in order to reduce systematic uncertainties. The red dots are the measured values, while the black curve represents a maximum-likelihood fit of the data, which is in excellent agreement with the quantum-mechanical prediction. The classical prediction yields a cosine function, the amplitude of which reaches only half of that of the measured curve.
(Courtesy of Gerald Eigen.)

came from a precise simulation of the experiment with the help of a Monte Carlo program which allowed a direct comparison of the measured energy distribution and the coincidence rates of the scattered photons with their expectation values. In this way it was possible to identify the errors that had troubled earlier experiments and led to the above-mentioned inconsistencies with quantum-mechanical theory.

Another outcome was the demonstration of the requirement of accuracy in order to allow the experimental checking of Bell's criteria for excluding the existence of "hidden variables". Both students had visited Bell and discussed their results with him. The late John Stuart Bell, who was involved in the build up of the CERN laboratories at Geneva/Switzerland has first quantified the criteria bearing his name in 1964. Since the 1980s, experiments with increased accuracy, such as the famous one by Alain Aspect[70] and his coworkers at Orsay in France, have received much attention. On the basis of Bell's criteria they all clearly speak against a local theory involving any hidden variables.

Today, experiments with material particles are almost routine. World champions in demonstrating the wave nature of larger material particles are Anton Zeilinger at the University of Vienna, Austria, and Jan Peter Toennies at the Max Planck Institute for Dynamics and Self-organisation at Göttingen in Germany. Zeilinger's record[71] involves the "bucky ball" particle of 60 carbon atoms, being 720 times as heavy as a hydrogen atom, while Toennies' masterpiece[72] concerns a particle of large extension, the complex between a bosonic ^{4}He- and a fermionic ^{3}He-atom which forms at temperatures of liquid helium. Because of the very low energy of interaction one finds a diameter of this complex in the range of 50 Å.

In the interpretation of quantum mechanics, modern schools have departed far from the one proposed by the Copenhagen school. Our world was operating quantum mechanically long before human observers appeared. Hugh Everett[73] came up with his "many worlds" interpretation, i.e. the parallel existence of numerous worlds, as many as are compatible with the fine-grained quantum-mechanical structure. If these many worlds cannot communicate with one another, we just happen to be in the one world which allows our existence. Richard Feynman modified this (somewhat bizarre and still rather anthropocentric) picture to one of many *potential* alternative histories.[74] James Hartl and Murray Gell-Mann considered the resulting history as having loss of coherence from the fine-grained quantum-mechanical world of probabilities into the coarse-grained quasi-classical structure we experience.[40] Losing coherence does not require an intelligent observer. Schrödinger's cat would die the moment the quantum phenomenon lost its coherence at the device triggering the macroscopic events and we, if we don't want to open the box, would smell it. A light quantum loses coherence the moment it is absorbed by a chromophore molecule which, in photosynthesis, uses this energy to trigger the formation of a sugar molecule from carbon dioxide and water. It does not wait for the human child who experiences the particular sugar molecule as a constituent of a sweet. Gell-Mann comments on the Copenhagen interpretation with the words: "Those of us working to construct the modern interpretation of quantum mechanics aim to bring to an end the era in which Niels Bohr's remark applies: 'If someone says that he can think about quantum physics without becoming dizzy, that shows only that he has not understood anything whatever about it.' "

For me, the replacement of an interfering (intelligent) observer by a mere entanglement with external events – that is to say, a disentanglement by "drowning" in a "sea" of wave functions – sounds more convincing. The "hard work" on the modern interpretation of quantum mechanics involves – besides Hartle and Gell-Mann – various names of the younger generation, such as Dieter Zeh and Wojtek Zurek, all quoted by John Wheeler.[75] Background decoherence is not just a way of making things plausible. However, it does not remove the strangeness of quantum-mechanical entanglement. Acting over large distances, it is something that defies our imagination, no less than it did Einstein's. How can one of the particles instantaneously "know" what happens to its partner although it is separated by a large distance? We simply have to acknowledge the experimental fact, which is so real that Charles Bennett bases exciting expectations for its technical realisation as a tool of "quantum telecommunication" upon it.

Theory *per se* cannot claim as much divinity as the first two syllables of the word (coincidentally) may suggest. It has to pay due respect to experimental facts. So far, quantum mechanics has never failed, and this – more than its internal consistency as a theory – is responsible for its claim to represent a final tenet of physics. Our brain is an adaptive organ which is able to filter out inconsistencies. What remains, after selective fixation, is what we call "understanding". The rest – as Shakespeare tells us – is "fantasies that apprehend more than cool reason ever comprehends".

1.5. What Are We to Call Ultimately Elementary?

The word element (from the Latin *elementum*, primary matter; German *Urstoff*) – as the chemist uses it – does not live up to original expectation. It shares this fate with the word "atom" (from the Greek *atomos*, indivisible). Chemical elements are represented physically by atoms, which can be split in various ways, such as by stripping off the electrons from the shell of the nucleus, which itself – with some more expenditure of energy – can be split into fractions down to single protons and neutrons. Explaining the highly ordered structure of the atom as determined by the electromagnetic interaction between protons in the nucleus and electrons in the shell – and especially among the electrons within the shell – required the full exploitation of quantum mechanics.

Yet the periodic order of the elements was recognised on empirical grounds long before the science of quantum mechanics came into existence. The German chemist Julius Lothar Meyer in 1864 wrote a book with the title *Modern Chemical Theory* in which he proposed a scheme for the arrangement of the chemical elements in relation to their atomic weights. At about the same time the Russian chemist Dimitri Ivanovich Mendeleyev came up with similar ideas from which he condensed his "final version" of the periodic table of the elements in 1871. He correctly left gaps for those elements that were not yet discovered. For some of these he even predicted their properties. Yet the discovery of the noble gases required the addition of another column in the table. Argon was found in 1894 by William Ramsay and Lord Rayleigh, followed by the isolation of neon, krypton and xenon in 1898 in Ramsay's laboratory. Also helium, although seen spectroscopically in the sun's atmosphere as early as 1865, was identified in the laboratory only in 1895, again by Ramsay, and it took him until 1910 to find the last member of the series, now called radon. In Figure 1.5.1 a present-day version of the table of elements is shown, in a representation that differs from the one usually found in textbooks of chemistry. It reflects more closely the theoretical interpretation based on the shell structure obtained from an independent-particle approximation combined with Pauli's exclusion principle. Table 1.5.1 contains the names of the elements 89 to 112, most of which were identified and named in recent years.

The independent-particle approximation is analogous to the picture of independent planets orbiting the sun, disregarding gravitational interactions among the planets. This approximation, of course, is much better in celestial than it is in atomic mechanics. The sun's mass exceeds by far the masses of the planets, whose mutual gravitational interaction is almost negligible. In contrast, what matters in an atom are the electric charges of nucleus and electrons, whose pairwise interactions are all of the same order of magnitude. This approximation, supplemented by Pauli's principle, leads to a shell model that is quite well substantiated by experimental results. The representation in Figure 1.5.1 refers to a shell structure as imposed by the main quantum numbers n (defining the shell) and the orbital angular momentum quantum number l, which runs from 0 to $n - 1$. The Pauli principle limits the maximum number of electrons that have a given main quantum number n to $2n^2$ (cf. the number of elements in each horizontal row). However, the instability of elements with very large atomic numbers limits the relation to s, p, d and f subshells. Each column in Figure 1.5.1 refers to a value of l. The order of filling the states is indicated in Figure 1.5.2 (note the exceptions). The main difference with respect to the representation favoured in chemistry textbooks is that the latter stress the similarities between outer electron configurations, which determine the chemical nature of each element: columns represent elements with corresponding chemical properties.

In this section I shall not comment further on atomic structure and its role in the theoretical foundation of chemistry, which now is textbook knowledge.[76] However, I must not neglect to mention three great pioneers of modern atomic theory who have provided the physical basis of chemistry: Friedrich Hund, Robert Saunders Mulliken and Linus Carl Pauling.

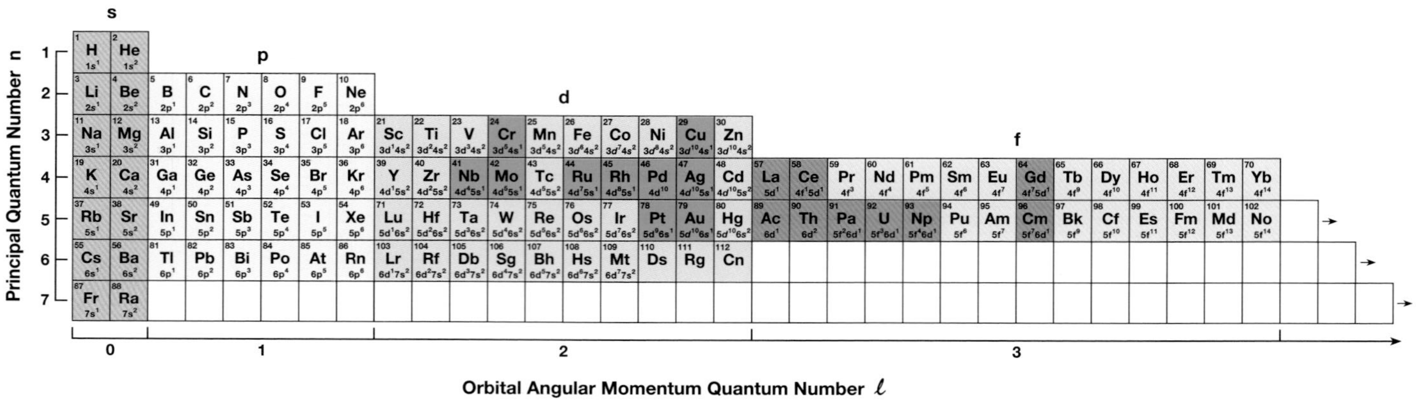

Figure 1.5.1

The periodic system of chemical elements

The energy state of an electron in the atomic shell depends on both its interaction with the nucleus and its interactions with other electrons in the shell. Four quantum numbers – the principal quantum number n, the angular momentum quantum number l and the two magnetic quantum numbers m_l (orbital) and m_s (spin) – determine the wave function of an electron. The Pauli principle forbids any two electrons in an atom to have identical sets of these quantum numbers.

The "independent-particle approximation" assumes that each electron in a multi-electron atom may be described by a potential function intermediate to a Coulomb potential which for large distance corresponds to that of a hydrogen atom (nuclear charge +1) and which at short distance approaches the Coulombic potential of a single electron with a nucleus of charge +Z. The "identical-particle approximation" together with Pauli's principle, then results in a "shell structure", dominated by the principal quantum number n, defining the "shell" (K, L, M, N for quantum numbers n = 1, 2, 3, 4), and the orbital momentum quantum number l, defining subshells (s, p, d and f electrons for l = 0, 1, 2, 3). This yields for n = 1 two s states with l = 0, for n = 2 eight states, i.e. 2 s and 6 p states, for n = 3 the two s and 6 p states plus ten d states, i.e. a total of 18 states, and for n = 4 the 18 states (as for n = 3) plus 14 f states. The maximum number of states for any value of n is $2n^2$, which equals the trivial series $2\sum_{l=0}^{n-1}(2l+1)$. The above representation of the periodic table of the elements is based on this relation. It emphasises the logical build-up of atomic shells.

The usual representation of this table in chemistry textbooks stresses more the outer electron configurations, which determine the chemical reactivity. In this representation each horizontal row corresponds to a shell (K, L, M, N...), while each vertical column refers to a given subshell defined by the l value. Each box has the atomic number Z in its upper corner. The shaded boxes indicate irregularities in the filling order, such as those filling an energetically close s or a d instead of an f state, as it were to be expected according to the serial order. Limits of the approximate model are also expressed by the empty boxes found for $n \geq 5$. These states are allowed by the model, but not possessed by any known element. The large density of electrons makes them extremely unstable. Table 1.5.1 provides the names of the elements 89–112.

Table 1.5.1 The Elements 89 to 112

Name	Symbol	Atomic Number
Actinium	Ac	89
Thorium	Th	90
Protactinium	Pa	91
Uranium	U	92
Neptunium	Np	93
Plutonium	Pu	94
Americium	Am	95
Curium	Cm	96
Berkelium	Bk	97
Californium	Cf	98
Einsteinium	Es	99
Fermium	Fm	100
Mendelevium	Md	101
Nobelium	No	102
Lawrencium	Lr	103
Rutherfordium	Rf	104
Dubnium	Db	105
Seaborgium	Sg	106
Bohrium	Bh	107
Hassium	Hs	108
Meitnerium	Mt	109
Darmstadtium	Ds	110
Roentgenium	Rg	111
Copernicium	Cn	112

Let me return to my question: "What is ultimately elementary?" In order to reach the next higher level of resolution we need another 100,000-fold "magnification"; the nucleus has a diameter of the order of magnitude of 10^{-15} m. It thus fills only the 10^{-14}th to 10^{-15}th part of the volume of the atom. In an examination a student expressed this in the words "atoms consist for the most part of air" (*sic*). We know already from the structure of atoms that their nuclei must be composite rather than elementary. Their building blocks, the nucleons, are protons and neutrons. The proton is positively charged while the neutron, as its name suggests, carries no charge. The neutron (940 MeV) is slightly heavier than the proton (938 MeV; note that I express the masses as their relativistic equivalent of energy, a common practice in particle physics). The isolated proton appears to be a very stable particle. Experiments suggest that it has a lifetime longer than 10^{33} years. The free neutron, on

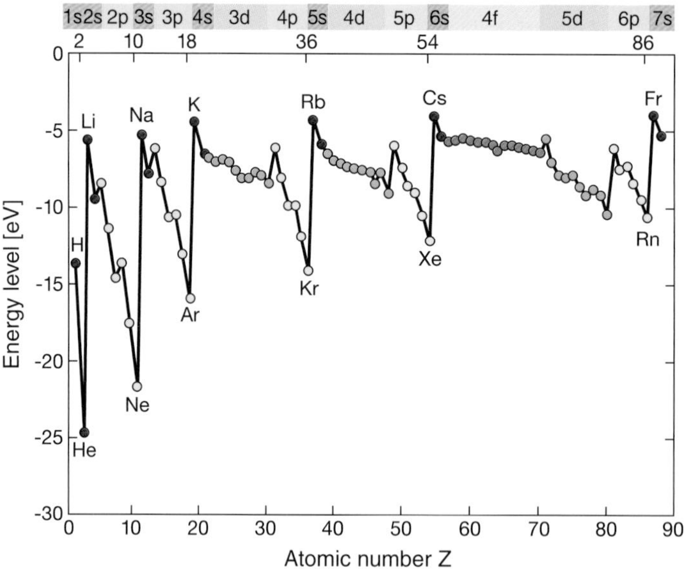

Figure 1.5.2

Energy levels of the highest-energy electron in the neutral atom as a function of atomic number Z

Note the repeating regularity which characterises the noble gases (filled p subshell) and alkali metals (one single electron in the s state) by extreme energy values.

the other hand, is rather short-lived. It decays with a half-time of only about 1000 s into a proton and an electron, at the same time producing – as we saw in Section 1.1 – a neutrino or, to be precise (see below), an antineutrino.

Inside the nucleus the neutron is stable; but how are protons and neutrons held together? There are no net electrostatic forces to keep them together; in fact there is strong repulsion between the protons in all nuclei larger than the isotopes of hydrogen. Here we encounter a new stabilising force, the "strong nuclear interaction". Unlike electrostatic or gravitational force, it does not extend out into space, but rather is entirely confined to the dimensions of the nucleus. It is many times stronger than the electrostatic force, and thus prevents nuclei that contain several protons from flying apart.

The property that distinguishes a proton from a neutron in the realm of strong interactions plays – in a formal mathematical treatment – a role similar to that of spin in electromagnetic interactions. It is therefore called "isospin" (Heisenberg, 1932),[77] and each nucleon is assigned a corresponding quantum number, in addition to its orbital and spin quantum numbers that refer to angular momentum. Apart from isospin, protons and neutrons are entirely equivalent in their strong interactions and therefore assigned to the same class.

Two models have been proposed to explain the structure of the nucleus. They represent – so to speak – limiting cases, and a combination of them might come closer to reality.

At first – in the early 1930s – the nucleus looked to physicists like a drop of liquid (although there was not yet really a good theory of liquids in existence). Carl Friedrich von Weizsäcker[78] came up with an empirical formula just by summing the terms referring to interactions that are expected to be involved. The main contribution was assumed to result from strong forces among the nucleons mimicking the Van der Waals forces active in a true liquid of neutral particles. (Van der Waals forces are of quite different, i.e. electromagnetic, nature, and are many orders of magnitude weaker than nuclear forces. However, what is of importance in this picture is that their range of action is restricted to small distances.) This contribution is proportional to the number of nucleons (i.e. to the atomic mass number A), corrected by a surface term that takes into account the fact that particles at the surface have fewer neighbours. Furthermore a Coulomb term that considered the repulsion between protons was added, along with several terms that took account of Pauli's exclusion principle, depending on a balance between even and odd numbers of protons and neutrons. The Weizsäcker formula could be adjusted numerically so as to yield quite satisfactory values for the binding energies that stabilise the various nuclei.

A more specific model considers the nucleus as being structured in a fashion similar to that of the electron shell of an atom. It is based on the observation that nuclei with certain compositions of neutrons and protons are particularly stable. Generally, nuclei with even numbers of neutrons or protons are found to be more stable than those with odd numbers. A particular stability is observed for neutron (or proton) numbers of 2, 8, 20, 28, 50, 82 and 126, referred to as "magic numbers" (Maria Goeppert-Mayer[79]), and this reminds us of the especial chemical stability of the noble gases, with electron numbers of 2, 10, 18, 36, 54 and 86 (cf. Figure 1.5.3). Examples of nuclei in which both neutron and proton numbers are "magic" are ${}^{4}_{2}He$, ${}^{16}_{8}O$, ${}^{40}_{20}Ca$, ${}^{48}_{20}Ca$, ${}^{208}_{82}Pb$ (the upper number being A, i.e. the sum of neutrons and protons, the lower being Z, the number of protons in the nucleus, called the "atomic number").

I remember a visit to Chicago in the mid-1950s, when I met Maria Goeppert-Mayer and her husband Joe. Maria was born in Göttingen, where her father was medical director of the university children's hospital. She had been working on her physics thesis on non-linear optics (two-photon merger) when she met Joe, who, like many colleagues from the USA, was visiting Göttingen to witness the birth of quantum mechanics. Joe, as the founder of "cluster theory" in statistical mechanics, was one of the great contributors to the physics of condensed matter. He was also famous for his Martini cocktails. The glasses in which he served them carried certain

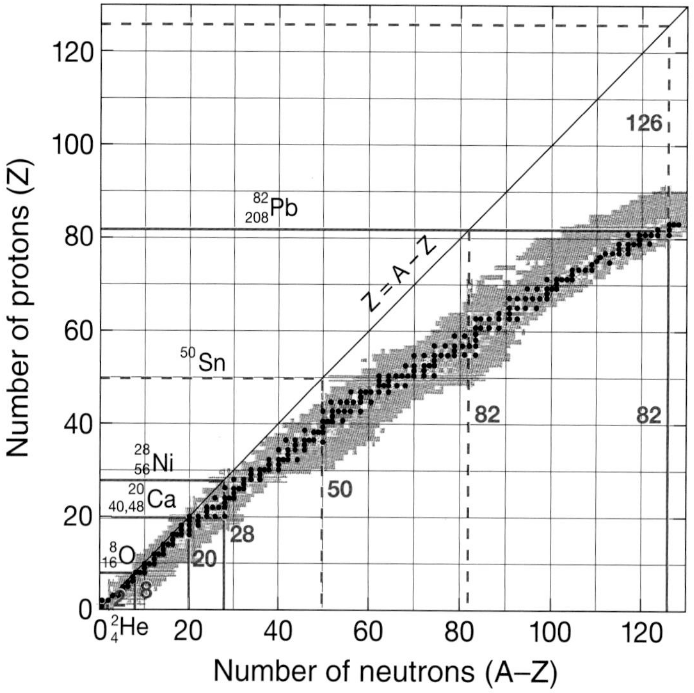

Figure 1.5.3

Number of protons versus number of neutrons in stable nuclei

The number of protons in an atomic nucleus determines its charge and is given by the atomic number, Z. The sum of the numbers of protons and neutrons that make up a given isotope is called the atomic mass number or (better) the nucleon number, A. Hence the abscissa in the diagram is A–Z and the ordinate is Z. The deviation from the straight line is caused by the increasing electrostatic repulsion among protons with increasing Z values.

numbers, the meaning of which was obscure to me at that time. They were the "magic numbers" introduced above, for the discovery of which Maria later shared the 1963 Nobel prize for physics with Hans Daniel Jensen and Eugene Wigner. Other important contributors to the nuclear shell model were Otto Haxel and Hans Suess.[80]

While the liquid-drop model assumes strong coupling among all nucleons, the nuclear shell model, like the electron shell model of an atom, uses independent wave functions for each nucleon. As for the atom, the total angular momentum involving orbital and spin contributions also has to be considered. In this way a nomenclature for energy levels similar to the one used for the atomic counterpart was introduced. It is again Pauli's principle, forbidding any two states to have identical quantum numbers, that appears to be the main source of order in the nuclear shell model.

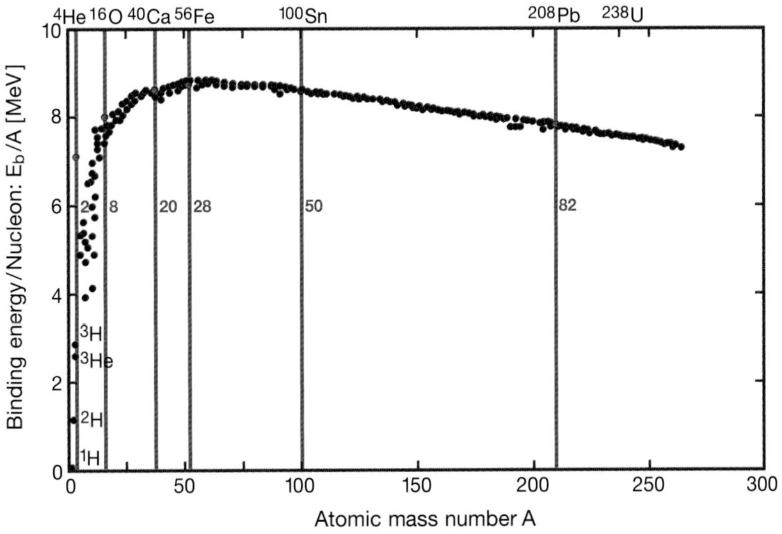

Figure 1.5.4

Binding energy per nucleon, E_b/A, as a function of nucleon number A

Note the flat maximum at ^{56}Fe. For A values < 55 fusion may prevail, while for A values > 56 fission will be the preferred process of instability.

Figure 1.5.4 shows the binding energies per nucleon (corresponding to the negative energy levels in Figure 1.5.3) for the various nuclei.

A comparison of Figures 1.5.3 and 1.5.4 reveals several important features:

1) The energy scale is about a million times larger, i.e. in the MeV range. Since thermal energies at normal temperatures are between a tenth and a hundredth of an eV, radioactive decay cannot be brought about by thermal excitation, but rather is a consequence of quantum-mechanical tunnelling.

2) Appreciable periodic fluctuations are confined to relatively low atomic mass numbers. The hydrogen nucleus $^1_1$H is a proton with a spin of $s = 1/2$. It can combine with a neutron of parallel spin, to form the deuteron $^2_1$H with total spin $S = 1$. The antiparallel state ($S = 0$) is not stable. This is a consequence of the fact that the strong force is mediated by an exchange of gluons with an intrinsic angular momentum $S = 1$. (In a way, this process can be seen as an analogue of that in which electromagnetic forces are mediated by an exchange of photons.) By the same token, states with either two neutrons or two protons do not exist because Pauli's principle would require such states to have $S = 0$. The nucleus $^4_2$He, which is called an α particle, is of particular stability. The α particle was first observed in 1896 by Antoine Henri Becquerel and identified by Ernest

Rutherford. Many nuclei with larger atomic mass numbers decay by emission of α particles. The fact that isolated protons or neutrons are never observed in radioactive decay is a consequence of the high stability of α particles and the comparatively unfavourable energy difference of states that differ by just one nucleon.

3) Nuclei with increasing mass favour an excess of neutrons over protons (as a consequence of the increasing electrostatic repulsion among the protons). Note that the energy curve passes through a flat minimum around the iron nucleus (Fe), meaning that for mass numbers below that of iron only fusion is energetically favoured, while for mass numbers above that of iron it is fission that may occur spontaneously. Among the approximately 560 known states only 254 are stable, the heaviest stable nucleus being ^{209}Bi, with $Z = 83$ protons and $(A–Z) = 126$ neutrons. The other nuclei are radioactive, with half-lives ranging from fractions of a millisecond to gigayears ($1\,\text{Gy} = 10^9$ years). Radioactivity includes β^-- and β^+-decay as well as electron capture. α-decay has been mentioned already. γ radiation is of electromagnetic nature, having wavelengths (usually) smaller than 10^{-12} m. It occurs when a decaying nucleus is left in an excited state. This corresponds to energies in the MeV range. A "periodic table" for energy states in the nuclear shell model is shown in Figure 1.5.5.

Despite the successes of this model, a nuclear shell structure that results from an assignment of independent wave functions to each nucleon represents just an idealised picture. Moreover, motion of the subnuclear particles could deform the

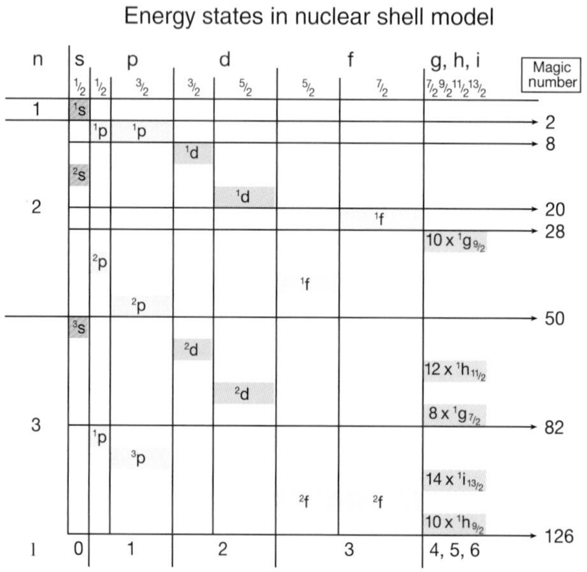

Figure 1.5.5
"Periodic system" of the nuclear shell model

The terminology indicates the analogies to the atomic shell model, but should not be taken too literally.

nucleus, which may be anything but perfectly spherical. With this terse remark on very important experimental and theoretical results[81] by three Nobel laureates of 1975, Aage N. Bohr, Ben R. Mottelson and Leo James Rainwater, I conclude that the atomic nucleus cannot be called "elementary", any more than a chemical element is really element-ary. This brings me to the question: How can we decide experimentally whether a particle is "composite" or "uniform"?

The answer is: probing by particle scattering. The experiments on which this answer is based and which, in fact, led to the first realistic model of an atom, the Bohr model, were the highlight of the life work of Ernest Rutherford. It was in his laboratory that Hans Geiger – who joined Rutherford (then at Manchester) shortly after completion of his doctorate at Erlangen in Germany – built the first particle counter, which was instrumental in the discovery of the atomic nucleus in 1912 by Rutherford, Geiger and Ernest Marsden.[82] The decisive observation was that α particles from radioactive decay, when passing through a thin metallic foil, are sometimes scattered at large angles ($> 90°$). This is possible only when a large force acts on the α particle, and for that the α particle has to approach a charged object within a distance much smaller than the size of the atom. Rutherford concluded that the atom possesses a positively charged nucleus.

In order to draw quantitative conclusions, theory had to be consulted. Rutherford derived a precise formula for scattering by electromagnetic interaction.[83] Important factors in this formula were the scattering cross-sections of a point-like projectile particle of given kinetic energy and a point-like target particle with finite charge. The formula has been thoroughly tested and verified. However, if the projectile energy is so large as to penetrate the target object, the formula breaks down. The list below is a record of studies with particles at ever-increasing energies, started by Rutherford and continued by Robert Hofstadter, using the Stanford Linear Accelerator in California (SLAC). It was this series of studies that opened the door to what we now call "ultimately elementary".

1) When bombarded by particles with energies up to 10 MeV, the nucleus appears as a point-like, and hence structureless, particle. This result was the basis of the liquid-drop model.

2) Around 100 MeV clear (negative) deviations from the model representing uniform particles indicated a nuclear structure.[84]

3) Between 100 MeV and 1 GeV, small-angle scattering agrees with point-like protons. Large-angle scattering, however, signalled the existence of a composite, i.e. non-point-like, structure of the proton (as was also suggested by its magnetic moment).[85]

4) Around 10 GeV, point-like substructures within the proton become visible. These were later identified as the quarks.[86]

5) Further increases up to 100 GeV continued to show point-like quark structures. When quarks with an energy of 50–100 GeV scatter one another (this is done by making protons collide with antiprotons) they behave like free particles within the protons that contain them. The present limit to which quarks can be considered to be "uniformly" point-like is 10^{-18} m.[87]

6) At even higher resolution, point-like uniformity is found for the electron and its anti-particle the positron. The electron was discovered as early as in 1897 by Rutherford's teacher and predecessor at the Cavendish chair of Cambridge University, Joseph John Thomson. Thomson determined the charge and the mass of this lepton from interactions with electric and magnetic fields. His first experiments were performed with cathode rays produced in X-ray tubes. The devices he used were the forerunners of the gigantic machines used nowadays in particle physics. Seven Nobel prizes were awarded to physicists who originally worked in his laboratory. One was Charles Thomson Rees Wilson (Nobel prize-winner in 1927), who in 1906 invented the cloud chamber for the detection of particles, the forerunner of today's bubble chamber, invented by Donald Glaser, who received the 1960 Nobel prize for physics.

So what is ultimately "elementary"? The electron (and its antiparticle, the positron) are certainly good candidates, at least on the basis of scattering experiments, but there are also sound theoretical reasons to consider them as potentially elementary. Moreover, as we know today, there is a whole family of leptons, consisting of the electron and some more massive, unstable homologues, which appear together with their neutrinos and which share certain properties that characterise them as the (at present) most elementary states of matter (see Section 1.6). As mentioned earlier, the electron can be traced back to Thomson (1897), and the positron – after having been predicted by Paul Dirac in 1927 – was found in 1932 by Carl David Anderson. The higher analogues mu and tau that eventually disintegrate into electrons or positrons (plus neutrinos) were discovered respectively by Carl Anderson and his coworker S. Neddermeyer[88] in 1937, and by Martin L. Perl and coworkers in 1975.[89] The electron neutrino was observed in 1956 by Clyde L. Cowan, Frederick Reines *et al.*,[90] and the muon neutrino in 1961 by Melvin Schwartz, Leon Lederman, Jack Steinberger *et al.*[91]

While scattering measurements showed that protons and neutrons are far from being "elementary", their constituents, the quarks, were shown to be as uniform as electrons – down to a resolution limit of 10^{-18} m. After Robert Hofstadter (Nobel prize-winner in 1961) had found a non-uniform structure of protons,[92] Jerome Isaac Friedman, Henry Way Kendall, Richard Taylor *et al.* pinned down their quark structure in 1967 by high-resolution scattering experiments.[93] The term "quark" was coined in 1964 by Murray Gell-Mann[94] who, apart from being one of the most imaginative physicists of our time, is also a linguistic genius and likes exotic wordings as they appear, for instance, abundantly in the work of James Joyce. Three quarks are

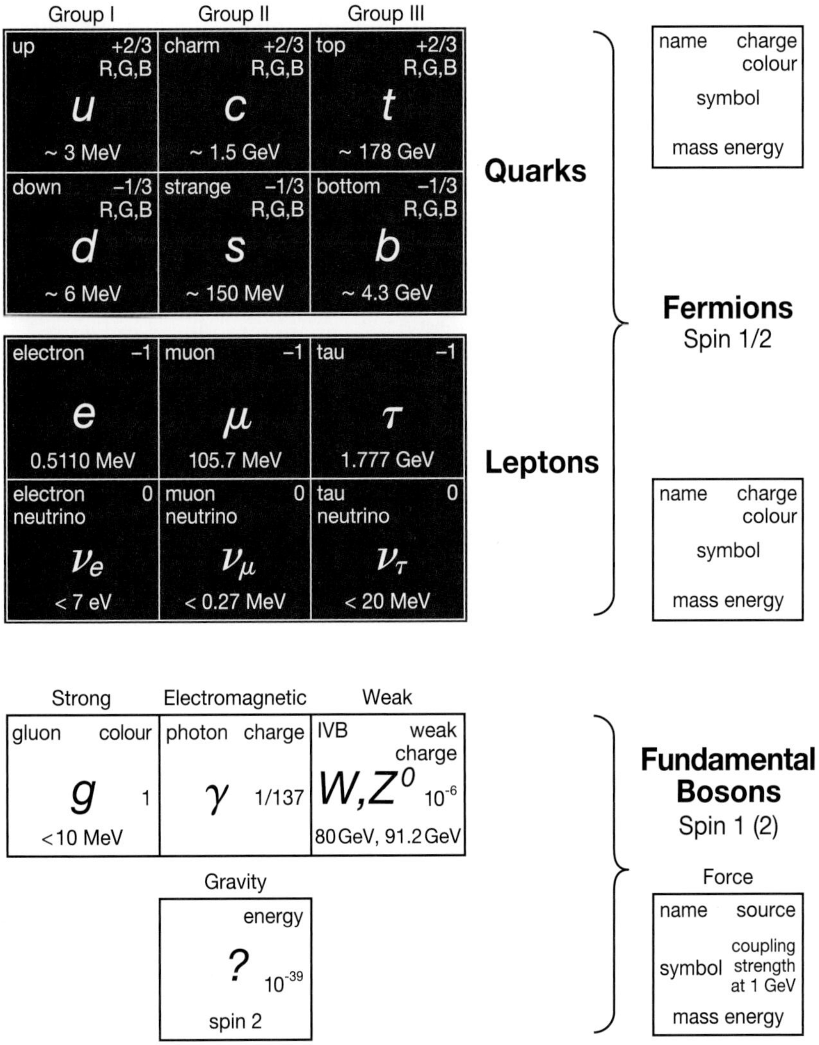

Figure 1.5.6
"Periodic tables" of quarks and leptons

More details are given in Section 1.6.

required to construct the baryons. "Three quarks for Muster Mark ..." is a quotation from *Finnegan's Wake*.

As in the case of leptons, quarks constitute a family, again including six members, of which all have been identified experimentally. Protons and neutrons are made up of "up" (u) and "down" (d) quarks. u carries an electric charge of +2/3 of an elementary charge, d of −1/3. Hence a proton involves the combination uud and a neutron udd. The higher homologues of u and d are c (for "charm") and s (for "strange") as well as t (for "top" or "truth") and b (for "bottom" or "beauty"). Of course, there is no logic in the names. It is just a continuation of the word game started by Murray Gell-Mann. Besides the three quarks originally required, i.e. "up", "down" and "strange", three more have been discovered experimentally, namely "charm" by Burton Richter, Samuel Ting *et al.* in 1974,[95] "bottom" (by Leon Lederman and his school in 1976)[96] and "top" (by CDF, the Collider Detector at the Fermi Laboratory, near Chicago, in 1995).[97] The "charm" quark was predicted theoretically by James Bjorken and Sheldon Glashow about 10 years before its observation.[98] The "periodic tables" of quarks, leptons and fundamental bosons are reproduced in Figure 1.5.6.

How the picture of these truly "elementary" particles was put together – in fact, how they arrange themselves in order to build up higher levels of structure – will be the subject of the following section. Looking at the tables of atoms, nuclei and their constituents, discussed in this section, we realise that, although we have now come down to a common basis, it does not appear to be much "simpler". Whatever that word may mean, any visualisation of this simplicity will uncover its strange aspect.

1.6. How Large is Zero?

The question: "Why is there something and not nothing?" has been accredited to the mathematician and philosopher Gottfried Wilhelm von Leibniz. Perhaps I should be a little less ambitious, and ask instead "Why is there something and not just anything?" Because we should first try to find out why a universe – characterised by certain natural constants – should build up certain singular rather than amorphous types of elementary structures. All we can do today is to start from present structures, looking top-down, although our material world is supposed to have come about bottom-up. However, that is certainly the way to build up theories, although our brain, being an adaptive organ, works neither one way or the other. We need both the top-down input and the bottom-up logical reflection. Even a mathematical proof is finally presented in a way different from the way in which it was found.

Our historical discourse has so far brought us down to units which with some justification might be assigned the quality of being "elementary" or – expressed with more caution – of being "non-composite". Figure 1.5.6 shows what are now called the "periodic tables" of elementary particles.

The quotation marks may indicate that the tables do not really uncover periodicities, apart from demonstrating a striking symmetry among the two "elementary" appearances of matter, leptons and quarks, a symmetry that may have deeper (more elementary?) origins. Leptons and quarks are called fermions, distinguished by fractional spins (e.g. 1/2, 3/2, etc.) The table also includes a row referring to particles that are the carriers of the basic physical forces, called bosons. They differ from the fermions by having integral spins (i.e. intrinsic angular momenta with quantum numbers of 0, 1, etc.). The "strong" interaction provides "glue" that allows quarks to form protons, neutrons, and a number of mesons, and also allows protons and neutrons to join together and form (more or less) stable atomic nuclei. The carriers of these interactions are correspondingly called gluons. Their mass-energy is < 10 MeV (assumed to be zero). Leptons are not able to sense this kind of interaction, but establish their contacts to quarks via so-called weak forces. These form the basis of processes such as the β-decay mentioned in Sections 1.1 and 1.2. The decay of a free neutron into a proton and an electron (with simultaneous production of an antineutrino) requires the exchange of "quanta" of weak interaction which are represented by "massive" particles such as W^+, W^- or Z^o. Because these quanta of interaction carry a relatively large mass, weak interactions are very short-ranged. We now have to return to theory; otherwise, the jargon used in Figure 1.5.6 will remain incomprehensible.

In fact, we departed from theory back in Section 1.4, after having discussed quantum mechanics, which have so far been an unquestioned verity of modern physics. Only a few years after quantum mechanics appeared, Paul Adrien Maurice Dirac of Cambridge University – a frequent visitor to the Göttingen quantum scene – came up with a relativistic version of Schrödinger's equation for the electron that incorporates its intrinsic angular momentum.[99] Dirac's theory predicted the correct size for the magnetic moment of the electron. The truly striking result, however, was a particular symmetry he discovered, namely the alternative appearance of an energy term with a negative sign (as the square root of one has the alternative solutions $+1$ and -1). This led Dirac to postulate an "electron hole" (as might be produced by removal of an electron from a "neutral sea"), equivalent to a state with positive charge. Five years later Carl Anderson detected the positron experimentally.[100] With Dirac's theory a new idea was born, "antiparticles", an idea that turned out to apply to any form of matter.

Attempts to apply Dirac's equation to the proton in order to find its antiparticle were initially unsuccessful, but this was due to the fact that the proton has a finite size and therefore possesses a correspondingly different magnetic moment. Nevertheless, the antiproton was found in 1955 by Owen Chamberlin, Emilio Segré and their coworkers Clyde Wiegand and Thomas Ypsilantis, by making protons collide with copper nuclei.[101] The existence of a mirror symmetry of matter and antimatter is today an accepted fact. Antiparticles have a charge (and baryon number) that is opposite to that of their fellow particles, but also neutral particles, such as the neutron, have an antipartner (with a negative baryon number). The symmetry between matter and antimatter is expressed in otherwise identical properties. A particle and its antiparticle, upon colliding, annihilate each other, and their energy is irradiated away in the form of γ-quanta, Z bosons or gluons. The transformation of a particle into its antiparticle is called charge conjugation (C-transformation), and I shall return to it in Section 1.9.

Dirac's equation was the starting point for an exciting development which led, via the quantisation of the electric field, to a generalisation called quantum field theory and, in particular, to a quantum theory of electromagnetic radiation, called quantum electrodynamics, which – following the modern trend of language mutilation – is invariably referred to as QED. Important steps toward this theory were a seminal paper by Pascual Jordan,[102] which appeared as early as 1927/28, and Heisenberg's S-matrix formulation[103] in the 1930s (S stands for the German *Stoss*, meaning collision, but it may also be expressed as *Streuung*, meaning scattering). The S-matrix is an operator, like the Hamiltonian, that applies to collision or scattering processes.

The true heroes of the development of quantum electrodynamics were Freeman Dyson, Richard Feynman, Julian Seymons Schwinger and Sin-itivo Tomonaga. The latter three were awarded the 1965 Nobel prize for their work, while Freeman Dyson fell victim to the terms of Nobel's bequest, which limit the number of a given year's laureates in a field to three. This does not mean that Dyson's contributions were of less importance. The reader who is interested in the whole story of the development of quantum electrodynamics is referred to the excellent monograph by Silvan S. Schweber, *QED and the Men Who Made It*.[104]

In the following paragraphs I shall focus essentially on the work of Richard Feynman, whose contribution opened an entirely new way of viewing particle interactions and which is most relevant for an understanding of the table of elementary particles. To the young physicist now, Richard Feynman is known by his best-selling books, above all by *The Feynman Lectures* (edited by his associates R.B. Leighton and M. Sands),[105] or by his more enchanting titles such as *Surely You're Joking, Mr Feynman!*[106] or *What Do You Care What Other People Think?*.[107] What made him most popular are his "diagrams" (see Figure 1.6.1). A young particle physicist, reflecting on them, got really excited: "It's amazing! You just look at the diagram and you can write down the equation." Indeed, the Feynman diagrams are much more than just pictorial descriptions of the processes they represent. Every symbol in the graph – depending on which type of gauge theory (see below) one is dealing with – has a clearly defined meaning.

The new idea expressed in those Feynman diagrams is that two interacting particles approaching one another exchange intermediate particles that are called "virtual" because they are not free but, rather, re-bind; these virtual particles are photons in the case of electromagnetic interaction or other fundamental bosons of interaction of the corresponding fields. Only some examples of Feynman diagrams are given in Figure 1.6.1. The (massless) quanta propagate with the velocity of light and transfer their momentum to their interaction partner. Thus, the action of force is reduced to a transfer of momentum, replacing the earlier mysterious forms of "instantaneous" action, be they nuclear, electromagnetic or gravitational fields.

Of course, theory involves more than such a trickery of representation. As mentioned before, what appears to be trickery does in fact have true physical meaning. It offers a way around difficult mathematical procedures, previously accomplished only by perturbation approximations. One of its important features was the "compensation" of infinities by a mathematical procedure, called renormalisation. A renormalisable theory nowadays is called a gauge theory. It is distinguished by a symmetry

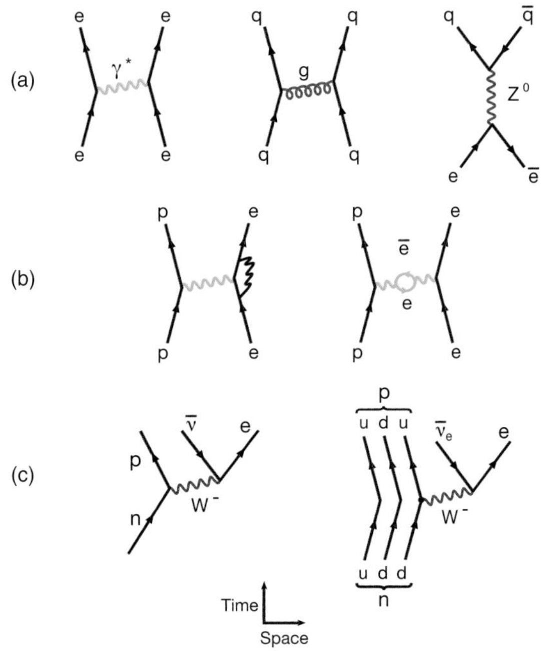

Figure 1.6.1
Feynman diagrams

Feynman diagrams are a visual code that is intended to aid calculations in particle physics. All symbols used in these diagrams, as well as their orientations (left and right, up and down etc.) have a clearly defined meaning, for instance straight lines are particles, where a letter at the beginning or end of a line indicates the kind of particle (e for electron, q for quark etc.); arrows pointing up or down may distinguish particles from antiparticles. Undulating lines represent bosons of interaction, such as photons, gluons, W or Z°. A vertex signifies interaction. The beginning and end of a line represent real particles at different points in either time or space. Interrupted undulating lines indicate "virtual" particles. Any diagram constructed represents a possible process, where conservation of energy and momentum is required at every vertex, lines entering or leaving represent real particles for which $E^2 = p^2c^2 + (mc^2)^2$ (see Section 1.2, E being total energy, p momentum and m mass). Each diagram is characterised by a definite amplitude which is a complex number. Feynman has given the rules for a quantitative evaluation of each diagram. Some examples of Feynman diagrams for processes treated in this book are given in the figure.

called gauge invariance. Quantum electrodynamics is such a gauge theory. Its roots reach back into the 1940s. In its final form it has never yet failed, and it may be considered the most accurate theory so far developed in physics.

As an example of the success of the theory, let us consider the so-called Lamb shift of the hydrogen atom. The hydrogen atom is the simplest member of the periodic table of the elements (Figure 1.5.1), just a proton "orbited" by an electron. The electron interacts not only with the proton but also with itself. According to

QED this can be described by steady emission and absorption of virtual photons, as depicted by one of the Feynman diagrams in Figure 1.6.1. The self-interaction of the electron causes it to "smear out" over a range of about a tenth of a femtometre, resulting in a slightly reduced attraction to the proton, especially when close to it. Quantum electrodynamics allowed a precise calculation of the resulting minute energy shift.

In order to measure this energy shift, Willis Eugene Lamb Jr. (Nobel prize-winner in 1955) used a beam of hydrogen atoms in the 2s state. The transition 2s $\rightarrow$ 1s is forbidden by a selection rule meaning, in classical terms, that a completely spherical charge and current distribution does not radiate energy. On the other hand, the transition 2p $\rightarrow$ 1s is allowed. Lamb therefore used a magnetic field to produce a Zeeman splitting and excited transitions with microwave radiation at a fixed frequency. By varying the magnetic field he could extrapolate the effect to zero magnetic field strength. The resonance frequency of the photon transition due to the "splitting" turned out to be 1057 MHz, which corresponds to the theoretical value with an accuracy down to many decimal places (Figure 1.6.2). This perfect agreement (now down to nine orders of magnitude) is the most convincing proof so far of the representation of electromagnetic interaction by the continual emission and absorption of virtual photons.

Returning to Figure 1.5.6, we now understand the relation between the electron and the particle that mediates its electromagnetic interactions, the massless photon. As a particle, the photon is a boson with spin quantum number 1. I remind the reader that "spin" is a quantum-mechanical ingredient that cannot be understood in classical terms (e.g. like a spinning-top). Otherwise, massless particles such as photons could not have a finite spin with quantum number 1.

The terms "boson" and "fermion" refer to the behaviour of the particles in statistics, i.e. Bose–Einstein and Fermi–Dirac statistics respectively, which is caused by the total wave function to be either symmetric (boson) or antisymmetric (fermion) with respect to the exchange of any two particles. Fermions, according to Pauli's exclusion principle, cannot occupy identical quantum states, while bosons have just the opposite, more sociable, tendency to "condense" into equal states with the loss of individuality.

The higher-mass sibling of the electron, the muon, was first discovered in cosmic rays, and at the time of its discovery it did not really make much sense. Having a mass intermediate between those of the electron and the proton, it was first called a "meson". "Who ordered that?" was Isidor Isaac Rabi's first reaction to its discovery in 1937, revealing how puzzled physicists were. The other, still more massive, sibling of the electron, the tau, was discovered in 1975. By this time it caused much less surprise. Furthermore, it was by now established that each of the leptons must be accompanied by its individual neutrino, as experimentally identified for ν_e (1956) and for ν_μ (1961). To understand this fact requires insight into the theory of weak interaction, in which both leptons and quarks are involved.

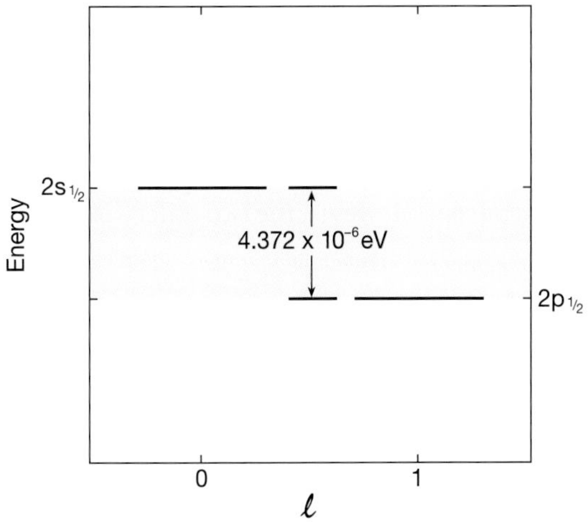

Figure 1.6.2
The Lamb shift

In order to measure this energy shift, Willis Eugene Lamb Jr. (Nobel prize-winner in 1955) used a beam of hydrogen atoms in the 2s state. The transition 2s $\rightarrow$ 1s is forbidden by the selection rule requiring $\Delta I = \pm 1$ (meaning in classical terms that a completely spherical distribution of charge and current does not radiate energy). In contrast, the transition 2p $\rightarrow$ 1s is allowed, Lamb therefore used a magnetic field to produce a Zeeman splitting, and induced transitions with microwave radiation at a fixed frequency of 2395 MHz. By varying the magnetic field, he could extrapolate the effect to zero magnetic field strength. The resonance frequency of the photon transition due to the "splitting" turned out to be 1057 MHz, which corresponds to the theoretical value in the above diagram with an accuracy of many decimal places. This perfect agreement (with an error now down to nine orders of magnitude) is so far the most convincing proof of the picture of electromagnetic interaction as the continual emission and absorption of photons.

The discovery of the quark family – although the non-uniform structures of protons and neutrons had long been appreciated – must be seen largely as a triumph of theory. There was quite a zoo of hadronic particles awaiting a unifying interpretation. (Hadrons are all particles subject to strong interactions. They include baryons, which are fermions with half-numbered spins, and mesons, which are bosons with integral spins.) The postulation of quarks came in the early 1960s, independently from Murray Gell-Mann[108] and George Zweig.[109] The particles were given this name by Gell-Mann (see Section 1.5). Superficially, this reduction of complexity looks like a typical "top-down" procedure, but its theoretical implementation is anything but simple. In 1960, Murray Gell-Mann and Yuval Ne'eman independently found that a symmetry concept based on the non-Abelian group SU(3) might provide an ordering principle for the multiplicity of hadronic particles. Quantum electrodynamics is similarly based on a simpler symmetry principle provided by the Abelian group U(l) (see Vignette 1.6.1). The classification of groups was done by mathematicians towards the end of the 19th century. SU(3) is one of the more advanced non-Abelian Lie groups, named after the Norwegian mathematician Marius Sophus Lie, which were classified by the French mathematician Elie Cartan in his thesis of 1894. Nevertheless, the theory that emerged from this concept, quantum chromodynamics[110] (or QCD), has a more complicated structure than its forerunner QED, which was based on the simpler Abelian group concept. However, the two otherwise show quite a number of important parallels.

Vignette 1.6.1

A group is called Abelian if multiplication is commutative, $A \cdot B = B \cdot A$. The groups U(n) are the groups of unitary operations A in n dimensions: $A^+A = 1$, where A^+ is related to A by transposition and complex conjugation. In one dimension ($n = 1$), the possible A are phase factors, $A = e^{i\zeta}$. In two dimensions ($n = 2$), the A are complex 2×2 matrices, while for $n = 3$ they are complex 3×3 matrices. As we know (from Section 1.4), matrix multiplication is, in general, not commutative. The S in SU(u) stands for the "special" requirement det (A) $= +1$, which excludes reflections. Hence, the elements of the group SU(n) correspond to complex rotations in n dimensions.

QCD theory was originally based on three kinds of quarks, the "up" (u), "down" (d) and "strange" (s) quarks, which were later complemented by the "charm" (c), "bottom" (b) and "top" (t) quarks, as listed in Figure 1.5.5. Gell-Mann called the symmetry he saw behind the eight-membered family of baryons the "Eightfold Way" according to a Buddhist term he was acquainted with from his linguistic and etymological hobbies, but SU(3) also allows for families of 10 or even 27 members. In Buddhism the eightfold path refers to the right way of understanding, motivation, speech, action, effort, livelihood, intellectual activity and contemplation, the eight pursuits for seeking enlightenment. The eightfold way of hadron formation includes combinations suggested by symmetry considerations. The original ideas had later to be revised with the introduction of "flavour" and "colour" (see below).

The construction of baryons required the combination of three quarks yielding integral charges, i.e. +1 for the proton (uud) or 0 for the neutron (udd). Each meson is made up of a quark and an antiquark, e.g. the three pions: π^- ($d\bar{u}$), p^0 ($u\bar{u}$) and π^+ ($u\bar{d}$). Accordingly, all baryons turn out to be fermions (spin 1/2 or 3/2), while all mesons are bosons (spin 0 or 1). The overall symmetry of the wave function of a fermion must be antisymmetric. This overall symmetry is the product of four symmetries: (spin) × (space) × (flavour) × (colour), which was used for justifying "colour". Quantum mechanics then requires certain selection rules, which, for instance, for an intrinsic angular momentum quantum number j of 1/2 forbid a state with three identical quarks. Such a state is allowed for $j = \frac{3}{2}$. The ω-particle (sss) represents such a case. It was predicted by Murray Gell-Mann and discovered in 1964 by V.E. Barnes *et al.*[111] from the Brookhaven National Laboratory and provided a triumphant confirmation of the quark model, which at that time was just evolving.

β-decay, mentioned in Section 1.1 (see also Figure 1.6.1), opened up the way to an understanding of weak interactions and – eventually – to a unification of weak and electromagnetic interactions. An example of "weak decay" is the disintegration of the particle π^+ into the pair $\mu^+ + \nu_\mu$, which has a half-time of about 10^{-8} s. The theory of Sheldon L. Glashow,[112] Steven Weinberg[113] and Abdus Salam[114] – all of whom had thought up the idea independently – was based on another non-Abelian group, namely SU(2). This theory, called quantum flavour dynamics, established the weak forces as a third form of intranuclear interaction. What type of symmetry breakage causes the splitting of the electro-weak forces into two separate components, the electromagnetic and the weak, is not yet known. Since the bosons of weak interactions have finite masses, the two forces can have equal strength only at very high energies, where the photons dominate (cf. below). "Why" the bosons W and Z_0 are so massive is not known.

Quantum flavour dynamics tells us that each lepton comes with a "flavour". They share this property with the quarks. They are all – like the quarks – fermions, having spins of 1/2. All three leptonic siblings, e, μ and τ, are married to individual neutrino partners, ν_e, ν_μ and ν_τ. The limitation to three such pairs has no apparent reason other than the fact that we refer to energies that are experimentally available to us. Charge and mass are assigned to only one of the flavours. The question of whether neutrinos do indeed possess a finite mass has been decided. This mass is extremely small and its precise value is not known. The bosons of weak interaction are the very massive W and Z particles that were predicted to exist and were named as early as 1956 by Tsung Dao Lee and Chen Ning Yang of Columbia University, New York,[115] and experimentally detected in 1982/83 by Carlo Rubbia, Simon van der Meer and their colleagues at the CERN laboratory at Geneva, Switzerland.[116] Otherwise neutrinos are dumb and deaf in any conversation that takes place by electromagnetic or strong interaction. The electron is the only stable lepton in isolation; both muon and tau decay with lifetimes in the ranges of micro- and picoseconds, respectively. Their electromagnetic interactions are mediated by the massless photons and (as mentioned above) these are described to an astonishing degree of accuracy by quantum electrodynamics.

There are far-reaching parallels between leptons and quarks. However, quantum chromodynamics, based on a more sophisticated kind of symmetry than quantum electrodynamics, involves more complicated behaviour.

Quarks also come in various flavours and carry fractional (electric) charges. In addition, there are three QCD analogues of electric charge, called "colours". Strong colour has nothing in common with what our eyes see, just as flavour in particle physics is not associated with what our taste-buds respond to. Again, quantum mechanical constraints require the three "colours" to complement each other and give something "integrally colourless", just as spectral colours can complement one another to give white. The three strong force charges are therefore assigned (arbitrarily) the colours R (red), G (green) and B (blue). An essential point is that the gluons, which mediate the strong force, are specifically associated with the two colours of the interacting quarks. There exist nine combinations of the three colours, but because of formal constraints only eight distinguishable gluons appear. Since gluons, unlike photons, can interact with one another, they cause a behaviour not found with photons. It is because of the gluon interaction that coloured quarks, antiquarks and their gluons are permanently confined to the separable "white" baryonic and mesonic particles, which we are able to experience.

Yes, this is quite a zoo of "elementary" particles, not much "simpler" than that in the periodic table of the 100-odd chemical elements. The species on display comprise no fewer than 18 quarks, three leptons (e, μ and τ) plus their neutrinos, all to be complemented by their antiparticles, yielding 48 fermions. In addition there are eight gluons, the photon and three intermediate bosons for the weak interactions (W_+, W_- and Z_0), bringing it altogether to 60 units,[117] possibly to be augmented by the recently observed Higgs boson. This complexity is ordered by the so-called standard model of particle physics, which, however, has to be fed with empirical parameters, such as the masses and charges of the particles, in order to allow quantitative estimates. Still, there is an essential qualitative difference between the two fundamental sets. In the energy range accessible to us, the particles mentioned are unique, and their schemes suggest the existence of a more fundamental and hence "simpler" yet entirely "strange" principle, while the chemical elements, within a very narrow range of interaction energies, can be combined, giving rise to the stupendous chemical "complexity" of living matter we are quite "familiar" with.

The basis of a unifying principle of particle physics might lie in the interactions (which are sketched again in Figure 1.6.3) and their symmetries.

Our universe is ruled by three fundamental physical constants: Planck's number (divided by 2π) $\hbar$, the velocity of light *in vacuo* c and the constant G of Newton's gravitation law (which as a limiting law is a consequence of general relativity). These constants, taken as unity, define elementary units for length (l_P), time (t_P) and mass (m_P) or energy (E_P). The latter, taken as kT_P (k being Boltzmann's constant) defines a temperature T_P. The index P refers to Planck, in whose honour these elementary units have been named (Table 1.6.1).

These units refer to a world which cannot be understood on the basis of any classical picture. For instance, the Planck mass corresponds to about 10^{19} proton masses, representing an energy many orders of magnitude above what could be reached with present-day accelerators. What is truly bewildering is the stupendous energy density one would obtain if the Planck mass were confined to a volume of Planck dimensions. It would exceed the mass density within an atomic nucleus (and

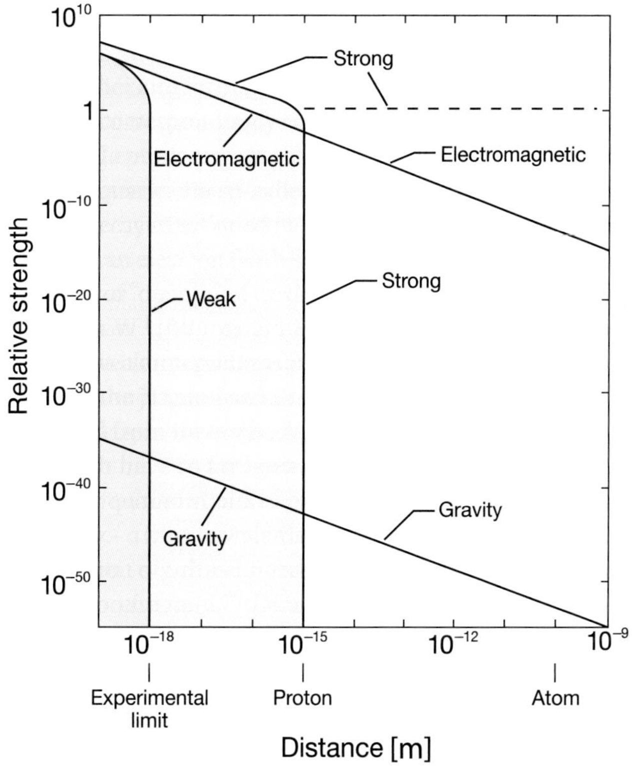

Figure 1.6.3

Fundamental forces and their variation with distance

We know four kinds of fundamental forces: electromagnetic, gravitational, strong nuclear and weak nuclear. At atomic and larger distances, only two of them are of relevant magnitude: the electromagnetic and gravitational forces. Applied to the interaction between proton and electron, the electromagnetic force prevails by 40 orders of magnitude. It is only large masses on astrophysical scales that could make gravitation a competitive factor. On the other hand, at distances of the size of a proton or neutron the strong force abruptly turns on and reaches a strength about 100 times the electromagnetic force, while at distances 1000 times smaller a similar behaviour is found for the weak force, which eventually becomes unified with the electromagnetic force. Quarks that make up the proton and the neutron do not exist in isolation. If they could, their mutual interaction would outweigh by far the electromagnetic force (dashed line) while electrons that respond strongly to the electroweak force are not affected at all by the strong force.

Table 1.6.1 The Planck Dimensions

Planck length	$I_P = 1.6 \times 10^{-35}$ m
Planck time	$t_P = 5.4 \times 10^{-44}$ sec
Planck mass	$m_P = 2.2 \times 10^{-8}$ kg
Planck energy	$E_P = 1.22 \times 10^{19}$ GeV
Planck temperature	$T_P \approx 10^{32}$ K

hence in a neutron star) by some 78 orders of magnitude. This does not make sense in our present picture of the universe – as little as the fact that the wavelength derived from Planck energy ($\lambda_P = hc/E_p$) comes out to be larger than the radius of the event horizon (ct_P) at Planck time t_P.

Let us now compare the relative force strengths of the fundamental interactions against the background of the Planck world. The relative strength of interaction can be characterised by a dimensionless parameter, called "coupling strength". It is generally denoted by the Greek letter α. Figure 1.6.4 shows the coupling strengths of the four fundamental physical forces as functions of energy.

The parameter α was first introduced for electromagnetic interactions by Arnold Sommerfeld in connection with the theory of line splitting in the spectrum of hydrogen, caused by the spin–orbital interaction of the electron. The dimensionless constant $e^2/\hbar c = 1/137.036\ldots$, was originally named the "fine-structure constant". Since it contains only fundamental physical quantities, such as the elementary charge e, Planck's number $h/2\pi$ and the velocity of light c, any theory which claims to explain the basis of physics should be able to reproduce this constant.

Arnold Sommerfeld was one of the great – and largely unsung – heroes of quantum physics. He was the first to apply special relativity to the quantum theory of Bohr's model of the hydrogen atom.[118] His physics department at the University of Munich was one of the leading educational centres of modern physics, and his pupils included, among many others, Peter Debye, Werner Heisenberg, Wolfgang Pauli and Hans Bethe. Sommerfeld's modesty was almost proverbial. Once asked what he considered to be his greatest discovery, he answered: Peter Debye.

Considering Sommerfeld's constant a "coupling strength", we realise from Coulomb's law that the square of electrical charge e^2 has the dimension energy × distance, or, if we choose the right units, $e^2 = 1.44\,\text{eV} \cdot \text{nm}$, with e being the elementary charge associated with Coulomb's law. Fixing a scale of energy and distance is a matter of definition, but in the case of electromagnetic interaction there is one choice prescribed by nature. It is the energy ($h\nu$) and wavelength ($\lambda = c/\nu$) of a radiation quantum (ν being the frequency of radiation), yielding for the product of energy and distance $\hbar c$, or for Sommerfeld's constant the above value, which we now call α_e, indicating by the index e that we are referring to the "coupling strength" of electromagnetic interaction.

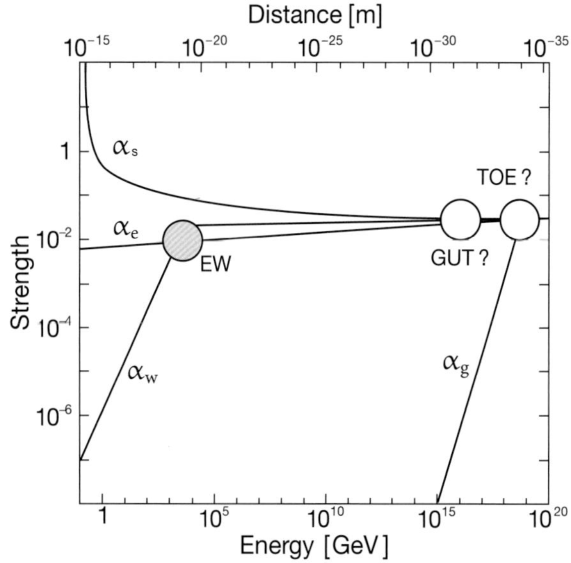

Figure 1.6.4

Dependence of the dimensionless coupling parameters α_i on the energy involved

The unification of the electroweak force (EW) with the strong force (GUT) and (possibly) eventually with the gravitational force (TOE) is still an unsolved problem.

In order to obtain the "coupling strength" of gravitation α_g we proceed in a similar way. According to Newton's law, the product Gm_1m_2 also has the dimensions of energy times distance, so that the ratio of the charge and mass products (i.e. e^2 and Gm^2) is dimensionless. Now we replace the masses according to special relativity by their energy equivalent, obtaining the dimensionless value for $\alpha_g = GE^2/\hbar c^5$, which is equal to $10^{-38}E^2$ if E is measured in GeV. The energy dependence ($\sim E^2$) distinguishes α_g from α_e. It tells us that gravitational and electromagnetic coupling strengths can become equal at energies of about 10^{18} GeV, which is one-tenth of the Planck energy.

For the coupling strength of weak and strong interactions I just report the results of theory (which is beyond the scope of this book). In the case of weak interactions we have a relation similar to that for the electromagnetic case, with electric charge e replaced by the weak charge. The weak charge is of a magnitude similar to that of the electric charge. The fact that at low energies the weak interaction is "weak" as compared with electromagnetic interaction has its origin in the large mass of the boson of interaction, which dominates the momentum transfer unless the energies are large. As a result, the weak coupling strength α_w is given by the electromagnetic coupling strength α_e, multiplied by the factor $(E/m_wc^2)^2$, where E is the energy, m_w the mass of the W-boson (≈ 80 GeV) and c the velocity of light. Hence at 1 GeV α_w is still four orders of magnitude smaller than α_e, but at energies of ~ 100 GeV it equals α_e. The weak coupling strength α_w, being proportional to E^2, shows an energy dependence qualitatively similar to that of the gravitational coupling strength α_g.

A more difficult picture emerges for the strong interactions because quarks inside hadrons show "asymptotic freedom". This means that quarks inside a hadron behave (asymptotically) like "free" particles. The quark–quark coupling strength decreases at short distances, but rises sharply at the size of the proton, while outside of hadrons quarks can hardly exist as free particles because they are powerfully forced to rebind. The coupling strength α_s is found to be close to 1, i.e. about 100 times larger than α_e, at energies around 1 GeV (the proton energy) with an inverse dependence on the logarithm of $(E/\Lambda)^2$, where E is the energy and Λ a constant being about 0.2 GeV. This means that the strong coupling strength α_s at energies below 1 GeV will strongly increase. At energies above 1 GeV, α_s will decrease towards α_e, yielding another coalescence (here of strong and electroweak interactions) at energies around 10^{16} GeV; that is one thousandth of the Planck energy.

The above description may assist the reader in understanding the graphs shown in Figure 1.6.4. These curves give us information as much as they stimulate new questions. If there is ultimately a unity of all forces, what causes them to divorce, to separate, condense or freeze out like phases?

Another question remaining open is that of how we are to envisage the origin of particles that have finite masses. The furnishing of particles with masses has been attributed to a field which was proposed, in connection with the Weinberg–Salam model of electro-weak charges, by the Scottish physicist Peter Higgs. The Higgs field is associated with a boson having zero spin and non-zero mass, to which Leon Lederman has paid tribute by raising it to the status of a "God particle".[119] As such, it would be an expression of the early universe's capacity to create matter. The existence of a Higgs boson has only just been experimentally verified. Its mass-energy is at the borderline of energy attainable with present-day machines.

Now that we are sure it exists, we can expect the Higgs boson to tell us in more detail why our material world is "something" and not "nothing" or some structureless "anything". According to Heisenberg's uncertainty relation "nothing" doesn't mean "anything". The quantum vacuum – if we are able to focus closely enough on it – could be called "anything" but I avoid calling it "zero", for in the physical universe it would then have no role other than that of a mathematical abstraction. More answers than questions!

Leon Lederman wrote:[120] "If the universe is the answer, what then was the question?"

1.7. Are We the Dust of Stars?

Up to the late Middle Ages it was a firm belief that the place where we live must be the centre of the world. This seemed obvious for reasons of faith as well as from undeniable "facts". There was firm religious conviction that *Homo sapiens* was the ultimate goal of creation, ergo, he must live at the centre of the world. And this view seemed to be confirmed by knowledge, which Webster defines as "acquaintance with facts". Look around! The horizon is a circle, the sun and moon appear in periodic cycles and (nearly all) stars are arranged in an orderly way in a celestial sphere *around* us. It was Nicolaus Copernicus who committed the first offence against the pride of mankind. He – followed by Tycho Brahe, Galileo Galilei and Johannes Kepler – moved us out of the centre of the solar system, which eventually, in turn, was "removed" from the centre of our galaxy according to Harlow Shapley. There, as we shall see, it would be quite uncomfortable to be. Nobody cares about these questions any more, although we still speak of the world "around us" and invent transcendental egocentrisms such as "anthropic principles". And if we live on the "surface" of a curved space, where might its centre be?

Isaac Newton was the first to bring lasting order into the way we perceive our world. He unified terrestrial and celestial mechanics.

Of all the forces we become acquainted with during our life, gravitation is the first we encounter. Things that are not supported fall down to earth. And we have learned the fact – more by abstraction than by experience – that the acceleration of things that fall is due not so much to their own properties as to the properties of the centre of attraction. It is true that the force exerted on a body is proportional to the body's heaviness, or its gravitational mass, but so is the inertia that resists acceleration. Newton's famous second law, actually first written in this way by Leonhard Euler, reads $F = m_i a$, where F is the force acting on a particle, m_i its inertial mass and a the acceleration. The law of gravitation states, among other things, that F is proportional to the particle's gravitational mass m_g. Newton knew that $m_i = m_g$ and took this fact as a mystery.

The equivalence of "gravitational" and "inertial" mass represents one of the principles on which Einstein's theory of general relativity is based. In a popular account of his theory, Einstein remarks that the equality of inertial and gravitational mass had long been noticed in classical mechanics "but it had not been interpreted" (Einstein, 1916).[121]

The story about Newton and the falling apple is well known. To tell the truth, I cannot see why an intellectual giant like Newton should have needed to take such a knock in order to start thinking about a force so obviously present on earth. Be that as it may, Newton came up with a gravitational law that represents a final and correct approximation for nearly all practical purposes in celestial mechanics. The law states that the force between two bodies is given by the product of their masses divided by the square of their distance apart. If the mass of one body, say that of the earth, is much larger than that of the other, say that of the object that falls to earth, then the acceleration is determined almost solely by the heavier attractant (the earth). In general, both bodies are accelerated towards their common centre of gravity. This is in particular true for a fixed star with an orbiting planet: both the star and the planet move around their common centre of gravity.

Among the other forces in physics, this behaviour is closest to that of two opposite charges, interacting according to Coulomb's law. The distance relationship in Coulomb's law is the same as that in Newton's law. This matching distance relationship allows us to compare the two forces, both of which, being inversely proportional to the square of distance, extend to an infinite range. The gravitational pull between proton and electron in a hydrogen atom is almost 40 orders of magnitude weaker than their mutual electrostatic attraction. Of course, here I am comparing two physically unrelated properties such as charge and mass, and that's where the analogy ends.

Newton's law is precise as a "limiting" law and as such of great utility in astrophysics. The gravitational constant, called G, which converts the product of two masses divided by the square of their distance (having the dimensions $kg^2\ m^{-2}$) into a force (having the dimensions $kg\ m\ s^{-2}$) is $G = 6.6726 \times 10^{-11}\ m^3\ kg^{-1}\ sec^{-2}$. This constant characterises a true physical relationship, rather than just converting a value from one measuring system into another using different definitions (cf. Boltzmann's constant). The gravitational constant G shares this property with Planck's constant h, which relates the energy of a photon to its frequency (and also provides the uncertainty limit in Heisenberg's relation) and to the velocity of light c, which appears in Einstein's relation establishing the equivalence of mass and energy. It can therefore be used together with h and c to calculate fundamental units of length, time and mass or energy (see Section 1.6).

So what is wrong with Newton's relation? Why is it not a final law? There is nothing wrong with Newton's gravitational law, except that it is based on a concept that cannot hold up generally and therefore fails under certain (admittedly quite extreme) conditions that were not known to Newton, or to anyone else at his time. If something falls down to earth we may ask: why does it do so? Newton's answer would have been: because of the presence of a gravitational field. However, physicists find it hard to remain satisfied with an explanation of "instantaneous action" at a distance. If the causes of the action are the two masses, then something must travel from one to the other. In the language of Feynman's diagrams (Section 1.6) we would say that the masses send out virtual quanta, gravitons, like the photons of electromagnetic interaction. These quanta, when absorbed by the partners, are responsible for a transfer of momentum. Newton himself felt the inconsistency of a picture in which both the space and the physical state of the space were assigned physical reality.

What does gravitation really do to bodies? What happens if two point-like masses approach one another very closely? Newton's law suggests that the force approaches an infinite value if the distance between the masses approaches zero. Will this lead to annihilation of matter in a "gravitational crunch"? In the same year that Einstein published his theory of general relativity (Section 1.8) the astronomer Karl Schwarzschild concluded from Einstein's theory for a non-rotating (almost point-like) star the possibility of such a gravitational crunch.[122]

Let us consider a star like our sun. The sun formed about 5000–6000 million (5–6×10^9) years ago from a gas cloud that consisted largely of hydrogen ($\sim 75\%$), helium ($\sim$ 25%) and only minute amounts ($<$ 0.23%) of heavier elements. This gas cloud contracted under the force of gravitation to its present size, with a concomitant increase in its temperature and pressure. Contraction and heating-up went on until thermonuclear reactions, i.e. the fusion of hydrogen to helium nuclei, commenced. The Bethe–von Weizsäcker cycle uses several nuclei related to carbon12 as catalytic agents (see Figure 1.1.1). The cycle is therefore also called the carbon cycle. The catalytic nucleus takes up successively four protons and releases – apart from a helium-4 nucleus (α particle) – two electrons together with their antineutrinos as well as three high-energy photons (γ rays). The amount of energy produced in one round is about 25 million electron volts (25 MeV).

The carbon cycle is the dominant source of energy above temperatures of 10^8 K. However, the temperature of the solar core is only a little over 10^7 K. Here, another three-step reaction scheme dominates, in which protons undergo straight reactions (the pp cycle):

1) Two protons fuse into a deuteron, releasing a positron and its neutrino.
2) A proton and a deuteron fuse to form a helium-3 plus a γ quantum.
3) Two helium-3 nuclei rearrange into one helium-4 nucleus and two protons.

There are further alternative reaction sequences, one leading from helium-3 and helium-4 to beryllium-7 and then, with another proton (via boron-8), to beryllium-8 (plus a positron), which forms two helium-4 nuclei. In all cases the overall reaction (through unstable intermediate nuclei) is four protons form one α particle (i.e. helium-4 nucleus), the excess charge being carried away by two positrons.

This cycle may be followed by further nuclear condensation processes, which require higher temperatures to complete. Beyond iron, no energy can be gained through the production of heavy nuclei by fusion (as was shown in Section 1.5). Heavier nuclei produce energy by nuclear fission, the best-known processes being those which take place in a uranium or thorium reactor.

The fusion reaction in the sun produces enormous amounts of heat and pressure, which counteract gravitational contraction and stabilise the sun at its present size. The sun will continue to shine upon us in this form for another 5000 million (5×10^9) years, although towards the end of this period it will get unpleasantly hot on earth. While in the early sun hydrogen-burning occurs exclusively in the (sufficiently hot) core, with increasing age the hot burning zone will move into layers closer to the surface. Eventually the nucleus will consist entirely of helium, yet (still) be too cool to allow further condensation of three helium nuclei to carbon-12. At the end of this phase the sun will swell up to a diameter that will include the present orbit of the earth. By then it will have become a *red giant*. Many such red giants can be observed in the heavens.

The final phases of red giants involve complex sequences of nuclear fusion reactions. At the centre of the decaying sun, a very dense core of matter will form. It will continuously gather hydrogen from its outer parts, whereby the burning zone will move towards the surface of what will by then have become a red giant. It will steadily lose material from its surface, forming what is called a planetary nebula. The core, by that time, will have shrunk to the size of the earth: the sun has now become a *white dwarf*, which after further cooling will turn into a *black dwarf*. About 10% of the stars of our galaxy are white dwarfs.

Further gravitational collapse of a star of this size and density is prevented by internal pressure exerted by the electrons. Its cause is the above-mentioned exclusion principle of quantum mechanics, named after Wolfgang Pauli, which forbids more than two electrons (which then must have opposite spins) to share the same region of space at the same time. The more one compresses the material, the more the electrons build up a counteracting internal pressure. Gravitational compression reaches a natural barrier when the electronic pressure balances the gravitational force. This is true only for a star like our sun. Subrahmanyan Chandrasekhar and – somewhat later, independently – Lev Landau have calculated that gravitational force requires only about 1.4 times the mass of the sun in order to overcome the electron degeneracy pressure implied by Pauli's principle. A star exceeding the mass of this Chandrasekhar limit will undergo further gravitational collapse. This occurs in a quite violent manner and entails the production of an enormous amount of heat. A *supernova*, as astronomers call such extremely luminous explosions, appears in the firmament. It finally reaches a density of a value found inside the nucleus of an atom, the enormous density of 10^{15} g/cm^3. At this density the whole mass of the sun would fit into a sphere less than 10 km in diameter. Such a star is called a *neutron star*: protons and electrons have combined, to form neutrons. The neutron star is stabilised by a pressure similar to that in the white dwarf, the difference being that it is the neutrons rather than the electrons that get claustrophobic because of Pauli's exclusion principle. A supernova explosion in a neighbouring galaxy, the Large Magellanic Cloud (160,000 light years away from our planet) visible to the naked eye, was observed and quantitatively tracked over a period of 300 days following the observation of the explosion on 23 February 1987.

Again, we have reached a natural barrier for the star's size and density, limiting further gravitational contraction. Yet this barrier can also be overcome if there is a sufficient amount of mass available. The limit is called – after its discoverers – the Landau–Oppenheimer–Volkov limit. It cannot be fixed as precisely as the Chandrasekhar limit, but there is agreement that it is reached at two to five sun masses. Note that our sun is not a particularly large star, and that we can find many stars in the cosmos that exceed this mass by as much as several hundred times. Their ultimate fate is to become a *black hole*. Here even the strong nuclear forces are outweighed. A black hole is matter so dense that even photons (caught by gravitational forces) cannot escape. This explains the name; the "hole" in space–time is "black", which means no light from there could reach us (although this turned out to be not entirely correct, as will be discussed in Section 2.9). Nevertheless, it is a bizarre, unimaginable world, separated from the rest of the universe by a so-called event horizon. There is observational evidence for the presence of black holes in the centres of galaxies, and our galaxy is no exception. Recent infrared studies at the centre of the Milky Way, performed by Reinhard Genzel and his group at the Max Planck Institute for Extraterrestrial Physics at Munich, Germany, have supplied clear evidence for the presence of a black hole of about 3 million sun masses. Reinhard Genzel kindly supplied the two pictures in Figure 1.7.1 and 1.7.2, and more details can be found in the legends to these figures.

Black holes can form spontaneously only from sufficiently large precursor stars, but there may be many smaller black holes scattered throughout the universe as left-overs of the origin of the universe, where they might have formed under the enormous pressure engendered by the limitation of space.

Black holes are perhaps the strangest objects in our physical universe. As was first shown by Karl Schwarzschild, they are imaginable in the scope of general relativity.

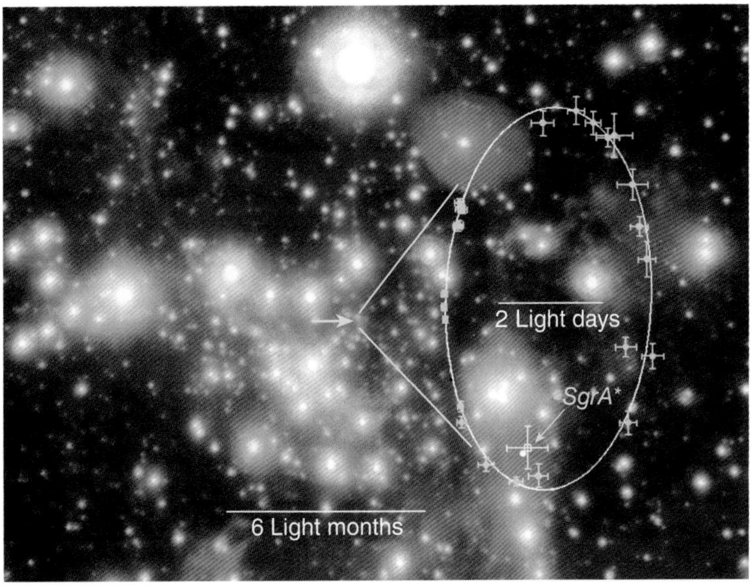

Figure 1.7.1

An infrared photograph of the Milky Way

This was taken in the summer of 2002 and shows the centre of our galaxy. The colours used here represent the stars' temperature: hotter stars are shown in blue, cooler ones in red. The "clouds" are interstellar dust. The arrow shows the position of the putative black hole close to the star SgrA* in Sagittarius. In this region a star called S2 has been observed; its orbit around the black hole is shown enlarged. See also Figure 1.7.2.
(Photograph courtesy of Reinhard Genzel.)

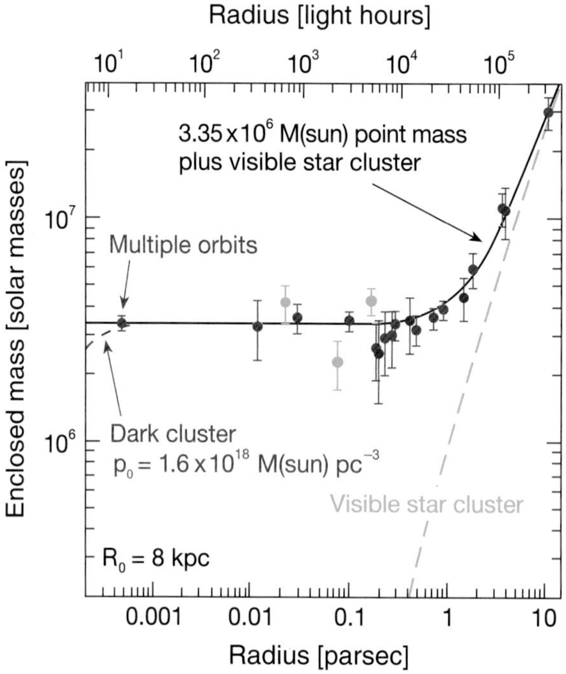

Figure 1.7.2

Demonstration of the presence of a black hole in our galaxy

This diagram, originally published in *Nature*, was kindly provided by the astrophysicist Reinhard Genzel. It demonstrates the existence of a black hole at the centre of our galaxy. The black line shows the total mass enclosed within various distances from SgrA* (Figure 1.7.1). This mass is measured by observing the movement of stars around it. For example, the mass measured within the orbit of the star S2 and others ("multiple orbits") is the same as that measured at greater distances, as the horizontal line indicates. The mass at the centre must therefore be concentrated in a body smaller than the orbit of S2. Above 0.5 pc the enclosed mass begins to rise because of the surrounding (visible) stars. The best fit to the measured curve is given by adding to the mass of these stars a central, invisible mass 3.35 million times that of our sun. This is the mass of the presumed black hole. The dashed lines show theoretical curves for other models that are rejected. (Courtesy of Reinhard Genzel.)

They are, so to speak, warps in space–time, so heavily curved that they are separated from the rest of the world. So do black holes still belong to "our world"? In order to make themselves detectable they have to send out at least gravitons. According to general relativity, the centre contains a singularity, which within the scope of quantum mechanics does not make much sense. If the existence of black holes can be proven, is there not another type of "claustrophobia" that prevents their disappearance in a singularity of space–time? Questions upon questions, which remain to be answered!

So far I have not considered what really is "up there". Being out in the desert now – far away from the dust we produce in our megalopolises – and looking at the sky at night, is for me, as for most of us, more an emotional than a scientific experience. To Immanuel Kant it gave as much confidence in the reign of divine powers as some inborn moral law he felt to be imperative for his doings. ("Der gestirnte Himmel über mir und das moralische Gesetz in mir" – "The starlit sky above me and the moral law within me".)

The main objective of astronomy – which is the most ancient among the sciences – is to provide quantitative information about the physical states (e.g. mass and temperature), motions and – last but not least – distances of material objects, such as stars or galaxies, in the vast and largely empty space of the universe. In order to fulfil this objective the astronomer (or astrophysicist) has to use the physical means that are available to him on planet earth. The most prominent among them are (i) geometrical methods, (ii) optics (in all ranges of electromagnetic radiation) and (iii) the consequences resulting from the theory of relativity.

Let me start, in this example, with the one mentioned first. The geometrical way of obtaining distances is triangulation. Figure 1.7.3 describes the principle. With the help of a telescope we can determine very precisely the angle at which an object appears in the night sky. Two angles (i.e. α_1 and α_2, with α_3 equal to $180°-\alpha_1-\alpha_2$) suffice to determine the shape of a triangle. In order to quantify distances we need in addition the precise length of a base line, which should be as long as possible. What is the longest base line that is realisable within our planetary limits? The answer

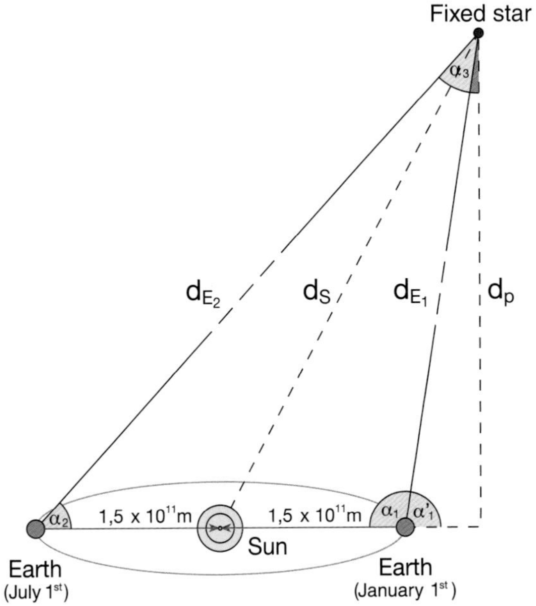

Figure 1.7.3

Triangulation method

This method is used to determine the position of stars at moderate distances.

is: about 300 million kilometres. How is this achieved, when the circumference of our planet is "only" about 40,000 kilometres? Figure 1.7.3 shows us the answer: we measure the angle at which the object is seen with the same instrument, located at the same position, twice, at an interval of half a year. The base line is then the diameter of the earth's orbit around the sun, which is precisely known to be $2D = 1.495979 \times 10^{11}$ m. D, the distance from the earth to the sun, is called the astrophysical unit.

The length of the base line, although huge, is still minute in comparison with the dimensions of the universe. Even though angle determinations can be very precise, the geometrical method does not lead us to the most distant objects in our universe. This fact is reflected in the definition of the most commonly used astronomical length unit, the parsec (pc). The term parsec is a hybrid of two words: parallax and second. In Figure 1.7.3 it is represented by the line d_P. Note that the proportions of this figure are quite unrealistic; the fixed star is in reality very much further away than the diameter of the earth's orbit.

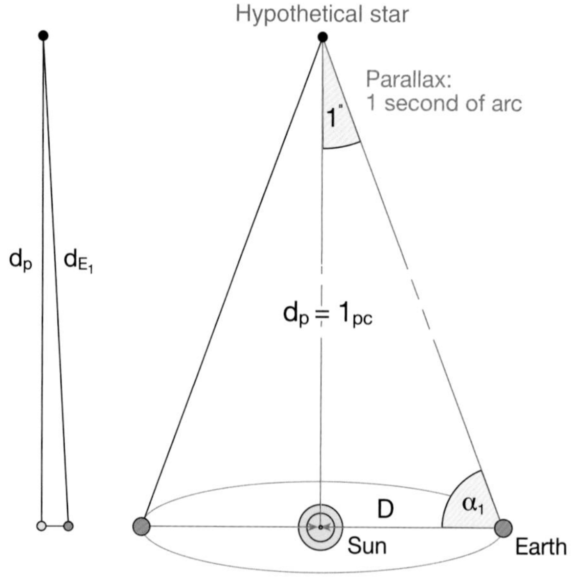

Figure 1.7.4
On the definition of the parsec

It is important to realise that an angle of one second of arc, being the apex angle of a triangle the baseline of which is the radius of the earth's orbit, produces the incredibly large distances of the parsec scale. In order to imagine this we have to abstract from the picture. The length is in reality about 200,000 times larger than the distance D. This causes the distance d_{E1} to be equal to d_P with a deviation of only about 10^{-9}%. At the same time the angle x_1 departs from 90° by less than a third of 1/1000 of 1%. From all this we realise how difficult it is to obtain distance data for other galaxies, which are separated from us by hundreds of Mpc.

Figure 1.7.4 shows what happens if we correct our representation to take account of this fact. The lengths of the lines d_{E1}, d_{E2}, d_s and d_p in Figure 1.7.3 then all become almost the same, and the fixed star is at a position almost precisely above the sun. Of course Figure 1.7.4 is still not to scale, considering how small one second of arc really is. The ends of the two sides enclosing the parallax of the second of arc are a distance D (one astrophysical unit) apart, and the hypothetical star is at a distance of 1 pc. In fact, for a d_p value of 1 pc, the distance from the star is so large compared with D (the radius of the earth's orbit) that the three lines d_{E1}, d_{E2} and d_p have nearly exactly the same length, the difference being only 2.5×10^{-11} dp. This is due to the large value of 1 pc, i.e. 1 pc $= 3.085677 \times 10^{16}$ m $= 3.26163$ light years (ly). In other words, 1 pc is larger than D by the factor $\frac{360 \cdot 60 \cdot 60}{2\pi} = 2.06 \times 10^5$ (the number of arc seconds in a radian).

Distances within our galaxy, the Milky Way, are of the order of magnitude of 1000 pc (kpc). The positions of all stars appearing in Figure 1.7.5 below have been determined by the parallax method. Intergalactic distances fall into the mega to giga parsec range. Examples can be found further below. In these cases the triangulation method would be far too insensitive.

Our access to material objects in space is based on electromagnetic radiation from radio waves to γ-rays. Frequency and intensity, or – projected into the optical

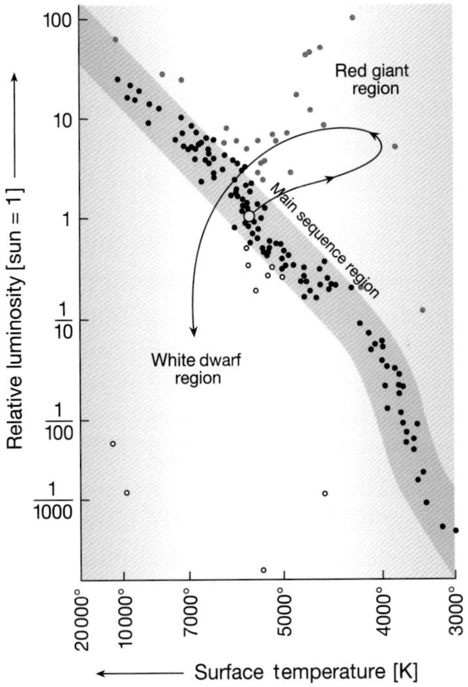

Figure 1.7.5

The Hertzsprung–Russell diagram

This diagram shows the sun-like stars in the galaxy of the Milky Way.

range – colour and luminosity, are our first-hand parameters for identification. Our knowledge about the evolution of stars, as described in the first part of this section, is based on an interpretation of optical data in terms of size, mass and temperature. Surface temperature, for instance, is reflected in the colour of a star: blue stars are hot, red ones relatively cool. More precise information comes from a detailed study of their spectra. If the distance of a star can be estimated, luminosity will tell us how much of its energy is radiated into space. Such correlations among optical parameters come to expression in the Hertzsprung–Russell diagram, an example of which is shown in Figure 1.7.5.

It was the Hertzsprung–Russell diagram that provided most of our knowledge about the evolution of stars. The diagram is named after the Danish astronomer Ejnar Hertzsprung, who was the first to classify the types of stars according to their colour in relation to their absolute brightness, and the US astrophysicist Henry Norris Russell who after focusing on interstellar distances used spectral properties in relation to luminosity in order to gain information about the temperature, mass and composition of stars. The use of the Hertzsprung–Russell diagram by astronomers all over the world enabled them to establish a luminosity scale that was later applied to a determination of intergalactic distances. Figure 1.7.5 shows the heavily populated band of "main-group stars" within our galaxy, whose masses vary from fractions to multiples of the mass of our sun, which appears in the middle of the diagram. It separates the (relatively rare) white dwarfs with their mass densities of thousands of kilogrammes per cubic centimetre (lower left corner of the diagram) from the red giants and supergiants, which have mass densities as low as a millionth of a gramme per cubic centimetre (upper right corner). The arrow line, originating at the position of the sun, indicates how to conjecture the evolutionary trajectory of our central star as to be expected some thousand million years from now. The background coloration of the picture refers (in a somewhat artistic way) to the colours in which the stars present themselves to our eyes.

Information about galaxies that are many mega parsecs away from the Milky Way comes mainly from the red shift of their spectral lines, which is caused by the Doppler effect. This effect is named after the Austrian physicist Christian Johann Doppler,[123] who lived in the 19th century. As an acoustic phenomenon, the Doppler effect can be observed in daily life. The siren of an ambulance sounds distinctly higher when it approaches than when it moves away from us, with a rather sharp drop when the ambulance is passing. When the distance to the moving ambulance is shortening, we receive more oscillations in a given time interval than when the distance is increasing. The temporal compression or dilation of oscillations is heard as an increased or decreased frequency. Doppler (correctly) suggested that the principle applies to any kind of wave motion, but (incorrectly) applied it himself to an interpretation of the colours of stars.

Special relativity offers a straightforward interpretation of the optical Doppler effect. The velocity of light in vacuum (c) is a universal constant and as such is independent of any motion of its source. The velocity of sound (being the reciprocal square root of the product of mass, density and volume compressibility, i.e. a thermodynamic property of the medium in which the sound wave is propagated) is also independent of the motion of its source or receiver.[124] The physical reason for the Doppler effect in both cases is that – depending on the relative velocity with respect to the source – the receiver records more (or fewer) vibrations in a given time interval. The interpretation of the optical Doppler effect becomes straightforward if we remember not only that c is a universal constant, but also that light quanta do not have a mass, but instead a finite momentum given by $E/c = h/\lambda$, where E is the energy, and λ the wavelength of the light quantum, and h is Planck's constant. Special relativity (see Section 1.2) then requires that systems moving with relative velocity u (towards or away from an observer) exhibit relative shifts ($\Delta\lambda/\lambda$) in the wavelengths of their spectral lines as is shown in (and explained in the legend of) Figure 1.7.5. Such Doppler shifts were first observed and quantitatively interpreted by Edwin Powell Hubble.

Spectral lines are narrow regions (their narrowness gives rise to the term "lines") in the rainbow-like spectrum of light that is obtained when white light is deflected by a prism or a grating into a horizontally extended "band". The positions in such a band indicate the wavelength of the light and are seen in the visible range with their characteristic rainbow colours. If the light consists of a mixture of light quanta of discrete wavelengths, the only regions appearing as illuminated lines are those that represent the wavelengths of the photons present in the light. Each wavelength corresponds to a fixed energy. Conversely, if light of all wavelengths (white light) passes through matter, light of the same fixed, characteristic energy is absorbed. In this case (an absorption spectrum) the lines appear as dark lines. This effect was first observed in 1802 by the English physicist William Hyde Wollaston, and was studied systematically by the German physicist Joseph von Fraunhofer[125] in the spectrum of sunlight, where the effect is caused by substances present in the gaseous atmosphere of the sun. Today some 25,000 Fraunhofer lines in the sun's spectrum are known.

Hubble, by measuring precisely the spectra of light from distant galaxies, found that the lines are red-shifted (i.e. shifted to a greater wavelength) relative to the corresponding light in the sun's spectrum. This is schematically represented in Figures 1.7.6 and 1.7.7, where the spectrum of the sun is used as a coloured background. The diagram records the red shift (ordinate $\Delta\lambda/\lambda$) as a function of the distance from the point of observation (distances in mega parsecs).

The experimental data obtained for several galaxies provide evidence for a linear relationship between distance and velocity (u) of recession, since according to special relativity for $u \ll c$ the red shifts are linearly related to $\beta = u/c$. However, the white curve and its deviation from the extrapolated linear relation (black curve) show what is to be expected for larger values of β. The origin of these curves has (arbitrarily) been placed at one of the clearly visible lines of the sun spectrum. In fact, Hubble

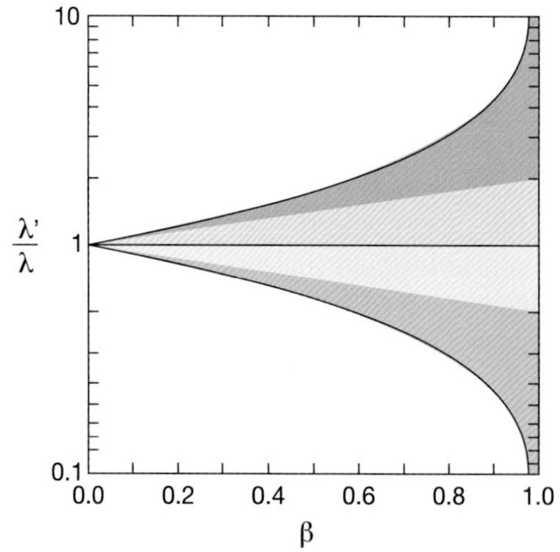

Figure 1.7.6
The relativistic Doppler shift of light

Light quanta have no mass, but they do have a momentum, which is equal to E/c, where E is their energy and c the velocity of light. The energy and momentum of the photons from a light source moving with a relative velocity (u) with respect to a resting observer, according to special relativity (cf. Section 1.2), is recorded as $E' = E\gamma(1 + \beta)$ with $\gamma = (1-\beta^2)^{-1/2}$ and $\beta = u/c$. The energy of a photon is $h\nu = hc/\lambda$ (with h = Planck's constant, ν = frequency and λ = wavelength). Then, using the relationship $(1-\beta^2) = (1-\beta)(1 + \beta)$, we obtain $\lambda'/\lambda = E/E' = [(1-\beta)/(1 + \beta)]^{1/2}$, where β is to be taken as positive if the light source is moving towards the observer (blue shift) and as negative when it is moving away (red shift).

used two well-defined lines in the near ultraviolet region caused by calcium ions in the atmosphere of the stars. Recent measurements detecting large red shifts, such as $\Delta\lambda/\lambda = 4.73$ for the distant quasar PC 1158 + 4635 (Figure 1.7.8), used the αLyman line of hydrogen ($\lambda = 121.4$ nm) shown in Figure 1.7.9.

For the galaxies represented in Figure 1.7.7 there is clear evidence for an initial linear relationship between $\Delta\lambda/\lambda$ and both speed and distance. This finding provided the basis for Hubble's law, according to which the velocity by which galaxies recede is linearly related to our distance from the galaxy: $u = H_o \times d$, with H_o being around $3 \cdot 10^4\ m \cdot s^{-1} \cdot Mpc^{-1}$. The difficulty in producing these experimental data lies in the indirect way astronomers have to determine distances by resolving luminosities. However, Hubble's law should not be formulated as a linear relationship between the red shift $\Delta\lambda/\lambda$ and either distance or speed. A linear dependence of $\Delta\lambda/\lambda$ on speed (except for small values of β) would contradict special relativity, while Hubble's law as such may well remain valid up to large values of speed and distance. The large red shifts, displayed in Figures 1.7.8 and 1.7.9, can be understood only on the basis of special relativity.

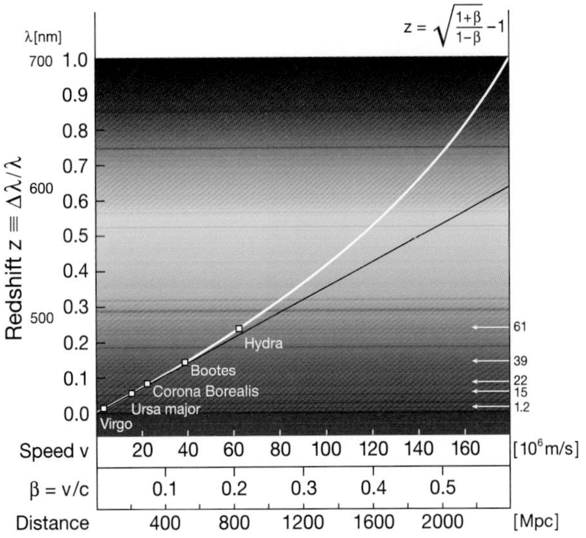

Figure 1.7.7

Red shift and Hubble's law (1)

According to Figure 1.7.6 the red shift $\Delta\lambda/\lambda$ of a receding light source is linearly related to its velocity u only for $u \ll c$. Hubble's law states a linear relation of distance d and velocity u, i.e. $u = Ho \times d$; Ho is called Hubble's constant. The white (relativistic) curve is used for a calibration of Hubble's law for known distances.

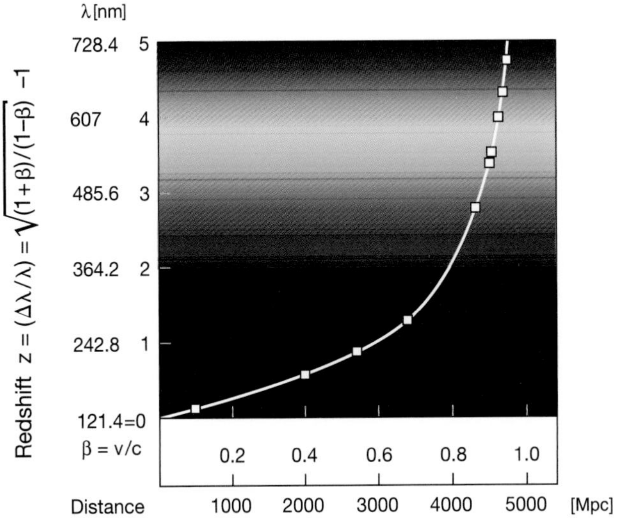

Figure 1.7.8

Red shift and Hubble's law (2)

A list of recent data was kindly provided by the astrophysicist Rudolf Kippenhahn, of which the results reproduced in this figure are selected examples. They cover a range up to (almost) light velocity and represent a wonderful proof of the relativistic nature of Hubble's law (see also Figure 1.7.9).

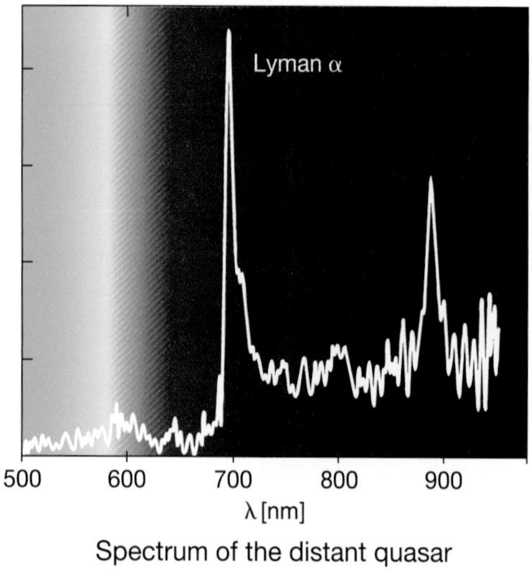

Figure 1.7.9

The extreme red shift of the quasar PC 1158 + 4635

The Lyman α line of hydrogen was used to determine the red shift z of the quasar PC 1158 + 4635. It amounts to $z = 4.73$, or a shift from the ultraviolet (121.4 nm) to the red (696.0 nm), corresponding to the extreme points in Figure 1.7.8.

Hubble's discoveries in the first half of the 20th century were revolutionary for our picture of the universe. They will be revisited in Section 1.8. A value of Hubble's constant between 30 and 100 km s^{-1} and per mega parsec suggests an age of our universe between 10,000 and 20,000 million years.

How many stars does the present universe comprise?

Astronomers today know this number only roughly, and if they were able to count it precisely they might find that quite a lot of matter, necessary to reverse the expansion of the universe, is missing. The present estimate is about 10^{11} masses of our sun per cubic mega parsec, or nearly 10^{23} solar masses (corresponding to 10^{80} proton masses) in the entire observable universe, the extent of which is some 10^{10} pc. General relativity (Section 1.8) will tell us that the critical density of a stable universe is 4 GeV m^{-3}, which is about ten times as large as the presently observed mass-energy density of 0.5 GeV m^{-3}. Black holes or other forms of "dark matter" such as "cosmic strings", finite masses of neutrinos and also dark energy (whatever that may be) may perhaps make up at least part of the difference.

Let's return to the present: Why is our sky dark at night?

Kant, in an early work of 1755 called *Theory of the Heavens*, postulated the universe to be infinite in space and time, and homogeneously filled with stars that shine for ever. The reason he gave was as simple as it was convincing for his contemporaries: to assume anything else would be an insult to the Lord. Heinrich Wilhelm Matthias Olbers, physician and astronomer in the town of Bremen, did not dare to doubt the validity of Kant's postulates, but in an article of 1823 he set out a problem (using arguments that the Swiss astronomer Jean-Philippe Loys de Chéseaux had formulated even before Kant, in 1743). The following description is in the words of my colleague and friend Peter Richter,[126] who once was associated with our group at Göttingen and now is professor of theoretical physics at the University of Bremen, where Olbers spent most of his life.

Olbers imagines standing in the midst of a large wood; any horizontal light ray that meets his eye must come from the surface of some tree. It is estimated that if the average thickness of a tree is d, and the average distance between two trees a, then the typical distance that the observer can look before his gaze meets a tree is a^2/d. By similar reasoning, if the average distance between stars is a, and their typical cross-sectional area F, then the horizon should appear filled with stars at a typical distance a^3/F. Thus, Olbers concludes, light ought to be coming towards us from all directions; the sky should everywhere be as bright as the sun. So there is a paradox, named the *Olbers paradox* by Hermann Bondi[127] in his book *Cosmology* of 1952. Olbers himself did not perceive it as a paradox, because he presented a solution: light may be attenuated on its way through space, and thus have only a finite range. But this solution turned out to be wrong, and the "paradox" was seen as a serious problem for any cosmological model.

Its resolution comes in different versions. The paradox disappears, of course, if any of Kant's postulates turns out to be wrong. We know today that the universe has an age of "only" some 15,000 million (10^9) years, so that light cannot meet our eyes from further away than some 10^{10} light years. Furthermore, stars do not shine for ever; the brighter they are, the shorter their life. So let us put numbers into the expression a^3/F. The average distance between stars in the universe is some 1000 light years (Olbers did not even know that within our galaxy, the average distance is 10 light years); the typical radius of a star is a million kilometres. This puts a^3/F in the order of 10^{23} light years. This is 10^{13} times more than the distance beyond which we cannot see because of the finite age of the universe. In his beautiful analysis of the problem, E.R. Harrison[128] shows that another way to express the same thing is to say there is not enough energy in the universe to fill it with radiation that would be in equilibrium with the stars' surface temperature, some 6000 K. Instead, the cosmic radiation is at a temperature of some 3 K (see Section 1.8), and given the Stefan–Boltzmann law which states that the energy density of radiation increases as the fourth power of the temperature, this temperature ratio of 2000 explains the missing factor of 10^{13}.

However, Olbers was not completely wrong if we take into account the history of our universe. The universe is expanding, and there was an epoch, some 300,000 years after the Big Bang, when it was 2000 times smaller and 2000 times hotter than it is now (this will be discussed in Section 1.8). During that epoch, the background radiation was indeed at a temperature of 6000 K; the universe was everywhere as hot and as bright as the present surface of our sun: Olbers' sky. But, of course, there were no stars, and there was nobody to observe them. Not even atoms had formed because nuclei and electrons cannot stay together at these temperatures. The universe was opaque, filled with charged matter – plasma – and light was confined therein. The formation of stars and galaxies could not start until expansion had cooled the cosmos to below 3000 K. Then atoms formed, the radiation escaped and gravitation could start to do its work of star and galaxy formation."

The process itself, i.e. why galaxies are clustered, has not by any means been answered satisfactorily yet. The primary material was nearly exclusively hydrogen and (possibly) some helium. There must have been certain density fluctuations that triggered nucleation and growth, favoured by gravitational attraction. Those stars heated up until fusion reactions, as discussed above, got going, keeping the gravitational pull in balance with further thermal expansion. The main product of such fusion reactions was (and is) helium-4, but as the British mathematician and astronomer Sir Fred Hoyle found out, there is some "resonance" that favours the unification of three helium-4 nuclei to form the carbon-12 nucleus, which – next to oxygen – is one of the important ingredients of the chemistry of terrestrial life. Heavier elements could come about only by nuclear reactions inside stars that had a sufficient lifetime and proliferated by gravitation-induced cosmic catastrophes. So I let one of the pioneers in the field, William A. Fowler answer the question posed in the title of this section:

> The first two elements and their stable isotopes, hydrogen and helium, emerged from the first few minutes of the early high-temperature, high-density stage of the expanding universe. ...Where did the heavier elements originate? The generally accepted answer is that all of the heavier elements from carbon (element six) up to long-lived radioactive uranium (element ninety-two) were produced by nuclear processes in the interior of stars in our own Galaxy. ...Let me remind the reader that our bodies consist for the most part of these heavy elements. Apart from hydrogen you are 65% oxygen and 18% carbon, with smaller percentages of nitrogen, sodium, magnesium, phosphorus, sulphur, chlorine, potassium and traces of still heavier elements. Thus it is possible to say that you, your neighbour and I, each one of all of us, are truly and literally a little bit of stardust.[129]

This answer may not be of great help to anyone, least of all to the philosopher who wants to understand what "we", living and thinking beings, are. But it does help to remind a humble biochemist what kind of chemical material he is working with. It had done a lot of space travelling long before it became transformed into its present complex state, in which it is now trying to travel back into space.

1.8. The Universe: A Finite Island in an Infinite Ocean of Space?

In 1916, shortly after he had concluded his theory of general relativity, Albert Einstein wrote a popular account of his work in which he said: "As regards space (and time) the universe is infinite. There are stars everywhere, so that the density of matter, although very variable in detail, is nevertheless on average the same."

Astronomical observations have essentially confirmed such a picture and have led to the formulation of the "cosmological principle" to which astronomers today still adhere: On scales greater than the distances between clusters of galaxies, the universe would look the same to an observer in any galaxy.

In the same book Einstein referred also to a "fundamental difficulty attending classical mechanics" which according to the best of his knowledge was first (1894) discussed in detail by the astronomer Hugo von Seeliger, who was the teacher and doctoral supervisor of Karl Schwarzschild, the greatest astronomer of the early 20th century. According to Newton's law, the universe should have a centre in which the density of stars is at a maximum. In that case, as von Seeliger saw it: "the stellar universe ought to be a finite island in the infinite ocean of space". However, this would cause a dilemma because the light emitted by stars would escape into space and never return. Matter therefore eventually should fade away. I will not go further into von Seeliger's arguments, in which he suggests modifications of Newton's law, which Einstein laconically dismisses as having "neither empirical nor theoretical foundation".

Einstein had actually been much more concerned with another fundamental conceptual difficulty of classical mechanics, the resolution of which was the central issue of his work after 1905. In fact, the problem had not escaped Newton, who saw it but could not resolve it. It was the Austrian physicist and philosopher Ernst Mach, who clearly identified it and, as a consequence, seriously considered eliminating the whole concept of space. In explaining his theory, Einstein accordingly paid due tribute to Mach, who had stated that the inertia of a mass must be related to some interaction of that very mass with the rest of the world. To him, inertia of a body in an otherwise empty world didn't make sense. Einstein's principle of equivalence, which he called "the happiest thought" of his life, bridged Mach's ideas with the mathematical formalism worked out half a century earlier by Georg Friedrich Bernhard Riemann. Expressed in daily language, the principle of equivalence says that for any observer in free fall there exists no gravitational force, or, in other words, there must be locally a complete equivalence between the effects of inertia and gravitation. This equivalence, when taken literally, would already explain the fact that "in general, rays of light are propagated curvilinearly in gravitational fields", the quotation marks referring to Einstein's own words.[130] As he points out, "a ray of light is no longer a straight line when we consider it with reference to an accelerated 'chest' ". (The "chest" was Einstein's favourite utensil, to enclose a person in order to let him or her experience the equivalence of gravitation and a force-driven acceleration.)

While Einstein's theory of special relativity was a stroke of genius, thought up and published within the same year, general relativity turned out to be a Herculean task, keeping Einstein busy for more than ten years. The mathematics were ready and waiting to be applied, but Einstein had first to be introduced to them – by his friend in Zürich, the mathematician Marcel Grossmann. The mathematical formulation culminates in the seemingly simple but strange form of an equation

Figure 1.8.1

Einstein's general-relativistic equation of gravitation

This equation provides an excellent example of "strange simplicity". The extreme compression of its information content (see Section 3.5) into a straightforward relationship between $G_{\mu\nu}$ and $T_{\mu\nu}$, $8\pi\kappa$ just being a constant, is of utmost "algorithmic simplicity". However $G_{\mu\nu}$ and $T_{\mu\nu}$, being tensors, are of a strange nature. Referring to four-dimensional space–time the compressed tensorial form of the above "master equation" represents ten different equations (see text). The tensor $G_{\mu\nu}$ is related to the curvature of space–time and hence describes the gravitational field, while the tensor $T_{\mu\nu}$ is associated with the density distribution of mass-energy involving stress–strain relations for the different coordinates.

shown on the blackboard in Figure 1.8.1. $G_{\mu\nu}$, the curvature of space, and $T_{\mu\nu}$, the mass-energy density, are so-called tensors. This equation is the kernel of general relativity. Let's see what it means and how one can arrive at this very condensed representation.

Riemann, one of the giants who held the chair of mathematics at Göttingen University in the succession of Peter Gustav Lejeune Dirichlet and Carl Friedrich Gauss, was the founder of a geometry that generalises Euclidean space by dispensing with Euclid's parallel axiom. In a general space of any dimensionality, the square of the distance between two points does not necessarily follow the Pythagorean rule of being equal to the sum of squares of its coordinate increments. If we consider small increments, as we do in differential geometry, we obtain in addition to the square terms $g_{\mu\mu}(dx_{\mu})^2$ several cross terms of the form $g_{\mu\nu}(dx_{\mu}dx_{\nu})$. For four-dimensional space–time the metric tensors include $4 \times 4 = 16$ coefficients $g_{\mu\nu}$, represented by a corresponding table which, because of the reciprocity $g_{\mu\nu} = g_{\nu\mu}$, reduce to 10 different positive numbers. Curvature of space and mass-energy density are represented by those tensors $G_{\mu\nu}$ and $T_{\mu\nu}$, yielding the above highly compressed representation of what in reality corresponds to ten equations. They replace the Pythagorean line element $ds^2 = dx^2 + dy^2 + dz^2$ or, in the case of Euclidean space–time, the Lorentz-transformable line element $ds^2 = dx^2 + dy^2 + dz^2 - c^2dt^2$ by $ds^2 = g_{ik}dx_idx_k + \ldots$, summed over all indices i, k from 11, 12, ...up to 44.

The equation above may be read from right to left, so it implies a certain density distribution of mass-energy as the cause of the curvature of space. It is important to note that this requires tensorial representation of both mass-density distribution and curvature. We have already become acquainted – in Section 1.2 – with the four-vector of energy-momentum. In a distribution of matter in space-time we have to allow for cross-terms. The energy-momentum tensor is the simplest second-rank tensor that applies to such a matter field. Reading the equation from right to left considers the density of mass-energy distribution as the cause of the resulting geometry.

One can just as well read the equation from left to right (as one usually reads an equation). Here the metric tensor $G_{\mu\nu}$ of curvature determines the energy-momentum tensor $T_{\mu\nu}$ and the principle of algebraic geometry establishes the existence of the sources. In this case, ten chosen functions of coordinates would determine the resulting tensor $T_{\mu\nu}$. In practice, however, this would rarely turn out to be useful.

The true problem is that we have to solve ten field equations, which together make up the constraints for a simultaneous choice of the ten components each of $G_{\mu\nu}$ and $T_{\mu\nu}$.

The two principles that guided Einstein's approach to general relativity were those of equivalence and covariance. Equivalence has been briefly described above. Covariance has already been introduced with special relativity, where it referred to an equivalence of all inertial systems. Here it is extended to include non-inertial observers, which requires the equations to have tensorial form. In fact, there is a third principle to be considered, which may be designated by "correspondence". It requires the convergence of general and specific relativity in the absence of gravitation and an approach to Newton's theory in the range of weak gravitational fields. The rest is Riemann's geometry, with the constraint of minimising the changes in the transition from simple vectorial to tensorial representation.

Meanwhile, we are in possession of several independent approaches to general relativity, from David Hilbert[131] (1915) to Andrei Sakharov[132] (1967) or Claudio Teitelboim[133] and his colleagues (1973). In fact, Hilbert presented his derivation, based on the principle of least action, five days before Einstein could expound his results to the Prussian Academy in Berlin. His results were identical with those of Einstein, yet he never challenged Einstein's priority. John Wheeler, in his address at Paul Dirac's 70th birthday,[134] discussed these various approaches in more detail. In Vignette 1.8.1 I present his compressed outline of the very elegant method used by Elie Cartan.

Vignette 1.8.1

Cartan's Principle: "The Boundary of a Boundary is Zero" and its Application to General Relativity

John Wheeler[134] introduces Elie Cartan's approach by quoting an analogy between geometry and mechanics. For mechanical systems, in order to be at equilibrium, two conditions have to be fulfilled:

i) At any point all forces have to add up to zero: $\sum F = 0$
ii) The sum of all "moments of force" has to vanish as well: $\sum (r - r_\rho) x F = 0$

The moment of force is the vectorial product of a relevant positional vector from an arbitrarily chosen point $\wp$ to the point of application of the force. Because of $\sum_{\text{all forces}} F = 0$ it does not matter, which point $\wp$ is used to define the moments.

According to Cartan, the analogue of the moment of force in geometrodynamics is the "moment of rotation". Let us first look at rotation of the underlying vector field, which is a measure of Riemann curvature and as such is associated with one of the faces of the cubes represented in Figure 1.8.3. The cube is one of the hypersurface elements (eight cubes) of space-time as introduced in Section 1.2 (Figure 1.2.4) referring to a given moment in time.

By placing a trivector (x, y, z) at the upper left hand corner of the cube we designate the various faces and their orientation. Focussing on the xz face in the foreground of the picture, we now transfer the vector parallel to itself around the whole cube in the two directions indicated (right/left and up/down). The vector undergoes a rotation which depends on the size of the faces and the relevant components of the Riemann curvature of the 4-dimensional geometry. We realise from the arrows drawn, that in this procedure we pass every edge of the cube twice in the two opposite directions indicated. Hence, the contributions of all edges cancel out: the one-dimensional boundary of the two-dimensional boundary of the three-dimensional volume is zero, or in Cartan's abbreviated formulation: $\partial\partial V \equiv 0$, where the symbol ∂ signifies "boundary of". Applied to Riemannian curvature of space-time, this is expressed in the form of the so-called Bianchi identity $\sum_{\substack{\text{all six}\\\text{faces}}} \binom{\text{rotation associated}}{\text{with each face}} \equiv 0$, named after the Italian mathematician Luigi Bianchi.

Now we come to the second condition of the mechanical analogy. Elie Cartan for this purpose introduced the "moment of rotation". It is defined in a similar way as the "moment of force", namely as the vectorial product of a positional vector and the rotation, associated with the relevant face.

Cont. ➲

➲ Cont.

The "moment of rotation" of a given face is represented by a trivector which comprises the bivector of rotation as a measure of curvature and the vector from $\wp$ to the centre of the face. It depends on the location of $\wp$, which can nevertheless be chosen arbitrarily, because it drops out when the sum of the trivectors is extended over all six faces yielding the "moment of rotation" trivector associated with the total elementary cube. It is this trivector of curvature which is related to 8π times the trivector representation of the amount of energy and momentum enclosed in the cube as the "content of source". This procedure is repeated for all space-like surface elements of space-time at constant t.

The question at stake is how this correlation of geometry with the source guarantees the conservation of the source as time goes on. Conservation of mass-energy is as important in geometrodynamics as is conservation of charge in electrodynamics. This conservation reveals itself automatically as a consequence of Cartan's principle $\partial\partial V \equiv 0$, now applied to four-dimensional space-time. It means that the "content of source", added up with due account of sign, comes out to be zero. The "content of source" in each cube, summed over all eight cubes constituting the surface of a 4volume element of space-time, which equals the sum, taken over all $6 \times 8 = 48$ faces of $1/8\pi$ times ($\wp$-dependent "moment of rotation" associated with each face), comes out to be identical to zero. The 48 faces, as Figure 1.8.3 shows, cancel out exactly for all 24 pairs of oppositely oriented faces that are shared by two different cubes. An analogous result is obtained for Maxwell's equations (see Vignette 1.3.1), demonstrating an inherent conservation of charge.

The gist of this table is not a derivation of the tensor equation in its most compact formulation, i.e., (curvature) $= 8\pi$ (density of mass-energy). There is a magnificent representation of the field by Misner, Thorne and Wheeler, which is both exhaustive and – though fun to read – exhausting. The purpose of this table is to show "how" general relativity creates a new insight into – what the authors call – "the mysteries and marvellous simplicities of this strange and beautiful Universe, our home".

Note: The propagation of sound requires a material medium. There is no sound propagation in a vacuum, and the velocity of sound is a thermodynamic property of the medium. The velocity of sound is therefore not a "fundamental" physical property like the velocity of light c, which appears to be a fundamental property of a vacuum. The value of c is given by the square root of the product of electric and magnetic permittivities, both being electromagnetic "properties" of the vacuum. It applies to the propagation of gravitational waves as well and is therefore of an "even more fundamental" nature than electromagnetism.

In order to express the relation between curvature of space–time and density of mass-energy, we have to rephrase the latter in terms of a length measure, which is possible by multiplying it by the "natural constant" $G/c^4 = 8.25 \times 10^{-45}\ \mathrm{m\,J^{-1}}$. The appearance of Newton's constant G, and of the velocity of light *in vacuo* c, underline the correspondence of general relativity with both special relativity and Newton's theory of gravitation in their respective ranges of validity. Newton's constant G is shown to be a fundamental natural constant like the vacuum light velocity and Planck's constant, in the sense in which I have used it in the definition of the Planck dimensions in Section 1.6.

Given the mass-energy density of the sun (amounting to about 10^{20} J m^{-3}) the curvature of space–time near the sun (measured in reciprocal square metres, i.e. mass-energy expressed in metres via G/c^4, divided by the volume given in cubic metres) comes out to be no more than 10^{-24} m^{-2}. The length of the light path of about 10^{11} m between sun and earth is necessary to make the deflection of photons, passing close by the sun, just observable on earth.

In fact, this happened very soon, right after the end of World War I, when the Royal Society equipped two expeditions – one to Sobra in Brazil and the other to the island of Principe in West Africa – in which several of the most prominent astronomers of Great Britain participated. Their task was to obtain precise photographic records of the solar eclipse on May 29th 1919. The magnitude of the effect expected according to Einstein's theory was hundredths of millimetres only, but the accuracy attained was sufficient for a convincing verification of the predictions of general relativity. It was essentially this event that established Einstein's stellar international fame.

Einstein himself at that time was already fully convinced of the correctness of his theory. One reason is that he was able to account precisely for an anomaly in the motion of the planet Mercury – a precession of its distance of closest approach to the sun, the so-called perihelion – which had no explanation within the framework of Newton's theory.

In our day, the fine and accurate technology of astronomical observation, especially in combination with a space telescope, allows the detection of the curvature of space–time by light from distant galaxies, thousands of billions of light years away, viewed through clusters of galaxies in the foreground.

What is more, even the minute acceleration of photons by the gravitational force of our planet can be studied in a laboratory on the surface of Earth. For sure, it requires the detection of a very tiny effect, namely the energy gained or lost by a photon travelling either towards or away from the surface of Earth. The relative change of its energy $\delta E/E$ is given by gl/c^2, where $g = 9.81 m/s^2$ is the acceleration furnished by the Earth's gravitational field (i.e. $g = GM/R^2$, with G being the gravitational constant, M the mass and R the radius of the Earth), l the length of the light path in the Earth's gravitational field and c the velocity of light in free space. (The energy term itself, $E = h\nu$, drops out because it appears also in the gravitational law as E/c^2, and this explains the appearance of $1/c^2$ in $\delta E/E$.) The value of $\delta E/E$ per metre of light path is only 1.09×10^{-16} m^{-1}. It is measurable as a Doppler shift (see Section 1.7) in the wavelength of the photons: a blue shift if the photon travels towards and a red shift if it travels away from the Earth.

In the visible or near ultraviolet range of light, such a small effect is far below any achievable resolution. Photon energies (E) are in the range of one or a few electron volts and the lifetimes of excited states in the electronic shells of atoms are usually between 10^{-7} and 10^{-9} s. Owing to quantum-mechanical uncertainty, this causes a finite bandwidth of the corresponding spectral "lines" in the range of $\delta E = 10^{-7}$ eV, yielding $\delta E/E$ values expressed by bandwidth and partial overlap of the emission and absorption lines (with resulting "resonance absorption") and exceeding by many orders of magnitude the relativistic effect to be expected.

The situation is quite different if we consider the excited state of an atomic nucleus (see Section 1.5) rather than processes that occur in its electronic shell. The photons emitted from the nucleus are gamma (γ) rays with energies some five orders of magnitude larger than those considered above. The lifetimes (τ) of excited nuclear states are of the same order of magnitude as those of the fluorescence states in the atomic shell, because they are also of electromagnetic nature. Since the energy (E) of the photons lies in the kilo to megaelectronvolt range, an unprecedentedly low width of the resonance lines (given by $h/\tau\ E \approx 10^{-12}$, h being Planck's constant divided by 2π) is to be expected in γ-ray spectroscopy. However, there was still a problem, associated with what is called "resonance absorption" and which is due to the law of conservation of momentum.

Conservation of momentum demands a recoil of the particle that emits a photon, just as it also implies momentum transfer from a photon that is absorbed. Hence photons, emitted in the decay of an excited state, have a lower energy than is required in order to produce that very state. In light optics (E being in the electron-volt range) the kinetic energy transmitted to an atomic particle by recoil is generally so small (e.g. 10^{-11} eV) that it can be disregarded. Hence the line shift is much smaller than the natural line width, and the overlapping of the lines of emission and absorption is the cause of resonance absorption. However, this is no longer true for nuclear excitation processes.

The kinetic energy obtained by recoil depends on the square of the momentum that is transmitted. We remember that kinetic energy E_k can be expressed as $p^2/2m$, where p is the momentum and m the mass involved. Let us consider a nucleus with an atomic weight of 50 and a γ quantum of an energy of 100 keV. If we multiply both the numerator and the denominator of the above ratio by c^2 we obtain $(pc)^2/2mc^2 = (10^5\ \text{eV})^2/100\ \text{GeV} = 10^{-1}$ eV (see Table 1.1 in Appendix A1.1). Emission and absorption lines are now separated by a distance that no longer allows "resonance absorption", a property needed for detecting the influence of gravity on electromagnetic radiation.

It was Rudolf Ludwig Mössbauer, a Ph.D. student in the physics department of the Technical University of Munich under Heinz Maier-Leibnitz, who used an ingenious trick to overcome the difficulty mentioned above. He found that fixing the atomic nucleus in a metal lattice has the effect of enlarging the recoiling mass to macroscopic dimensions and thereby making possible an almost recoil-free emission and absorption of the energy-rich γ quanta. The Mössbauer effect was recognised by the 1961 Nobel prize for physics, which the 32-year-old Mössbauer shared with Robert Hofstadter, who had explored the charge distribution inside the proton and the neutron by Rutherford scattering (see Section 1.5). There are many applications of Mössbauer's discovery, ranging from nuclear physics to chemistry. Mössbauer did his original experiments using as a radioactive source osmium 191, which is converted via β-decay (see Section 1.5) to excited iridium 191: $^{191}Os \rightarrow ^{191} Ir^{xt} + e^- + \bar{\nu}_e$. The excited iridium is the source of γ photons with an energy of 129 keV. Moving the emission or absorption source at rates of only a few centimetres per second produces Doppler shifts and hence allows the precise overlap of the sharply defined emission and absorption levels.

It was this discovery of "resonance absorption", with extremely sharp spectral lines, which was immediately used by another pioneer in the field, Robert Vivian Pound. Together with his coworker Glen Rebka Jr., in 1960, he was able to detect directly the attraction of photons in the gravitational field of the earth, as expected from general relativity.[135] Pound and Rebka used as a γ source the iron isotope ^{57}Fe in its excited state, produced from ^{57}Co via K capture. This is the capture of an electron by the nucleus according to $p + e^- \rightarrow n + \nu_e$, where the captured electron comes from the inner K shell of the atom and thereby leaves a vacancy that is filled by an electron from a higher energy level, with simultaneous emission of an X-ray in the keV range (see Figures 1.8.2 and 1.8.3).

The experiment, sketched in Figure 1.8.4, required careful preparation to ensure that emission source and absorber have closely matched energy levels. For this purpose the emission source, a thin sheet of iron, was electroplated with ^{57}Co (which

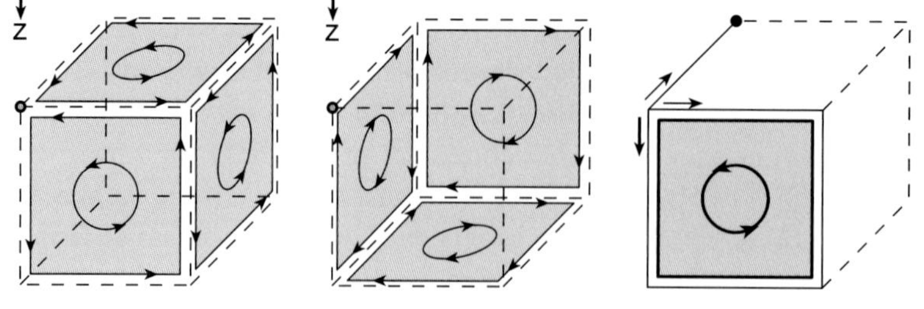

Figure 1.8.2

The three-dimensional hyperface of the four-dimensional hypercube of space–time

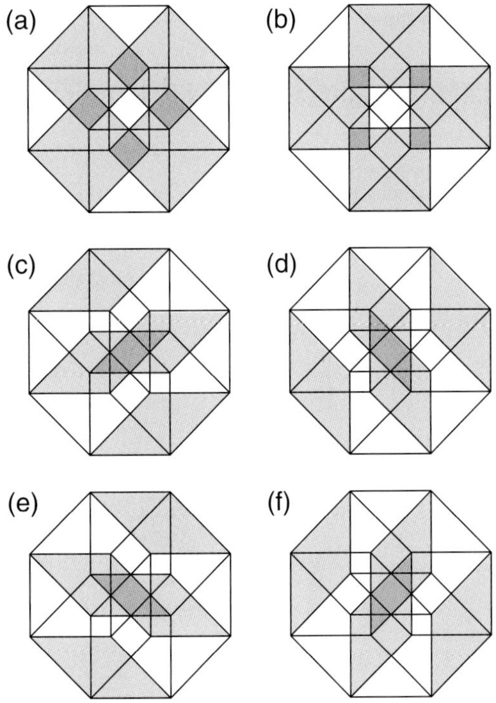

Figure 1.8.3
The eight hyperfaces in the hypercube of space–time

has a half-life of 271 days) and annealed, so that the cobalt could reproducibly penetrate the iron sheet, while the absorber consisted of pure ^{57}Fe. Source and absorber were mounted vertically 22.6 m apart in a five-floor building. The linewidth $\delta E/E$ amounts to about 10^{-12}, that is, about a hundred times larger than the expected gravitational shift. It therefore required a very exact determination of the shape of the absorption curve by averaging over large numbers of photons, with correspondingly long exposure times and under careful temperature control.

To find the exact position of the absorption maximum, Pound and Rebka used a Doppler-shift compensation procedure. The Doppler shift is produced by moving the source with a certain (very small) velocity υ towards or away from the absorber, which causes a line shift from an energy E to an energy $E(1 \pm \upsilon/c)$. The velocity υ has to be chosen so as to yield a maximum of resonance absorption, which means compensating exactly for the gravitational shift $\pm gL/c^2$, yielding $\upsilon = gL/c = 7.4 \times 10^{-5}\,\mathrm{cm\,s^{-1}}$. This was achieved by using a mechanical device that converted clock-driven rotation into slow linear motion (see Figure 1.8.4). By

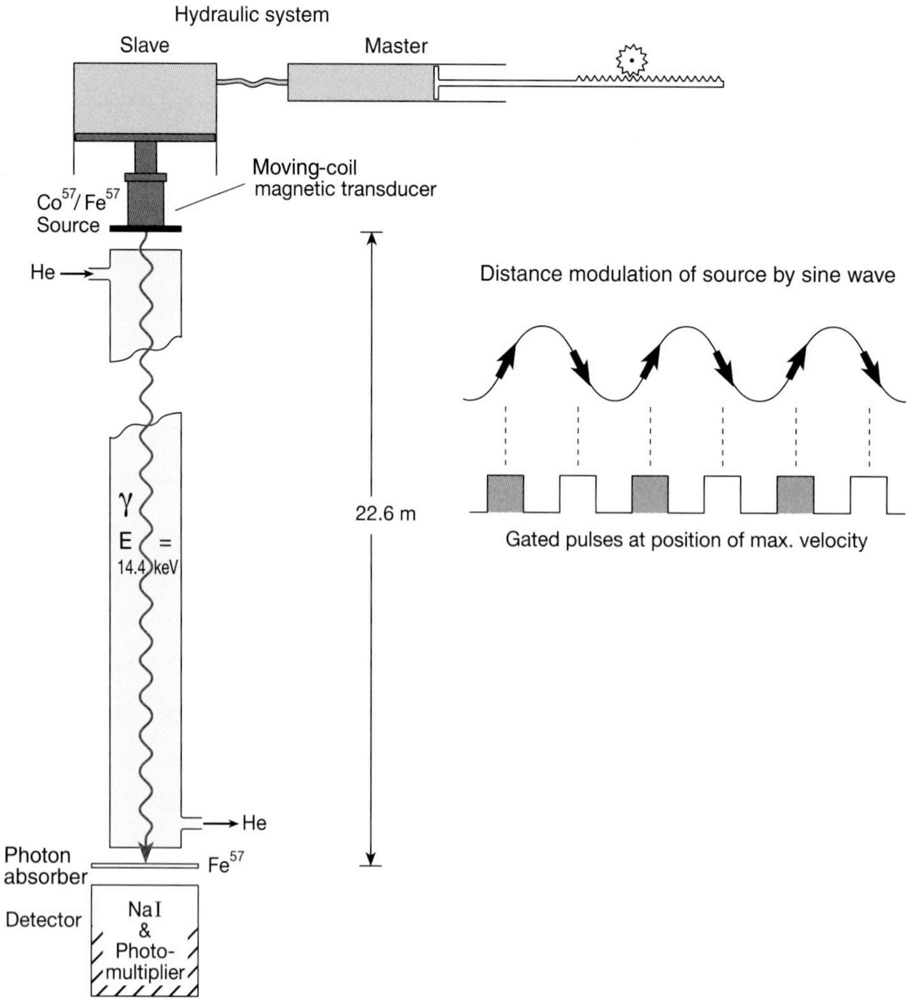

Figure 1.8.4

The gravitational attraction of a photon to the earth

This diagram shows the scheme of the Pound–Rebka experiment, by which the gravitational attraction of a photon to the earth was measured in the laboratory. The photon source is the 14.4 keV X-ray from ^{57}Fe decay. The required narrow width is obtained by moving the source slowly backwards and forwards, and utilising the Mössbauer effect. The Doppler-shifted photons fall over a distance $L = 22.6$ m and gain an energy $\Delta E = EgL/c^2$, where g is the acceleration due to the earth's gravity and E/c^2 is the mass equivalent of the photon's energy. The energy change due to the Doppler effect produced by the slow motion of the source is $E–E(1–\beta) = E\beta$. The trick is to choose a velocity of source motion (and hence a β value) for which β is equal to gL/c^2, thereby producing an optimal resonance condition for the absorption of the photons at the target. Moreover, the effect was measured in both directions separately, i.e. with the photons travelling both up and down the tower. The picture shows only the principle and not the (quite involved) technical detail of Pound and Rebka's experiment.

adjusting precisely the angle of the photons' motion, Pound and Rebka were able to confirm that the maximum effect was detected when the direction of the emitted photons was exactly normal to the surface of the earth. Measuring the shifts for both the upward and downward motion of the photons, the results obtained were $(\delta E/E)\text{down}-(\delta E/E)\text{up} = (5.1 \pm 0.5) \times 10^{-15}$. The theoretical value according to general relativity is 4.9×10^{-15}.

The experiment described in the paper of Pound and Rebka is a masterpiece of physical artistry, preceding our age of laser technology and femto- to attosecond pulse equipment. Experimental testing is complementary to producing new ideas, regardless of whether its result is affirmative or negative. Take the Michelson–Morley experiment (Section 1.2). The negative result preceded our understanding of it and hence provoked a new theoretical interpretation by special relativity. General relativity, on the other hand, was way ahead of any of its observational or experimental verifications. In fact, the proof of the existence of gravitational waves is still missing, although it may be close to realisation.

Theory and experiment have to go hand in hand, and we shall see in later parts of this book, dealing with physical reflections on life and thought, that the worth of a theory is as great as its amenability to experimental testing. The late Karl Popper would have preferred the word "falsification", but given the many feedback loops between trial and error the situation usually is too complex to allow a binary "right or wrong" decision.

In the rest of this section I shall discuss the consequences of all our theoretical and experimental insights into the structure and history of our universe.

General relativity has turned out to be the most relevant theory of the universe – final if we exclude very small distances or a possible singularity at very large energies. It is a dynamical theory, although Einstein himself wanted it to be static, which was the classical view of the universe. So he added to his equation a constant term, the so-called cosmological constant, in order to compensate for gravitational attraction by some "antigravitational" repulsion and thereby to force the system into a stationary state. Later, when the expansion of the cosmos became manifest and could be accounted for by non-static solutions of Einstein's equations, he called his cosmological constant "the biggest blunder of my life" (*die größte Eselei meines Lebens*). Apart from the fact that Einstein's theory is now generally accepted as a dynamical theory of the universe, the cosmological constant occupies an important position in present-day deliberations concerning a pre-existing negative curvature of space–time.

Einstein's non-linear field equations as such are very difficult to handle, and at first even Einstein himself did not expect that an exact solution would ever be obtained. All the greater was the surprise that soon after Einstein's first publication in 1915 the German astronomer Karl Schwarzschild, who made important theoretical and experimental contributions to 20th-century astronomy, produced an exact solution for the geometry in the neighbourhood of a point mass. He also laid the foundation stone for the theory of black holes. Schwarzschild died soon afterwards from an illness that he contracted while serving in the German Army in World War I at the Russian front.

The general dynamic solution of Einstein's theory was first provided in 1922 by the Russian meteorologist Alexandr Alexandrovich Friedmann.[136] It distinguishes two cases (with a third, intermediate, one in between) (Figure 1.8.5). The solutions depend critically on the mass-energy content of the universe and decide whether the presently observed expansion of the universe will go on for ever or whether the universe will finally collapse, the intermediate case being a critical mass density at which expansion slows down asymptotically and which is called the "Einstein–de Sitter universe"[137] (see Figure 1.8.5). Friedmann's equation correlates the density of mass-energy with a scaling parameter describing expansion. It was the epoch-making discovery of line shifts in the spectra of stars in the distant galaxies, made by the US astronomer Edwin Powell Hubble, that lifted Friedmann's equation to practical significance and finally convinced Einstein of the dynamic nature of the universe[138].

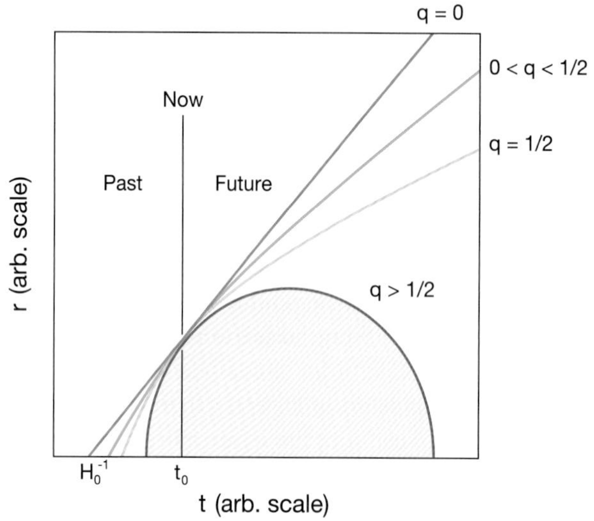

Figure 1.8.5
Will the universe expand for ever?

Friedmann's solution for the dynamics of a universe with uniform and isotropic mass density, based on general relativity, provides three possibilities, depending on the energy density of the universe expressed by its curvature parameter q. These are: unlimited expansion ($q < 1/2$), approach of steady conditions at critical mass density ($q = 1/2$; the "Einstein–de Sitter universe") and reversal of expansion, i.e. later contraction of the universe ($q > 1/2$; the "big crunch").

The cosmological principle states that whatever applies to "our" part of the universe should also apply for any other part. In particular, Hubble observed two well-defined lines in the near ultraviolet region, caused by calcium atoms in the stars' atmospheres. In the way already described in Section 1.7, Hubble was ultimately able to present an impressive diagram demonstrating a linear increase in the red shift with increasing distance of the galaxy from which the light comes. (The difficulty in producing such a diagram lies in the indirect way astronomers have to determine distances by resolving luminosities.)

One can interpret Hubble's revolutionary findings as a "swelling" of the universe much like the swelling of an inflated balloon. The relative expansion of the universe is described by Hubble's constant, which is a parameter in Friedmann's equation. Its present value is estimated to be 15–30 km s^{-1} per million light years, its reciprocal being equivalent to an estimate of the age of our universe of 10,000 to 20,000 million years.

There is now almost unanimous agreement among cosmologists that the universe has such a finite age. The assumption is that it originated some 10,000 to 15,000 million (i.e. $\approx 10^{10}$) years ago in a gigantic explosion referred to as the "big bang", as described in 1946 by the Russian-born US physicist George (Georgi Antonovich) Gamow, who revised an original (1927) idea of the Belgian cosmologist Georges Edouard Lemaître. (Gamow's idea became known through the famous "Alpher–Bethe–Gamow" paper of 1948.[139]) The phrase "big bang" was coined by Fred Hoyle, who used it to ridicule the model. Present-day acceptance of the big-bang hypothesis rests more on factual evidence than on deductive theoretical models. What are the empirical facts? Let me quote three pieces of evidence:

i) The first is the Doppler shift in spectra, as discovered by Hubble. In fact, the interpretation as a Doppler effect is not quite adequate. It is the expansion of space as a whole that causes what is called a "Doppler shift".

ii) The second fact is the distribution of elements, in which we find a pronounced preponderance of light elements such as hydrogen, helium-4, deuterium, helium-3 and lithium-7. This would be expected kinetically when these elements build up from protons and neutrons. Studies of this kind were pioneered by Fred Hoyle, who together with Hermann Bondi and Thomas Gold proposed an alternative "steady-state model" of the universe, but who later honestly admitted that the data on the origins of the elements in fact support the big-bang model.[140]

iii) The third argument is provided by the cosmic background radiation detected by Arno Penzias and Robert W. Wilson,[141] which refers to a radiation temperature of about 3 K and its ripple inhomogeneities, disclosed by the Cobe satellite experiment by George Smoot.[142] They are the echoes of what happened at the dawn of the universe (see Section 1.3).

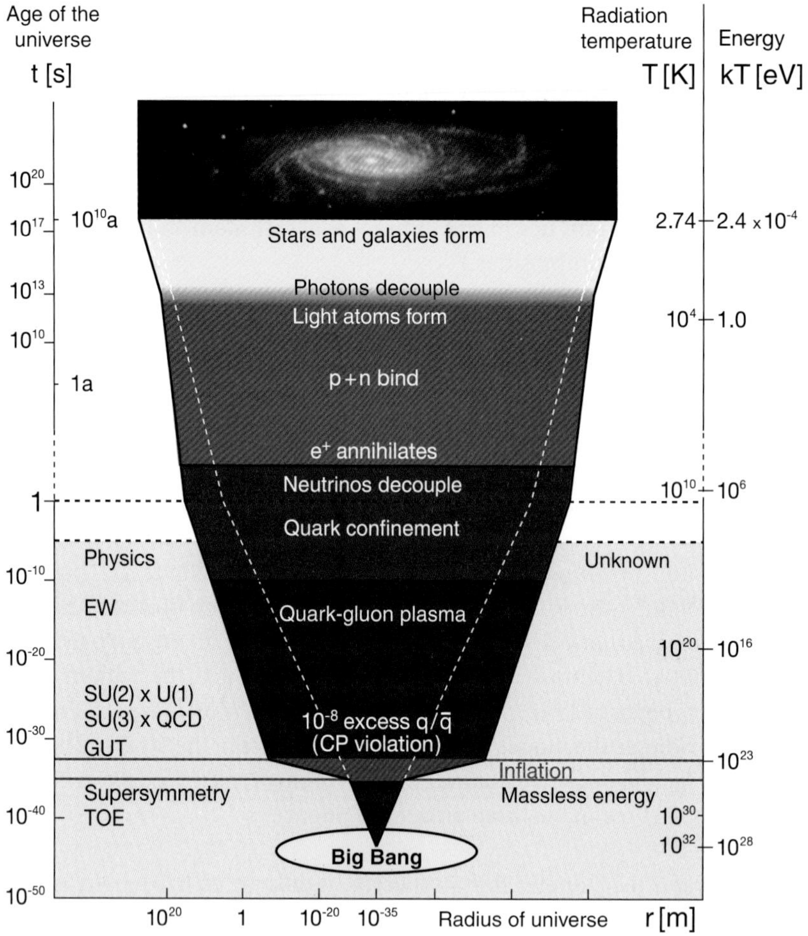

Figure 1.8.6

Origin and evolution of the universe

The line of reasoning from present-day cosmic background radiation back through the evolution of the cosmos leads the way sketched in Figure 1.8.6, which combines the results of Sections 1.6 to 1.8 (see also Vignette 1.8.2). When the expanding universe cooled down, it went through a phase in which atoms were still separated into nuclei and electrons. In such a hot and dense plasma, photons were scattered by free charges and, thereby, thermal equilibrium between matter and radiation was established. The universe at that time was radiation-dominated and opaque. Upon further cooling – to a temperature of about 3000 K – electrons and nuclei combined, and the universe became matter-dominated and transparent. In the absence of free charges, radiation was no longer confined by matter, and the thermal equilibrium between the two ceased to exist. The average distances between photons (and, concomitantly, their wavelengths) simply increased in proportion to the increasing size of the universe, exhibiting, so to speak, a steady red shift. A quantitative treatment of this scenario shows that the photons – although no longer in contact with matter – maintain an equilibrium energy distribution that is described precisely by Planck's formula, but referring to a temperature that drops in inverse proportion to the scale of expansion. The finding of Penzias and Wilson, i.e. a drop in the radiation temperature from about 3000 K to 3 K (the precise figure is 2.735 K), correlates well with an estimated expansion of the universe by a factor of about 1000 and – as has now been confirmed by subsequent studies at different wavelengths – also correlates well with Planck's formula. It may be seen as the reverberation of an early phase of evolution of the cosmos, referring to a time about 300,000 years after the big bang, and hence including a substantial part of the time span between the origin of the universe and the present, especially when events are viewed, as they correctly should be, on a logarithmic time scale.

Vignette 1.8.2 On the construction of the diagram in Figure 1.8.6

The diagram relates four parameters from the beginning of the universe to the present time: time (t) in seconds [s], size (R) in metres [m], temperature (T) in degrees Kelvin [K] and energy (kT) in electron volt [eV], k being Boltzmann's constant.

It makes use of two relations:

i) The law of Stefan and Boltzmann, in the specification due to Planck's law (see Figure I.3.3): $R = \sigma(kT)^4$ where R is the total power per area radiated and $\sigma = 1.03 \times 10^9\ W \cdot m^{-2} \cdot eV^{-4}$, and

ii) the Friedmann equation, which relates the square of the Hubble parameter (H) to the energy density $\rho(T) : 8\pi G\rho(T)/3c^3$, where G is Newton's gravitational constant and c the velocity of light. Since the reciprocal of Hubble's parameter yields the characteristic expansion time t_{exp}, we have $t_{exp} = \sqrt{3c^2/8\pi G\rho(T)}$. (The current expansion time $t_0 = H_0^{-1}$ is $5 \cdot 10^{17}$ s.) The present background radiation temperature is 2.74 K. We then have for the radiation-related energy density $\rho_r(T) = (0.4\ MeV/m^2) \cdot (T/2.74)^4$. Combination of these equations yields for the expansion time of the universe (as long as the energy of the universe is radiation-dominated): $t_{exp} = 3.77 \cdot 10^{20}/T^2$ s if T is measured in [K].

At temperatures below 10^4 K the energy density of the universe becomes dominated by matter. This contribution $\rho_m(t)$ is given by $\rho_m(T) = (0.5\ GeV/m^3)(T/2.74)^3$,

Cont. ➲

➲ Cont.

leading to an expansion time (as a function of temperature) of $t_{exp}(T) = (2.74/T)^{3/2} \cdot \sqrt{3c^2/8\pi G\rho_0}$, where ρ_0 is the present mass-energy density of the universe. The change from $t_{exp} \sim T^{-2}$ to $t_{exp} \sim T^{-3/2}$ at temperatures less than 3000 K is seen clearly in the slope of expansion in Figure 1.8.6. If in the above equation ρ_0 is assumed to be the critical mass-energy density required for a closed universe according to Einstein's theory of general relativity, the expansion time becomes equal to $5 \cdot 10^{17}$ [s].

When one looks at Figure 1.8.6, what catches the eye first is the transition from a light to a dark background in the upper quarter of the diagram. If we could have been around as observers during this phase of the history of the universe would have been as eye catching as it is in the diagram. However, this would have been quite uncomfortable for any observer at any place in the universe.

Essentially two events happened to coincide in this late phase of evolution: The energy content of the universe, so far dominated by radiation processes, switched to become mainly determined by the mass-energy content. There are about a million times more photons than material particles in the universe, but the expansion causes a continuous red shift which is equivalent to a continuous decrease in the energy of all photons, whose contribution to the total energy content was eventually reduced to a small fraction. The other effect caused the universe, which up to this point had been opaque because of the interaction of electromagnetic radiation with the electric charges, to become transparent. The ionisation energies of atoms, as was shown in Figure 1.5.1 for the outer electrons of the atomic shell, are of the order of magnitude of one or a few electron volts. In the temperature range between one thousand and ten thousand Kelvin the atoms become neutral and, since there is a meticulous balance between positive and negative charges, most of the free charges disappeared.

The other, no less dramatic changes in the composition of the universe, occurred under visually less spectacular circumstances. Before the formation of neutral atoms, when photons decayed to the present 3 K background radiation level (see Figure 1.3.4), the lighter nuclei formed from free protons and neutrons. This phase was preceded by the formation of protons and neutrons from a confinement of quarks, as was described in Section 1.6. At the same time the neutrinos decoupled, freeing electromagnetic interactions of electrons and allowing the "clumping" of matter necessary for the formation of stars. At the same time, and without any obvious causal relation, radiation lost its dominant rôle. Weinberg estimated about a million photons per nucleon. Their energy density, being proportional to the fourth power of temperature (see Section 1.3) exceeded by far that of the mass-energy of particles. However, the expansion of the universe enlarges the wavelength of

Cont. ➲

Cont.

a photon. Since the numbers of both photons and particles stays about constant, the energy content of the universe eventually became dominated by matter rather than by radiation. This happened progressively after the temperature had reached a few thousand degrees Kelvin, where the energies of the photons become about a million times smaller than the mass-energy of an electron. Since ionisation energies of neutral atoms such as the hydrogen atom (see Figure 1.5.1) are of the order of magnitude of a few eV, the electrons recombine with the nuclei to form neutral atoms. The neutralisation of charges not only made the universe transparent, but also was an important prerequisite for allowing gravitational dumping of matter and hence star and galaxy formation.

At the other end of the evolutionary time scale, i.e. towards higher temperatures, other important phases had to be traversed until atomic nuclei could form. Let me start right at the beginning. The big bang, the occurrence of which is rendered highly probable by the three arguments discussed in this section. Apart from the likelihood of the event as such, we have no idea how it started. If it was a quantum fluctuation in the range of Planck dimensions, its occurrence in Planck time (10^{-43} s) was limited to an amount of energy of some 10^{19} GeV (i.e. Planck energy). This is almost nothing in comparison with the energy content of the present universe, but it is unrivalled in energy density. With a Planck length of $l_p = 10^{-34}$ m this comes to the order of magnitude of 10 eV/m^3, as much as the mass-energy density of a neutron star. We do not know anything about the physical laws that operate under such conditions. But it is the range in which all forces must have been unified and in which gravitation, as the first of the four forces, veered away from the others.

At times 5–10 orders of magnitude longer, the universe is supposed to have undergone a period of exponential growth in which its present mass-energy content was almost reached. This was followed by separation of the strong forces, again as a result of some physics still unknown, that left electromagnetic and weak interaction still united. Under these conditions the universe can be imagined as a plasma of quarks, gluons, leptons, weak-interaction bosons, and γ-quanta.

How the universe is thought to have come about is vividly described by Steven Weinberg in his book *The First Three Minutes*, which concentrates on the time interval for which the physics are nowadays sufficiently well understood.[143] However, physicists and cosmologists also like to speculate about earlier phases of our universe, and this brings us back to the results (and the open questions) of Section 1.6.

All our questions are certainly rooted in our present way of thinking. They may become irrelevant once we have reached a new platform. With this proviso we may now look again at Figure 1.8.6.

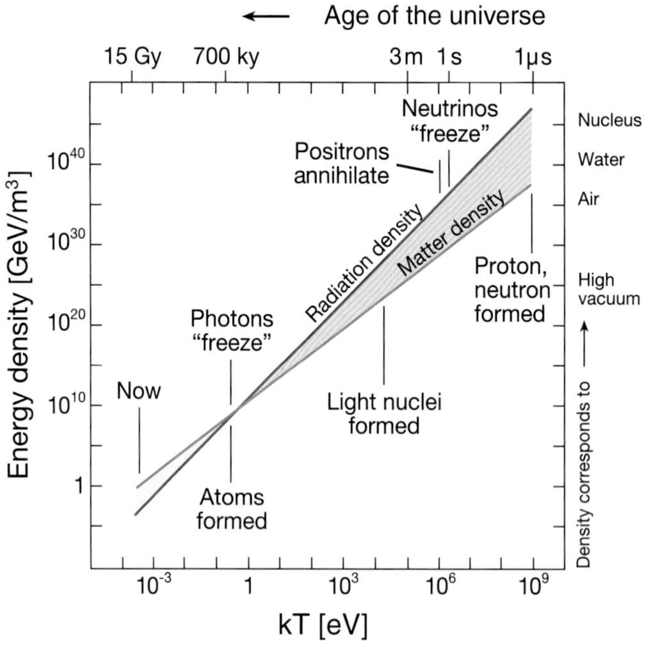

Figure 1.8.7

The main events in the evolution of the universe – as far as they can safely be reconstructed

The energy density versus temperature after the hadrons have frozen out.

We are on solid physical ground now only for times longer than a millionth of a second. This means that more than half of the logarithmic time scale (shaded in Figure 1.8.6) still – more or less – lies in the dark. Figure 1.8.7 records the main events in the better-charted range of $> 1\,\mu s$. For quite some time (on the logarithmic scale), the universe was a radiation-dominated plasma, more than 10^{12} K hot and less than 1 km in diameter. The subsequent era is described neatly in Steven Weinberg's book. I shall talk instead about the preceding era, in order to show how physicists imagine possible solutions for a largely unsolved problem.

What worries physicists most is the "how" of the phase transitions that decoupled the various forces. The earliest of the transitions, separating gravitation from the other forces, must have happened immediately after the Planck era. It was followed by what has been called the period of inflation, i.e. an exponential growth period during which the original energy fluctuation gained its mass, probably associated with "condensation" of the strong forces. Such an exponential growth means the doubling of the mass-energy content after the elapse of a certain time interval. Imagine a doubling time of (say) 10^{-35} s. After a time of 2×10^{-33} s the universe would then have grown by a factor $2^{200} \approx 10^{60}$, i.e. from its original Planck size to almost its present mass equivalent of about 10^{80} proton masses.

There is indeed good reason for such an assumption – or should I rather say "Guth reason", because it was the MIT physics theorist Alan Guth[144] who put forward the theory of inflation. This theory also includes the possibility of branching into a multitude of different universes. Inflation flattened the originally quite inhomogeneous universe and made it appear uniform. Moreover, it filled it with matter up to about its critical density (probably in association with the Higgs mechanism discussed in Section 1.6). Since inflation started from a single speck of dense material in the wildly fluctuating original universe, there may have been ample chances for generating a multitude of parallel universes.

What about my original question?

"A finite island in an infinite ocean of space" may be a nice metaphor to describe the universe, but it has little physical background in what we know so far. What do we mean by "space" if the word refers to a possible "multiverse" of expanding universes each of which is creating its own space? We cannot deal with such a question in classical terms that originated in our experiences and conceptions. Comprehending a dynamic cosmos requires us to retreat to the smallest dimensions that we can think of, given the known parameters of our physical universe. It requires the unification of cosmology with particle physics. Modifying Lederman's question slightly, I may ask, in response to this section's title: If the existence of our universe, of which we are the only (?) witnesses, is an answer, then we must formulate our question in a new language.

1.9. Symmetry: *A Priori* or *A Posteriori*?

"The essential uncertainty of quantum behaviour might create the uneasy suspicion that anything goes in this world. But since the universe has (on all observed scales of length, time, mass and so forth) a very definite structure, it seems extremely unlikely that everything is allowed. There must be at least some things that are more allowed than others." I found this sentence in a book entitled *The Force of Symmetry*, by the Dutch physicist Vincent Icke.[145] It is true, in our world most things are forbidden, in the sense that they are unlikely in the extreme. But then, is symmetry an *a priori* recipe for how to organise the universe, or is it the final "optimum" result in the sense of Leibniz, so to speak a manifestation *a posteriori* of its genesis? What I mean is perhaps best shown in Figure 1.9.1 by the beautiful symmetry we encounter with radiolaria, planktonic protozoans, shown in a masterly drawing by the German zoologist and philosopher Ernst Haeckel, whose talents were not limited to these two fields, as his drawing shows. These creatures certainly represent the optimum achievement of a functional organisation, being an *a posteriori* result of biological evolution.

Figure 1.9.1

Symmetry in unicellular life

A representative (*Sagenoscena stellata*) of the class of radiolaria, which are unicellular protozoa, exhibiting fascinating symmetries. Taken from Ernst Haeckel, who described systematically and illustrated hundreds of those species in his monumental work *Kunstformen der Natur* (Bibliographisches Institut, Leipzig, 1904).

The beauty of symmetry (see Figure 1.9.2) has fascinated mathematicians and physicists ever since Plato, as is documented in the writings of many of the leading figures in these fields. Nevertheless, beauty as such is a phenotypic category, a result of "expression". The symmetries that govern physics and have dominated our universe since its dawn have always been considered to reflect a universe subject to symmetries that were given *a priori*.

First, there are the elemental, continuous symmetries associated with space and time. We recall Noether's theorem, according to which continuous symmetries are inherently linked to conservation laws. The whole of classical physics, including the statistical foundation of thermodynamics and chemical kinetics, is based on the conservation laws for energy, momentum and angular momentum, which are in turn

Figure 1.9.2
The beauty of snow crystals

In his novel *The Magic Mountain* (*Der Zauberberg*) Thomas Mann calls them, "Little jewels, insignia, orders, agraffs – no jeweller, however skilled, could do finer, more minute work ... And among these myriads of enchanting little stars ... there was not one like unto another." The picture shows a small selection from 2453 photographs of natural snow crystals assembled by W.A. Bentley and W.J. Humphreys in their book *Snow Crystals* (Dover, 1931).

consequences of the invariance of the Lagrangian or Hamiltonian with respect to time and to translation and rotation in space. Relativity and quantum mechanics have modified these laws, but not repealed them. Another classical conservation law – although its origin is less clear – is concerned with electrical charge. No way is known of either creating or destroying electrical charge; hence, its conservation (not to be confused with conservation of charge conjugation, described below) is taken to be of universal validity.

Symmetry is described mathematically by group theory. A group $G = (g_1, g_2 \ldots)$ is a set with an associative binary operation called multiplication, where $g_i \times g_k$ always exists and is itself an element of the group. A group is called "Abelian" if the commutative law $g_i \times g_k = g_k \times g_i$ is fulfilled for all its elements. All one-dimensional groups are Abelian, and so is translation in space and rotation in a plane (i.e. around one axis). An object that remains the same after some manipulation by the group elements is called "symmetric" under the particular group. Mathematicians have classified all groups (cf. Elie Cartan[146]), distinguishing point groups having a finite number of elements from continuous or Lie groups with an infinite number of continuous elements. All groups in which any two elements do not commute with each other are called "non-Abelian".

A good example of non-Abelian group behaviour is rotation in three-dimensional space, while rotation in a plane is represented by an Abelian group. Rotation in space with three mutually perpendicular axes of rotation requires "memory" of the order of successive rotations around the single axes. Anyone who has played with a Rubik's cube[147] knows how difficult it is to reproduce a certain pattern. Given the number of possible moves, operating by random walk can be quite frustrating. This is due to the non-Abelian nature of the group of three-dimensional rotations, where the non-commutative symmetry of the elements of the group makes it impossible to return to the initial state by simple repetition of an operation. The quantum theory of the angular momentum of atoms, where "intrinsic" contributions are to be included, is an example where higher symmetries are to be taken into consideration.

In particle physics we encounter a whole variety of non-Abelian symmetries. One of the fundamental discrete symmetries of matter is the existence of antimatter, as first revealed by Dirac's theory of the electron (Section 1.5). However, something appears to be wrong with this symmetry, because we immediately ask: Why does our universe essentially consist of matter? Where is the antimatter? Of course, a trivial answer would be: if both were around we would not exist, because owing to their symmetry they would annihilate one another, thereby producing a stupendous amount of energy, which would make life quite uncomfortable. Well, the answer is certainly not wrong, but it does not help us much to understand the situation; in fact it heightens the problem.

Particle physics indeed gives us reasons to believe that particles come about in pairs, in order to fulfil a symmetry requirement. Take a look at Figure 1.9.3. The mirror image of a rotational ellipsoid would be an identical body. The mirror image

(a)

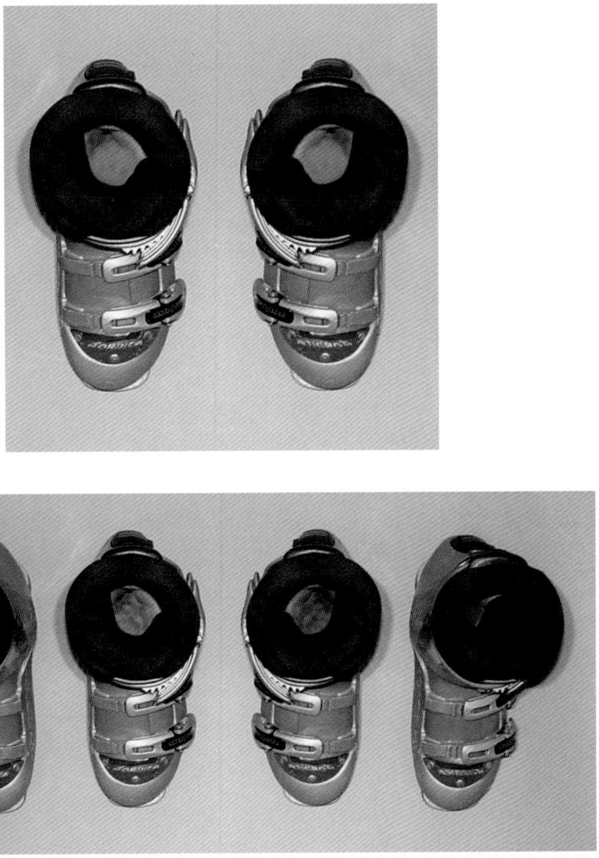

(b)

Figure 1.9.3
Symmetry of chiral objects

A chiral object, such as a right boot, is reflected and thereby converted into its mirror image – the left boot. A complete pair of such objects (a right plus a left boot) when mirrored, remains the same.

of a right shoe, however, is a left shoe, which no manner of rotation can make congruent with the right shoe. The difficulty disappears if we take a pair of shoes – which, of course, has to include a right and a left one. Its mirror image is just such a pair of shoes (supposing that the two shoes, apart from their "handedness", are indistinguishable). Closer to fundamental physics is the example of a mirrored spiral (Figure 1.9.4).

"Handedness" is an important property at the particle level, where it is called "parity" (abbreviated to P). Parity is expressed by a quantum number that can be either +1 or −1. It describes the transformation of the wave function if the spatial

$u\bar{u}$ π^0,η										
$d\bar{u}$ π^-	$d\bar{d}$ π^0,η	$\bar{d}u$ π^+								
$s\bar{u}$ K^-	$s\bar{d}$ $\bar{K}^0$	$s\bar{s}$ η,η'	$\bar{s}d$ K^0	$\bar{s}u$ K^+						
$c\bar{u}$ $\bar{D}^0$	$c\bar{d}$ D^+	$c\bar{s}$ D_s^+	$c\bar{c}$ η_c	$\bar{c}s$ D_s^-	$\bar{c}d$ D^0	$\bar{c}u$ D^-				
$b\bar{u}$ B^-	$b\bar{d}$ $\bar{B}_d^0$	$b\bar{s}$ $\bar{B}_s^0$	$b\bar{c}$ B_c^-	$b\bar{b}$ η_b	$\bar{b}c$ B_c^+	$\bar{b}s$ B_s^0	$\bar{b}d$ B_d^0	$\bar{b}u$ B^+	B-physics	
$t\bar{u}$ T^0	$t\bar{d}$ T_d^+	$t\bar{s}$ T_s^+	$t\bar{c}$ $\bar{T}_c^0$	$t\bar{b}$ T_b^+	$t\bar{t}$ η_T	$\bar{t}b$ T_b^-	$\bar{t}c$ T_c^0	$\bar{t}s$ T_s^-	$\bar{t}d$ T_d^-	$\bar{t}u$ $\bar{T}^0$

Figure 1.9.4
Table of mesons

This shows the $5^2 = 25$ possible combinations of each one of the five quarks (u, d, s, c and b) with each of their antiquarks. The last (colourless) row refers to the top quark (t), the energy of which is too high for present-day experimentation and must therefore be reserved for the future. For details of B-physics see text.

coordinates x, y and z are reversed, i.e. replaced by their negative values −x, −y and −z. If the wave function changes its sign, one speaks of negative parity; if it does not do so, parity is conserved owing to the wave function's inherent symmetry.

The electromagnetic and strong interactions proved to be invariant under parity transformation. It was assumed that this should also hold for the weak interaction. Therefore, it came as a great surprise when in 1956 Tsung Dao Lee and Chen Ning Yang showed that the weak force causes K mesons to decay in a way that disobeys parity conservation.[148] In 1957 Chien-Shung Wu and her collaborators at Columbia University, New York,[149] demonstrated experimentally the violation of parity in weak interactions by aligning cobalt-60 nuclei in a magnetic field and measuring the directions of the emitted electrons produced by β-decay: ${}^{60}Co \rightarrow {}^{60}Ni + e^- + \bar{\nu}_e$. The result was that more electrons were emitted in the direction opposite to the magnetic field, which is opposite to the direction of the nuclear spins. If parity were conserved, the numbers of electrons emitted parallel and antiparallel to the magnetic field would have been equal.

Let us return to antimatter. There is another parameter, called charge conjugation (represented by a quantum number C), that characterises symmetry with respect to matter–antimatter transformation. Charge conjugation involves several properties, one of which may be charge. Electron and positron carry opposite charges, as proton and antiproton do. Neutron and antineutron, however, do not differ in charge, but in their magnetic moment, which for neutron and antineutron have opposite signs relative to their spins. Charge conjugation, again, turned out to be a symmetry for electromagnetic and strong interactions, but – like parity – failed for weak interaction. The decay of the pion (π^-) into the muon (μ^-) and the muon anti-neutrino ($\bar{\nu}_\mu$) has been detected, while the corresponding decay of the pion's antipartner (π^+) (into the anti-muon (μ^+) and the neutrino ν_μ) would occur only when both C and P are conserved at the same time. (The pion π^- is a meson, π^+ its antipartner. The muon μ^- is the heavier analogue of the electron, its antipartner μ^+ that of the positron. Neutrino ν_μ and antineutrino $\bar{\nu}_\mu$ are the muon-associated neutrinos; see Figure 1.5.5.)

It was long believed that the symmetry of joint transformation CP is conserved, but its violation was observed by Val Logsdon Fitch and James Watson Cronin in 1964 with the K-meson system.[150]

If together with reversal of space coordinates we include a time reversal (denoted T) for which (itself) in weak interactions a small dissymmetry has also been observed, a new situation arises. CPT invariance represents a fundamental requirement of relativistic quantum field theory. No violation of this has been experimentally observed so far; but if there were any such violation it would alter drastically the structure of relativistic quantum field theory. The CPT theorem was first derived by the theoretical physicist Günter Lüders of Göttingen University.[151] This theorem states that all physical processes are invariant under CPT reversal.

Symmetry violation may provide an answer to the question of why matter is in great excess over antimatter in our universe. There is a small favouring of quarks over antiquarks of the order of magnitude of 10^{-8} involved in processes in the very early phase of the universe (Figure 1.8.6) caused by CP violation. In subsequent phases matter and antimatter annihilated one another up to the "small" excess of matter that was left. The energy produced by the annihilation appeared as a large excess of radiation. However, the standard model does not reproduce the observed baryon/photon ratio, being in error by about nine orders of magnitude. CP violation is therefore currently the object of active experimental research. It is not the fact as such which is now in the foreground of interest, and which is anyway no longer in doubt. It is the quantitative aspect that could throw light on the nature and origin of the effect, and it is quite likely that various kinds of CP violation exist. In order to study them more quantitatively, experiments have to be extended to energies appreciably higher than those involved in the original discovery of the effect. The energy of kaons lies well below 1 GeV. To do quantitative research on CP violation one requires higher energies. Good candidates for this are the so-called B mesons, which derive their name from the b = bottom quarks involved, and which have an energy of about 5 GeV. As shown in Figure 1.9.4, mesons are combinations of a quark and an antiquark, where u, d, c and b refer to up, down, charm and bottom (see Figure 1.5.6). Figure 1.9.5 shows a recent detection of CP violation by B mesons

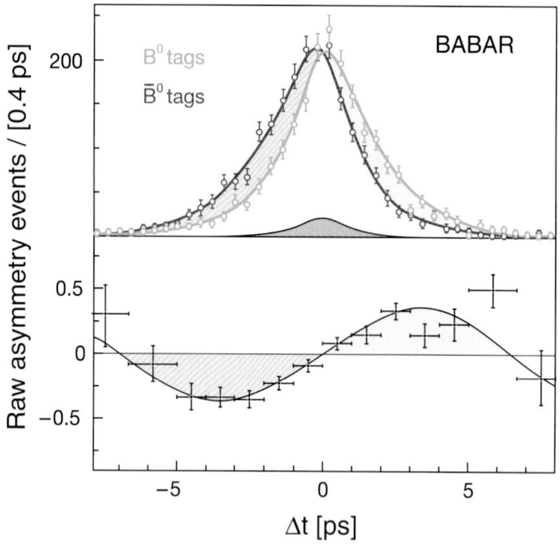

Figure 1.9.5
Distribution of the decay of the B^0 meson and its antiparticle

The B^0 meson consists of one bottom antiquark ($\bar{b}$) and one down quark (d). Its antiparticle ($\bar{B}^0$) correspondingly contains the bottom quark (b) and the down antiquark ($\bar{d}$). The $B^0/\bar{B}^0$ pair is produced in a decay of a higher-energy state, called the $\Upsilon(4S)$, of the ($b\bar{b}$) meson. Both the B^0 and the $\bar{B}^0$ meson can decay to the same final state called a "CP eigenstate". In order to find out whether it was the B^0 or the $\bar{B}^0$ meson that decayed, the remaining meson ($\bar{B}^0$ or B^0) has to be "tagged" (by its specific decay properties). The red curve in the figure corresponds to the decay of the B^0 meson, while the green curve corresponds to that of the $\bar{B}^0$ meson. The asymmetry is emphasised by the lower curve. The points in the upper curves are measured values from the "BABAR" experiment. (The dark-shaded curve indicates the background.) The curves were obtained by a sophisticated fitting procedure. If CP were not violated, the two curves would lie precisely on top of one another. The observed asymmetry is in agreement with the prediction of the standard model of particle physics.
(Courtesy of Gerald Eigen.)

carried out by the BABAR collaboration at the Stanford Linear Accelerator in California. (The term BABAR was proposed by the particle physicist David Hitlin of Pasadena because of the onomatopoeic kinship of the B meson (b, b-bar) with the comic-strip figure of Babar the elephant.) B-physics is a promising branch of research in present-day experimental particle physics,[152] while T-physics, involving the top quark, is reserved for the future.

Apart from matter–antimatter correspondence there are other discrete symmetries in particle physics which I mention here only in brief. All quarks have an intrinsic angular momentum of 1/2 and a baryon number of 1/3. The up and down spins form an isospin doublet of $I = 1/2$. The associated magnetic quantum numbers

m_I are $+1/2$ and $-1/2$. The strange quark behaves quite differently. It has a much larger mass than both the up and the down quark, and its analogue the charm quark is heavier still. The strange quark is an isospin singlet ($I = 0$) with $m_I = 0$. However, it also has a new property, called "strangeness" (S). Strangeness is a quantum number that takes care of the fact that certain particles are produced only in association with other particles. The quantum number assigned to strangeness (s quark: $S = -1$) is additive, with $\Sigma S =$ constant. The antiparticles have S values of opposite sign (examples: $\Omega^- = 955$ has a strangeness of -3; $K^+ = u\bar{s}$ and $K^o = d\bar{s}$ are assigned S values of $+1$). The law of strangeness conservation holds for electromagnetic and strong interactions, but not for weak interactions. The electric charge of quarks is related to m_I, S and B by $q/e = m_I + (S + B)/2$. The quantum numbers of the hadrons (which include baryon numbers, electric charge, spin, isospin and strangeness) are equal to the sums of the quantum numbers of their constituent quarks.

Particle symmetries stress the pair nature of the property to be conserved (as was referred to in Figure 1.9.3). This is obvious for matter and antimatter. A γ quantum may produce an electron–positron pair, but not any of the isolated single particles. Isospin symmetry classifies the proton and the neutron as being one state differing in its isospin quantum number m_I, but otherwise belonging to the same nucleon state. The (not exactly self-explanatory) term isospin or isotopic spin was chosen in the early days of particle physics because of its analogy to the "up and down" spin states of the electron. The term "strangeness" is certainly no more lucid, but again we are dealing with some pair or additive property that is conserved, as is baryon number in strong interactions. In this respect there is one major symmetry still missing, namely that between pairs of bosons and fermions, which is called "supersymmetry". It made its appearance on the stage of particle physics only lately.[153] According to supersymmetry, every boson has a corresponding fermion partner and *vice versa*. It must be said at once that none of the known bosons and fermions are superpartners, and it is left to the future to find any of them experimentally. However, the introduction of supersymmetry proved useful in reducing the infinities that trouble present-day relativistic quantum field theory. The symmetries among bosons and fermions cause such infinities to compensate one another and hence to cancel out. It is possible that any theory able to absorb the standard model will be of a supersymmetric nature.

Physicists these days are at a crossroads, and it is not clear how they are to proceed further. Any quantum field theory has to fulfil the gauge postulates, i.e. it has to exhibit a continuous symmetry called gauge symmetry (which involves transformations of the fields and potentials). If electrodynamic, weak and strong interactions are to be accounted for by a single, all-embracing theory, the symmetry group should include U(1), SU(2) and SU(3), i.e. it might be SU(n) where n is at least 5. Following this road to SU(5) has led to some interesting new discoveries, for instance a rule for the quantisation of the electric charge that is in accordance with what is observed.[154]

However, at the same time it uncovers difficulties which hint at higher symmetries; but where is one to stop along the road from S(5)?

Instead, another procedure has led to more promising results: "superstring" theory. (The prefix "super" in this word refers to a compliance with the requirements of supersymmetry.) The basic idea is to replace the extreme approximation of a "point-like" (i.e. zero-dimensional) particle by a one- or higher-dimensional construction, such as a string or loop.[153] Superstring theories of the fermion, such as an electron, require a space of ten (including one temporal) dimensions. Instead of approximating the electron by a point, one considers structures in which all dimensions but one are shrunk to (almost) zero. This definition is not to be taken too literally. A string, like a rope, is of course a three-dimensional structure in which just one dimension greatly exceeds the two others. The same is true for (heterotic) superstrings, for which one (or more than one) of the ten dimensions are extended beyond the Planck length. String theorists then also speak of "branes", in analogy to membranes in which two of their three dimensions are extended beyond molecular sizes. "Branes" in string theory refer to structures in which several of the ten spatial dimensions are curled up to the size of the Planck length, while one or few of them appear to be extended. One may have problems imagining those high-dimensional structures, but physicists can deal with them in mathematical terms. It is assumed that the normal modes of vibration of those string-like structures represent elementary states of matter.

The premise of superstring theory is the postulate that the physical universe involves dimensions beyond the generally accepted ones, i.e. the four dimensions of space and time on which general relativity is based. A first step in this direction was taken, soon after the advent of Einstein's theory, in 1919 by the then hardly known mathematician Theodor Kaluza[155] (see Figure 1.9.6). In 1926, the Swedish mathematician Oskar Klein[156] refined Kaluza's prescient idea in a way that stated explicitly: "The spatial fabric of our universe may have both extended and curled-up dimensions". However, Klein's conjecture of dimensions that are curled into circles of a size in the range of Planck dimensions, in the absence of any possible experimental support, did not find many subscribers. Only when physicists started to combine general relativity with the Yang–Mills field[157] as well as with the concept of supersymmetry did the general idea provoked by Kaluza begin to take up momentum again.

Let me, at this point, intersperse my story with a personal reminiscence. Theodor Kaluza, at the time he had these ideas, was at the University of Königsberg, the birthplace and lifelong home of Immanuel Kant. After World War II Königsberg became part of the former Soviet Union and is now called Kaliningrad. Kaluza was later professor at the University of Kiel and, from the mid 1930s, at the University of Göttingen, where I attended his lectures in mathematics while I was a student of physics after the war in the 1940s. David Hilbert, who had died in 1943, had left his hallmark on the standard of mathematics at Göttingen. I particularly cherish

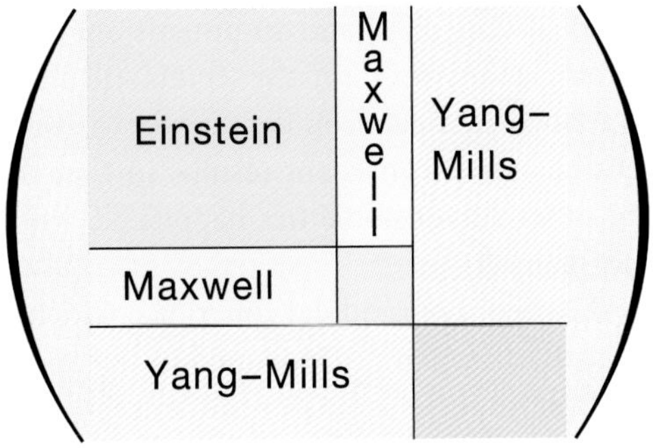

Figure 1.9.6

Generalisation of Kaluza's unification of relativity and Maxwell's theory

Quotations from Michio Kaku's book *Hyperspace*[162]:
If we go to the Nth dimension, then the metric tensor is a series of N^2 numbers that can be arranged in an $N \times N$ block. By slicing off the fifth and higher columns and rows, we can extract the Maxwell electromagnetic field and the Yang–Mills field. Thus, in one stroke, the hyperspace theory allows us to unify the Einstein field (describing gravity), the Maxwell field (describing the electromagnetic force), and the Yang–Mills field (describing the weak and strong force). The fundamental forces fit together exactly like a jigsaw puzzle.

Apparently, one of the first physicists to perform this reduction was University of Texas physicist Bryce DeWitt, who has spent many years studying quantum gravity. Once this trick of splitting up the metric tensor was discovered, the calculation for extracting the Yang–Mills field is straightforward. DeWitt felt that extracting the Yang–Mills field from N-dimensional gravity theory was such a simple mathematical exercise that he assigned it as a homework problem at the Les Houches Physics Summer School in France in 1963. [Recently, it was revealed by Peter Freund that Oskar Klein had independently discovered the Yang–Mills Field in 1938, preceding the work of Yang, Mills and others by several decades.]

my recollection of my final test in mathematics, where Kaluza was my examiner. His questions (about Taylor's and MacLaurin's theorems) were well chosen, and the charm and dignity of his personality made the test a most pleasant and memorable event. Unfortunately, none of us knew Kaluza's early work on five-dimensional field theory, which in the physics of that time seemed to have passed entirely into oblivion. Michio Kaku's book (see below) contains many entertaining stories about the later fate of Kaluza's ideas.

The pioneers of superstring theory are John Schwarz of the California Institute of Technology at Pasadena, Michael Green of Queen Mary's College in London[158] and last but not least Edward Witten[159] of the Institute for Advanced Study at Princeton. It is mainly the latest version – called heterotic superstring theory, to which another group[160] (including the Princeton physicist David Gross and his coworkers Emil Martinec, Jeffrey Harvey and Ryan Rohm, called the Princeton string quartet) made essential contributions – that seems to promise lasting success. The name "heterotic" refers to the fact that there are two kinds of vibration, clockwise and anticlockwise, which belong to compact spaces of 10 and 26 dimensions which, however, can be combined to produce a single theory. One of the secrets still unrevealed by this theory is the question of where those mystical-seeming numbers 10 and 26 come from. Only spaces of these dimensions yield consistent results, in that the nonsensical terms occurring in spaces of other dimensionalities happen to drop out. Why this is so remains at present unexplained.

There are hints that the theory of modular functions, which the legendary Indian mathematician Srinivasa Ramanujan[161] developed in the last year of his short life, may offer some explanation. In these functions the "miraculous" numbers 10 and 26 play a role similar to the one they play in heterotic superstring theory: they produce identities which lead to a cancellation of all the unwanted terms, thereby demonstrating symmetries that are unique for spaces that have the two magic numbers of dimensions.

Excellent descriptions of the exiting aspects of string theory can be found in two recent monographs by Michio Kaku[162] and by Bryan Green,[163] both of whom are active players in the game. My reason for referring to this work in this book is its relation to a, possibly ultimate, aspect of symmetry in the physical universe.

This brings me back to the question in the title of this section. Given our universe, the fundamental symmetries appear to be of an *a priori* nature. They do not allow alternative consequences, and the consequences do not have any feedback that would adapt the symmetries in order to produce an optimal outcome. However, my answer contains an implicit caveat in the phrase "our universe". How did "our" universe come about? Any exponential growth phase is based on a process that is of an inherently autocatalytic nature. Autocatalysis involves feedback, from the results back to their cause. It is thus inherently linked to selective fixation of an optimal result. The latter point may not be obvious in the present context, but I shall return to it at the end of Chapter 5. This point is relevant in connection with questions in later parts of this book. Why are the fundamental symmetries of our universe so well adapted to allow our existence? In the evolution of life, all symmetries (and all seemingly logic-based principles) are *a posteriori* manifestations of a single *a priori* requirement, namely, the optimal facilitation of existence. Life as such came about by adaptation to environmental conditions. It is not the physical state that is so well adapted to our form of life, it is rather our form of life which is adapted to "our" physics. Is it not reasonable to suppose that similar requirements of optimality may have governed the evolution of a sustainable cosmos?

The basic symmetries of physics, according to the present view, seem to be of an axiomatic nature. Nevertheless, in a lecture presented at the Einstein Centennial Celebration at the Institute for Advanced Study in 1979, John Archibald Wheeler[164] reminded us that "laws derived from symmetry considerations hide the machinery underlying physical law".

1.10. Why "All That"?*

Two friends from New York, travelling through Israel, got invited to a party in Tel Aviv. They were immediately welcomed and treated very hospitably. After dinner, their hosts started to tell jokes, in Hebrew of course, which the two friends did not understand. While one of them got pretty bored after a while, the other burst into loud laughter after every joke. His companion asked him: "What do you keep on laughing at? You understand as little Hebrew as I do!" The other replied: "You're right – but I trust these people!" And if I was asked now why I – not being a particle physicist myself – tell all "their stories", my answer, too, is: I trust these people. And why do I care about their work? My final answer to this question has to await the final section (5.10) of this volume. Meanwhile let us comfort ourselves with the following answer.

The four central themes of this book are "matter", "information", "life" and "thought". Our world is made of matter and we ourselves consist of matter. So we are pondering over the question: How does the world have to be made up in order to provide a place where intelligent life could come about?

The British cosmologist Martin Rees argues that there are *Just Six Numbers* (the title of his recent book[165]) that constitute "the recipe for a universe that could engender sustainable conditions suitable for an evolution of life and intelligence" (see Table 1.10.1). The truly astonishing outcome of his thoughts is that the fabric of this universe, the properties of atoms, their sizes and masses as well as the forces that link them together, are extremely sensitive to the values of these numbers.

In preceding sections I have discussed these numbers, which include the following:

1) The ratio of the two coupling strengths of the electrical and gravitational forces α_e/α_g (cf. Section 1.8), which for proton masses ($E \approx 1$ GeV) yields a value of about 1000,000,000,000,000,000,000,000,000,000,000,000, or 10^{36}.

2) The fraction of mass converted to energy in the sun (cf. Section 1.1), which is about 0.007 and expresses how firmly atomic nuclei are held together.

3) The ratio of observable and "critical" mass, which decides whether the universe will always expand or recontract, at present thought to be approximately 0.04, but which may be closer to unity if one accounts for "dark" matter (cf. Section 1.8).

4) The finite, but very small, cosmological constant (cf. Section 1.8), indicating some antigravity force, which was observed for the first time in 1998.

* "All that" is a term used nowadays by physicists for "matter plus energy", including all that they know about and all that they do not.

Table 1.10.1 Rees' Six Numbers

1	N	~1 000 000 000 000 000 000 000 000 000 000 000 000	Strength of electrical over gravitational force
2	ε	~0.007	Fraction of mass converted to energy in the sun
3	Ω	~0.04	Ratio of visible to critical mass density in the universe
4	λ	≥0	Cosmological constant
5	Q	~0.000001	Fraction of mass-energy required to break up stars
6	D	3	Number of dimensions of the visible world

According to Martin Rees, these six numbers "constitute a recipe for our universe ...If any of them were to be untuned, there would be no stars and no life".

Ad 1: Since Newton's and Coulomb's laws have the same distance relationship the two forces, e.g. between the electron and the proton in an hydrogen atom, can be compared directly.

Ad 2: The mass defect corresponds to the energy released in the fusion reaction (cf. Section 1.1, page 3: $0.0294/4.032 = 0.0073$). This number defines how firmly atomic nuclei are held together.

Ad 3: According to special relativity, the critical mass density decides whether the universe will always expand or will later shrink again. Dark matter – such as black holes, cosmic strings or finite neutrino masses – raises this number closer to one.

Ad 4: The (finite, but very small) cosmological constant is related to an original curvature of the universe. It is a sign of a finite vacuum energy and – as the biggest surprise in 1998 – of antigravity.

Ad 5: The amplitude of ripples in the background radiation as detected by the Cobe experiment (see Section 1.8), in cheating an original inhomogeneity in the distribution of matter eventually leading to the formation of stars, galaxies and clusters.

Ad 6: Why are the dimensions favoured to build up our physical world, while other dimensions (as suggested by string theory) appear to be "curled up"?

5) The amplitude of ripples in the background radiation, as detected by the Cobe experiment (cf. Section 1.8), indicating an original inhomogeneity in the distribution of matter that may have nucleated the formation of stars, galaxies and clusters. and last but not least

6) The number of dimensions which build up our visible three-dimensional world, while other dimensions according to superstring theory appear to be "curled up" within the Planck range (cf. Section 1.9).

I will not pursue the question of whether it is only these six numbers that shape our universe. Michael Rowan-Robinson, in another recent book,[166] quotes "nine numbers of the cosmos", while the standard model of particle physics (see Section 1.6) has to be tuned by no less than 61 adjustable parameters, as added up by Murray Gell-Mann in his book *The Quark and the Jaguar.*[55] As all the authors concede, it is unlikely that these parameters will ultimately turn out to be independent of one another. The unexpectedly huge entropy of a black hole rather suggests a more uniform constitution of the earliest forms of matter referring to the

Planck dimensions of space, time and mass-energy (cf. Section 1.8). Let us therefore dig a bit more for the deeper mysteries in our present picture of the physics of matter.

The first and perhaps most conspicuous mystery is the quantum itself. Here I do not mean its theory, quantum mechanics. I rather mean the results of this theory as they present themselves to our imagination.

Quantum mechanics as a theory has been subjected to the most scrupulous experimental testing (see Section 1.6). The Einstein–Podolsky–Rosen experiment, originally conceived as a *Gedankenexperiment*, has been realised in various forms in recent years. The criteria for evaluating such experiments, as codified by John S. Bell, allow an unprejudiced interpretation, according to which quantum-mechanical theory does not involve any "hidden" variables. The solutions of Schrödinger's equation yield only probabilities for space/time behaviour. Nevertheless, the theory is deterministic in terms of the resulting wave functions. These yield a delocalised behaviour, with its inevitable consequence of "entanglement" of quantum states over (possibly) large distances. A quantum state characterised by its wave function is in principle unknowable: any measurement or other kind of "fixation" destroys it and hence interferes with (or decoheres) its previous entanglement.

Let me stress again: it is not quantum mechanics as a theory that is the mystery. We can easily adapt our imagination to any self-consistent theory. What remains unimaginable are the resulting effects, the inherent uncertainty of which, for instance, makes it impossible to "clone" quantum states, such as light quanta in their respective states of polarisation. The same is true for entanglement, which involves instantaneous action over finite distances. This does not contradict special relativity, according to which we cannot transmit any signal at a speed greater than that of light. The incomprehensibility of quantum-mechanical entanglement lies in the simultaneity of instantaneous action over distances, with the fundamental impossibility of using it for transmission of information – the major obstacle in utilising entanglement for the purpose of telecommunication.

Henning Genz, in a recent book,[167] stated that one of the greatest challenges of physics is to derive quantum mechanics solely from its principles. Any apparent discrepancy, such as the one mentioned above, should then tell us what is wrong with our common way of reasoning. He is convinced that this will involve basic changes in our conception of time and space.

The second fundamental mystery, the universe, expresses itself in a certain incompatibility between quantum theory and relativity. It, too, results from our incomplete understanding of time and space on both their smallest and largest scales. "Quantum vacuum" and "black hole" are both terms that conceal our ignorance about time and space, including the beginning and the possible end of our world. The discrepancy between the two theories becomes visible only in the extremes of dimensions, most conspicuously in the range of Planck dimensions. However, there are ranges of dimension where quantum mechanics and relativity not only peacefully coexist but indeed complement one another, to produce an unprecedented precision in reproducing experimentally established facts. Quantum electrodynamics bears witness to their ability to represent physical reality.

One of the phenomena poorly understood today is the unification of forces that occurs with decreasing spatial separation. Electromagnetic and weak nuclear forces coalesce to the electro-weak force at distances between 10^{-18} and 10^{-20} m, corresponding to interaction energies of 10^2 to 10^5 GeV. This is theoretically well understood.[168,169] Its pendant, dealing with unification of the electro-weak with the strong nuclear forces, at separations of 10^{-30} m, equivalent to energies around 10^{15} GeV, called grand unification theory (GUT), still is a vision, perhaps not too far from realisation. What I really mean when I refer to them as "poorly understood" is expressed by the question: what do these unifications mean? They occur like phase transitions caused by certain symmetry breakages. But, as John Wheeler says, symmetries hide the underlying mechanisms. If the unification of two forces is characterised by approaching equal strength, what causes the unified force to diverge with increasing separation?

This question is of even greater relevance for a unification of the three forces with gravity expected at dimensions near the Planck range. Relativity predicts a singularity, while quantum mechanical uncertainty does not allow us to pinpoint space dimension down to zero tolerance. The four-dimensional continuum of space-time in any theory of quantum gravity is to be replaced by some "super-space", with dramatic consequences for both particle physics and cosmology.

Here physics is at a crossroads. The optimists hope for a final theory (or at least dream of one). Optimism is justified if – as previous attempts at unification demonstrate – convergence to a final theory can be reached. Whether or not super-string theory represents such a final theory is an open question. In this connection one word of clarification is necessary. An expression like "theory of everything" (TOE) for such a "final" approach is an unfortunate choice. It makes sense only in relation to the unification of forces. Gerard t'Hooft (Nobel prize for physics 1999), who became famous as a student when (in 1971) he solved the problem of infinities occurring in the electro-weak theory of Weinberg, Salam and Glashow[111–113] by renormalisation, said with regard to a final theory that should unify all forces: "Evolution according to these laws will give rise to a nearly infinite complexity, a complexity sufficiently extensive to include the marvellously perplexing wonders abounding in our universe – the emergence of life and intelligence being only one of these." There is nothing wrong with these optimistic expectations as far as they concern the material prerequisites of life and thought, being just somewhat more than "the dust of stars". However, the ladder from the dust of stars to life and intelligence is a "Wittgenstein ladder", of which we have to dispose once we have reached the steps that represent the chemical complexity of even the most primitive stages of life. Living matter does not recall any principles of a "theory of everything", as little as our brains are inherently reminiscent of quantum gravity. I do not want to be misunderstood: these theories take care of the material basis of life and culture. But the complexity underlying life and thought is organised according to quite different principles.

The pessimists, on the other hand, think that there may be no end, and that we are merely moving from one layer to the next. Their view is represented by those physicists who do not believe that the existing "mysteries" can be resolved by any "final" theory, and who rather favour a radical rethinking of the approach. Perhaps it isn't quite fair to call them pessimists; this pejorative term certainly does not apply to their most prominent representative, John Archibald Wheeler. "Law without law" is his creed, expounded in a recent book[170] in the series *Masters of Modern Physics.* It is essentially based on a catch-phrase by Niels Bohr: "No elementary phenomenon is a phenomenon until it is a registered phenomenon". (Note the use of the two words "phenomenon", which refers to the Greek expression for "appearance", and "registered", which replaces Bohr's original word "observed", in order to stress that it is not conscious observation, but rather a mere decoherence by fixation providing some record that is required.) This Bohr quotation establishes the role of the observer as a "participator in defining reality". John Wheeler symbolised the meaning of this phrase by a story, which I let him tell us now in his own words:

> We had been playing the familiar game of Twenty Questions. Then my turn came, fourth to be sent from the room, so that Lothar Nordheim's other 15 after-dinner guests could consult in secret and agree on a difficult word. I was locked out unbelievably long. On finally being admitted, I found a smile on everyone's face, the sign of a joke or a plot. I nevertheless started to attempt to find the word. "Is it animal?" "No." "Is it mineral?" "Yes." "Is it green?" "No." "Is it white?" "Yes." These answers came quickly. Then questions began to take longer in the answering. It was strange. All I wanted from my friends was a simple yes or no. Yet the one queried would think and think, yes or no, no or yes, before responding. Finally I felt I was getting hot on the trail, that the word might be "cloud". I knew I was allowed only one chance at the final word. I ventured it: "Is it 'cloud?'" "Yes," came the reply, and everyone burst out laughing. They explained to me there had been no word in the room. They had agreed not to agree on a word. Each one questioned could answer as he pleased – with the one requirement that *he* should have a word in mind compatible with his own response and all that had gone before. Otherwise if I challenged he lost. The surprise version of the game of Twenty Questions was therefore as difficult for my colleagues as it was for me.

The normal version of the game of "Twenty Questions" should in principle, if played intelligently, have a deterministic solution. With 20 hierarchically ordered "yes or no" questions one should be able to screen a multitude of $2^{20} \approx 10^6$ alternative words (cf. Chapter 3). The large dictionaries proudly announce that their contents include several hundred thousand entries. Hence, a systematic narrowing-down should allow the player to pinpoint any word in the dictionary. Well, that may be true in principle, but it is not in practice. When we play the game we do not usually bring along a dictionary. Indeed, we rather try to encircle the word using logical, and not lexicographic, arguments. There are no criteria that would allow a precise transformation of one into the other. In fact, the arbitrariness of language, especially in the assignment of words, makes any deterministic solution futile. This holds in particular for the version of the game described by John Wheeler. It requires an evolutionary outcome based on a logic that is accessible to all players. The final phase of the game described was more of a random guessing game than a strictly logical procedure.

So far, so good, but why is this game an example of "law without law"? Yes, the final result was unpredictable; it took shape only with the progression of the game. A biologist would say that the game is representative of the process of biological evolution: yes and no! The result definitely is not predetermined in the early phases. So is it really "law without law"? There is certainly no law that predicts the specific outcome, although the process as such is not at all a random one. There is a law that we may call a Darwinian law, and this law is as physical as the Newtonian law of classical mechanics, and like Newton's law it can be formulated in mathematical terms. In the game of evolution, no living being is a being until that very being has been promoted to reality by the choice of questions (mutations) and answers (given by natural selection). This sentence is taken almost verbatim from Wheeler's paper except that "word" has been replaced by "living being". And he translates it to Bohr's statement about elementary phenomena, admitting: "This comparison between the world of quantum observations and the surprise version of the game of 20 questions misses much, but it makes the central point." Very true, so does the game of evolution.

The difficulties, of course, rest in the fact that – as Wheeler says – "all the game is missing", which indeed is more than the game explains. In evolution, the guiding principle is "to save existence". The mechanism is reproduction. In connection with mutation it produces selection and adaptation. The result is not predictable, since the sequence of appearance of mutations, although biased by the populated states, does not obey any predetermined order. Hence the result is truly deterministic only in retrospect. In the universe an analogous situation may have been present in the "autocatalytic" phase of inflation. Again, I have to trust the cosmologists that our universe is a special outcome of countless possibilities of multiversal realisations. Then, our symmetries would in fact be of an *a posteriori* nature, and the fundamental constants would be (in Wheeler's words) "artefacts of history" which determine our space time-dimensions. If this analogy makes any sense for an evolving universe, we must enquire not only about the mechanisms "hidden in symmetries", but also about the reproductive feedback that secures existence. This means that the evolving laws are not just one accidental choice out of a huge diversity, but rather that they belong to a (possibly extremely) narrow choice of "evolvable" laws yielding a sustainable world. Lee Smolin, in a recent book *The Life of the Cosmos*,[171] has compared the situation of the early cosmos with that of biological evolution. The situations, however, are too different to allow them to resemble each other in any straightforward way.

Nevertheless, let me come back to this discussion after I have analysed in more detail the nature of evolutionary processes (see Section 5.10).

When earlier in this book I modified Leibniz' question to: "Why is there something and not just anything"? I had in mind the problem of the very narrowly defined physical parameters that are so remarkably well adapted to our existence. Some cosmologists and physicists assign primary importance to this fact by asserting that the universe is the way it is because we do exist. This is the anthropic principle, first expressed by Robert Dicke[172] in 1961. It was developed into a formal principle by Brandon Carter,[173] who in 1974 introduced the term "anthropic principle". John D. Barrow and Frank J. Tipler wrote a book[174] with the title *The Anthropic Cosmological Principle*; it was embraced by philosophers and theologians because they argued that theology may have a role in both astrophysics and biology.

Carter formulated the anthropic principle (which he later conceded to be misnamed) in two versions, called the weak form and the strong form. His weak formulation reads: "What we can expect to observe must be restricted by the conditions necessary for our presence as observers", a statement that can hardly be contradicted. What aroused controversy was the strong form, which in Carter's words reads: "The universe necessarily has the properties requisite for life – life that exists at some time in its history". The first formulation – perhaps as an appendix to Greek philosophy – states what "is the case", i.e. the world is as it is, and if it were different we could not be here to observe it. What is perturbing in the strong formulation is the word "necessarily". In other words, the universe "must" have certain properties in order to allow the evolution of intelligent life at some time and, one may add, at some place(s). Is this not a re-assertion of the Ptolemaic scheme, placing us back at the centre of the universe, now in a causal rather than a spatial sense? No wonder the principle has been embraced by (some) philosophers and theologians as a basis for a "holistic" world view. As such, it is of an ideological nature and may be neither provable nor falsifiable. Whether or not we call our universe a "designer universe"[162] does not change the nature of the laws of physics. The question of a designer – in any case *ultra vires* – is thus pursued *ad absurdum* as being a scientific question. The business of physics is not personal belief, but rather a consistent description of nature.

If I understand Martin Rees correctly, the multiverse saves the freedom of chance for our existence and thereby safeguards cosmology against the determinism of an anthropomorphic world view. However, the tremendously narrow limits of the six parameters put us for the existence of our universe into a similar situation as for the existence of human life. The assumption of an unconstrained origin of our universe may then be as false as the belief that within a thousand million years the $10^{\text{one thousand million}}$ possible arrangements of nucleotides in the human genome could have come about by a mere process of throwing dice.

Even if we leave aside the "anthropic principle" I still do not understand why, in this discussion, cosmologists insist that their parameters are responsible for the existence of intelligent life, the origin of which was, after all, even more at the mercy of chance than that of the universe. Would it not be more appropriate to state that the cosmological parameters allow the existence of stars in which a chemistry beyond hydrogen could come about? That is what the six numbers identified by Martin Rees actually do. As mentioned above, and as pointed out by Rees himself, it is not clear whether these parameters are really independent of one another. They are probably not, but how they are related is one of the great unsolved problems of physics. And it certainly makes a big difference whether these parameters came about by random choice or by evolutionary adaptation to a state that favours its own existence.

Whatever the true number of independent parameters, they are a precondition for chemistry and, as such, they are only a necessary prerequisite of life. By no means, however, are they sufficient to allow life to flourish or to produce intelligent beings. If that were so, we would long have been able to reproduce the *Urform* of life in the laboratory. We indeed ask why life appeared on our planet about 4000 million years ago (as has been shown to be true for the genetic code on the basis of a comparative sequence analysis of the various code adaptors (tRNAs)),[175–177] which means that life followed quickly once its chemical prerequisites had been fulfilled. And we also ask why it then took a time almost comparable with the age of our planet before multicellular and eventually intelligent life evolved. This suggests that primitive life should be quite abundant in our universe, while intelligent life may well be much harder to find within distances allowing communication. Up to now we have no evidence for extraterrestrial life anywhere in our universe.

All this demands that we revisit the above questions after we have confronted the strangeness of simple matter with the enormous combinatorial complexity which it may achieve under conditions more familiar to our experience (see Section 5.10). Let me, in conclusion, quote once more from John Wheeler:[178] "It will take the power of all thought together if we are ever to understand why we have 'something rather than nothing'. We can believe that we will first understand how simple the universe is when we recognise how strange it is."

LITERATURE AND NOTES

1. Bethe, H. A. (1967). "Energy Production in Stars". Nobel Lecture. December 11; Bethe, H. A. (1939). "Energy Production in Stars". *Phys. Rev.* 55: 434–456.
2. Von Weizsäcker, C. F. (1937). "Über Elementumwandlungen im Innern von Sternen". *Phys. Z.* 38: 176–191 and 39 (1938): 633–646.
3. Feynman, R. P., Leighton, R. B. and Sands, M. (1964). "The Feynman Lectures on Physics", Vol I.: p. 4-1, Addison-Wesley, Reading, M.A.
4. Mayer, J. R. v. (1851). "Bemerkungen über das mechanische Aequivalent der Wärme". Heilbronn. Compt. rend. t. 29. 1849; Mayer, J. R. v. (1867). "Die Mechanik der Wärme". *Verlag der J. G. Cotta'schen Buchhandlung.*

5. Noether, E. (1918). "Invariante Variationsprobleme". *Nachr. Kgl. Ges. Wiss. Gött. Math.-phys. Kl.* 1918: 235–257.
6. Pauli, W. E. (1930). "Offener Brief an die Gruppe der Radioaktiven bei der Gauvereins-Tagung zu Tübingen", "Liebe Radioaktive Damen und Herren", 4.12.1930, Zürich. Published in: Pauli, W. E., Hermann, A., Weisskopf, V. F. and v. Meyenn, K. (Publisher). (1979). "Wissenschaftlicher Briefwechsel mit Bohr, Einstein, Heisenberg u.a.", Band II: 1930–1939. Springer, Berlin.
7. Fermi, E. (1934). "Versuch einer Theorie der Betastrahlen". *Z. Phys.* 88: 161–177.
8. Cowan, C. L. Jr., Reines, F., Harrison, F. B., Kruse, H. W. and McGuire, A. D. (1956). "Detection of the Free Neutrino: A Confirmation". *Science* 124: 103–104; Reines, F. and Cowan, C. L. Jr. (1956). "The Neutrino". *Nature* 178: 446–449.
9. Carnot, N. L. S. (1872). "Réflexions sur la puissance motrice du feu et sur les machines propres à développer cette puissance". Annales scientifiques de l'École Normale Supérieure Sér. 2, 1: pp. 393–457. Paris.
10. English version of Ref 9: Carnot, N. L. S. and Thurston, R. H. (ed.) (1890). "Reflections on the Motive Power of Heat and on Machines Fitted to Develop That Power". New York: J. Wiley & Sons.
    German version: Carnot, N. L. S. and Ostwald, W. (ed.) (1892). "Betrachtungen über die bewegende Kraft des Feuers und die zur Entwicklung dieser Kraft geeigneten Maschinen". Leipzig, W. Engelmann.
11. According to W. Nernst absolute zero temperature due to the Third Law of Thermodynamics can never be reached: Nernst, W. (1906). "Ueber die Berechnung chemischer Gleichgewichte aus thermischen Messungen". *Nachr. Kgl. Ges. Wiss. Gött.* 1: 1–40.
12. "It is said that there's no such thing as a free lunch. But the universe is the ultimate free lunch." Guth, A. H., quoted by Hawking, S. (1988). "A Brief History of Time: From the Big Bang to Black Holes". p. 129. Bantam Press, London.
13. Clausius, R. (1865). "Über die Wärmeleitung gasförmiger Körper". *Ann. Phys.* 125: 353–400.
14. Carathéodory, C. (1909). "Untersuchungen ueber die Grundlagen der Thermodynamik". *Math. Ann.* 67: 355–386.
15. Einstein, A. (1905). "Zur Elektrodynamik bewegter Körper". *Ann. Phys. Chem.* 17: 891–921.
16. Poincaré, H. (1905). "Sur la dynamique de l'électron". *Comptes rendus hebdomadaires des séances de l'Académie des sciences* 140: 1504–1508; Poincaré, H. (1906). "Sur la dynamique de l'électron". *Rendiconti del Circolo matematico di Palermo* 21: 129–176.
17. Michelson, A. A. (1881). "The Relative Motion of the Earth and the Luminiferous Ether". *Am. J. Sci.* 22: 120–129.

18. Michelson, A. A. and Morley, E. W. (1887). "On the Relative Motion of the Earth and the Luminiferous Ether". *Am. J. Sci.* 34 (203): 333–345.
19. Lorentz, H. A. (1899). "Simplified Theory of Electrical and Optical Phenomena in Moving Systems". *Proc. KNAW* 1: 427–442.
20. Newton, I. (1687). "Philosophiæ Naturalis Principia Mathematica". Publisher: Jussi Societatus Regiae ac typis Josephi Streater; prostat apud plures bibliopolas. Place: Londini.
21. Einstein, A. (1905). "Ist die Trägheit eines Körpers von seinem Energieinhalt abhängig?" *Ann. Phys.* 18: 639–641.
22. Einstein in a letter (19.6.1948) to L. Barnett, cited after Okun, L. B. (1989). *Phys. Tod.* 42 (6): 31–36. (Courtesy to Albert Einstein Archives at the Jewish National and University Library, Hebrew University of Jerusalem, Israel.)
23. Rohlf, J. W. (1994). "Modern Physics from α to Z_0". John Wiley & Sons. New York.
24. Minkowski, H. (1909). "Raum und Zeit". Jahresbericht der Deutschen Mathematiker-Vereinigung, Vol. 18. B. G. Teubner, Leibzig and Berlin. According to a lecture at the 80. Naturforscher-Versammlung, Cologne, Sept. 21, 1908.
25. Cartan, E. (1952–1955). "Oeuvres completes". 3 volumes. Gauthiers-Villars. Paris. Reprinted in Edition du CNRS 1984.
26. Hawking, S. (1988). "A Brief History of Time: From the Big Bang to Black Holes". Bantam Press, London.
27. Kaufmann, W. (1901). "Die Entwicklung des Elektronenbegriffs". *Phys. Z.* 3(1): 9–15; Kaufmann, W. (1901). "Die magnetische und elektrische Ablenkbarkeit der Bequerelstrahlen und die scheinbare Masse der Elektronen". *Gött. Nachr.* 2: 143–168; Kaufmann, W. (1902). "Über die elektromagnetische Masse des Elektrons". *Gött. Nachr.* 5: 291–296; Kaufmann, W. (1902). "Die elektromagnetische Masse des Elektrons". *Phys. Z.* 4 (1b): 54–56; Kaufmann, W. (1903). "Über die "Elektromagnetische Masse" der Elektronen". *Gött. Nachr.* 3: 90–103.
28. Abraham, M. (1902). "Dynamik des Electrons". *Gött. Nachr.* 1902: 20–41.
29. Russell, B. (1921). Preface. In: Wittgenstein, L. "Logisch-Philosophische Abhandlung". Ostwald, W. (ed.). *Ann. Naturphil.* 14: 186–198.
30. Wittgenstein, L. (1921). "Logisch-Philosophische Abhandlung". *Ann. Naturphil.* 14: 185–262.
31. Hilbert, D. (1900). "Mathematische Probleme". Lecture at the International Congress of Mathematicians at Paris.
See also: Hilbert, D. (1900). "Mathematische Probleme". *Nachr. Kgl. Ges. Wiss. Göttingen, Math.-Phys. Kl. (Gött. Nachr.)* 3: 253–297.
32. Cantor, G. (1895). "Beiträge zur Begründung der transfiniten Mengenlehre I". *Math. Ann.* 46: 481–512; Cantor, G. (1897). "Beiträge zur Begründung der transfiniten Mengenlehre II". *Math. Ann.* 49: 207–246.

33. Faraday, M. (1832–1857). "Experimental Researches in Electricity". *Collection of scientific papers or transcriptions of lectures*. Vols i and ii. 1839, 1844. Taylor, R. and J. E., London. Vol. iii. 1855. Taylor and Francis, London.
34. Ampère, A. M. (1827). "Mémoire sur la théorie mathématique des phénomènes électro-dynamiques uniquement déduite de l'expérience". *Mémoires de l'Académy Royale des Sciences* 6: 175–388.
35. Maxwell, J. C. (1865). "A Dynamical Theory of the Electromagnetic Field". *Phil. Trans. R. Soc. Lond.* 155: 459–512.
36. Boltzmann, L. E. (1909) "Wissenschaftliche Abhandlungen", Vols I, II and III. Hasenöhrl, F. (ed.). Barth, Leipzig, reissued by Chelsea, New York, 1969.
37. Gibbs, W. J. (1876–1878). "On the Equilibrium of Heterogeneous Substances". *Trans. Connecticut Acad.* III: 108–248, 343–524.
38. Rayleigh, L. (1900). "Remarks upon the Law of Complete Radiation". *Phil. Mag.* 49: 539–540.
39. Jeans, J. H. (1905). "On the Partition of Energy between Matter and Aether". *Phil. Mag.* 10: 91–98.
40. Planck, M. (1900). "Zur Theorie des Gesetzes der Energieverteilung im Normalspectrum". *Verh. Deutsch. Phys. Gesell.* 2 (17): 237–245.
41. Stefan, J. (1879). "Über die Beziehung zwischen der Wärmestrahlung und der Temperatur". S.-B. Akad. Wiss. Wien 79 (2): 391–428.
42. Boltzmann, L. (1884). "Ableitung des Stefan'schen Gesetzes, betreffend die Abhängigkeit der Wärmestrahlung von der Temperatur aus der electromagnetischen Lichttheorie". *Ann. Phys. Chem.* 22: 291–294.
43. Smoot, G. F. (1994). "*COBE observations of the cosmic background*". *Front. Sci. Ser.* 8: 115–136.
44. Wilson, R. W. and Penzias, A. A. (1967). "Isotropy of Cosmic Background Radiation at 4080 Megahertz". *Science* 156 (3778): 1100–1101.
45. Einstein, A. (1948). "Message for Max Planck Memorial Service". Courtesy of the Albert Einstein Archives, The Jewish National and University Library, The Hebrew University of Jerusalem, Israel. Translated from German to English by P. Woolley.
46. Planck, M. (1900). "Entropie und Temperatur strahlender Wärme" *Ann. Phys.* 1(4): 719–737; Planck, M. (1901). "Ueber das Gesetz der Energieverteilung im Normalspectrum". *Ann. Phys.* 4: 553–563; Planck, M. (1901). "Ueber die Elementarquanta der Materie und der Elektricität". *Ann. Phys.* 309(3): 564–566.
47. Lenard, P. E. A. v. (1894). "Ueber Kathodenstrahlen in Gasen von atmosphärischem Druck und im äussersten Vacuum". *Ann. Phys.* 287 (2): 225–267.
48. Einstein, A. (1909). "Zum gegenwärtigen Stande des Strahlungsproblems". *Phys. Z.* 10: 185–193.
49. See the following section Ref. 50.
50. De Broglie, L.-V. P. R. (1925). "Recherches sur la théorie des quanta". *Ann. Phys.* 10(3): 22–128.

51. Compton, A. H. (1923). "A Quantum Theory of the Scattering of X-Rays by Light Elements". *Phys. Rev.* 21(5): 483–502.
52. Bohr, N. H. D. (1913). "On the Constitution of Atoms and Molecules, Part I". *Phil. Mag.* 26: 1–24.
53. Rutherford, E. (1911). "The Scattering of α and β Particles by Matter and the Structure of the Atom". *Phil. Mag.* 6 (21): 669–688.
54. Mehra, J. (ed.) (1973). "The Physicist's Conception of Nature". Proceedings of a Symposium on the Development of the Physicist's Conception of Nature, held at Trieste, September 18–25, 1972, in honour of the 70th birthday of Paul Dirac. Reidel, Dordrecht.
55. Gell-Mann, M. (1994). "The Quark and the Jaguar: Adventures in the Simple and the Complex". Little, Brown Book Group, London.
56. Schrödinger, E. (1926). "Quantisierung als Eigenwertproblem (Erste Mitteilung)". *Ann. Phys.* 79: 361–376.
57. Bohr, N. H. D. (1977). "Collected Works, Volume 4: The Periodic System (1920–1923)". Ed. Nelson, J. R. North-Holland. Amsterdam, New York and Oxford.
58. Wiles, A. (1995). "Modular Elliptic Curves and Fermat's Last Theorem". *Ann. Math.* 141(3): 443–551.
59. Heisenberg, W. (1925). "Über quantentheoretische Umdeutung kinematischer und mechanischer Beziehungen". *Z. Phys.* 33: 879–893.
60. Born, M., Heisenberg, W. and Jordan, P. (1925). "Zur Quantenmechanik II". *Z. Phys.* 35: 557–615.
61. Hamilton, W. R. (1834). "On a General Method in Dynamics". *Phil. Trans. Roy. Soc.* II: 247–308.
62. Heisenberg, W. (1927). "Über den anschaulichen Inhalt der quantentheoretischen Kinematik und Mechanik". *Z. Phys.* 43: 172–198.
63. Weinberg, S. (1995, 2005). "The Quantum Theory of Fields", Vol. iii. Cambridge University Press, Cambridge.
64. Thomson, G. P. and Reid, A. (1927). "Diffraction of Cathode Rays by a Thin Film". *Nature* 119: 890–890.
65. Davisson, C. and Germer, L. H. (1927). "Reflection of Electrons by a Crystal of Nickel". *Nature* 119: 558–560.
66. Breit, G. and Wigner, E. (1936). "Capture of Slow Neutrons". *Phys. Rev.* 49: 519–531.
67. Einstein, A., Podolsky, B. and Rosen, N. (1935). "Can Quantum-Mechanical Description of Physical Reality be Considered Complete?". *Phys. Rev.* 47: 777–780.
68. Pais, A. (1994). "Einstein Lived Here". Oxford University Press, Oxford.
69. Wu, C. S., Ambler, E., Hayward, R. W., Hoppes, D. D. and Hudson, R. P. (1957). "Experimental Test of Parity Conservation in Beta Decay". *Phys. Rev.* 105: 1413–1415.

70. Aspect, A., Grangier, P. and Roger, G. (1982)."Experimental Realization of Einstein-Podolsky-Rosen-Bohm Gedankenexperiment: A New Violation of Bell's Inequalities". *Phys. Rev. Lett.* 49(2): 91–94; Aspect, A., Dalibard, J. and Roger, G. (1982). "Experimental Test of Bell's Inequalities Using Time-Varying Analyzers". *Phys. Rev. Lett.* 49(25): 1804–1807.
71. Arndt, M., Nairz, O., Vos-Andreae, J., Keller, J., Van der Zouw, G. and Zeilinger, A. (1999). "Wave–Particle Duality of C60 Molecules". *Nature* 401: 680–682.
72. Kalinin, A., Kornilov, O., Schollkopf, W. and Toennies, J. P. (2005). "Observation of Mixed Fermionic-Bosonic Helium Clusters by Transmission Grating Diffraction". *Phys. Rev. Lett.* 9511(11): 3402.
73. Everett, H. (1956). "Theory of the Universal Wavefunction". Thesis, Princeton University, Princeton, pp. 1–140; Everett, H. (1957). "Relative State Formulation of Quantum Mechanics". *Rev. Mod. Phys.* 29: 454–462.
74. Gell-Mann, M. loc. cit. 55.
75. Wheeler, J. A. (1994). "At Home in the Universe". American Institute of Physics Press, Woodbury, N.Y.
76. Pauling, L. (1939). "The Nature of the Chemical Bond and the Structure of Molecules and Crystals: An Introduction to Modern Structural Chemistry". Cornell Univ. Press. Ithaca, NY.
    Pauling, L. (1949). "General Chemistry". W. H. Freeman, San Francisco, Dover Publications. New York.
77. Heisenberg, W. (1932). "Über den Bau der Atomkerne". *Z. Phys.* 77: 1–11. The discovery is due to Heisenberg, but the name "Isospin" was coined by Wigner: Wigner, E. (1937). "On the Consequences of the Symmetry of the Nuclear Hamiltonian on the Spectroscopy of Nuclei". *Phys. Rev.* 51: 106–119.
78. Weizsäcker, C. F. von (1935). "Zur Theorie der Kernmassen". *Z. Phys.* 96: 431–458.
79. Goeppert-Mayer, M. (1948). "On Closed Shells in Nuclei". *Phys. Rev.* 74: 235–239.
80. Haxel, O., Jensen, J. H. D. and Suess, H. E. (1948). "Zur Interpretation der ausgezeichneten Nucleonenzahlen im Bau der Atomkerne". *Die Naturwissenschaften* 35: 376–376.
81. Bohr, A. N. (1975). "Rotational Motion in Nuclei". Nobel Lectures, Physics 1971–1980. Ed. Stig Lundqvist. World Scientific Publishing Co., Singapore. 1992.
    Mottelson, B. R. (1975). "Elementary Modes of Excitation in the Nucleus". Nobel Lectures, Physics 1971–1980. Ed. Stig Lundqvist. World Scientific Publishing Co., Singapore. 1992.
    Rainwater, J. (1975). "Background for the Spheroidal Nuclear Model Proposal". Nobel Lectures, Physics 1971–1980. Ed. Stig Lundqvist. World Scientific Publishing Co., Singapore. 1992.

82. The Geiger–Marsden experiment was done in 1909 and led in 1911 to the Rutherford atomic model. Geiger, H. and Marsden, E. (1909). "On a Diffuse Reflection of the α-Particles". *Proc. R. Soc.* A 82: 495–500.
83. Rutherford, E. (1911). "The Scattering of α and β Particles by Matter and the Structure of the Atom". *Philos. Mag.* 6(21): 669–688.
84. Mott, N. F. (1929). "The Scattering of Fast Electrons by Atomic Nuclei". *Proc. R. Soc. Lond. A* 124: 425–442.
85. McAllister, R. W. and Hofstadter, R. (1956). "Elastic Scattering of 188 MeV Electrons from Proton and the Alpha Particle". *Phys. Rev.* 102: 851–856.
86. Breidenbach, M., Friedman, J. I., Kendall, H. W., Bloom, E. D., Coward, D. H., Destaebler, H., Drees, J., Mo, L. W. and Taylor, R. E. (1969). "Observed Behavior of Highly Inelastic Electron-Proton Scattering". *Phys. Rev. Lett.* 23(16): 935–939; Bloom, E. D., Coward, D. H., de Staebler, H. C., Drees, J., Mo, L. W., Taylor, R. E., Breidenbach, M., Friedman, J. I. and Kendall, H. W. (1969). "High-Energy Inelastic e^- p Scattering at 6° and 10°". *Phys. Rev. Lett.* 23: 930–934.
87. Gross, D. J. and Wilczek, F. (1973). "Ultraviolet Behavior of Non-Abelian Gauge Theories". *Phys. Rev. Lett.* 30: 1343–1346. See also: Gross, D. J. (2004). "The Discovery of Asymptotic Freedom and the Emergence of QCD". Les Prix Nobel. (2005). "The Nobel Prizes 2004". Ed. Tore Frängsmyr. Nobel Foundation. Stockholm. Wilczek, F. (2004). "Asymptotic Freedom: From Paradox to Paradigm". Les Prix Nobel. (2005). "The Nobel Prizes 2004". Ed. Tore Frängsmyr. Nobel Foundation. Stockholm.
Politzer, H. D. (1973). "Reliable Perturbative Results for Strong Interactions". *Phys. Rev. Lett.* 30: 1346–1349. See also Politzer, H. D. (2004). "The Dilemma of Attribution". Les Prix Nobel. (2005). "The Nobel Prizes 2004". Ed. Tore Frängsmyr. Nobel Foundation. Stockholm.
88. Neddermeyer, S. H. and Anderson, C. D. (1937). "Note on the Nature of Cosmic-Ray Particles". *Phys. Rev.* 51: 884–886.
89. Perl, M. L., Abrams, G. S., Boyarski, A. M. et al. (1975). "Evidence for Anomalous Lepton Production in e+e- Annihilation". *Phys. Rev. Lett.* 35 (22): 1489–1492.
90. Cowan, C. L. Jr., Reines, F., Harrison, F. B., Kruse, H. W. and McGuire, A. D. (1956). "Detection of the Free Neutrino: A Confirmation". *Science* 124: 103–104; Reines, F. and Cowan, C. L. Jr. (1956). "The Neutrino". *Nature* 178: 446–449.
91. Danby, G., Gaillard, J.-M., Goulianos, K., Lederman, L. M., Mistry, N. B., Schwartz, M. and Steinberger, J. (1962). "Observation of High-Energy Neutrino Reactions and the Existence of Two Kinds of Neutrinos". *Phys. Rev. Lett.* 9: 36–44.
92. Hofstadter, R. (1964). "The Electron Scattering Method and its Application to the Structure of Nuclei and Nucleons". Nobel Lectures. Physics 1942–1962. Ed. E. Fues, pp. 560–581. Elsevier Pub. Co., Amsterdam-London-New York.

93. Friedman, J. (1991). "Deep Inelastic Scattering: Comparisons with the Quark Model". *Rev. Mod. Phys.* 63: 615–627; Kendall, H. W. (1991). "Deep inelastic scattering: Experiments on the Proton and the Observation of Scaling". *Rev. Mod. Phys.* 63: 597–614; Taylor, R. E. (1991). "Deep Inelastic Scattering: The Early Years". *Rev. Mod. Phys.* 63: 573–595.
94. Gell-Mann, M. (1964). "A Schematic Model of Baryons and Mesons". *Phys. Lett.* 8(3): 214–215.
95. Aubert, J. J., Becker, U., Biggs, P. J. et al. (1974). "Experimental Observation of a Heavy Particle J". *Phys. Rev. Lett.* 33 (23): 1404–1406.
SLAC-SP-017 Collaboration. Augustin, J.-E., Boyarski, A. M., Breidenback, M. et al. (1974). "Discovery of a Narrow Resonance in e + e- Anihilation". *Phys. Rev. Lett.* 33: 1406–1408.
96. Herb, S. W., Hom, D. C., Lederman, L. M. et al. (1977). "Observation of a Dimuon Resonance at 9.5 GeV in 400-GeV Proton-Nucleus Collisions". *Phys. Rev. Lett.* 39: 252–255.
97. CDF Collaboration. Abe, F., Akimoto, H., Akopian, A. et al. (1995). "Observation of Top Quark Production in Anti-p p Collisions". *Phys. Rev. Lett.* 75: 2626–2631.
98. Bjorken, B. J. and Glashow, S. L. (1964). "Elementary Particles and SU(4)". *Phys. Lett.* 11: 255–257.
99. Dirac, P. A. M. (1928). "The Quantum Theory of the Electron". *Proc. R. Soc. Lond.* 117(778): 610–624.
100. Anderson, C. D. (1933). "The Positive Electron". *Phys. Rev.* 43: 491–494.
101. Chamberlain, O., Segrè, E., Wiegand, C. and Ypsilantis, T. (1955). "Observation of Antiprotons". *Phys. Rev.* 100: 947–950.
102. Jordan, P. (1927). "Über eine neue Begründung der Quantenmechamik II". *Z. Phys.* 41: 797.
103. Heisenberg, W. (1943–1944). "Die beobachtbaren Größen in der Theorie der Elementarteilchen". Part 1: *Z. Phys.* (1943). Vol. 120: 513–538. Part 2: *Z. Phys.* (1943). Vol. 120: 673–702. Part 3: *Z. Phys.* (1944). Vol. 123: 93–112.
104. Schweber, S. S. (1994). "QED and the Men who Made it; Dyson, Feynman, Schwinger and Tomonaga". Princeton University Press, Princeton.
105. Feynman, R. P., Leighton, R. and Sands, M. (1964, 1966). "The Feynman Lectures on Physics". 3 volumes. Addison-Wesley Pub Co. Reading (Massachusetts), Palo Alto, London.
106. Feynman, R. P., Leighton, R. (contributor) and Hutchings, E. (ed.) (1985). "Surely You're Joking, Mr. Feynman!: Adventures of a Curious Character". W. W. Norton & Co. New York.
107. Feynman, R. P. (1988). "What Do You Care What Other People Think?" W. W. Norton. New York.
108. Gell-Mann, M. (1964). loc. cit. 94, pp. 214–215.

109. Zweig, G. (1964). "An SU(3) Model for Strong Interaction Symmetry and its Breaking". CERN Report No.8181/Th 8419 and CERN Report No.8419/Th 8412. Genéve, Switzerland.
110. Barnes, V. E., Connolly, P. L., Crennell, D. J. et al. (1964). "Observation of a Hyperon with Strangeness Number Three". *Phys. Rev. Lett.* 12 (8): 204–206.
111. Glashow, S. I. (1961). "Partial-Symmetries of Weak Interactions". *Nucl. Phys.* 22: 579–588.
112. Weinberg, S. (1967). "A Model of Leptons". *Phys. Rev. Lett.* 19: 1264–1266.
113. Salam, M. A. (1968). "Weak and Electromagnetic Interactions". In: Elementary Particle Theory. (ed. N. Svartholm): 367–377. Stockholm: Almqvist & Wiksell. Salam, M. A. (1979). "Gauge Unification of Fundamental Forces". Nobel Lectures. Physics 1971–1980, Editor Stig Lundqvist. World Scientific Publishing Co., Singapore, 1992.
114. Lee, T. D. and Yang, C. N. (1956). "Question of Parity Conservation in Weak Interactions". *Phys. Rev.* 104: 254–258.
115. Rubbia, C., Van der Meer, S., Arnison, G. et al. (UA1 collaboration) (1983). "Experimental Observation of Isolated Large Transverse Energy Electrons with Associated Missing Energy at $\sqrt{s}$ =540 GeV". *Phys. Lett.* 122B: 103–116.
116. Sommerfeld, A. (1919, 1929). "Atombau und Spektrallinien", two volumes. Braunschweig, Vieweg.
117. Lederman, L. M. and Teresi, D. (1993). "The God Particle: If the Universe is the Answer, What is the Question?" Houghton Mifflin Company, Boston.
118. Einstein, A. (1917)."Über die spezielle und die allgemeine Relativitätstheorie". Springer, Berlin, Heidelberg and New York. The quotation from this work on special and general relativity in Chapter 19 "The Gravitational Field" reads: "If now, as we find from experience, the acceleration is to be independent of the nature and the condition of the body and always the same for a given gravitational field, then the ratio of the gravitational to the inertial mass must likewise be the same for all bodies. By a suitable choice of units we can thus make this ratio equal to unity. We then have the following law: The gravitational mass of a body is equal to its inertial mass.

It is true that this important law had hitherto been recorded in mechanics, but it had not been interpreted. A satisfactory interpretation can be obtained only if we recognise the following fact: The same quality of a body manifests itself according to circumstances as "inertia" or as "weight" (lit. "heaviness"). In the following section we shall show to what extent this is actually the case, and how this question is connected with the general postulate of relativity."
119. Genzel, R. et al. (2002). Personal communication.
Schödel, R., Ott, T., Genzel, R., Hofmann, R., Lehnert, M., Eckart, A., Mouawad, N., Alexander, T., Reid, M. J., Lenzen, R., Hartung, M., Lacombe, F., Rouan, D., Gendron, E., Rousset, G., Lagrange, A.-M., Brandner, W., Ageorges, N., Lidman, C., Moorwood, A. F. M., Spyromilio, J., Hubin, N. and Menten, K.

M. (2002). "A star in a 15.2-year orbit around the supermassive black hole at the centre of the Milky Way". *Nature* 419: 694–696.

120. Lederman, L. and Teresi, D. (1993). "The God Particle: If the Universe Is the Answer, What Is the Question?". Houghton Mifflin Company, Boston.
121. Schwarzschild, K. (1916). "Über das Gravitationsfeld eines Massenpunktes nach der Einsteinschen Theorie". *S.-B. Kgl. Akad. Wiss.* 1916 (part 1). pp. 189–196. Deutsche Akademie der Wissenschaften zu Berlin. Berlin.
122. Schwarzschild, K. (1916). "Über das Gravitationsfeld einer Kugel aus inkompressibler Flüssigkeit nach der Einsteinschen Theorie". *S.-B. Kgl. Akad. Wiss.* 1916 (part 1). pp. 424–434. Deutsche Akademie der Wissenschaften zu Berlin. Berlin.
123. Buys-Ballot, C. H. D. (1845). "Akustische Versuche auf der Niederländischen Eisenbahn, nebst gelegentlichen Bemerkungen zur Theorie des Hrn. Prof. Doppler". *Ann. Phys. Chem.* 142: 321–351.
124. Wollaston, W. H. (1802). "A Method of Examining Refractive and Dispersive Powers, by Prismatic Reflection". *Phil. Trans. R. Soc.* 92: 365–380.
125. Fraunhofer, J. (1817). "Bestimmung des Brechungs- und Farbenzerstreuungs-Vermögens verschiedener Glasarten, in Bezug auf die Vervollkommnung achromatischer Fernrohre". *Denkschriften der koeniglichen Akademie der Wissenschaften zu Muenchen für die Jahre 1814 und 1815*, 5: 193–226 (also in *Ann. Phys. (Gilbert)* 56: 264–313).
126. Richter, P. H. (1995). "Das Olberssche Paradoxon". *Sterne und Weltraum* 34: 804–809.
127. Bondi, H. (1952). "Cosmology". Cambridge University Press, Cambridge.
128. Harrison, E. R. (1965). "Olbers' Paradox and the Background Radiation Density in an Isotropic Homogeneous Universe". *Mon. Not. R. Astr. Soc.* 131: 1–12.
129. Fowler, W. A. (1993). "Experimental and Theoretical Nuclear Astrophysics; the Quest for the Origin of the Elements". Nobel Lectures. Physics 1981–1990. Editor-in-Charge Tore Frängsmyr, Editor Gösta Ekspång. World Scientific Publishing Co., Singapore.
130. Einstein, A. (1920). "Relativity: The Special and General Theory". Chapter XXII. "A Few Inferences from the General Theory of Relativity". Methuen & Co Ltd. London.
131. Hilbert, D. (1915). "Die Grundlagen der Physik". *Kgl. Ges. Wiss. Nachr., Math.-Phys. Kl.* 8: 395–407.
    Hilbert, D. (2007). "The Foundations of Physics". Chapter 9 in Janssen, M., Norton, J. D., Renn, J., Sauer, T. and Stachel, J. "Genesis of General Relativity". Springer Netherlands.
132. Sakharov, A. D. (1967). "Violation of CP Symmetry, C-Asymmetry and Baryon Asymmetry of the Universe". *Pisma Zh. Eksp. Teor. Fiz.* 5: 32–35, translation in *JETP Lett.* 5: 24–27 (1967); Sakharov, A. D. (1967). "Quark-Muonic Currents and Violation of CP Invariance". *Pisma Zh. Eksp. Teor. Fiz.* 5: 36–39, translation in *JETP Lett.* 5: 27–30 (1967).

133. Teitelboim, C. alias Bunster, C. (1973). "Hamiltonian Gravitation: Are Constraints Redundant?" *Phys. Rev. D* 8: 3266–3270; Regge, T. and Teitelboim, C. (1974). "Role of Surface Integrals in the Hamiltonian Formulation of General Relativity". *Ann. Phys.* 88: 286–318.
134. See Ref. 54: Wheeler, J. A. "From Relativity to Mutability". Part 1–9. pp. 202–250. In: Mehra, J. (Ed.) (1973). "The Physicist's Conception of Nature". Proceedings of a Symposium on the Development of the Physicist's Conception of Nature, held at Trieste, September 18–25, 1972. In honour of the 70th birthday of Paul Dirac. Dordrecht: Reidel.
135. Pound, R. V. and Rebka Jr., G. A. (1959). "Gravitational Red-Shift in Nuclear Resonance". *Phys. Rev. Lett.* 3(9): 439–441; Pound, R. V. and Rebka Jr., G. A. (1960). "Apparent Weight of Photons". *Phys. Rev. Lett.* 4 (7): 337–341.
136. Friedman, A. (1922). "Über die Krümmung des Raumes". *Z. Phys.* 10 (1): 377–386. English translation in Friedman, A. (1999). "On the Curvature of Space". *Gen. Relativ. and Gravitation* 31: 1991–2000.
137. Einstein, A. and de Sitter, W. (1932). "On the Relation between the Expansion and the Mean Density of the Universe". *Proc. Natl. Acad. Sci. U.S.A.* 18: 213–214.
138. Hubble, E. (1929). "A Relation between Distance and Radial Velocity among Extra-Galactic Nebulae". *Proc. Natl. Acad. Sci. U.S.A.* 15 (3): 168–173. See also Lemaître, G. (1927). "Un Univers homogène de masse constante et de rayon croissant rendant compte de la vitesse radiale des nébuleuses extra-galactiques". *Ann. Soc. Sci. Bruxelles* 47: 49–59, English translation: Lemaître, G. (1931). "A Homogeneous Universe of Constant Mass and Growing Radius Accounting for the Radial Velocity of Extragalactic Nebulae". *Mon. Not. R. Astron. Soc.* 91: 483–490.
139. Alpher, R. A., Bethe, H. and Gamow, G. (1948). "The Origin of Chemical Elements". *Phys. Rev.* 73(7): 803–804.
140. Bondi, H. and Gold, T. (1948). "The Steady-State Theory of the Expanding Universe". *MNRAS* 108: 252–270.
Hoyle, F. (1948). "A New Model of the Expanding Universe". *MNRAS* 108: 372–382.
141. Penzias, A. A. and Wilson, R. W. (1965). "A Measurement of Excess Antenna Temperature at 4080 Mc/s". *Astrophys. J.* 142: 419–421.
142. Smoot, G. F., Fixsen, D. J., Cheng, E. S. et al. (1994). "Cosmic Microwave Background Dipole Spectrum Measured by the COBE FIRAS Instrument". *Astrophys. J.* 420(2): 445–449.
143. Weinberg, S. (1977). "The First Three Minutes: A Modern View of the Origin of the Universe". Basic Books. New York. (Updated with new afterword in 1993.)
144. Guth, A. H. (1981). "The Inflationary Universe: A Possible Solution to the Horizon and Flatness Problems". *Phys. Rev. D* 23: 347–356.

145. Icke, V. (1995). "The Force of Symmetry". Cambridge University Press, Cambridge.
146. Cartan, E. (1952–1955). "Oeuvres completes", three volumes. Paris. Reprinted by Edition du CNRS (1984).
147. Hungarian patent HU170062 for the Magic Cube in 1975; Rubik, E., Varga, T., Keri, G., Marx, G. and Vekerdy, T. (1987). "Rubik's Cubic Compendium", English translation by A. Buvös Kocka, with an afterword by D. Singmaster. Oxford University Press, London; Frey, A. H. and Singmaster, D. (1982). "Handbook of Cubik Math". Enslow Publishers, Hillside, N.J.
148. Lee, T. D. and Yang, C. N. (1956). "Question of Parity Conservation in Weak Interactions". *Phys. Rev.* 104 (1): 254–258.
149. Wu, C. S., Ambler, E., Hayward, R. W., Hoppes, D. D. and Hudson, R. P. (1957). "Experimental Test of Parity Conservation in Beta Decay". *Phys. Rev.* 105: 1413–1415.
150. Christenson, J. H., Cronin, J. W., Fitch, V. L. and Turlay, R. (1964). "Evidence for the 2π Decay of the K_2^0 Meson". *Phys. Rev. Lett.* 13: 138–140.
151. Lüders, G. (1954). "On the Equivalence of Invariance under Time Reversal and under Particle-Antiparticle Conjugation for Relativistic Field Theories". G. *Kgl. Danske Videnskab. Selskab, Mat.-fys. Medd.* 28(5): 1–17; Lüders, G. (1957). "Proof of the TCP Theorem". *Ann. Phys.* 2: 1–15.
152. Aubert, B., Boutigny, D., Gaillard, J. M. *et al.* BABAR Collaboration. (2002). "Measurements of Branching Fractions and CP-Violating Asymmetries in B0$\rightarrow \pi + \pi$-, K+π-, K+K- Decays". *Phys. Rev. Lett.* 89, 281802-1–281802-7.
153. Gross, D. J., Harvey, J. A., Martinec, E. J. and Rohm, R. (1985). "The Heterotic String". *Phys. Rev. Lett.* 54: 502–505.
154. Georgi, H. and Glashow, S. (1974). "Unity of All Elementary-Particle Forces". *Phys. Rev. Lett.* 32:.438–441.
155. Kaluza, T. (1921). "Zum Unitätsproblem in der Physik". *S.-B. Preuss. Akad. Wiss. Berlin, Math. Phys. Kl.* 1921: 966–972.
156. Klein, O. (1926). "Quantentheorie und fünfdimensionale Relativitätstheorie". *Z. Phys.* 37(12): 895–906.
157. Yang, C. N. and Mills, R. (1954). "Conservation of Isotopic Spin and Isotopic Gauge Invariance". *Phys. Rev.* 96: 191–195.
158. Green, M. B. and Schwarz, J. H. (1984). "Anomaly Cancellation in Supersymmetric D=10 Gauge Theory and Superstring Theory". *Phys. Lett. B* 149: 117–122.
159. Witten, E. (1986). "Non-Commutative Geometry and String Field Theory". *Nucl. Phys. B* 268: 253–294; Witten, E. (1986). "Interacting Field Theory of Open Superstrings". *Nucl. Phys. B* 276: 291–324.
160. Princeton String Quartet.
161. Berndt, B. C. (1991). "B. Ramanujan's Notebooks", Parts I–V. Springer-Verlag, New York.

162. Kaku, M. (1994). “Hyperspace: A Scientific Odyssey Through Parallel Universes. Time Warps, and the Tenth Dimension”. Oxford University Press, Oxford.
163. Greene, B. (1999). “The Elegant Universe: Superstrings, Hidden Dimensions, and the Quest for the Ultimate Theory”. W. W. Norton. New York.
164. Wheeler, J. A. (1980). “Beyond the black hole”. pp. 341–375. In: Wolff, He. ed. “Some strangeness in the proportion. Centennial symposium to celebrate the achievements of Albert Einstein”. Addison-Wesley Educational Publishers Inc. Reading, Massachusetts.
165. Rees, M. J. (1999). “Just Six Numbers: The Deep Forces That Shape the Universe”. Weidenfeld & Nicolson. London.
166. Rowan-Robinson, M. (1999). “The Nine Numbers of the Cosmos”. Oxford University Press, Oxford.
167. Genz, H. (2002). “Wie die Naturgesetze Wirklichkeit schaffen. Über Physik und Realität”. Carl Hanser Verlag, München.
168. Susskind, L. (1997). “Black Holes and the Information Paradox”. *Sci. Am.* 272: 52–57.
169. Susskind, L. (2005). “The Cosmic Landscape: String Theory and the Illusion of Intelligent Design”. Little, Brown & Co. New York.
170. Wheeler, J. A. (1994). See Ref. 75.
171. Smolin, L. (1999). “The Life of the Cosmos”. Oxford University Press, Oxford.
172. Dicke, R. H. (1961). “Dirac’s Cosmology and Mach’s Principle”. *Nature* 192: 440–441.
173. Carter, B. (1974). “Large Number Coincidences and the Anthropic Principle in Cosmology”. In: “IAU Symposium 63: Confrontation of Cosmological Theories with Observational Data”, Longair, M. S. (ed.). Reidel, Dordrecht, pp. 291–298.
174. Barrow, J. D. and Tipler, F. J. (1988). “The Anthropic Cosmological Principle”. Oxford University Press, Oxford.
175. Eigen, M. and Winkler-Oswatitsch, R. (1981). “Transfer-RNA: The Early Adaptor”. *Naturwiss.* 68: 217–228; Eigen, M. and Winkler-Oswatitsch, R. (1981). “Transfer-RNA, an Early Gene?”. *Naturwiss.* 68: 282–292.
176. Eigen, M., Gardiner, W., Schuster, P. and Winkler-Oswatitsch, R. (1981). “The Origin of Genetic Information”. *Sci. Am.* 244: 88–118.
177. Eigen, M., Lindemann, B. F., Tietze, M., Winkler-Oswatitsch, R., Dress, R. and von Haeseler, A. (1989). “How Old is the Genetic Code? Statistical Geometry of tRNA Provides an Answer”. *Science* 244(4905): 673–679.
178. Wheeler, J. A. (1979). “The Quantum and the Universe”. In “Relativity, Quanta and Cosmology”, Vol. II, Pantaleo, M. and de Finis, F. (eds). Johnson Reprint Corp., New York.

2 Energy and Entropy

2.1.	Who's Afraid of . . . ?	139
2.2.	Gibbs' Paradox: How Equal is "Equal"?	153
2.3.	How Real is the Microstate?	161
2.4.	Probability: Expectation, Frequency or Fuzziness?	166
2.5.	How Real is the Macrostate?	172
2.6.	How Many Trees Make a Wood?	178
2.7.	Can Arbitrarily Complex Chemical Systems Ever Reach Detailed Balance?	188
2.8.	How is Entropy Related to Order?	197
2.9.	Who Keeps Our Clocks Running?	207
2.10.	Entropy: What Does It Mean?	214

2.1. Who's Afraid of . . . ?

Clearly, I do not mean Virginia Woolf, in fact, as little as Edward Albee, the author of the award-winning play of 1964 "Who is afraid of Virginia Woolf?" really meant the English novelist, critic and feminist who lived from 1882 till 1941. He took the title from a graffito he found on some men's-room wall. No, what I mean here is entropy, and I have every reason to ask: who is afraid of it? Even nowadays chemistry students still get nervous if an examination question in physical chemistry includes the word "entropy".

The term "entropy" was coined by Rudolf Clausius in his famous 1865 paper in "Poggendorf's Annalen"[1]. Clausius wanted to express by a phenomenological term the capacity of a macroscopic system for internal microscopic alteration. He called it the "Verwandlungsgehalt" (changeability content) of matter. As was customary, he derived the term from Greek. The verb entrepein (ἐντρέπειν) means to change, to alter or to modify, but Clausius did not explicitly refer to this verb. Instead, he quoted the noun tropae (ἡ τροπή = the change) and added the prefix "en" (= "in"), to make it sound similar to energy (ἔργον) = work), stressing "because the two terms are closely related in their physical importance". In the same paper, Clausius formulated the two principles of thermodynamics as a set of laws applying to the world as a whole.

Entropy, despite its more than 130 years of existence as a defined variable of state, is still considered with a mixture of uncertainty, awe and suspicion by the public at large. The cartoon from the *New Yorker* (Figure 2.1.1) expresses vividly what I mean. I am sure that Clausius would have felt hurt by the comment made in the caption of the cartoon. No one has scruples about using the word "energy" in daily conversation – notwithstanding the fact that we do not know what energy really is. Yet the word

Figure 2.1.1

"In my opinion, Mrs. Wendell – and I believe Dr. Steinmuth will concur – if you can live with entropy you can live with anything."

From *The New Yorker*

"entropy" remains largely confined to scientific textbooks. Even consulting a dictionary is not very rewarding. One is usually offered a collection of terms that associate entropy with "useless energy", "lack of order" or "measure of probability", none of which matches the true meaning of a term that is physically well defined. The definition given by Clausius equates the increase of entropy with the "reversible" supply of heat at a given temperature, divided by that temperature (see Vignette 2.1.1). I admit that this formulation does not give a lay person much help in grasping what entropy is really about. First of all, the words "reversible supply of heat" and "temperature" require some further elucidation. Let me start with temperature.

Vignette 2.1.1 Definition of entropy according to Clausius

The entropy is considered as a function of volume (V), temperature (T) and particle number (N). Q_{rev} is the reversibly added heat.

$$S_2 - S_1 = \int_1^2 \frac{dQ_{rev}}{T}$$

At constant N we have:

$$S(V_2, T_2, N) - S(V_1, T_1, N) = \int_1^2 \left(\frac{\delta S}{\delta V}\right)_T dV + \left(\frac{\delta S}{\delta T}\right)_V dT$$

For an ideal gas the equation of state reads:

$$PV = NkT;$$

The heat capacity at constant volume is:

$$C_V = \frac{3}{2}Nk$$

The equation:

$$S(V_2, T_2, N) - S(V_1, T_1, N) = Nk \log V_2/V_1 + 3/2\, Nk \log T_2/T_1$$

is satisfied by:

$$S(V, T, N) = k\,\{N \log V + 3/2\, N \log T + f(N)\}$$

Everyone has some intuitive understanding of temperature. However, anyone who really understands what temperature is should have no trouble with entropy either. Why is it that temperature is so familiar to us, while entropy remains so strange?

Probably because we have a sensory awareness of "hot" and "cold". Furthermore, we know how to measure temperature. You will find a thermometer in every household. We knew how to measure temperature long before we had any idea what temperature really is. If we had known better, we might have given temperature the dimensions of energy (per amount of matter).

In contrast, in order to measure entropy we need a calorimeter. This instrument does not even belong among the standard equipment of a chemical laboratory, let alone a household. Temperature is, so to speak, the "intensity" of heat or of the underlying thermal motion, a measure of the average kinetic energy related to a degree of freedom of motion. Entropy is the "extensive" complement which could be expressed in dimensionless units, as long as reference to the amount of matter involved is made.

A point-like material particle has three translational degrees of freedom, which means that it can move freely in all three dimensions of space. A body with spatial extension may also rotate about any of its three axes. A diatomic molecule with two point-like atoms separated by a finite distance has only two of these axes available for rotation. Point-likeness, of course, is an approximation to reality, but quantum theory tells us why, in this case, it is a realistic approximation. The diatomic molecule then – among its maximally possible six degrees of freedom – has one internal degree of oscillatory motion of its two atoms with respect to each other. Molecules, made up of many atoms, likewise have many oscillatory degrees of freedom, associated with the normal modes of coupled internal motion, and hence are accompanied by correspondingly large entropic increments per molecule. Quantum theory (Section 1.3) establishes how energy is distributed among the various degrees of freedom. In the limit of high temperature, each degree of freedom of oscillatory motion (including potential and kinetic energy) reaches a limiting energy value of kT, where k is Boltzmann's constant (1.38×10^{-23} J K^{-1}) and T is the absolute temperature.

These insights into the nature of the "thermal fluid", called heat, were gained in the late 19th and early 20th century. By that time, the scale of temperature, referring to the thermal expansion of a test body, and measured in variously defined "degrees" (Réaumur, Fahrenheit, Celsius), was long fixed. Otherwise, Boltzmann's constant could have been set to a value of 1, which would have meant measuring temperature in units of kT (as indeed was done by Gibbs, who was unaware of Boltzmann's work). Since T as an "intensity" is independent of the amount of matter, entropy should be its "extensive" complement. Other examples of complementary variables are pressure (intensive) and volume (extensive), electric or magnetic field density (intensive) and dielectric polarisation or magnetisation (extensive). Thermodynamics – at least as far as it applies to systems at equilibrium – is a wonderful, logically closed edifice that rests on two axioms: the conservation principle of energy (the first law) and an extremum principle for entropy (the second law). One may perhaps add a "zeroth" law, usually tacitly assumed, namely the existence of equations of state, i.e. defined relations among the extensive and intensive variables. (These relations define quantities such as specific heats at constant volume or pressure, isothermal or isentropic compressibilities, coefficients of thermal expansion, dielectric permittivities or magnetic permeabilities.) Thermodynamics then consists of an application of variational principles that carefully distinguish between intensive and extensive variables.

The phenomenology of thermodynamics can be put onto a rigorous axiomatic basis. The roots of this approach are found in the works of Josiah Williard Gibbs (1839–1903). The father of the axiomatic theory of thermodynamics is the Greek-born German mathematician Constantin Caratheodory, whose 1909 paper[2] *Untersuchungen über die Grundlagen der Thermodynamik* (Studies on the foundations of thermodynamics) is a landmark of classical thermodynamics (see also the clear account of Caratheodory's work in English by Robert Eisenschitz[3]). A more recent rigorous approach to axiomatic thermodynamics is that of Gottfried Falk[4] and Herbert Jung.

It may come as a surprise that Clausius' definition of entropy as a state variable (meaning that it characterises a state, independent of the system's previous history) does not necessarily imply that entropy is extensive (cf. Table 2.1.1). The way Clausius introduces entropy is to add to a system minute portions of heat, piece by piece, at their respective temperatures in a reversible way. Since heat itself is not a variable of state it has to be added in small portions, with carefully controlled book-keeping of the amounts added and their respective temperatures. Integration then yields a finite difference in entropy between the system's initial state, before the addition of the first portion of heat, and its state after all the portions of heat have been added to it. As a function of state, entropy then requires an integration constant, as yet undefined. As long as the extent of the system does not change, the integration constant deserves its name, i.e. it is constant. Clausius did not make any statement to this effect. However, if no assertion about the amount of matter involved is made, then the integration "constant" may well turn out to be a function of the amount of matter. Hence, in respect of the extensive character of entropy, Clausius' definition is incomplete. In particular, the entropy of an ideal gas (Table 2.1.1) according to Clausius' definition is not (necessarily) an extensive function.

This deficiency was noted explicitly by Wolfgang Pauli.[5] In his *Lectures on Physics* he emphasised this incompleteness of Clausius' definition and proposed to complement it by a scaling law, to ensure extensiveness. The scaling law states explicitly that any increase in the size of the system requires an increase in its entropy by the same factor (cf. Vignette 2.1.2).

The above deficiency, of course, had been noted from early on. However, we learned that this was a shortcoming of classical statistical mechanics and could be rectified only by quantum mechanics, which explicitly require indistinguishability of particles of the same kind. The entropy constant thus calculated – apart from providing a numerical value – becomes a function directly proportional to the

Vignette 2.1.2 Pauli's analysis of the extensivity of entropy

Pauli introduces scaling by using the undetermined f(N) to correct the term N log V, which does not comply with the scaling condition, which reads:

$$S(qV, T, qN) = qS(V, T, N)$$

$$0 < q < \infty$$

with

$$f(N) = -N \log N + f(1)$$

As f(1) now is a constant, we obtain

$$S(V, T, N) = Nk\,\{\log V/N + 3/2 \log T + f(1)\}$$

This causes $S = f(V, T, N)$ to be an "extensive" function.

amount of matter in the system, thereby establishing the extensiveness of entropy. Thermodynamics as a logically closed phenomenological theory should not depend on its mechanistic background. Pauli's method of supplementing Clausius' definition by a scaling law appears to be the most satisfactory way. Moreover, it leaves open the possibility for entropy not to be extensive wherever the scaling law does not apply (which is the case for certain long-range interactions as well as for surface phenomena). It was the late Edwin Thompson Jaynes[6] who re-opened the discussion about the nature of entropy, and we shall return to this later in the chapter.

The original idea of introducing entropy came out of Carnot's abstraction of the process that converts heat to work, which refers to a system at equilibrium. Entropy is a measure that identifies the part of energy which cannot be turned into work. This part can never become negative or even zero, otherwise one would have a perpetual-motion machine. Hence, in any non-equilibrium process – and especially in an order-producing process of self-organisation (as discussed later in this book) – the net change of entropy will always be positive. This is the lesson of the second law of thermodynamics, which, like the first law, is an axiom and cannot be deduced from anything else. However, while energy is a naturally "given" property that we can measure (even if we do not know what it "really is"), entropy is a construct of the human mind. Entropy is "explained" by a definition that in its detail depends on what knowledge we have about the system. If the system does not exchange heat, work or matter with its surroundings – in which case we call it an "isolated system" – entropy eventually reaches a maximum value, corresponding to the establishment of equilibrium. Entropy for its statistical definition does not require the establishment of equilibrium, while temperature (appearing in statistical thermodynamics as an "integrating denominator") makes sense only when the system has reached thermal equilibrium, at least for the degrees of freedom under consideration.

The latter point may require some comment. The term "thermal equilibrium" has to be explained in more detail. Energy distributes itself among all available degrees of freedom, such as translational, rotational, vibrational or spin states in magnetic fields. However, it does so with different characteristic rates, so that, for instance, we may talk about a vibrational or a spin temperature, the latter being comparatively slow in establishing its equilibrium. In a sound wave of high frequency, even vibrational equilibration is unable to follow rapid adiabatic compression and dilation, with their associated changes of translational and rotational temperature. Vibrational relaxation occurs in a characteristic frequency range in which the velocity of sound undergoes a dispersion with concomitant absorption of sound energy, as was first noted and explained by Albert Einstein.[7]

As Table 2.1.1 shows, entropy can be expressed as a function of the extensive variables (internal) energy and volume. (Reciprocal) temperature can then be defined as the partial derivative of entropy with respect to internal energy at constant volume. The more complex the system, i.e. its composition and the processes of chemical

Table 2.1.1 Thermodynamic Functions

State Function	Independent Variables	Partial Derivatives
Energy U	$S, V, N_1 \ldots N_n$	$dU = TdS - PdV + \sum_{i=1}^{n} \mu_i dN_i$
Entropy S	$U, V, N_1 \ldots N_n$	$dS = \frac{1}{T}\, dU + \frac{P}{T} dV - \sum_{i=1}^{n} \frac{\mu_i}{T} dN_i$
Helmholtz Free Energy A ($A = U - TS$)	$T, V, N_1 \ldots N_n$	$dA = -SdT - PdV + \sum_{i=1}^{n} \mu_i dN_i$
Enthalpy H ($H = U + PV$)	$S, P, N_1 \ldots N_n$	$dH = TdS + VdP + \sum_{i=1}^{n} \mu_i dN_i$
Gibbs Free Energy G ($G = U - TS + PV$)	$T, P, N_1 \ldots N_n$	$dG = -SdT + VdP + \sum_{i=1}^{n} \mu_i dN_i$

P = pressure, V = volume, T = temperature, μ_i = chemical potentials, N_i = particle numbers. S, V, N_i and U (and hence A, G) are extensive, T, P and μ_i intensive variables, where $\mu_i \equiv (\partial G/\partial N_i)_{PTN_{k \neq i}}$

Examples: $dU(S, V, N_i) = (\partial U/\partial S)_{V,N_i} \cdot dS + (\partial U/\partial V)_{S,N_i} \cdot dV + \sum_{i=1}^{n} (\partial U/\partial N_i)_{S,V,N_i} \cdot dN_i$

or: $1/T = (\partial S/\partial U)_{V,N_i}$ or: $S = (\partial G/\partial T)_{P,N_i}$

equilibration among its constituents, the more variables, such as chemical potentials and individual particle numbers, may be explicitly taken into consideration when writing down entropy as a function of extensive variables. Table 2.1.1 also lists corresponding relations for other thermodynamic "potentials". Systems that are in a heat bath at constant temperature with which they can exchange thermal energy are best described by so-called "free" energies. At constant temperature the change of free energy corresponds to (usable) work that is performed on, or by, the system. Whether or not this work is in fact utilised (see Section 2.2) depends on our knowledge, on the basis of which we can construct machines that make use of it. At constant temperature we can write the change in internal energy as a sum of two terms, one being the change of (usable) free energy, the other being the non-usable part expressed by the change of entropy (multiplied by temperature). The only physical change in the system would be in its internal energy. Even if this is zero, we still have the freedom of compensation between free energy and entropy change. Any realisation of work is compensated for by a heat flux from a thermal bath. This somehow "anthropomorphic" freedom can lead to seemingly paradoxical behaviour, which will be discussed in Section 2.2.

In this context let me say a few words about the chemical variables μ_i and N_i, which appear in Table 2.1.1, because they are good examples of our knowledge about internal parameters of a system, which may or may not be accessible. Moreover, complex chemistry is a hallmark of biological systems, which constitute a major topic in this book. In the early 1950 I developed a method, called chemical relaxation spectrometry, which is based on the perturbation of a chemical equilibrium by means of rapid changes of external parameters, such as pressure, temperature or electrical field density. The response of the system not only allowed the determination of the speed of fast chemical reactions down to the nanosecond range, but also the analysis of complex reaction mechanisms in terms of elementary steps of chemical change.

In Table 2.1.1 I referred to chemical states, thermodynamically defined by their particle numbers (N_i) and their attributed chemical potentials μ_i, as defined in the table. In a closed system, i.e. in the absence of any transport of particles to and from the environment and with defined temperature and pressure, we may look at the internal transformations by referring to each particular reaction j occurring in the system, rather than to each single molecular state of particles involved. For this we introduce as the extensive variable which responds to an external perturbation the "extent of reaction" ξ_j, which for all participants of a given reaction j are defined by $dN_{ij} = \nu_i d\xi_j$, where ν_i is the stoichiometric coefficient of a particle i as a participant in the reaction j. In other words, all participants of a given reaction change simultaneously in stoichiometric proportion, so that their changes can be expressed by one extensive reaction parameter ξ_j.

The term $\sum_{i=1}^{n} \mu_i dN_i$, referring to single chemical states i, now can be rewritten as a sum, referring to all (m) reactions j taking place in the system: $\sum_{j=1}^{m} A_j d\xi_j$, where A_j is called the affinity of reaction j: $A_j = \sum_{i=1}^{n_j} \nu_{ij}\mu_{ij} - \nu'_{ij}\mu'_{ij}$, the indices ij referring to all participants i in reaction j, while the apostrophe distinguishes reactants (which have a negative sign) from reaction products (with a positive sign). At detailed balance both A_j and ξ_i become zero because no net reaction occurs. For any state near equilibrium both A_j and ξ_j are finite and their change – irrespective of the order (or molecularity) of the reaction – can be described by a system of linear differential equations.

The method can be applied to any arbitrarily complex system near equilibrium. Because of the reactive couplings among different chemical states we have a system of coupled differential equations which has to be transformed to normal co-ordinates, as is known also for systems of coupled oscillators. A general thermodynamic theory of such "linear systems" was developed by Joseph Meixner[8], a descendent of Sommerfeld's school. As in oscillatory systems, the eigenvalues of relaxation refer to normal modes of reaction rather than to single molecular states. They define a spectrum of relaxation times, which are the time constants that are experimentally detectable in response to an external perturbation. Since the solutions in terms of normal modes appear to be exponential functions, their resolution is not as sharp as we are accustomed to in the case of resonance phenomena. But it was this technique of chemical relaxation spectrometry, which for the first time allowed resolution of the complex reaction mechanisms of enzymes, that is demonstrated with the example of an allosteric enzyme at the end of this section (Vignette 2.1.3).

After this little excursion into chemistry, I return to the question in the title of this section. It is important to note that entropy, unlike energy, is a defined construct of the human mind, but in its detailed interpretation depends on knowledge that we are in possession of. Edwin Jaynes therefore has called it "anthropomorphic" in character. All we have done so far is to provide a phenomenological definition. Clausius, who was the first to give such a definition, left open the question of extensiveness, which is most relevant for systems that exchange matter with their environment. As an extensive variable (thermodynamic) entropy represents a bulk figure about the "extent" of material states among which (thermal) energy is distributed. As such, it expresses more the extent of (combinatorial) complexity of the system than it expresses – as we are often told – the system's particular state of "order" or "disorder" (which anyway would require an adequate definition).

The title of this chapter is "Energy and Entropy" and, indeed, entropy reflects the sponge-like ability of matter to store energy in its multiple individual atomic and molecular degrees of freedom. However, as was shown first by Josef Stefan and Ludwig Boltzmann[9], even a hollow space, the walls of which are at non-zero temperature, and which therefore is "filled" with black body radiation, are characterised not only by a finite energy, but also by an entropy (see Section 2.10).

Vignette 2.1.3

Allosteric enzymes may be considered as "intelligent" catalysts. Beside their active (catalytic) site they have another (*allo-*), spatially separated (*-steric*), binding site that provides for regulation of their catalytic or binding activity. In this way they can respond intelligently to "supply and demand".

As an example, consider the haemoglobin molecule (which gives red blood cells their colour). Its task is the transport of oxygen molecules from the lung to the tissue, where the oxygen is used to burn a metabolite, or the reverse transport of the ultimate product of burning (CO_2) from the tissue to the respiratory organ. As such it is not really an enzyme, but owing to its sophisticated enzyme-like binding properties it may be awarded the "honorary title" of an enzyme. This was suggested by the late Max Perutz, to whom we owe most of our knowledge about this amazing molecular machine.

The haemoglobin molecule includes four protein subunits, each equipped with a binding site for O_2. The binding of an oxygen molecule at one site exerts an effect on the binding affinities of the other three subunits, thereby regulating the binding activity according to supply and demand. In the lung, where oxygen is plentiful, the binding affinity is high, so that the haemoglobin is saturated with oxygen. In contrast, at the metabolic sites, where oxygen is needed, it is all given up and replaced by CO_2, which in turn is exhaled in the lung. This is a brief description of the system. What we want to explore is the mechanism of such a sophisticated reaction system.

In Figure 2.1.2 two possible reaction schemes are shown that could explain the co-operative behaviour of such a system. Co-operative binding expresses itself in the particular form of the equilibrium binding curve, as is demonstrated for oxygen binding by myoglobin (involving only one subunit) and haemoglobin (involving four subunits) in Figure 2.1.3. Equilibrium properties, on the other hand, reflect the interactions involved only in a cumulative way and therefore are not well suited for drawing conclusions about the mechanism. The allosteric binding curve of haemoglobin can be described – in principle – by a two-parameter equation. A three-parameter equation almost suffices to describe "half an elephant" (Figure 2.1.4). What I want to say by this is that we need a different approach in order to distinguish between mechanisms such as the two schemes of allosteric binding illustrated in Figure 2.1.2.

The relaxation technique, which I mentioned in the text of this section, provides such a tool. A chemical equilibrium, with all its steps in detailed balance, is perturbed by a sudden change of an external parameter, such as pressure,

Cont. ➲

Cont.

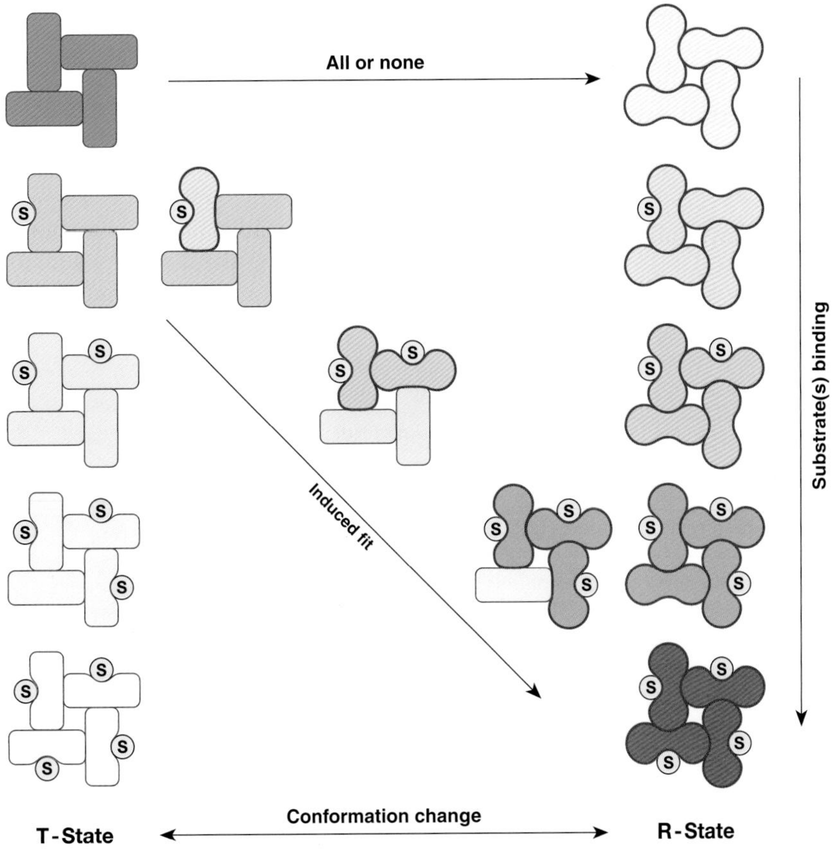

Figure 2.1.2

Reaction schemes for allosteric control with the example of a four-subunit enzyme

The two vertical columns refer to two different conformational states of different affinity for a substrate S (yellow): the T state (blue; lower affinity for S) and R state (red; higher affinity for S). According to the model of Monod, Wyman and Changeux all subunits – by reason of symmetry – can only be in one or other of these two conformations. There is an "all or none" conformational change that includes all subunits. In contrast, Koshland *et al.* assume that substrate binding involves an "induced fit" of the particular protein subunit. The Koshland mechanism then involves only states along the diagonal. In the simplest case the Monod mechanism yields only three relaxation times: (1) binding of the substrate to R subunits, (2) binding of the substrate to T subunits and (3) an all-or-none conformational change R $\leftrightarrow$ T. The Koshland mechanism includes only substrate binding by induced fit. However, this takes place in four steps, which for strong co-operativity should differ considerably in their rates.

Cont.

➲ Cont.

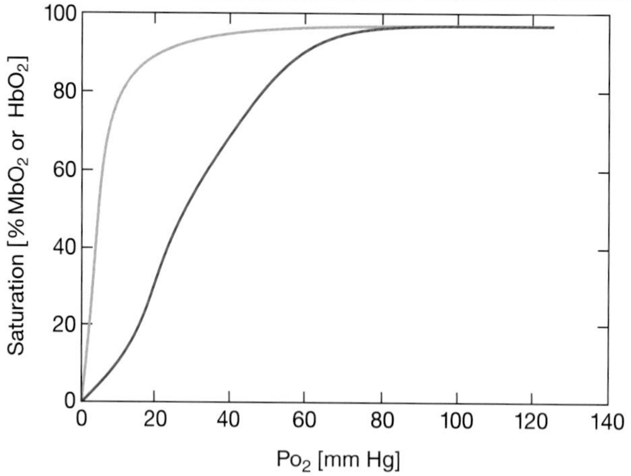

Figure 2.1.3

Oxygen-binding curves for myoglobin (light red) and haemoglobin (deep red)

The hyperbolic curve for myoglobin is what one expects on the basis of the law of mass action. The curve starts with a linear term and the slope steadily decreases towards saturation. The allosteric curve for haemoglobin is S-shaped. It starts with a low slope which then increases before approaching saturation. Such a curve was established by Adair in 1925. Adair measured four different binding constants, but the shape of the curve could be simulated with the help of only two constants.

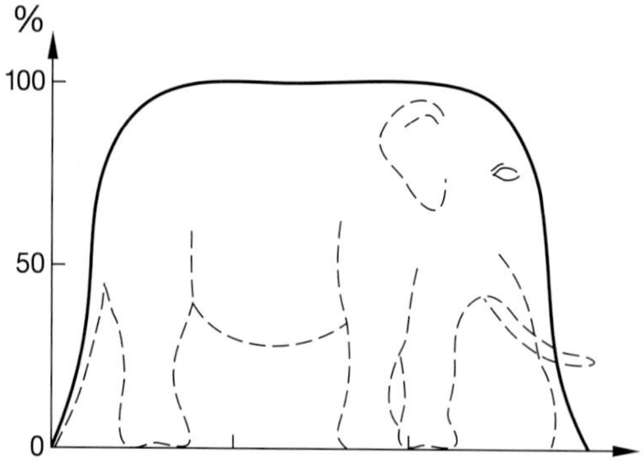

Figure 2.1.4

No comment!

Cont. ➲

Cont.

temperature or electrical field strength. All single steps then respond with characteristic rates, eventually restoring detailed balance. Owing to the coupling among all the steps, re-equilibration proceeds along the "normal modes" that result from solving the system of linear differential equations of the variables describing the deviations from equilibrium. The rate parameters appear in the relaxation times in a way defined by the reaction couplings and the transformation of the corresponding rate equations into a diagonal form. Instead of going into the detail of the theory, I will describe some experimental results that allowed the identification of the reaction mechanism of allosteric control for a particular case.

The system studied was the four-subunit enzyme glyceraldehyde-3-phosphate dehydrogenase. The experiments were done by Kasper Kirschner, who had joined our group in the early 1960s, coming from Feodor Lynen's laboratory in Munich. I had just worked out the complete normal-mode theory of chemical relaxation of enzymes which could be directly applied to the different possible allosteric mechanisms. This was done together with Georg Ilgenfritz, who at that time carried out those studies for his doctoral thesis.

Figure 2.1.5 shows the experimentally detected relaxation spectrum, consisting of three well-separated processes in the time ranges down to microseconds.

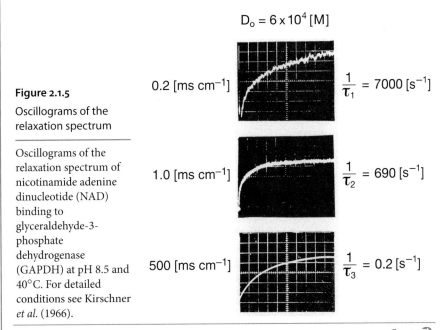

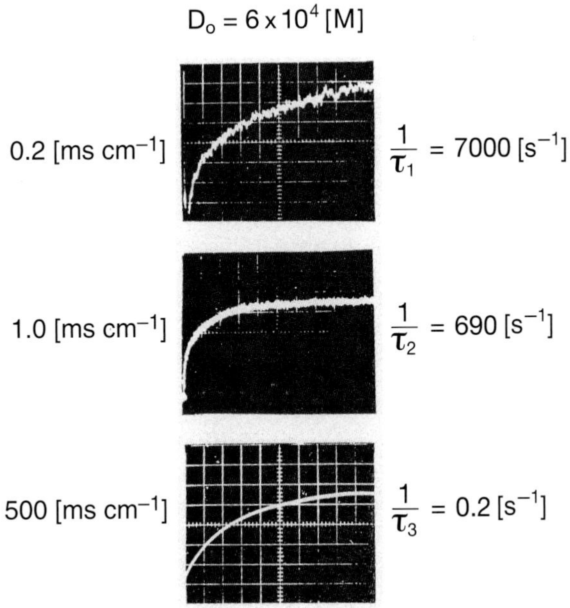

Figure 2.1.5

Oscillograms of the relaxation spectrum

Oscillograms of the relaxation spectrum of nicotinamide adenine dinucleotide (NAD) binding to glyceraldehyde-3-phosphate dehydrogenase (GAPDH) at pH 8.5 and 40°C. For detailed conditions see Kirschner *et al.* (1966).

Cont.

Cont.

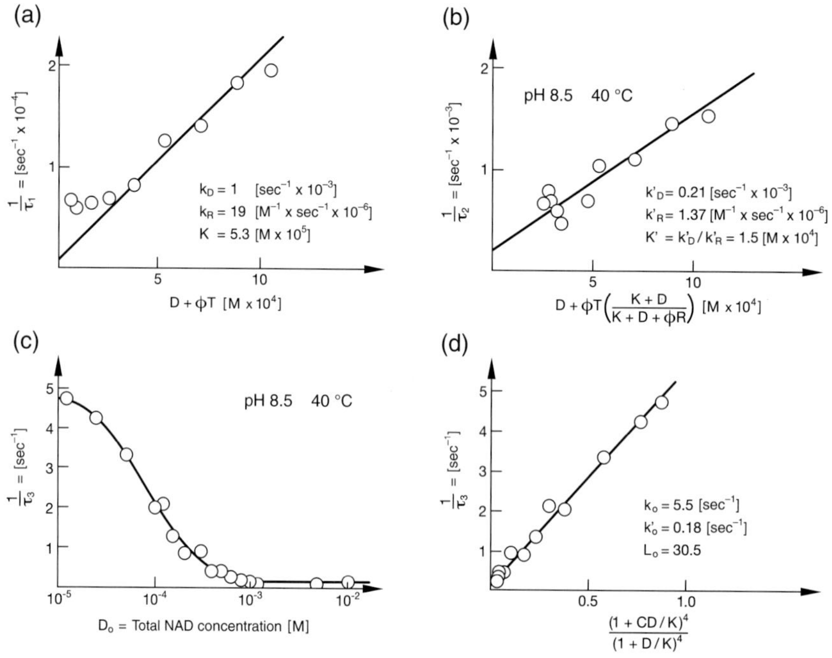

Figure 2.1.6

Dependence of reciprocal relaxation times on concentrations

Each point on each graph shows a relaxation time taken from an oscillogram such as those shown in Figure 2.1.5.

(a) $D + \Phi_R = S +$ free R sites
(b) $D + \Phi_T = S +$ free T sites
(c) $D_0 = \log S_{total}$
(d) $(1 + S/K_T)^4/(1 + K_R)^4$, yielding $n = 4$

$1/\tau_3$ is independent of enzyme concentration

An interpretation of these curves yields the following respective conclusions:

1) Substrate binding in R state, all subunits behaving very similar
2) Substrate binding in T state (as in 1, but no coupling between R and T)
3) All-or-none conformation change of all R and T states
4) Proving conclusion 3

This is exactly the Monod mechanism in its simplest form. The Koshland mechanism would have yielded four relaxation times, which might have overlapped partially, but which would all have had dependences similar to $1/\tau_1$ and $1/\tau_2$, the values of which increase with NAD concentration as expected for second-order binding reactions. The fact that $1/\tau_3$ shows a contrary behaviour is a conclusive piece of evidence in favour of the "all or none" mechanism (see Figures 2.1.8 and 9).

Cont.

➲ Cont.

The concentration dependences shown in Figure 2.1.6 allowed a clear identification of the "all or none" relaxation scheme, proposed by Jacques Monod together with Jeffries Wyman and Jean-Pierre Changeux. The assignment is explained in the legends of those figures.

Later studies with other allosteric enzymes (including haemoglobin) have revealed that both reaction schemes, i.e. Monod's "all-or-none symmetry" and Koshland's "induced fit", do appear in nature and can be identified. Jacques Monod, who was very enthusiastic about our results, drew very far-reaching conclusions about symmetry in nature, which I do not share. We have no evidence other than symmetry in life as an *a posteriori* result of evolution. Recently Jean-Pierre Changeux has drawn exciting conclusions about allosteric control mechanisms in neural networks.*

* The studies described here were carried out at the same time (i.e. 1965–66, when Monod and Koshland published their theories).

2.2. Gibbs' Paradox: How Equal is "Equal"?

Is this another version of my earlier question (Section 1.6) "How large is zero?"? Not quite, but I am certainly not going to digress into politics or social sciences. Neither will I try to answer the question of "how" to classify (subjects or) objects. The question I am asking refers to material complexity and to what extent this may show up in entropy. The following apparent paradox, first thought up by Josiah Willard Gibbs,[10] serves as an excellent example. It was discussed in Arnold Eucken's classical textbook of chemical physics[11] and, more recently, in a very instructive essay by the late Edwin Jaynes.

Two volumes, V_1 and V_2 are separated by a diaphragm, as indicated in Figure 2.2.1. The volume V_1 contains N_1 molecules of an ideal gas of type 1. Likewise V_2 is filled with N_2 molecules of another ideal gas of type 2. The term "ideal" gas means that the molecules do not interact in any way with one another, as though they were mass points of zero size, which neither collide nor exert any force upon each other. However, being "points" of finite mass they posses a finite momentum (and kinetic energy) which they can transfer to the walls of the vessel in which they are contained, or to a piston that can thereby "sense" their finite pressure. More realistically, an ideal gas is a gas at such low pressure that collisions among the molecules are negligible compared with those between the molecules and the walls (where momentum and energy can be exchanged). Correspondingly – by placing the vessels in heat baths – the gases can be given defined temperatures T_1 and T_2, and thus build up finite

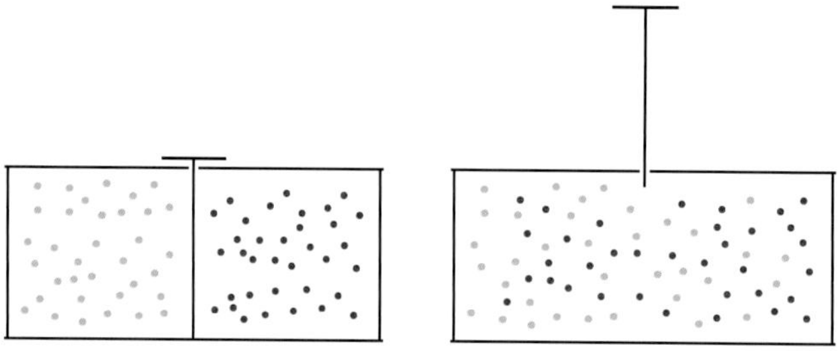

Figure 2.2.1

Entropy of mixing

Two ideal gases, occupying equal volumes V_1 and V_2 and comprising equal particle numbers N_1 and N_2, are separated by an impenetrable partition wall. Removing this wall will cause the two gases to mix. The final pressure is $P = P_1 = P_2$, while the final volume is $V = V_1 + V_2$. If both gases consist of identical (or indistinguishable) particles there is no entropy change connected with the mixing, as we have ideal gases and the number of particles of each gas per unit volume remains constant. If, however, the particles in the two compartments at the start differ (and therefore can be distinguished), the mixing entropy will have a finite value (cf. the experiment described in Figure 2.2.2 and Table 2.2.1).

pressures p_1 and p_2. Dilute gases, such as air containing mainly nitrogen and oxygen, behave like ideal gases, as evidenced by their compliance with the ideal-gas law of Boyle and Mariotte (as long as the temperature does not become too low). This law states that the product of volume and pressure is universally constant for a given temperature and number of molecules involved. For the sake of simplicity let us assume that the two volumes V_1 and V_2, the molecule numbers N_1 and N_2, the temperatures T_1 and T_2, and (hence) the pressures in both vessels are the same, while, however, types 1 and 2 still denote different classes of molecules.

We shall now carry out two different operations and then compare their results. In the first experiment we simply remove the diaphragm and allow the two gases to diffuse freely into both vessels and mix completely. To characterise the situation we note that after removal of the diaphragm the total volume $(V_1 + V_2) = 2V$, the total number of molecules $N_1 + N_2 = 2N$, the temperature $T_1 = T_2$ and the pressure $p_1 = p_2$ remain the same as they were before. What has changed is the fact that the two gases, which before were separated, are now mixed, meaning that each type of gas has expanded to twice its former volume. The total volume is the sum of both volumes. If gases distribute themselves in $V_1 + V_2$, the total volume has remained constant, i.e. Total volume = $V_1 + V_2$ (see Fig. 2.2.1). Dealing with ideal gases, on the other hand, means that the molecules of the two gases do not interact with one another. “Mixing” them simply means that they share the same volume.

Vignette 2.2.1 Gibbs' paradox

A thermodynamic "cycle" is a sequence of changing thermodynamic states in which the final state is identical to the initial one. The Carnot cycle is a prominent example of this and has been used in the derivation of many of the relations following from the first and second laws of thermodynamics.

The most puzzling result is perhaps that obtained in the first step of the second experiment (Figure 2.2.2) where, despite an increase in the total particle number and (correspondingly) of pressure to twice its initial value, the entropy did not respond with any change. According to Table 2.1.1 entropy is a function of (internal) energy (U), total volume (V) and individual particle numbers (N_i). However, what is shown in Figure 2.2.2 is only the first step in the experiment. The two experiments should show the same result if their initial and final states are identical. This is true for the initial situation, where we have two separated compartments 1 and 2, having equal volumes, pressures and temperatures in both experiments, but it is not true for the final situation (shown in Figure 2.2.2), where the total volume (V_1) is half of that in Figure 2.2.1 (which is $V_1 + V_2$) and the pressure correspondingly twice as large in the mixture-sharing volume V_1 as that distributed over a total volume of $V_1 + V_2 = 2V_1$.

Now let us calculate the entropy change in both experiments. The mixing of the two gases in the first experiment can be considered in two steps. First we dilate both separate compartments V_1 and V_2 to twice their values, i.e. $2V_1 = 2V_2$, so the work

$$2 \cdot \int_{V_1}^{2V_1} PdV = 2RT \int_{V_1}^{2V_1} dV/V = RT\ln 2$$

(using the ideal-gas law $PV = RT$ and assuming $\sum N_i = N_A =$ Avogadro's number). The temperature decrease due to dilation of the gases is compensated for by a flow from the heat bath, accounting for an entropy increase of $2R\ln 2$. The second step of this mechanism consists of unifying the two dilated compartments, reducing their total volume $2(V_1 + V_2)$ to $V_1 + V_2 = 2V_1$. This requires a device such as the one described for the second experiment, which would work only for two "distinguishable" gases, involving no entropy change (as in the first step of the second experiment). If the two gases were indistinguishable, the unification would have to be effected by simple compression from $P_{1/2}$ to P_1, reversing the work gained in the first step (i.e. the net entropy balance would be zero).

Now to the second experiment. In order to arrive at the final situation obtained in the first experiment, we have to dilate the mixture from its volume V_1 to a volume $(V_1 + V_2) = 2V_1$, whereby the pressure is released from twice its initial

Cont. ➲

Cont.

value (i.e. 2P) and falls to a value of P. Here we dilate only one volume, but at twice its pressure as compared with the first experiment. That means that the work is again

$$\int_{V_1}^{2V} PdV$$

but this is now performed at twice the pressure compared with the first experiment, i.e. $\Delta S = 2R\ln 2$.

No paradox is left! Both experiments yield an explicable result.

What remains puzzling is the interpretation of "mixing entropy". "Mixing" in this case means sharing the same space. There is no other way in which non-interacting substances, as represented here by ideal gases, can be mixed. The entropy does not come in with the unification of the two compartments, but rather with the expansion, as both substances share the same volume of space. If interactions occur, as is the case in real gases and in liquids, one can interpret these effects as changes in the nature of space (potentials) brought about by the presence of matter. Any calculation of these effects requires knowledge of the internal interactions. Only to this extent does entropy depend on our knowledge.

The paradox that Gibbs noticed (see Vignette 2.2.1) was that while the (internal) energy remains constant during the mixing, the entropy increases by a certain amount, called "mixing entropy". However, this is true only if the two types of molecules can be distinguished as being "different", whatever that means. If they had been classified as "identical", there would have been no entropy change whatsoever. But how are we to classify them, if they are assumed to be ideal gases? Are isotopes of a given element "equal" or "different"? How "equal" is equal? Apparently, some paradox was involved.

Now let us look at the second experiment, also done with ideal gases. Here we use the device shown in Figure 2.2.2. The two pistons are made up of two semipermeable membranes, one (green) that allows the passage of type-1 molecules (red) only, and another green one that allows the passage of type-2 molecules. Moving the pistons allows the two gases to penetrate one another and mix reversibly, whereby the final volume remains the same as the initial volume of each of the separated gases ($V_1 = V_2$). This means that the total volume, which initially was the sum of the two separate volumes, $V_1 + V_2 = 2V_1$, in the second experiment is now reduced to $V_1 = V_2$. Since (at constant temperature) the reduced volume then is shared by both types of molecules, i.e. by a total number $N_1 + N_2 = 2N_1$, the pressure has risen to twice its former value. There is no doubt that the overall state of the mixture

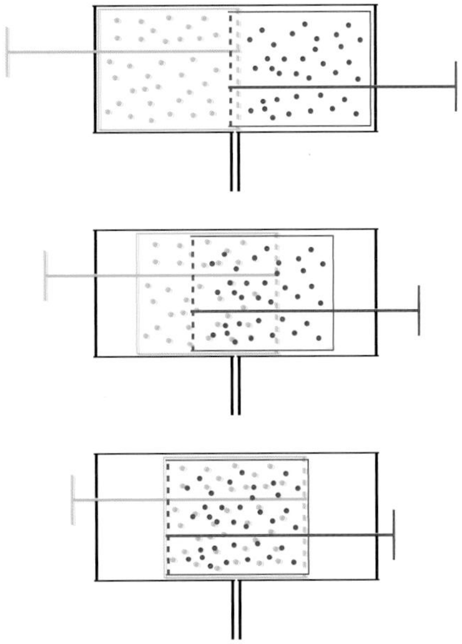

Figure 2.2.2
Gibbs' paradox

There are two compartments with two different gases, as in the former experiment, but now separated by two semipermeable diaphragms. The green diaphragm blocks the path of green particles but is open for red ones, while the red diaphragm stops red particles but is permeable to green ones. Both diaphragms can be moved by pistons within the cylindrical vessel, as indicated in the two lower pictures. Mixing of the two gases can now be performed by mutual penetration of the two chambers, so that the total gas will be compressed, yielding a final volume of $V = V_1 = V_2$ and a pressure $P = P_1 + P_2$. In this case the entropy of mixing for two different (and distinguishable) gases is zero. Note that the two gases must be distinguishable, as is brought about by the two semipermeable membranes, which are able to distinguish between the two gases. For an explanation of this experiment and the one in the previous figure, one has to be aware that ideal gases are point-like particles that are able to transfer momentum to the walls of their vessels (and hence to a piston) but do not show any interaction whatsoever with one another).

of the two ideal gases exhibits a more drastic change in the second than in the first experiment. However, since ideal gases can be imagined as point-like, featureless material particles, which take no notice of one another, there is no detectable change in entropy such as was seen in the first experiment, even though both pressure and volume have changed visibly in the second experiment.

If I had been asked to guess about the entropy change in the two experiments, the answer would have most probably been in conflict with reality. So my advice is not to guess, but rather to calculate, as is done in Vignette 2.2.1. The result is: entropy increases in the first experiment, where nothing seems to change, while it remains constant in the second experiment, despite the increase in pressure.

Let's try to understand these results without going into the details of calculation. I gave some hints above when I stressed the nature of an ideal gas. For the second experiment, this tells us immediately that each gas remains exactly in the compartment in which it was before, and since ideal gases do not "notice" one another, no entropy changes could occur – even though, after merging the two volumes, twice as many molecules bounce against the walls of the joint volume (which is half as large as the sum of both volumes in the initial state and also half of what the final volume had been in the first experiment). We can rectify the situation immediately by expanding the final volume in the second experiment to its initial size ($V_1 + V_2$). The work, when compensated from the heat bath, accounts exactly for the entropy increase associated with the mixing (see Vignette 2.2.1) in the first experiment.

So there is no real "paradox", but by now, if not earlier, the reader might object: how could we base our arguments on two types of gases assumed to be "ideal", i.e. made up of point-like, featureless and hence indistinguishable particles? My reply is: treating the two gases as "ideal" only means that the "ideal-gas law" is applicable. But here lies the crux of the matter: if the particles were really entirely identical, i.e. not distinguishable by any means, then the first experiment indeed would have been associated with zero entropy change. Individual molecules in an entirely uniform gas "mix" all the time, so why should two compartments filled with identical molecules at identical pressure and density show any effect upon unification? Even then I could expand the two compartments separately, but I would have to "pay back" the work gained, since any unification of both compartments could be effected only by compression. Now we have a paradoxical situation, which provokes the question: how equal is "equal"? Clearly, oxygen and nitrogen with present-day analytical techniques are distinguishable, although not easily in the framework of these two experiments. Or, can isotopes of the same element be considered "equal"? We would certainly not be able to record any change of entropy in such experiments. It is sufficient to know that the two gases are different in order to assign them a different free energy when they are separated. This free energy difference is equivalent to potential work which, some day, might be realisable if some hypothetical material could be found that exhibits different solubility for the two isotopes – today a crazy notion. Nonetheless, this free energy difference would be annihilated upon plain mixing. Macroscopically, we would not observe anything because the internal energy does not change. All that happens, happens in our mind, where we compensate the loss in free energy by $T\Delta S$, i.e. by an increase in entropy. In other words, if we were able to realise the free energy difference by some work, then there must be a corresponding change of entropy. The outcome depends on the availability of appropriate equipment.

The above problem was discussed by the late Edwin T. Jaynes, who in 1992 had revisited Gibbs' paradox (which has survived in textbooks for over 100 years). Jaynes believes that Gibbs himself had the precise answer to the question involved as early as in 1875. He quotes a passage from Gibbs' text *Heterogeneous Equilibrium*. However, that passage is written in a ponderous, nearly unintelligible, style so that – as Jaynes remarks – "only his faith in Gibbs" made him persist in grappling with it.

Entropy is a function of the "thermodynamic state", which is characterised by macroscopic variables. This macrostate must be contrasted with any "real physical state" which is always a microstate (Section 2.3) that changes steadily as time goes on. Each thermodynamic state is compatible with a very large number of microstates and it makes no sense to assign entropy to a microstate. Moreover, we might define different entropies, depending on which macroscopic variables are at our disposal. Entropy as a function for a given macrosystem may show different values, depending on how much we know about, for example, chemical or isotopic composition and possible reactions among the constituents which express themselves in the structure of the microstate. This is what Jaynes calls the "anthropomorphic" character of entropy. In this respect, entropy differs from energy, which is clearly specified for both the micro- and the macrostate.

Jaynes demonstrates his point with hypothetical elements "so rare that they have not yet been discovered". He jokingly calls them "superkalic" and cooks up names such as "Whifnium", "Whafnium" or "Whofnium". These "elements" are imagined to serve as semipermeable materials that are even capable of separating isotopes of noble gases in the way I described above in the second experiment. He then shows that, depending on one's knowledge about this material, one may play all sorts of tricks using the second law. He shows that apparent violations an experimenter may see in fact only signify ignorance about the relevant macrovariables involved.

Jaynes' arguments have not remained uncriticised, in particular from the camp of physical chemists who do real experiments with real chemicals rather than thought experiments with hypothetical, ideal systems. They claim that the result of an experiment, if done correctly, cannot depend on the experimenter's subjective knowledge. If two students do the same experiment and get different results, then one of them (or both) must have done something wrong. The point is: a given experiment will have a given result, independently of the subjective knowledge of the experimenter. However, a given experiment is done with "given equipment". The latter is chosen by the experimenter and hence depends on his subjective knowledge. Only in this sense are Jaynes' arguments to be considered.

I have to admit that there is indeed a danger in making one's statements too pithy. Expressions like "anthropomorphic" or "subjective" are too easily misunderstood. However, for a logically closed framework such as thermodynamics, it is important to define the premisses clearly. They do not come from "within" the framework; rather they are dependent on what information is available to us and they make sense only if the information is realisable. To make this clear, Jaynes intentionally used simple and idealised systems, such as ideal gases.

The fact that "real" gases do not follow the ideal gas law was found soon after Boyle and Mariotte, who lived in the 17th century. Molecules do attract one another and

– if they approach too closely – repel each other as well. In 1873 Johannes Diderik van der Waals[12] introduced a semi-empirical correction of the gas law by adding a volume-dependent term to pressure, accounting for attraction among the molecules (which indeed can be envisaged as some additional force pressing them together), and by subtracting from the volume available for the total system that part which is occupied by the molecules themselves because of their finite size. Van der Waals' equation was most successful in representing exactly (small) deviations from the ideal gas law as was substantiated by a more rigorous "virial theory". Moreover, it mimicked a phase transition to a more condensed state, which occurs when the gas is liquefied. (The problem will be looked at in more detail in Section 2.8.)

Whether or not such deviations from ideal behaviour are caused by less specific general interactions (such as van der Waals' forces) or even by very specific chemical changes is an additional piece of information to be gained from a source outside thermodynamics. Two "real" gases (as they are called, emphasising their non-ideal behaviour), on being mixed at constant temperature, may show entropy as well as internal-energy changes. If one also knows that chemical transformations take place, one can express the thermodynamic function by additional variables such as affinities (intensive) and extents of reaction (extensive), as was exemplified in Section 2.1. The merit of thermodynamics is that it holds independently of the degree of complexity of the system involved. The more additional information one has about this complexity, the more detail can be expressed using additional variables in the thermodynamic relations. Entropy, which expresses the extensive complexity of the system, depends on how much information is available. What Jaynes tried to express by his idealised examples is that one can measure this entropy, but how much of the information available can be utilised depends on the measuring device and thus on our knowledge for designing the device correspondingly.

After this short digression into reality let me come back to the question posed in the title of this section. Equality in a mathematical sense is an abstraction. Our very early ancestors certainly realised that each hand (usually) has five fingers, and they used these five fingers to learn how to count. This abstraction of numbers long preceded any mathematical formalisation, and it included equality: 5 equals 5. Yet your five fingers are not equal to my five fingers, neither do they involve equal numbers of cells, nor are they arranged in exactly the same way, nor do your cells resemble mine in every detail. Does this mean that equality does not exist, that ultimately everything is fuzzy? Wrong question! Equality is the result of a certain classification. Your fingers differ from mine in every respect but one: the numbers of the units we classify as "fingers" do agree – in most cases – from person to person. How detailed a classification can be depends on the precision with which we can detect (or measure) the features of the material to be classified. Fuzziness in many cases presents a true problem, and I shall return to it in Section 2.4.

How equal is "equal"? We see that a sharp borderline between equal and unequal is a matter of definition and hence – again – a construct of our mind. Practically speaking, it is as sharp as the distinction allowed by the devices we can construct. Particles, such as bosons and fermions, apparently distinguish sharply between equal and unequal, but equality of bosons means loss of individuality. This is the lesson of this section, which was meant to contrast the mental construct of entropy with the physical fact of energy.

2.3. How Real is the Microstate?

In the foregoing discussion I have already used the term "microstate" and referred to its "physical reality". However, I haven't yet specified what the physicist means by it, especially when he contrasts it with the word "macrostate". Let me now make up for this omission.

The world we view with our eyes is a world of macroscopic objects. It is the world of our experience, to which our intuition is adapted and from which we take examples in order to model phenomena that evade our direct sensual perception. The terms "microstate" and "macrostate" refer to systems consisting of the same physical objects and, in particular, to the same amount of matter. In other words, the microstate is not a microscopic "clipping" from a macroscopic object, on which we focus with the help of some "magnifying glass". It is rather a "snapshot", which resolves at the atomic level the positions and momenta of all microscopic particles present in the macroscopic sample. Hence, the terms micro- and macro- do not refer to the size of the object under study, but rather to the temporal and spatial resolution with which we view the object. The microstate describes what we would see with each single snapshot (if we had suitable equipment), and the macrostate would then result as a superposition of all possible snapshots, taken at any instant. The single snapshot yields a clear picture of each atom's position and velocity (never mind how the snapshot is taken), while the superposition must yield something quite fuzzy. At any instant it is the microstate that represents physical reality, whatever we may mean by that. Yet it is not all that simple, as the reader might already have guessed from the reservation just expressed. If I speak of reality, then I must also be prepared to state how I could grasp this reality. And here I find several obstacles in my path.

First of all, the quantum-mechanical picture of the microstate in abstraction is a blurred picture anyway. In order to establish the microstate firmly, I would have to interfere quite drastically with the system and thereby destroy the microstate under examination. Therefore, we should avoid getting side-tracked by questions of practical realisation. The microstate of any moderate-sized system is in any case too complex to allow us to "see" it or to calculate, and thereby identify, the co-ordinates and momenta of all the particles involved. We therefore have to assume that in principle the system can be described by superposition of all the wave functions of its constituent particles, again a platonic assumption, the realisation of which would include overcoming many hurdles. At least we may assume that owing to the existence of molecules and atoms there should be something like physical reality associated with the microstate, a "reality" that we can postulate, but not observe in practice.

The complexity of the microstate is the main obstacle to its realisation. We can, of course, try to simplify the situation by reducing the system to a few states, in the extreme focussing on just one single molecule. Systems consisting of only a few molecules are called clusters. They enjoy much popularity in present-day experimental research. I shall revisit them in Section 2.7.

Another flourishing subject is the study of single molecules. Modern laser techniques allow the focusing of light to dimensions comparable with its wavelength, i.e. to the submicrometre scale. This corresponds to volumes below 1 femtolitre (10^{-15} l). Labelling of substances with suitable fluorescent dyes makes single molecules of these substances detectable whenever they enter the laser focus. With the help of technical tricks such as confocal optics, time-resolved recording and registration of fluctuations by means of auto- or cross-correlation techniques, the stochastic nature of single molecular events can be made visible.[13] Preferred objects of those studies are biomolecules, and applications reach up to biotechnology and medical diagnostics (e.g. screening of pharmaceutically relevant material, expressed in molecular mutant spectra).

Likewise, the chemical kinetics of single molecules in the gas phase can be studied with crossed molecular beams and optical or mass-spectrometric recording. This all yields information about "physically real" states.

However, the great problem of the microstates of macroscopic systems is their exponential combinatorial complexity, which being exponential in nature shows up already in systems of quite moderate size (cf. the games described in Section 2.6). This complexity is of the order of magnitude of N^N, where N is the number of particles involved. Imagine what this means if N is of the magnitude of Avogadro's number (i.e. $\sim 10^{24}$).

Ludwig Boltzmann asked the following question: how long would it take for all the 10^{18} molecules contained in one cubic centimetre of air at about 1/30 of atmospheric pressure to reproduce their initial positional coordinates within 10 Å and their velocities (or momenta) within 0.2% of their original values (of about 500 m s^{-1})? The answer he gave was: one would have to wait for about $10^{10^{19}}$ years. Yes, you read correctly: ten to the ten to the nineteenth, a one not "only" with nineteen, but rather with ten to the nineteenth zeros! The age of our universe is only a tiny fraction of that time, namely about 10^{10} years, i.e. ten to the tenth, or ten thousand million years.

The time that elapses until the recurrence of a particular microstate is called the Poincaré recurrence time, after the French mathematician Jules Henri Poincaré (1854–1912). It is so large because of the huge combinatorial number of possible microstates. It is to be distinguished from the recurrence time for a particular equilibrial macrostate, which in fact is shorter the larger the number of microstates that make up the macrostate in question.

The microstate changes steadily as a consequence of thermal motion of all the molecules or atoms involved. Hence, a system spends only a tiny fraction of its time (of the order of a thousand millionth of a second) in the neighbourhood of any given microstate. It changes and changes and changes, from one instant to the next, without any end. How can one suitably describe this complex panorama of changing microstates, each of which refers to a particular combination of space and velocity co-ordinates for every particle in the system. The answer is "phase space".

The phase space is a 6N-dimensional construction of Euclidean space where each of the N particles involved is assigned six co-ordinates, three for its co-ordinates in geometrical space and three for the corresponding velocities or momenta. In this co-ordinate system, the complete microstate is represented by a phase point which, because of the continuous temporal change of the positions and momenta of all particles, describes a trajectory in phase space. Since the Hamiltonian function depends on the squares of the co-ordinates, energy is expressed by the Pythagorean metric, i.e. the total energy is obtained from the sum of the squares of all the co-ordinates. Constant energy in this space limits the motion of the phase point to a given hyperplane, which is the analogue of the surface of a sphere in three-dimensional space. Furthermore, we note that the product of a space and its conjugate momentum variable yields an action, of which Planck's constant "h" is the smallest unit. Hence the elementary cell of phase space in quantum statistics has a volume of h^{3N}.

Let me summarise the microstate concept, which essentially goes back to Boltzmann and Gibbs. The phase point is representative of the total system, specifying the three position co-ordinates and their conjugate momentum co-ordinates of all the N particles that comprise the system. Since these co-ordinates change steadily for all N particles, the phase point will describe a trajectory in phase space characterising the temporal evolution of the total system. If the system does not exchange energy with its environment, the trajectory of the phase point is restricted to a fixed hyperplane of constant energy. In 1887, Boltzmann hypothesised that the phase point must, in the long run, reach every point of the hyperplane. This, of course, cannot be true if the phase point is a point in the mathematical sense. Boltzmann's example of recurrence of the microstate in an ideal gas, as demonstrated above, shows also that exact ergodicity, as this behaviour is called, would be quite unrealistic. The Polish–Dutch physicists Paul and Tatjana Ehrenfest[14] therefore replaced Boltzmann's conjecture by what is called the "quasi-ergodic hypothesis", which states that the trajectory may come arbitrarily close to any point of the hyperplane if one cares to wait correspondingly long. The question of ergodic behaviour will be revisited later because it is of importance in the evolution of information-gathering systems.

Ergodicity will also be revisited in the discussion of macrostates because it yields one of the reasons why the steadily changing microstates in the end lead to some almost deterministically fixed macroscopic behaviour, rather than to a totally blurred appearance of the system as observed over larger periods of time and in macroscopic dimensions of space. At first sight, this involves another puzzle because – as the above example has shown – during any decently realistic span of time, or within any realisable space, only a really minute fraction of the vast total number of microstates would turn out to be populated. How could such a minute fraction of populated states yield clearly defined moments (i.e. averages and variances) that are representative of the total system? I shall come back to this question when discussing the macrostate in Section 2.5.

There is another apparent incongruence which troubled Boltzmann greatly. Boltzmann[15] tried to find a kinetic foundation for the fact that in an isolated system entropy can only increase, as required by the second law of thermodynamics. With the help of a kinetic collision approach (German: *Stossansatz*) he derived a function (denoted by H) which in an isolated system was supposed to change in one direction only – in this case decreasing in value with increasing time. However, the concept of the H-theorem stirred up controversies among physicists. Boltzmann's ansatz for the collisions among molecules is based on the laws of mechanics, which are inherently symmetric in time. In other words, the laws are the same if the time co-ordinate t is inverted, i.e. replaced by –t. Hence, if x(t) is a solution for the motion in the system, x (–t) should be a solution too. This accords with our expectation that there will be fluctuations in the form of deviations from equilibrium. By using distribution functions for calculating the collision numbers, Boltzmann introduced in addition a statistical element. This causes his H-function on the whole to behave correctly in the above sense, albeit not distinguishing clearly between fluctuations and true irreversibility.

Among Boltzmann's adversaries, apart from his Austrian colleagues Joseph Loschmidt[16] and Ernst Zermelo[17], was Henry Poincaré[18], who with his own work laid the ground for a late rehabilitation of Boltzmann's idea, as Ilya Prigogine[19] recently pointed out (see also Section 2.9). Poincaré had found that the solution of the three-body problem of astronomy cannot be given in explicit terms. It must be, as we say today, of a chaotic nature, as shown in the example depicted in Figure 2.3.1.[20] A chaotic solution (as will be discussed in more detail in Section 5.1) is deterministic, but extremely sensitive to its initial conditions and to the smallest perturbations, which are certainly present in any multiparticle environment. Hence, time reversal, even in mechanics, would not mean a true and exact reversal of a fluctuation.

This brings me back to my question about the physical reality of microstates. The aforementioned problem underlines the necessity of treating the microstate on a probability basis, as will be discussed in Section 2.5, after elucidating the nature of probability in the next section. I may seem to have lost track of the problem of entropy. However, that is not so. A microstate has no entropy – despite the fact that it has a defined energy! Entropy is solely associated with the macrostate, hence I shall discuss it in Section 2.5.

(a)

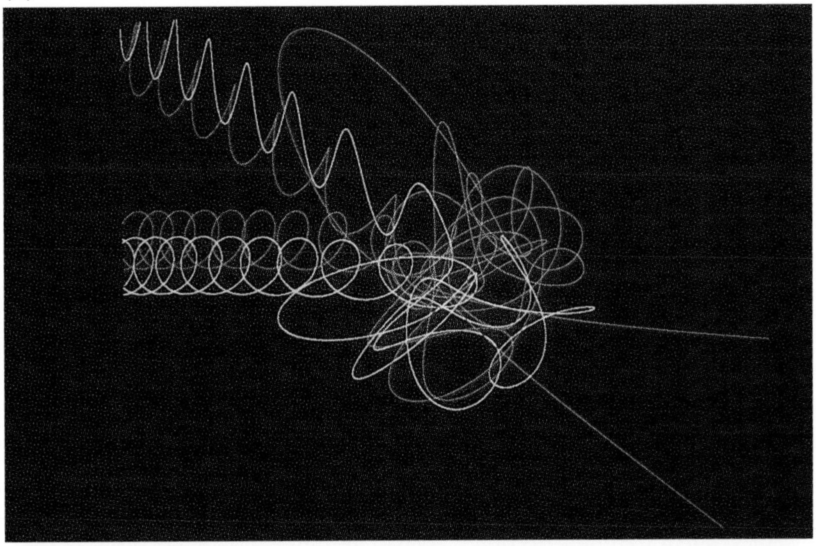

(b)

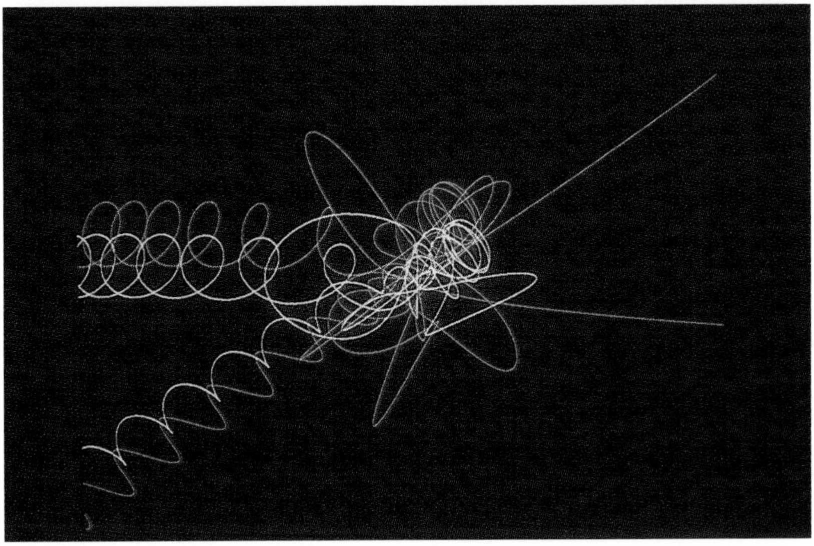

Figure 2.3.1

Chaotic three-body problem

A double star (coloured red and yellow) meets a single star (blue) and engages in chaotic three-body interactions. The final result of the second example (b) is an exchange of partners.
(Courtesy of Manfred Schroeder.)

Is the microstate physically real? The true question is more of a semantic nature: what do we mean by "physically real"? Isn't any physical reality only a reflected reality? The concept of the microstate as such is supported by all the physical knowledge in our possession, so we can construct it in our mind. But we cannot realise it by observation, as little as nature can manifest all its possible constellations.

2.4. Probability: Expectation, Frequency or Fuzziness?

The word "probability" has its root in Latin. The verb *probare* means "to test". Whatever is capable of withstanding a test should have an objective basis. However, to us the word also has some subjective connotations. It refers to something we have to guess. The German word for probability is *Wahrscheinlichkeit*, meaning something like "apparent truth". The two interpretations caused a long-lasting controversy among mathematicians and philosophers of science, one that even today is far from being settled. In fact, it is enriched today by contrasting probability with a new term that is currently in fashion: "fuzziness". I shall first discuss the semantics of the word "probability" before trying to remove our fuzziness about "fuzziness".

Jumping into history, and leaving aside the Italian Renaissance as well as some forefathers such as Blaise Pascal, Pierre de Fermat and Christian Huygens, let me start right away with Jacob Bernoulli (Figure 2.4.1) and what he called the "law of averages". In his book *Ars Conjectandi*, published in 1713, Jacob Bernoulli stated a formal principle which defined probability as resting on our state of knowledge. The probability p(A) that a proposition A applies is defined as M/N, where M is the number of favourable cases among a total number N of equally likely outcomes. This rule, of course, can only be applied to situations where all possible cases can be enumerated. Bernoulli realised that this assessment severely limits the practical application of his principle merely to certain games of chance, where the possible outcomes are defined by the inventor of the game. In Nature, we cannot usually be sure of knowing all the numbers that define M/N. In particular, Bernoulli asked: "What mortal will ever determine the number of illnesses", a question that, against the background of AIDS and (more recently) of prion diseases, still bothers us a lot.

Suppose one assumes a value for M/N and calls it the probability, p. In order to test the hypothesis that p is constant, one has to carry out successive trials and record the frequency of having m successes in n trials. If the assumed value of p was correct, then it should be matched by the frequency $f = m/n$ in the limit of large values of n. The accuracy of the coincidence between observed frequency f and supposed probability p was quantified by Abraham de Moivre and later by Pierre Simon de Laplace.

Figure 2.4.1

Cartoon by Sydney Harris.

From *The New Yorker.*

We now ask the inverse of the above question: given m and n or the frequency m/n, can we find a relation that tells us what M (or p) is likely to be for a given size of N? A formula that answers this question was given by an amateur mathematician, a clergyman, named Thomas Bayes. His name nowadays stands for a school of analogical or inductive reasoning in scientific inference. Modern Bayesianism has as many protagonists as it has detractors. Bayes' paper "An Essay Towards Solving a Problem in the Doctrine of Chances" was published posthumously in the 1763 volume of the Royal Society's *Philosophical Transactions*. Bayes' theorem, which was reformulated by Laplace, received little attention otherwise, and it had little influence on the foundations of statistics in the late 18th and 19th centuries, being largely ignored up to the 1940s, when it was revived.

At that time the predominant view regarded frequency as the only objective basis of a mathematical theory of probability. One of the main adepts of this school (which is represented by prominent mathematicians of the early 20th century), Richard von Mises, writes in a book[21] with the title *Probability, Statistics and Truth* (first published in German in 1928): "The theory of probability deals exclusively with frequencies in long sequences of observations; it starts with certain given frequencies and derives new ones by means of calculations carried out according to certain established rules". The last part of the sentence is especially instructive. If one looks into the subject index of von Mises' book, one will not find the word "conditional probability". And if one looks up what is meant by "initial" (*a priori*) and "final" (*a posteriori*) probability, one finds a repeated, staunch defence of the view that such expressions are dispensable whenever one is able to carry out operations with defined frequencies according to established rules. Von Mises refers to four fundamental operations called "selection", "mixing", "partition" and "combination". In other words, not the subjects as such are missing in treating probability, but rather access to the subjects is prescribed. Guesses and inferences are considered unreliable requisites of the human mind and should be kept out of mathematics. Probabilities of microstates can be inferred, but they can never be identified as "frequencies in long sequences of observations". And what about the abstractions on which all mathematics rest? Aren't they also (idealised) constructs of the human mind?

Objective probability was axiomatised, as late as in 1933, by Andrei N. Kolmogorov[22]. He introduced three self-evident (and universally accepted) properties of probability (cf. Ghahramani). The entire theory of probability is today rigorously based on these axioms, which also apply to subjective probabilities. This Bayesianism as a "quantitative approach in terms of degrees of belief regimented according to the principles of probability calculus" has also been put on to a solid axiomatic basis in recent years. As such, it has superseded early work on formal logic from Ludwig Wittgenstein to Thomas S. Kuhn. What is referred to as Bayes' formula (which never was written down explicitly by Bayes himself) rests on the notions of prior and final probability, establishing a relation between assumption and confirmation through learning modelled as conditionalisation. It has become fashionable on the basis of its quasi-philosophical implications, but – in its simple form – it is mathematically trivial. It is based on the symmetry of the way in which the joint probability for two events A and B can be expressed. The probability $p(A,B)$ that both events will occur can be found by combining the probability $p(A)$ that event A will occur "unconditionally" with the "conditional" probability $p(B|A)$ that event B will occur given the occurrence of A. However, the same argument can be made with A and B exchanged. It is this symmetry that leads to Bayes' formula. The problem of conditional probabilities will be revisited in subsequent chapters, mainly in connection with information-transfer processes (Section 3.3). Edwin T. Jaynes, who is an advocate of the Bayesian approach, has used the concept to characterise entropy by a maximum principle. Jaynes' enunciation[23] of the principle of maximum entropy in 1957 is considered by insiders as one of the most important recent breakthroughs in statistical physics.

What does maximum entropy mean? In characterising the principle, Jaynes is not referring to the (long-known) fact that systems in certain states, e.g. equilibrium states, can be characterised by a maximum entropy or a minimum free energy (Section 2.7). He is looking for the correct way to characterise entropy as a state of knowledge about the system. This explains why he refers to Bayesian inference. Maximum entropy starts from Shannon's entropy concept (which in its mathematical formulation is equivalent to Boltzmann's expression, see Table 2.5.1), basing probabilities on Laplace's principle of insufficient reason. One may, for instance, define the expectance for every realisable state as equi-possible and then add constraints from observation that limit their frequency of occurrence. Constraints in a physical system may be "moments" of the probability distribution such as averages (first moments), mean square deviations or variances (second moments), and so on. Among all distributions that are consistent with prior probabilities and constraints, maximum entropy selects those which utilise no more and no less information than available. Any larger entropy would not have made use of all the information given, and any entropy smaller than maximum entropy would have violated the principle of insufficient reason by making unjustified assumptions. Kapur and Kesavan[24] characterise this choice (somewhat emphatically) as consistent with two principles of science (and common sense):

Speak the truth and nothing but the truth

and

Make use of all the information that is given, and scrupulously avoid making assumptions about information that is not available.

One may add for the particular case of the maximum-entropy principle: make sure that the probability distribution on which your reasoning is based is sufficiently rich and that the states you invoke can be populated. We shall re-encounter the principle frequently in subsequent sections (without explicitly referring to it). Here it may serve as a further characterisation of entropy. Its applications are numerous. For instance, the Maxwell–Boltzmann distribution is derived from assuming uniform *a priori* probabilities and defined expectation values of energy using the method of Lagrangian multipliers. In an analogous, but quantitatively different, way Bose–Einstein or Fermi–Dirac statistics are obtained.

While physicists in general apply the concept of probability without questioning its philosophical background, and while in mathematics the frontiers between the camps seem to be softening, the concept as a whole has come under fire from the supporters of a new "paradigm", fuzzy logic, or, to quote one of its protagonists Bart Kosko[25]: "What Einstein did with gravity was to eliminate it. That's what we do here: eliminate probability". At a conference at Los Angeles in 1985 Peter Cheeseman[26], a strong opponent of Kosko, offered "$50 to anyone who could solve the fuzzy logic of any problem that Bayesian inference could not". Admittedly $50 is not a very risky investment, yet I would be cautious, not so much because of the scientific problem at stake, but rather because of the "fuzziness" of the language involved. What is it all about?

In 1964 Lotfi Zadeh, an immigrant to the USA, born in Soviet Azerbaijan and by then professor of electrical engineering at the University of California in Berkeley, presented for the first time to a critical audience of his scientific colleagues his ideas about a theory of fuzzy sets. His aim was to unify set theory with multivalued logic, which is at the basis of a philosophy of vagueness, the main exponent of which was Max Black. Zadeh had made a name in "system theory", a term he coined in 1951. His new theory on fuzzy sets did not make many friends at the time it was proposed. It neither met with enthusiasm among theorists, nor did it find any major industrial application in an otherwise blooming era of computer technology in his country. Zadeh, who has written many papers on his brain-child, "Fuzzy Sets and Applications" (the title of a volume of selected papers of his)[27], was certainly not an expert in public relations. It was left to his younger, less modest followers to proclaim the "scientific revolution". However, their big enemy is fuzziness itself, an undoubtedly necessary requisite of human language, without which our culture could not have evolved.

Ludwig Wittgenstein, at least in his earlier work, had tried to confine language to logical determinacy. His advice was to say only things that can be said.

The literature on "fuzzy thinking" is full of examples that are simply due to the fuzziness of language. Let's take a look at some of them, starting with: "Is the Rhine a long river?" If a child were to ask me this question, my answer would probably be a clear "yes". In saying so I make the mental reservation that this answer corresponds entirely to the standards of imagination of a child. If Kosko himself were to make the same inquiry I would reject it as an improperly defined question.

Another example discussed by Kosko is concerned with finding an apple in his refrigerator. One may possibly find a partly eaten apple, say half of one. Is it then "an apple" or is it "not an apple"? Again, an improperly specified question with regard to this point. (If I knew the person in question, and her or his patterns of behaviour, I could make a guess about whether apples in the refrigerator are half-eaten or not). These are simply bad examples if I want to prove or disprove the usefulness of probabilities. Yet the claim of "fuzzy thinking" is to replace the binary "black and white" world of probability by the grey world of fuzziness.

That is fuzziness of language! Why is the world of probability black and white? Let's take the microstate. We assign to it a certain probability p = "black". However, the remaining (1 – p) is "white" because it is a mixture of all other "colours" that are "non-black". By taking into consideration the whole "spectrum" of microstates I can calculate the entropy that belongs to the macrostate and that I can test experimentally. Entropy is not a probability. It is a deterministic quantity. It is the logarithm of a defined volume in phase space. I can quantify its degree of precision, which usually is so high that I cannot record deviations experimentally. I can calculate average deviations and may call them "fuzziness" (language is patient). However, the point is that probability theory quantifies what otherwise is vaguely postulated.

On the other hand, a good case can be made for "fuzzy thinking". I shall not talk in this paragraph about its most natural product, fuzzy logic. That is a subject in itself. I want to restrict my discussion to macrostates, which are objects of our consideration, both in nature and in technology. The question: "What is the probability that it will rain tomorrow?" has no answer in terms of probability theory that applies to microstates. It is true that meteorologists can guess pretty well if they have data on the air pressures, wind strengths, temperatures and humidities of today, but their non-linear dynamic equations may contain instabilities and chaotic forms of behaviour that make a precise forecast impossible (see Section 4.8). The reason is that the evolution of the system – although deterministic – depends so critically on initial and boundary conditions that it becomes practically (and in the end truly) unpredictable. Similar cases are encountered in technical systems. The prime example quoted by fuzziologists is the underground railway system of Sendai, a city in central Japan. Engineers have found means to make the trains move with uncanny precision, to start and stop so smoothly that none of the passengers have to hang on to the straps. In their construction they had taken into account the fact that both input of perturbations and control of reaction are based on fuzzy data, too complex to be handled by pre-programmed devices. So they found a solution that responded to fuzziness under the sole constraint of optimal smoothness.

There are innumerable examples of a similar kind, in which the notion of fuzziness can be very valuable. These usually refer to macroscopic states that are too complex to be analysed on a microscopic scale but can be characterised in a multidimensional parameter space by regions of fuzziness. I therefore would dare to consent to the extreme probabilist's view that "Bayesian inference can perform every task fuzzy logic can, and therefore fuzzy logic is unnecessary". My counter-argument is: they are trying to answer different questions (for different purposes). That will become especially obvious when we come to treat phenomena of life and thought. Fuzziologists should reduce the fuzziness of their language, for otherwise they do their idea a disservice. It is – even self-contradictory – ideology to think that the only way of solving problems of uncertainty is "fuzzy thinking". Eliminating probability would be a form of one-track thought.

How is entropy related to measures of probability? Entropy is based on the assumption of a probabilistic distribution of microstates. However, referring to the macrostate, entropy itself is a deterministic measure that can be quantified within defined limits of fluctuation. Jaynes' formalism chooses out of all probability distributions that are consistent with given constraints (e.g. given moments such as averages and their mean variances) the one distribution that corresponds to maximum entropy. It is the one that just uses all the information available. Any smaller entropy would require the use of additional information that was not contained in the constraints. In this approach, the dichotomy of subjective and objective probability (expressed as defined frequency) has been overcome by objectivation of the Bayesian approach. The term "probability" for characterising the expectance of singular macroscopic events should be handled with caution. Information not available or fundamentally unobtainable may then be expressed as fuzziness. Taking full account of fuzziness may lead to the invention of devices that show reliable performance on the basis of irregular single events.

2.5. How Real is the Macrostate?

This question, of course, takes up the discussion of Section 2.3. The macrostate is the state of a macroscopic material system, depending on certain macroscopically measurable variables. In relation to the microstate it is a construction of our mind; it does not exist as such and therefore cannot be "observed".

Again, I am not very happy with this statement. What is it that we really can observe? Is it not ultimately just something that takes place in our mind?

The macrostate is the totality of all microstates and, as such, it can be perceived by measuring its properties. One of these properties is entropy. Entropy in thermodynamics is a deterministic variable, statistically characterised by its mean with an often exceedingly small variance. It is not a probability itself, in the sense of frequency or guessed expectance. And here, as mentioned already in Section 2.3, comes one of our true surprises: the macrostate is obtained by taking account of all possible microstates. As we shall see, entropy as a function of the macrostate is given by the logarithm of the total phase volume (defined below). In reality, in particular during the relatively short time-span of measuring entropy values, in fact even during the whole life-time of our universe, only a tiny fraction of the total number of microstates over which the averaging procedure is extended has been – or, as Boltzmann's example of a recurrence time shows, could possibly have been – populated. And when I say "tiny fraction", I mean just that. It's not just a tenth, or a millionth; no, it is near to N^{-N} where $N \approx 10^{24}$ is not more than one mole of substance, i.e. a few grams, that is required. Given the fact that only a negligibly small fraction of microstates is "sampled", how can we nevertheless get a defined value of entropy that is more precise the larger the system?

Expressions for entropy were first derived independently by Ludwig Boltzmann[28] and Josiah Willard Gibbs[29]. When I now refer to Boltzmann I don't mean his work on the H-theorem but rather the work in which he studied the energy distribution to (yet unspecified) cells in phase space. Later in quantum statistics the volume of such elementary cells was identified as h^{3N}. Since the treatment of Hamiltonian mechanics is based on so-called canonical transformations, one speaks of a microcanonical ensemble if the available phase volume is limited to an energy shell between E and $E + \delta E$. Canonical ensembles are those in which exchanges of energy with the environment (e.g. a heat bath) are also allowed. Correspondingly, the integral or sum is to be taken over the whole "phase space", i.e. including all energies from zero up to infinity but weighted with the Boltzmann factor $e^{-\beta E}$. A grand canonical ensemble additionally allows for exchange of matter and is therefore not confined to a constant particle number. I shall not digress into these details here. What is more important in the context of this chapter is the notion of the hugeness of the phase volume. If the particle number is of the magnitude of Avogadro's number, we are dealing with a 10^{24}-dimensional space. The volume of such high-dimensional space is almost entirely confined to a thin surface layer in this high-dimensional space (cf. also Section 4.5). Therefore, it makes no major difference if the microcanonical entropy is calculated as the "volume" or the "surface" of the phase space sphere.

The entropy is obtained by introducing into Clausius' concept the statistical expressions for energy and work. Entropy is then given by the logarithm of the phase volume "ϕ", normalised to a dimensionless form. Gibbs arrived at such an expression without knowing about Boltzmann's work. His entropy was dimensionless (as it is used nowadays in information theory). Temperature then would have to have the dimension of an energy, corresponding to kT (where k is Boltzmann's constant). Today we write both Boltzmann's and Gibb's expressions, resulting from the phase integral, as:

$$S = k \ln \phi$$

where k is Boltzmann's constant and ϕ is the phase volume.

Boltzmann arrived at his expression from straight probability considerations. In his work the phase volume is expressed by an – entirely equivalent – expression that he called W (for the German word *Wahrscheinlichkeit*, probability), which, like ϕ, is a huge number obtained from a combinatorial expression of the "statistical weights" of all states. Probabilities, according to their normalised mathematical definition, are small numbers, the sum of which – extended over the entire distribution – is 1. In this respect the allusion to probability was not a fortunate choice, since it may be misleading. However, one may easily take the reciprocal of Boltzmann's W by introducing for all single states true probabilities p_i, with $\hat{A}_i p_i = 1$. Then W^{-1} would become $\prod_i p_i^{p_i}$ and $S = -k \hat{A}_i p_i \ln p_i$ would be an always positive expression which

Table 2.5.1 Entropy Expressions

Boltzmann	$S = k \ln W$	W = probability function $W^{-1} = \Pi p_1^{p_i}$ (see Shannon)
Gibbs	$S = k \ln \Phi$	Φ = phase volume
Planck	$F = U - TS = kT \ln Z$	$Z = \sum_i e^{-U_i/kT}$: partition function U_i = energy eigenvalues of states (discrete quantum states U_i)
von Neumann	$S = -k\,\mathrm{Tr}(\rho \ln \rho)$	Tr = trace of density matrix (variables ρ)
Shannon	$S = -\sum_i p_i \ln p_i$	p_i = probability of individual states

formally agrees with Shannon's independently derived formula for informational entropy (see Section 3.2). In fact, the concept of entropy is much too broad to be limited to thermodynamics only. Another point is that the concept of entropy can be extended to certain non-equilibrium situations, while temperature makes sense only for equilibrated systems, i.e. those in which energy is distributed among all the states available.

Quantum theory replaced the phase volume integral by a discrete sum, which is called the partition function and often denoted by Z (German: *Zustandssumme*, sum of states). We could interpret it as a sum taken over all exponentials $e^{-E_i/kT}$, where the E_i values refer to the non-degenerate energy levels (i.e. the eigenvalues of the Hamilton operator) of the total system. Quantum theory introduced a conceptual simplification due to the fact that energy can be taken up by the different degrees of freedom only in a discontinuous fashion. Only at sufficiently high temperatures do quantum states become so densely occupied that these sums can be treated in the continuous classical way, as integrals. Furthermore, owing to the indistinguishability of identical particles, quantum theory simplifies considerably the "counting" of discernible states.

The statistical derivations of entropy in terms of the logarithms of W, ϕ or Z, as they were given by Boltzmann, Gibbs, Planck[30], Shannon[31] and von Neumann[32] (Table 2.5.1) are entirely congruent. I shall not enlarge further on what is supposed to be the stuff of textbooks, but will rather highlight those problems that I find relevant in the context of this book.

One of the main problems to be revealed in later parts of this book is that of complex systems. All living systems are structures of huge combinatorial complexity. The statistical treatment of material systems as described in these sections also introduces us to the problem of complexity. However, we shall see that the "problem" that manifests itself in the incredibly large number of microstates and that seems to be "tamed" in the macrostate is of an entirely different nature from the problems we shall be facing later in this book. Understanding this difference, and understanding

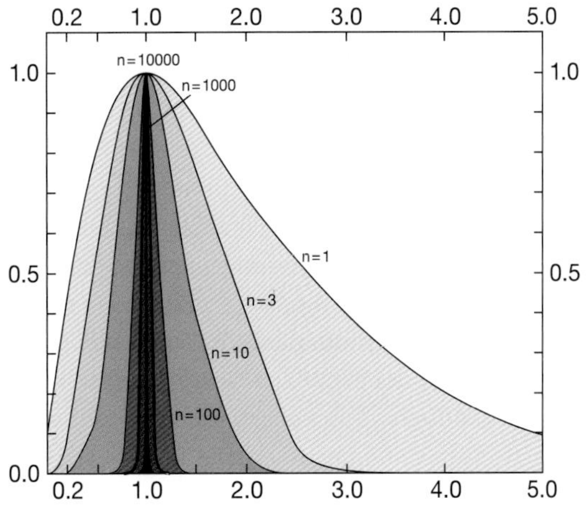

Figure 2.5.1

The function $f(x) = (xe^{1-n})^n$, as explained in the text (see also Table 2.5.2).

why and how approaches to their solutions have to differ, will be of great help for our discussions in Chapters 4 and 5.

I opened the discussion of the macrostate by pointing out that only a very small fraction of the microstates that constitute the macrostate are (or have ever been) populated. Nevertheless, the macrostate yields a statistical definition of entropy (and other thermodynamic variables) with a degree of precision that is the higher the larger the size, and hence the combinatorial complexity, of the system in question. The complexity is of an exponential nature, which means that the size of the system, expressed by the particle number N, appears in the exponent of the function that characterises the macrostate. It is this very fact of exponential complexity that also contains the key to the solution of the problem.

Consider, for example,[33] the function $f(x) = (xe^{1-x})^n$, which is represented in Figure 2.5.1. It consists of two factors, one of which (x^n) increases sharply (particularly for large n) with increasing x, while the other ($e^{n(1-x)}$) approaches zero exponentially. The function f(x) is chosen so as to yield, for all values of n, an upper limit of 1, which appears when x is equal to 1. What is remarkable about this function is the incredible steepness of the maximum appearing for large values of n. This is shown in Figure 2.5.1 for n values of 1, 3, 10, 100, 1000 and 10000. However, the n values I have in mind are much larger, for instance of the magnitude of Avogadro's number (10^{24}), which would yield curves that could not be distinguished from an invisibly thin bar at $x = 1$.

Let me demonstrate this in Vignette 2.5.1 for values in the neighbourhood of $x = 1$, by introducing $x = 1 + \varepsilon$ in the function f(x), which then can be expanded for ε values that are much less than 1. According to the results obtained in Vignette 2.5.1, at n values of 10^{24} the function would have decayed to $1/e$ ($\sim 1/3$) of its maximum value at x values that differ only by 10^{-12} from $x = 1$. In other words the main decay (i.e. by 2/3 of its value) occurs between x values of 0.999999999999 and 1.000000000001.

Vignette 2.5.1 The sharpness of thermodynamic distributions

In his book *Theory of Heat* (*Theorie der Wärme*) the Göttingen theoretical physicist Richard Becker[33] (another confrère of Sommerfeld's famous school) emphasised the sharp maximum that characterises Maxwell's velocity distribution which is typical of many thermodynamic functions. The mathematical form of this distribution is closely related to the function $f(x) = (x\,e^{-x})^n$. The function xe^{-x} has also a maximum at $x = 1$ where it assumes the value e^{-1}. It will change if we look at different powers as is done for $f(x) = (x\,e^{-x})^n$ where n usually is an extremely large number. If we change this function by multiplying it with the constant e^n to $(x\,e^{-x+1})^n$ the maximum value becomes one for all values of n (which may have the order of magnitude of Avogadro's number $N_1 \approx 10^{24}$). It can easily be seen from Figure 2.5.1, that for values of x other than one, all curves quickly converge to zero.

The details are mathematical routine. This can be best seen if in the neighbourhood of $x = 1$ we replace x by $1 + \varepsilon$. Then the function decays from 1 to $1/e$ already for very small values of n, e.g. for $\varepsilon = \sqrt{2/n}$.

This is demonstrated in Figure 2.5.1 for n values of one, three, ten, one hundred, one thousand and ten thousand. However macrostates in thermodynamics are usually characterised by much larger numbers close to or larger than $N_A \approx 10^{24}$ where our representation would have yielded a very thin and hardly visible line (e.g. with a relative line width of $\sqrt{N_A}/N_A \approx 10^{-12}$. It is one of the most important conclusions of statistical thermodynamics that macrostates (and their thermodynamic functions, such as entropy) are deterministically well defined, usually within immeasurably small error limits. In fact, the ultimate limit for $n \to \infty$ is called the "thermodynamic limit".

A typical example of the behaviour described above is provided by the binomial distribution represented by the binomial coefficients $\binom{n}{k} = \frac{n!}{(n-k)!k!}$, with $0 \le k \le n$. This plays a part in mixing problems, chemical distributions and mutant spectra in molecular genetics. Consider a straightforward chemical transformation $A \leftrightarrow B$, in which both states (A and B) are isomers of equal chemical energy content. Macroscopically the system can be described by the law of mass action $n_A/n_B \equiv K_{AB}$, where n_A and n_B are the particle numbers and K_{AB} is the mass action constant, which in the present case is assumed to be one. How precise is such a law?

If the total number $n_A + n_B = n$ is constant, we might call $n_A = k$, $n_B = n - k$. Since single particles are subject to thermal fluctuations n_A and n_B may assume any value between 0 and n, but with the constraint $n_A + n_B = n =$ constant. Such a situation is described by the binomial law where each state is populated

Cont. ➲

Cont.

according to the binomial coefficient $\binom{n}{k} = \frac{n!}{(n-k)!k!}$. We see immediately (and can easily prove) that the distribution yields a maximum for $n - k = k = n/2$. Using Stirling's approximation $n! \approx n^n e^{-n}\sqrt{2\pi n}$ (which already for $n \geq 10$ yields values that are more than 99% accurate) we calculate the maximum value of $\binom{n}{k}$ to be $2^n \cdot \sqrt{2/\pi n}$. Moreover, we can now expand the function around its maximum value (as we did above for f(x)) and obtain for $k = n/2 \pm \kappa \cong n/2(1+\varepsilon)$ with $\varepsilon = \kappa \cdot 2/n \ll 1$ the same decay law as in the above case: $e^{-\kappa^2 \cdot 2/n}$, i.e. a decay to e^{-1} of the maximum value for $\kappa = \sqrt{n/2}$ or for $\varepsilon = \sqrt{2/n}$.

Note that in the case of chemical equilibrium n is a huge number, e.g. for a few grams of material it is of the order of magnitude of N_A (1 mole = M grams). Hence the natural error in the law of mass action would be of the order of magnitude of one in 10^{12}. This is what we observe on the macroscopic scale, which involves about 10^{24} particles and $2^n = 2^{10^{24}} \approx 10^{10^{23.5}}$ possible states of the distribution. Microscopically, the decay to 1/e involves about 10^{12} states around $k = n/2$. The decay law is of the form e^{-ax^2} involving erf $(x\sqrt{a})$, i.e. the Gaussian error integral $\frac{1}{2}\sqrt{\frac{\pi}{a}}\int_0^{x\sqrt{a}} e^{-t^2}dt$ or a decay range from 1 to 1/e of the extension of $\sqrt{n}$ involving some 10^{12} states. Almost the whole population is located in this range of extremely low relative extension.

The binomial distribution will be revisited in connection with problems of selection and evolution in later parts of the book. But there we shall encounter quite different aspects of complexity, which involve the specification of single individual states in complex systems, which is not possible in chemically equilibrated systems and which requires a confrontation of binomial complexity with single states requiring a sharp identification (selection) and amplification of (chemical) microstates.

We find this type of behaviour in all functions that characterise the macrostate and hence yield definite values for the thermodynamic variables. A single molecule in a gas at a given temperature has a very broad energy distribution, and the temperature does not tell us anything about its energy. It is only the large magnitude of N that, in the case of the ideal gas, establishes the relation $E = \frac{3}{2}$ NkT, i.e. which correlates temperature in a unique way with thermal energy. This answers our question of why the macrostate is so well defined, despite the fact that an overwhelming majority of the microstates is never populated. Since the distribution is so incredibly sharp, the states that are populated at any instant are predominantly the ones that represent the sharply defined first moment of the distribution. Hence, the complexity of the absurdly large multitude of microstates is "suppressed" by the dominance of the sharply defined maximum. In other words, the exponential complexity of the problem is never really expressed. The laws for the ideal gas, for instance, are superficially very simple laws, despite the fact that they are backed by a hyperastronomical combinatorial complexity of individual states, which describe the physics going on at the microlevel.

In fact, physicists do not really get bothered by the combinatorial complexity involved in the statistical mechanics of equilibrium states. The complexity that we shall encounter in life processes, and which will accompany us through large parts of this book, is of a quite different nature. There, just one single state may bring along particular properties, which become determinant for the "living state", and our problem will be to find the analogues of such particular single microstates. Averaging in the above way would never allow us to find them. What we need is a "selective" procedure. These remarks should suffice for the moment, but it is important to realise how the problems of equilibrium statistics differ from those we have to solve for the statistical dynamics of life processes.

In this chapter we have been introduced to a complexity that is expressed by a stupendous number of possible microstates. However, it is a manageable complexity, one that leads to a well-defined numerical value of entropy, and therefore it does not bother us in this case. For a sufficiently large system, the "fuzziness" in entropy is negligible. Complexity in this sense is even the basis of determinate robustness. The individual microstate as such, although "physically real", is of minor importance for the appearance of the system in ascertainable reality. The incredibly long recurrence times render the individual microstates relatively unimportant; indeed, an overwhelmingly large fraction of these are never realised within times as long as the age of the universe. Despite this fact, the macrostate (even at relatively low particle numbers), owing to the extremely sharp distribution of microstates, is well established.

So what does the word "real" mean? The microstate is a state we have no access to; it is always the macrostate that we encounter in reality. Entropy as an extensive variable in its explicit form depends on how much we know about the system that establishes its macrostate. The fact that the population of only a minor fraction of the microstates determines quite well their mean representing the macrostate may intuitively be astonishing, but is a mere consequence of the Gaussian character of the binomial distribution. Our experimental information about a system is nearly exclusively based on its macrostate rather than on any individual (physically "real") microstate.

2.6. How Many Trees Make a Wood?

Think of a single molecule that can exist in either of two defined chemical states, say A and B. Examples are legion in all branches of chemistry, but using an abstract notation here I want to make sure that I do not get lost in discussing any particular chemistry. For simplicity, let's assume that both states refer to isomers that do not differ in their chemical energy content. But at any time only one molecule is present. It can switch back and forth between its two energetically equivalent chemical states A and B, residing in each with a finite life-time, which as such is defined only as an average. Can we say that such a molecule at any instant is in chemical equilibrium? If we define "chemical equilibrium" as a statistically fixed state then the answer must be "no", because in any single case the probability of making an error due to fluctuations is as large as that of the predicted value.

One could suggest that "balance" requires at least two states that exist simultaneously. Very well, let's consider two such molecules. But this does not really help because while one molecule is in one of the two states the other molecule need not necessarily be in the inverse state. The case of one molecule may be simulated by repeatedly tossing a coin, which results in an unpredictable series of "heads" and "tails" (i.e. A and B). The case of two molecules is not very different; it corresponds to a game of chance in which we use two coins simultaneously. Equal energies of states A and B are equivalent to "fair" coins. I don't think we can claim that such a system is at "chemical equilibrium", in the same way that we cannot call two trees a wood.

It is true that, for a sufficiently long series of trials, states A and B will eventually show up with, on average, equal frequencies, so that we may call that long-term average a "balance". However, knowing the exact history of outcomes does not help us at all in predicting which of the two states will emerge at the next moment. Similarly, in tossing two coins or "looking at" two molecules, a simultaneous appearance of head and tail (or of A and B) – observed over a long series of trials – will result from only one-half of the trials.

The example of tossing two coins is popular in probability theory as a demonstration that the assumption of "equally likely outcomes" is a level of judgement that cannot be fixed by mathematical arguments alone. Consider the three possible outcomes of tossing two coins. They are: two heads, one head and one tail, and two tails. Assuming that these outcomes are equally likely would lead one to predict a relative frequency of one-third for each outcome in a sufficiently long series of trials. Such a result may be obtained in certain situations, for instance if individual states have to be assumed to be indistinguishable. Coins, however, like other macroscopic physical objects, are distinguishable, resulting in two possible realisations of "one head and one tail". Hence in the long run the balanced solution "head and tail" will appear twice as often as either of the extremes of two heads or two tails. Chemical compounds, in their reactions at normal temperatures, generally behave in this manner.

Anyway, chemists do not usually "play" with single molecules. What they call "chemical equilibrium" is the result of a microscopic lottery viewed at a macroscopic level, yielding a deterministic "law of mass action" that describes the chemical balance. Paul and Tatjana Ehrenfest[34] devised a model that demonstrates the nature of such a statistically based law. They used two urns (representing states A and B) filled

Figure 2.6.1

The glass bead game designed by Ruthild Winkler-Oswatitsch

The design and rules of the game are explained in the text.

with numbered stones that were to be chosen by some lottery and then exchanged between the two urns, thus simulating the effects of chance and law. Some physicists disrespectfully called Ehrenfest's urn game the "dog and flea model", supposing two dogs running side by side and exchanging one (or many) fleas.

Ruthild Winkler-Oswatitsch[35] has fashioned an elegant "glass bead game" (Figure 2.6.1), which can be adapted to any types of rules and for which they demonstrate *ad oculos* the transition from stochastic to deterministic behaviour, depending on the rules applied. In the following I shall replay Ehrenfest's urn model, using two sorts of coloured glass beads, yellow and blue ones, which are placed on a board. The board has co-ordinates by which we can identify each individual square, as on a chessboard, from 1 to 8 and from A to H. Accordingly, we use two octahedral dice, the eight faces of each corresponding to the co-ordinate markings. (One can equally well use cubic dice and play the game on a 6 × 6 board.) Rolling the pair of dice identifies the co-ordinates of any square on the board to which a particular rule, depending on its occupancy and specified by the process under study, is then applied.

In the Ehrenfest model we start by arbitrarily placing a blue or a yellow bead on each of the 64 squares of the board. For the eventual result it really doesn't matter how they are arranged at the start; it may be any of the possible mixed compositions of

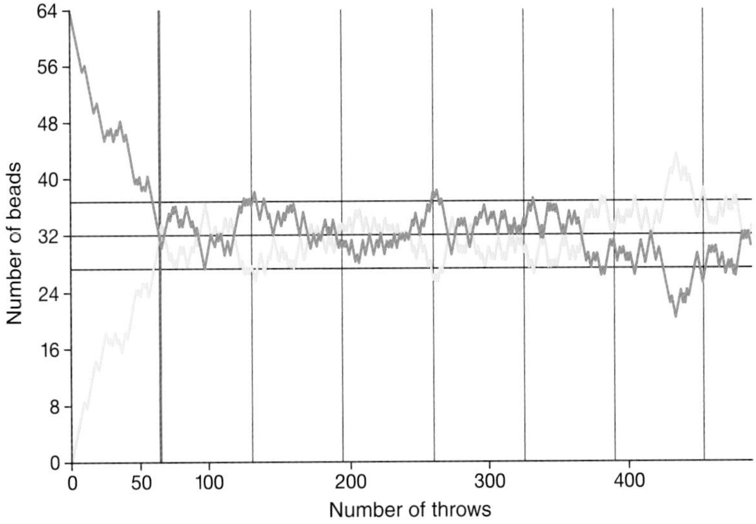

Figure 2.6.2
A typical run for Ehrenfest's urn model

For details see text.

blue and yellow beads. However, the observable course is most illustrative if we start with all 64 squares occupied either by blue or by yellow beads. Now we roll the pair of dice and execute the following rule: the bead on the identified square is removed and replaced by one of the opposite colour. That's all! It simulates the reaction A to B or B to A. We can continue throwing the dice as long as we want.

Admittedly, this game is not very exciting. However, we should try not to get bored before we have completed at least 64 moves. Figure 2.6.2 illustrates the typical course of a game, starting from a situation in which the board is covered uniformly with blue beads. We may call 64 moves one generation because each square has, on average, one chance to change the colour of the bead occupying it. "Equilibrium" between blue and yellow is almost reached within one generation.

How do we define "equilibrium"? A suggestive answer – since chances for blue and yellow are equal – would be: equipartition, that is, 32 blue and 32 yellow beads. Looking at Figure 2.6.2 this answer seems "almost" correct. We see that with 64 beads the indeterminacy of tossing one or two coins has been replaced by a fairly well-established determinacy. Only some "fuzziness" of the deterministic solution is left. Let's do a more quantitative analysis in terms of micro- and macrostates, partition functions and entropies as we have learned in the preceding sections.

In our game, a microstate is any of the individual distributions of blue and yellow beads on the board. We distinguish squares by their co-ordinates in order to specify individual species. Removing and replacing a bead means reaction of the particular individual from one to the other state (A to B or vice versa). Spatial separation in this game does not refer to true geometric space. In modelling a chemical reaction the individual molecules in the game, symbolised by beads and individually identified by their co-ordinates on the board, are supposed not only to move around freely but also to be kept at "constant temperature". All we distinguish is their chemical states; this is justified if chemical and thermal equilibration occur on different time-scales. Since each square can be occupied in two ways, blue or yellow, there are altogether $2^{64} \approx 2 \times 10^{19}$ "chemical" microstates. That's a lot, if we think of a small "cluster" as comprising not more than 64 "molecules".

Think of playing this game by carrying out a move every second. In order to visit every microstate just once (on average), we would have to play the game for appreciably longer than the time that our universe has been in existence. (We recall from Chapter 1 that the "big bang" happened some 15 thousand million years – about 5×10^{17} s – ago.) The truly surprising fact is that we reach equilibrium (as we shall define it) within less than 100 s (supposing one move per second). That means that only a tiny fraction of the microstates has to be tried out in order to reach the defined macrostate of "chemical equilibrium". At the level of microstates, no two games would look alike or even similar. The Poincaré recurrence time, introduced in Section 2.3, is of the order of magnitude of $10^{19}\tau$, where τ is the elementary time of changing the "microstate".

The ratio of blue and yellow beads on the board is to be called the macrostate, each being represented by a (usually) very large number of microstates. In chemical thermodynamics the macrostate is defined by macroscopic variables, such as "affinity" and "extent of reaction". Both these variables become zero at "chemical equilibrium", which appears to be a "force-free" state. Since in nature all molecules present would change in parallel (with some stochastic distribution), the relaxation time of the macrostate tells us something about the average reaction time (or life-time) of a single molecule. In fact the reciprocal (macroscopic) relaxation time is given by the inverse of the sum of the two rate constants $A \rightarrow B$ and $B \rightarrow A$, since both processes contribute to equilibration (in the present example with equal average rates).

This is a simple example of what we really mean by micro- and macrostate. In any particular microstate the colour of each individual square on the board is uniquely defined, while the macrostate refers only to net population numbers of occupation. A chemical macrostate is defined by the molar fractions or concentration ratios of the reactants and products. Let us take a look at those macrostates. There are altogether 65 such states, ranging from black-to-white ratios from 64:0 to 0:64. The two extremes of these macrostates are each represented by only one single microstate. The respective numbers of microstates representing each macrostate are given by the binomial coefficients, the sum of which here adds up to $2^{64} \approx 2 \times 10^{19}$ microstates. The binomial that describes this distribution has its maximum at equipartition, i.e. 32 blue versus 32 yellow beads (Figure 2.6.3). In our game, the number of microstates that represents this macrostate is about 10% of the total number of microstates, i.e. larger than 10^{18}. This huge maximum number of representations by microstates causes the distribution to settle quickly around equipartition. Each of the microstates

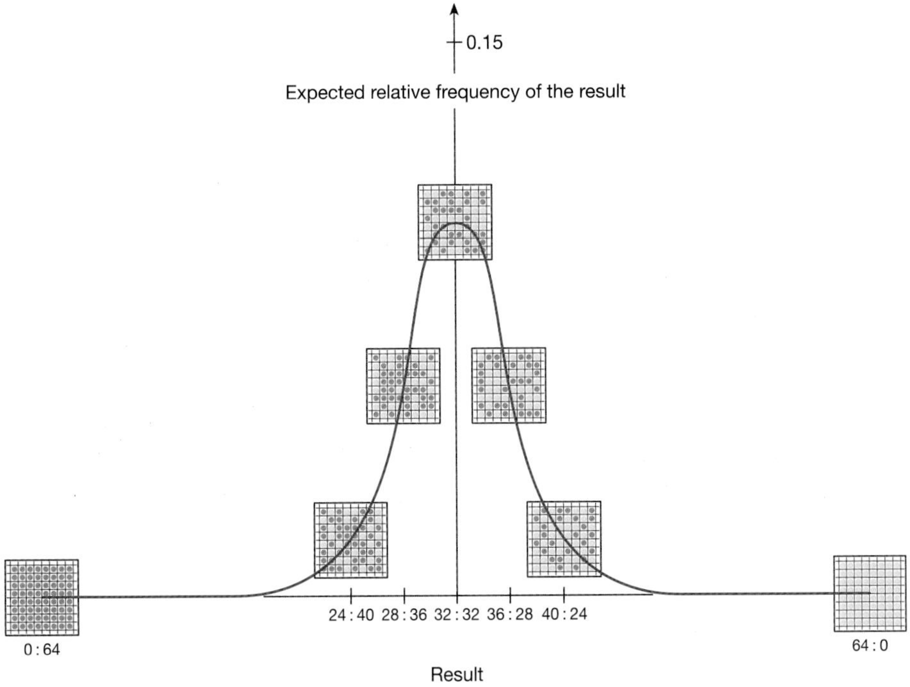

Figure 2.6.3
Distribution curve of the Ehrenfest model

For details see text.

(in this example) has exactly the same *a priori* probability of occurring. In other words: the extreme states, with all squares occupied either by blue or by yellow beads, are just as likely to occur as any particular one of the huge number of single microstates involving a balanced ratio (i.e. 32:32). The reason why the equilibrium settles around equipartition is of a purely statistical nature (see Vignette 2.5.1); it is "enforced" by the large number of microstates being associated with the macrostate of equipartition. Here we truly see "mass"-action at work. On the other hand, small deviations from equipartition, according to Figure 2.6.2, do occur relatively frequently. According to Figure 2.6.3 each of the two macrostates neighbouring equipartition (i.e. 31:33 and 33:31) is represented on average by almost as many microstates as the exactly balanced macrostate. However, most of the 2^{64} microstates fall into the narrow range around equipartition, having a half width of $\sqrt{N}$ (N being the total number of particles or beads involved). Fluctuations in this range of $\sqrt{N}$ are very frequent. Recurrence times for macrostates are inversely proportional to the number of microstates, hence the balanced state reproduces itself most frequently.

What then is the correct physical description of the equilibrated macrostate? In fact, none of the macrostates alone, defined by a ratio of integral population numbers, represents the "equilibrium state". The stable equilibrium state refers to the complete binomial distribution that comprises the total number of (equivalent) microstates, and shows a sharp peak at a certain ratio (for equal energies this ratio is 1:1). Maintaining a population precisely at this fixed ratio, called the equilibrium ratio, would require continuous external interference, which in practice could not easily be materialised.

The above results suggest that we need many more trees to create the impression of a wood than we need molecules to define a "macroscopic" system. The term "macroscopic" does not really refer to the number of molecules involved, but rather to the number of physical microstates that establish macroscopic behaviour. Sixty-four pieces in the Ehrenfest game produce a total of 2^{64}, i.e. more than 10,000,000,000,000,000,000, possible microstates, yielding a macrostate of quite deterministic structure. Sixty-four trees cannot even be called a grove – notwithstanding the possible objection that a grove, in order to become a wood (with typically thousands of trees), need not extend over miles, as was impressively described in John Galsworthy's short story with this title[36].

A system consisting of only a few particles is called a cluster. It is the relatively sharp transition from micro- to macroscopic behaviour that makes clusters a preferred object of study in present-day research into the condensed states of matter. I agree that the beautiful complexity of woods is, at the very least, associated with the number of trees that make up the wood. And the same is true if we envisage the beautiful complexity of matter. Our world is not made up of two-coloured pieces, but rather of a vast variety of different chemicals, the number of possible combinations of which by far exceeds the capacity of our universe to accommodate all of them. What is going on at the microscopic level is largely hidden to us. It includes chaotic processes similar to the one depicted in Figure 2.3.1. It is therefore in principle impossible for us to follow in any detail the molecular mechanics, which are based on reversible laws, up to the macroscopic level of statistics that includes irreversible processes.

Another issue is how to describe chemical equilibration at the macroscopic level. In the experiment represented by Figures 2.6.2 and 2.6.3, we started with the extreme state of a uniform occupation of the board by blue beads and then watched its "chemical relaxation"[37], i.e. the exponential change of the population numbers approaching the equilibrium state. The change of population numbers in this case is solely the consequence of a "statistical force". Equilibration here can be described as the "loss of memory" of the initial microstate. The state of maximum entropy we call equilibrium is a state of "no memory".

In order to describe chemical equilibration in macroscopic terms let us introduce a new kind of space that we may call "population space". In Figure 2.6.4a, I drew such a space diagram for the Ehrenfest game. The number of yellow beads is plotted against the number of blue ones. Since the rules of the game are such that the total number of beads, i.e. the total population of the board, is always 64, the possible occupation states of the game board all appear on a straight line that connects the numbers 0 and 64 on both co-ordinates. Hence the constraint of constant population

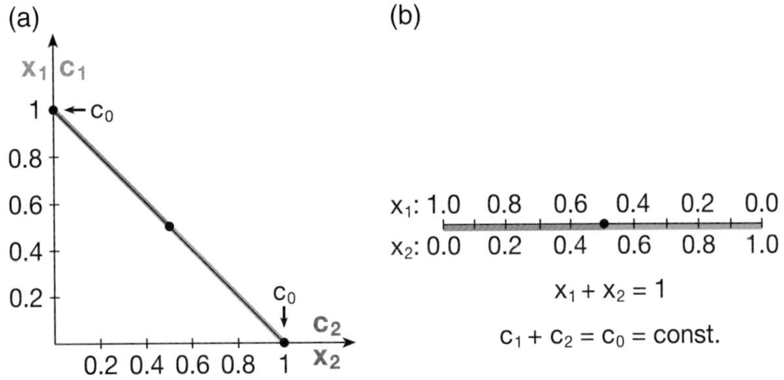

Figure 2.6.4

The two-dimensional simplex

(a) A two-dimensional concentration space diagram: $c_1 + c_2 = c_0 =$ constant. (b) All possible states of distribution appear on a line connecting the co-ordinate points $c_1 = c_0$ and $c_2 = c_0$, which is the simplex for a two-component system.

reduces the dimension by one, and instead of a planar co-ordinate system we can use a one-dimensional (i.e. linear) "simplex" (Figure 2.6.4b). All possible macrostate situations are found along this line; in case of equal *a priori* probabilities for blue and yellow they tend towards a "fixed point" at the midpoint of that line. Chemical equilibration is indeed guided by a gradient that we could call a purely "entropic force", which tries to reduce any deviation from the equilibrium point wherever it is situated along the line. This "force" becomes stronger the larger the deviation. If the *a priori* probabilities for blue and yellow are not equal – which in chemistry means that the two states differ in their energies – then the equilibrium will settle at some other position along the simplex-line, according to the law of mass action, which refers to a minimum of free energy.

The simplex formalism can be extended to concentration (or population) spaces of any dimension. It is shown in Figure 2.6.5 for a ternary reaction system. All points now lie on a triangle. The simplex in this case is a two-dimensional diagram. The corners of the triangle describe the states in which only one of the three reaction partners constitutes the total population. Equilibrium is again characterised by a stable fixed point. If three equivalent partners are involved, the fixed point is situated at a defined position within the plane of the triangle, as shown in Figure 2.6.5b. The more one of these compounds dominates, the closer the fixed point comes to the corresponding apex. If the equilibrium involves only two partners (i.e. if the third state is unstable) then the fixed point will lie on the line connecting the (relative) population numbers of both partners that are in existence. In Figure 2.6.6 we see the same simplex in a slightly different representation, emphasising the ranges of dominant influence of one of the three reaction partners.

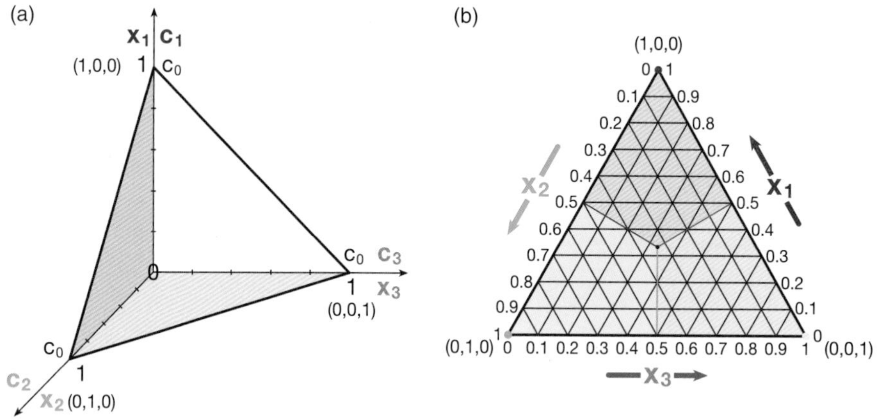

Figure 2.6.5

The three-dimensional simplex

For a three-dimensional concentration space, all possible points are on a triangle, which is the simplex for a three-component system.

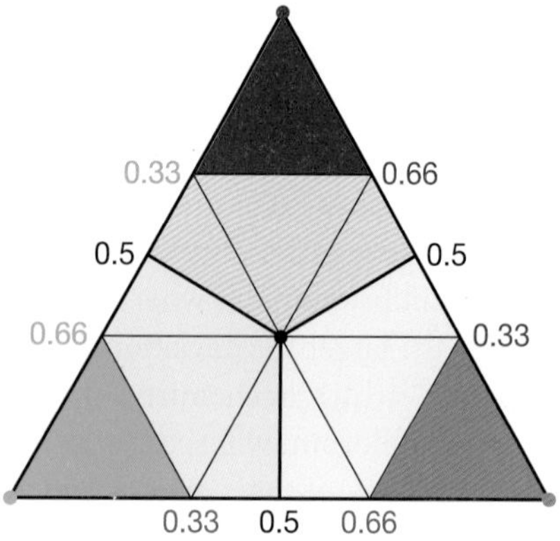

Figure 2.6.6

Dominance regions of the three-dimensional simplex

The coloured areas indicate the regions in which one of the three components is present at a concentration higher than the other two combined (dark shading) or at a concentration greater than either of the others (light shading).

The triangle simplex, representing the chemical changes relating three independent chemical compounds with the constraint of conservation of the sum of their concentration variables, is an illustrative example demonstrating a mathematical concept that holds for an arbitrary number of reaction co-ordinates. A pictorial representation of the general case, however, is hampered by our inability to imagine spaces of more than three dimensions. The simplex that applies to four reaction partners is a regular tetrahedron. This follows from the simple fact that all four compounds require an equivalent representation, which is materialised by a (platonic) body faced by four equilateral triangles, i.e. a regular tetrahedron. Each of the four apices represents a situation in which one of the four compounds has a relative concentration of one (the others being zero). Any point along an edge means finite concentrations of only two of the compounds, i.e. the ones whose apices are connected by the edge. Likewise, the presence of only three compounds confines fixed points to one of the four triangular faces of the tetrahedron. If all four reactive states are populated, the corresponding fixed points must lie inside the tetrahedron.

The simplex referring to five reacting compounds is a four-dimensional hypertetrahedron (the term being chosen in analogy to the hypercube). It is representable in terms of its subspaces: tetrahedra, equilateral triangles, lines and points. In the first

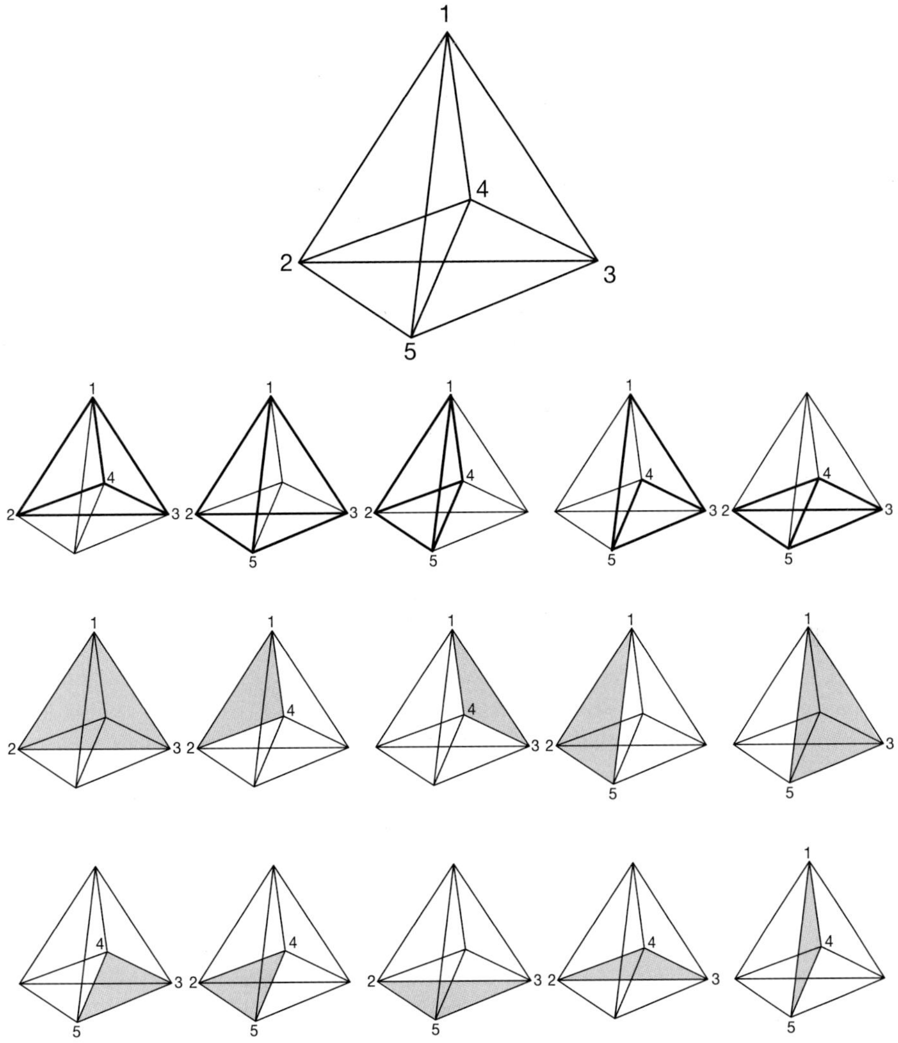

Figure 2.6.7

The five-dimensional simplex

The simplex is a four-dimensional hypertetrahedron in which the various triangular subspaces are shaded.

row of Figure 2.6.7 I have emphasised the next five subspaces by thick lines. In the two lower lines the subsequent subspaces, equilateral triangles, are marked by shading. Since all co-ordinates r (lines) run from 0 to 1, each of the five vertices represents one pure compound $i(x_i = 1, x_{k \neq i} = 0)$. The phase points of two coexisting species lie at one of the ten edges, those of three coexisting species at one of the ten triangular faces and those of four coexisting ones inside one of the five tetrahedra. Complete coexistence of all five compounds yields a phase point inside the hypertetrahedron. With analogous rules we may build up any higher-dimensional simplex construction.

Since such a generalisation leads to ever more complex constructions, I was curious how the word "simplex" came into existence. Pioneering work on such problems was done by Henri Poincaré who, however, did not use the term "simplex". It was his contemporary, the Dutch mathematician Pieter Hendrik Schoute, who introduced the term "simplex"[38]. Schoute argued that the most "simple" geometries are those that have no diagonal. This is true for the equilateral triangle, the tetrahedron and its higher analogues, and that is exactly what is required for this type of representation. Schoute decided that a word like "simplicissmus" was too clumsy and proposed instead the shorter term "simplex".

The temporal evolution of a reaction behaviour among a fixed number of compounds is described by a trajectory inside the system's representative simplex. Equilibration yields a point-like, stable focus as the "fixed point" eventually reached, and there are many other cases of analogous stationary reaction systems that show such behaviour. If the trajectory approaches a reproducible, cyclically closed curve, such a system is called a "limit cycle" and indicates oscillations which in chemical systems are of a non-sinusoidal shape (it generally can be observed in certain non-linear systems involving feedback terms). A spiral ending up in a point, likewise, indicates an oscillation with decaying amplitude. An overwhelming majority of solutions are of a "chaotic" nature. Although they result from deterministic solutions of their rate equations, they never reproduce their shape, which makes them extremely sensitive to their initial conditions. They may be considered to be "irregular" oscillations, showing no real periodicities, but with trajectories usually confined to a certain region of the simplex, in which they never reproduce one another in their temporal evolution. At this point I shall not go into further detail, but I shall revisit those reaction systems in more detail in Section 4.8.

2.7. Can Arbitrarily Complex Chemical Systems Ever Reach Detailed Balance?

A rash answer would be: yes! If the external conditions allow equilibration, and if internal states are free to change correspondingly, why shouldn't entropy reach its maximum, or free energy its minimum, indicating that total equilibrium has been established? Well, such an answer is at best premature. What we really need to know is: what is it that we consider balanced at equilibrium? The preceding section has provided a partial answer, but the system we considered was deliberately chosen to be simple. Reality is not a board game with two sorts of glass beads. In this section I shall be referring to an arbitrarily complex system which, I think, resembles reality more closely.

The Ehrenfest game we played is not necessarily limited to two players. We could, for instance, play the game with any number of participants using a corresponding number of differently coloured pieces. We have then to introduce, in addition, a "roulette wheel" to decide what colour should replace the piece removed by the throw of the dice. Suppose we use in this game more than 64 different colours. Then each colour would be represented, on average, by (less than) one copy and we have a situation as described at the beginning of Section 2.6, in which the large relative size of fluctuations precludes a deterministic manifestation of equilibrium. We can make no meaningful statement about this macrostate other than to say that it corresponds to a fully occupied board. Distinction of colours in the sense of equilibration no longer makes sense, especially if the number of different colours to choose between is arbitrarily large compared with the numbers of squares on the board. If all colours are equally probable we still can calculate a deterministic entropy value that refers to the "fully occupied board" of equivalent states.

Returning to chemistry in macroscopic systems, one could in principle have an unlimited number of substances B, C, D etc., all being in equilibrium with A, i.e. $A \rightleftarrows B$, $A \rightleftarrows C$, $A \rightleftarrows D$ etc. All these substances then are in "detailed balance" with each other, too. Describing such a system by corresponding thermodynamic variables such as affinities and extents of reaction, referring to the "normal modes" of the system (resulting from a diagonalisation of the matrix of the rate coefficients associated with all coupled chemical forward and backward processes involved) makes sense only if these modes have an identifiable macroscopic representation.

If we recall the results of Section 2.2, the thermodynamic parameters can be defined as functions of variables that depend on our state of knowledge. Entropy (S) can be written as a function of the two extensive variables U (internal energy) and V (volume), i.e. S(U, V). In addition – depending on our knowledge about ongoing reactions within the system – we can specify reaction variables such as ξ_i (the extent of the i-th reaction) and its complementary affinity A_i, which adds additional terms to entropy now being a function of more than two extensive variables, such as $S(U, V, \xi_k)$. Either representation describes the system correctly. (In the first case the total differential dS(U,V) would have two terms, i.e. $(\partial S/\partial U)_V\, dU$ and $(\partial S/\partial V)_U\, dV$, but in the second case it contains (at least) a third term, $(\partial S/\partial \xi_i)_{U,V}\, d\xi_i$, with the two first terms referring explicitly to constant ξ_i.)

Am I now talking about our real world? Look around at the surface of the earth. Except in certain deserts and in the oceans (and sometimes even there), the appearance of our planet, including all the artifices produced by human interference, is derived largely from the chemical diversity of living structures and their artefacts. As shown in the first chapter, our material world up to the level of atoms appears to be of a limited combinatorial complexity. This is mainly a consequence of the existence of fundamental symmetries, and is due partly to the fact that interactions at that level soon become saturated. Atoms are made of protons, neutrons and electrons. Electrons belong to the one class of elementary matter that (at present) cannot be reduced further. Protons and neutrons are made of quarks and antiquarks, which, like electrons, carry the trademark of being elementary. The glueing-together of protons and neutrons in atomic nuclei is of limited adhesive quality. Up to iron, the combination yields more energy than it requires in order to come about. Above iron, fission rather than fusion liberates energy, resulting finally in highly unstable, radioactive, nuclei. The combination of electrons in the atomic shells is also highly constrained, as is reflected in the periodic structure of the table of elements. In short, the combinatorial complexity of matter up to the level of atoms is limited to the hundred-odd known elements and their isotopes. The burst of complexity takes place at the level of chemistry, and from there up it seems to be open-ended. In order to take place, it had to await the cooling of the original universe down to temperatures not far from absolute zero. Only then could life organise its unbounded chemical complexity and give the earth its present-day appearance.

This is not the place to delve more deeply into the complexity of chemistry (see Section 4.1). However, in the present context I have to explain the term "arbitrary complexity" in the title of this section in order to support my view that the chemical complexity of life defies an organisation that might be achievable solely through chemical equilibration.

It is true that the chemicals responsible for the organisation of life, and hence for the ultimate richness of organic matter, belong to only relatively few classes of substances. Take the two classes responsible for the legislative and executive tasks of biomolecular organisation, the nucleic acids and the proteins. DNA makes use of only four monomers, known by their chemical abbreviations A, T, G and C, while proteins utilise 20 different "natural" amino acids. Complexity enters with polymerisation. The DNA molecule representing the genome of a bacterial cell, e.g. of *E. coli*, comprises four million monomers; hence, it is an (almost) unique choice among $4^{4\ \text{million}} \approx 10^{2,400,000}$ (a one with 2.4 million zeros) alternative copies. And even the smallest single protein molecules are unique choices out of more than 10^{100} alternative polymeric chains, most of which would chemically be as stable as the ones selected on the basis of their functional properties.

In the case of microstates, we were able to comfort ourselves with the thought that our world keeps going without requiring all these states to be populated. In the same way, not all the possible alternative states of a gene could possibly be populated either. For that, too, our world is far too small. However, each single gene counts as an individual, which has to fulfil its particular function, and as such it has to be a reproducible entity. So let me try and restore our respect for large numbers, which might have been lost in the preceding sections.

Numbers like "a million million million million" do not yet make chemists feel uneasy because they resemble the numbers of molecules (e.g. Avogadro's number) that they are used to working with in their laboratories. In Chapter 1 we learned that the (visible) matter in the whole universe is the equivalent of about 10^{80} masses of the lightest atom, the hydrogen atom, and the volume of the entire universe could not accommodate more than the equivalent of about 10^{108} closely packed hydrogen atoms. Hence, a tiny fraction of all possible (small) protein molecules would suffice to fill the entire universe. So "arbitrarily complex" chemical systems are indeed part of our world.

In preparative chemistry, one way of achieving such complexity is combinatorial synthesis, i.e. synthesis by chance, with subsequent artificial selection. However, for full-sized protein molecules, this does not get us very far. Even a decapeptide already has about 10^{13} alternative sequences.

We are really in trouble if we want to achieve true chemical equilibrium in a world of proteins or nucleic acids. The energies of the primary structures, involving the covalent linkages of the peptide bond in proteins or the phosphate diester linkage in nucleic acids, are fairly uniform in magnitude. They are modulated by the energies of the secondary and tertiary structures formed, but they still remain within a relatively narrow range. Equilibrium would require any two sequences i and j to be populated according to their relative differences $\Delta_{ij}G$ in free energies, i.e. in proportion to a Boltzmann factor $e^{-\Delta_{ij}G/RT}$. Hence, nearly all of them should be present. But they aren't! They cannot be! They would form once, if at all, and those lucky enough to be around in just one copy would soon clutter up the entire space available.

In spite of this, in biological reality the sequences that are required *are* present, and not just in one single copy. On the contrary, they exist in macroscopic amounts, so that their reactions can be described by macroscopic rate equations. Why are precisely those sequences present in large abundance? The answer is: Because the system is not at equilibrium, cannot be, and "doesn't want to be". Let's immediately clear up one (possible) misunderstanding. Thermally, these systems are at equilibrium throughout, so they can be described by their thermodynamic functions. What is (far) out of equilibrium is their chemistry, i.e. their formation and their turnover. A DNA chain forms by template-instructed polycondensation from energy-rich nucleoside triphosphates, whereas it decomposes into energy-deficient nucleoside monophosphates.

So let us play a game which resembles more closely a reaction, the product of which is brought about by replicative instruction. At first glance a simple reversal of Ehrenfest's rule seems to do the job and yield the correct result. A square, specified by the throw of dice, amplifies its colour (via self-reproduction), i.e. it introduces another bead of its own colour at the expense of a bead of the other colour. Ruthild Winkler-Oswatitsch has designed such a game in analogy to the Ehrenfest game. A typical course is shown in Figure 2.7.1. It simply shows a reversal of the curve of Figure 2.6.2, the result being the sharp selection of one colour. In other words:

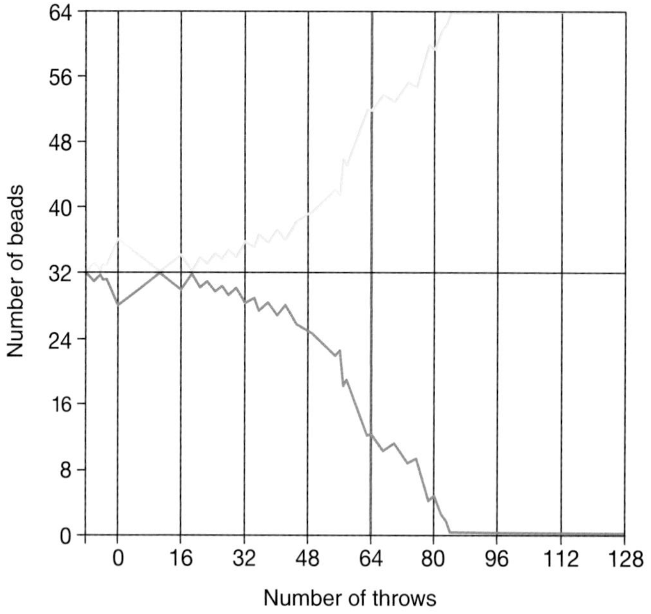

Figure 2.7.1
Reversal of the Ehrenfest game

The rules of Ehrenfest's model are reversed. A bead, the co-ordinates of which have been determined by a throw of the dice, is doubled at the expense of another bead of the other colour, i.e. any bead of the opposite colour is removed from the board. The game yields "all or none" selection within almost one generation (64 moves).

the equilibrial "fraternisation" of both types of beads is replaced by an "all or none" decision. Which of the two colours becomes selected remains unpredictable until the process has passed its first phase. Initially, both types have exactly the same chance of being selected, and in a long series of experiments both results appear (on average) with equal frequency. The Gaussian distribution curve of Figure 2.6.3 is now replaced by two sharp peaks, which become singular lines at the extreme values of both bead numbers, as shown in Figure 2.7.2.

However, something is wrong with this version of the game, or rather: something appears to be unrealistic. Doubling of beads means self-reproduction, which certainly is realisable in Nature. In fact, it will turn out to be a necessary prerequisite of "natural selection" (see Chapters 4 and 5). However, linking self-reproduction with the removal of the competitor (which is responsible for the acceleration in the later phase of the selection process) sounds artificial. In the Ehrenfest model the coupling of formation of a blue bead with the removal of a yellow one (and *vice versa*) can be understood as a chemical transformation between both partners. It introduces microscopic reversibility, the hallmark of chemical equilibration. For the selection

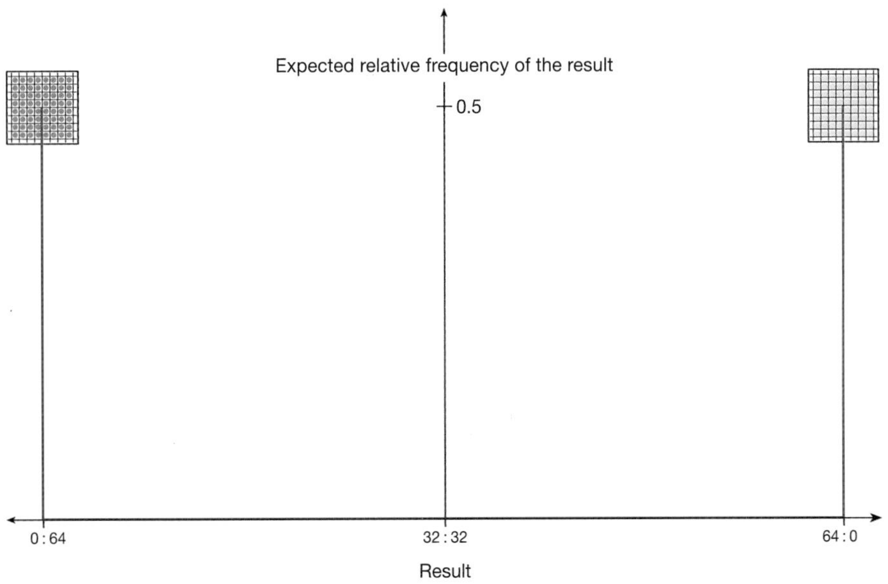

Figure 2.7.2

The distribution curve for "all or none" selection

The "curve" shows two singular lines for the population ratios 64:0 and 0:64 and no possible populations in between.

game such a coupling does not make any sense because we are far from chemical equilibrium in the process described.

What we need to do is to uncouple formation and removal (or decomposition). This can be easily done by throwing the dice twice for each step. The first throw decides which colour (namely that on the square that is hit) is to be removed. The square hit in the second throw then decides which colour is to replace the one removed. Reproduction and decomposition (removal) have thereby become independent of one another, while both are made proportional to their respective population numbers. If these two throws of the dice are performed in strict alternation, the total population is kept constant (at 64 beads).

The result of such a game is "selection" as well, and it is therefore also represented by the final distribution depicted in Figure 2.7.2. However, the process as such appears to be quite different from that represented in Figure 2.7.1. First of all, it takes much longer than one generation. Then, upon closer inspection, it reveals the square distance-over-time characteristic typical of a random-walk process (see Section 4.6), before eventually leading to clear-cut selection.

Let us take a look at a straightforward random-walk process and see how it is expressed in our glass bead game. Ruthild Winkler-Oswatitsch chose for simulation the simple process of linear diffusion. A particle undergoing linear diffusion has only two choices, to go forward or to go backward, and it does both with equal probability. Such a process can be most easily simulated by tossing a coin. So the rules of the game are:

Heads: Replace any yellow bead by a blue one.
Tails: Replace any blue bead by a yellow one.

The course of this game is shown in Figure 2.7.3. What is represented in this figure is not an average expectance, which requires the mean square displacement $<x^2>$ to be proportional to 2τ, where τ is the time during which the displacement occurs (in our case the number of throws of dice required to reach $<x^2>$, the square root of which is called the variance). No, what is seen in Figure 2.7.3 is a true curve of an experiment, which includes all fluctuations that are innate to a diffusion process.

We may interpret the experiment as demonstrating selection whenever it first reaches one of the extreme bead ratios 64:0 or 0:64. That would definitely be the end of the experiment, since the board would then be completely filled with one sort of bead, which would mean that the other sort had died out and could not reappear (as this appearance would require the reproduction of a bead no longer present). Hence this part of the game shows exactly the distribution depicted in Figure 2.7.2. It turns out to be a fairly good approximation to "natural selection", but it would be premature at this point to start a discussion of possible differences due to an oversimplification

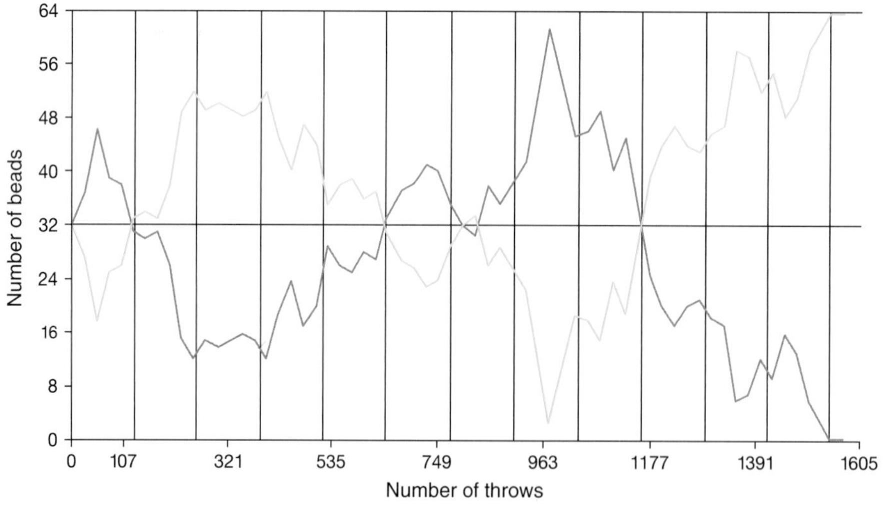

Figure 2.7.3
The random-walk game

The game is based on tossing a coin. The rules are: for "heads" a bead of one colour (e.g. a blue bead) is replaced by a bead of the other colour (here a yellow one); for "tails" the opposite move is made.

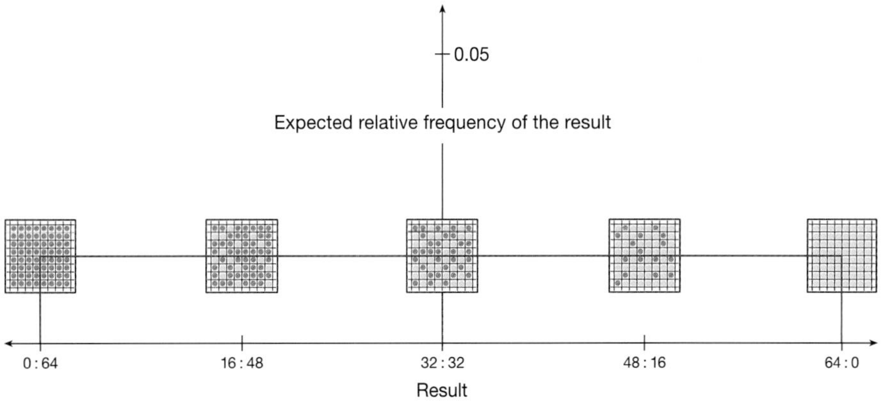

Figure 2.7.4
Distribution of the random-walk game

As Figure 2.7.3 suggests, each ratio of colours may occur with equal frequency if the game goes on "for ever".

of the rules. This is done in Section 5.3 in connection with the theory of "neutral selection" as proposed by Motoo Kimura.

The true coin-tossing game does not stop at the first selection point in Figure 2.7.3, but rather may go on "for ever". As is seen in the figure, after a long time has passed, every bead ratio will appear with equal frequency. Therefore the distribution curve will look rather like the one drawn in Figure 2.7.4, i.e. a rectangular distribution in the whole range of existence, as one would expect for a diffusion process that after a sufficiently long time fills a certain volume homogenously with matter.

Let me come to a conclusion. The reader may have noticed that I have circumvented the problem of Darwinian selection as expressed by the term "survival of the fittest". Likewise, he may have been surprised that a "selective advantage" is not even necessary for achieving natural selection. These problems will be discussed in detail in Chapters 4 and 5. Here I was mainly concerned with the effect of statistics. As for chemical equilibrium, we get a sharp ordering by the law of mass action if only entropic "forces" are at work, i.e. if both states reaching equilibrium have the same internal energy or enthalpy (ΔU or $\Delta H = 0$); we still get selection if there is no "selective advantage". It is not deterministic, i.e. one cannot predict which of the two alternatives becomes selected, but when the selection happens it is sharply all or none. Only if there is sufficient selective advantage among competitors does it become deterministic, as the example in Figure 2.7.5 shows, which at the moment should serve just to emphasise the difference between this kind of selection and "neutral selection". The latter requires, as prerequisites, just "replicability" and conditions "far from equilibrium". At or near equilibrium the system would be solely

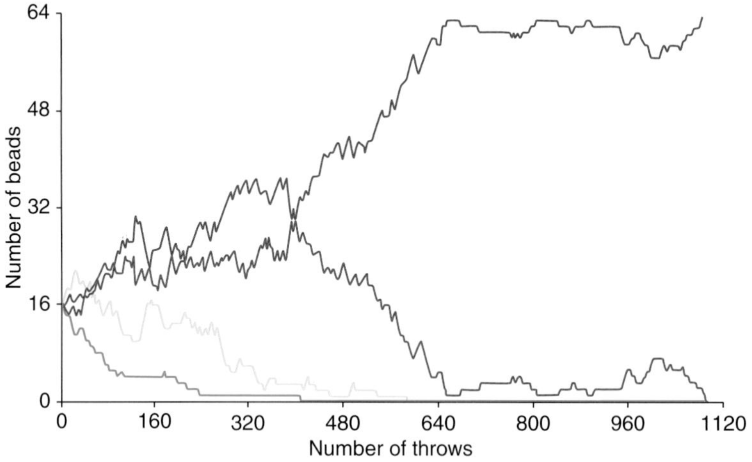

Figure 2.7.5
The Darwinian selection game

Here we have true "survival of the fittest" among the four colours, i.e. the blue one. The game uses a different "probability" of reproduction and/or removal for each colour.

determined by a difference between free energies and react only by relaxational decay to equilibrium, irrespective of the complexity of the reaction mechanism involved.

In this chapter I am almost exclusively concerned with questions of the statistical behaviour of multiparticle systems, and the title of this section asks whether complexity in its ultimate limit excludes equilibration. The answer, now, is that there is an internal discrepancy between the actual and the possible in arbitrarily complex systems. In many cases a system, because of its complexity, can never reach a definite composition and instead is bound to drift randomly through all reachable states. However, this cannot be the complete answer. In biology we encounter an unlimited complexity of material states. Yet the states we encounter are anything but orderless. They do not show any order of the kind we are used to in physics. It may be better characterised by the term "functional order", which is highly complex, but not "orderless".

The three prototypes of material arrangement (Figures 2.6.3, 2.7.2 and 2.7.4) that I have discussed in this section will turn out to be instrumental for a physical understanding of our world. Equilibrium obeys strictly the tyranny of structural stability. Life and thought are examples of Nature's power to overcome the limits set by the dictate of structural organisation. Life represents a kind of organisation that is primed by optimal functionality and as such only able to exist far from equilibrium. Erwin Schrödinger, more than 50 years ago, realised this fact when in his book *What is Life?* he stated as one of the characteristic features of life: "living matter evades the decay to equilibrium".

2.8. How is Entropy Related to Order?

In mathematics, the term "order" is clearly defined. A set is called an "ordered" one if for any pair of elements a and b the following three conditions hold:

1) a is either smaller or larger than b, or the two are equal.
2) The three alternatives, listed under (1), are mutually exclusive. No two of them can be valid at the same time.
3) The relations must be transitive, i.e. if a is smaller than b, and b is smaller than c, then a must be smaller than c, too.

What appears to be entirely intelligible for numbers does not necessarily correspond to daily experience. Football team A in a certain game may turn out to be superior to team B, while in another game B beats team C. This does not necessarily mean that in the next game C will lose against A. And it is not limited to competitions only. Who, for example, would want to put Bach, Mozart and Beethoven into a particular order? We may do it individually, by taste, but even that may change with time and mood. What is wrong in all such cases is that condition (1) is not fulfilled. The sets under consideration may contain incomparable elements. Conditions (2) and (3) then only hold for those elements of the sets that are comparable according to condition (1). In mathematics such a system is called "partially ordered".

We may admit that the above cases involve multifaceted properties, each of which is hard to express in a quantifiable way. But our subject in this chapter is entropy, a thermodynamic state function that is clearly definable in quantitative terms. Yes, anything that can be expressed by numbers should lend itself to being ordered according to the three criteria quoted above. However, consider the following experiment.

A box, filled with a clear liquid, is injected in one of its eight corners with a droplet of ink. The dye molecules then diffuse across the whole box. The entropy of mixing will steadily increase, in accordance with the second law of thermodynamics. The ink will spread in the most direct possible way, which means that some dye molecules will reach the centre of the box before they fill up the remaining seven corners and, in particular, the one diagonally opposite the injected corner. However, one could alternatively conceive of a (non-realisable) path – again with steadily increasing entropy – in which all the corners are visited first, as though one-eighth of the original droplet had been injected into each of the eight corners. That all sounds trivial, but it shows that there are different ways of arriving at a state of maximum entropy that correspond to incomparable situations.

The German theoretical chemist Ernst Ruch[39] developed a scheme by which he could quantify the consequences of partial order in the process of mixing, calling the quantity that replaces entropy "mixing character". Two situations in his scheme are characterised by a "mixing distance", rather than by a mere entropy difference, thereby providing a stronger version of the second law. Ruch developed his scheme with the help of so-called partition diagrams, which were introduced by the British mathematician Albert Young, in order to characterise the complete order of numbers.

Consider the upper row of nine boxes (which may be numbered correspondingly) in Figure 2.8.1. Let us "socialise" this ranking order by giving all boxes equal ranks. This means we transform the one horizontal row into nine vertical columns, each containing initially one box, all boxes being in equivalent positions. In order to do this we need certain rules: In a stepwise procedure only boxes from longer rows can be moved to the next available shorter one, whereby no subsequent row is allowed to become longer than any one preceding it in the diagram. I call this process "socialisation" because shorter rows are exclusively favoured at the expense of longer rows. For four boxes there is only one way of doing this. However, with nine boxes the process not only yields alternative routes, but within them certain intermediate configurations occur for which no comparable ranking can be assigned (see Figure 2.8.1). The numbers we are dealing with in statistical physics are not four or nine, but rather of the magnitude of Avogadro's number. For any realistic physical system, the partition diagrams will turn out to be tremendously complex. Ernst Ruch has applied his thoughts to a study of chirality and he has worked out a general theoretical frame. So far his work has not found the recognition that it deserves.

Many physicists call entropy a measure of order or (better) of disorder. However, we have to bear in mind that its statistical definition refers to a macrostate in phase space rather than in geometrical space. This means that we would have to define carefully what we mean by "order", and we shall see that this does not always match our intuition. Let me start with some examples where the correlation with an order parameter seems to be obvious. One of those examples is the phenomenon of phase transition (see also Appendix 2).

In order to characterise a phase transition, let me start from a straightforward chemical example, for instance the transformation of a biopolymer between two conformational states. The polymer might be a peptide or nucleotide chain that can exist in either of two states, an ordered helix or a random coil. A formal representation is given by the example in Section 2.6, i.e. the two-state equilibrium: $A \rightleftharpoons B$. Deviating from the above discussion, we now assume that both states A and B differ in their energy or enthalpy content by ΔH°, where ΔH° is the enthalpy difference between A and B under some standard conditions of temperature and pressure. The blue curve in Figure 2.8.2 describes the temperature dependence of the relative concentration of B, called x_B, assuming a reaction enthalpy five times as large as the thermal energy (RT) at room temperature. $x_B = 0$ means that all of the substance is in state A, while $x_B = 1$ indicates that it has all been transformed to state B. We see that the transition range (from 10% to 90%) is fairly spread out over about 300 K. The relatively low value of ΔH° corresponds to an involvement

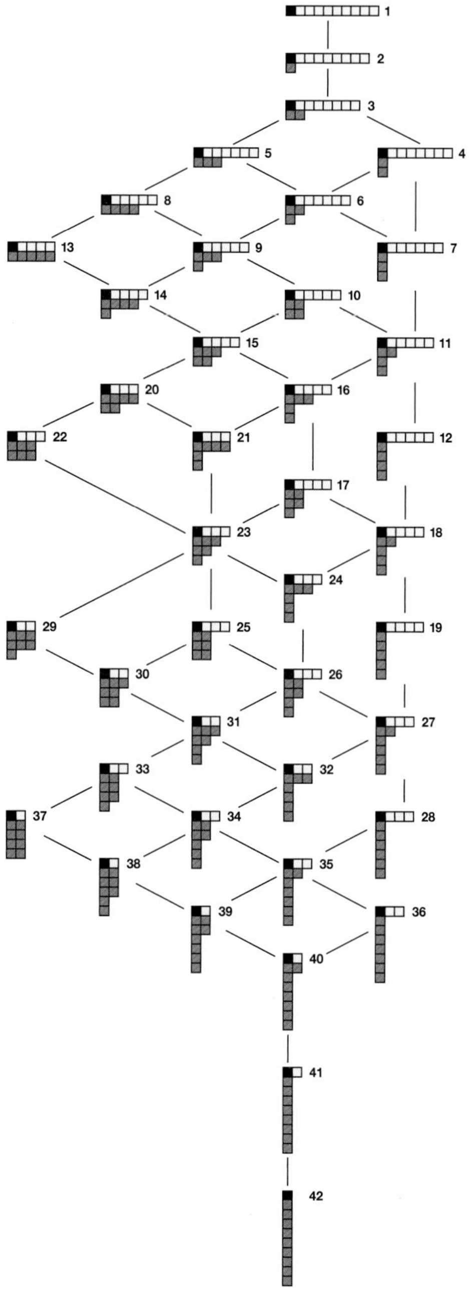

Figure 2.8.1

Ernst Ruch's partition diagram

The "game" is described in the text.

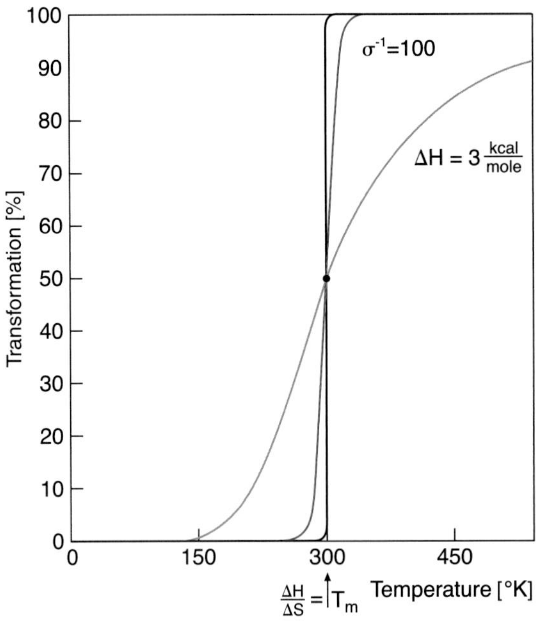

Figure 2.8.2

Co-operative versus non-co-operative chemical transformations

The blue curve describes the extent of transformation according to the temperature dependence of the constant K of the law of mass action (d lnK/dT $= -\Delta H/RT^2$). This corresponds to a simple chemical equilibrium A $\leftrightarrow$ B with $\Delta H = 3$ kcal mol^{-1}, while the red curve refers to the co-operative transformation of a polymeric state where for each monomer ΔH is of the same value (3 kcal/mol^{-1}), but the transformation next to a state already reacted is favoured by a factor 100. The polymer has to contain substantially more than ten monomers (see text). The black line describes a transformation involving very many monomers: this transition is still continuous, but because of the large number of monomers it is now very steep and thus appears like an “all or none” transition.

of only one hydrogen bond in the formation of the ordered structure (in aqueous solution).

The example so far is not particularly exciting, and it was chosen only for the sake of comparison. In the 1950s and 1960s we studied the kinetics of the helix–random coil transition of a number of oligo- and polypeptides as well as of polynucleotides of various lengths. The above example would be typical for the initial step of helix formation “nucleating” a hydrogen-bonded structure. The further steps of helix formation proceed in a co-operative fashion. In other words, the stability (and rate) of any subsequent bond formation next to a bond already formed appears to be enhanced by some factor. The red curve in Figure 2.8.2 corresponds to a 100-fold increase in bond strength. Theory tells us that the steepness of the transition curve is proportional to the square root of this enhancement factor, which in case of the red curve means a ten-fold increase in steepness. The curve behaves as if the enthalpy difference ΔH° has increased by a factor of ten as compared with the blue curve. To return to reality: in our experiments with nucleic acid helices we found enhancement factors up to 10,000, meaning that the co-operative block includes on average as many as a hundred (i.e. $\sqrt{10000}$) monomeric units. The α-helices of polypeptides behave in a similar fashion: they form and melt as co-operative units. However, the word "melting" is not quite correct, as will be seen below. Ice, for instance, melts in a discontinuous fashion at 0°C, showing a true singularity at its melting temperature in a sufficiently extended system, fulfilling the "thermodynamic limit". The black line in Figure 2.8.2 approximates such a behaviour by a continuous curve valid in finite systems or for very large ΔH values.

Co-operative behaviour has been studied by physicists, in particular in connection with the phenomena of ferromagnetism or antiferromagnetism. In his doctoral thesis, 1925 Ernst Ising[40] tried to describe ferromagnetic behaviour with the help of a simplified model. He assumed a linear chain of elementary magnets, the spins of which are in one of two states ("up" or "down"), influenced by the respective state of their nearest neighbours. From his model Ising concluded (correctly) that a linear chain can undergo co-operative changes, but no true phase transitions. For this to happen, the enhancement factor would have to become infinite, which is not possible for interactions of finite strength. Ising extrapolated his finding further, and inferred (incorrectly) that nearest-neighbour models for any dimensionality greater than 1 (being more realistic for ferromagnets) should likewise fail to show true phase transitions. However, in 1944 Lars Onsager[41] proved that even a two-dimensional model of this kind may undergo true phase changes. Onsager's exact computation of the partition function of the two-dimensional Ising model was a mathematical tour de force, unprecedented and surpassed only by his solution of the Ising model in the presence of a magnetic field (1948).

I take this opportunity to slip in some personal remarks. Lars Onsager was the most ingenious theoretical chemist I knew. I met him first at Yale University in 1954 at a celebration of Peter Debye's 70th birthday. He talked to me about a new idea of his, and (as always) before he came to the end he got excited and started to laugh, so that I could not work out what he actually intended to say. I must have put on a quite discontented face. Suddenly somebody tapped my shoulder. It was Peter Debye (cf. Section 1.8, p. 83), who said: "Don't try to understand him, but be sure that he is always right." Later we – that is Ruthild Winkler-Oswatitsch and I – often went on walks with Lars when we visited each other at various places in the USA or Germany. Lars could explain everything, whether it was weather, botany, zoology, the arts or history, not to mention physics. So Ruthild, whenever I could not answer a question of hers, said, "I shall ask Lars, he knows everything."

The one-dimensional Ising model had a late and unexpected renaissance with the helix–coil transition found in linear biopolymers. What is it that distinguishes their co-operative behaviour from a true phase transition? Co-operativity is one obvious prerequisite. However, co-operativity as such is not sufficient because it cannot be of an infinite strength, and hence in a linear model it does not lead to an infinite correlation length. What is missing is the instability that causes co-operativity to propagate through the whole phase. True infinity, of course, could be achieved only in systems of infinite extent, i.e. in the so-called "thermodynamic limit". The "infinite extent" of a system is something which – as an approximation – is physically realisable, other than "infinite strength" of local interaction.

Let me expand on this point somewhat further by looking at a model that was successful in describing the phase transition associated with evaporation and condensation. As early as 1881, van der Waals came up with a two-parameter equation that reproduces remarkably well the phase transitions between liquids and gases. In his doctoral thesis (1873). "On the Continuity of a Liquid and Gaseous State", he had already dealt with the subject, knowing that the ideal gas law could be derived from kinetic theory under the assumptions of point-like molecules and an absence of intermolecular forces of attraction. The two parameters he introduced were meant to take care of those simplifications: the molar volume V was corrected by subtraction of a constant hard-sphere volume term b, and the (external) pressure by adding an "internal" pressure term a/V^2. This latter term can be explained by the assumption that the free energy (F), due to attractive forces among molecules, is negative and inversely proportional to the volume. At constant temperature (T) the pressure p equals $(\partial F/\partial V)_T$, yielding a positive contribution to pressure from internal interactions that is inversely proportional to V^2. Figure 2.8.3 shows the pressure–volume relation according to the van der Waals equation. In this figure two regions are emphasised, namely the one designated to a coexistence of two phases (light shading) and the one forbidden by stability criteria for homogenous phases, i.e. $(\partial V/\partial p)_T > 0$

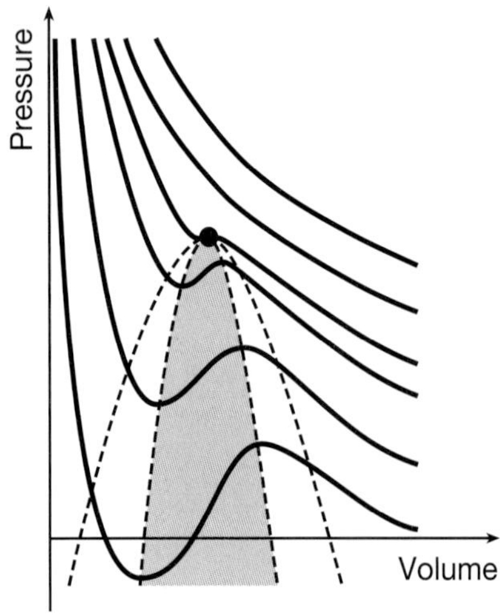

Figure 2.8.3

Van der Waals isotherms

Pressure–volume diagram for a liquid–gas transition according to the law of Van der Waals. (For a more extensive discussion see Appendix A2.)

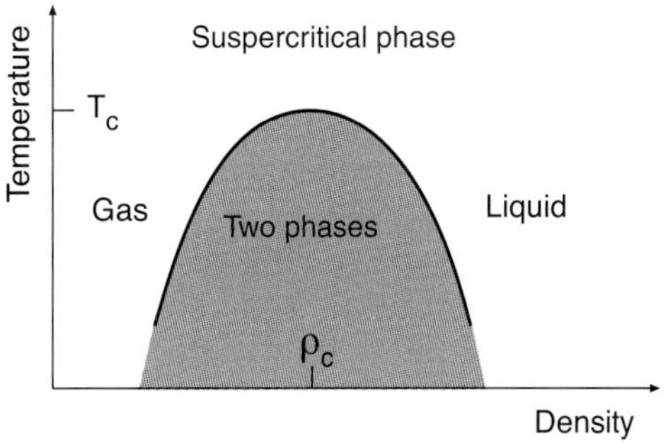

Figure 2.8.4
Temperature–density diagram near the critical point

This example describes the liquid–gas transition in the neighbourhood of the critical point, as described in more detail in Appendix A2.

(darker shading). The two regions coincide at the critical point, and it is on this that I shall now focus.

However, let me first say that it is not my intention to reproduce in this book standard knowledge from textbooks. For that purpose I refer to the excellent text by Berry, Rice and Ross[42].

The critical point to which I referred above appears in the temperature (T)–density (ρ) diagram, shown in Figure 2.8.4, or in an analytical representation as a power law:

$$(\rho_{\text{liquid}} - \rho_{\text{gas}}) \sim (T_c - T)^{\beta}$$

with β being universally close to 0.33. The most accurate measurements suggest that β – although approximately one-third – may not be a rational number. The universality behind the correct power law is demonstrated by the fact that the same critical coefficient is found for entirely different kinds of critical phase transitions, for instance in the behaviour of ferromagnets or antiferromagnets at their critical temperature (i.e. the Curie temperature for ferromagnets, or the Néel temperature for antiferromagnets).

Earlier theories had predicted a power law different from the one described above. The most successful classical approach was that of Lev Davidovich Landau,[43] who predicted a value of 0.5 for the exponent in the power law.

So what is wrong with Landau's theory? In fact, nothing is "wrong" with it. Wherever mean-field approaches apply, his theory is correct and almost unbeatable. In particular, it indicates universality of critical phenomena and in many cases yields an excellent representation of experimental data. Moreover, it tells us where it is prone to fail. Being a "mean field" approach it ignores fluctuations. All microscopic phenomena include fluctuations, but these usually cancel out if macroscopic dimensions are involved. This is not the case for critical phenomena. We all know the phenomenon of critical opalescence, which results from fluctuations of density and hence of the refractive index at all scales of length. The same is true for magnetic phenomena near the Curie temperature, or for superfluidity phenomena in ^{4}He. Fluctuations on any length scale require an iterative reconsideration of the partition function at all length scales. This is accomplished by the renormalisation group, applied by Kenneth G. Wilson.

Renormalisation was originally introduced in relativistic quantum field theory in order to remove the infinities that occur in higher terms of the solutions obtained by perturbation theory. Wilson did his thesis work with Murray Gell-Mann and Francis Low, who both came up with a renormalisation scheme for perturbation theory. Wilson's version of the renormalisation group was based on quantum field theory, but it was then given a more general physical interpretation. His colleagues at Cornell, the physical chemists Ben Widom and Michael Fisher, whose influence Wilson specially acknowledges in his 1982 Nobel lecture[44], had prepared the ground by the notion of "scaling laws", which generally apply to processes in the neighbourhood of a critical point. According to a classification by Ehrenfest, critical phenomena are called second-order phase transitions. This classification assigned an order to a discontinuity of the i-th derivative of free energy at the point of phase transition. According to the modern view, this definition fails except for the first-order process, which involves a generation of latent heat. The failure of Ehrenfest's definition is the failure of mean field assumptions. "Second-order phase transitions" involve non-analytical behaviour rather than true discontinuities. A λ-point in the temperature dependence of heat capacity is a cusp, rather than a true divergence. It was Wilson's method that showed the universality of such different phenomena as ferromagnetism near the Curie temperature and liquid–gas transitions at the critical point. It yielded the correct (presumably irrational) number for the critical exponent, to which mean field theory assigned a rational number, such as 1/2 in Landau's theory. A profound discussion of these quite recent advances can be found in Goldenfeld's[45] *Lectures on Phase Transitions and the Renormalization Group*.

Let me summarise the results of my (purely descriptive) account in three statements:

1) Phase transitions in physical space involve co-operative interactions.
2) The correlation length at the phase transition in the thermodynamic limit ($N \rightarrow \infty$) is infinite.
3) Critical phenomena involve fluctuations on all length scales. Renormalisation leads to the correct (experimentally observed) critical coefficients.

This short digression will be amended in Appendix 2 because phase transitions, caused by broken symmetries, have played an important role in the first part of the book and will play an even greater one in later parts (Chapters 4 and 5), where they will apply to entirely different spaces. Therefore it is important to know their general physical character. My point of departure was the correlation of entropy with order, for which the different types of phase transitions mentioned provide excellent examples. In the physics of phase transitions, a well-defined so-called "order parameter" has proved to be a useful descriptive tool. It characterises the state of a phase of the system below its transition temperature in terms of a non-zero quantity. Above the phase-transition point, this parameter becomes zero. Despite the fact that the term "order" defies a clear-cut, general definition, the "order parameter" in the case of phase transitions makes sense. It is a variable in the theory of phase transitions that is characteristic of the phase, and it reflects how sharply the order in the phase (whatever that may be) changes with temperature or any other analogous variable. Phase transitions and the order parameters that characterise them are related to a "broken symmetry" of the system. In this terminology, the ideal gas has a higher symmetry than the liquid. In the (ideal) gas the movement of all particles is virtually free in all directions of space. The liquid, in turn, has a higher freedom of motion and hence higher symmetry than a solid, in particular a crystal, where freedom of motion is highly constrained by gradients of potential energy that establish the bond order within the crystal.

Order here is linked to restrictions of freedom (e.g. by forces), which in the case of the crystal as compared with the liquid do not correspond to our intuition, which stresses the spatial symmetry of crystal structures versus the chaotic arrangements in liquids and gases. Broken symmetry, in an abstract formulation, refers to a situation where the state of a many-body system has a symmetry lower than suggested by the Hamiltonian that describes the system. In solid-state physics, other typical cases of broken symmetries apart from ferro- and antiferromagnetism are superconductivity and Bose-Einstein[46] condensation. In particle physics, the electroweak theory (Section 1.6) provides an example of a relativistic quantum theory with broken symmetry. The concept can also be applied to abstract spaces. In Section 4.8 it will be shown that generation of meaning can be described by a phase transition in information space: this too can be exactly described by a physical order parameter.

The entropy difference associated with a material phase transition is always related to the energy difference between the two phases. This is most obvious for so-called first-order phase transitions (Section 5.4) such as melting or boiling. First-order phase transitions involve a "latent heat" (e.g. ΔH), which means that enthalpy H (at constant pressure) changes discontinuously at the transition point. If T_{tr} is the corresponding temperature (e.g. the melting or boiling temperature), then the entropy change is given by $\Delta S^0 = \Delta H^0/T_{tr}$.

However, since entropy refers to phase space while our intuitive concept of order is focused on physical space–time, it may be misleading to call entropy a measure of "order". Let me now give some examples that run counter to the above intuitively obvious cases. They are taken from an article by Stephen Berry[47] called "Entropy, Irreversibility and Evolution".

The spontaneous crystallisation of sodium thiosulphate from a supersaturated solution is accompanied by an increase of entropy (McGlashan[48]). However, what is being observed is the increase of spatial order that occurs when the molecules of solute and solvent are separating. Of course, what we do not see is the concomitant change in water structure that contributes to the entropy balance. Another example of this kind, which is relevant in biological organisation, is the problem of hydrophobic interaction, which results mainly from exclusion of water structure. The formation of (spatially highly ordered) biological membranes is an entropy-driven assembly of lipid molecules from their previously "disordered" state in aqueous solution.

In particular, Berry mentions the complete breakdown of the simplistic correlation of entropy with disorder when gravitation comes into play. A star shows a radial density gradient that can be thought of as a kind of spatial ordering; but it refers to the state of highest entropy. Uniform density would not represent a stable state for a star, since then its entropy would be lower. A very spectacular case, the black hole, is discussed in the next section.

Of course, one could define the order and disorder of a particular state by referring to its amount of entropy. However, defining order through entropy and then saying that entropy expresses order would entail a circular argument. The crucial point is that entropy is defined for the phase space that is spanned by positional and their conjugate momentum co-ordinates, while our intuitive comprehension of order is usually limited to geometrical space. To say that ordering is correlated with reducing and disordering with increasing the "phase volume" is neither very poetic nor very illuminating. But it is correct, even if it is not as charming (and vague) as a sentence I recently read in a text on energy: "Entropy is the physical scientists' way of quantifying chaos."

How grotesque this "quantification" of chaos can be is shown in the following story of a well-meant attempt to illustrate entropy. I found it in an otherwise excellent physical text. I quote:[78] "As an example, suppose that we have a room containing air and a few crumpled-up newspapers. The air and the paper together contain less entropy than they would have if the paper were burned in the air to form carbon dioxide, water vapour and a bit of ash." So far, so good! The howler comes with the conclusion of the story, which reads: "That (i.e. the increase in disorder) is why the paper burns naturally... Entropy increases during burning; entropy would decrease during unburning; thus burning occurs and unburning does not." Physicists sometimes like to flaunt their contempt for chemical details, so let us look at these. Newspaper is indeed a chemical compound, albeit not very well defined, with cellulose as its basic constituent. Cellulose is a polymer of glucose, $C_6H_{12}O_6$. When the latter burns, the products of combustion indeed have a higher entropy than the starting materials ($\Delta S° = 259.0\ \text{J}\cdot\text{mol}^{-1}\ \text{K}^{-1}$ under standard conditions). This contributes $-T\Delta S° = -77.2\ \text{kJ}\cdot\text{mol}^{-1}$ to the combustion reaction. However, the enthalpy change on burning, i.e. the heat evolved, is about 40 times greater ($\Delta H° = -2801.6\ \text{kJ}\cdot\text{mol}^{-1}$, adding up to $\Delta G° = -2878.8\ \text{kJ}\cdot\text{mol}^{-1}$). The numbers under conditions other than standard ones (within realistic limits) are a little different, but they change neither the order of magnitude nor the conclusion to be drawn. Thus, our physicist friend has correctly considered the contribution of entropy to driving the reaction, but by omitting a "minor" chemical detail he has considerably exaggerated its importance. It is the large enthalpy content rather than the "increase of disorder" (which after the heat has been absorbed is hardly visible) which explains "why the paper burns naturally".

Erwin Schrödinger could be accused of a similar neglect, when he says in his book *What is Life?* that we maintain the order of life by feeding on "complicated organic compounds" representing an "extremely well-ordered state of matter". He might have been thinking of the proteins in meat. However, if we are ill, then an infusion of glucose solution will do the job perfectly. The "currency" of our metabolism is ATP (adenosine triphosphate), which is replenished mainly from the energy content of chemical bonds. Of course, Schrödinger knew this, and the charge that he neglected free energy is not justified. As was shown in Table 2.1.1, the use of Gibbs free energy G(T,p) is the most convenient form (and the safest way) to deal with reactions that take place at constant temperature and pressure, but an explicit account of enthalpy and entropy would do as well. Schrödinger, who lectured on statistical thermodynamics, was well aware of this. It was his exaggeration of the entropy rather than energy content of food, or – as he put it – the necessity "to extract order from the environment" which met with opposition. Hence Schrödinger added a note in a later edition in which he acknowledged the criticism, in particular by his colleague Sir Franz (Eugen) Simon, with the words: "Simon is quite right in pointing out to me that actually the energy content of food *does* matter". The French biologist (and Physiology Nobel Laureate of 1965) André Lwoff[49] was more outspoken, commenting: "If fed with pellets of negentropy ... even a physicist would succumb."

2.9. Who Keeps Our Clocks Running?

In his lectures on physics Richard Feynman[50] asks the question: "What is time?" One of the answers he offers is: "Time is as long as we wait", and surely, Mr Feynman was joking. In order to find out "how long to wait" we need a clock. Everybody knows: "A watched kettle never boils"! That's why Feynman continued: "What really matters anyway is not how we define time, but how we measure it."

A definition of time is as hard to produce as a definition of energy. In fact, the two quantities are complementary to one another. Quantum mechanics relates the uncertainty of time to the uncertainty of energy, the product of the two being greater than or equal to $\hbar$, Planck's constant divided by 2π. I am now asking: who (and I really mean "what") keeps our clocks running? And in particular: "running in one direction"! The dynamical laws that underlie the mechanism of any clock are symmetrical with respect to time. If we replace t by $-t$, they don't change.

Stephen Hawking, in his popular book *A Brief History of Time* specifies three "arrows" of time. The "psychological arrow" is the direction in which we remember the past but not the future. It is based on the conscious processing of what we perceive and remember. Newton's characterisation of time, namely an absolute time, which flows continuously and independently of any external event, was an abstraction from what he could measure using a clock. On the other hand, mechanics proved to be invariant with respect to the direction of the "flow of time". Relativity rejected the absolute character of time, but maintained its invariance against reversal, just as quantum mechanics never challenged the latter.

Hawking distinguishes the psychological from the "thermodynamic arrow of time". The only physical law proposing an explicit arrow of time is the second law of thermodynamics, in Clausius' provocative formulation: "The entropy content of the world tends to a maximum."Since entropy is based on "molecular mechanics" this statement – apart from its premature extrapolation to the world as a whole – seemed to involve a paradox: how could a mechanics that is invariant with respect to the direction of time produce a law which implies a direction of time?

Boltzmann, who tried to resolve this difficulty by showing that thermodynamic irreversibility resides in the statistical nature of molecular dynamics, could not convince his critics although, in principle, he knew that he was right. Entropy rise is linked to an increase in the number of populated microstates, but it was hard to put up a clear demarcation between fluctuation and irreversibility. Today, the matters that worried Boltzmann are less pressing. Dynamical chaos (cf. Figure 2.3.1) makes most trajectories practically irreversible (Prigogine[51]). To put it more clearly, it is not chaos that saves the thermodynamic arrow of time from being abolished by the symmetry of the laws of mechanics. It is rather the steep gradient in the distribution curve of microstates that produces the dissymmetry and forces fluctuations to decay towards the state of maximum entropy. What chaos does is to invalidate the argument that time reversal, which leaves the laws of mechanics unchanged, but never occurs in reality, must produce a reversal of all possible trajectories. It does not do so for any extended span of time because of the dissymmetry of statistical interference taking place in real time. Hence, the naturally irreversible trend towards equilibrium is caused by "statistical forces", and these are responsible for establishing the "thermodynamic arrow of time". An isolated system moves towards equilibrium, while an equilibrated system (without external interference) never "de-equilibrates". At equilibrium there is complete equivalence of past and future, based on the inherent statistical reversibility of all microscopic processes.

Clausius' courageous statement regarding entropy, in fact, anticipates Hawking's third arrow, the "cosmological arrow of time". In Clausius' time it was interpreted as the "heat death of the universe", according to Carl Friedrich von Weizsäcker, a "scene of devastation with ruins and skeletons", rather than some "homogeneous mash". The modern view differs quite a bit from this "classical" picture: the ultimate state of the universe may be a universal black hole. General relativity leaves three possibilities for a cosmological time arrow, depending critically on the total mass content of the universe and on the initial curvature of space. The three are: expansion for ever, stationary existence and one, or several, cycles of expansion and contraction (see Section 1.8).

General relativity does not allow the universe to be static. It rather predicts that space–time had a beginning at the singularity of the "big bang" and, if its mass–energy is "supercritical", comes to an end in the singularity of a "big crunch". In other words: a dynamic universe according to Einstein's theory must include singularities in the form of points and finite boundaries. This was proven in a paper by Roger Penrose and Stephen Hawking.[52] A singularity at the big bang, for instance, would mean that the universe at its beginning would have been infinitely dense and infinitely hot, having zero entropy. Those conditions at the beginning or end of the world would have caused all physical laws to break down. Hawking, however, has also pointed out that there is no need for true reversibility. The universe may as well end up at zero temperature and with a stupendous amount of entropy, as associated with a universal black hole.

If those states were to be describable by laws of quantum gravity, a description of the history of the world would require a knowledge of its initial boundaries, even if it is only describable by a superposition of all possible histories, as was proposed by Richard Feynman according to the requirements of quantum mechanics.

According to Stephen Hawking and Jim Hartl,[53] a quite different situation may arise if the time axis can be assumed to be imaginary, i.e. if t can be replaced by $i \times t$ ($i = \sqrt{-1}$, the imaginary unit). We remember (see Section 1.2) that Hermann Minkowsky, in his seminal paper "Space and Time", mentioned a similar idea, arguing that the greater symmetry of such a space might simplify the physical laws to be expected. In particular, such a substitution could restore an Euclidean metric for which there is no "beginning" or "end". Hawking refers to such a world as being governed by "no boundary boundary conditions". He quotes the example of the surface of our planet which, although being finite ($4\pi R^2$ with R being the radius), has no boundary. One never reaches any edge, just as the north and south poles do not represent point-like singularities.

In a similar way, three-dimensional space could be the hyperface of a four-dimensional "sphere" of volume $\pi^2 R^4/2$ (see Section 4.5). Such a "sphere" would be demarcated with respect to the surrounding space by its three-dimensional surface, which itself has a finite size of $2\pi^2 R^3$ and no boundary. The same is true for a four-dimensional space bent into a fifth dimension, in which a "sphere-like" volume has the size of $\frac{8}{15}\pi^2 R^4$ with a finite, but limit-less hyperface of $\frac{8}{3}\pi^2 R^4$. Such finite spaces could be dynamic in (imaginary) time, but they would never include internal singularities in the form of edges or point-like infinities. Any being that was able to sense all but the radial dimensions could advance in this space without ever reaching a boundary.

Unfortunately, such spaces will remain "imaginary" to us unless we find experimental means to test them. Misner, Thorne and Wheeler[54] in their monumental documentation *Gravitation* (p. 51) include a (black-framed) obituary under the title "Farewell to 'ict' " (ict being the imaginary coordinate of time).

Is the cosmological arrow related to the entropic arrow of time? The Australian philosopher Huw Price[55] argues that what deserves attention is not the increase in entropy in the future, but rather the low entropy in the past. He calls future and past human prejudices. The question would become particularly relevant if the universe at some time were to stop expanding and start to contract.

Price accuses Hawking of failing "to apply an atemporal view in a consistent and thorough way". He alleges that Hawking (and other contemporary cosmologists) make the "mistake" of using a double standard of arguments when looking into the future and the past. According to Price, physics must be symmetric with respect to past and future.

But why should it be? Why should there be a symmetry, typical of reversible processes as are present on average, in equilibrated systems? Hawking has made clear that even in a closed universe there need not be such a symmetry. And the kaon (see Section 1.9) seems to "know" the difference between past and future – for our present understanding still a mystery. Certainly, this is not a playground for philosophical contemplation (however philosophy may understand itself today). It is a matter of serious physical discernment.

The most persuasive argument for Hawking's "irreversible" scenario derives from the entropy of a black hole. Stephen Hawking had realised that the area of the event horizon of a black hole could never decrease. Sucking in matter or radiation energy (or merging with another black hole) would always cause an increase in its surface area. More important, Hawking found out that the light rays forming the event horizon have to run parallel to or away from one another, which prevents the black hole from contracting because doing so would violate the conditions for at least some of these light rays.

A physical quantity's property of never decreasing is well known from the second law of thermodynamics, which states that in certain systems, on a macroscopic scale, entropy can never decrease spontaneously. Hawking concluded that, if there was a relation between the surface area of a black hole and its entropy, then there must also be a finite temperature associated with the black hole's surface, and hence it must radiate away some heat. Hawking's eventual conclusion was: "Black holes aren't so black". The resulting surface temperatures, depending on the size of the black hole, turned out to be of the order of magnitude of some millionths of a degree Kelvin or less. Since the microwave radiation background of the universe corresponds at present to 2.7 K, the black hole absorbs more energy than it radiates out. What I am presenting here in purely descriptive form is the outcome of solid calculations. The mechanism of black-hole radiation is fed by quantum fluctuations, as described in Chapter 1.

Meanwhile (i.e. before Hawking could satisfy himself of the existence of black-hole radiation) a graduate student of John Wheeler's at Princeton, Jacob Bekenstein,[56] was wondering what happens to matter that falls into a black hole. Does its entropy get lost? Inside a black hole there is nothing that could distinguish any particles of matter from one another. So would this mean a violation of the second law of thermodynamics? Bekenstein had the right idea. He came across Hawking's finding of the increase in the black hole's surface area as a consequence of the uptake of matter, and he correlated it with the increase in the black hole's entropy.

The numbers that result for the black hole's entropy are gigantic if we compare them with conventional entropy data. Roger Penrose[57] calculated the entropy of the universe (including a mass equivalent to about 10^{80} protons), given the presently measured temperature of background radiation of 2.7°K, to be about 10^{88} in absolute entropy units (i.e. setting Boltzmann's constant to 1). If the total mass of the universe were to form a giant black hole its entropy would be 10^{123} (same units). That is an increase of 35 orders of magnitude. If such a black hole were the final state of our universe, the second law of thermodynamics would certainly have held true: "The entropy *of the world* approaches a maximum."Bravo, courageous Clausius! However, such a fate of the universe is only one of the possibilities discussed by astrophysicists.

Entropy here appears in a new light. We remember that Edwin Jaynes warned us to call it simply a measure of "disorder", because otherwise we would have to define carefully what we mean by order. He further emphasised its anthropomorphic nature, argued also by Eugene Wigner (Section 2.2). In order to interpret entropy we must specify the variables of which entropy is to be considered a function. Is a black hole, in which matter is downgraded into its most elementary constituents, and of which we just know the mass, charge and spin, a less "ordered" state than our present universe? And what is the meaning of the Bekenstein–Hawking entropy? It is the surface area of the black hole divided by four times the Planck area, i.e. something like the number of little squares of Planck size that fit the black hole's surface or, converted to bits, the number that would be required to specify all elementary constituents of which the black hole consists. Does this have any real meaning, as suggested by Figure 2.9.1, or is it just an analogy? Beware of anthropomorphic interpretations before one can account for them in terms of calculable numbers!

When entropy was first interpreted in terms of statistical mechanics, an ideal gas could be regarded as composed of microscopic billiard balls. With molecules, there came contributions from rotations, which at room temperature do not completely "get going" in the classical sense, while vibrations are not excited at all. This changes with rising temperature, and in a plasma even electrons contribute to the specific heat. What are the states that are responsible for the huge entropy values? We didn't know – until January 1996!

Andrew Strominger and Cumrun Vafa[58] considered a special type of black hole, one that is endowed with charge and with just the minimum content of mass required to carry the charge. They showed that these black holes involve configurations, the

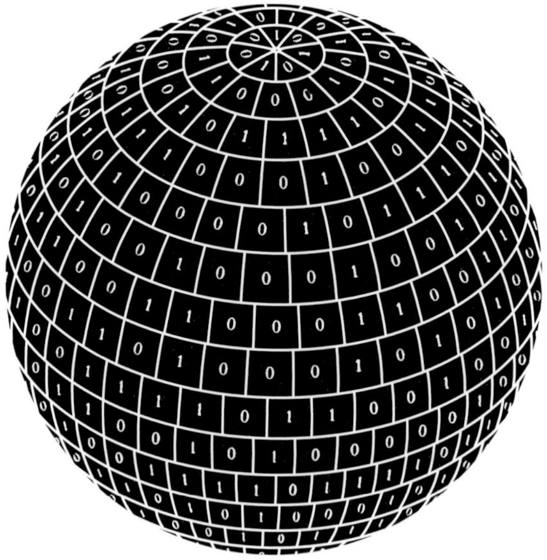

Figure 2.9.1
The entropy of a black hole[54]

According to Bekenstein and Hawking, the surface of a black hole expresses its entropy. In this picture this is symbolised by partitioning the surface area of the event horizon of a black hole into 2^N domains, each having a size of 2.77 times the Planck area (see Section 1.8), being equal to $4\hbar$ ($\hbar$ is Planck's number divided by 2π). Wheeler draws far-reaching conclusions from this representation in elementary binary units and we shall revisit this in Section 5.10.

properties of which are uniquely determined by arguments rooted in supersymmetry (Section 1.9). Strominger and Vafa "constructed" such black holes "starting from scratch" by adding together all strings with defined states of specified dimensions (cf. Greene[59]), in the same way as one would construct an atom starting with quarks and electrons, and fitting them together up to the level where electrons orbit a nucleus. In the black hole, the constituents are called "branes", a word derived from "membrane". An abstract membrane is a planar object in which two dimensions are expanded while the third dimension is of (practically) zero extension, analogous to a line in one dimension. "Branes" in the superstring theory are membrane-like states of higher spatial dimensions up to ten (see Section 1.9). These so-called BPS-states in supersymmetric theory are configurations having properties that can be exactly determined from symmetry considerations. BPS refers to the names of the authors of the BPS-theory, B to Eugéne Bogomol'nyi from Moscow/Russia, P to Manoj Prasad and S to Charles Sommerfield, both from Yale University, New Haven/USA.

The results were obtained independently by the Russian and the American Groups. Strominger and Vafa designed their black holes from those states and counted the number of possible rearrangements that leave the overall properties unchanged. These numbers could be correlated with the black hole's surface area and yielded the Bekenstein–Hawking entropy (see Section 5.10).

What about Hawking's two other "arrows of time"? While the laws of mechanics are symmetric, there is one law, based on motion, that explicitly postulates an arrow of time. It is the second law of thermodynamics, which states that all spontaneous macroscopic processes proceed with an increase in entropy. Our ability to distinguish between past and future has been associated with a "psychological arrow of time". It is based on the faculty of the brain to store and memorise information that applies exclusively to the past. From all we know about our brain (which will be discussed in later parts of the book) we can say that this psychological arrow cannot be seen independently of the thermodynamic arrow. Apart from all its miraculous qualities, the brain is also a "thermodynamic system". As such, it cannot possess properties that are incompatible with the laws of physics and chemistry.

On the other hand, the psychological arrow of time establishes a "strong" temporality that is not necessarily required by the thermodynamic arrow. A system in steady state shows a constant entropy production, and near equilibrium entropy change varies linearly with the deviation from equilibrium. In these cases one could easily compensate for entropy production by external entropy flows. The resulting "temporality" I call "weak" in contrast to non-linear situations where instabilities or other irreversible phenomena are involved. The popular example of the broken pieces of a plate or cup fallen to the ground, which cannot (spontaneously) reconstitute themselves, belongs in this category of non-linear "instabilities". The precise neural mechanism of recording the past is not yet well known. The brain indeed creates at every instant a new "picture" of our world (Donald McKay), the "instant" being an amount of time between a millisecond and a second. Apart from this, we inherit a diurnal clock that is adapted to (and synchronised by) our planetary cycle.

Another biological time arrow is that of evolution by natural selection. As will be seen in Chapters 4 and 5, natural selection introduces instabilities that cause the "strong" temporality of life, reflected by irreversible changes in the time course of evolution. An evolutionary tree, with its fan-shaped topology, represents an irreversible divergence of the underlying genetic process. This does not exclude convergent adaptation, for instance certain proteolytic enzymes which had to adapt to similar functions and thereby developed similar active centres. However, their evolutionary histories, as documented by the DNA sequences of their genes, appeared to be entirely unrelated. On the other hand, the compound eye of insects and the lens eye of mammals, often quoted as "witnesses" of an adaptive, non-Darwinian evolution, were recently shown in a series of exciting papers by Walter Gehring and his group[60] to have descended from a common precursor.

So, "who" keeps our clocks running? After reading more than just this question one might ask: "Who is who?" Since I am not a theologian, let me modify my question: whoever "who" is, what does he have to do in order to keep our clocks running? Or in short: "what" keeps our clocks running?

We have – according to Stephen Hawking[61] – distinguished three clocks: the "cosmological", the "entropic" and the "psychological". From what I say nearer the end of this section, the strong temporality of the "psychological" clock is certainly based on the non-equilibrium nature of all processes of life and thought, as will become clearer in subsequent chapters of this book. But what about the "cosmological" clock?

I see only one common possible physical cause for all three clocks: the cause that is expressed in the entropic law, to be interpreted in the almost tautological way of saying: every material system tends to approach its "most probable" state. "Most probable" is that macrostate that is represented by a maximum number of microstates.

However, there is a caveat. One might initially conclude that time must eventually stop when complete equilibrium (including "detailed balance") is reached. But what about large-scale fluctuations? Boltzmann once thought that such a large-scale fluctuation might explain our existence. It is true: the larger the scale of such a statistical fluctuation the less probable it is. But we cannot exclude it. A completely reversible world with all the "negative conclusions" of general relativity discussed in Section 1.8 would end up in the "foam" of fluctuations of quantum vacuum. And that is where our present-day wisdom ends.

2.10. Entropy: What Does It Mean?

This book started with the question "What is energy?" Some readers might have been puzzled by the question, and even more so by the answer, which was: "We don't know". Isn't energy familiar to everyone in daily life? Even politicians talk about energy, how precious it is, and that we shouldn't use up too much of it. Physically, that is not possible, because of the first law of thermodynamics. So what they mean is that we should not produce too much entropy. But if they were to say that, not many people would understand them, and they would have no chance of being re-elected.

Energy is given to us by the reality of Nature. Entropy, on the other hand, is a human construct, albeit a necessary one if we want to understand Nature. Hence, since we are able to define entropy, we must know what it is. Entropy (S), according to Clausius' phenomenological definition, is a function of the internal energy (U) and the volume (V) of the system, i.e. S(U, V). It is closely associated with its complementary variable, temperature (T). The product of temperature and entropy has the dimensions of energy. When complemented by a scaling law (Section 2.1), entropy expresses an "extent", or amount, of (thermal) energy, while temperature is a measure of its "intensity". The work of Clausius introduced the new qualities taken on by matter when it appears in the form of macroscopic ensembles. The statistical definition of entropy tells us about the capacity of matter for taking up energy into all its different quantum levels, while temperature is a measure of how much energy on average is contained in every available individual degree of freedom. The idea of temperature makes sense only for an equilibrated system, while entropy, representing an average that refers to the macroscopic systems, can also be defined for non-equilibrium systems. This statement suggests that a deeper understanding of the qualities of macroscopic matter requires a statistical inspection of its microscopic mechanical properties.

The two laws of thermodynamics – again, underivable principles – were formulated explicitly by Rudolf Clausius. They reflect the two characteristic traits of energy, namely, to conserve strictly its amount in a closed system (first law) and to disperse itself to all available internal degrees of freedom (second law). The latter property causes entropy to approach the greatest possible value in the equilibrated system. Clausius' formulation of these laws referred to the world as a whole, which must have seemed to him the only system that can exist in true isolation. Apart from this, he could not have known much about our universe. The second law – with the premise of the first law – then suggested to him that the universe will eventually end up in a timeless equilibrium state, called "heat death". Modern cosmology has revivified this question and replaced "heat death" by a "frozen" phase of matter, the black hole, in which time (and space) would cease to exist. I call it "frozen" because its surface, representing a huge entropy, is extremely "cold". I refrain from speculating further about the yet undecided future of our universe (i.e. whether it will keep on expanding for ever, or re-contract to a black hole, or whether it has an eternal steady or oscillatory existence). Rather, I wind up now by reviewing various aspects of entropy and their consequences. In order to do this I have to digress first to its statistical basis of entropy, as clarified essentially by Ludwig Boltzmann and Josiah Willard Gibbs.

Entropy in dictionaries is often related to "randomness", "disorder" or "chaos". These interpretations are "not even wrong". They include pitfalls, which have been demonstrated particularly in Section 2.8. The statistical interpretation of entropy, given a macrostate, is based on the probability distribution for all possible microstates of the system (Section 2.3). The width of the probability distribution is $\sim\sqrt{N}$, where N in simplest cases is the number of particles (which define degrees of freedom) involved. The relative width $\sim \frac{1}{\sqrt{N}}$ then decreases as N increases, and hence entropy measures the statistical weight of the macrostate (Section 2.5) in terms of microstates. In the thermodynamic limit (i.e. for an infinitely large system) entropy is not of a probabilistic nature, but is a well-defined, deterministic variable of state.

An entropy can be assigned to any probability distribution:

$$P = (p_1, p_2, \ldots p_N) \text{ with } 0 \le p_i \le 1 \quad \text{and} \quad \sum_{i=1}^{N} p_i = 1.$$

Then, entropy is nothing but the average logarithm of all the probabilities in a (possibly very steep) distribution. In fact, we shall encounter entropy in many other situations, ones which are hardly related to thermodynamics. In thermodynamics entropy is related to an energy distribution among quantum states, which, for instance, differs from the distribution of letters in a language. This means that we find entropy both in statistical thermodynamics and in information theory. Temperature, on the other hand, is uniquely characteristic of thermodynamics and makes sense only for a probability distribution of energy that has settled into equilibrium.

What I have just said may become more convincing when we look at the distributions themselves, and when we realise how closely the systems that we are used to dealing with in practice resemble the "thermodynamic limit".

We have been impressed by the vast numbers we obtain in counting microstates, numbers that exceed any practical realisation in our perceivable world. Ehrenfest's urn game, played with just 64 glass beads, as described in Section 2.6, involved some 10^{19} microstates. In order to populate them all by throwing dice we have to live for a longer time than our present universe. Does then the statistical aspect of entropy provide us with an access to complexity? The staggering answer is "no"! The larger the complexity of the distribution, the simpler its appearance. The whole distribution coalesces to a sharply defined line representing entropy. How come? Think of playing the game, with not 64 but a million beads. We would again get a Gaussian distribution curve with a certain half-width. In absolute terms, the half-width is a thousand times larger than in the game we played, but in relative terms the distribution is a thousand times sharper. In our daily experience with matter, numbers are of the order of magnitude of Avogadro's number, i.e. 10^{24}. The distribution curve is now a sharp line, its relative sharpness being one in 10^{12}, or one in a million million. These numbers defy our imagination. What is one cent for somebody who owns (or owes) ten thousand million (10^{10}) dollars? Only a few people in our world are that rich. But the physical world is extremely fine-grained, and that is why entropy usually has an extremely sharp value.

Why did I say that those sharp distributions also remove complexity? Isn't $\sqrt{N}$ for very large N still a large number in absolute terms? Yes, it is, but the microstates that get populated in such a distribution are nearly all of the same macroscopic appearance. In an Ehrenfest game, played with two sorts of glass beads, almost all microstates involve approximately equal numbers of both beads, all randomly distributed on the board. And that, translated into chemistry, would cause them to behave identically. The only evidence entropy would provide in this game is contained in the length of the sharp line into which the distribution curve has coalesced.

It tells us about the number of beads involved (or positions on the game board). This game has 2^N equivalent "microstates". Hence the entropy in this simple case would amount to N bits. What this means in terms of information will be the subject of the next chapter.

It is important, of course, that we do not draw too many conclusions from too simple models. If we were to play more realistic games, for instance if we in addition introduced interactions among the beads, or if we allowed co-operativity with varying correlation length, then we would obtain more sophisticated results, which would be reflected in the values obtained for entropy. Additional constraints, reducing the number of possible microstate configurations, would be reflected in lower entropy values. Nevertheless, in such a situation entropy is characterised by a single number that is an average representative of a possibly extremely complex situation. This number does not tell us how complex the situation is, and it does not tell us anything about the even more complex, chaotic dynamics that may be involved. Its information content is so low as just to remind us of how much we do *not* know. It is like asking a telephone operator for a number and getting as an answer "23 bits". I do not say this in order to denigrate the importance of the entropy concept in statistical physics or in communication theory; rather, I want to demonstrate what it tells us and what it cannot tell us. This will be very important for later parts of the book. Only the precise definition of entropy can tell us where it applies and where it does not. By no means does it offer any true access to complexity as such. It is the compression of complexity into one single number, an average characterising a macrostate. It does not allow us to identify any individual microstate among the vast number of possible ones.

One of the main themes of this book is the marvel of life. Life is a phenomenon that requires penetration into the complexity that is represented by individual states of matter. Life is specified complexity. However, it also has general prerequisites. Life cannot exist in a system that does not exchange energy and matter with its environment. Any living system in the thermodynamic sense is an open system, feeding on free energy. Hence, the existence of life does not contradict the second law of thermodynamics. This means that the biosphere steadily produces entropy, which, at stationary temperature conditions, must be compensated for. In Section 2.8 we learned that the major part of this entropy comes from an uptake of chemical bond energy. The primary source of this energy is sunlight, harvested by microorganisms and by plants. Let's find out how.

A hollow space, the walls of which are at temperature T, contains electromagnetic radiation – so to speak a "quantum gas" – of an energy density u(T), which is uniquely dependent on T. As was emphasised in Sections 1.2 and 1.3, light quanta do not have any mass, but they do possess a finite momentum, and hence they exert a pressure p on the walls of the hollow space. The phenomenon as such is most intriguing. We may remember that the pressure wave set off by the blast of an atomic bomb is one of the main sources of its destructive power. What I want to emphasise is the pressure that is always present, even in the complete absence of matter within a hollow space, i.e. in a perfect vacuum, meaning that a true void does not exist in our world; it cannot exist on the premises of quantum physics.

In the present context there is another non-trivial consequence of the finite pressure of black-body radiation. The energy density involves a finite work term – to create a volume V at finite pressure p requires work (i.e. $\int_0^V pdv$). Hence there is a corresponding (negative) free energy $\left(-\frac{1}{3}V{\cdot}u(T)\right)$ and a finite entropy term $\left(-\frac{4}{3}\ \frac{Vu(T)}{T}\right)$. Since the latter contains temperature in the denominator, the entropy of high-temperature radiation is lower than that of low-temperature radiation of the same energy: $V{\cdot}u(T)$. In other words, at balanced total energy, more light quanta of lower energy are required, representing a higher entropy value (cf. Vignette 2.10).

This is the situation we find on earth. We receive high-energy, or low-wavelength, quanta from the sun, which correspond to the sun's surface temperature of 5770 K. The maximum of the Planck curve (Section 1.3) at this temperature lies in the visible range of the spectrum, at wavelengths around 0.4 μm, corresponding to energies of about 3 eV. The earth, on the other hand, emits electromagnetic radiation into space, which has its maximum at a wave-length near 10 μm, i.e. in the far infrared range, corresponding to the average surface temperature on earth of somewhat below 300 K. The corresponding energy of radiation quanta is only about 1/10 eV. Hence we receive high-energy light from the sun and emit low-energy light into space, which at constant energy represents a correspondingly higher entropy. This is a rough estimate, and it rests on the assumption that our planet is on average in balance with respect to the exchange of radiative energy. Yet it corresponds to a continuous net export of entropy which, considering the surface area of the earth, is of the order of $1\ J\ m^{-2}s^{-1}\ K^{-1}$.

The conclusion is that the export of entropy through radiation compensates for the average production of entropy and thus allows steady, but non-equilibrium, conditions on earth. These conditions are a prerequisite for the evolution of life. Life itself represents an entropy-producing, non-equilibrium state of matter. I am not saying that it is the mere conversion of sunlight to infrared (heat) radiation that maintains life on earth. What I have considered so far is just the net balance of energy conversion, which applies to any planet, whether or not it is endowed with life. Moreover, it is a balance that is subject to large fluctuations, expressing themselves in long-term climate variations. Life, the existence of which is adapted to such overall environmental conditions, enabling it to come about and to persist, depends on many other prerequisites, of which energy conversion is only one. Solar energy is directly utilised in photosynthesis, where it is primarily converted into chemical bond energy. There follows a long chain of interconversions of free energy involving entropy production until it ends up in the form of thermal motion, which is the source of the export of its radiant energy into space. Figure 2.10.1 reproduces an artist's reflection upon all these facts caused by sunlight and ending up in the complex forms of life.

Vignette 2.10.1

Integrability: The function $Y(x_1, x_2 \ldots x_n)$ of n variables x_k changes if an x_k changes by dx_k.

Then $a_k = \partial Y/\partial x_k$ for all coefficients a_k yields

$$\frac{\partial a_k}{\partial x_j} = \frac{\partial a_j}{\partial x_k}$$

This is the integrability condition for $Y(x_1 \ldots x_n)$

Independent variables: T and V:

$$dU = \frac{\partial U}{\partial T}\, dT + \frac{\partial U}{\partial V} \cdot dV$$

$$dS = \frac{dU - \delta A}{T}; \quad \delta A = -pdV$$

$$dS = \frac{1}{T}\frac{\partial U}{\partial T}\, dT + \frac{1}{T}\left(\frac{\partial U}{\partial V}\, dV + pdV\right)$$

Integrability: $$\frac{\partial}{\partial V}\left(\frac{1}{T}\frac{\partial U}{\partial T}\right) = \frac{\partial}{\partial T}\frac{1}{T}\left(\frac{\partial U}{\partial V} + p\right)$$

or $$\frac{\partial U}{\partial T} = T \cdot \frac{\partial p}{\partial T} - p$$

Boltzmann's derivation of the radiation law: Electrodynamics yields $p = \frac{1}{3}u(T)$

Here : $u(T) = \frac{U(T)}{V}$ = radiation power density.

Hence with $U = V \cdot u(T)$ and $p = \frac{1}{3}u(T)$

we get $\frac{1}{3}\left(T \cdot \frac{\partial u}{\partial T} - u\right) = u$

or $T\frac{\partial u}{\partial T} = 4u$ and integration yields: $u = CT^4$

where C is an integration constant.

This is the Stefan–Boltzmann law.

U = (internal) energy; u = energy-density; T = temperature; S = entropy; V = volume; p = pressure; δA = work ($\int pdV$)

Figure 2.10.1

Alexander Calder, "Evolution" (Author's collection.)

More detailed estimates of the processes mentioned above have been pursued by several research groups of which I name but three: those of the late Ilya Prigogine[62] in Belgium, the late Mikhail Volkenstein[63] in Russia and Werner Ebeling[64] in Germany. I am stressing these questions in this summary of Chapter 2 as further examples that shed light on the nature of entropy. First of all, it is surprising that radiation involves an entropy at all. The "how" is a consequence of the law of Stefan and Boltzmann, according to which the energy depends on the fourth power of temperature.

Formally, the Stefan–Boltzmann law can be derived from Planck's radiation formula by integrating over the whole spectrum of wavelengths. In particular, the precise form of the Stefan–Boltzmann law can only be obtained in this way, revealing it to be a quantum law. However, the fact that the radiant energy depends on the fourth power of temperature did not have to wait for Planck's results. Boltzmann derived this dependence as early as in 1889, after Stefan had discovered it experimentally in 1879. It is a pure consequence of thermodynamics, a relation that follows from the fact that entropy is a state variable, to which integrability conditions apply. The details are explained in Vignette 2.10.

Here we encounter another example of the superficial relation between entropy and "order" (Section 2.8). In this case it would read: high-energy radiation corresponds to lower entropy than low-energy radiation of the same total energy. The latter point, i.e. constant total energy, is the important one because $u(T) \sim T^4$ and $S(T) \sim T^3$ both strongly increase with temperature. The "low order" of low-temperature radiation then corresponds to the larger number of quanta required – really not a world-shattering conclusion about "order". The more exciting lesson is that the entropy concept applies even to cases which are not obviously related to material order.

We discovered a similarly surprising fact about the entropy of a black hole. The huge gravitational energy produces a "mash" of Planck units of matter (if we can still call them that), yielding its finest achievable subdivision.

Another correlation is that of entropy with information. A maximum entropy is also a maximum of unknowingness, which in this case is a sort of unknowableness. Entropy is therefore correlated with "information", as suggested by the formal agreement of Shannon's and Boltzmann's expressions in Table 2.5.1. Boltzmann himself had already argued that entropy is equivalent to "lack of information". But then we have to ask: information for whom? If we were to answer: "for us", we might run into trouble if we do not carefully distinguish the amount that we need in order to be "informed" from what its content means to us. But that is a topic for Chapters 3 and 4, and some problems that ensue if we try to "get informed".

LITERATURE AND NOTES

1. Clausius, R. (1865). "The Mechanical Theory of Heat – with its Applications to the Steam Engine and to Physical Properties of Bodies". Ed. Hirst, T. A. Pub. John van Voorst, 1 Paternoster Row, London.
2. Caratheodory, C. (1909). "Untersuchungen über die Grundlagen der Thermodynamik" [Studies on the Foundations of Thermodynamics]. *Math. Ann.* 67: 355–386.
3. Eisenschitz, R. (1955). "The Principle of Caratheodory". *Sci. Prog.* 45: 246–260.
4. Falk, G. and Jung, H. (1959). "Axiomatik der Thermodynamik". Ed. S. Flügge. *Handbuch der Physik*, III.2 *Prinzipien der Thermodynamik und Statistik*. Springer-Verlag, Berlin, pp. 119–175.
5. Pauli, W. (1973). "Lectures on Physics". Vol. 3. MIT Press, Cambridge, MA and London.
6. Jaynes, E. T. (1992). "The Gibbs Paradox". In "Maximum Entropy and Bayesian Methods". Ed. Smith, C. R., Erickson, G. J. and Neudorfer, P. O. Kluwer Academic Publ., Dordrecht, Holland, pp. 1–22.
7. Einstein, A. (1920). "Schallausbreitung in teilweise dissoziierten Gasen". *Sitzungsber. K. Preuss. Akad. Wiss., Physik.-math. Kl.* 1920: 380–385.
8. Meixner, J. (1953). "Relaxationsverhalten der Materie". *Kolloid Z.* 134: 3–14.
9. Boltzmann, L. (1896, 1898). "Vorlesungen über Gastheorie". 2 Volumes – Leipzig 1895/98. English version: "Lectures on Gas Theory". Translated by Stephen G. Brush (1964) Berkeley: University of California Press; (1995) New York: Dover.
10. Gibbs, J. W. (1902). "Elementary Principles in Statistical Mechanics". Yale University Press, New Haven, CT. Reprinted in Gibbs, J. W. (1928). "The Collected Works of J. Willard Gibbs", Ed. W. R. Longley and R. G. Van Name. Longmans, Green & Co., New York, London, Toronto.

11. Eucken, A. (1944). "Lehrbuch der Chemischen Physik". II. Band "Makrozustände der Materie" in Gemeinschaft mit Dozent Dr. Klaus Schäfer. 2. Teilband "Kondensierte Phasen und heterogene Systeme". Akademische Verlagsgesellschaft Becker & Erler, Leibzig.
12. Van der Waals, J. D. (1910). "The Equation of State for Gases and Liquids". In: "*Nobel Lectures, Physics 1901-1921*". Elsevier Publishing Company, Amsterdam, 1967.
13. Eigen, M. and Rigler, R. (1994). "Sorting Single Molecules: Applications of Diagnosis and Evolutionary Biotechnology". *Proc. Natl. Acad. Sci. U.S.A.* 91: 5740–5747.
14. Ehrenfest, P. and Ehrenfest, T. (1911). "Begriffliche Grundlagen der statistischen Auffassung in der Mechanik". *Encykl. d. Math. Wiss.* 4 (32): Anm. 89a and 90.
15. Boltzmann, L. (1872). "Weitere Studien über das Wärmegleichgewicht unter Gasmolekülen". *Wien. Ber.* 66: 275–370. English translation: Brush, S. G. (1966). "Further studies on the Thermal Equilibrium of Gas Molecules". *Kinet. Theor.* 2: 88–174.
16. Loschmidt, J. (1876). "Über den Zustand des Wärmegleichgewichts eines Systems von Körpern mit Rücksicht auf die Schwerkraft". *S.-B. Kais. Akad. Wiss. Wien, Math. Naturwiss. Cl.* 73: 128–142.
17. Zermelo, E. (1896). "Über einen Satz der Dynamik und die mechanische Wärmetheorie". *Ann. Phys.* 57: 485–494.
Zermelo, E. (1896). "Über mechanische Erklärungen irreversibler Vorgänge". *Ann. Phys.* 59: 793–801.
Boltzmann's reply:
Boltzmann, L. (1896). "Entgegnung an die wärmetheoretischen Betrachtungen des Hrn. E. Zermelo". *Wied. Ann.* 57: 772–784.
Boltzmann, L. (1897). "Zu Hrn. Zermelos Abhandlung 'Über die mechanische Erklärung irreversibler Vorgänge.' " *Wied. Ann.* 60: 392–398.
18. Poincaré, H. (1889). "Sur les tentatives d'explication mecanique des principes de la thermodynamique". Comptes rendus de l'Académie des sciences (Paris) 108: 550–553. An English translation is found in the Appendix in Olsen, E. T. (1993). "Classical Mechanics and Entropy", Found. Phys. Lett. 6: 327–337.
19. Prigogine, I. (1949). "Le domaine de validitè de la thermodynamique des phénomènes irréversibles". *Physica* 15: 272–284.
20. Courtesy M. R. Schroeder (Personal Communication).
21. Von Mises, R. (1928). "Wahrscheinlichkeit, Statistik und Wahrheit". Springer, Vienna. English translation: "Probability, Statistics and Truth". London: Hodge 1939; London: Allen & Unwin/ New York: Macmillan 1957; New York: Dover / London: Constable 1981.
22. Kolmogorov, A. N. (1933). "Grundbegriffe der Wahrscheinlichkeitsrechnung". Springer, Berlin.
23. Jaynes, E. T. (1957). "Information Theory and Statistical Mechanics". *Phys. Rev.* 106: 620–630.

24. Kapur, J. N. and Kesavan, H. K. (1992). "Entropy Optimization Principles with Applications". Academic Press, New York, London and San Diego.
25. Kosko, B. (1986). "Fuzzy Entropy and Conditioning". *Information Sciences* 40: 165–174.
    Kosko, B. (1990). "Fuzziness vs. Probability". *Int. J. Gen. Syst.* 17 (2–3): 211–240.
26. Cheeseman, P. C. (1985). "In Defense of Probability". Proc. Ninth International Conference on Artificial Intelligence, Los Angeles. pp. 1002–1009. Morgan Kaufmann. San Mateo, CA, USA.
27. Yager, R. R., Ovchinnikov, S., Tong, R. M. and Nguyen, H. T. (1987). "Fuzzy Sets and Applications: Selected Papers by L. A. Zadeh". John Wiley & Sons, New York.
28. Boltzmann, L. (1877). "Über die Beziehung zwischen dem zweiten Hauptsatz der mechanischen Wärmetheorie und der Wahrscheinlichkeitsrechnung respektive den Sätzen über das Wärmegleichgewicht" [On the relation between the second fundamental law of the mechanical theory of heat and the probability calculus with respect to the theorems of heat equilibrium]. *Wien. Ber.* 76: 373–435. Reprinted in *Wissenschaftliche Abhandlungen* Ludwig von Boltzmann, Vol. II; Chelsea: New York, 1968: pp. 164–223.
29. Gibbs, J. W. (1902). "Elementary Principles of Statistical Mechanics". Yale University Press, New Haven, pp. 206–207; Gibbs, J. W. (1948). "The Collected Works of J. W. Gibbs", Vol. 1. Yale University Press New Haven.
30. Planck, M. (1901). "Über das Gesetz der Energieverteilung im Normalspektrum". Drudes Annalen 1901: 553–562. Reprinted in Ostwald, W. (1923). "Die Ableitung der Strahlungsgesetze". In: Ostwalds Klassiker der exakten Wissenschaften, Band 206: 65–74. Akad. Verlagsges. Leipzig.
    Planck, M. (1906). "Theorie der Wärmestrahlung". J. A. Barth: Leipzig. Translated into English by Morton Masius in M. Planck, "The Theory of Heat Radiation". Dover, New York, 1991; p. 119.
31. Shannon, C. E. and Weaver, W. (1949). "The Mathematical Theory of Communication". University of Illinois Press, Urbana, IL.
32. von Neumann, J. (1955). "Mathematische Grundlagen der Quantenmechanik" (Mathematical Foundations of Quantum Mechanics). Springer. Berlin.
33. Becker, R. (1955). "Theorie der Wärme" [Theory of Heat]. Springer, Berlin, Göttingen and Heidelberg.
34. Ehrenfest, P. and Ehrenfest, T. (1907). "Über zwei bekannte Einwände gegen das Boltzmannsche H-Theorem". *Phys. Z.* 8: 311–314.
35. Eigen, M. and Winkler-Oswatitsch, R. (1982). "Glasperlenspiele mit dem Zufall". *Natur*(August) 91–99; Eigen, M. and Haglund, H. (1976). "Life-game, with Glass Beads and Molecules, on Principles of Origin of Life". *J. Chem. Educ.* 53(8): 468–470.
36. Galsworthy, J. (1935). "Selected Short Stories"/Volume 2 of "Heritage of literature series: Section B". Longmans, Green & Co. London, New York, Bombay.

37. Eigen, M. and Winkler-Oswatitsch, R. (1975). "Chemical Relaxation: Das Spiel" see 'Laws of the Game'; loc. cit.
38. Schoute, P. H. (1902). "Mehrdimensionale Geometrie", Vol. 1. G. J. Goschen, Leipzig.
39. Ruch, E. (1975). "The Diagram Lattice as a Structural Principle". *Theor. Chim. Acta* 38: 167–183.
40. Ising, E. (1925). "Beitrag zur Theorie des Ferromagnetismus". *Z. Phys.* 31: 253–258.
41. Onsager, L. (1944). "Crystal Statistics I. A Two Dimensional Model with Order-Disorder Transition". *Phys. Rev.* 65: 117–149.
42. Berry, S., Rice, S. and Ross, J. (2000). "Physical Chemistry", 2nd edn. Oxford University Press, New York and Oxford.
43. Landau, L. D. and Lifshitz, E. M. (1980). "Statistical Physics", Part I, 3rd edn, Course of Theoretical Physics, Volume 5. Pergamon, New York. (See also Ref. 42, p. 674 ff; Breakdown of Landau Theory, see Ref. 45).
44. Wilson, K. G. (1982). "The Renormalization Group and Critical Phenomena". In: Nobel Lectures, Physics 1981–1990. Editor-in-Charge Tore Frängsmyr, Editor Gösta Ekspång. World Scientific Publishing Co., Singapore, 1993.
45. Goldenfeld, N. (1992). "Lectures on Phase Transitions and the Renormalization Group". Addison-Wesley, Reading, MA and New York.
46. Bose, S. (1924). "Plancks Gesetz und Lichtquantenhypothese". *Z. Phys.* 26: 178–181. English translation: (1976). Bose, S. "Planck's Law and the Light Quantum Hypothesis". *Am. J. Phys.* 44 (11): 1056.
Einstein, A. (1925). "Quantentheorie des einatomigen idealen Gases - Zweite Abhandlung". *S.-B. Preuss. Akad. Wiss*: 3–10.
47. Berry, S. (1995). "Entropy, Irreversibility and Evolution". *J. Theor. Biol.* 175(2): 197–202.
48. McGlashan, M. L. (1966). "The Use and Misuse of the Laws of Thermodynamics". *J. Chem. Educ.* 43: 226–232; (1965). *The Cambridge Quarterly* I(1): 64–68.
49. Lwoff, A. (1962). "Biological Order". MIT Press, Cambridge. Concluding phrase: "Even a physicist must succumb if fed with pellets of negentropy".
50. Feynman, R., Leighton, R. and Sands, M. (1964, 1966). "The Feynman Lectures on Physics". 3 volumes, Vol I, Chapter 5. Addison Wesley, Reading, MA.
51. Prigogine, I. (1949). See Ref. 19.
52. Penrose, R. (1965). "Gravitational Collapse and Space-Time Singularities". *Phys. Rev. Lett.* 14: 57–59; Hawking, S. W. and Penrose, R. (1970). "The Singularities of Gravitational Collapse and Cosmology". *Proc. R. Soc. Lond. Ser. A* 314: 529–548; Hawking, S. and Penrose, R. (1996). "The Nature of Space and Time". Princeton University Press, Princeton.
53. Hartle, J. and Hawking, S. W. (1983). "Wave Function of the Universe". *Phys. Rev.* D28: 2960–2975.

54. Misner, C. W., Thorne, K. S. and Wheeler, J. A. (1973). "Gravitation". W. H. Freeman, San Francisco.
55. Price, H. (1996). "Time's Arrow and Archimedes' Point". Oxford University Press, Oxford.
56. Bekenstein, J. D. (1973). "Black Holes and Entropy". *Phys. Rev. D* 7(8): 2333–2346.
57. Penrose, R. (1979). "Singularities and Time-Asymmetry". In: "General Relativity: an Einstein Centenary", Hawking, S. W. and Israel, W. (eds). Cambridge University Press, Cambridge.
58. Strominger, A. and Vafa, C. (1996). "Microscopic Origin of the Bekenstein-Hawking Entropy". *Phys. Lett. B* 379: 99–104.
59. Greene, B. (1999). "The Elegant Universe". Vintage Books, Random House, New York.
60. Gehring, W. J. (2009). "Neue Perspektiven über die Entwicklung und Evolution der Augen". Mitt. Naturforsch. Ges. Basel 11: 215–230.
Gehring, W. J., Kloter, U. and Suga, H. (2009). "Evolution of the Hox Gene Complex from an Evolutionary Ground State". A contribution to *Hox Genes*, a volume of *Curr. Top. Dev. Bio.*, edited by Olivier Pourquie. Vol. 88: 35–61.
Blanco, J., Girard, F., Kamachi, Y., Kondoh, H. and Gehring, W. J. (2005). "Functional Analysis of the Chicken δ1-*Crystallin* Enhancer Activity in *Drosophila* Reveals Remarkable Evolutionary Conservation between Chicken and Fly". *Dev.* 132: 1895–1905.
Gehring, W. J. (2005). "New Perspectives on Eye Development and the Evolution of Eyes and Photoreceptors". *J. Hered.* 96: 171–184.
61. Hawking, S. (1988). "A Brief History of Time", Chapter 9. Bantam Press, London. The quotation reads:
"The increase of disorder or entropy with time is one example of what is called an arrow of time, something that distinguishes the past from the future, giving a direction to time. There are at least three different arrows of time. First, there is the thermodynamic arrow of time, the direction of time in which disorder or entropy increases. Then, there is the psychological arrow of time. This is the direction in which we feel time passes, the direction in which we remember the past but not the future. Finally, there is the cosmological arrow of time. This is the direction of time in which the universe is expanding rather than contracting."
62. Prigogine, I. (1997). "The End of Certainty: Time, Chaos and the New Laws of Nature". Free Press, New York.
Prigogine, I. (2002). "Advances in Chemical Physics, Dynamical Systems and Irreversibility: Proceedings of the XXI Solvay Conference on Physics: 122". John Wiley & Sons, New York.
63. Volkenstein, M. H. and Volkenstein, M. (1994). "Physical Approaches to Biological Evolution". Springer, New York.

Volkenstein, M., Shenitzer, A. (translation) and Burns, R. G. (translation). (2009). "Entropy and Information". Progress in Mathematical Physics 57. Birkhäuser, Basel.

Volkenstein, M. (1970). "Molecules and Life: An Introduction to Molecular Biology". Plenum, New York, London.

64. Ebeling, W. (1991). "Models of Selforganization in Complex Systems – MOSES (Mathematical Research)". Wiley-VCH. Weinheim.

Ebeling, W. and Feistel, R. (1988). "Evolution of Complex Systems: Selforganisation, Entropy and Development (Sovietica)". Kluwer Academic Publishers, Dordrecht.

Ebeling, W. and Ulbrich, H. (1985). "Selforganization by Nonlinear Irreversible Processes". Proceedings of the Third International Conference Kuhlungsborn, GDR. Springer Series in Synergetics. Springer-Verlag, Berlin.

3 Entropy and Information

3.1.	Who Informs the Demon?	227
3.2.	Information = Entropy?	232
3.3.	Whose Information Is It?	243
3.4.	Why Coding?	249
3.5.	Do We Live in a Markovian World?	269
3.6.	How Much Information is in Mathematics?	272
3.7.	Can a Turing Machine Create Information?	283
3.8.	Whose Information is in Our Genes?	287
3.9.	How Far is it from Shannon to Darwin?	299
3.10.	Where is the "Temperature" of Information?	306

3.1. Who Informs the Demon?

In his book *Theory of Heat* James Clerk Maxwell[1] conceived "a being whose faculties are so sharpened that he can follow every molecule in his course, and would be able to do what is at present impossible to us". "Maxwell's Demon" – as this hypothetical being has been named – was supposed to fool the second law of thermodynamics by means of intelligent action. In the language developed in Chapter 2, the demon is able to perceive (or guess about), and possibly influence, individual microstates.

Think of a box enclosing two compartments that are separated by a wall containing a hole, just large enough to let single molecules pass through and, in addition, equipped with a flap that can be opened and closed (Figure 3.1.1). The two compartments contain equal numbers of molecules of the same kind, thermally equilibrated

Figure 3.1.1
Maxwell's demon

Maxwell's demon tries to outwit the second law of thermodynamics. He quickly opens a flap when a fast-moving molecule (red arrow) approaches from the left. He keeps the flap closed when such a molecule comes from the right. He does the opposite for slowly moving molecules (blue arrow).
Figure designed by Claus-Peter Adam.

at some given temperature. The demon is sitting at the opening, watching all the molecules as they move with different speeds according to Maxwell's distribution law. Whenever a "fast" molecule approaches the hole from one side, the demon quickly opens the flap and lets it pass, while keeping the flap closed if such a high-speed molecule comes from the other side. For "slow" molecules he does just the opposite. In other words, he lets them pass from the second to the first compartment, but blocks their way if they come in the other direction. By doing so, the demon cheats the second law of thermodynamics because one compartment cools down while the other warms up. Spontaneous demixing of "hot" and "cold" molecules (excuse the sloppy terminology) is associated with a decrease of entropy, thereby violating the second law. This may only happen by a thermal fluctuation (usually of quite small size, and not as suggested in Figure 3.1.2).

The question is: can one outwit nature by intelligent action, i.e. by clever use of information? Or better: What are the costs of doing so? It was Leó Szilárd who brought up this question in his doctoral thesis[2], which he wrote in 1929 while working in Albert Einstein's physics department at Berlin University. Szilárd's conclusion was that the information the demon requires – supposing that no frictional forces were involved in opening or closing the flap – is equivalent to the decrease in entropy

'Holy entropy! It's boiling!'

Figure 3.1.2

Mr Tompkins in Wonderland

Mr Tompkins experiences the extremely rare moment of a giant thermal fluctuation which causes part of his cold drink to boil. Illustration by John Hookman, from George Gamow's delightful book, *Mr Tompkins in Paperback* (1993, Cambridge University Press, Cambridge).

involved in the process of demixing "hot" and "cold" molecules. He emphasises that his finding presented a "radical departure from [current] thinking" because it based a thermodynamic quantity not on the underlying mechanics but rather on "information which is not really in his [i.e. the demon's] possession, because he guesses it". Szilárd is speaking of information that would be required by the demon in order to circumvent the second law of thermodynamics. We recall that Ludwig Boltzmann had already related entropy to "missing information", namely the knowledge of all microstates actually populated, which contribute to the robust average represented by the entropy of the macrostate. In more recent times, the IBM scientists Rolf Landauer and Charles Bennett[3] discussed a device that can use information acquired and recorded from one physical system to decrease the entropy of another, a process that works most efficiently close to equilibrium.

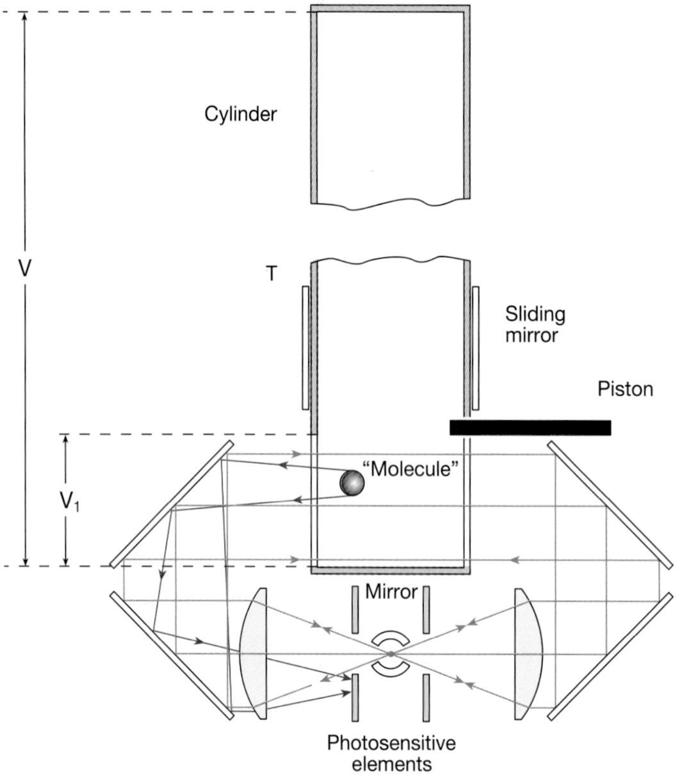

Figure 3.1.3

Sketch of Dennis Gábor's fictitious machine[4]

This machine resembles a "well-informed heat engine". The presence of a molecule in V_1 is traced by a light beam. The scattered photons are recorded by a photosensitive element, which triggers a relay to slide a piston into the cylinder and at the same time slides a mirror to cut off the light beam. The molecule then transfers momentum to the piston, moving it upwards.

Redrawn after Figure 13.5 on page 180 of Léon Brillouin's book[7] *Science and Information Theory*.

Szilárd's ideas were taken up by Dennis Gábor[4] who received the 1971 Nobel prize in physics for his epoch-making work on holographic photography. He constructed a "machine" that directly demonstrates the close relation between entropy and information. The principle of its construction is shown in Figure 3.1.3 and explained in the legend.

Gábor's machine, of course, is an abstract machine, based on a *Gedankenexperiment*. It would hardly work in reality – not only because it makes use of frictionless motion, a single molecule lifting a piston upon detection, and not to speak of the metabolic energy of the nerve cells involved in thinking up and operating such a machine. Yet we too are part of nature, and we have to ask how the generation of information, which is always connected with a decrease in entropy in some parts of the total natural system and hence must be compensated by an increase in other parts (often called the environment), can occur through natural self-organisation.

The generation of information is ultimately a "rectification" of microscopic fluctuations, which are often hard to realise as such in the macroscopic behaviour of a system. In this way we can "fool" Nature by intelligent action, which nevertheless requires access to some source of (free) energy, the use of which involves the production of entropy. The "intelligent action" can be delegated to a machine, but in performing this delegation we cannot outmanoeuvre physical law.

It is in this sense that we must understand Szilárd's and Gábor's efforts. The fact that Brillouin's interpretation[5] (to which I refer in the legend of Figure 3.1.3) leads to the result that information requires twice the value of the entropy involved is not because account was taken of realistic working conditions. As I said before, the machine is based on a *Gedankenexperiment* and would not work under any realistic conditions. It is true that today we are able to detect single molecules, but the devices needed for this are quite involved, and the momentum of a single molecule would hardly suffice to set a macroscopic piston into motion. Brillouin's analysis is based on ideal assumptions. The involvement of twice the entropy value is a consequence of an interpretation of the technical term "information".

The machine produces mechanical work, which has to be compensated for by entropy production, e.g. in the heat required to keep the temperature – and thereby the energy and momentum – of the single molecule constant. For this, the single molecule is localised in V_1 and the triggered sliding of the piston can be regarded as a (negative) jump of entropy which subsequently increases to its original value, assuming that the molecule is anywhere in the total volume V. But a knowledge of the temporal behaviour, i.e. the ratio of τ and t_0, is an additional piece of information required for the machine to work. Not knowing τ in relation to t_0 would leave us uncertain when to slide the piston in and out. But wouldn't this introduce a semantic – although physically interpretable – element of information?

Brillouin defines "information" by probabilities based on the number of all possible situations, e.g. for a molecule that is within the total volume $V(P_0)$ and, after information is introduced, for all those cases where it is within the restricted volume V_1. The definition of information, based on prerequisites of this kind, is the main subject of the present chapter. However, this is not what we usually mean by "being informed". It is semantic information, based on selective evaluation, which also represents the basis of life, a main subject in this book. It includes the question of how life and intelligence originated in nature, and how life was able to organise itself into its presently accessible, highly sophisticated forms.

Life appears to be quite wasteful in its energy consumption, and the same is true for the metabolism of the nerve cells involved in thought. This holds for the macroscopic as much as for the microscopic level. All food production on earth is ultimately based on the sun's energy, but which however cannot be used in its primary form. Moreover, it does not just pass through the photosynthetic cycle. We enjoy it as much in the form of a steak – ultimately also a product of photosynthesis, one that has been "refined" in the rumen of cattle, which is a quite wasteful metabolic process. And on the microscopic level: all life is based on the synthesis of nucleic acids from energy-rich triphosphates that decompose into energy-deficient products under conditions that are far distant from those states, near chemical equilibrium, that are the most economic for information transfer. Of course, life is adaptable and tries to optimise as much as possible the mechanisms that were fixed in the early phases of evolution by the environmental constraints that then existed.

This chapter will deal with "information" as something given, characterised by some definite amount, without asking how it was generated; to ask that would have to include the question of its meaning or value (see Chapter 4). However, I shall repeatedly confront these two issues in order to prepare the ground for our later discussion of a new theory that includes the aspects of the "meaning" and "value" of information.

3.2. Information = Entropy?

The founder of what is nowadays called "information theory" was Claude Shannon[6] (see Table 3.2.1). He arrived at this concept during the late 1940s, on abstract mathematical grounds, while working as a research scientist at the Bell Telephone Laboratories. Accordingly, the problems he was concerned with were closely connected with telecommunication. The ground-breaking paper he wrote in 1949 together with Warren Weaver had the title "The mathematical theory of communication". In fact that is precisely what it is. The term "information theory" was adopted some years later, fostered mainly by Léon Brillouin in his seminal book *Science and Information Theory*[7].

The theory of communication is dominated by a mathematical expression that Shannon at first called "uncertainty". There is a story that John von Neumann,

Table 3.2.1 The Constant of Shannon's Entropy

Shannon's entropy involves a constant that depends only on the base of the logarithm. For "natural logarithms" ($\log_e$) it is set to one, and as such replaces Boltzmann's constant in thermodynamic entropy. Correspondingly for any base other than e the conversion factors are:

$\log_e 10 = 2.30259\ldots$	$\log_{10} e = 0.43429\ldots$	$\log_2 e = 1.44270\ldots$
$\log_e 2 = 0.69315\ldots$	$\log_{10} 2 = 0.30103\ldots$	$\log_2 10 = 3.32193\ldots$

For binary sequences one includes the conversion factor in the unit "bit" and expresses entropy simply in units of $\log_2 2 = 1$ bit (i.e. $p = 0.5$).

to whom Shannon introduced his work, immediately realised the mathematical equivalence between Shannon's "uncertainty" and entropy as it appears in statistical mechanics. So he asked him: "Why don't you call it 'entropy' too?" And he added that it would give Shannon a lead over opponents in discussions because "only a few people in the world understand what entropy really is".

Claude Shannon, who had done his graduate work at the Massachusetts Institute of Technology (MIT), eventually (1956) returned there as Donner Professor of Science. MIT is the true home of information and computer sciences. One of Shannon's teachers was Norbert Wiener, the father of cybernetics. The concept of information plays an important role in Wiener's work and is reflected in his book *Cybernetics*[8], which appeared in the same year as Shannon's theory. Among the many legendary stories about Norbert Wiener is the one related by Jeremy Campbell[9]: "In 1947, a year before the publication of Shannon's theory, Robert Fano was working on his doctoral dissertation at MIT. At intervals, Wiener would walk into Fano's room, puffing at a cigar, saying: 'Information is entropy'. Then he would turn around and walk out again without another word." Robert Fano, by the way, is well known to information theorists through his work on coding theory (e.g. Shannon–Fano coding and the Fano bound). He was also involved in getting Shannon appointed to MIT.

At that time the concept of information had not yet been formulated in quantitative terms, but Wiener certainly had the right idea, which was already in the air. The goal was to free "information" from the anthropomorphic connotations that the term has in everyday language. To "inform" somebody involves the transfer of some quantity. If one does it through a machine, this quantity has to be protected against noise during the process of transmission. Hence, there must be something referring to the amount of information that can be quantified and that is independent of all the qualities of "content", such as meaning and value, which eventually cause one "to be informed" and to respond accordingly. Shannon came up with a new mathematical concept for this quantification, with far-reaching consequences for telecommunication and particularly computer science.

What is the role of entropy in this theory?

Entropy was introduced in Chapter 2 as an important concept in statistical thermodynamics. What does it have to do with problems of communication? Frankly: very little – as long as we have in mind its physical aspect only. First of all, entropy is an average characterising a macrostate that usually involves contributions from a vast number of microstates, while information refers to a defined situation including only one or a few physically real microstates. Moreover, letters or digital units have little in common with atoms or molecules – physically! Yet both digits and molecules confront us with probability distributions, regardless of whether we have to guess how the various letters are arranged in sequences that form words, sentences and texts, or how atoms and molecules distribute themselves along the position and momentum co-ordinates of phase space.

In both cases we have a probability distribution $P = (p_1, p_2 \ldots\ldots p_N)$ where in the most general case p_i is a real number denoting the probability of an outcome among the total number N of microstates involving a total of n particles or a possible symbol combination of a message of length ν. The probabilities are normalised, that is to say, each p_i value is greater than or equal to 0 and less than or equal to 1, while the sum over all p_i values equals 1.

Every such statistical distribution has an entropy $H(N) = K\sum_{i=1}^{N} p_i \log(1/p_i)$ (cf. Boltzmann's expression in Table 2.5.1). In this general case, N is the total number of possible states, i.e. all allowed microstates or symbol combinations, while K is a constant that relates entropy to the statistical problem involved (by proper choice of variables, K can in many cases be set to 1). Since the probabilities p_i contain weighting factors, entropy can be considered an average of the logarithms of the p_i values. Remarkably, this implies that entropy exhibits an additive property that characterises an "amount". Indeed, the fact that the logarithm transforms products into sums implies that, given any two systems with probability distributions $P = (p_1, p_2, \ldots, p_N)$ and $Q = (q_1, q_2, \ldots, q_M)$, the composite system whose probability distribution $P \times Q$ is given by the products $p_i q_j$, $(i = 1, \ldots, N; j = 1, \ldots, M)$ has an entropy that is the sum of the entropies of P and Q, i.e. we have $H(P \times Q) = H(P) + H(Q)$.

However, there is an important difference between the use of entropy in statistical mechanics and its use in information theory. In statistical mechanics, entropy refers to the distribution of energy, for which a conservation law applies. There, entropy is complemented by temperature (which we can express in energy units by including Boltzmann's constant in the temperature). The product of entropy and temperature has a direct physical meaning, describing the thermal component of energy. In information theory, there is no analogue of temperature. Why not? Because we still have to decide what to call "information". We have just identified its "amount" with entropy, but this covers only one aspect of what we mean by the word "information".

In order to make what I have just said easier to understand, let us look at an example. If I call a telephone information service I usually want to get a particular subscriber's number. An answer such as: "23 bits" (which would imply a seven-digit decimal number) would be of no help. It would just tell me that I am looking for one of several million equally probable numbers. All I know now is how much information I am short of. The semantics are missing, which in this case means that the phone number I want to know refers to the one and only one telephone subscriber I want to speak to. Could a second quantity, such as temperature, be of any help? No, it could not!

The product of entropy and temperature, although it refers to a macrostate because it results from averaging over many microstates, makes sense because of energy conservation. Moreover, temperature can only be defined in thermally equilibrated systems. "Semantically" equilibrated information doesn't make any sense. An average phone number is no phone number! I shall come back to this problem in Section 3.10.

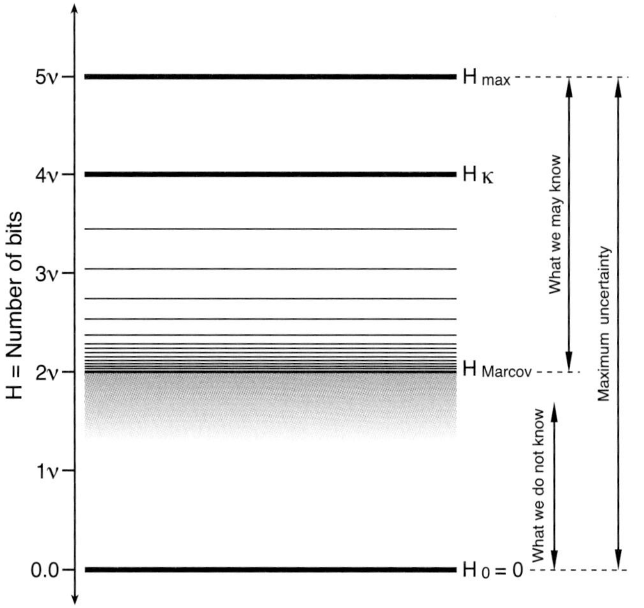

Figure 3.2.1

Schematic representation of the levels of (entropic) information

The upper level H_{max} represents "maximum uncertainty" referring to a probability distribution in which all possible outcomes are equally likely. The next level H_K, considers the different probabilities of the occurrence of letters in the English language (according to Table 3.2.1). The subsequent lines refer to higher concatenations expressing word statistics and other constraints (e.g. syntactical and grammatical) on language which converge to a limit demarcating all structural influences delimiting the information capacity required for expressing meaning. The precise borderline between structural and semantic information is not known. The remaining capacity is sufficient to represent any facts in terms of language in order to provide us with (meaningful) "information", i.e. with facts we otherwise could only guess. The total amount of information submitted is proportional to the length of a message, i.e. to the number ν of symbols contained in the message. Every constraint in the probability distribution defining the entropy ($\Sigma p_i \log p_i$) reduces the uncertainty.

Shannon's theory treats only one aspect of information, namely, its "extensive" aspect, and I should like to explain this in further detail with the help of Figure 3.2.1.

The upper line in the schematic diagram represents maximum uncertainty about a message given as a sequence of ν symbols. If the alphabet includes κ different symbols, then there are altogether κ^{ν} possible sequences of length ν (i.e. κ possibilities for each of the ν positions). If all κ symbols of the alphabet were to have the same prior probability of occurrence (i.e. $p_i = 1/\kappa$), and if no further constraints limit their combination into sequences (which means that each of the κ^{ν} possible sequences has the same probability of occurring), then the entropy counted for ν positions would be simply $\nu \log_r \kappa$. If κ were equal to 2 (binary system), we would choose a logarithm to the base $r = 2$ and call the "amount" of informational entropy ν bits. If κ corresponds to the 27 characters of the English alphabet (i.e. 26 letters plus one space symbol), then the information entropy for each letter comes out as $\log_2 27 = 4.76$ bits, i.e. it would require slightly less than a five-bit code length to represent each letter. Multiplying by the number of symbols (ν), we obtain 4.76ν as the highest entropy that a sequence of ν letters (of the English language) can achieve. Any additional constraint will lower this value.

Needless to say, the above assumption of equal probability for all symbol combinations is entirely unrealistic. It would be the root of a Babelian confusion of languages, because if two languages using the same characters (as most European languages approximately do) were to assign meaning to all possible combinations, then the semantics of any combination of letters would either have to be identical in all languages or any meaning would be completely lost. According to the Book of

Figure 3.2.2
"Tower of Babel"

By Pieter Bruegel the Elder. Used with permission from the Kunsthistorisches Museum, Wien.

Genesis (chapter 11, verses 1–9), Yahweh decided to stop the building of the Tower of Babel (Figure 3.2.2), which was intended to reach up to Heaven. So He went down to Earth and confused the formerly universal language of the people in order to make it impossible for them to communicate. This brought a stop to any common activity, including the building of the Tower, and the people scattered in confusion across the face of the Earth.[10]

Back to Figure 3.2.1. The second line takes into consideration the most obvious constraint, namely, the one imposed by individual *a priori* probabilities of the different symbols. The most frequent "symbol", probably in every language, is the space or blank symbol. In English it appears with a probability of about 20%. The most frequent letter in English is "e", which has a probability of 10.5%, while the least frequent letters "j", "q" and "z" have probabilities that amount to only 0.1% (Table 3.2.2). In the absence of other constraints, taking account of individual symbol frequencies would reduce the average amount of information per symbol to $\sum_{i=1}^{k} p_i \log_2(1/p_i)$, that is to say from 4.76 to 4.03 bits/symbol. Further constraints of language will bring about additional compression of information entropy. These constraints reach from nearest-neighbour relations among letters in words, through more subtle peculiarities of language structure, up to frequencies of usage of particular words. This may eventually lead to separate frequency ratings of all the κ^{ν} sequences of length ν.

The entropies associated with higher-order correlations are called Markov entropies H_M^k, where the order k can run from 1 to ν (ν being the total length of the symbol sequence under consideration). They are named after the Russian mathematician Andrei Andreyevich Markov because they arise by considering the

Table 3.2.2 The probability of occurrence for the letters of the English language

Symbol	Probability, p	Symbol	Probability, p
Space	0.2	L	0.029
E	0.105	C	0.023
T	0.072	F, U	0.0225
O	0.0654	M	0.021
A	0.063	P	0.0175
N	0.059	Y, W	0.012
I	0.055	G	0.011
R	0.054	B	0.0105
S	0.052	V	0.008
H	0.047	K	0.003
D	0.035	X	0.002
		J, Q, Z	0.001

sequence as resulting from a Markov process. The probability of any symbol occurring in a message now becomes a "conditional probability", i.e. the probability that a symbol occurs, given that it is preceded by a chain of k other symbols, in a particular message. Markov processes are explained in more detail in Section 3.5. Let me here quote some numerical examples that refer to the English language.

We have seen that the maximum informational entropy, as obtained for uniform character frequencies (Markov order zero), is reduced from 4.76 bits to 4.03 bits per letter, if individual character frequencies according to Table 3.2.2 are taken into consideration. The sum in the entropy expression is extended over the 27 probability terms $(-p_i \log_r p_i)$. If the $27^2 = 729$ possible nearest-neighbour combinations are rated as well, then the informational entropy per symbol is reduced further, to 3.32 bits/letter. Using all data for groups of three letters leads to yet another reduction, to 3.1 bits/letter. To get this far, the frequencies of 19683 (i.e. 27^3) three-letter groups had to be evaluated. For any symbol sequence of finite length, the input data now become increasingly uncertain.

Another source of redundancy is word statistics, as was first treated by Benoit Mandelbrot in his PhD thesis of 1952.[11] If statistics for the usage of complete words are compiled, one finds for the 8727 most abundant words in the English language (such as "the", "of", "and") an entropy value of 11.82 bits/word. Since the average word length (including blanks) is 5.5 characters/word, the above number means a reduction of the informational entropy to 2.15 bits/character (which is an overrated reduction, since the most frequently used words contain fewer than five or six characters). Leaving meaning and value aside, there are more constraints, introduced by syntax and grammar, that bring about further compression of the *amount* of information.

The reduction from 4.76 to 2.16 bits per letter is an enormous reduction. A script with a length of 100 letters would have only $2^{216} \approx 10^{65}$ instead of $2^{476} \approx 10^{143}$ possible arrangements when re-coded in binary form. As large as the reduction factor of 10^{78} is, the remaining "tiny" residue is still sufficient to represent much more than what is needed for the capacity of any language. In other words, most of the remaining letter combinations still have no semantic value.

Shannon tried to estimate empirically the reduction of information per symbol, as caused by all sorts of correlations or redundancies. To do this he got his students to play the following game.[12] A sentence is selected by one individual and has to be guessed, letter by letter, by another individual who does not possess any initial information about the sentence. His guesses (in which he tries to use whatever information about language is available to him) are answered by "yes" or "no". The questions, of course, have to obey a rational hierarchy, which reduces systematically the possible outcomes, as in the game of Twenty Questions referred to in Section 1.10. The number of guesses per letter is recorded, summed up and divided by the number of letters in the sentence. Since the answers are "binary", one obtains in this way the number of bits per letter. The impressive result of several such experiments was a reduction in the amount of information down to values of 1.4 bits/letter. There is, of course, no clear demarcation line between constraints imposed by language structure and those suggesting some possible meaning.

Summarising the content of Figure 3.2.1, we may define informational entropy as "what we still do not know after we have taken into account everything that we do know". It is the latter part of this statement that constitutes the substrate for Shannon's "information theory". If all symbols were uniformly distributed (H^o), then the best we could do would be to state our total ignorance about any symbol used. The more we know about languages in general and the language of the message in particular, the more economically we can communicate by taking into consideration higher-order concatenations of symbols. The information theorist calls an encoding of such concatenations or blocks of symbols "the kth extension of the code" (see below).

All the subtleties of Shannon's theory rest on the "ladder" of the H_M^k values ($k = 1, 2, \ldots \kappa^\nu$) in Figure 3.2.1, for which $H_M^k \leq H_M^{k-1}$ holds, an inequality that was rigorously proven by the Russian mathematician Alexander Khinchin[13]. This "ladder" of H_M^k values converges to a (fictitious) limit H_M that utilises all possible constraints imposed on the symbols of the language used. The total space of information 2^ν then is reduced to $2^{\nu H_M^k}$, where H_M^k is the kth-order Markov entropy per symbol. Owing to the exponential relation, the true space occupied is usually only a tiny fraction of the total "space" (see below).

Sequences pertaining to the reduced space $2^\nu H_M$ ($\ll 2^\nu$) are the ones we actually meet in reality. This is true not only of languages, but also of the sequences of proteins that we encounter in living organisms. If the length ν of a sequence is large enough, it will contain most of the highly probable (k-fold) concatenations of symbols. This would make the Markov entropies H_M^k for all long sequences converge eventually to the same value, and all long sequences of length ν would have the same probability of appearance, i.e. $2^{-\nu H_M}$. All other sequences, belonging to the much larger group of $2^\nu - 2^{\nu H_M} \approx 2^\nu$ sequences, would appear with a probability of (almost) zero. The fact that we have two types of sequences, a relatively small group of sequences all occurring with (nearly) the same "high" probability ($2^{-\nu H_M}$) and a much larger group of sequences having an *altogether* (almost) zero probability of occurrence, is explained by a theorem named after Shannon[14], McMillan[15] and Breimann[16], and was stressed by Hubert Yockey[17] as applying to proteins as well.

The above conclusion brings me to the question I actually want to ask in this section: why is entropy an important quantity in a theory that deals with the communication of specific messages? Entropy in statistical thermodynamics is the measure of a macrostate that results from the superposition of a huge number of microstates, all of which are compatible with all constraints that delimit their probability distribution. As was shown in Section 2.7, equilibration of the distribution means loss of recollection of any particular microstate within that distribution. If we denote a particular message, i.e. the particular arrangement of symbols in a sequence that we may call a "microstate", then the issue of communication theory is "how to get the message from the sender to the receiver without any loss of information" or "how to 'conserve' this very 'microstate'". The average that characterises the macrostate does not play any obvious role, a problem that is foreign to (equilibrium) thermodynamics. However, note that communication theory must apply to *any* message sent through the transmission line.

The problem of "loss of memory" is definitely an important one if we think of the "messages" that are stored and processed in computers that handle thousands of millions of bits every second, including all kinds of iterative procedures over greater lengths of time. As we shall see, this requires a sophisticated means of error correction. However, "equilibration" as such is not an issue of Shannon's theory. Errors certainly do occur, and have to be corrected before they can "equilibrate" to total nonsense. So the question remains: why does entropy play such an important role in information theory? There are essentially two reasons:

1) If a message is long enough, then on average the symbols will occur according to the statistical constraints of the language, the same way as they are reflected in the entropy. Hence the theory applies to *any* sufficiently long message, regardless of its content, as long as it complies with the rules of that language. This is the reason for the existence of high- and low-probability groups of sequences, as was shown above.

2) Entropy is related to the length (ν) of the message (Figure 3.2.1), and as such is ultimately a measure of distance. It is appreciably smaller than the unconstrained length ν, but it is proportional to this length, i.e. it is equal to νH_M, where H_M is the average Markov entropy per symbol, a quantity less than 1.

The first reason is immediately intelligible from the preceding discussion. It can best be seen by looking at some (meaningful) sentence written in a language that we do not understand. Another good example is the nucleotide sequence of a gene. Despite the fact that the gene represents a well-defined function, and it is expressed in a well-defined structure of a protein molecule, the nucleotide sequence as such looks pretty much like a random sequence of the four letters A, T, G and C. Although highly organised with regard to functional purpose, it does not on the whole show any particular structural order, other than the order resulting from general constraints and therefore found in any natural sequence.

The second reason stated above is related to the first one. However, it is not so obvious that information entropy is related to distance or length. In order to demonstrate this, we have to consider a suitable space for representing information. Richard Hamming (one of the pioneers of communication theory) introduced this concept for the purpose of encoding messages.[18] It is modelled along the idea of multidimensional phase space in statistical mechanics; however, it differs from this in some important respects. Here I shall give a brief introduction to this concept. A more general treatment of "information space", which forms the basis of evolutionary theory, will follow in Sections 4.4 and 4.5.

In brief, information space is *a discrete point space* (to be introduced in Section 4.4) that has a *non-Euclidean metric*. Consider the simplest case of a binary sequence of ν positions, for which each of the 2^{ν} possible alternatives is assigned an equal probability of $2^{-\nu}$. The space we construct provides one and only one point for each of the 2^{ν} sequences. The important issue is how to arrange the points in (a visualisation of) this space so as to reflect their neighbour relationships. The rule is to distribute the points in such a way that distances between any two points reflect the kinship between the sequences involved. In other words, any two binary sequences having different symbols (0 or 1) in only one position should be nearest neighbours, i.e. separated by a distance of one unit. Likewise, two sequences that differ in k positions should be separated by a distance k, and the same must apply to any two sequences represented in this space.

It is immediately clear that this concept can be realised only in a space that (1) is a discrete space and (2) has a high dimensionality. As an example, consider the three-dimensional model of the point space in Figure 3.2.3. The lines do not represent co-ordinates, but just connect adjacent points. Hence only the eight vertices of the cube represent the states denoted by binary symbols. The volume of a point space is just the total number of its vertices. Generalisation of this model leads to a hypercube of dimension ν, where ν is the length of the binary sequence under consideration. We see from Figure 3.2.3 that a fourth position cannot be accommodated by the three-dimensional diagram, but rather requires the addition of a fourth dimension. Figure 3.2.4 represents the projection of a six-dimensional hypercube. The metric of this space is well defined. It is non-Euclidean, since a hypotenuse, which would connect two non-adjacent points, passes through non-existent regions of this space and therefore makes no sense. The distance between any two points is simply the shortest sum of edges that connect the two points. This distance is

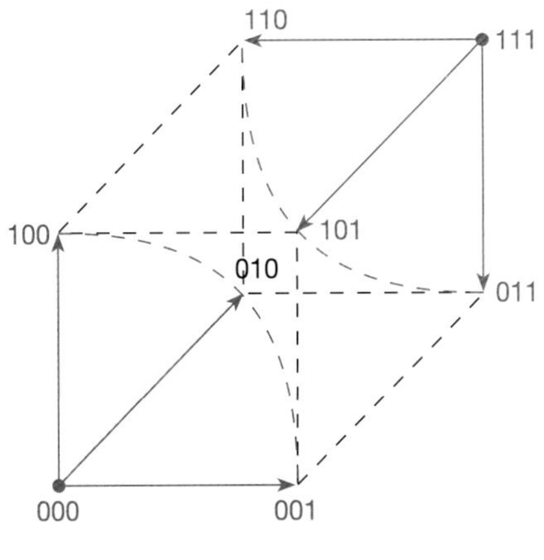

Figure 3.2.3
Hamming space

The example shows triplets of binary symbols defining a three-dimensional point space. All points having the same distance with respect to one of two reference points (000 and 111) define the surface of two "spheres" as shown in this simplest possible diagram (for more details see Sections 4.4 and 4.5).

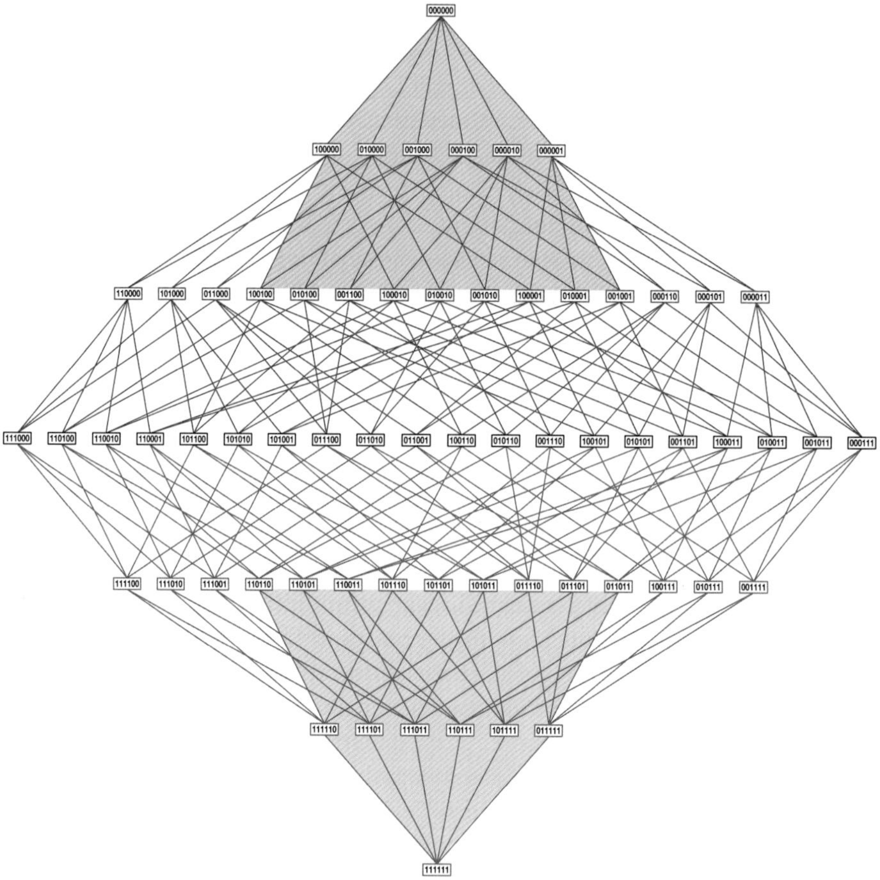

Figure 3.2.4

Two non-overlapping Hamming spheres in six-dimensional sequence space

See also Figure 3.4.3.

named the Hamming distance in honour of Richard Hamming (for more details, see Section 4.4).

We now see how information entropy is related to the length of a message. We have a relation similar to the one in statistical mechanics, namely, that entropy is the logarithm of a "phase volume". If all symbols are independent of one another and occur with the same probability $p = 2^{-\nu}$, the phase volume is simply the volume of our point space. The logarithm of this volume is $\nu \log_r 2$ or, for binary logarithms ($r = 2$), ν bits, i.e. the largest Hamming distance in the hypercube. As is true for physical phase space, constraints upon the probability distribution – up to individual probabilities for all 2^{ν} possible sequences – will lower the phase volume and hence the information entropy. The "available" volume of information space will then be $2^{\nu H_M}$ rather than 2^{ν}, where H_M is the Markov entropy per symbol, i.e. a quantity ≤ 1.

We are now ready to answer the question in the title of this section. Yes, the equality is justified if we mean the amount of information. It is the predicate of Shannon's first theorem, which identifies the lower bound of the length of an encoded message with its entropy. Let me quote the theorem in Richard Hamming's words: "Using a sufficiently large (kth) extension of the code we can bring the average length of each code word for the symbols of the primary source as close as we please to its entropy."

This theorem links entropy to the length of the sequence of the code symbols that represent a message in the most economical fashion. It does not yet assume the presence of any noise. It just deals with the representation of an amount of information.

3.3. Whose Information Is It?

In our times we communicate almost as much with machines as we do with people. We exchanged and still exchange information by phone, fax, telex and e-mail, listen to radio and watch television, store and retrieve data in computers and let them do our homework and, increasingly, our housework. This all involves compression, storage, transmission, transformation, retrieval and representation of information, requiring its protection against all forms of noise. If we think of a computer that carries out thousands of millions of operations every second, we can only guess what this means in terms of reliability and economy.

When Shannon developed his theory, the main issue was sending information from one place to another without loss of its "content". The general scheme of telecommunication is shown as a block diagram in Figure 3.3.1 (see also Hubert Yockey[19]). In Figure 3.3.1 I have deliberately represented the source and the sink as black boxes. They produce and use the information that is sent through the transmission line. How they do it does not concern us in the present section, but will be a main issue in later parts of the book. If, in connection with the diagram, I talk here about gain and loss of information, then it has little to do with what is going on in those black boxes. Nevertheless, the problems discussed in this chapter will provide important prerequisites for our later discussion, which we could characterise as "how to generate information from noise". The present question is rather "how to prevent the loss of information caused by noise" during the process of communication.

The scheme in Figure 3.3.1 obviously refers to an exchange of information between a sender and a receiver; but *whose* information is the scheme dealing with? It was Edwin Jaynes[20] who raised this question and supplied the answer: it is neither the sender's nor the receiver's actual "information" that is involved. Why should the

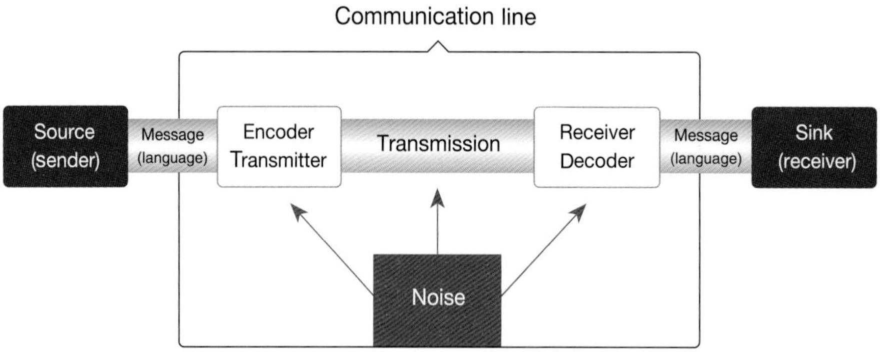

Figure 3.3.1

Schematic diagram of a transmission channel

This is the standard diagram of a signalling system. Source and sink have been drawn as "black boxes", to indicate that we have no knowledge of the components necessary for producing or utilising the information to be sent as a message. In fact, we do not need such knowledge for the theory presented in this section, which therefore properly should be termed "communication theory".

sender conjecture about probabilities and entropy? He knows exactly what he is going to send out. The message originated in his head – goodness knows "how", but here it is. By the same token, the receiver knows it, at the latest when he has received the message in its correct form.

So *whose* information is it?

There remains only one other person who is (or has been) involved in this process. It is the communication engineer who has to design the transmission line for his telephone company. He doesn't care about the content of any of the messages to be sent through this line. He just has to make sure that *any* message released by a source will arrive at the receiver in an undistorted form. Proper encoding is one reason why entropy, a measure of the "capacity" of information, plays such a decisive part in communication. It is the property to which the transmission system has to be adapted, therefore the main player involved is the engineer. It is "his" information that we are dealing with in the present section.

The central element in the transmission scheme sketched in Figure 3.3.1 is the channel through which the signals carrying information are to be sent. Every such channel has a physically defined capacity for transmission of signals, which consist of quantities (or quanta) of electromagnetic energy. The line may be a microwave cable, a radio link, a magnetic tape or some glass-fibre optics. A cable, for instance, has a finite frequency band width (Δf) that limits the duration (Δt) of any signal passing along it to $\Delta t \geq 1/\Delta f$. Fibre optics are becoming increasingly the medium of choice because of their inherent physical advantages. Be that as it may, the statistical problem involves a finite set of discrete symbols, each having a finite length l_i or time of duration t_i. From now on we may ignore the physical nature of the signals. The quantities that matter are their numbers and durations. The total duration of a message of length L_ν is simply the sum of the durations of all ν symbols contained in the message, yielding a total time T_ν. Its average per symbol (T_ν/ν) is defined in the same way as the average symbol length (L_ν/ν), which was introduced with Shannon's first theorem.

Now consider the set $N(T_\nu)$ of all possible sequences of symbols having a given total duration T_ν. The individual messages in this set differ in selection and order of their symbols, but all have the same duration T_ν, which is the sum over all individual time lengths. Very long sequences will – for a given probability distribution and length – eventually converge to the same total time T_ν. In other words: the set $N(T_\nu)$ will include all "allowed" sequences of length ν, and $\log N(T_\nu)$ will define its information entropy. The rate at which information is transmitted through the channel is $\log N(T_\nu)/T_\nu$. This rate, which may vary considerably for shorter messages, tends asymptotically to a constant limiting value (for a given length) as the length L_ν gets larger and larger. Shannon defined the channel capacity as the limit of $\log N(T_\nu)/T_\nu$ for T_ν approaching infinity (i.e. $C \sim \lim_{T_\nu \to \infty} \frac{\log_r N(T_\nu)}{T_\nu}$). The limit of a constant ratio $\frac{N(T)}{T}$ is reached when the Markov entropy H_M^k reaches its limiting value. Hence we can replace the above logarithmic term by the Markov entropy, for which all symbol interactions are taken into account and which for sufficiently long sequences becomes constant.

Of course, the capacity of a channel is defined by its physical properties, which are constrained by technical limitations. Optimising an information channel therefore means matching the code and its maximum rate of transmission with those limiting physical properties, i.e. adapting the length and duration of the source symbols in an optimal way. Note that although time enters through the channel capacity we are not applying any dynamical theory. Time is "compensated" for by matching the firing rate with the limiting bandwidth. The rest is probability theory dominated by information entropy.

In Section 3.2 we considered the coding problem only for noise-free transmission. Noise, however, is omnipresent and hence cannot be prevented. It will "blur" any message sent through the channel. All one can do is try to correct errors that may be introduced by the noise. The presence of noise means in particular that the probability distribution of the symbols at the output of the channel may not agree with the input probability distribution of the source symbols.

In an abstract way, the problem of noisy transmission can be described by the scheme represented in Figure 3.3.2.[21] The left-hand side of the diagram refers to the input by the sender, the right-hand side to the output at the receiver. In between is the channel, which is characterised by a transition matrix, the elements of which are "conditional probabilities".

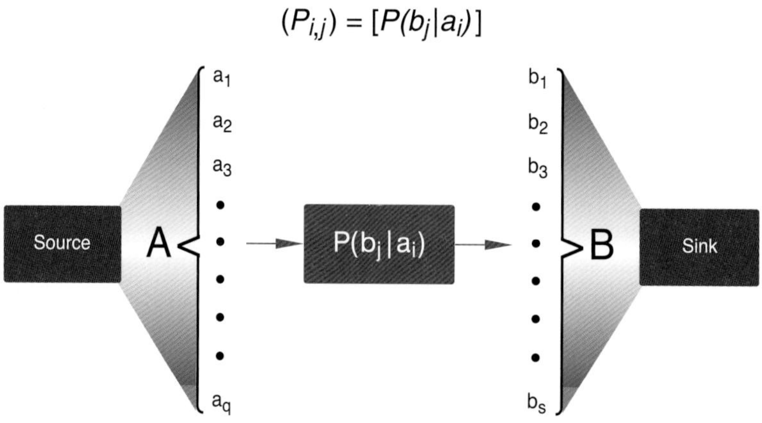

Figure 3.3.2
The mathematical representation of the transmission channel

A channel is described by a set of conditional probabilities $P(b_j|a_i)$, expressed in the form of a matrix. The single terms $p(b_j|a_i)$ refer to the single letters a_i of an input alphabet and the corresponding ones of the received message b_j. The sizes of the two alphabets need not be the same. Error-correcting devices will require a larger alphabet of letters to be received than sent, while other requirements may necessitate more input than output symbols. Hence the matrix is usually not symmetrical. The ith row of the matrix corresponds to the ith input symbol a_i and the jth column to the jth output symbol b_j. The sum of all elements in a given row is always one. Such conditions ensure that for each input something will come out, and that whenever something is put into the system, then something comes out. This model has been considered in detail by Richard Hamming in his book *Coding and Information Theory*[18]

Now we are back to one of our original questions: how does entropy enter the problem? The symbol sequence representing the message has to be conserved during transmission; it must not be permitted to "equilibrate" to a mixture of any of the $2^{\nu H_M}$ copies in the high-probability group of symbol arrangements represented by entropy. The message may be altered somewhat by noise, but not to an extent that would make it irreparable by error-correction mechanisms. It is precisely this problem to which the conditional probabilities of the channel transition matrix are ascribed. $p(b_j|a_i)$ means the probability that the symbol b_j appears at the output of the line, given that symbol a_i entered at its input. Likewise, $p(a_i|b_j)$ looks at the same process from the other direction: it is the probability that the input symbol was a_i, given that a particular symbol b_j appeared at the receiver end. In addition, we introduce the probabilities for the occurrence of any pair of symbols a_i and b_j with

$$p(a_i, b_j) = p(a_i) \times p(b_j|a_i) = p(b_j) \times p(a_i|b_j)$$

This general equality is known as Bayes' theorem of conditional probabilities (Section 2.4).

The total numbers of symbols a_i and b_j need not be the same. Hence, the transition matrix does not have to be square. Since different redundancies in both alphabets may be involved, and since error correction requires a more extensive source code (cf. Section 3.4), the input and output alphabets usually differ. Otherwise, the model is completely described by the matrix of conditional probabilities $p(b_j|a_i)$. In this matrix, the ith row shows all the terms for symbols b_j that are received as a consequence of sending a symbol a_i. Their sum must equal 1, since every input a_i must have some output, the individual terms showing the distribution of probabilities for the outcome b_j. Each column j then refers to the received symbol b_j, giving the contributions to the occurrence of the output b_j induced by the various inputs a_i. Finally, the product of the input probability $p(a_i)$ with the conditional probability $p(b_j|a_i)$, summed over all indices i and j must again equal one, since the total input must correspond to the total output.

We may now use the various probability distributions for formulating their entropies according to the general scheme $H = -\Sigma p_i \log p_i$. We call these entropies H(A), H(B), H(A|B), H(B|A) and H(A,B). The capital letters A and B again refer to all a_is and b_js in an obvious way. This leads us to what is called the "mutual entropy" or "mutual information" I(A;B), which is the central issue of Shannon's most important theorem. It comes out as $I(A; B) = H(A) + H(B) - H(A, B)$, which is identical to $I(A; B) = H(A) - H(A|B) = H(B) - H(B|A)$ (for more details see Hamming's monograph (Ref. 18)).

Although the above paragraph deals mainly with definitions, I must apologise for violating my principle of keeping mathematical formalism to a minimum. I shall now do penance for my negligence and explain in words what it is all about. Mutual entropy provides a measure of the information gain of the whole system. On the basis of its mathematical structure it is always greater than or equal to zero. "Equal to zero" means that A and B are completely independent of one another. In other words: none of the information in A can be found in B, or the message A has been made completely meaningless by superposition of noise. According to the expression quoted above $I(A; B) = 0$ means that $p(b_j|a_i) = p(b_j)$ holds for all i,j. Of course, $I(A; B) = 0$ makes no sense for a "transmission" channel. On the other hand, the maximum of mutual information is given by the channel capacity, where the input symbols have to be chosen to fit the channel capacity. Generally speaking, the system's mutual entropy is the extent of information common to both input and output. This explains the name "mutual information", which is used in communication theory but has no obvious counterpart in statistical mechanics (because we do not usually deal with single microstates there). If we express mutual information as "gain of information", we must be aware that the "gain" is of purely communicative, rather than generative, nature. Gain means gain in sharing, not in creating (an aspect that is important to note with respect to later parts of this book).

Shannon's main theorem is largely based on the concept of mutual entropy. The theorem states: "There exist codes which are arbitrarily reliable and which can signal at rates arbitrarily close to the channel capacity."[22]

The proof of this, even for the special case of a binary symmetric channel, for which it was first conceived, is beyond the scope of this book. It involves theory of coding based on general Markov processes, their extensions and error correction of extended codes; it includes quite a number of originally strange ideas such as random encoding and averaging over all random codes, which eventually led Shannon to success in proving that for sufficiently long messages the mutual entropy I(A;B) can be adapted to the channel capacity C, involving the assumption that whatever works for *an average of random codes* must work for at least one special code. I shall not go into all these details, but prefer to focus on some of the steps involved, which will turn out to be important for later parts of the book. What is relevant in this respect is the conclusion to be drawn from the results, rather than the details of the theory, which in any case can be found in textbooks.

It was shown in the preceding section that a suitably large extension of a code in the absence of noise can provide an optimal code structure, matching arbitrarily closely the entropy of the source alphabet. Then, after introducing channel capacity, Shannon could prove the fundamental result that even in the presence of noise one can send at a rate as close to the channel's capacity as one pleases, without losing reliability of transmission. However, Hamming cautions us that this is an "in principle" result rather than a practical one. Let me quote Hamming: "That in the past we have had great trouble finding those good codes is another thing entirely. The reason is not far to seek – in the proof we have regularly assumed that we could take the block length k long enough. This 'long enough' is, in fact, very long indeed." He concludes: "Thus the theorem sets bounds but does not suggest much other than that very good codes must be very long, rather impractical ones."

Despite its somewhat critical undertone, reflecting the view of an expert in computer science, this remark clearly expresses the nub of Shannon's concept. The main objective is to raise the vocabulary to a maximum size by concatenations of the input symbols, e.g. an alphabet of 27 letters with its "meagre" five-bit positions to the kth power for the much more readily available combinations of k letters differing drastically in their abundance. This allows us to utilise the full extent of entropy constraints that the intricate structure of language can offer, in order to make communication as precise and economical as possible. As in statistical mechanics, it is found that in order to obtain good averages only a minority of the states, in particular those of high abundance, have to be populated.

Shannon's classical theory does not deal with the content of information (i.e. its semantic aspect) or with its dynamics of generation and conservation. It assumes that "information" as such is given as a defined sequence of symbols that does not require any specification. All that is required is that the sequence belongs to the "high probability group" of statistical expectance, for which the constraints of language are defined. There are no populations of sequences (of the kind that, as we shall see, are required in evolutionary theory). The probabilistic character is not based on distributions of interchangeable microstates, as is the case for statistical thermodynamics. It refers to one and the same message, which has to be long enough to allow the statistics to establish a macrostate reflecting clearly the constraints of language, as does any other (possibly meaningless) sequence in this group defined by H_M. In fact, a sufficiently long sequence can deal with any possible situation in our material and spiritual world. This underscores the fact that Shannon's "extent of information" is totally independent of its semantic content.

Richard Hamming says in the introduction to his book: "In this abstract theory there is no transmission, no storage of information, and no 'noise is added to the signal'. These are merely colourful words used to motivate the theory. We shall continue to use them, but the reader should not be deceived; ultimately this is merely a theory of the representation of symbols." He continues: "Information theory sets bounds on what can be done but does little to aid in the design of a particular system." This indeed is the work of the engineers who utilise the information provided by theory. The result is the giant post-modern edifice of cyberspace.

The theory itself became an active branch of mathematical research, integrating various tools such as group theory and the theory of finite fields (both pioneered by Évaríste Galois). In my representation I want to emphasise the suggestive power of Shannon's theory without overstressing its range of applicability. Its limits are that it deals with a properly defined amount of information and cannot in itself be extended to include value and meaning. To achieve this, we need a complementary dynamical approach.

3.4. Why Coding?

When in the 1830s the American painter Samuel Finley Breese Morse thought up the code that is named after him, he wanted to use the newly invented electromagnet to convert the letters of the alphabet into electrical signals, in order to use them for telecommunication. In 1835 he produced a workable electric telegraph, but Morse was neither the only nor the first person to have had such an idea. In Göttingen, in 1833, Carl Friedrich Gauss – the "princeps mathematicorum", as he was called by his patron, the Duke of Brunswick – co-operated with the physicist Wilhelm Weber to build an electromagnetic telegraph. His first message (see Section 5.10) was sent from the astronomical observatory to the physics institute.

Certainly, one of the answers to my question, "why coding?", is: coding accomplishes a conversion of abstract "logical" symbols into transmissible physical signals. These can be electrical, optical or acoustic impulses of different duration, frequency or amplitude, or they could be of a chemical nature, as many signals in biological systems are.

However, the problem of coding is of a much more fundamental nature. It lies at the root of human knowledge. We experience the world around us through our sensory organs; in other words, we perceive physical signals which our brain has to convert into a context that can be reflected in words. In the first chapter, I mentioned the desire of physicists to include the observer in a "participatory universe", but any research in this area – both in physics and in neurobiology – is still in its infancy. I shall revisit this question in various parts of the book. In the present section I shall have to restrict myself to the subject of coding in communication theory. Even in this restricted area, coding is much more than a simple matter of converting language patterns into sequences of physical signals.

Coding has to be seen within the context of precise and economical representation of a certain information content. As such, it includes both compression and expansion of texts. It is encoding that allows a practical realisation of Shannon's first theorem, according to which a message of length ν can be compressed to its entropy $\nu \cdot H_M^k$, where H_M^k is the Markov entropy of kth order (per letter), a quantity appreciably smaller than 1. Optimal compression of texts for the purpose of transmission is of economic value. Moreover, it tells us how much information is really involved. On the other hand, there are also good reasons to expand the length of a message. This is to be seen in connection with Shannon's second theorem, according to which the encoded message can be sent at a rate arbitrarily close to the channel capacity, which is adapted to the Markov entropy H_M^k. However, owing to the omnipresence of noise, coding now requires additional symbols to be used for the automatic correction of errors.

Accordingly, the main purpose of coding is the practical realisation of both economical and error-free representation of the messages that arrive at the receiver. In addition, there might be a need for "encrypting" the message in order to make it available only to a receiver who is able to decipher it. I shall treat the latter problem at the end of this section; this part may be omitted by the reader because – despite its present-day relevance – it is not required for an understanding of subsequent sections.

How do we construct a code?

I once had a discussion about the nature of information with my colleague and friend, the Swiss chemist Albert Eschenmoser. Albert asked: "Isn't a spark in a mixture of hydrogen and oxygen 'information'?" My spontaneous answer was: "No, it's just a signal"; but I had to admit that such a signal can trigger the release of (or destroy) quite a lot of information. Yet Albert insisted: "Isn't the signal itself already 'information'?" Yes, one bit is the limiting case of information. Nevertheless, the main part of the information is contained in the instability of the oxygen–hydrogen mixture. In such a case the semantic content of a single signal can be devastating. Think of Hiroshima!

Of course, answering a question with "yes" or "no" provides just one bit of information; but there again, a great deal of information is already contained in the question.

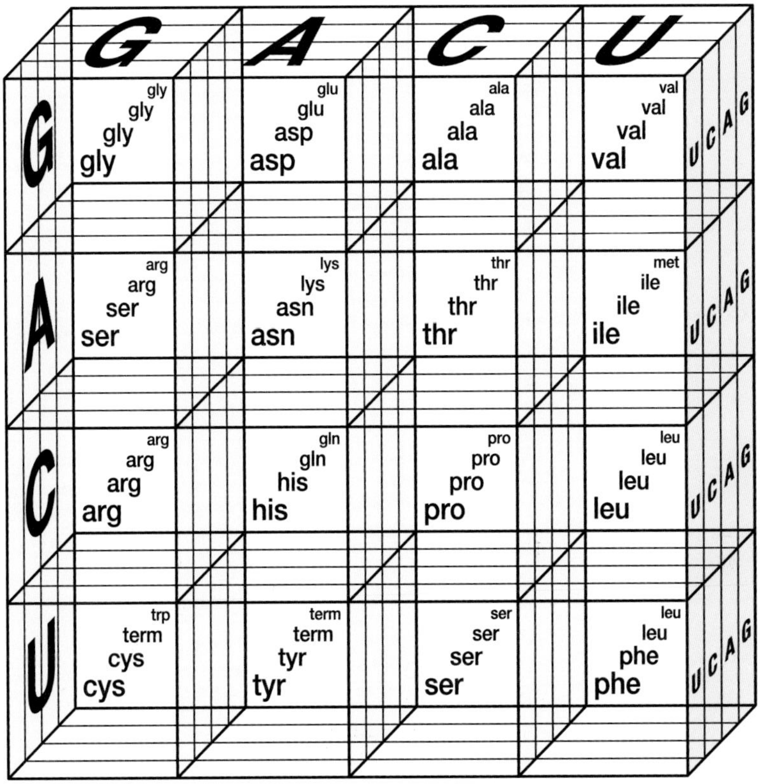

Figure 3.4.1

The genetic code

This is a typical example of a block code. It consists of 64 triplets of a quaternary alphabet and encodes the 20 natural amino acids found in proteins, including termination (term) and initiation (met) symbols. The amino acids are, in their three- and one-letter codes of our language: ala (A) = alanine, arg (R) = arginine, asp (D) = aspartic acid, asn (N) = asparagine, cys (C) = cysteine, glu (E) = glutamic acid, gln (Q) = glutamine, gly (G) = glycine, his (H) = histidine, ile (I) = isoleucine, leu (L) = leucine, lys (K) = lysine, met (M) = methionine, phe (F) = phenylalanine, pro (P) = proline, ser (S) = serine, thr (T) = threonine, trp (W) = tryptophan, tyr (Y) = tyrosine, val (V) = valine.

Usually, the coding problem concerns texts in a given language. Codes can then form "words" of fixed or of variable length. Both methods were already used by Gauss (see Figure 5.10.2). A five-bit code for an alphabet that contains $32 = 2^5$ letters (if we include some punctuation marks) seems to be perfectly matched to quintets of binary symbols. Such a code is called a block code, but it is not necessarily an optimal code. The genetic code (Figure 3.4.1) is a block code of nucleotide triplets. It may not

Table 3.4.1 The Morse Code

Letter	Code
A	·—
B	—···
C	—·—·
D	—··
E	·
F	··—·
G	— —·
H	····
I	··
J	·— — —
K	—·—
L	·—··
M	— —
N	—·
O	— — —
P	·— —·
Q	— —·—
R	·—·
S	···
T	—
U	··—
V	···—
W	·— —
X	—··—
Y	—·— —
Z	— —··

be the best choice on purely logical grounds, but it came about under evolutionary constraints, which put physical interaction before logic. These constraints are still reflected by the present code.

Morse code, mentioned above, differs from block codes by making use of variable-length code words. It is a ternary code, using dots, dashes and spaces (Table 3.4.1). In the international version, agreed upon in 1851 and still current, the shortest symbol, a single dot, is assigned to the most frequent letter in the English language, E, while the infrequent letters J and Q are encoded by the longest symbols, which consist of one dot and three dashes. The four-dash symbol and three more four-digit codewords are not used at all for encoding letters. Apart from its economical symbol assignment, the Morse code is not particularly efficient. A dash lasts three times as long as a dot; the space between dots and dashes corresponds to one dot, the space between letters within a word to three dots, and that between words to six. The Morse code was later replaced by the van Duuren code, which uses seven positions for each letter, of which three are always filled with 1 and the other four with 0. This code has a capacity of $7!/3!4! = 35$ code words. The receiver is thus able to detect several types of error by simply counting the 1s in every codeword.

In the computer world there is a need for a much larger source alphabet, including lower-case and capital letters, punctuation marks, numbers and mathematical symbols, frequently used commands for computer operations and other abbreviations. A typical example is the (now already "classical") ASCII code (Table 3.4.2), which has a source alphabet of $2^7 = 128$ characters. They are encoded inside the computer in a code called octal, because it uses eight symbols (0 to 7), each encoded by a triplet of binary digits. Hence it runs from 0 to 7, 10 to 17, 20 to 27, and so on till 70 to 77, and then 100 to 107 till 170 to 177, including altogether $8 \times 8 \times 2 = 2^7$ characters. Since computers usually work with bytes, which comprise bits, there is an eighth position free to be used for parity checks (cf. below).

A variable-length code can usually be made more economical than a code with fixed lengths. One can optimise such a variable-length code to achieve maximum efficiency, which means using a minimum number of digits for representing a given amount of information. If the probabilities of occurrence of the various source symbols differ appreciably – as is the case in our language – variable-length encoding can offer significant advantages over block encoding.

However, variable-length encoding brings along other problems, namely, how to recognise the end of a given code word and distinguish it from the beginning of the next one, i.e. how to guarantee unique decoding. The problem is best explained by an example, given by Hamming. Take a source alphabet with four symbols and the following coding rules: $s_1 = 0$, $s_2 = 01$, $s_3 = 11$ and $s_4 = 10$. If the message received reads 00110, it can be translated to either $s_1s_1s_3s_1$ or $s_1s_2s_4$. If the code had read $s_1 = 0$, $s_2 = 10$, $s_3 = 110$, $s_4 = 111$, then decoding would have yielded the unique answer $s_1s_1s_3$ because every 0 indicates the end of a code word, while three 1s or any multiple of three 1s after a 0 mean s_4 or a multiple of s_4.

One calls a code that is uniquely decodable right from the beginning of the message "instantaneous". To see what this means, replace the above code by $s_1 = 0$, $s_2 = 01$, $s_3 = 011$, $s_4 = 111$. The message is then also uniquely decodable, but it is not "instantaneous" because in order to know the end of a symbol one may first have to go to the end of the message, as would be the case for the sequence 0111...1111. Here one has to count from the end the triples of 1, arriving at the first codeword, which is s_1, s_2 or s_3. Optimally adapted instantaneous codes, taking into account the probabilities of the various symbols sent, are called Huffman codes. They make use of the variability of word length, ordering them inversely to the probabilities of the respective source symbols, i.e. $l_1 \leq l_2 \leq l_3 \ldots \leq l_\kappa$ according to $p_1 \geq p_2 \geq p_3 \ldots \geq p_K$.

Shannon and Fano introduced another way of coding which is (for long messages only slightly) less efficient than Huffman coding, but which has the advantage of

Table 3.4.2 Seven-Bit ASCII Code

Octal code	Char.	Octal code	Char.	Octal code	Char.	Octal code	Char.
000	NUL	040	SP	100	@	140	“
001	SOH	041	!	101	A	141	a
002	STX	042	“	102	B	142	b
003	ETX	043	#	103	C	143	c
004	EOT	044	$	104	D	144	d
005	ENQ	045	%	105	E	145	e
006	ACK	046	&	106	F	146	f
007	BEL	047	'	107	G	147	g
010	BS	050	(	110	H	150	h
011	HT	051	)	111	I	151	i
012	Lf	052	*	112	J	152	j
013	VT	053	+	113	K	153	k
014	FF	054	,	114	L	154	l
015	CR	055	–	115	M	155	m
016	SO	056	.	116	N	156	n
017	SI	057	/	117	O	157	o
020	DLE	060	0	120	P	160	p
021	DC1	061	1	121	Q	161	q
022	DC2	062	2	122	R	162	r
023	DC3	063	3	123	S	163	s
024	DC4	064	4	124	T	164	t
025	NAK	065	5	125	U	165	u
026	SYN	066	6	126	V	166	v
027	ETB	067	7	127	W	167	w
030	CAN	070	8	130	X	170	x
031	EM	071	9	131	Y	171	y
032	SUB	072	:	132	Z	172	z
033	ESC	073	;	133	[	173	{
034	FS	074	<	134	\	174	|
035	GS	075	=	135	]	175	}
036	RS	076	>	136	^	176	~
037	US	077	?	137	–	177	DEL

starting directly from the probability p_i and going from there to the word length I_i, building upon the fact that for each probability p_i there is an integer l_i lying between $\log(1/p_i)$ and $\log(1/p_i) + 1$. Summing l_i over all probabilities p_i (the sum of which is 1) yields what is called Kraft's inequality, which is necessary and sufficient for the existence of an instantaneous code. This fundamental inequality states that the average length l; of such a code lies between the entropy H(S) of the symbol system and H(S) + 1. It was the point of departure for Shannon's derivation of his first theorem.

The second part of my answer to "Why coding?" refers to error correction in connection with Shannon's second theorem for noisy messages (Section 3.3). No modern computer, carrying out billions of operations within a short time-span, could work without using error-correcting codes.

The essence of error correction is to find a way to add some coding devices that allow the tracking-down and correcting of errors without impairing the whole economy of coding. An easy but uneconomical way would be to send every message more than twice and decide by "democratic vote" which symbol is the correct one. Hamming's method, by contrast, is appreciably less costly and much more elegant. It consists of using some extra space in the message for employing rules by which the correctness of the transmitted message can be checked. These rules imply certain parities among the symbols which must be fulfilled. If, for instance, we feel reasonably sure that there cannot be more than one error within a block of digits of a certain length, then we add to each such block another digit which adjusts the number of 1s, making their total always even. Hence, in each block of n digits to be transmitted, the sender uses n – 1 positions for encoding the message and the nth position for a parity check. The receiver then counts the number of 1s in each block. An odd number indicates an error. This method cannot detect two (or any other even number of) errors in a block, nor does it identify the position of an error within a block. In order to achieve this, one has to employ further independent parity checks.

For finding the parity in a string of 0s and 1s, we can use two-state finite automatons, such as the one shown in Figure 3.4.2. Starting at state 0, each 1 in the message will cause a change of state. The final state at the end of the message gives the parity count. Computers are usually equipped for doing such counts of the numbers of 1s in the accumulator register.

Hamming worked out an error-detecting code based on such parity checks. He used an algebraic method and demonstrated its function in the geometric space model discussed in Section 3.3. I have two reasons to discuss his geometric approach in this book:

(1) The geometric scheme is more lucid and provides an illustrative physical representation.

(2) The generalised scheme can be used to demonstrate the effect of mutations in the evolutionary generation of information (see Sections 4.4, 4.5 and 4.7).

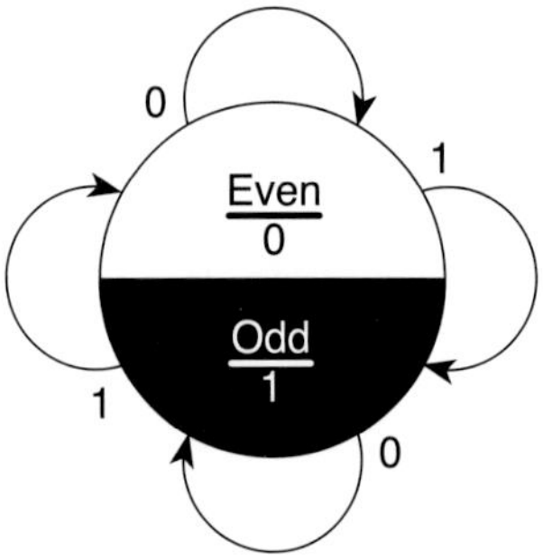

Figure 3.4.2
Error detection by parity checking

A message is written in (sufficiently long) blocks of n symbols (in binary notation), of which only (n – 1) symbols refer to the message, while the nth symbol makes possible a parity check. It is a "0" if the block includes an even number of 1s, and a "1" if it contains an odd number. In this way it can be found out whether a block contains one error. This holds for the original message, which may undergo changes in transmission. "Parity" then refers to the agreement of the original message with the one that arrives at the receiver. This is a procedure (mod 2), as explained later in this section. This method distinguishes even from odd numbers. Hence additional checking is required in order to distinguish one single error from any odd number of errors as well as the possible presence of an even number of errors. For this purpose a suitable length of block should be chosen. Methods that determine the presence and position of errors have been worked out by Richard Hamming[18]. The two-state, finite automaton, shown in the picture, can carry out such a parity check. The two states (white and black) remain unchanged for agreement of 0s or 1s, while they change for any disagreement. The final stage at the end of the message gives the parity count.

Let's go back to Figures 3.2.3 and 3.2.4. We see immediately how to define "spheres" in n-dimensional information space. The surface of a sphere in three-dimensional Euclidean space is the geometric locus of all points that have an equal distance from a given point, which marks the centre of the sphere. This holds as well for a Euclidian space of any dimension and, in particular, for Hamming's space, where all points at an equal distance from a given point define the surface of a Hamming sphere. Figure 3.2.3 has been drawn to emphasise the role of the points 000 and 111 as references, but such a diagram could be drawn for any point in Hamming space and for any number of digits, in particular those representing the string of symbols representing a message.

Now let us consider the role of such spheres in Hamming's theory of error-correcting codes. For this we subdivide the whole point space into non-overlapping spheres of equal volume, as shown in Figure 3.4.3 (see also Table 3.4.3). The total volume of the space, which is 2^{ν}, divided by the volume of a subsphere, must be larger than or equal to the maximum number of subspheres. This relation can be fulfilled as an equation only in very special cases. The codes derived for such a case are called "perfect codes". Perfect codes for the correction of more than one error per sequence are, in fact, very rare (see Section 4.5). If equality does not hold, the maximum volume is just an upper limit for the number of subspheres that can be accommodated in the total volume of space.

Let us try to understand Hamming's concept in more detail. As an example, a total set of non-overlapping Hamming spheres with radius 1 in seven-dimensional sequence space is shown in Figure 3.4.3. The centres of all neighbouring spheres in this case have to have a mutual distance of ≥ 3. The volume of each "radius-one sphere" includes eight points, of which seven are the "vertices" of its surface. For each point in Hamming space, a minimum distance must be 1 in order to guarantee the uniqueness of the code. Among spheres, a distance of 2 is sufficient for detecting the presence of one error, as explained in the legend of Figure 3.4.3, while the localisation of the error necessitates a minimum distance of 3. For the correction of two or more errors we have to use Hamming spheres of radius 2 (with a minimum distance of 5) or larger, allowing multiple parity checks. The construction of higher error-correcting codes can be based on Hamming's geometrical model. A full theory, which eventually becomes quite complex, has been developed over the years.

In this representation we can clearly identify the centre sequence as the only one resulting from all possible one-error changes in a set. Since the ν-dimensional point space – unlike its two-dimensional representation – does not distinguish any single point we need to know the full neighbourhood in order to find its respective centre. This is true in a similar way for all spheres with radii $k > 1$, having $\binom{\nu}{k}$ points at their surface (ν being the length of the sequence). The method becomes highly efficient for ν values much larger than 7, as shown in Table 3.4.3. In its most highly developed form the method includes tools of pure mathematics, such as group theory, Galois' theory of finite fields, linear programming and other topics.[23,24,25,26]

Hamming's geometric approach to error-correcting codes goes beyond the scope of Shannon's theory of information because it focuses on an identification of individual sequences rather than on their statistical representation only. Coding theory originated at (about) the same place and during the same period as information theory. Its publication, however, was delayed until April 1950, when it appeared in the Bell System Technical Journal.[27] Today, the idea of an information space as a new arena for performing physics in biology proves to be most useful (see Sections 4.4 and 4.5). In later parts of this book I shall revisit (overlapping) Hamming spheres as sites of mutant distributions in the information space of genes, or as excitation patterns in networks of neurons. This concept will demonstrate its relevance to reality when we try to understand the origin of semantic information or meaning in biology, which transcends an interpretation in terms of structural properties of matter.

Another approach to coding is to be found in cryptography. If we consider coding as "representation of symbols" for the sake of communication, then we may, for a moment, also think of its opposite role, namely coding for the purpose of hiding information. Strictly speaking the word "code" should then be replaced by the word "cipher", but the two are often confused and used as synonyms in an inconsistent way.

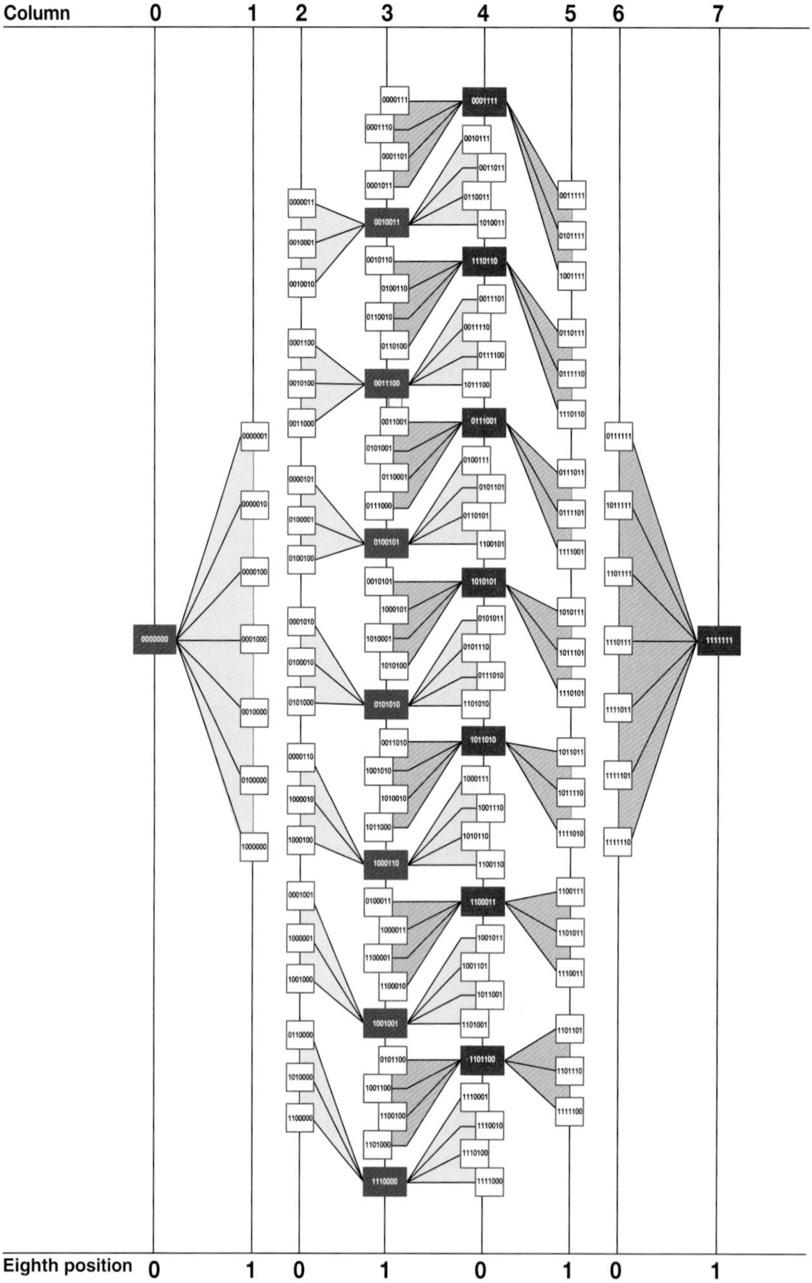

Figure 3.4.3 (Cont.)

Yet a message can be both encoded and enciphered, or "encrypted". Th
represent the message in a most economical form and make it available u
to anyone using the equipment involved, which may be a telecommunication n
computer. The device is usually constructed by an electrical engineer for comme
purposes and often includes automatic procedures for coding, decoding and erro
correction. Enciphering and deciphering of messages that are to be kept secret is a task left to the user; this is a problem in its own right. In the following pages I shall briefly digress into this topic of coding information.

Cryptology is a fascinating discipline, and not only because of its obvious relevance in our chip-card age. It is so exciting because it leads us to the limits of today's mathematics. The classical method is to apply a secret code to a message and to utilise the dissymmetry of being, or not being, in possession of the right "key" for deciphering it in order to protect it from illegitimate access. However, if the dissymmetry rests solely on personal knowledge, and not on fundamental dissymmetries, there

---

Figure 3.4.3

Hamming spheres and error-correcting codes

We consider a message consisting of seven digits (0 and 1). An eighth digit to be used for parity checks is added before transmission. A "0" in the eighth position indicates that the original message contains an even number of 1s, while "1" indicates an odd number. The message received is one of the $2^7 = 128$ possible combinations of 0s and 1s at the seven message positions, possibly containing one (and in the present model only one) mistake. The whole scheme of 128 points can be subdivided into 16 "spheres", each containing eight (= one central and seven surface) points. The central points are shown in dark colours (blue or red). The whole scheme distinguishes eight vertical rows, numbered 0–8 according to the number of 1s in all sequences of a given row. The number at the bottom of each row (0 or 1) indicates whether the number of 1s is even or odd (eighth position). The scheme, indicating a "perfect code", is used to explain Hamming's concept of error-correcting codes.

The geometrical model immediately shows that the central point of a sphere, if it fulfils the parity check, must be the correct message because any of its one-error mutants in the original message would have given the opposite result. Likewise, if the parity check – which was introduced with the original message – does not agree with the received message, the latter must contain one error (or an odd number of errors). Since *a priori* no point is distinguished, it is possible to find out which alternative scheme would cause one of the seven one-error mutants of the recorded sequence to be a central sequence that fulfils the parity check. For this, all seven one-error mutants have to be known. This explains why just "detecting" one error requires only a distance of two among spheres, since any of the mutants in a given sphere cannot be at the same time a central sequence of another sphere within the same scheme. The fact of the non-fulfilment of the parity check, however, is sufficient for indicating the presence of one error. On the other hand, identifying the error requires non-overlapping spheres and therefore a minimum distance of three. This concept can be generalised to any larger number of errors using additional parity checks, but carrying out this generalisation involves quite a bit of mathematics.

Table 3.4.3 Hamming Spheres

Radius	Surface $= \binom{\nu}{k}$			Volume $\sum_{1}^{k} \binom{\nu}{k}$			Percentage of symbols to be used for communication		
k	$\nu = 10$	100	1000	$\nu = 10$	100	1000	$\nu = 10$	100	1000
1	10	100	1000	11	101	1001	67	93	99.0
2	45	4950	499500	56	5051	500501	43	81	$\sim$ 98.1
5	252	7.5×10^7	8.3×10^{12}	638	7.9×10^7	8.3×10^{12}	10	73	$\sim$ 95.7

Symbols with only two places are approximate. The last column has been obtained in the following way: The total volume of space 2^ν (for $\nu = 10$, 100 and 1000 ≈ 1024, $\sim 10^{30}$ and $\sim 10^{301}$) is divided by the volume of a sphere in order to yield the number of spheres filling the total volume. These numbers are expressed as 2^x, where x yields the number of message symbols available for communication (with correction of 1,2, or 5 errors).

As the table shows, the method works best for large values of ν. For $\nu \gg k$ the surface is simply given by $\nu^k/k!$ The volume of the sphere is then practically given by this surface term, a relation we have found also for the multidimensional Euclidean phase space of statistical mechanics (Section 2.5). In particular, for large values of ν the extra digits to be used for error corrections are in the range of one to a few percent. A value of $\nu = 1000$ corresponds to no more than three lines of a typescript encoded in binary digits. A modern computer with its billions of operations could not work in the absence of error-correcting codes.

is always a way of reconstructing the key from the regularities of th message. That is how Alan Turing and his team cracked the Enigma code War II. The mathematician Jan Stewart[28] describes the problem in the fo way. A message M is encrypted with the help of some function f to become f Decryption then means finding an inverse function f^{-1}, such that $f^{-1}f(M) = M$. In many cases, knowing how to encrypt a message automatically imparts a knowledge of how to undo the encryption.

However, this symmetry is not present in all cases. The dissymmetry brought about by "knowing or not knowing" may be partly transferred into the mathematics of decryption by making the unaided finding of f^{-1} very time-consuming and thereby practically intractable.

Jan Stewart calls such functions, where f is very easy but f^{-1} practically impossible to compute, "trapdoor functions". Trapdoor functions as such are not yet very useful because they present the same problem to the legitimate as to the illegitimate recipient of the message. The way out of the dilemma is to find some extra piece of (secret) information that makes the determination of the deciphering function f^{-1} easy for the informed decipherer, but nearly impossible for any wire-tapper. In fact the method and all quantities involved – except this "extra piece" of information – may well be publicly known, and automatic devices that are able to run the decryption program on providing the (otherwise secret) piece of information are all around us. Such a system, called a "public-key cryptosystem", is ideally adapted to our chip-card age, in which we can do our banking through automata.

The encryption–decryption principle described is the basis of the RSA system, named after its inventors Ronald L. Rivest, Adi Shamir and Leonard Adleman,[29] who opened a new epoch of cryptography. (In fact this system was invented independently by James Ellis, Clifford Cocks and Malcolm Williamson, who were working for the British Secret Service and, accordingly, were not allowed to publish their results.) As a "trapdoor" function, the RSA system utilises the factorisation of large numbers, that is, their decomposition into primes. In order to compute the trapdoor function f one just multiplies two sufficiently large primes. Of course, very large prime numbers, with hundreds or thousands of digits, cannot just be taken from reference books. Such a library would exceed the spatial bounds of our universe. Therefore, it is important to be able to identify a given large number as being a prime. The largest prime number known at the time I write this is $2^{43112609} - 1$, which runs to about 13 million decimal places[30]. With present-day computers, the primality of a 1000-digit number can be tested within one or two days.

Multiplying two arbitrarily chosen large primes p and q in order to obtain the trapdoor function f is a trivial task. However, the reverse process, i.e. decomposing the product pq into p and q, is a formidable problem. Any systematic approach with pq comprising hundreds of places would make negligible progress. A pq value of 500 places would require some 10^{40} steps, which would render even our best ntirely helpless.* (Remember that the age of the universe amounts to ne 10^{17} s.)

RSA method requires three numbers. One is the above-mentioned product et us call it R. The second number is an encoding key E, and the third a decoding ey D. Only the last key is to be kept secret, while R and E, both more or less arbitrarily chosen, can be made publicly known. In order to determine D, one must know p or q individually. How to get to D is the non-trivial part of the problem, and this requires two auxiliary ideas, which I am going to discuss now.

The first idea leads us back to the foundations of arithmetic and number theory as established by Carl Friedrich Gauss in his *Disquisitiones Arithmeticae* (1801). Jan Stewart explains the essential idea with the example of a clock running periodically through 12 hours. The clock arithmetic knows only the numbers 0 to 11 (with 12 equivalent to 0). Three hours after 10 o'clock is 1 o'clock, regardless of how many cycles the clock has completed before. According to Gauss this can be described as an arithmetic "modulo 12". "Modulo" (abbreviated mod) simply means "according to the modulus" which tells us that every cycle starts again with 0 at 12 o'clock. So we have $13 \equiv 1 \pmod{12}$ or $14 \equiv 2 \pmod{12}$ and so on, simply ignoring all multiples of 12. Instead of using for the above relations the equality sign "=", one uses the sign "≡", which is read as "is congruent with". One may, of course, choose any number n as the modulus. The number (mod n) then means the remainder left after subtraction of any multiple of n. A numerical example is $5^2 \equiv 1 \pmod{12}$. It means $5^2 = 25 = 2 \times 12 + 1$, one being the remainder, which for mod 12 runs from 1 to 11 for numbers between 25 and 35 while 12, 24 and 36 are taken as equivalent to 0. This odd-sounding arithmetic has proven to be of fundamental importance in dealing with questions of divisibility and number theory. The remainders (mod n) will play the main role in coding and decoding using prime numbers.

The second auxiliary idea goes back even further in the history of mathematics, namely to a theorem proposed by Pierre de Fermat. It is called Fermat's "little" theorem, in order to distinguish it from his "last" theorem which, however, remained a conjecture until its proof was recently achieved by Andrew Wiles (see Section 3.6). Fermat's (little) theorem (see Vignette 3.4.1) proposes that, if P is a prime and a is some (positive) integer that is not divisible by P, then the (P – 1)th power of a minus 1 (i.e. $a^{P-1} - 1$) is divisible by P (i.e. is equal to nP, n being an integer). Fermat, living over 100 years before Gauss, certainly did not know of "modulo arithmetic", which allows an illustrative access to his problem. In this terminology, Fermat's theorem simply reads $a^{P-1} \equiv 1 \pmod{P}$. This is the equivalent of saying that a clock, the full

* I have been asked whether this does not contradict the last sentence of the preceding paragraph. However, these are two different questions. In the first, we are asking whether a given sequence of digits makes up a prime number; in the second, we are seeking to identify its prime factors p and q, which is as far more laborious task.

Vignette 3.4.1 On Fermat's "Little" Theorem

Group theory is certainly the most general and elegant branch of mathematics for dealing with problems like Fermat's "little" theorem – yet neither its terminology nor its logical depth are easily accessible to anyone not trained in mathematics. Nonetheless, Fermat's problem is transparent enough to allow a thorough inspection of some numerical examples, providing the essential clues for a general proof. For this purpose, we consider what I call a "prime clock" (Figure 3.4.4). This is a clock with a dial the complete round of which includes the numbers 1 to P.

In the example below I chose P equal to 17. According to Ian Stewart, who in his book *From Here to Infinity* introduced the analogue of a clock for problems of modulo arithmetic, the prime clock is defined by P (mod P) $\equiv$ 0 (mod P). P (in our example equal to 17) is congruent with zero and the positions of the hands of the clock are remainders of full rounds: r_i (mod P).

In this terminology Fermat's "little" theorem states: $a^{P-1} \equiv 1$ (mod P), where a is any integer not divisible by P. In order to analyse this relation let me verbally describe the scheme I am using below where a^{P-1} is explicitly written as a (P–1)-fold product – i.e. $a \cdot a \ldots a$; P times – for all a values $1 < a < P$. The integer a must not be divisible by P, and problems for $a > P$ (with $a = nP + r_a$, n being an integer ≥ 1 and r_a being a remainder $< P$) can be reduced to the former case.

In this scheme each horizontal series represents the exponential a^{k-1} as the k-fold product $(a \cdot a \cdot a \ldots)$ written in terms of remainders r_{k-1} (mod P). This splitting of the exponential a^{k-1} into a multiple product is equivalent of saying that each subsequent term r_k in this series, starting with $r_{k-1} = 1$ for $k = 1$, follows from its precursor r_{k-1} by multiplication by a. It turns out that the last term in this series for k = P always equals the first term, and that is what Fermat's theorem proposes. But how does it come about?

First, we realise that in such a series each term r_i is unambiguously determined by its precursor r_{i-1} (through multiplication by a) and hence (again after multiplication by "a") fixes the subsequent remainder term r_{i+1}. This constraint is important for obtaining a well-defined periodic structure in terms of remainders in which the maximum period includes P terms, but in addition – as the example shows – periods which may contain P/k terms, depending on k being a divisor of P–1. P–1 is always an even number and hence divisible by two. If (P–1)/2 is another prime (example $P = 23$; $(P - 1)/2 = 11$), only three kinds of periods occur, namely those of (P–1) terms, those of (P–1)/2 terms and one with only two terms. (The scheme for $P = 23$ yields 10 sequences each of the two first-mentioned categories and one sequence of alternating 1 and 22.) Another interesting case is the prime 31, showing periods with 2, 3, 5, 6, 10 and 30 symbols. If the sequence contains P–1 symbols, they all are different from one another yielding $\sum_{k=1}^{P-1} k = P \cdot (P-1)/2$, i.e. a divisibility by P or

Cont. ➲

ont.

$a^{P-1} = 1$ (mod P), which corresponds to Fermat's "little" theorem $a^{P-1} - 1$ being divisible by P. It is true that all fractional periodicities upon repetition reach the value 1 (mod P) for $k = P$; we must nevertheless ask the question why other periodicities that do not do so (e.g. those with $k > P/2$) cannot appear. One reason is that for any given a the sequence is uniquely fixed. The other reason is that this unique fixation includes symmetries which become apparent when one looks at the dial of the "prime clock". As for any clock there is a symmetry for going clockwise and anticlockwise. We could call the numbers that have equal distance from P "complementary", because they cancel out when added, as complementary colours in a colour cycle "cancel out" to white. We can arrange the prime clock like an ordinary clock, with the prime P at the top of the dial. Then P–k and k (for any $1 > k \leq P{-}1$) cancel out to $P \equiv 0$ (mod P) (see Figure 3.4.4). The only difference from an ordinary clock is that P/2 does not exist (as a full "hour"). This symmetry is present in all periods with an even number of terms. It is obvious for those series in the scheme which have a period of P–1 and Gauss' formula for the sum of numbers 1 to P–1 which is divisible by P tells us that it conforms with Fermat's theorem. But the same symmetry also tells us that at the middle term of such a series must be ($r_{(P-1)/2}$) equal to P–1. And that is true also for fractional periods which come out to be at the term for $k = (P-1)/2$ either with $a = 1$ or $a = P-1$. The addition of the fractional periods then takes care of the fact that a^{k-1} for $k = P$ always yields a remainder $r_P = 1$ (mod P).

There is much more detail involved in the symmetries of prime clocks than I wish to show here. All I wanted to demonstrate is that a proof of a theorem with such far-reaching consequences in cryptology as Fermat's "little" theorem can be performed using mathematical operations that are intelligible for a layman and therefore may indeed be of value for teaching in high schools. Take, for instance, a puzzling assertion such as "9999999999999999 is divisible by 17", which in the above scheme just takes one line ($a = 10$; $P = 17$; $10^{16} = 1$ (mod 17)), meaning that $10^{16}-1$ is divisible by 17 leaving an integer as large as 588235294117647. Certainly a proof performed in the way outlined above takes many words. The task of mathematics is to make proofs as short and as general as possible, and for the problem in question – as mentioned above – group theory is the appropriate language. The problems I have treated in this table involve relatively small primes, such as $P = 13$, 17, 23, 31 or 43, leading already to quite respectable numbers for a^{P-1}. The primes used in cryptology, in order to construct a "trapdoor function", contain many more than just two digits, in order to paralyse a computer in any attempt to reconstruct the code used.

round of which includes P hours, starting at 1 o'clock will always arrive o'clock after a series of P – 1 moves of its hand, each of which produces a re $p_i = ap_{i-1}$ from the remainder p_{i-1} of the preceding move. Both p_{i-1} and p_i be taken as (mod P). They are positions on the clock which may have been reac through one or several rounds of the clock. The index i runs from 1 to P, the firs term being $p_1 \equiv 1$ and the last resulting term $p_P \equiv 1$ (mod P). In this procedure the total number of rounds increases exponentially with i as a^i. As an example, take P = 13 and a = 3. The second last term ($a^{P-1} = 3^{12}$) amounts to 531441. This number equals $1 + 40880 \times 13$, which is congruent to 1 (mod 13).

What I have shown so far is an illustration of Fermat's theorem in terms of modulo arithmetic rather than an explanation, let alone a proof. The first proof of the theorem was provided by Leonhard Euler. The modern version is a typical case for group theory, which is almost designed for problems of this kind. There is no need to reproduce it here. Instead, I shall try to explain the process again for the reader in Vignette 3.4.2 and in the caption to Figure 3.4.4. Such examples use relatively small primes, such as 13, 17, 23, 31 or 43, involving only two decimal places. The largest primes known today have millions of places. The trick of the method, described below, is to construct trapdoor functions by decomposition of numbers into large primes, a reconstruction of which would burn out even a modern computer.

Vignette 3.4.2

In order to encipher and decipher a message we express each letter of the message (M) by a number, e.g. by its place in the alphabet (a = 01, b = 02, ...j = 10...z = 26; spacing = 00). The message is now a sequence of digits and can be split into blocks represented by their numbers B_i, e.g.:

1105 0516 0020 0809 1411 0914 0700

Apply the encoding key E according to the formula EB_i (mod n), where n is the product of two large primes p and q. The encoded message C consists of the blocks $C_i \equiv EB_i$ (mod n). Each C_i is the remainder r_i if EB_i is expressed as a multiple of n, according to $EB_i = a_i n + r_i$, where a_i is an integer and r_i is the remainder with $0 < r_i < n$. An example for the above message with E = 5 and n = 11 × 13 = 143 is:*

0 0 7 8 0 1 2 0 0 0 8 9 0 0 8 7 0 0 8 9 0 0 2 3 0 0 9 8

If the decoding key D is known, decoding according to $DC_i \equiv B_i$ (mod n) is an analogous procedure.

Cont. ➲

ont.

In order to determine the decryption key D apply condition $E \times D \equiv 1$ (mod z), where z is (p – 1)(q – 1). Start with $z \equiv r_1$ (mod E), i.e. determine the remainder r_1 if z is expressed as a multiple of E. This remainder is then the modulus for expressing E in terms of r_1, yielding a second remainder r_2, i.e. $E \equiv r_2$ (mod r_1). The procedure is to be iterated until a remainder r_k appears that is equal to 1, or $r_{k-2} \equiv 1$ (mod r_{k-1}). This yields the equation $r_k = 1 = r_{k-2} - b_{k-2}r_{k-1,}$, where b_{k-2} is an integer that reveals how the remainder r_{k-2} can be expressed in multiples of the remainder r_{k-1}, leaving the remainder $r_k = 1$. Now one can reverse the iterative procedure and substitute the r_i values by the preceding relation until one arrives again at the first relations in which only E and z occur. In other words: the equation $1 = r_{k-2} - b_{k-2}r_{k-1}$ can be traced back to a relation 1 = multiple of z – remainder ($\sim$ E). The coefficient of E in this remainder can be identified with D because it is identical to the decryption key according to the condition $1 \equiv DE$(mod z).

A numerical example may illustrate the procedure for calculating D. We choose as parameters $n = p \times q = 490,048,499$, being the product of two primes $p = 1009$ and $q = 48611$. n is made publicly known while p and q are kept secret. One of them has to be known for calculating $z = (p-1)(q-1) = n + 1 - (p+q)$. With the above values we have $z = 48,998,880$. As encoding key we choose the prime $E = 73$. We evaluate the equations: $z \equiv r_1$(mod E), $E = r_2$(mod r_1) $\ldots r_{k-1} = r_k$ (mod r_{k-2}) and each $r_k = 1$ for $k = 5$. The series of equations reads:

$$z = 48998880 = 671217 \times E + 39$$

$$E = 73 = 1 \times r_1 + 34$$

$$r_1 = 39 = 1 \times r_2 + 5$$

$$r_2 = 34 = 6 \times r_3 + 4$$

$$r_3 = 5 = 1 \times r_4 + 1$$

$$r_4 = 4; r_5 = 1$$

Starting from the last equation $1 = r_3 - r_4$ we re-substitute iteratively r_4, r_3, r_2 and r_1 and arrive at the equation:

$$1 = 15z - 10068263E$$
$$(\text{with } 15z = 734,983,200 \text{ and } 10,068,263E = 734,983,199)$$

Cont. ➲

Cont.

establishing the congruence: $10,068,263E \equiv 1 \pmod{z}$ and hence identifying $D = 10,068,263$, satisfying the condition

$E \times D \equiv 1 \pmod{z}$

* The numbers for E and N were intentionally chosen to be fairly small, so that the reader can reproduce the computation steps with his or her pocket calculator. The block size determining the number B must be smaller than (and should be close to) n because the remainders of the operation EB (mod n) are smaller than (but possibly close to) n. In other words if large primes and hence n values with many places are involved, B values should be correspondingly large. The values of B and E, given that block sizes are (arbitrarily) chosen to be 4, and have even been (again intentionally) chosen too small, as the many zeros in the encoded message C indicate. This has been done in order to show the constraints of arbitrariness in the choice of parameters. Moreover, with EB in the above example, numbers with up to 18 places could possibly occur, which already require some tricks to be handled by a pocket calculator. An adapted value of n for the above choice of E would be $n = 97 \times 89 = 8633$.

The RSA system is based on this strange-looking, yet nonetheless internally entirely consistent, type of arithmetic. Three numbers play a dominant role: (1) the product of two large primes p and q, providing the "trapdoor function" $pq = R$, (2) the enciphering key E, which can be chosen arbitrarily and (3) the deciphering key D, which must not be easily accessible. For enciphering, the message must be converted into a sequence of digits and then broken up into blocks of numbers B_i of length $<N$. These blocks are to be enciphered according to the formula C (= encrypted message) $\equiv EB_i \pmod{N}$. In order to decrypt the message the deciphering key D (containing the "trapdoor function") is required, subject to the condition $DE \equiv 1 \pmod{z}$ where $z = (p-1)(q-1)$. Once D has been determined, one may conclude the decryption procedure by re-establishing the message M according to the formula $DC \equiv B \pmod{N}$, a relation that follows from a generalisation of Fermat's theorem by Leonhard Euler. The above relation $DE \equiv 1 \pmod{z}$ is symmetrical in D and E and allows D and E to be exchanged. However, it puts one constraint on E, namely, that it must not have a common divisor with z. Since p and q are primes, $p - 1$ and $q - 1$ are even numbers, and their product consequently is divisible by 4. One is therefore on the safe side if one chooses an arbitrary prime number for E and makes sure that z is not a multiple of E. More details can be found in Vignette 3.4 and here are a few notes:

(1) The main point is that while *en*cryption involves only the publicly known keys E and N, *de*cryption requires knowledge of the individual primes p and q from which the product N was formed. This is expressed in the relation $DE \equiv 1 \pmod{z}$, which involves $z = (p-1)(q-1)$. For the "trapdoor function" to

Figure 3.4.4
The prime clock

A prime number is, by definition, one that is not divisible by any integer except 1 and itself. The "hours" of the prime clock introduced in this figure, being 1/P-fractions of a total round, therefore look somewhat strange. The prime clock, however, is a lucid means of illustrating problems of modulo arithmetic. The case presented in Table 3.4.3 reveals certain interesting symmetries. All numbers smaller than P form $(P - 1)/2$ complementary pairs, the sum of which is equal to $P \equiv 0 \pmod{P}$. The two numbers of each complementary pair have equal distance from P, if measured clock- versus anticlockwise. Hence when added they yield 0 (mod P). In fact, they behave like complementary colours on a colour wheel. Examples for $P = 13$ are $12 = -1$, $11 = -2$, $10 = -3$, $9 = -4$, $8 = -5$, $7 = -6$, with 13 being 0 (mod 13).

be effective, the numbers p and q – and hence N and z – have to be very large.

(2) Because N and z are very large, the intermediate computational steps also involve very large numbers. This calls for the involvement of adequate computation equipment and of tricks to aid calculation. Note that E enters the coding procedure as an exponent. One therefore keeps it as small as possible, which, as the example in Vignette 3.4 shows, causes D to become very large if z is very large.

(3) The fact that the method works is due to its internal mathematical consistency, for which no proof is given in this representation (cf. also the more detailed monographs by L Bauer[31] and A. Beutelspacher[32]). Jan Stewart's excellent book provides the mathematical background, using a language that non-mathematicians can understand, while Rudolf Kippenhahn's[33] and Stephen Singh's[34] very lucid popular accounts are treasure troves for finding details about the history of cryptography and its present-day applications.

Are primes the final solution of cryptography? We can't say, because we are at the limits of what is known and we therefore can hardly predict the future. There seems to be a competitive method coming out of physics, which seeks to utilise both quantum-mechanical uncertainty and entanglement of wave functions to produce a trapdoor situation[35,36]. In this case, the encrypting is not due to the enciphering of a message but rather to its physical processing.

3.5. Do We Live in a Markovian World?

A Markov process is usually described in a dictionary (if at all) as a random process for which the rate of change depends only on the state immediately preceding the change, rather than on its earlier history. Since the same condition applies to any of the preceding states, the overall process links the system's presence quite consistently to its more or less distant past. The process is named after the Russian mathematician Andrei Andreyevich Markov, who understood it in a more general probabilistic way, linking the probability of the occurrence or existence of a given situation to its (possibly) causative conditions. The process is called a Markov chain if it involves discrete, rather than continuous, states.

Markov himself originally did not think of sequences of events in time. He looked for examples in literature rather than in physics. So he calculated the probabilities of the occurrence of vowels and consonants in Aleksandr Sergeyvich Pushkin's *Eugene Onegin*. In this spirit, in fact, Markov's idea has been applied by Shannon in his theory. In computer language, a "zero memory source" refers to symbols the appearance of each of which is totally independent of the symbols preceding it in the chain. In this case, each symbol is characterised by its *a priori* probability. A jth-order Markov process then is described by a conditional probability for symbol s_i as $p(s_i|s_{i1}, s_{i2}, \ldots s_{ij})$, i.e. the probability that the symbol s_i occurs, "conditioned" by the fact that it was preceded by the sequence of symbols $s_{i1}, s_{i2} \ldots s_{ij}$. A first-order Markov process thus describes nearest-neighbour interactions, and this represents an example that is frequently encountered in both language and physics. A typical case is the conditional probability $p(u|q)$, which is nearly equal to 1, since in dictionaries of European languages one will hardly find any word of non-foreign origin or non-artificial construction that contains a q without a subsequent u (like "Iraq").

In this general way, Markovian behaviour is the basis of Shannon's statistical information theory where it does not refer to stochastic processes (that is to say, to statistical time series), nor is it limited to the spatial analogue of the immediate past regarding a temporal sequence as a sequence of nearest-neighbour interactions. On the other hand, the Markov process may be subject to other restrictions, such as ergodicity. Ergodicity means that from any state one can get, in one way or another, (arbitrarily close) to any other state. In the long run – or, what is equivalent in this case, for sufficiently long sequences – the system can be described by a limiting distribution, independent of its initial state. Ergodic behaviour is clearly expressed in Figure 3.5.1, and its message is best seen if we contrast it with the diagram for non-ergodic behaviour in Figure 3.5.2. Depending on which path is taken, one ends up in certain traps, from which no return to the starting point is possible. As we shall see later, non-ergodic behaviour is typical of biological evolution, and also of our way of thinking. In physics it occurs only in systems far from equilibrium where, owing to some non-linear dynamic behaviour, instabilities may occur that prevent a return to the initial state. Systems at or near equilibrium do behave ergodically (see Section 2.5).

Markov processes in general may appear to be quite messy. Symbol correlations in a message, such as we have considered in the preceding sections, are usually ergodic

Figure 3.5.1

Transition graph of an ergodic Markov process

The term "ergodic" was initially introduced in statistical mechanics (see Section 2.5). In the present context it means that any state in the system can be reached from any other state, so that the system can settle to a final distribution that is independent of the initial state. Chemical states at equilibrium, exhibiting "detailed balance", are excellent examples of ergodicity.

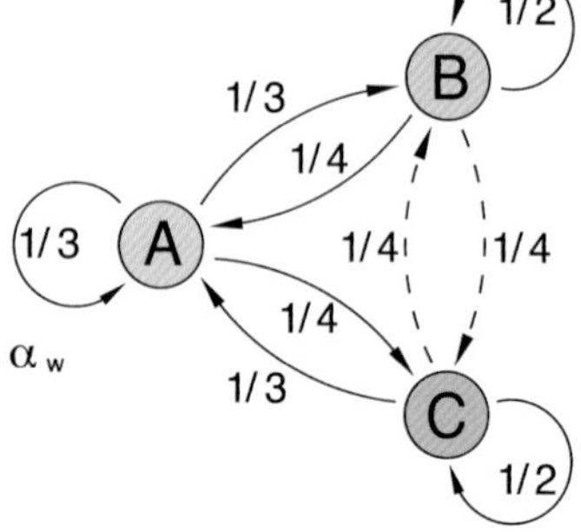

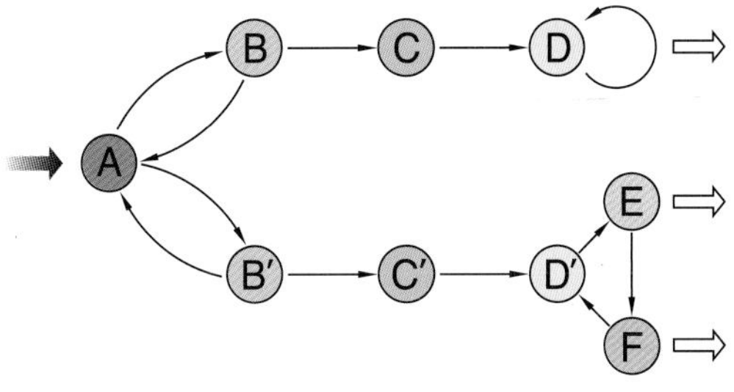

Figure 3.5.2
Example of a non-ergodic Markov process

Evolutionary processes involving "natural selection" are usually non-ergodic. We shall find examples later in this book.

and do not show any kind of capricious or erratic behaviour, despite their Markovian behaviour being of a higher order. This means that the occurrence of a given symbol depends not so much on nearest neighbours, but rather on the specific appearance of whole blocks of letters. It is this property of the ergodic Markov process, expressed by its entropy, that is used to improve the efficiency of encoding messages before they are sent through the transmission line. Transmission, on the other hand, is a process in time, so we may well use the transition probability structure of the message for predicting the next symbol in a stream of data running through the transmission line. In a more abstract sense, the rationale behind a Markovian structure is the attempt to base probabilistic predictions of the future behaviour of a system on its present (probabilistic) structure. Does this mean that we are trying to describe the relations between past and present as those of cause and effect?

This question is related to the one posed in the title of this section. However, we should be reminded that Markov processes do not allude to the "real" world but, rather, to how we may model physical reality in probabilistic terms. They tell us what we may or may not expect, be it for diffusion in physical space or for the ups and downs of the stock market. All this does not interfere with my question of whether or not our world is to be seen as Markovian, a question exclusively concerned with how to model the processes that occur in it. Cause and effect are usually linked by deterministic relations. We are used to formulating probabilistic events by differential equations, such as the Fokker–Planck or Kolmogorov's master equation, that describe expectations and their higher moments. My question is really to what extent Markovian processes prevail in the physical universe. Another way of phrasing this question is: can we name typical physical phenomena that are non-Markovian in nature? I put this question to several colleagues who were concerned with problems of this kind. The result was not very satisfactory.

Generally, the first response was that a Markov process is mathematically well defined, and all I had to do was to look at processes that do not fit this definition. That is certainly true for cases that involve a complex entanglement of different events. But how is it if we get down to the level of elementary stochastic events? Do we have here any examples of non-Markovian behaviour? Let's take the example of a first-order Markov process that requires the conditional probability of an arbitrary future event $X(t_o + \tau)$ to depend only on $X(t_o)$. In other words, for making any probabilistic statement about future behaviour it is of no advantage to know the entire history of the process X(t) for $t \leq t_0$; just to know its current state $X(t_o)$ is sufficient. Another question is: do we really know everything about the present state? It may well be that the history has left hidden variables that are not accessible to us and therefore cannot be used in Markovian modelling.

Take the case of rubber elasticity or other phenomena that involve hysteresis. A high resolution on the molecular scale is often needed in order to characterise a given state, without knowledge of its past beyond the immediately preceding states. In such a case one could always still refer to Markov processes of higher than first order. This is also true for evolutionary processes in biology or for all kinds of mental phenomena. I remember Donald McKay once saying that our mind creates, every moment, a new "world". A "moment" on mental time scales may be of the order of milliseconds, but the resulting new "world" certainly depends on the whole history of preceding "moments" stored in short-term and long-term memory in a fashion much more complex than that of simple hysteresis phenomena. Yet this whole "past" might be compressed in a very complex fashion in our present state of mind. So far, I have not been able to find anyone who could name an example of non-Markovian phenomena on the elementary level, supposing that the structure of the "present state" can be analysed with sufficient resolution.

Markov modelling applies to areas as diverse as genetics, neurobiology, engineering, planning of the manpower economy and others. It relates the world of "being" to the world of "becoming" (Prigogine[37]), and as such is the basis of phenomena such as the adaptation and optimisation involved in genetic evolution and neural learning.

When I ask the question: "Do we live in a Markovian world?" I do not just refer to the Markovian character of information theory, rather, I mean the time-dependent stochastic processes with which the term is usually associated in physics. The true importance of my question lies in the specific form of my answer, which is an assertion or conjecture: "There is no elementary form of memory." How do I know? My answer is: I don't, but I can't *imagine* any elementary form of memory. In cases where it seemed to be present, such as in rubber elasticity or in magnetic hysteresis, it turned out to be reducible to more elementary phenomena that are entirely devoid of memory. Neural memory, anyway, is a very complex phenomenon, and it will concern us in later parts of the book. "Elementary" memory would be of a revolutionary character, and so far has never shown up, but: "never say never".

3.6. How Much Information is in Mathematics?

Let me start with a fairy tale. The scene is a town in central Germany which the poet Heinrich Heine, who lived there as a student for several years, characterised[38] as "famous for its sausages and its university", the inhabitants of which he classified into four ranks: students, professors, philistines and "livestock" (note his ranking!). Since then, the town has become a real city, with more than 100,000 inhabitants, of whom more than a quarter are professors and students. Its name is Göttingen (see Figure 3.6.1). The name seems to involve two German words: *Göttin* (goddess) and *Gen* (gene). This interpretation of its name will be a central issue in my tale, although I have been convinced by experts that it is all nonsense – or let's say "pure fairy tale" – because the name in reality derives from "gutingi", which means some (ancient) settlement along a creek. This creek flows into the Leine, which Heine called a big river (because his friend "had to take a run-up in order to jump across"). But let's now focus on its university, the site of the scene where our tale is played out.

Compared with the age of the city – and by European standards – the Georgia Augusta, as the University of Göttingen is called, is not particularly old. While the town, once a member of the Hanseatic League, can look back on more than 1000 years of history, during which it once even housed the palace (pallatium) of the king of what later became the Holy Roman Empire, its university was founded only in 1734 by George II, King of England and Elector of Hanover, the kingdom in Germany to which the town of Göttingen belonged. By the way, the same king, in the same decade, also founded the University of Princeton. While still fairly young, the University of Göttingen grew up to a respectable size, and for some time it housed the largest scientific library in the world. Benjamin Franklin stayed for a while at Göttingen in order to study…guess what? Not any particular discipline of science, but rather the structure of this very library.

The fame of the University of Göttingen, which is considered one of the birthplaces of the Enlightenment (thereby attracting many students – although often for reasons not quite in keeping with the pleasure or the intention of its royal founders), includes an unparalleled tradition in mathematics. This goes back to Heinrich Heine's contemporary, Carl Friedrich Gauss. Among Gauss' followers were such eminent names as P.G. Lejeune Dirichlet (who was married to Rebecca, the younger sister of the composer Felix Mendelssohn-Bartholdy), G.F. Bernhard Riemann, Felix Klein, Hermann Minkowsky, David Hilbert and his contemporaries Edmund Landau, Emmy Noether, Hermann Weyl, Richard Courant and lately Carl Ludwig Siegel, who returned to Göttingen after World War II from his exile in Princeton.

In the early part of the 20th century, Göttingen's fame in mathematics was paralleled by that in physics. Thanks to the unique atmosphere, great names were attracted to chairs in the physics department, among them Peter Debye, Max Born and James Franck. In the 1920s Göttingen became the birthplace of quantum mechanics, and the world élite of physics met there, including Niels Bohr, Werner Heisenberg, Wolfgang Pauli, Paul Dirac and others, as I have already described in Section 1.4. When Hans Kopfermann (the mentor of two 1989 physics Nobelists, Wolfgang Paul and Hans Dehmelt) visited the USA after the war he came back and reported in

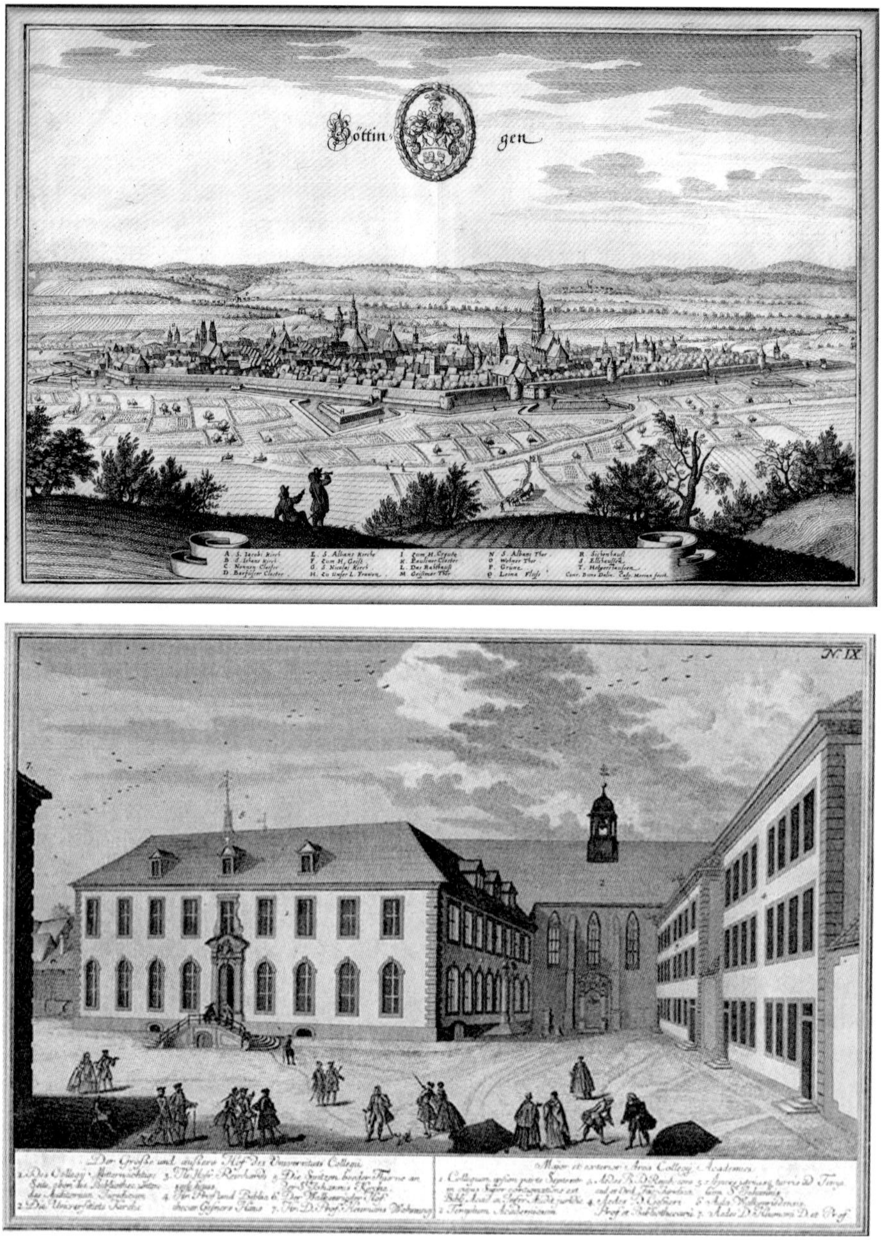

Figure 3.6.1

Göttingen in the 18th century

The upper picture is an engraving of the old town, and the lower one shows some original University buildings in the town centre.
(Both published by Merian.)

the famous Monday colloquium of physics: "I was surprised to find out in what high esteem Göttingen is still held. Let us be very careful not to squander this reputation."

When Gershom Scholem, the great scholar of the Kabala, once visited me he asked, after I had shown him around at the Institute: "What is worth seeing in Göttingen?" My spontaneous answer was: "The cemeteries." So we went looking for graves, and Scholem, who at school in Berlin was a classmate of (the late) Carl Ludwig Siegel, could tell some story at almost every grave we visited. (Among them there was a whole row including the graves of Walter Nernst, Max Planck, Max von Laue, Otto Hahn and, nearby, Max Born.) From this visit to the Göttingen cemetery I returned with many take-home lessons about my alma mater.

Among all the dignitaries mentioned I should not forget the Grimm brothers, Jacob and Wilhelm, who were professors at Göttingen University and are considered to be the founders of systematic German philology. Being upright citizens (they were not yet called democrats at that time), they both participated in the protest of the *Göttinger Sieben* (seven professors of the University) when the king tried to take back some of their rights guaranteed by the constitution – and were promptly dismissed from their posts. To most people the Grimm brothers are better known for the fairy tales that they collected by interviewing peasants in the countryside around Göttingen.

Why am I recounting all this? Because it gets me back to the promise I made at the beginning of this section, namely to tell a fairy tale. Don't worry! I hadn't forgotten, I was just preparing the scene. So I do not have to start: "Once upon a time, there was a town . . ." My tale starts in heaven – and ends up in hell. The story begins thus:

The Lord always looked with some suspicion on what was going on in a town, the name of which claimed its origin from the gene of a goddess. And when he realised the advancements of all those mathematical geniuses of Göttingen, he became afraid that they really were inheriting special genes that are usually reserved for deities. Then along came David Hilbert who, at the second international congress of mathematicians[39] in 1900 posed 23 trend-setting problems, of which the tenth dealt with the foundations of mathematics, which he thought could be provided by mathematics itself. He conjectured that "by proceeding from the postulates of arithmetic, it should prove impossible to reach contradictory results by means of a finite number of logical deductions". Hilbert himself, in accordance with the formal logic (later) developed in the Principia Mathematica[40] by Alfred N. Whitehead and Bertrand A.W. Russell, and philosophically reflected by Ludwig J.J. Wittgenstein[41], firmly believed that the efforts in proving the consistency of arithmetic (*Beweistheorie*) and of mathematical logic in general would some day turn out to be successful. This belief is reflected by the inscription on his gravestone in the Göttingen cemetery: "Wir müssen wissen, wir werden wissen" ("We must know, we shall know").

In the eyes of the Lord this was pure blasphemy. To make mathematics entirely self-explanatory would render Him obsolete. That was enough! Mathematics, the "queen of science", had to be saved from becoming a soulless automatic device that would run by itself.

And God sent (His) Gödel!

The tragic end of the tale is that Göttingen mathematics was destroyed – not by Gödel, who would have been embraced, if not by Hilbert himself, then at least by his colleagues Landau, Noether, Weyl and Courant. No, it was the devil, who at about the same time decided to send a countryman of Gödel, named Adolf Hitler, to Germany. He quickly disposed of Göttingen mathematics by driving all these brilliant colleagues of Hilbert into exile. When Hilbert was asked in 1934 by a minister of the new lords: "How is mathematics at Göttingen now?" He looked at him: "Mathematics at Göttingen? It doesn't exist any more." Hilbert, after losing all his colleagues, retired and lived in bitterness until his death in 1943. Two of his younger colleagues, Franz Rellich and Arnold Schmidt, looked after him and helped him with his writings. They became professors of mathematics at Göttingen after the war, and both were among my teachers. Gödel himself had to escape from Vienna and found refuge at Princeton, when that devilish man over in Europe started to occupy and destroy the world before, eventually, he was stopped. I am afraid that most of my tale happens to have been bitter reality. If you want to learn more about it, read the two wonderful books about Hilbert and Courant by Constance Reid.[42,43]

But let's now get back to the reality of mathematics. What did Gödel find out? Take a mathematical structure like that of real numbers. Hilbert's intent was to find within the framework of this structure a set of rules that could prove every possible mathematical truth that is implied in this structure. For this to work, the structure must not contain any statement of which the negation is equally true. Consider the sentence: "This statement is false" where "this" is reflexive, i.e. referring to "this" selfsame sentence. If the sentence is correct, it asserts at the same time that it is incorrect. Hence one finds oneself inescapably entwined in a loop. This sentence is just a simplified example of the classical paradox: "All Cretans are liars", supposed to have been stated by Epimenides (6th century BC), who himself was Cretan. Douglas R. Hofstadter[44] has provided a number of instructive examples both in science and in the arts, the latter usually representing pseudo-loops that could be disentangled by going off the dimensions of the system. If, however, we demand that the structure be internally consistent, as we do in logic and mathematics, these examples become highly relevant. For any given mathematical structure, such as natural numbers, we can choose rules to find out and prove the truths involved in the structure. What Gödel showed is: whatever rules we choose, there remain unprovable truths, or – in more general terms – mathematics using its rules as a formal system cannot provide the means of generating all the truths contained in it because there remain undecidable problems that cannot be resolved by the axioms or rules of the formal system. Unlike the "lying baron" von Münchhausen, it cannot pull itself up out of the mud by tugging on its own pigtail.

In order to comprehend such a general statement about mathematics, we have to ask not only *what* Gödel showed but also *how* he did it. His essential trick was to transform the problem into statements about natural numbers, a process afterwards called "Gödel numbering". A self-referential statement, like the sentences quoted above, after having been translated into the language of number theory, can then be shown to be equivalent to a polynomial equation that has no solution in the form of integers. The results of Gödel's work were two theorems, one called the "incompleteness"[45] and the other the "consistency"[46] theorem. They were originally related to natural numbers, but could be shown to be relevant for any formalisation of arithmetic.

Why are Gödel's results relevant to my question in the title? Let us return to Hilbert. If his expectations had come to fruition, then the answer to my question would have read: there is zero information in mathematics. In other words, its "entropic" information would have to be zero, because everything is constrained by rules. It is important to realise that I do not speak about "ourselves being informed by mathematics". Our brain, as an information-gathering system, first has to adapt to the wealth of information that has been, and is being, accumulated by mankind as a whole.

I mean mathematics that is "out there" and in itself is entirely constrained by rules, regardless of whether or not we know them all (which we certainly do not, for otherwise mathematics would be complete and nothing new could be found). Gödel's theorem "saves" us from this situation by leaving mathematics an open system with "in principle" unlimited amount of information to be generated.

Gödel's legacy is generally accepted today, and it is a tragedy that the greatest mathematician of his time was so shocked when Gödel's work appeared that he found it impossible to accept. Gödel never came to Göttingen. Why not? Because he was not invited by Hilbert; indeed, there was only a very short period left in which such an invitation would still have been possible.

Intimately linked to the problem of the "amount of information contained in mathematics" is the question of how to express this amount in mathematical terms. It leads to the formulation of a new version of information theory, called "algorithmic information theory". One of the fundamental questions addressed by this theory is how to define randomness.

Consider the following three rows of natural numbers equal to or less than 6, each of which can in principle be obtained by rolling a (fair) cubic die:

6 6 6 6 6 6 6 5 2 6 6 3 5 4 4 3 6 1
1 5 3 6 5 4 3 1 2 1 4 3 6 4 5 2 6 2
6 6 4 2 1 4 4 1 5 3 6 5 5 2 3 1 2 3

By all physical criteria the sequences should be called "random" because they were obtained using a random physical procedure. However, simply viewing each of these sequences, without knowing how they were obtained, does not tell us whether or not they really are the outcome of random processes. Of course, these sequences were constructed, because I did not have the patience to try out all the $6^{18} \approx 10^{14}$ throws in order to select them according to any given criteria. For the first row, the criterion was a clear preponderance of the 6, as one would expect for a correspondingly loaded die. The second and third rows certainly correspond more closely to what one would expect to be random. Comparing the two rows more closely tells us that they agree in making exactly the same use of all six numbers, i.e. three times each, which should in general hold true only for very long sequences. Nevertheless, each of the sequences has the same *a priori* expectation value, a result that we have already encountered in a different form in the Ehrenfest game, discussed in Section 2.6, where the occupation of the chessboard with differently coloured pieces yields the same prior probability for uniform occupation by one colour as for any individual "random-looking" representation of all colours.

Algorithmic information theory was invented independently by three authors: Ray Solomonoff[47] (1964), Andrej N. Kolmogorov[48] (1965) and Gregory Chaitin[49] (1966). Evidently, this theory makes little sense for sequences as short as those quoted above, just as Shannon's theory requires sufficiently long messages in order to allow a recovery of many of the higher correlations or redundancies that are reflected in entropy. The magnitude of (specific) numbers we want to look at might exceed a million or a billion, it may have no definite limit, just as is the case for irrational numbers. The algorithmic information content of a string of digits is defined as the number of bits making up the shortest computer program that could reproduce the original string. This takes account of the fact that any string of digits that can be generated by a program containing fewer bits can be "compressed" to the bit length of the program.

We see already that this variant of information theory has an objective quite different from that of Shannon's theory. Not only does it deal preferentially with numbers and thus lend itself more readily to mathematical analysis. Numbers, like letters, can be represented in binary symbol notation, and the role of constraints that express themselves in the higher-order terms of Markov entropy is taken over by algorithms. The main difference, however, is that Shannon's theory is adapted to the purposes of optimal encoding and transmission of arbitrary messages, constrained by common rules of a language, while algorithmic information theory deals with individual messages such as given numbers that may be reconstructed using the specific algorithms involved.

Compare the four sequences:

0 0 0 0 0 0 0 0 0 0 0 0 0 0 0 0 0 0 0 0 0 0 ....
0 1 0 1 0 1 0 1 0 1 0 1 0 1 0 1 0 1 0 1 0 1 ....
0 0 1 0 1 1 1 0 0 1 1 0 1 0 0 0 1 1 1 1 0 1 ....
1 1 0 0 1 0 0 1 0 0 0 0 1 1 1 1 1 1 0 1 0 1 ....

Regardless of its length (n) the first sequence is most economically expressed by a program stating "Print zero and repeat it n times". The program for the second sequence would be only slightly longer, i.e. "Print zero-one and repeat this n times". The two last cases look like random sequences for which no shorter program than the sequence itself exists. That is probably (at least in the way I produced it) true for the third sequence. However, the fourth sequence is the beginning of the number π, encoded in binary digits. Its computer program is almost as short as that for the first two sequences, if we base it on $\pi = 4(1 - 1/3 + 1/5 - 1/7 + 1/9)$ and so on (for more examples see John Casti[50]).

The question of how to define a sequence of digits as being random can now be answered most concisely by algorithmic information theory: "A random sequence does not allow any description more compact than spelling out all its digits." In terms of program length or computer running time this would place its algorithmic complexity at the upper end of the scale. The term "algorithmic complexity" then seems to be somewhat better adapted to the situation than "algorithmic information content" because an entirely random sequence does not really contain any "algorithmic information", unless one interprets the word information again as "missing information". The scale of algorithmic complexity thus starts with the simplest possible program – such as "print out only ones [or zeros]" – and ends at the top with a program that spells out each single digit.

Murray Gell-Mann, in his book *The Quark and the Jaguar*[51], proposed an additional term: "effective complexity". The absence of any algorithm or regularity at the upper end of the scale of algorithmic complexity is assigned zero effective complexity, while the most simple, e.g. egalitarian, algorithm at the other end of the scale should also have minimal (in this case still finite) effective complexity. A plot of effective complexity versus algorithmic information content (Figure 3.6.2) will then yield some (generally) convex plot with one or several maxima somewhere between the two extreme ends.

Charles Bennett, a member of the think-tank at IBM laboratories, whose name we encounter in several places in this book (e.g. when talking about informational entropy, quantum computation, or cryptography) suggested supplementing the concept of effective complexity by two more numbers called "depth" and "crypticity". Depth is a measure that characterises how difficult (or time-consuming) it is, starting from a highly compressed program, to spell out the complete sequence of digits that embodies the algorithm. Crypticity, on the other hand, quantifies the minimum time the computer requires for the reverse process, i.e. deriving the shortest program starting out from the actual bit string. The latter reflects all the effort a theorist may have put in to find the algorithm, which might be simple. Yet think of the number π as represented by Leibniz's formula comprising a series of consecutive reciprocal odd numbers with alternating signs. In order to reach a good approximation one has to include quite a number of terms in the series. Such a procedure is much more involved than a simple repetition, although the lengths of the programmes do not

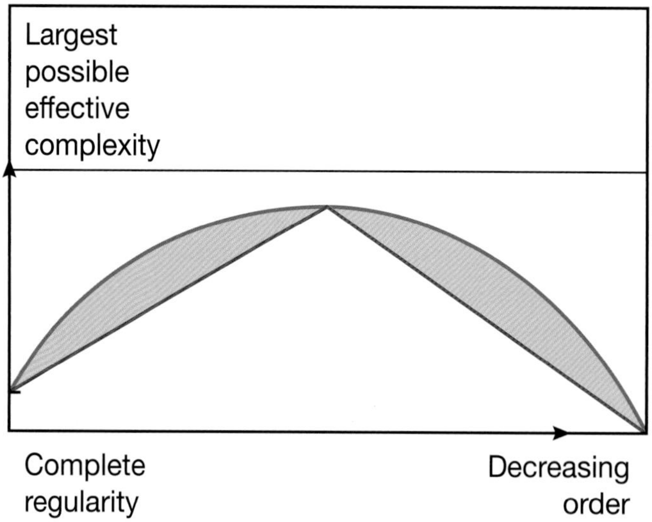

Figure 3.6.2

Schematic plot of "maximum effective complexity" as a function of "algorithmic information content" (according to Gell–Mann[51])

See text. The minimum at the left-hand end of the curve is close to, but not equal to, zero, while the curve at the right-hand end approaches zero for entirely random behaviour.

differ greatly. Conversely, the task of quoting the largest number up to which the Goldbach conjecture* consistently is of an "all or none" nature depends on whether or not the conjecture has been proven. In case of Fermat's last theorem the proof took about 350 years. The definition of depth, as given by Bennett, starts from a whole set of programs that are able to compute the actual string and determine for each the time required to reach the goal. The average procedure in which shorter programs are given higher weights yields the depth measure. One might characterise depth as a procedure to get "from the possible to the actual". Crypticity reflects the effort of the scientist even more than it does that of the computer.

Both depth and crypticity still require algorithms to be computable, a task that finds a fundamental limit in Gödel's incompleteness theorem. A plot such as that proposed by Murray Gell-Mann therefore has to be taken with some caution. It may not reflect practicability very well, although it is justifiable in principle on the basis of the definitions of algorithmic complexity wherever they can be quantified.

* Christian Goldbach (1699–1764): Every even number greater than 2 can be expressed as the sum of two primes.

A similar attempt, undertaken in the early 1970s by Christine and Ernst Ulrich v. Weizsäcker[52], lacks this consistency (see also B.-O. Küppers[53]). Their concept goes under the label "pragmatic aspect of information" and practically follows a definition of "aesthetic information" as was proposed by the mathematician George David Birkhoff as early as 1931. It essentially states that, in order to acquire new information, one needs both prior knowledge, which is to be confirmed, and something that one did not know before, i.e. "novelty". It is certainly true that without confirmatory information we cannot understand anything. This confirmatory information includes constraints on language, as accounted for by Shannon entropy, as well as subjective expectances based on whereabouts and circumstances. On the other hand, without any novelty everything is constrained and even Shannon entropy must be zero.

I intentionally use in this connection the term Shannon entropy instead of Shannon information because the authors show a diagram in which the abscissa ranges from 0% confirmation (= 100% novelty) to 100% confirmation (= 0% novelty). The convex curve they draw refers to an ordinate marked. It starts at the co-ordinate origin, increases, passes through a maximum and returns to zero at the far right end, where it approaches the Shannon curve asymptotically. It is not the form of the curve, nor is it the question of whether pragmatic information can really be quantified in any such way; rather, it is the missing congruence between two different concepts of information which makes the curve they draw, not the idea as such, meaningless. At the far right end, it is identical with Shannon entropy, which is clearly defined and represents "potential" information, i.e. knowledge we do not have. At the far left end it is knowledge we gain by undefined means. If Shannon information increases, knowledge decreases. In other words, the curve lumps together two quantities that are defined in different ways, producing a curve shaped like the one in Figure 3.6.2, which represents a well-defined function. As a concept it only makes sense as a qualitative representation of what is meant by "effective complexity".

Otherwise, algorithmic information theory, although being subject to differently defined constraints, proved to be equivalent to Shannon's theory. For communication processes the simple entropy concept is replaced – more adequately – by "mutual entropy" (based on conditional probabilities), also called the Kullback–Leibler measure of "cross-entropy".

So let us sum up and compare algorithmic information theory with Shannon's probabilistic theory. Although Gregory Chaitin showed the equivalence of these two theories as early as in 1973, there remain notable differences that become obvious if we ask what purpose(s) the two theories serve. Shannon's theory, although designed for the optimisation of message transmission, does not deal with particular single messages; rather, it deals with the constraints of language, expressed as the average measure of entropy. The single message and its content (to be distinguished from its amount) does not play any essential role, except that it has to be long enough to reflect a good deal of the statistical properties of language. The correctness of sentences is rated solely by criteria of language, not by criteria of truth and falsehood, i.e. being right or wrong. The sentence "the moon is made of green cheese" is perfectly acceptable by the criteria of language. It is just this property that enables intelligent users of language to deal with any situation, even if the meaning of individual sentences gets trapped in loops of undecidability.

Algorithmic theory, on the other hand, deals with the language of mathematics and, when applied to single "messages", it is indeed sensitive to their "semantic" content. This can be better expressed as: the rules of the mathematical language are so stringent that they dominate the truth or falsehood of any mathematical assertion – at least as Hilbert had hoped. Gödel's incompleteness theorem constrains this statement: "Mathematical assertions are either true or false – or not decidable." Apart from this, the algorithms exert a similar influence on the amount of information as other constraints of language, but: the constraints of language apply to any (sufficiently long) sentence, those of mathematics apply to individual sentences. Only in the limit of "no constraint" are the two theories exactly equivalent. An unconstrained probability distribution, where every outcome has the same probability, yields a maximal entropy, which is equivalent to total randomness of a sequence and the absence of any algorithm.

This comparison will be useful if we proceed from mathematics and language to biological information which, besides genetic information, includes neural information and eventually all the results of thought that are based on it. For messages written in human language, we have to distinguish entropic from semantic information. "Entropic" information is what can be expressed by the (average) amount of what we called information entropy. Semantic information was strictly separated from it, as something generated in a "black box" (Figure 3.3.1). Since semantic information, a product of our mind, is expressed in the form of language within this black box, the precise demarcation line between semantic and entropic information is not easy to define. In algorithmic theory there is no such demarcation line – apart from the "incompleteness" of the language of mathematics. Otherwise it is just the "semantics" of mathematics that determine the constraints in algorithmic information. Genetic information and its expression in the life process include both amount and its semantics in terms of chemical reactions.

The question of whether algorithmic information could serve as a "bridge" to an understanding of semantic information in the physical (including the biological) world will be a main issue in later parts of this book. Gregory Chaitin[54], in a paper of 1979, goes as far as to conjecture that a precise concept of complexity would contribute to the definition of a living organism. This is a projection which – if it were true – would reach very far into the future. As I see it today, such a view may fall victim to the impossibility of describing the outcome of an evolutionary (or learning) process in biology by any "static" information concept using statistically fixed probability distributions or defined mathematical algorithms – just as "formal" mathematics, according to Hilbert's original view, did in fact fall victim to Gödel's incompleteness theorem. This may not seem entirely plausible on the basis of what has been described in this book so far, but we may catch a hint of insights to come by linking Gödel's theorem with the impossibility of creating new information in any formally closed system.

3.7. Can a Turing Machine Create Information?

Only a short time after Gödel delivered his proof, Alan M. Turing came up with the concept of an idealised machine that brings the abstract results of the preceding section into a more pragmatic perspective and closer to our intuition. Algorithmic information theory, in referring to the most condensed "programme" by which a sequence of digits can be represented, can indeed be based on such an abstract machine. Any real machines – and we are now in possession of tremendously efficient computing devices (as foreseen by Alan Turing and John von Neumann) – are necessarily finite in respect of size, speed and sophistication. As a basis for theory we need an idealised machine that is free of these limitations, but accordingly it can only work "in principle".

I have never read any posthumous description of Turing's machine that would make sense in terms of any possible technical realisation, which simply means that such a device does not (and cannot) exist in reality. Therefore I do not pretend that it could do, by using cyber-age jargon and talking about tapes of infinite length having an infinite storage capacity and that can move in space, about a scanning head that can read and write and simultaneously change its internal state (whatever that may be). The important thing is that it is an ideal and not a real machine, that it has a huge memory in which all of the known axioms and algorithms of mathematics are stored, that its scanning head has access to this storage and uses it in order to respond to the information presented in finite bit strings of input data, and that this indefinitely long tape runs and runs and runs, until it eventually stops – or keeps running for ever.

The last point, i.e. whether or not the tape eventually comes to a halt, is the most important in the whole story because it tells us if the machine has solved a problem or if it is unable to do so. Again, we don't care to wait until we get tired and switch off the machine ourselves. If the problem had been Fermat's last theorem and the machine had worked at human pace, our mathematician colleagues would have had to wait for about 350 years before the machine stopped. But since it is an abstract machine, having an infinite tape, it may as well be supposed to run infinitely fast. Charles Bennett's "depth" and "crypticity" would then pose no problems for it.

What we really need in order to understand how such an ideal machine works is not a pseudo-technical description of the machine itself, but rather a concise outline of its way of handling mathematical operations. The language the machine uses can be assumed to be encoded in binary symbols (0 and 1). The programmes that instruct the mathematical operations by all known algorithms are written in such a code. In order to make themselves executable by the machine they need a program, called the Turing–Post program, that involves seven commands:

(1) Print 1
(2) Print 0
(3) Move one position to the right on the tape
(4) Move one position to the left on the tape
(5) If the tape shows 1: Go to step x of the program
(6) If the tape shows 0: Go to step y of the program
(7) Stop

What Turing showed in his 1936 paper is that this simple list of commands, when supplied to the various programs containing all known mathematical algorithms, enables the machine to execute any type of computing feasible at any attainable level of mathematics.

The name of the program refers to Alan Turing[55] and Emil Post[56], who – both in the same year (1936) – came out with the idea that anything "algorithmic", i.e. following from applying mechanically the rules of the system, could for that very reason be handled by a mechanical device. In fact, this idea is associated with a third name, Alonzo Church[57], who – also in 1936 – arrived at it by a route entirely different from that of Turing and Post. Church developed a very elegant formal scheme, called λ-calculus. Turing and Church showed independently that their schemes were entirely equivalent: Church's formally more economical scheme refers exactly to what an ideal Turing machine with unlimited storage capacity could handle. Roger Penrose[58] has proposed that one might distinguish Church's from Turing's thesis on the basis that Turing himself might have had more in mind than what is covered by the purely mathematical statements of Church. By "more" he means any physical device, including the "human calculator". However, since the equivalence of the two theses has been demonstrated, it is futile for us to ponder about what one or the other mathematician might have had in mind.

One may have doubts about whether a concept as simple and straightforward as Turing's universal machine is subtle enough to provide all the operations necessary to cover the whole potentiality of mathematics. Turing showed that this is indeed the case – as long as one doesn't bother about time. The development of modern computers, moreover, demonstrates vividly how powerful the technological realisation of such a straightforward principle can be.

The crux of the Turing concept is the "halting problem". If we identify an algorithm with a programme that eventually stops with a definite answer, then the decisive question is which input does or does not cause the machine to stop after a finite time. Turing had the brilliant idea of making the "halting problem" itself a problem to be addressed by the Turing machine.

A common method of proving a claim in mathematics is to disprove its negation by showing that it contains logical assertions that contradict one another. In this way Georg Cantor proved the uncountability of transcendentals by his "diagonal argument" (cf. Appendix 3.1). In Turing's case the question was: How does the machine "know" when to stop if there is not a particular algorithm that can tell it to do so? Turing showed that certain input strings that represent the codes for a Turing–Post program may come out with inconsistent instructions for the machine to stop, and that therefore the halting problem is not generally decidable. We again encounter "logical loops" of the kind discussed in Section 3.6, such as the two sentences

"The following statement is true."
"The preceding statement is false."

The proof, as Jan Stewart emphasises, is another "application of Cantor's diagonal argument [see Appendix 3.2], like many in undecidability theory". The undecidability of the halting problem can be considered a (more technical) expression of Gödel's incompleteness theorem.

Roger Penrose[59], in his book *Shadows of the Mind*, calls this result the "Turing–Gödel conclusion". He analyses carefully 20 possible technical objections which could be brought up against the conclusion in order to demonstrate that human thought cannot be simulated by a Turing machine because "conscious thinking must indeed involve ingredients that cannot be even simulated adequately by mere computation; still less could computation, of itself alone, evoke any conscious feelings or intentions. Accordingly, the mind must indeed be something that cannot be described in any kind of computational terms."

This conclusion contains an answer to the question posed in the title of this section, an answer to which I would fully subscribe. Of course, so far I have said nothing about human consciousness, and at this stage we are not yet ready to discuss the problem of conscious thinking in any meaningful way. Therefore I am not going to discuss now how I think the problem should be approached. Neither am I willing at this point to comment on how Roger Penrose thinks the problem of conscious thinking may one day be solved.

It suffices at this stage to conclude that a closed and formal system cannot generate any information that is not inherent in its algorithms or, expressing it in Gregory Chaitin's (1982) conclusion about algorithmic information theory: "I would like to say that if one has ten pounds of axioms and a twenty-pound theorem, then that theorem cannot be derived from those axioms." That sounds almost like a conservation law, which can be true only for a closed formal system and may then be formulated as: "A Turing machine generates as little information as a Carnot engine produces energy." Needless to say that in any open system, with fluxes of energy and matter, the situation is entirely different. Using "learning algorithms" a computer (which is an open system) is indeed able to generate information if we provide it with adequate learning algorithms and selection criteria. But that is not our concern in this section.

Let me rather use the preceding conclusions to say a few words about the nature of mathematics. They are provoked by a recent "conversation" between the (molecular) biologist Jean Pierre Changeux of the Institut Pasteur in Paris and the mathematician and Field's medallist Alain Connes of the Collège de France, published under the title *Conversations on Mind, Matter, and Mathematics.*[60]

A key issue of the conversation was the question (as phrased on the jacket of the English edition): "Do numbers and the other objects of mathematics enjoy a timeless existence independent of human minds, or are they products of cerebral invention?" In a paper following up the conversation between the two scholars, Alain Connes[61] gave a brief account of this view under the title *On the Nature of Mathematical Reality*, which, in particular, is related to the context of this section (whereas details of the biologist's view refer more to later parts of the book).

There was agreement on the above conclusion, namely that mathematics is "more" than what can be simulated by a Turing machine, even given that it is fed with the latest version of known algorithms. From the biological point of view, the human brain is an adaptive organ, inheriting very little *a priori* information at birth (apart from preprogrammed "instinctive" behaviour), but equipped with a tremendous capacity to learn. Learning in the beginning is dominated by reproductive adaptation, which at later stages decreases in qualitative efficiency, replacing pace by sophistication that involves reflection and contemplation. This is the time during which our brain adapts itself to the level of knowledge that has been accumulated by the joint efforts of the whole human race. It is also the period in the human life cycle where the individual brain starts to contribute to this general endeavour of mankind.

Mathematics, providing a special playground for this joint human activity, cannot simply be characterised as the "established sequence of logical deductions in formal systems which have exclusively been constructed by the human mind". Alain Connes illustrates these (his) words with a short excursion into the reality of mathematical research. Without following him now into the details of the examples he discussed, I may supplement his view by quoting Richard Courant[62], who said: "Mathematics as an expression of the human mind reflects the active will, the contemplative reason, and the desire for aesthetic perfection. Its basic elements are logic and intuition, analysis and construction, generality and individuality. Though different traditions may emphasize different aspects, it is only interplay of these antithetic forces and the struggle for their synthesis that constitute the life, usefulness, and supreme value of mathematical science." And he concludes: "Fortunately, creative minds forget dogmatic philosophical beliefs whenever adherence to them would impede constructive achievement. For scholars and layman alike it is not philosophy but active experience in mathematics itself that alone can answer the question: 'What is mathematics?' "

Alain Connes contrasts the formalistic view of mathematics with that which he calls the view of a Platonist, who (in the extreme) considers himself as the "discoverer of an objective mathematical world, the existence of which for him is beyond any doubt." Reference is then made to Kurt Gödel who (in Connes' elegant formulation) said: "In every sufficiently complex and consistent formal system there is a true statement about natural numbers that is not provable using the algorithms of the system." It is clear that this is a denial of Hilbert's idea of a complete internal consistency in formal mathematics. And what is meant here is the mathematics that comes out of our mind. But it also says that this denial is of a final nature: any possible approach of the human endeavour to the "objective mathematical world of the Platonist" leaves a "finite gap". Connes quotes Wilhelm von Humboldt who, referring to science in general, calls all possible results "a truth that has not been and never will be completely uncovered"[63].

In the previous paragraph, I distinguished the information sited in the individual brain from the higher-level information that is the property of mankind, deposited in our libraries. In the case of mathematics, the latter is the present formal system, with all its open questions. This system is strong enough to remove mistakes, but at no stage will it be free of questions that are undecidable by the formal rules. The formal system may be steadily broadened by new ideas requiring intuition, yet the formalist is cautioned by Gödel that there always will remain questions that cannot be answered by any formal system. On the other hand, the arguments of the Platonists are subject to the same limitations. The third level of an "objective mathematical reality" can never be reached by any of us mortals. Reality does not require proof or disproof, it is just "out there", not tolerating any inconsistency. If both the formalist and the Platonist heed these constraints, their dissent might someday become a pseudo-problem.

3.8. Whose Information is in Our Genes?

This question is certainly not intended to open a discussion about the patent rights of biotechnology companies on any of the 30,000-odd genes we carry in our hereditary pool, called "the human genome". My question is rather to be understood in a manner similar to the one raised in the title of Section 3.3. In other words, I am still concerned with the problem of how to transmit messages in an optimum way. This time, however, the messages are genetic programmes, contained in DNA or RNA molecules rather than in tele-texts or computers. As long as the problem is "communication", Shannon's theory should apply equally well in these cases. And, indeed, it is one of the theory's later successes that it really does so.

The messages laid down in our genes are texts written in a molecular language. For the moment, it will suffice to know that genetic messages – like the verbal messages transmitted in telecommunication – are sequences of symbols that encode a molecular language of rich expressiveness. Chemical communication, like verbal exchange of information, requires a "rich phraseology", here in terms of physical interactions, which are inherent to the complex three-dimensional structure of protein molecules with its distribution of physical force centres. The proteins, with their 20 classes of monomeric units (natural amino acids), have developed a repertoire of "articulation" that is comparable to that of human language, e.g. of English with its 27 characters (26 letters plus one blank). As chemical effectors, involved in all sorts of biological tasks, individual proteins are made up of some 100–1000 monomeric units. They are able to adapt to any conformation required by their functional assignment.

For comparison, nucleic acids, unless they are involved in executive functions, such as processing of messengers or translation of messages, behave more like code tapes securing their sequential read-out. They differ from bit strings in making use of a quaternary (A, T, G, C), rather than a binary (Y, R) alphabet. This is because the system had to organise its read-out by making use of exclusive complementary binding between pairs (A–T and G–C) of the quaternary digits. Nature solved the problem by employing two classes of organic bases: purines (A and G, referred to collectively as R) and pyrimidines (T and C, referred to collectively as Y). Each pair consists of a purine plus a pyrimidine, and the exclusive pairing provides both pairs with the geometry required for the polymer strand to match the active site of the enzymic code-reading machinery. That's all I want to say for the moment. More of the details will be explained in the biological sections in a second volume. Let me now return to communication theory.

In analogy to Figure 3.3.1, I start with a block diagram for the transmission line that applies to the chemical communication between the genetic legislative and the functional executive in the living cell. As is stressed in Figure 3.8.1, the analogy between the two models is obvious, except that the new diagram is a bit more detailed. The two black boxes of "source" and "sink" in the first diagram, as well as the intermediate stages, are replaced in the second diagram by the names of the well-known molecular compounds involved in the transfer of genetic information in a living cell.

In the case of the telecommunication line, the encoding of the message, the adaptation of the channel capacity and the decoding – including correction for possible errors introduced by noise – are problems to be taken care of by the engineer who designed the equipment, guaranteeing conservation of the message sent from the "source" to the "sink", regardless of its content. Nothing, in this case, is known about the origin and fate of any individual message, which result from (largely unknown) processes occurring in the heads of some people.

On the other hand, the genetic transmission line was constructed at the same time as all the messages that are processed with its help. However, that is not really important with regard to the problem discussed now. The number of different messages sent through the line is still vast; for instance there are messages of some 30,000 genes in the case of the human genome. They are not all read out in the same cell, but the transmission lines in all the cells of an organism are constructed according to the same blueprint, the general structure of which is the same even for widely differing organisms. Hence, they accept all sorts of genetic messages. So we have a problem similar to that in telecommunication, and it doesn't really matter "who" constructed the transmission line.

So the analogy between the tele- and the genetic communication line – at least in this respect – is fairly complete, and (small wonder!) the latter has adapted

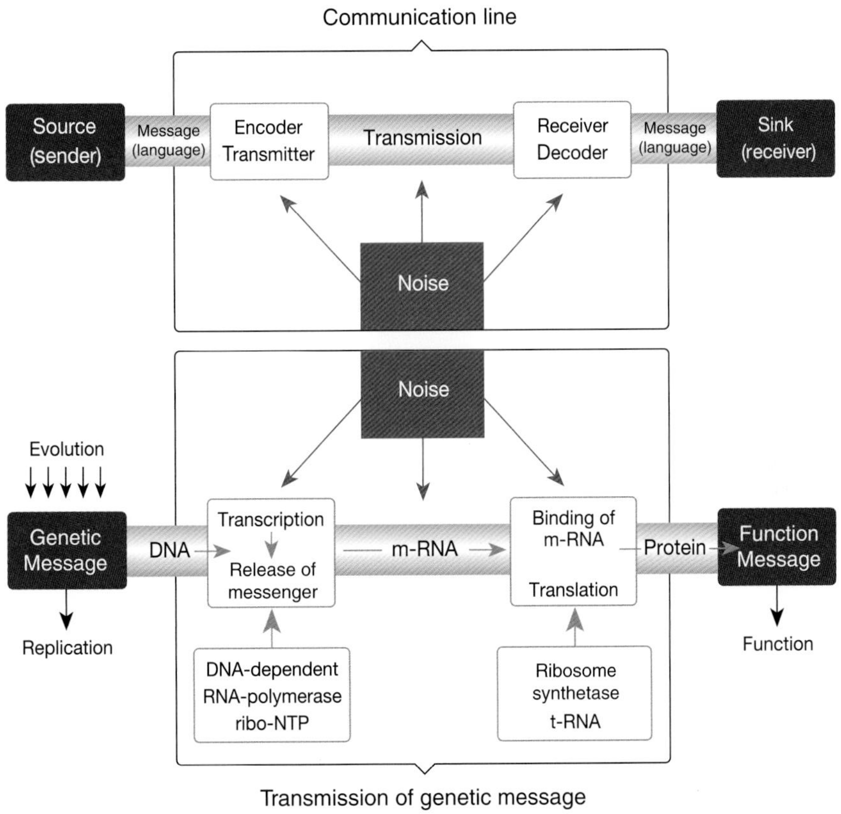

Figure 3.8.1

Diagram for the transmission of genetic information into biological function

A comparison is made with the diagram for the transmission of messages according to Shannon's theory (see Figures 3.3.1 and 3.3.2).

itself during evolution to the criteria of Shannon's theory. Yes, the code is perfectly matched to the capacity of the natural transmission channel, and errors, which unavoidably occur during transcription and translation, are steadily corrected. Sources of "noise" include the mispairing of some nucleobases in the transcriptive replication of the messenger, the misloading of the code adapter by incorporating a wrong amino acid and the mispairing of the anticodon, situated at the adapter, with a wrong codon of the messenger RNA. All these processes may change the phenotype, but they have no effect on the genotype of the organism.

The replacement of the black boxes of Figure 3.3.1 by the concrete structures in the second diagram of Figure 3.8.1 should not mislead us into assuming that we really know the source of information that is being sent through the horizontal transmission line of the cell. In other words, an answer to the question "Whose information is it?" cannot be different from that given in Section 3.3. Here, as there, what we mean by "information" is just "informational entropy". As in Section 3.3, this is the only aspect of information that is involved in "constructing" an optimal communication line. The origin of whatever is being sent through the horizontal line is not at stake. Nevertheless, the vertical arrows in the present diagram should indicate that we cannot simply ignore this question, particularly since it is not only the genetic message that somehow had to come into existence, but also the communication apparatus itself. It owes its very existence to the same mechanism that brought about the messages that it processes. It is this fact that I wanted to indicate by the vertical arrows at the left and right ends of the lower diagram in Figure 3.8.1.

While in this chapter I must leave open any question regarding the nature of the vertical arrows, I should emphasise that, whatever the answer may turn out to be, the vertical process includes information transfer too. Not only is the single cell in a multicellular organism a product of ontogeny, which starts with the multiplication and stepwise diversification of the fertilised egg cell, but the latter is itself an individual descendent within the progeny of a whole species, which in turn is the offspring of a phylogeny, ultimately rooted in the origin of life, dated around 4000 million years ago. And we have to keep in mind that in as far as these "vertical" processes involve generation and proliferation of (semantic) information, they cannot be described by a theory that is based solely on informational entropy, to which qualities such as meaning or value are foreign.

In this section, however, I am concerned with the quantitative and not with the generative aspect of information. In other words, I do not care "how" the information that we inherit in our genes came about. So I can start from the fact that it does exist, and that it includes the information for constructing a highly efficient transmission line for the expression of genetic information in present-day living organisms. If in Section 3.3 the answer to the question "whose" information we were dealing with was "whoever constructed the transmission line", we can now start from a similar premise and try to analyse how well Shannon's criteria are fulfilled in biology. For this purpose I am going to consider four examples which I regard as representative of the work done at many laboratories around the globe.

(i) The Markovian nature of genetic information

The Berkeley campus of the University of California has proved to be one of the strongholds in the field of biological information. The aura of this place was largely determined by the late Thomas H. Jukes, one of the great theoretical biologists of our times, and by the deceased biomathematician Hans Bremermann. As a result of their encouragement, in the 1960s Lila Gatlin wrote a monograph, *Information Theory and the Living System*,[64] in which she presented an early application of Shannon's theory to the results of a chemical analysis of DNA (frequencies of nucleotides and their combinations) in various organisms. Her interest was mainly in demonstrating the Markovian nature of nucleotide sequences.

In order to explain what is meant by this, I refer to Figure 3.2.1. It shows informational entropy, as it depends on constraints that apply to the probability distribution. The maximum entropy (^0H_M) is obtained if all probabilities are equal. Different *a priori* probabilities of the various symbols reduce the entropy to ^1H_M, e.g. for the symbols of the English language – as we may remember from Section 3.3 – from 4.76 to 4.03 bits per letter. The next constraint applies to pairs of letters. Different nearest-neighbour frequencies for all symbol pairs lead to a further reduction of entropy to ^2H_M, and that continues for higher-order symbol concatenations until a limiting value ^nH_M, called the Markov entropy of nth order, is reached (see Section 3.3).

In her book Lila Gatlin analysed a large number of data for various organisms, including bacteriophages, mammalian and plant viruses, bacteria, protozoa, invertebrates, plants and vertebrates. She could clearly verify the effect up to the second-order Markov entropy, establishing clear differences for various species. Higher-order effects turned out to be too small to be seen in the experimental data. I shall not go into further detail of her evaluations, which she presented in terms of redundancies, just as one may express the entropy reductions that are caused by the various constraints. Instead, I refer to data for singlet and doublet base frequencies, obtained by J. Josse, A.D. Kaiser and A. Kornberg[65] for *Micrococcus phlei.*

I mention Lila Gatlin's monograph mainly because it was an early documentation of one of the basic problems of molecular biology; i.e. DNA sequences and their relation to information theory. The experimental work on single-base and nearest-neighbour frequencies was initiated by the epochal work on DNA polymerisation by Arthur Kornberg (Nobel prize-winner in 1959, shared with Severo Ochoa) and his Stanford school. Beside Josse and Kaiser, whom I mentioned above, M.N. Swartz[66] and T.A. Trautner worked in Kronberg's laboratory, while J.H. Subak-Sharpe[67] (then at Göttingen) and many others also contributed to the experimental data. I remember the time in the early 1960s, when we worked at the Max Planck Institute in Göttingen on the dynamics of base-pairing and conformation changes in polynucleotides. In those years Friedrich Cramer came to Göttingen, where he founded a laboratory of nucleic acid chemistry. He brought along J.H. Subak-Sharpe. We used to hold a weekly joint tea meeting for our two departments, called the "nucleic-acid tea-hour". Once when I invited a visitor to come along to that event, he hesitated for a moment and then asked with some trepidation: "What does nucleic-acid-tea taste like?"

(ii) Shannon's theorem and errors in information transfer from DNA to protein

Shannon's theorems, treated in Sections 3.2 and 3.3, prove that codes can be found that allow us to send messages from a source to a sink with rates arbitrarily close to the channel capacity, including as few errors as we wish. In a recent book entitled *Information Theory and Molecular Biology*[19] Hubert Yockey presented a survey of the applications of information theory to the field of molecular biology. Apart from its occasionally quite polemic style, the book contains an excellent review of the foundations of Shannon's theory, and it presents convincing examples of its relevance to the transfer of information from DNA through mRNA to protein. In the following I shall discuss an example taken from Yockey's own work.

The genetic code is a triplet block code (Figure 3.8.2). According to Crick's central dogma, information can flow only from DNA to protein, not *vice versa.* (In this formulation I have used the RNA symbols, as if information could flow only from RNA to protein, while it can flow – in the reverse sense – from RNA to DNA.) Since the probability space of the source refers to an input alphabet differing from that of the receiving probability space, a corresponding matrix of conditional probabilities has to be applied (cf. Section 3.3). More plainly: the 20 classes of natural amino acids plus a termination signal in a block code together employ 64 triplets of nucleotide bases, yielding a 64 by 21 matrix of transition probabilities. In the absence of any noise the elements of this matrix would be either 0 or 1. The conditional entropy $H(y|x)$ for a receiving alphabet B with elements y, given a source alphabet A with elements x, would then be 0 (since $p \log p$ is 0 both for $p = 0$ and for $p = 1$). The mutual entropy, calculated according to the results of Section 3.3, would be $I(A; B) = H(x) - 1.7915 - H(y|x)$. In other words, 1.7915 bits of the original entropy $H(x)$ per base triplet are lost simply because (even in the absence of noise) the information of 64 codons is reduced to 20 amino acids plus one termination symbol. In the presence of noise, according to a certain misreading-probability per nucleotide base, Yockey calculates for the cytochrome c molecule (see Figure 4.1.5) an "information content" of 374 bits, as compared with 438 bits in the absence of noise-induced errors.

As Yockey noted, there is no deleterious error threshold, a violation of which would cause the disintegration of the entire information. As he puts it, only "smooth" error terms (such as linear, quadratic or logarithmic terms) occur, leading to only slightly noisy data. Any loss of message content could be easily compensated for by error-correction mechanisms. However, Yockey's conclusion that an error catastrophe cannot occur at all has no general logical foundation. Indeed, it cannot occur for information sent through a transmission line where the message of the "source" is replaced by a (possibly erroneous) message arriving at the "sink", without including any duplication or multiplication of the information at intermediary stages. But Yockey's argument cannot be generalised because he fails to distinguish between the transfer and the generation of information. In a recent paper (see Ref. 68) he claimed that in our work we had found such an error threshold to apply to "the transfer of information from DNA through mRNA to protein". We have never worked on that problem and have not said anything like that. What we find is an error threshold in the *generation* of information in DNA or RNA in multiplying populations. And these are not – as he phrased it – "mechanist-reductionist speculations" of ideologists, rather they are (theoretically well-founded) experimental facts. But, as I said before, I do not intend to talk about the *generation* of information in this section. Yockey's ideas about the origin of information in living systems are quite quixotic. Nevertheless, the relevance of Shannon's theory for the *transmission* of genetic messages is unquestioned and Yockey has performed a valuable service in demonstrating this.

(iii) Replaceability of amino-acid residues in functional proteins

An unconstrained base triplet at the genetic source corresponds to an information quantity of six bits, of which about 1.8 bits do not appear at the receiver end, namely, the protein with its alphabet of 21 units. The functional proteins we study nowadays have been selected during the evolutionary process according to their fitness. The amino-acid residue we find at any given position represents an optimum choice and therefore cannot be exchanged easily with another residue. Yet we observe the phenomenon of "neutral drift", which means that even these molecules, which have been selected for their efficiency, steadily undergo changes in their amino-acid sequences without losing their functional competence. Hence, there should be a mutual resemblance among certain of the members in the repertoire of natural amino acids.

The late Margaret O. Dayhoff is the *grande dame* of protein evolution. Her *Atlas of Protein Sequence and Structure*[69] is a standard source and her article in volume V of the *Atlas*, "A Model of Evolutionary Change in Proteins", is a classic of protein chemistry. It was Ludwig Hofacker of the Technical University of Munich who, starting from the Dayhoff matrix of mutation frequencies, tried to solve the problem of functional equivalence of amino acids and to achieve a theoretical understanding of the effect of "neutral evolution". The dynamics of the process itself, with its consequences for a steady "neutral drift" in systems that have already been optimised in evolution, is a subject of non-linear stochastic theory, which was solved in a masterly way by the late Motoo Kimura (see Section 5.4).

Earlier work by Ludwig Hofacker[70], together with his co-worker R. Coutelle and his colleague R.D. Levine from the Hebrew University in Jerusalem, had shown that evolutionary changes within a family of proteins favoured an optimal strategy, in the sense that it maximises the number of potentially functional sequences generated by "accepted point mutations" within a pool of neutral mutants, subject to functional constraints. In a later paper, Hofacker, together with his Slovenian co-worker Branko Borštnik,[71] worked out the fundamental aspects of the neutral patterns in protein evolution. They started from a 20-dimensional characteristic space spanned by the eigenvectors of a "property preservation" matrix that is closely related to Dayhoff's mutation frequency matrix. What turned out to be preserved in the mutations was a subspace related to a combination of three physical properties of the amino-acid residues (polarity, hydrophobicity and residue volume), rather than to any of these properties on its own. The crucial point is that the (20-dimensional) space of eigenvectors contains the three properties, preferentially expressed as a three-dimensional subspace. The genetic code was shown to be optimally adapted to favour certain mutants with property-preserving control, all of which were located within a sphere in this subspace. The diameter of this sphere yields an average "distance" among the equivalent residues, which is called the Borštnik–Hofacker distance, and which is to be contrasted with the Hamming distance among the amino acids involved.

A fascinating aspect of Hofacker's work is the application of rigorous methods of mathematical physics to a relevant biological problem. We cherish Ludwig as a participant of many years' standing in our winter seminars at Klosters, where he enriches our discussion and guides us on challenging skiing tours in the Swiss mountains.

(iv) Is the genetic code optimised?

In his book *Coding and Information Theory*, Richard Hamming made it quite plain that proper encoding of the source symbols is the essence of information theory. The examples discussed above stress this point. It is proper encoding that makes it possible for a genetic message to be transmitted safely to the executive machinery of the cell, for errors to be corrected and for some potential of functional flexibility to be preserved among the mutants that are "acceptable" and therefore prevail.

Vignette 3.8.1 provides a short sketch of the early evolution of the genetic code which is represented in its three-dimensional Hamming space configuration of Figure 3.8.2. The amino acids are spelled out together with their percentage frequencies as they appear in present-day proteins (according to M.O. Dayhoff el at.[72]). In this kind of representation the code reveals itself clearly as a three-dimensional block code. As we have seen in earlier sections, a block code is less efficient than a Huffman code of variable length, but the example of the Shannon–Fano code (Section 3.4) showed that its average length is not very different from that of a corresponding Huffman code. The slightly lower efficiency by no means offsets the advantages gained through having an "instantaneous" readout of the message, which is a property inherent in all block codes.

Vignette 3.8.1 The Initial Code

This vignette shows the average frequency of appearance for each of the 20 natural amino acids in present-day proteins, as evaluated from a pool of 314 sequences, in which each one comes from a different family of proteins. They were published by Margaret O. Dayhoff and her co-workers in Supplement 3 of her famous *Atlas of Protein Sequences and Structures*[72].

Present-day structures are primordial products of evolution, which at this basic level encompasses a period of about 4000 million years. This is a very long time – and we were all the more surprised at how well the present-day sequences still reflect early evolutionary events. The code had to start from internal "interactions" only. The most stable base pair is that of G and C. It is ten times stronger than A and U. Hence a codon GGC or GCC would bind to its complementary anticodon about 1000 times more strongly than the codon AAU or AUU. Its logic then was a result of evolution. Considering a triplet code from early on, such a code must have started with the most abundant nucleotides present and their stablest partners. From our experimental studies, we soon concluded that a code rich in G-C pairs would be very favourable. Moreover, a code such as GNC or CNG (for which the complementary codon, read in the opposite direction, would also be GNC or CNG) would be the one best able to make a message readable.

In any case a triplet code – in the absence of special reading devices – requires differences in its first and third positions, which can be used for reading a message free of overlapping alternatives. Our choice for an initial code model best adapted to such a requirement was GNC. Since the GC pair (according to our experimental

Cont. ➲

Cont.

studies) is the most stable base pair, this is the optimum model for providing an initial code. N can be any of the four bases (G, C, A, U), so this model would start by encoding only four of the amino acids present in the primordial soup. The available evidence suggests that the two most abundant original amino acids were glycine and alanine. The next ones should include a negative charge (such as aspartate and glutamate) and a highly hydrophobic side chain (such as valine and leucine). The negatively charged carboxyl group could also bind a divalent metal ion, such as Mg^{++} or Ca^{++}, and in this way also provides a positive charge in the original code.

Figure 3.8.2 shows that glycine, alanine, valine/leucine and aspartic/glutamic acid are – even in present-day proteins – still among the amino acids that are most frequently encountered.

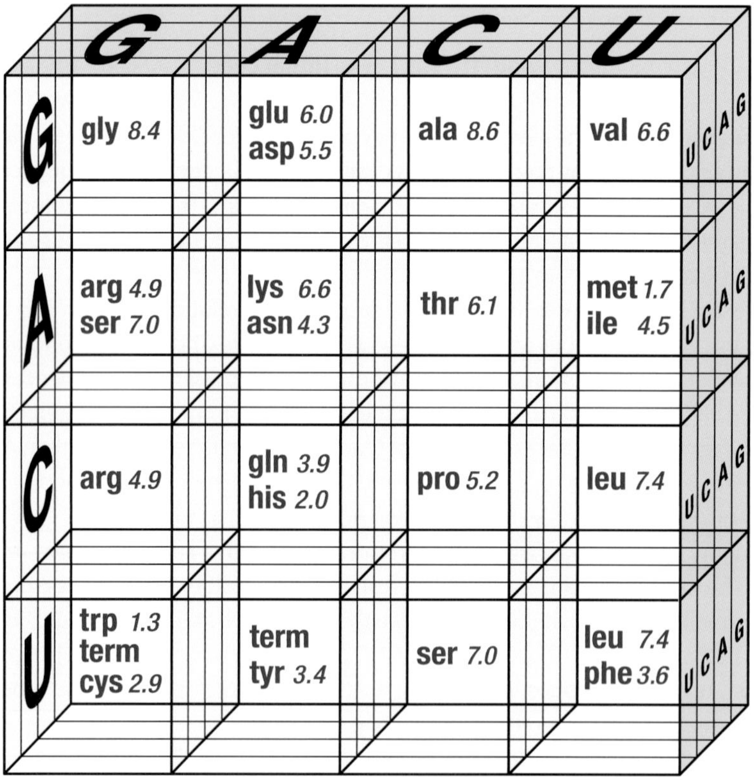

Figure 3.8.2 (Cont.)

Cont.

➲ Cont.

Figure 3.8.2

Average percentage of amino acids in present-day proteins according to Margaret O. Dayhoff

This figure uses my scheme for the genetic code introduced in Figure 3.4.1. The numbers are the average percentages of amino acids occurring in present-day molecules. One can derive from these numbers a sequence of the 20 amino acids in decreasing order of occurrence, i.e. Ala (8.6), Gly (8.4), Val/Leu/Ile (6.6/7.4/4.5) Asp/Glu/Asn/Gln (5.5/6.0/4.3/3.9), Ser/Thr (7.0/6.1), Lys/Arg (6.6/4.9), Pro (5.2), Phe (3.6), Tyr (3.4), Cys(2.9), His (2.0), Met (1.7), Trp (1.3). The order found today includes original influences, such as abundances of primordial amino acids, as well as requirements arising out of the evolutionary process. In the above series I have combined several cases which reflect their similarity in both structure and evolution of function. Perhaps the most surprising feature is the role played by the original abundances of the amino acids in the primordial pool. This is especially true of the chemical sophistication of an amino acid located at the active site of an enzyme. It is needed for the enzyme's catalytic function, but is only a minute portion of the total sequence. A much larger part of the sequence is required for determining the globular structure of the enzyme, which could have been optimised in later phases of evolution, but which originally had to start from the abundances then prevailing (e.g. glycine). For this we developed a model that agrees well with the available evidence (see Edward N. Trifonov (1999), "Elucidating Sequence Codes: Three Codes for Evolution", *Ann. NY Acad. Sci.* 870: 330–338). In order to be readable without external machinery, the original code had to emerge from "internal" interactions as described in this legend.

The strongest base pair is GC. In order to yield a read-out of information, free of overlapping, the first code scheme should have included GNC, which can be read in both the 3' → 5' and 5' → 3' directions. N could have been any of the four bases (A, U, G, C), so the original code can have included only four amino acids, starting with the (by far) most abundant ones Gly and Ala, but adding two more functions: electric charge (Asp and Glu) and hydrophobic interaction (Val and Leu). By extension to RNY (R = purine, Y = pyrimidine) the system could build up a machinery that makes use of more sophisticated mechanisms. All this has been described and found to be in excellent accordance with historical facts.

It was Thomas Jukes[73] who proposed an adaptive development of the code through several extensions. As Francis Crick[74] pointed out, there should be no alteration in the size of the reading frame in the course of evolution because this would ruin all assignments made earlier. Furthermore, the initial code must be non-overlapping, fixing a frame that can be read in the absence of enzymes (which in early phases of life were not yet available). Ruthild Winkler-Oswatitsch and I suggested that the initial code should be based on physical rather than logical arguments, and we proposed a scheme GNC, where in the beginning only the middle nucleotide N was assigned to an amino acid[75] (more details will be given in Chapter 6).

What do I mean when I say "physical rather than logical arguments"? In the beginning there is nothing but physical interaction. Logic comes in with evolution because of the advantages it offers. G–C pairing, in the absence of enzymes, is about ten times stronger than A–U pairing[76]. Hence for a sufficiently stable, non-overlapping triplet frame that can be read in the same way in both the plus and the minus strand of RNA, the two choices are GNC or CNG. We chose GNC on the basis of sequence data which clearly show the traces of an early scheme of this kind. For an assignment of N, the primordially most abundant amino acids are primary candidates, an argument clearly supported by the present code: GGC = glycine; GCC = alanine; GAC = aspartic acid and GUC = valine (see also Figure 3.8.2). I shall not go into more detail at this point because all I need here is a starting condition that cannot be taken from evolutionary arguments.

According to Jukes' evolutionary model there was a "first extension" of the four-letter alphabet, which yielded a doublet code, relegating the role of the third position in the frame to that of a mere spacer. The doublet code already allowed an assignment of 14 to 15 amino acids, but it then got into a bottleneck, requiring a second extension, which led to the triplet code now in use. All the later phases (which, again, are not discussed in detail here) allowed the evolution of proteins with efficient catalytic properties, making the initially obligate frame scheme (GNC followed by RNY) increasingly obsolete.

In this section I want to answer the question: is the modern genetic code optimal in terms of coding and information theory? This problem was studied in detail by two French scientists, G. Cullman[77] and J.M. Labouygues[78]. They came to the following conclusions:

First: the genetic code is instantaneous and must have been so at all stages of extension. There is a mathematical relation, called Kraft's inequality (see Section 3.3) that provides the necessary and sufficient condition for a code to be instantaneous. This condition is fulfilled for any block code, and hence for the genetic code, allowing the direct read-out of messages without prior knowledge of the complete message.

Second: the genetic code is optimal with respect to the most economical use of the four nucleotides. This means that under the constraint of being instantaneous, the code has a minimum word length.

Third: the genetic code is optimal with respect to the avoidance of translation errors. It favours mutants with small Borstnik–Hofacker distances from the wild type. In the second extension of the code, amino acids were assigned to base triplets that derived from preceding doublets of the first extension, involving Hamming distances of "one". The fact that nowadays more than 20 different amino acids can be found in proteins has not required a third extension of the code. Instead, any "newcomers" needed are produced by special enzymes after translation, by the modification of individual amino acids wherever required.

To conclude this section, let me come back to my original question: "Whose information?" All I have dealt with so far is "communication" of information, not "information itself". This may sound paradoxical: didn't I also say something about products of evolution, such as the genetic code, or about evolution itself, such as the neutral changes observed primarily at later stages of evolution, after some optimal adaptation had already been achieved? Yes, that is true, but I did not really speak of any mechanism that could tell us how the (semantic) information, representing functional order, was generated. The table of the genetic code, unlike the periodic table of the chemical elements, does not reflect any *a priori* order caused by physical law. It represents an *a posteriori* order due to adaptation to function and purposefulness. What we observe today is what has been achieved by the adaptive mechanisms of evolution. Distances are therefore weighted distances, such as entropies, and "errors" are established changes that show up in present-day alignments of sequences, i.e. after they have been evaluated with respect to their functional equivalence. The order of life is a functional order, and it does not manifest itself (at least not in any obvious manner) as an algorithmic order in genotype sequences. There is in general no computer program deducible from DNA sequences that would lead to a substantial compression of the length of the encoded program. Nevertheless, a gene does carry highly relevant "meaning".

Hubert Yockey writes[79]: "It is a curious fact, highly relevant to the origin of life, that the genetic code is constructed to confront and solve the problems of communication and recording by the same principles found both in the genetic information system and in modern computer and communication codes." And: "The existence of a genome and the genetic code divides living organisms from non-living matter. 'Mechanism' applied to life is a hopeless category."

I disagree with his conclusion expressed in the last sentence! The logic of life is an *a posteriori* outcome of a dynamical process that started from interactions among "non-living" particles of matter. There is no *a priori* division between "living" and "non-living" matter. The secret of life is rooted in the origin of adaptiveness. If we want to understand the *a posteriori* order of living matter, we must track down the roots of adaptiveness. The hallmark of biology is its exponential complexity, a problem that nature has obviously solved in polynomial time (see Section 4.3).

3.9. How Far is it from Shannon to Darwin?

A brief answer could be 8880 miles or 14,288 kilometres[80]. This is the distance between the international airport near Limerick on the west coast of Ireland, named

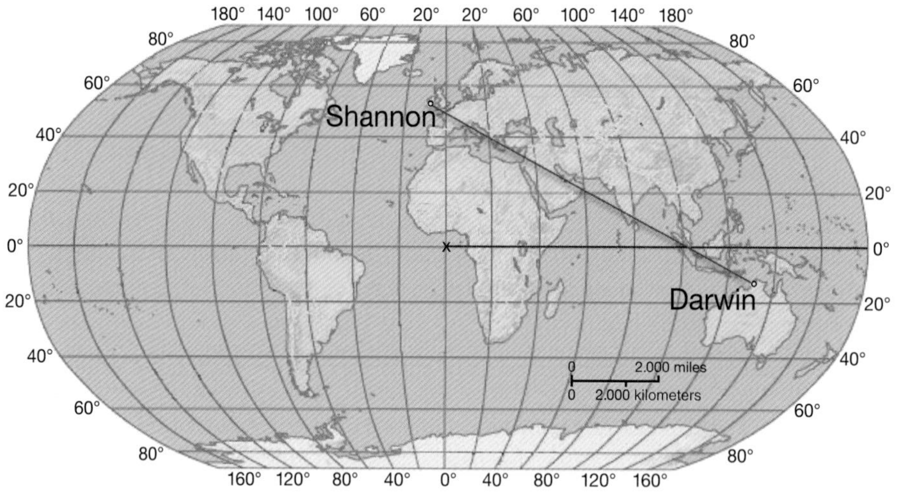

Figure 3.9.1
The distance from Shannon to Darwin

The direct route from Shannon airport (Ireland) to Darwin on the northern coast of Australia. I am grateful to Jack Dunitz for calculating the exact distance of this route: 14,288 km or 8880 miles.

after the river Shannon, and the seaport in Northern Australia (Figure 3.9.1). The latter owes its name to a first mate of the HMS Beagle, who, having participated in the now legendary voyage of the Beagle (1831–1836) was so impressed by his (not yet widely known) fellow-traveller Charles Darwin that on a subsequent trip to Australia he proposed christening one of the nameless coastal stations after him[81]. That's an interesting story, but not really what I wanted to enlarge upon. My question refers of course to the theories of Claude Shannon and Charles Darwin. The two places mentioned are almost as widely separated as two points on the Earth's surface can be – which makes them a good illustration of the distance separating their respective namesake theories. Why are they so far apart?

Of course, one could ask: why should there be any connection at all between telecommunication and evolutionary biology? So let me approach the problem in a different way. Information, if we have in mind its semantic aspect, is something that has to be generated. Messages sent through communication lines have originated – no matter how – in the heads of people. Likewise, the information contained in our genes had also to come about. How? Somehow during biological evolution! Oh yes, that's a cue: could information not be generated by a suitable evolutionary mechanism? Darwin said: Biological evolution is based on natural selection. One of the two aspects of information, i.e. its quantitative amount, is described by Shannon's entropy. Is its other aspect – namely "quality", "meaning" or "value" – perhaps related to Darwin's concept of evolution? It is therefore reasonable to ask whether there is a connection between the theories of Shannon and Darwin, and, if so, what this connection is.

Our question is thus: how was information, characterised by meaning and value, able to come into existence? This question cannot be answered by Shannon's theory. Can, perhaps, Darwin help us?

There is one difficulty in answering this question. Shannon's theory is a mathematical theory applied to a physical problem: transmission of a message through a noisy communication line. Darwin's theory is of a different nature. Let me call it (following Ernst Mayr) a "biological" rather than a "physical" theory. By this I do not mean that biology lies outside the limits of what is accessible to physical theory. No! "Biological theory" refers to the complex reality of life, as we observe it. Darwin came to his conclusions by what we might call a "top down" process, that is, by observing biological reality in all its detail. His main desire was to explain adaptability to any environmental constraint. He never formulated his theory in mathematical terms. Physical theory, on the other hand, is based largely on abstraction from reality, an abstraction that focuses on recurrent regularities. Its results are "if–then" answers, ones that can be cast into a mathematical form.

It is interesting to note that Darwin himself, in fact, thought that there might be some general (and that can only be a "physical") law behind his principle, and he expressed this clearly in a letter to a colleague (see Section 5.1). A general law of this kind will be the major topic of Chapters 4 and 5. As a physical law – in accordance with Eugene Wigner's postulate[82] – it does not deal with processes themselves, and it by no means deals with all the complex processes we observe in the reality of life. Rather, it treats – as Wigner formulated it – the "regularities among processes" which then could explain how reality is shaped. The regularity of evolutionary biology is the ability of a material system to adapt optimally to a complex reality, allowing life to come about. The theory of life is a theory for the generation of information which, in fact, is the nub of Darwin's "dangerous idea"[83].

Darwinian evolution can only occur in populations. A single gene copy cannot evolve, as little as an individual fish can become a bird. Populations are necessary in order to provide a spectrum of alternatives for selection. In thermodynamics, populations are also necessary for the equilibration of alternative chemical states, although this is a situation totally different from natural selection, which does not occur in equilibrated systems. Evolution is based on function and – what is really new in physics – the "purposefulness" of the function involved. Purpose requires a feedback from the effect to its cause. Genetic feedback is based on an inherent property of nucleic acids, that of producing (more or less) identical copies by replication. "Repli"-cation is a fitting term because a reply represents feedback in the form of an answer to a question, which can open a dialogue between cause and effect. There is no other chance for a given sequence copy – because of its vast number of alternatives – to come about other than by replication. This, being an inherent property of a whole class of material entities, engenders novel physical properties and, as will be seen later, is the basis of generating (genetic) information. An "answer" given by (mutation-prone) replication produces new "questions", answered by selective response. "Genetic information" is generated in this way in evolving populations of mutants. A similar kind of process is active in our central nervous system. Here we have populations of excitation patterns built up by firing nerve cells that interact with the memory, fixed by synaptic connections within the neural network. That – likewise – causes communication through selective response, although quite different molecular mechanisms are involved.

In Section 3.7 we discussed a mathematical reality "out there", to which mathematical theory adapts and which is of an *a priori* nature. Biological reality, on the other hand, being itself the result of an adaptive process, occurs as an *a posteriori* phenomenon in our material universe – at least in the way as it appears now. As Theodosius Dobzhansky once said, nothing in biology makes sense except in the light of evolution.

Evolutionary theory dealing with genetic messages is a dynamic theory that applies to extended populations of mutants. A singular genetic message doesn't make sense in Darwinian theory. On the other hand, information theory, which is a probability theory of symbol representation, refers to singular messages that are sent through a transmission channel. I do not see any common root connecting the two theories. The "distance" between them is indeed very, very large. In fact, we would have to depart into a new dimension, as we do when we travel "as the crow flies" from Shannon airport in Ireland to Darwin in Australia. Otherwise we are forced to make a large detour, sailing around the Cape of Good Hope as the first Irish settlers did.

This is as much about Darwinian theory as I want to relate at this point. However, I should like to conclude this section with a little story that may demonstrate that what I have termed "Darwinian theory" is not so foreign to contemporary physics, even if the latter deals with non-viable material systems.

In 1969 I attended a meeting of the Institut de la Vie at Versailles in France. I was invited to talk about my theory of the evolution of self-replicating molecules. When I finished, the blackboard was filled with equations, but only some of them were in my handwriting: more than half had been left by the speaker who gave the lecture before mine. It was the physicist Hermann Haken, the founder of a branch of non-linear dynamic theory that he has termed "synergetics".[84] His talk was on the theory of laser – the word being an abbreviation for light amplification by stimulated emission of radiation. At that time the laser was still a novel physical device, which had been invented only a few years earlier and had been recognised by the award of the 1964 Nobel prize to Charles H. Townes, Nikolai G. Basov and Aleksander M. Prokhovov[85]. In the discussion following my lecture, we realised that my equations describing molecular evolution were similar – and some almost identical – to those left over on the blackboard by Hermann Haken. Could it be that things as different as laser modes and DNA resemble one another so closely in their dynamical behaviour? The self-replicating molecule could just as well be a virus, i.e. an entity at the borderline of "real" life. A laser, on the other hand, is as "inorganic" as a physical system could be – but, as it turned out, it must possess certain properties characteristic of a viable system. So let us look a bit more closely at what goes on inside a laser, in order to understand its magic behaviour, which makes it such a useful device in our cyber-age where it is used for optical communication, reading of digital information stored on compact discs or as an instrument for surgery, welding, holography and many other purposes.

A laser is a light source that produces (nearly) *parallel* beams of *monochromatic* and *coherent* radiation in a wide range of wavelengths, from microwaves to X-rays. Any classical (i.e. non-laser) light source sends out photons in all directions, as a result of spontaneous emission processes. These photons are neither coherent (i.e. they do not oscillate in phase with one another) nor can they be bundled to form perfect parallel beams, with a width down to the dimension of their wavelength. In the laser this is achieved by enclosing the "light medium" in a resonator. This is a cylindrical transparent chamber, the ends of which are closed by mirrors, the mirror at one end being semi-transparent. Light generated in the chamber is reflected back and forth many times, until it eventually leaves the cavity through the semitransparent mirror at one end as a highly parallel beam. Photons with small deviations from the cylinder axis would soon have escaped laterally during the many back-and-forth reflections of the beam (Figures 3.9.2 and 3.9.3).

The important issue, however, is the generation of monochromatic coherence which, in a way, may be seen as "generation of information". The laser medium consists of excitable atoms, ions or molecules. They have to be excited by shining in light from an external source, usually a flash-lamp or a second laser. The process is called pumping because the excited entities in the laser medium, when sending out a light quantum, fall back into their ground state. The excited states therefore have to be steadily replenished by "pumping in" optical energy. The photons produced by the laser medium are originally out of phase, like the light quanta produced by conventional sources.

But now comes the crucial point. If the number of excited entities produced in the laser medium is sufficiently large, they will increasingly be hit by photons produced in the laser and thereby be stimulated to emit their energy as a light quantum in phase with the photon that causes the event. What is critical in this "stimulated emission" is the population of excited states. At a certain threshold value, where the majority of relevant states are excited, "population inversion", as it is called, sets in and eventually forces the whole population into complete in-phase oscillation. Population inversion is like a critical (non-equilibrium) phase transition. However, "phase" in this context does not refer to the fractional period of an oscillatory cycle, but rather to a characterisation of the conformation of a state, in the sense in which the physical chemist uses the term "phase transition" for the melting of ice or boiling of water (see Section 2.8). It means that the phase of those entities that emit photons

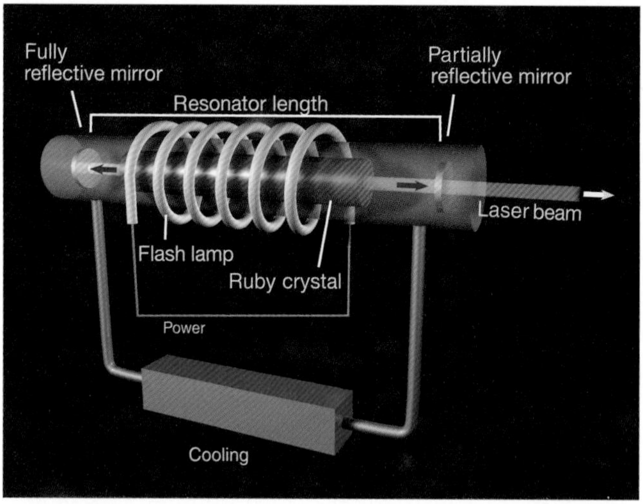

Figure 3.9.2

Schematic representation of a laser

The principle of an optically pumped solid state laser, as used in early experiments by Charles H. Townes and Arthur L. Schawlow, is described in the text. The first device of this kind was built and put to use in 1960 by Theodore H. Maiman. The laser is illuminated by a high-intensity flash of light having a frequency higher than that emitted by the laser. The distance between the two mirrors determines the resonance frequency of the oscillator producing the amplified laser beam. While Charles Townes shared the 1964 Nobel prize for physics with Nicolay Gennadiyevich Basov and Aleksandr Mikhailovich Prokhorov for his early ideas on the laser principle, Arthur Schawlow was co-recipient – together with Nicolaas Bloemberger (USA) and Kai Manne Siegbahn (Sweden) – of the 1981 physics prize.[85] Drawing by Claus-Peter Adam.

in an uncorrelated way becomes unstable with respect to the phase of those entities that, owing to stimulated emission, behave in a correlated way. Note the presence of a co-operative action with unlimited correlation length, which is typical of a phase transition (see Section 2.8). Starting from a fluctuation, the light field serves as an order parameter for the synchronous motion of electrons in the excitation centres, while the excited entities, by virtue of their joint in-phase emission of quanta, create the coherent light field. Hermann Haken called this process "enslavement". The field enslaves the excitable atoms, which in turn generate the exciting field.

Let me now come back to the biological analogue. The "cross-talk" between enslaving and being enslaved in order to produce coherent emission of light in a laser corresponds to the cross-talk between cause and effect in replication, which I alluded to in the first part of this section. DNA or RNA replication is at the basis of the processes

Figure 3.9.3
Modern tuneable dye laser equipment

Theodore Haensch, in his work which was recognised with the 2005 Nobel prize for physics, used more modern equipment. The picture, showing an optically pumped continuously tuneable laser using the dye Rhodamine 6G, was kindly provided by Theodor Haensch, who wrote in his letter that the more modern solid state and fibre lasers are not as spectacular because they hide their working principles in elegant boxes. Another reason to show this picture is that the principle of the tuneable dye laser was pioneered by my colleague Fritz Peter Schäfer of the Göttingen Max Planck Institute.

that produce genetic information through selection. We shall see (Section 4.8) that the correspondence even goes as far as to include an analogy between "population inversion" in laser function and "natural selection" in the evolutionary process. Both are "phase transitions", taking place in different "spaces", but according to analogous physical criteria. But that is as far as the analogy reaches. Otherwise, it misses more than it explains.

The competing phase fluctuations at the onset of population inversion in a laser do not originally differ in their "fitness"; their potential to be selected changes increasingly as a consequence of a sort of "mass action". Are they therefore equivalent to neutral mutants? Hardly, because they lack the property of mutation. The absence of this property and of "fitness" in any case excludes the main consequences that replication has in biology. So my aim in discussing the above example was not to recruit it as a model, but rather to explain an important prerequisite of Darwinian selection.

Another characteristic of light quanta is that they cannot be "cloned". The laser selects only from wave trains that differ in their phases. Light quanta also possess a spin, or polarisation. Spin is a property that can be physically explained only in terms of quantum mechanics. Heisenberg's uncertainty relationship would be violated by synchronising all individual spins of an ensemble of photons. Hence, it is impossible to produce identical copies of light quanta.[86]

3.10. Where is the "Temperature" of Information?

One could reply to this question by posing a counter-question: why should information have a temperature at all? Information is not related to energy, and temperature expresses an "intensity" of thermal energy. Well, what about an analogue of temperature, an analogue that expresses some "intensity" of information and thereby clearly sets it apart from informational entropy? This is by no means a trivial question! An answer to it might tell us what information really is, but not necessarily. After all, temperature did not tell us what energy really is. Information may be something that cannot be reduced further, at least in a world that is reflected in our minds.

The quantity that Shannon originally described as "uncertainty", and which is often simply defined as representing "information", should adequately be termed "(informational) entropy", a term that was also adopted by Shannon. The customary choice of the letter H – whether or not it was supposed to be reminiscent of Boltzmann's H function (Section 2.3) – is at least appropriate to differentiate it from thermodynamic entropy, the latter usually being denoted by the letter S. Boltzmann's constant relates his statistical expression to energy. This constant could just as well have been associated with temperature, as originally proposed by Gibbs, characterising temperature unequivocally as an increment of the intensity of energy. Moreover, it would have yielded a complete formal agreement between Boltzmann's and Shannon's entropy (see Table 2.5.1). Yet the agreement is a formal one only, and therefore anything but complete.

Any statistical distribution can be assigned an entropy. In physics, one wants entropy to be an extensive variable. Accordingly, it refers to the extent of matter, a macrostate that includes all possible excitable quantum states among which energy can be distributed. In communication theory entropy stands for all possible symbol representations that are necessary in order to conserve the content of a message during the process of transmission. If the sequences are sufficiently long, all concatenations yield for entropy an expression that holds for all possible distributions. Mutual entropy then allows one to compare source and sink, which is necessary for conserving the content of any defined sequence. That is as far the correspondence between Boltzmann's and Shannon's entropy stretches. In principle, a sequence of symbols – such as the text of a given novel – represents a microstate, while the macrostate is given by the entirety of all texts in that language.

In spite of these formal coincidences, there is quite a difference in the physical background of the two cases, and here I do not mean only the obvious difference between quantum states and symbol arrangements. Thermodynamic entropy refers to a macrostate of an ensemble of material particles that represents an average of all statistically possible microstates at a given temperature. For informational entropy one may well ask: where is the distribution? The answer is: there is only one fixed message, but its long-range order, which is responsible for its meaning, is of no importance to the communications engineer who has to construct a device to serve for all possible symbol sequences. If the message is long enough, it will contain all sorts of symbols and symbol concatenations characteristic of the particular language, wherever in the message they may occur, and it is this distribution of symbol combinations allowed by the language that determines its entropy. The consequence is: for a given language every sufficiently long message will converge to the same normalised entropy value because all code extensions of higher Markovian order will eventually show up with defined probabilities. The entropy just sums them up, regardless of where in the long-range order they occur. The true concern is that the message is really long enough to allow the utilisation of the results of Shannon's second (and main) theorem, which says that "codes exist which are arbitrarily reliable and which can signal at rates arbitrarily close to the channel capacity". Richard Hamming adds to this – and I can almost hear him groaning – "the theorem sets bounds, but does not suggest much other than that very good codes must be very long, rather impractical ones" (see Section 3.3).

It is this statistical behaviour that establishes the analogy between informational and thermodynamic entropy. This can be seen even better if we cut up our messages into pieces, the lengths of which include of all code extensions taken into account. This, of course, destroys the very meaning of the message, but the (normalised) entropy, obtained by averaging over all pieces, remains the same. If the length L of the message tends to infinity ($L \to \infty$), the number of pieces N also will do so. In statistical mechanics this corresponds to the "thermodynamic limit", which is reached for sufficiently large particle numbers. The statistical laws hold exactly only in this limit, and the correspondence between informational and thermodynamic entropies refers to such a limit. A particle number $N = 10^{24}$ (Avogadro's number) is a natural order of magnitude in thermodynamics. A system as small as a bacterial cell (one femtolitre) still contains some 10,000 million water molecules. A message containing 10,000 million letters would correspond to some 1000 books, each as long as Tolstoy's *War and Peace*. However, in Section 2.6 we have seen that these statistical laws are still good approximations at much lower particle numbers, as in the game where N was only 64.

What is more important in the present discussion is the fact that informational entropy only rates "structural" properties of our languages as far as they are known to us, but nothing that is related to the particular content of a message. As I said in Section 3.3: informational entropy tells us how much we do *not* know after we have taken into account everything that we do know. This, of course, is meant for messages with novel semantic content. A good example is the gene of an unknown protein, which at first sight looks pretty much like a random sequence.

Another example, more relevant for my original question about an analogue of temperature in Shannon's theory, is a given "piece of information", say a book, that has been translated from one language into another, e.g. from German into English. Here the semantic content is fixed, and the difference in informational entropy merely reflects differences in the structures of the two languages. In this case one could define an analogue of temperature, i.e. an average "intensity", as a property of language, which is complementary to the "extent" rated by the informational entropy of that particular language. It is a fact that on average (not necessarily for any particular phrase) an English text after translation comes out to be somewhat shorter than its German counterpart. Hence the "intensity" of English must be correspondingly larger than that of German, or English is "hotter" than German.

However, if we are dealing with semantic information as such, our problem should not depend on the choice of language. There is an "intensity" of information content, which is related to the length of the message. Running through a hotel at midnight and yelling "fire" represents a maximum of information intensity. Just compare it with the answer to an invitation to some meeting, reading: "I am sorry to be unable to attend your meeting because I cannot come" (Sydney Brenner, personal communication).

By the same token, I have some qualms about calling informational entropy the "amount of information". After all, thermodynamic entropy is not called an "amount of thermal energy". It is *entropy*, and as such it describes in a clearly defined way the extensive aspect of energy. As far as semantics have been "exorcised" from information theory (I borrow the term from Hermann Haken[87]), a "temperature" of information, or its analogue, simply doesn't exist. The word "information" instead of "informational entropy" would be justifiable only by arbitrary definition.

In the rest of this book I shall abandon the practice, quite conventional among physicists (following Léon Brillouin), of using the word "information" if its entropy is meant. I prefer to identify the word "information" with the meaning it has in common use in English. Informational entropy H and thermodynamic entropy S cannot simply be equated (or even added together (as has been done)), unless a normalisation has been applied to create comparable conditions.

In the first section of this book I stressed the point that we do not know what energy really is. However, I immediately qualified my statement by adding that we nevertheless know how to quantify it, regardless of which form it appears in. And we know how to measure it. This is due to two properties that were first expressed by the two laws of thermodynamics: (1) in a closed system the amount of energy is strictly conserved and (2) energy distributes itself among all available degrees of freedom, yielding a maximum value of entropy. For an isolated system this means that any exchange of energy or matter with the environment is excluded. If we now try to make analogous statements about information, we get into a precarious situation. We have not yet decided what information (apart from informational entropy) really *is*, but we know for sure that neither a conservation nor a dissipation law exists. Information can be generated as well as destroyed, and the most common way of destroying information is to distribute it randomly among all possible symbol states, in other words: to "equilibrate" it. The definition of temperature requires (at least partial) equilibration. Hence it is not possible to assign any analogue of temperature to information. The true reason for this is the absence of laws analogous to the first and second laws of thermodynamics. If we want to understand the semantic aspect of information we must be able to answer the question: how does information originate? Instead of any conservation and dissipation law we shall rather require an extremum principle that involves evaluation and optimisation. This is the subject of Chapters 4 and 5.

The above difference was the reason why we failed to find a route "from Shannon to Darwin" in the preceding section. Let me illustrate it again with an example, the evolutionary adaptation of an enzyme molecule to its functional efficiency, which nowadays can be performed experimentally in an automatically controlled "evolution machine" (see Section 5.9). In such a process the informational entropy of the total polypeptide sequence hardly changes during the entire experiment, although at each stage of the process we find a different long-range order – or, put better, organisation – of the monomers in the whole sequence. What is the "force" that changes the long-range organisation?

Here we are brought back to our quest for the difference between informational and thermodynamic entropy. It is not only the absence of temperature (or its analogue) in information theory. In addition to the absence of temperature there is also the replacement of entropy by "mutual entropy" (Section 3.3), which establishes the difference between the entropy concepts in the two theories, as was pointed out clearly by Hubert Yockey. Mutual entropy is based on conditional probabilities: given a source A with an alphabet of s_A letters from which a symbol a_i with probability $p(a_i)$ has been sent, and a receiving end B with an alphabet of s_B letters, then the conditional probability of the appearance of the symbol b_j at B is $p(b_j|a_i)$. Mutual entropy does not make much sense in statistical thermodynamics, since it would require an identification of individual microstates, but it does make sense in communication theory that deals with individual messages which in transmission get (slightly) changed by noise.

For years I had thought that conditional probabilities might yield some access to a "generative principle" of information. However, what in information theory is called "gain" of information, describing the change of informational entropy involved in the transition from a given to a new probability distribution, rated by the new distribution, always comes out to be positive, simply for mathematical reasons. It does not help in implementing a new distribution that leads to an entropy compression, such that in the end only one sequence (or a few degenerate copies) fulfils the purpose. What is required is a "force" that brings into being the new distribution instead of the former one. It has to involve dynamic instabilities, which underlie "selection".

This can materialise only in populations where it is possible to establish new probability distributions, and this may occur in relatively small steps. In this process "error" should not be characterised simply as noise that has to be eliminated. It is the source of new information and as such has to be maintained, albeit within certain limits. The final outcome, i.e. the selection of a unique new sequence that is optimally adapted to some functional purpose, is just the opposite of equilibration. None of this can be achieved by communication theory, which is designed for optimally representing and conserving a message sent through a transmission line.

In conclusion of this whole chapter I would like to point out the problem associated with the nature of "information", which is not yet fully appreciated by many physicists. Yes, we do have a theory that bears the name "information theory". It is indeed a valuable theory and has given us many new insights. However, it only covers the aspect of "potential information", i.e. the quantitative aspect of information: how much do we have to know in order to become completely informed? "Completely informed" in our language means having the ability to distinguish between different alternatives, or, in physical language, to discern between different microstates.

In Chapter 2, I stressed the relation between micro- and macrostates. In statistical thermodynamics entropy is a measure of macrostates, while true information is always associated with microstates. In language, the microstate refers to a single and unique message, while the macrostate is an average of many messages, a property of language. The non-existence of temperature and the replacement of entropy by mutual entropy demarcate Shannon's theory from statistical thermodynamics. These two approaches meet in their mathematical formation concept, which includes meaning. They differ in respect of equilibration among alternative microstates (in the one case) and selection of one particular microstate (in the other). Both are governed by extremum principles for equilibration: within a macrostate, entropy is maximised (or free energy is minimised) and one particular microstate is established in a kind of "natural selection". In other words: entropy refers to an average of (physical) states, information to a particular (physical) state.

In the scientific literature we can find much confusion in the use of the word "information". The following example illustrates this. I read it in a very interesting article by Leonard Susskind in *Scientific American* (1997) dealing with "black holes and the information paradox". It describes the huge entropy that Hawking and Bekenstein associate with black holes, a problem mentioned in Sections 2.9 and 5.10. So far so good, but how did the editorial staff of the journal interpret the results of the article? A comment on the cover of the journal reveals the answer: "Data lost in collapsed stars may not be gone for ever". If data are broken down to the Planck scale they will still yield a maximum of their "entropy", which indeed is suggested by the

Schwarzschild model, but they have certainly lost their "semantic" information, like a novel that has been broken down to a jumble of single letters.

LITERATURE AND NOTES

1. Maxwell, J. C. (1871). "Theory of Heat". Greenwood Press, Westport, CT. Later edition: Longman, Green and Co., London, 1902.
2. Szilárd, L. (1929). "Über die Entropieverminderung in einem thermodynamischen System bei Eingriffen intelligenter Wesen". *Z. Phys.* 53 (11–12): 840–856.
3. Landauer, R. (1961). "Irreversibility and Heat Generation in the Computing Process". *IBM J. Res. Dev.* 5: 183–191.
   Bennett, C. H. (2003). "Notes on Landauer's principle, Reversible Computation and Maxwell's Demon". *Stud. Hist. Philos. M. P.* 34: 501–510.
4. Gábor, D. (1951). "Lectures on Communication Theory – Machine Demonstrating Relations between Information and Entropy". M. I. T. Lectures. Technical Report No. 238. Research Laboratory of Electronics. M. I. T. Cambridge, Massachusetts. See Ref. 7 pp. 179–181.
5. Brillouin, L. N. in Ref. 7 pp. 176–182.
6. Shannon, C. E. and Weaver, W. (1979). "The Mathematical Theory of Communication". University of Illinois Press, Urbana, IL. Based on two papers by C. E. Shannon in the *Bell Syst. Tech. J.* Reprinted in Slepan, D. (ed.) (1974). "Key Papers in the Development of Information Theory". *IEEE Press*, New York.
7. Brillouin, L. N. (1959). "La Science et la théorie de l'information". Masson & Cie, Paris. English translation: Brillouin, L. N. (1962). "Science and Information Theory". Academic Press, New York.
8. Wiener, N. (1948). "Cybernetics Or Control and Communication in the Animal and the Machine". John Wiley and sons. New York. See also Wiener, N. (1956). "What is Information Theory?" *IRE Trans. Inform. Theory*, Vol. I T-2: 48.
9. Campbell, J. (1982). "Grammatical Man: Information, Entropy, Language and Life". Simon and Schuster, New York.
10. Pennock, R. T. (1999). "Tower of Babel: The Evidence Against the New Creationism". MIT Press, Cambridge, MA.
11. Mandelbrot, B. (1952). "Contributions á la théorie mathématique des jeux de communication". Ph. D. Thesis. Paris. Institute of Statistics, Univ of Paris 2: 1–124.
12. Shannon, C. E. (1951). "Prediction and Entropy of Printed English". *Bell Syst. Tech. J.* 30: 50–64.
13. Khinchin, A. I. (1957). "Mathematical Foundations of Information Theory". Dover, New York (translated by R. A. Silverman and M. D. Friedman).

14. Shannon, C. E. (1948). "A Mathematical Theory of Communication". *Bell Syst. Tech. J.* 27: 379–423, 623–656.
15. McMillan, B. (1953). "The Basic Theorems of Information Theory". *Ann. Math. Statist.* 24 (2): 196–219.
16. Breiman, L. (1957). "The Individual Ergodic Theorem of Information Theory." *Ann. Math. Statist.* 28: 809–811.
17. Yockey, H. P. (1977). "A Calculation of the Probability of Spontaneous Biogenesis by Information Theory". *J. Theor. Biol.* 67: 377–398.
18. Hamming, R. W. (1980). "Coding and Information Theory". Prentice Hall, Englewood Cliffs, NJ.
19. Yockey, H. (1992). "Information Theory and Molecular Biology". Cambridge University Press, Cambridge.
20. Jaynes, E. T. (1979). "Where do we Stand on Maximum Entropy?". In: "The Maximum Entropy Formalism", Levine, R. D. and Tribus, M. (eds). MIT Press, Cambridge, MA, pp. 15–118; Jaynes, E. T. (1989). "Where Do We Stand on Maximum Entropy?" In: "Papers on Probability, Statistics and Statistical Physics", Rosenkrantz, R. D. (ed.). Kluver, Dordrecht, p. 233 ff.
21. Hamming, R. W. (1980). "Coding and Information Theory". Prentice Hall, Englewood Cliffs, NJ, p. 130 ff.
22. Shannon, C. E. (1949). "Communication in the Presence of Noise". *Proc. IRE* 37: 10–21.
23. Berlekamp, E. R. (1968). "Algebraic Coding Theory". McGraw Hill, New York.
24. MacWilliams, F. J. and Sloane, N. J. A. (1977). "The Theory of Error Correcting Codes". Addison-Wesley, Reading, MA.
25. McEliece, R. J. (1977). "The Theory of Information and Coding". Addison-Wesley, Reading, MA.
26. Wakerly, J. (1978). "Error Detecting Codes, Self-checking Circuits and Applications". North-Holland, Amsterdam.
27. Hamming, R. W. (1950). "Error Detecting and Error Correcting Codes". *Bell Syst. Tech. J.* 29: 147–160.
28. Steward, I. (1996). "From Here to Infinity: A Guide to Today's Mathematics". Oxford University Press, Oxford and New York.
29. Rivest, R. L., Shamir, A. and Adleman, L. (1978). "A Method for Obtaining Digital Signatures and Public-Key Cryptosystems". *Comm. ACM* 21(2): 120–126.
30. Edson Smith, George Woltmann, Scott Kurowski *et al.* of the Mathematics Department, University of California at Los Angeles. See e.g. www.mersenne.org.
31. Bauer, F. L. (1997). "Decrypted Secrets: Methods and Maxims of Cryptology". Springer-Verlag, New York.
32. Beutelspacher, A. (1990). "How to Communicate Efficiently". *J. Comb. Theory, Ser. A* 54(2): 312–316.

33. Kippenhahn, R. (1997). "Verschlüsselte Botschaften: Geheimschriften, Enigma und Chipkarte", 1st edn. Rowohlt, Reinbek at Hamburg.
34. Singh, S. (1999). "The Code Book: The Secret History of Codes and Codebreaking". Fourth Estate, London.
35. Fraser, G. (2006). "The New Physics for the Twenty-First Century". Cambridge University Press, Cambridge and New York.
36. Zeilinger, A. (2006). "Essential Quantum Entanglement", Chapter 10 (Part III). In: "The New Physics for the Twenty-First Century". Fraser, G. (ed.). Cambridge University Press, Cambridge and New York, pp. 257–267.
37. Prigogine, I. (1980). "From Being to Becoming: Time and Complexity in the Physical Sciences". Freeman, New York.
38. Heine, H. (1826). "Die Harzreise". In: Heine, H. (1826–31). "Reisebilder", Vol. 1. Hoffmann und Campe, Hamburg.
39. Hilbert, D. (1900). "Mathematische Probleme". 2nd International Congress of Mathematicians. Paris. In: Nachr. Königl. Ges. Wiss. Gött. Math.-Phys. Kl. 1900: 253–297.
40. Whitehead, A. N. and Russell, B. (1910). "Principia Mathematica". Cambridge University Press, Cambridge.
41. Wittgenstein, L. J. J. (1922). "Tractatus Logico-Philosophus". Routledge and Kegan, London.
42. Reid, C. (1970). "Hilbert". Springer, New York (with an appreciation of Hilbert's mathematical work by Herman Weyl).
43. Reid, C. (1976). "Courant in Göttingen and New York. The Story of an Improbable Mathematician". Springer Verlag, New York.
44. Hofstadter, D. B. (1979). "Gödel, Escher, Bach: An Eternal Golden Braid". Basic Books, New York.
45. Gödel, K. (1931). "Über formal unentscheidbare Sätze der Principia Mathematica und verwandter Systeme I". *Monatshefte Math. Phys.* 38(1): 173–198.
English translation: Gödel, K. (1962). "On Formally Undecidable Propositions of Principia Mathematica and Related Systems." Translation B. Meltzer. Basic Books.
46. Gödel, K. Corollary to Incompleteness Theorem: The consistency of such a system cannot be proved.
47. Solomonoff, R. J. (1960). "A Preliminary Report on a General Theory of Inductive Inference". *Tech. Rept. ZTB*-138, Zator Company, Cambridge, Mass.
Solomonoff, R. J. (1964). "A Formal Theory of Inductive Inference, Part I". *Inform. Control* 7(1): 1–22.
Solomonoff, R. J. (1964). "A Formal Theory of Inductive Inference, Part II". *Inform. Control* 7(2): 224–254.
48. Kolmogorov, A. N. (1965). "Three Approaches to the Quantitative Definition of Information". *Probl. Inform. Transm.* 1(1):1–7.

49. Chaitin, G. J. (1966). "On the Length of Programs for Computing Finite Binary Sequences". *J. ACM* 13(4): 547–569.
Chaitin, G. J. (1969). "On the Length of Programs for Computing Finite Binary Sequences: Statistical Considerations". *J. ACM* 16: 145–159.
50. Casti, J. L. (1994). "Complexification: Explaining a Paradoxical World Through the Science of Surprise". Harper Collins, New York.
51. Gell-Mann, M. (1994). "The Quark and the Jaguar: Adventures in the Simple and the Complex". Little, Brown & Co., London.
52. Weizsäcker v., E. and Weizsäcker v., C. (1972). "Wiederaufnahme der begrifflichen Frage: Was ist Information?". *Nova Acta Leopoldina* 37/1: 535–555.
Weizsäcker v., E. and Weizsäcker v., C. (1998). "Information, Evolution and 'Error-friendliness' ". *Biological Cybernetics* 77(6): 501–506.
Weizsäcker v., E. (1974). "Erstmaligkeit und Bestätigung als Komponenten der pragmatischen Information". In: E. von Weizsäcker (ed.) "Offene Systeme, Zeitstruktur von Information, Entropie und Evolution". Klett-Cotta, Stuttgart, Germany. pp. 82–111.
53. Küppers, B.-O. (1990). "Information and the Origin of Life". MIT Press, Cambridge, MA.
54. Chaitin, G. J. (1970). "To a Mathematical Definition of 'Life' ". *ACM SICACT News* 4: 12–18; Chaitin, G. J. (1979). "Toward a Mathematical Definition of Life". In: "The Maximum Entropy Formalism", Levine, R. D. and Tribus, M. (eds). MIT Press, Cambridge, MA, pp. 477–498.
55. Turing, A. M. (1936). "On Computable Numbers, with an Application to the Entscheidungsproblem". *Proc. London Math. Soc. Ser.* 2(42): 230–65; Turing, A. M. (1937). "On Computable Numbers, with an Application to the Entscheidungsproblem. A correction". *Proc. London Math. Soc. Ser.* 2(43): 544–546.
56. Post, E. L. (1936). "Finite Combinatory Processes – Formulation 1". *J. Symbolic Logic* 1: 103–105.
57. Church, A. (1936). "An Unsolvable Problem of Elementary Number Theory". *Am. J. Math.* 58(2): 345–363.
Church, A. (1941). "The Calculi of Lambda Conversion". *Ann. Math. Stud.* No. 6. Princeton University Press.
58. Penrose, R. (1994). "Shadows of the Mind: A Search for the Missing Science of Consciousness". Oxford University Press, Oxford.
59. Penrose, R. (1989). "The Emperor's New Mind: Concerning Computers, Minds, and the Laws of Physics". Oxford University Press, New York.
60. Changeux, J. P. and Connes, A. (1998). "Conversations on Mind, Matter, and Mathematics". Princeton University Press, Princeton.
61. Connes, A. (1992). "Sur la nature de la réalité mathématique". ["On the Nature of Mathematical Reality"]. *Elem. Math.* 47(1): 19–26.

62. Courant, R., Stewart, I. and Robbins, H. (1996). "What Is Mathematics?: An Elementary Approach to Ideas and Methods". Oxford University Press, Oxford. (Revision of the earlier text of Courant and Robbins (1941) by I. Stewart.)
63. Paper by ref. to W. von Humboldt.
64. Gatlin, L. (1973). "Information Theory and the Living System". Columbia University Press, New York.
65. Josse, J., Kaiser, A. D. and Kornberg, A. (1961). "Enzymatic Synthesis of Deoxyribonucleic Acid. VIII. Frequencies of Nearest Neighbor Base Sequences in Deoxyribonucleic Acid". *J. Biol. Chem.* 236(3): 864–875.
66. Swartz, M. N., Trautner, T. A. and Kornberg, A. (1962). "Enzymatic Synthesis of Deoxyribonucleic Acid. XI. Further Studies on Nearest Neighbor Base Sequences in Deoxyribonucleic Acids". *J. Biol. Chem.* 237: 1961–1967.
67. Subak-Sharpe, J. H., Bürk, R. R., Crawford, L. V., Morrison, J. M., Hay, J. and Keir, H. M. (1966). "An Approach to Evolutionary Relationships of Mammalian DNA Viruses through Analysis of the Pattern of Nearest Neighbor Base Sequences". *Cold Spring Harb. Sym.* 31: 737–748.
68. Yockey, H. O. (2002). "Information Theory, Evolution and the Origin of Life". *Inf. Sci.* 141(3–4): 219–225.
69. Dayhoff, M. O. (ed.) (1961). "Atlas of Protein Sequence and Structure". National Biomedical Research Foundation, Washington, DC.
70. Coutelle, R., Hofacker, G. L. and Levine, R. D. (1979). "Evolutionary Changes in Protein Composition – Evidence for an Optimal Strategy". *J. Mol. Evol.* 13(1): 57–72.
71. Borštnik, B. and Hofacker, G. L. (1985). "Functional Aspects of Neutral Patterns in Protein Evolution". In: "Structure and Motion, Membranes, Nucleic acids and Proteins", Clementi, E., Corongiu, G., Sarma, M. H. and Sarma, R. H. (eds). Adenine Press, Guilderland, pp. 277–292.
72. Dayhoff, M.O., Schwartz, R. M. and Orcutt, B. C. (1978). "A Model of Evolutionary Change in Proteins". In: "Atlas of Protein Sequence and Structure", Vol. V, suppl. 3, Dayhoff, M. O. (ed.), pp. 345–352. National Biomedical Research Foundation, Washington, DC.
73. King, J. L. and Jukes, T. H. (1969). "Non-Darwinian Evolution". *Science* 164 (3881): 788–798.
74. Crick, F. H., Barnett, L., Brenner, S. and Watts-Tobin, R. J. (1961). "General Nature of the Genetic Code for Proteins". *Nature* 192: 1227–1232.
75. Eigen, M. and Winkler-Oswatitsch, R. (1981). "Transfer-RNA: The Early Adaptor". *Naturwiss.* 68: 217–228; Eigen, M. and Winkler-Oswatitsch, R. (1981). "Transfer-RNA, an Early Gene?". *Naturwiss.* 68: 282–292.
76. Eigen, M. and Pörschke, D. (1970). "Co-operative Non-enzymic Base Recognition. I. Thermodynamics of the Helix-Coil Transition of Oligoriboadenylic Acids at Acidic pH". *J. Mol. Biol.* 53(1): 123–141.

77. Cullman, G. and Labouygues, J. (1983). "Noise Immunity of the Genetic Code". *Biosystems* 16: 9–29.
78. Cullman, G. and Labouygues, J. (1987). "The Logic of the Genetic Code". *Math. Model* 8: 643–646.
79. Yockey, H. P. (2000). "Origin of Life on Earth and Shannon's Theory of Communication". *Comput. Chem.* 24(1): 105–123.
80. Dunitz, J. Personal communication. Courtesy: Calculating distance Shannon - Darwin.
81. Wills, C. Personal communication.
82. Wigner, E. P. (1957). "The Structure of the Nucleus". Proc. 11th Conf. Robert A. Welch Found. Chem. Res. I: 86. Houston, Texas.
83. Dennett, D. (1995). "Darwin's Dangerous Idea: Evolution and the Meanings of Life". Simon and Schuster, New York.
84. Haken, H. (1983). "Synergetics. An Introduction: Nonequilibrium Phase Transitions and Self-Organization in Physics, Chemistry, and Biology". Springer, Heidelberg.
85. Townes, C. H. (1964). "Production of Coherent Radiation by Atoms and Molecules". Nobel Lectures, Physics 1963–1970, Elsevier, Amsterdam, 1972.
    Basov, N. G. (1964). "Semiconductor Lasers". Nobel Lectures, Physics 1963–1970, Elsevier, Amsterdam, 1972.
    Prokhorov, A. M. (1964). "Quantum Electronics". Nobel Lectures, Physics 1963–1970, Elsevier, Amsterdam, 1972.
86. Wootters, W. K. and Zurek, W. H. (1982). "A Single Quantum Cannot be Cloned". *Nature* 299: 802–803; Bussey, P. J. (1983). "On Replicating Photons". *Nature* 304: 188; Wootters, W. K. and Zurek, W. H. (1983). *Nature* 304: 188–189.

4 Information and Complexity

4.1.	How Complex is Chemistry?	317
4.2.	How Does Nature Tame Chemical Complexity?	333
4.3.	An Unsolved Mathematical Problem: P = NP?	344
4.4.	Are We Points in Hilbert Space?	354
4.5.	Hyperspace: Trick or Treat?	372
4.6.	How Does Matter Move in Physical Space?	387
4.7.	And How to Get from Here to There in Information Space?	404
4.8.	Can a Simplex be Complex?	423
4.9.	What Does "Meaning" Mean?	436
4.10.	Pure Thought = Poor Thought?	449

4.1. How Complex is Chemistry?

There is an old German nursery rhyme that asks the child: "Do you know how many starlets there are under the blue tent of the sky?" Of course, the children do not know, but they are comforted by the fact that they don't have to because the Lord has counted "that huge whole number" exactly for them, not missing a single one.

If you ask a chemist how many molecules our universe may harbour, his answer would be as imprecise as that of a modern astronomer requested to give an estimate of the number asked for in the nursery rhyme. In fact, the chemist and the astronomer would probably argue on a similar basis. They would both start from an estimate of the total mass content of the observable universe, which is something above 10^{80} proton masses (see also Chapter 1). "Something above" means a (logarithmic) uncertainty due to open questions such as the precise mass of neutrinos, or how many forms of "dark matter" – such as black holes or cosmic strings – are in existence and how much they contribute to the total mass of the (detectable) universe. Leaving aside an answer to these questions, both the chemist and the astronomer would encounter various other difficulties in making their estimates.

The astronomer has to assume some mass average for all the stars in the universe. This, however, is not straightforward, and neither does it make much sense because of the tremendous diversity of the sizes of stars. A mass of 10^{80} protons would be the equivalent of 3×10^{28} earth or about 10^{23} sun masses; but we know that there are fixed stars much larger than our sun, and that there are planets, moons and asteroids much smaller than the earth.

For the chemist, the difficulties are even worse. How many different types of molecule should he take into account? What would a correctly weighted average size be? How could such a number be representative of all the matter in the universe, given the many places that may show totally different chemistry? How many places are entirely hostile to chemistry? The binding energies of outer electrons in atoms, essentially responsible for chemical behaviour, fall into the range of 1–10 eV. A thermal energy of 1 eV corresponds to a temperature of nearly 12,000 Kelvin. The inner electrons of heavy atoms have energies reaching the scale of kilo-electron volts, corresponding to temperatures in the interior of our sun. As was discussed in Chapter 1, at such temperatures nuclear fusion gets going. Around 10 million Kelvin, as in the sun, this fusion consists essentially of the condensation of four protons, involving reactions with light nuclei, such as helium-3, lithium-7, beryllium-7 and -8, and boron-8. There are many stars that reach temperatures of 100 million Kelvin and higher, where the carbon cycle shown in Figure 1.1.1 dominates.

It is a matter of definition to what extent we want to include fully or partially ionised plasma in this discussion. Under these conditions the answer may turn out to be quite trivial. It would just include some or all of the 100-odd states of chemical elements, possibly including some of their combinations. However, all this is not what I am actually looking for. It is not even interesting, unless I want to know what is going on at specific places. As far as the general question is concerned I could retreat to the answer given by the astrophysicists, i.e. quoting the total content of matter in terms of proton masses.

What I really want to discuss in this chapter is the complexity of chemistry under fairly hospitable conditions, i.e. the conditions we encounter on our planet. And here we are not interested in just counting numbers of molecules present. About a 0.1% of the total mass of the earth (6×10^{24} kg) is water, which means about 10^{47} H_2O molecules. They establish the medium in which the chemistry of life got going. But what we really want to know right now is the *potential* complexity of chemistry: How many *different* molecules are possible under these conditions? Again, this is not a simple question! We might ask how many kinds of molecules have been synthesised in our laboratories. That number is recorded regularly by the journal *Chemical Abstracts*. It amounts to more than 10 million organic versus less than 1 million inorganic compounds. In the "club" of recognised compounds only those substances that fulfil the *Beilstein* requirements can become members. (The *Beilstein*, named after its founder, was the first multivolume dictionary to list all the known organic compounds that meet certain demands of purity and yield of synthesis.)

The distinction between organic and inorganic compounds is a bit arbitrary. What is called "organic" is essentially the chemistry of carbon compounds, chains or rings, containing hydrogen with or without oxygen, nitrogen, phosphorus, sulphur or other elements. However, carbonic acid ($CO_2 + H_2O \rightleftharpoons H_2CO_3$) and its salts – such as soda (Na_2CO_3) – or cyanides (derived from hydrogen cyanide, HCN) and cyanates (containing the radical CNO) are assigned to inorganic chemistry, which is otherwise the province of all non-carbon compounds. Originally, the term "organic" referred to substances existing in or derived from microbes, plants or animals, i.e. living organisms.

Why is the question regarding chemical complexity itself of a complex nature? Because we don't know where to put the limit! Let us look at a very simple defined class of compounds, the saturated hydrocarbons, which have the general composition C_nH_{2n+2}. The constitutions of the first three members, occurring in natural gas and widely known as methane, ethane and propane, can only be arranged in one way. This is not so for the higher members, starting with butane (Figure 4.1.1).

Butane has only one alternative constitutional isomer, called isobutane. In Figure 4.1.1(b) a second class of isomer is introduced, called "conformational" or "stereo-" isomers, terms that refer to the different spatial arrangements that a given constitution can assume, usually by rotation about a single bond. The whole field is excellently reviewed by Gerhard Quinkert, Ernst Engel and Christian Griesinger[1]. Since the focus of this section is the complexity of chemical compounds, I shall restrict myself in the following paragraphs to a discussion of constitutional isomerism.

For butane (four carbon atoms) there are only two possible isomers (Figure 4.1.1). The hydrocarbon with 20 carbon atoms ($C_{20}H_{42}$) has more than 300,000 isomers, the one with 30 carbon atoms ($C_{30}H_{62}$) has over 4000 million constitutional isomers. $C_{40}H_{82}$ has more than 10^{13} isomers, in fact precisely 62,491,178,805,831.[2] This is just one class of hydrocarbons; others are those including double or triple bonds, such as ethylene ($H_2C{=}CH_2$) or acetylene ($HC{\equiv}CH$), again forming chains with many combinations. There are cyclic compounds, such as cyclohexane (C_6H_{12}), which has only single bonds, or those involving rings containing double as well as single bonds. Furthermore, there are condensed ring systems, and all sorts of chains and cycles (and combinations thereof) that include other atoms such as oxygen, nitrogen, phosphorus and sulphur. An example from Nature – at the same time a masterpiece of synthetic chemistry[3] – is shown in Figure 4.1.2.

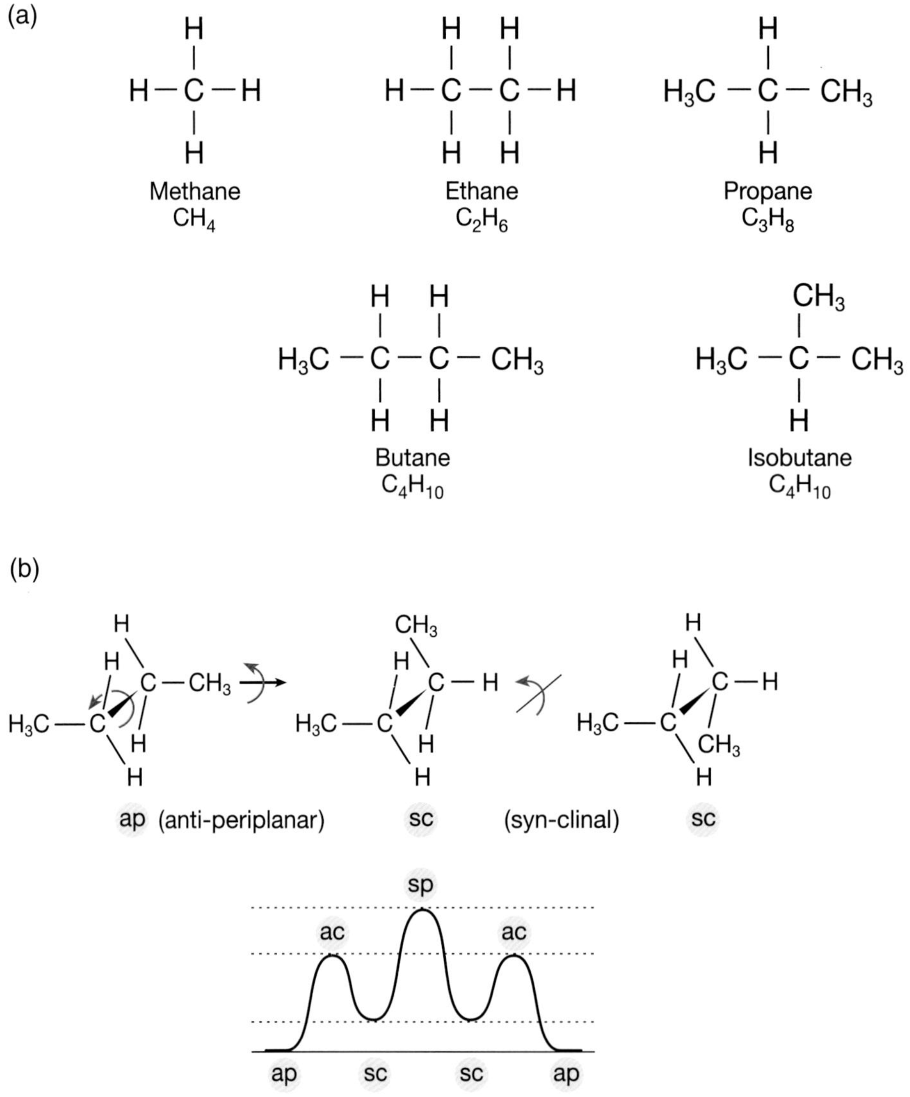

Figure 4.1.1

The hydrocarbons C_nH_{2n+2} for $n = 1$ to 4

(a) "Constitutional isomers" appear first for $n = 4$ with isobutane. For $n > 4$ their number increases drastically (see text). Two constitutional isomers differ in their bond structure. In other words, a constitutional transition involves the breaking of a chemical bond. (b) Apart from this isomerism there exists a "configurational isomerism", where the transition results from the rotation of a group around a given bond between two more or less stable positions. The lower part of the figure describes such a process. Time constants for rotational transitions range from "very slow" to "very fast" (such as nanoseconds).

(a)

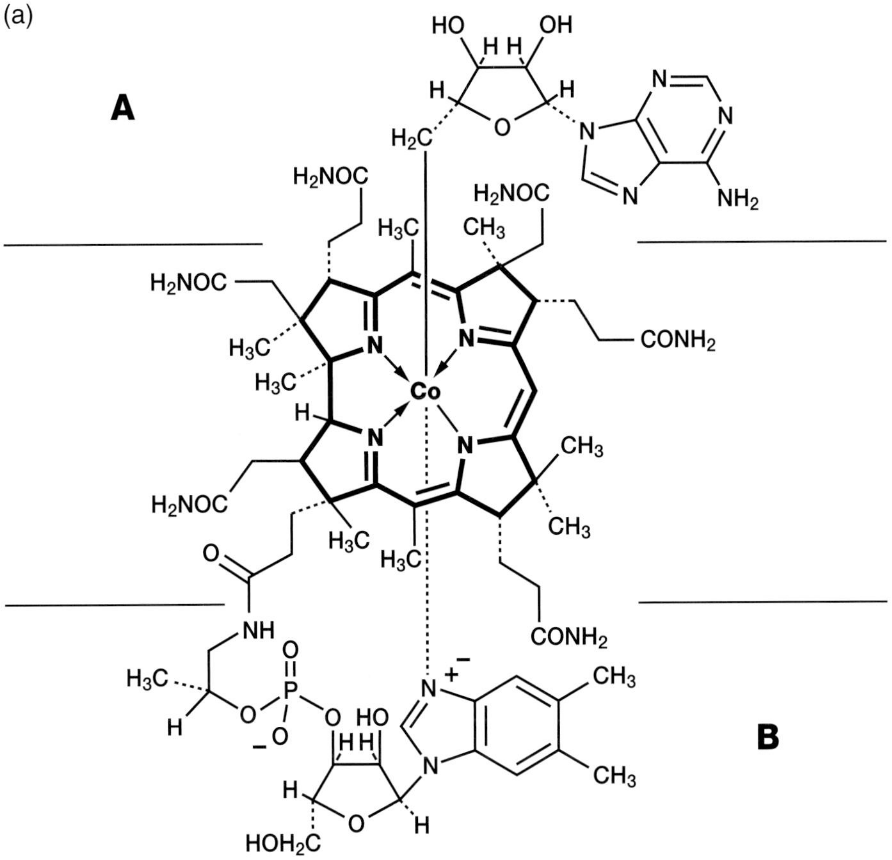

(b)

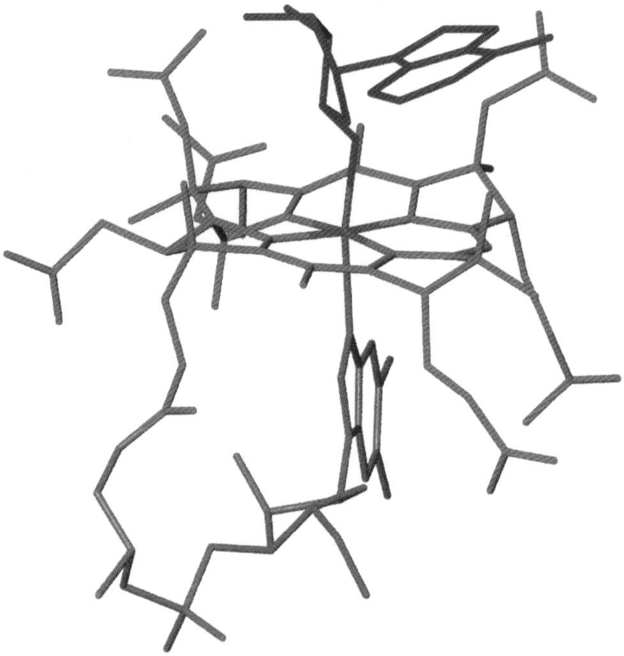

Figure 4.1.2

(Cont.)

Does all this matter?

Don't most of the chemicals that we are exposed to have a fairly simple structure, like the O_2 we have to breathe, like the H_2O we drink (more or less enriched) or like the CO_2 that plants take up together with H_2O in order to convert light into metabolic energy in the form of sugar? Likewise, low-chain hydrocarbons drive our automobiles, and aspirin pills ($CH_3COOC_6H_4COOH$) comfort us if we are plagued by headache. For the chemist, the latter substance is still a fairly simple compound, although researchers in medicine do not really know in detail how it works.

Organic nature does not shrink back from complexity, but chemical complexity is so rich that actually only a very tiny fraction of all possible compounds is used. The situation becomes truly complex if we ask "which fraction?" and "why?" Take as an example a substance like $C_{20}H_{40}O$ (see Figure 4.1.3), one of the some 300,000 hydrocarbon isomers of $C_{20}H_{42}$ in which two hydrogens have been replaced by an oxygen bridge. It is the sex pheromone of the gypsy moth[4] and only in its specific configuration can it be perceived by the sensory organ of this insect. The word pheromone is a contraction of the words "pherein" (Greek for "to transfer") and "hormone". Female insects send the pheromone out to signal that they are ready to mate. The male has extended feather-like antennae (Figure 4.1.4) to sense minute amounts of pheromone (down to single molecules). Another example is the sex attractant of the female silkworm moth *Bombyx mori*, shown in Figure 4.1.4. The molecule has to be in the *cis–trans* configuration (I) in order to be perceived by the sensory organ of the male insect. Adolf Butenandt and his coworkers[5] first isolated the molecule, determined its structure and synthesised it. Since then, many such examples have become known. Certain insects also use similar substances to mark trails or to warn their fellow creatures of dangers encountered.

Figure 4.1.2

Molecular structure of vitamin B_{12}

(a) The chemical formula of coenzyme B_{12} (adenosylcobalamin). The molecule consists of a central ligand system, the "corrin" ring, the chromophore of which is responsible for the molecule's beautiful red colour. In its central hole, the corrin ligand is complexed by its four nitrogen atoms to a cobalt ion, which is further linked to an adenosyl residue **A** by a covalent metal–carbon bond above the corrin ring, and complexed to another nucleoside **B** below the ring. The structural feature essential for the coenzyme's biochemical function is the (weak) cobalt–carbon bond: during reactions catalysed by the coenzyme, this bond is cleaved homolytically to furnish the highly reactive adenosyl radical. In vitamin B_{12}, a cyano group is bound to cobalt in place of **A**, and in cobyric acid (a natural product, as well as an intermediate in the chemical synthesis of vitamin B_{12}), **A** is replaced by a cyano group and **B** by a water molecule.
Courtesy of Albert Eschenmoser.
(b) The three-dimensional shape of the coenzyme B_{12} molecule.
Courtesy of Albert Eschenmoser, Zürich and Bernhard Kräutler, Innsbruck.

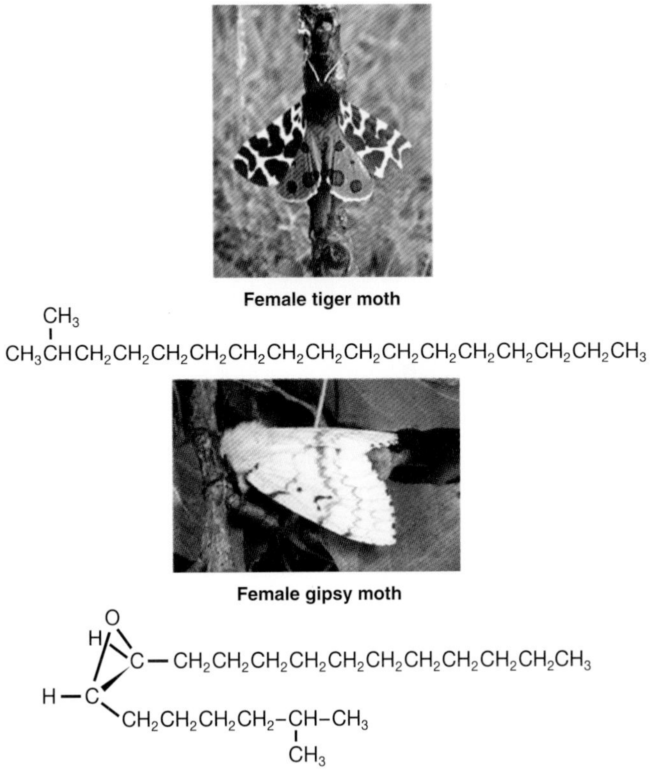

Figure 4.1.3

Two examples of pheromones of female insects

Courtesy of Robert J. Ouellette[2].

To get back to my question: 300,000 isomers, for a chemist, is not a large number. Three hundred thousand molecules still make up an invisibly small spot, and each one of the 62 million million isomers of $C_{40}H_{82}$ together make up just 58 ng, but no individual in its particular configuration is the same as any other. The fact that we encounter individual copies of the complex organic repertoire of Nature in larger, reproducible amounts is in fact a consequence of the existence of life on earth, and as such a consequence of the reproductiveness of living beings.

Where is the limit of chemical complexity? How large can a single molecule be? How large would it have to be for all of its combinatorial alternatives (each represented by a single copy) to take up all the space available in the whole universe? These questions are not a problem for macromolecular chemistry! As we shall see, these molecules do not even have to be unreasonably large. They are much smaller than what we are used to in biochemistry.

I. (bombykol)

II.

III.

IV.

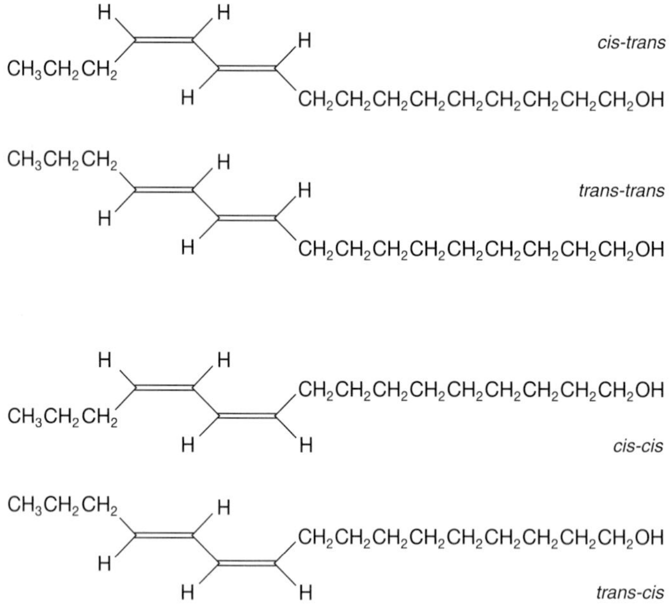

Figure 4.1.4

The four isomers of bombykol

Only the *cis–trans* structure (I) appears to be active as a sex attractant. It is detected by the male insect through its feather-like antennae. (For details see text.)

Figure kindly provided by Karl Ernst Kaissling.

If we take our universe to be a sphere with a radius of 15,000 million light years, its volume would amount to about 10^{79} m^3. About 10^{105} protein molecules of length 100 amino acid residues (i.e. roughly like the enzyme cytochrome c shown in Figure 4.1.5) would neatly occupy such a volume in closest packing. However, there are some 10^{130} possible alternative protein sequences of that very length. Less than the tiny fraction of a quadrillionth (10^{-24}) of these 10^{130} different alternatives would completely fill the whole universe, yet our universe is – essentially – empty. All

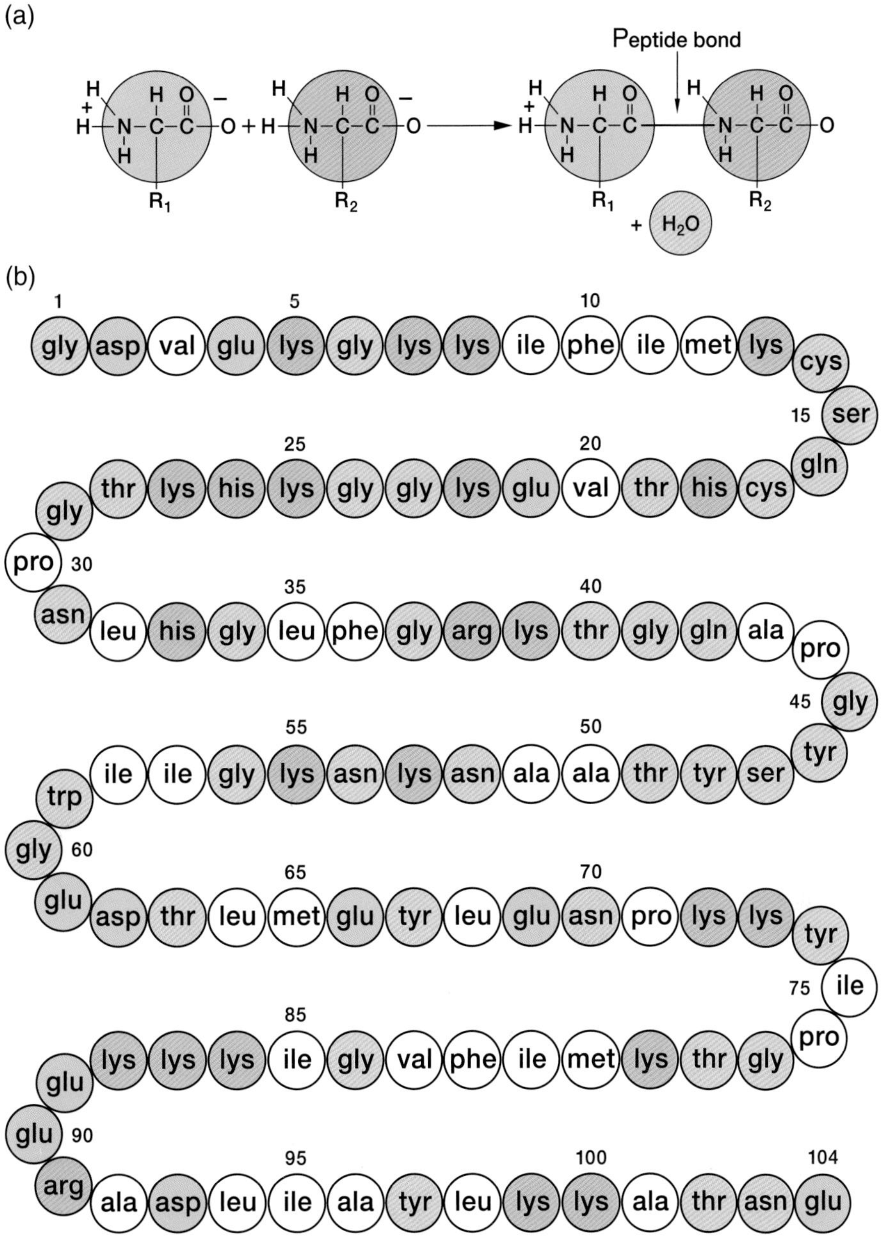

Figure 4.1.5

(Cont.)

(c)

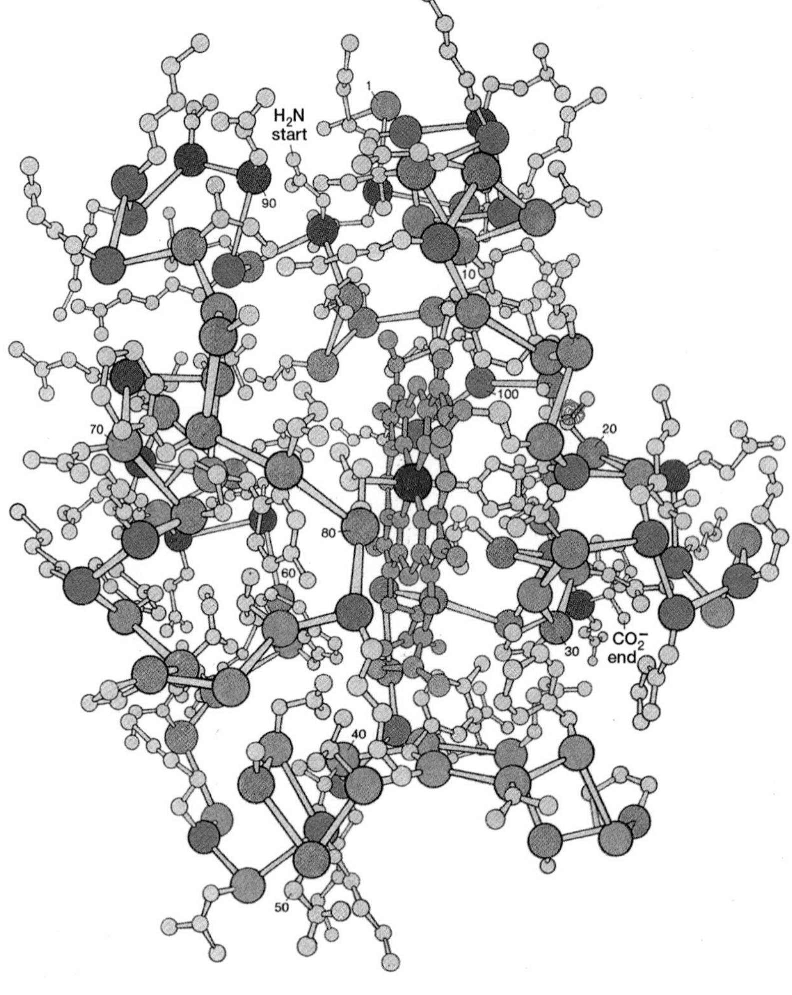

Figure 4.1.5

Cytochrome c

The polypeptide chain contains 104 amino acid residues, linked together by condensation (a) and (b). The three-dimensional structure (c) (after R.E. Dickerson (1972), The Structure and History of an Ancient Protein, *Scientific American*, 226: 58–72) contains an iron complex at its active centre. The catalytic activity involves an electron transfer $Fe^{2+} \rightarrow Fe^{3+}$ enhanced by the side chains of amino acids at the centre (resolution ~2Å).

this means is that any systematic inspection of the total set – in order to find one specified sequence – is entirely outside the bounds of possibility, given the spatial (and temporal) limits of our universe. Put differently: the potential complexity of chemistry surpasses any reasonable limit within our universe.

One might raise the objection that a protein molecule is the product of life. Yes, but we have the chemical means to synthesise it. Since all living matter is in some chemical state, the (potential) complexity of chemistry is, by definition, far beyond the complexity of viable matter and we shall see that the taming of this complexity was one of the most important prerequisites for the origin of life. An unanswered question, however, is whether we can tame this complexity in the laboratory as well. The chemist, when he synthesises a compound, wants to get it in a high yield – and as pure as possible. Chemical complexity is diametrically opposed to this desire. It requires all sorts of skills and tricks of preparative chemistry, such as used in the examples shown in Figure 4.1.6[6].

After having read about micro- and macrostates in Chapter 2, we may well no longer be too impressed by large numbers. The numbers there went up as high as N^N, where N itself was as large as 10^{20} or more. However, the population of relatively few of these microstates suffices to specify well-defined averages that are

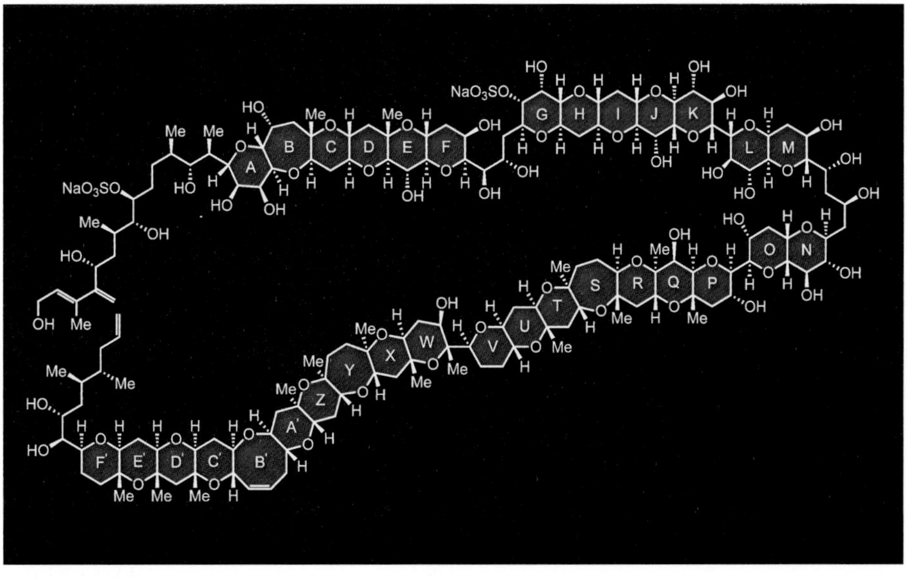

Figure 4.1.6
Proposed structure of maito toxin[6]

This molecule is the largest and most toxic non-protein natural product known today. It is produced by the dino flagellate *Gambierdiscus toxicus* and activates voltage-dependent Ca^{++} channels (LD_{50} = 50 ng/kg; 1 mg can kill a million mice).
Courtesy of Kyriacos C. Nicolaou.

expressed in thermodynamic variables. We have seen that this is a consequence of the particular form of the distribution function (see Figure 2.5.1). In Chapter 3 we saw how the function that characterises the complexity of a macrostate, entropy, can be used to define a certain meaning for the concept of information. However, that definition yielded a negative complement of what we usually call information. It tells us how much we would have to know in order to be "exactly informed". The chemist who prepares a pure compound in high yield "produces" an example of this type of specified information. How? In most cases, that is still really an art. Otherwise the result would be some useless hotchpotch of substances, for instance all those that would coexist under the conditions of complete chemical equilibration. Under those conditions they are by no means all equally probable: the mixture as such may even be well defined, given that free energies are known, but to the chemist a random mixture as such doesn't make any sense, even though we may guess that life had to start from a hotchpotch of chemicals. What I want to make clear is that in chemistry we encounter a new individual quality behind the large numbers, something we cannot express by averages. Each individual state counts in itself.

Today we are witnessing some promising attempts to penetrate chemical complexity by what is called combinatorial chemistry.[7] Some call it irrational chemistry, but it is not at all "irrational"; otherwise it would indeed yield the hotchpotch I referred to. Combinatorial chemistry has become most prominent in huge protein libraries, which have become an almost priceless treasure of the pharmaceutical industry. The rationale behind combinatorial chemistry is the artificial selection of uniform samples from mixtures of randomly varied combinatorial composition, based on a limited class of chemical monomers. Such a technology demands parallel processing using nano-structured devices, as well as high-throughput screening and selection.

What is the source of all this plentifulness and richness of chemistry? All chemical interaction is electromagnetic in nature. However, different kinds of electromagnetic interaction are encountered in chemistry.

What we usually call the "chemical bond" is perhaps the hardest to understand. It was only quantum mechanics that could clarify its nature. Interactions at atomic distances in any case require quantum-mechanical treatment, independently of whether or not we have some kind of "classical" understanding of them. Yet it is the lack of any classical analogue that causes our difficulties in understanding the covalent bond.

Covalent chemical bonds may have "lifelong" stability, and – unlike the Coulombic force that acts between electrically charged partners – they are saturable and have a precise bond length. This imparts a combinatorial character to chemical complexity, a complexity that increases exponentially with the number of atoms involved.

The chemist uses the term "radical" for a group of chemically linked atoms that possess at least one unpaired electron, causing it to be – sometimes extremely – reactive. In fact, it can be added (as we are acquainted with Pauli's principle) that the term "pairing" of electrons hints at the cause of covalent bonding. Atoms as such are electrically neutral. The positive charge of the protons in the nucleus is exactly balanced by the orderly arrangement of the negatively charged electrons in the shell (see also Section 1.5). How precisely quantum mechanics and, in particular, quantum electrodynamics can handle this problem has been shown in Chapter 1. The Lamb shift represents one of the most accurately accounted-for phenomena in physics. However, the exact treatment of chemical problems has so far been limited to relatively simple structures. Exact solutions can hardly be found for systems that involve more than three interacting particles. Difficulties – although of a different kind – had already been encountered with the three-body problem in classical mechanics (see Figure 2.3.1). On the other hand, there are excellent approximative methods available, which in our computer age have led theoretical chemistry far beyond its original limitations, which confined it to being more of an explanatory than an inventive nature. Nowadays theoretical methods are being used to build up structures such as full-sized protein molecules (see the monograph by Andrew R. Leach[8]). The discovery of new so-called "lead" compounds in the chemical and pharmaceutical industry is often based on such methods of molecular modelling. (A lead compound is a substance that shows a certain desired biological activity, e.g. as an agonist or antagonist, and is considered to be a basis for starting the development of a new drug.)

I do not intend to go into further details of this branch of theoretical chemistry. It started out from the treatment of atomic structure by Max Born and Robert Oppenheimer[9] and by Douglas R. Hartree and Vladimir Aleksandrovich Fock[10] leading to the orbital theory, made available to chemistry by Erich Hückel[11] or the valence-bond theory of Walter H. Heitler and Fritz W. London[12]. They provided the basis of modern models of force fields, energy landscapes, molecular mechanics and dynamics, all well documented in the literature. Chemists in their synthetic work make frequent use of rules derived from theory. A prominent example in organic and theoretical chemistry is the Woodward–Hoffmann rule,[13] which is based on the concept of conservation of orbital symmetry.

Another part of chemical complexity is due to several weaker types of electromagnetic or electrostatic interactions that I have hardly touched upon so far. The tendency of electrons to pair, which makes helium so stable as compared with hydrogen (see Figure 1.5.2), or which causes hydrogen atoms to form the H_2 molecules, is also responsible for the formation of stable ionic states. In the crystal lattice of salt (NaCl), the individual partners are present as Na^+ and Cl^- ions. The name "ion" is derived from the Greek word for "to go", referring to the fact that, because of its charge, an ion migrates in an electric field. The ionic state, of course, results from the fact that the lone electron in the s shell of the sodium atom has moved to the unfilled p shell of the chlorine atom, producing two particles with complete outer shells. The interaction between the two ions is thus largely Coulombic in nature, the potential energy of interaction being proportional to the reciprocal of their distance apart. At atomic distances, the Coulombic interaction energy is of the order of magnitude of 10 eV. This is still a quite strong type of chemical interaction.

In water, however, sodium chloride is easily soluble. Here, the ions surround themselves with one to two shells of water molecules. The process is called hydration, and it is based on the strongly dipolar nature of the water molecules. This allows the ions to exist as separable species. The Coulombic force between two ions is reduced by a factor of almost 80 (which is the magnitude of the dielectric constant of water). The reduction of force is due to the shielding of the ionic charge by the dipolar water molecules. At a distance of about 5 Å the interaction is reduced already to the level of thermal energy at room temperature. The distance at which these energies become equal is called the Bjerrum distance, named after the Danish physical chemist Niels Bjerrum[14]. The still active Coulomb force causes the ions to form so-called ion clouds, in which the average distance between oppositely charged ions is slightly smaller than that between equally charged partners. This superimposes an exponentially decaying shielding effect of the ion cloud, which was first calculated by Peter J.W. Debye and Erich Hückel.[15,16] The shielding effect of an ion cloud can be very influential for highly charged particles, such as the polyionic strands of DNA. Hückel, as mentioned above, was also one of the early pioneers of theoretical chemistry, where he was the first to provide a molecular-orbital approximation for describing π-electron systems as we find them in partially double-bonded aromatic structures such as benzene rings.

The water molecule H_2O is a relatively strong dipole. Its strength is equivalent to the energy to be expected for a separation of one positive and one negative elementary charge by a distance of nearly a half (0.385) of anÅngstrøm unit (10^{-10} m). The OH bond is itself just 1 Å in length. Again, the primary cause of the non-homogeneous charge distribution within the H_2O molecule is the fulfilment of Pauli's requirement for an asymmetric wave function. The two hydrogen atoms, sharing their electrons with those in the p shell of the oxygen atom, become slightly positive, providing a residual negative excess charge at the two remaining lone electron pairs of the oxygen atom. The water dipoles interact fairly strongly with one another, making water a quite anomalous liquid. Compared with the less polar ammonia or the non-polar methane – both of these molecules being of about the same size as H_2O – the melting and boiling points of water are shifted to comparatively high values, allowing the liquid state to exist between 0 and 100° C, a temperature range most suitable for the evolution of life on earth. Moreover, water shows a density anomaly, i.e. a maximum in the temperature-dependence of its density, at 4° C. Without this, our lakes would freeze from the bottom up and possibly never thaw completely in summer. The dipole–dipole interaction of water molecules is of a specific nature (referred to as a "hydrogen bond") and reaches strengths that exceed normal thermal energy at room temperature by about tenfold. These specific bonds generally occur between protons of polar molecules and lone electron pairs of electronegative partners (F > O > N). Hydrogen bonds in water form a bulky structure similar to that found in ice crystals. The bulky structure of ice causes its density at the melting point to be lower than that of liquid water between 0 and 4° C, so ice floats on water.

Hydrogen bonds (H bonds) are important structural elements of biopolymers such as nucleic acids and proteins (see below). Hydrogen bonds between isolated partners do not easily form in water because the partners also undergo hydrogen bonding (sometimes more strongly) with water molecules. However in non-polar solvents quite stable pairs can form, which for the nucleic acid base pair GC are held together with an energy ten times higher than thermal energy at 300 K (0.026 eV). Hence, in the interior of biopolymers, where there are no competing water molecules, these base pairs contribute appreciably to the stability of the conformation of the macromolecule. In nucleic acids, where base pairs are lined up like the teeth of a zip, they provide the basis for fidelity of copying in information transfer.

Another stabilising effect of importance in biopolymers should be mentioned here. It is called hydrophobic bonding, but it has more to do with the exclusion of water structure from the interacting hydrophobic partners. I can best demonstrate it with the following model. Think of two separate bubbles in water. They can be regarded as essentially empty. In order to produce such "holes" in the liquid one has to work against the (appreciable) surface tension of water. The amount of energy required to create a surface inside the liquid is proportional to the area of the surface produced. If the two bubbles coalesce to form one larger bubble, the total surface area is reduced and energy is gained. Hence, the two bubbles "attract" one another, as if there were some force of interaction between them. However, apart from the few gas molecules within each bubble there is nothing that could really interact. Even so, there is a "net reaction" between two "pieces of nothing", which give all the appearance of having a finite affinity for one another. Now, one might fill up the two bubbles with some hydrophobic substances. This will introduce some additional attractive interaction between the two substances (cf. below), but the main effect resulting from the unification of two "holes" in the water structure remains unchanged. Hydrophobic free energies, owing to their dependence on water structure, also have relatively large entropic components.

This brings us to that "additional attractive interaction" which is effective among all neutral substances. Where does it come from?

First of all, dipolar or multipolar substances – as a whole – are neutral. However, we sense their partial charges whenever we get close enough to them. The force that acts between two dipolar molecules depends not only on their distance apart, but also on their orientation relative to one another, and hence may be either positive or negative. In polar liquids, such as water in the absence of external electric fields, there is no prevailing net orientation of the freely mobile dipoles, although closely interacting dipoles favour orientations with attractive interaction. In fact there is a Boltzmann distribution that describes the balance between the attraction just described and the thermal motion that interferes with an alignment of the dipoles. The average potential energy comes out to be inversely proportional to the sixth power of distance, i.e. confining interactions to a quite small "neighbourhood" of mutual approach. The reason for this strong distance dependence is that, in addition to the interaction among the fixed dipoles, there is also an effect of thermal motion on the mutual orientation of the dipoles.

It is interesting to note that exactly the same distance relationship exists for entirely neutral molecules which, by coming close together, can mutually induce dipole moments by "polarising" the (negative) electronic shells relative to the (positive) nuclei of the atoms involved. Those interactions are called van der Waals' interactions (Section 2.2). They arise in all forms of condensed matter. One does not need to be a chemist to guess that there must be some sort of attractive interaction even among entirely neutral molecules, such as the noble gases or other molecules such as nitrogen or hydrocarbons, because all these substances, at sufficiently (sometimes very low) temperatures liquefy and most of them even solidify.

The British physicist John E. Lennard-Jones devised a more realistic potential function, consisting of two terms that describe the van der Waals forces. The (negative) term representing attraction varies in a manner inversely proportional to the sixth power of distance ($1/d^6$) and hence has quite a short range. The other (positive) term, representing repulsion, varies with the 12th power ($1/d^{12}$) of distance, which almost mimics a "hard-sphere" potential. The complete potential function then shows a minimum at a "distance of closest approach" which characterises the condensed liquid state. As the low boiling temperatures of noble gases such as helium and neon indicate, the interaction energies there are pretty low.

The final explanation of Lennard-Jones' terms was left to quantum mechanics. The attraction term is explained by the dispersion, or London, forces, named after Fritz London,[17] who devised the quantum-mechanical theory of these in 1930. It is due to the induction of dipole moments by polarisation of the electron clouds at close distances. Since this electromagnetic type of interaction proceeds with finite velocity (i.e. that of light), this may play a role for larger molecules, changing the distance relationship slightly, as was shown by the Dutch physicist Jan Hendryk Casimir[18]. The "very steep" repulsion term, which according to R.A. Buckingham and Terrel L. Hill may be written in exponential form, is another expression of Pauli's exclusion principle that does not allow overlapping electron probability densities. It is the type of force discussed in Section 1.7 that withstands the gravitational collapse of matter. Using a very delicate and ingenious way of recording, the Dutch physicist Marcus J. Sparnaay[19] provided a direct experimental demonstration of the attractive term of van der Waals' forces. I had the pleasure of being introduced personally to Sparnaay's excellent work in its very early stages.

So how complex is chemistry?

First of all, the covalent chemical bond gives rise to a potentially unlimited complexity. A polymer is like a sequence of natural numbers. To any terminal position one may add a further monomer. Of course, in three dimensions there are branchings, cyclic closures and other sorts of termination, and in practice there are clear limits of stability. However, the molecule of inheritance, DNA, has shown how far the exponential complexity can reach. The single giant DNA chain of the *E. coli* genome comprises some 4 million monomers. Chemically this means a combinatorial complexity of about $10^{1\ \text{million}}$ (a 1 followed by a million zeros) alternative sequences. The human genome contains several such chains, each comprising many million nucleotides.

The basis of combinatorial chemical complexity is the covalent chemical bond together with superimposed weaker interactions. The periodic table of the elements is still fairly simple, explained by the "strange simplicity" of quantum-mechanical symmetries. While quantum mechanics also explains reactivity as rearrangements in the electronic structure of the combining atoms, it now leads to an unbounded potential for extension. The expression of unlimited complexity, and therefore of chemical diversity, requires a new ordering principle. The ordering principle will turn out to be an information-generating principle, the emergence of which is the topic of this chapter.

4.2. How Does Nature Tame Chemical Complexity?

If we look around on earth, our first impression is certainly not that of some random jumble of chemicals. Yes, all we see is made of chemicals, but the chemical order we find on the surface of the earth is largely influenced by biological organisation. If we look more closely at each chemical structure we indeed discover complexity of the highest degree, but it appears in an orderly fashion, it is a "tamed" complexity. And here we encounter a paradox. All these "biological molecules" are chemicals, so they can certainly be only a relatively small subset of all possible chemical compounds. On the other hand, if we dig out any matter not belonging to the biosphere, for instance collecting it from somewhere in the universe where life is absent, we observe a chemistry that is in no way comparable with the degree of complexity typical of the chemistry of life. So the paradox lies in the fact that chemical complexity could unfold only by first being reduced to a subset of molecules, i.e. essentially to two or a few classes. These are the nucleic acids, which made complexity reproducible, and the proteins, which made all kinds of reactions possible by catalytic control.

The total set of chemical complexity is of a purely hypothetical nature and, as we have seen, its extent is hard for us to estimate. Without organisation it had no chance of ever being realised. The only – or better, the simplest – organisation to bring about unlimited complexity was life. Without this organisation the process of "complexification" would have stopped in a very early phase, for several reasons. It would have run out of supply long before it would have been able to equilibrate to any truly complex state. Production rates would never have matched decay rates. Moreover, the whole process would have interfered with itself by producing inhibitors or even "killer" substances. The production of "everything possible" would necessarily include such compounds.

Yes, it is a true paradox: extended complexity, as we encounter it today in the living world, could only come about by first reducing (potential) chemical complexity. In this process complexity changed its nature. "Chemical" complexity became "informational" complexity. Any reaction desired now became available through a programmable system of suitable catalysts, i.e. enzymes. Only in this way was chemical complexity able to unfold until it was eventually raised into a supramolecular organisation. Let me underpin this by some examples.

All proteins, the molecular executive of life, are made up of just 20 types of monomers, the natural amino acids (see Figure 4.1.5). "Small" protein molecules (with molecular weights around 10,000) comprise "only" 100 or so of these monomers, large ones (with molecular weights of hundred thousands and above) correspondingly more. Hence the combinatorial "complexity", again only a potential one, lies between 20^{100} and 20^{1000} – i.e. 10^{130} and 10^{1300} – possible sequences. The magnitude of these numbers is in any case incomprehensible for us. We have seen (Section 1.8.) that the total matter content of the observable universe amounts to an equivalent of about 10^{80} proton masses, equalling no more than 10^{75} or 10^{76} protein masses, something completely negligible in comparison with the potential complexity of the protein world.

We can repeat such a thought-game with the other main class of biological macromolecules, the nucleic acids, which – in the form of genes – encode the protein molecules. Nucleic acids are made of only four types of monomer (see Figure 4.2.1(a)), using triplets of them in the genetic code for the 20 natural amino acids, which was shown in Figure 3.8.2. A typical gene, encoding a given protein molecule, comprises about 1000 nucleotide monomers, a typical (RNA) virus (which encodes several proteins) 10,000, a simple bacterium several million and the human genome several thousand million nucleotide monomers (only a small fraction of which represent genes). Such a complexity, again, cannot be realised in material form; only a minute fraction of all possible sequences – given the limits of our universe – could possibly ever have been tried or tested in nature. The total volume of the universe amounts to "only" something like 10^{108} cubic Ångstrøms and the universe that has allowed the chemistry of life to arise is not much older than 10^{17} seconds. On the other hand, it takes – even using a well-adapted enzyme machinery – more than 1 s to synthesise or decompose a single protein molecule in the laboratory.

I do not produce these numbers in order to demonstrate the statistical "improbability" of life. That has often been done before, but it does not make much sense. Those numbers would have statistical significance imputed to them only if the molecules of life had come about by throwing dice. We shall see that the statistics we have to use for any system subject to natural selection are of an entirely different kind; hence these numbers have no meaning at all with regard to the origin or evolution of life. Otherwise neither you nor I would have been around to reflect on

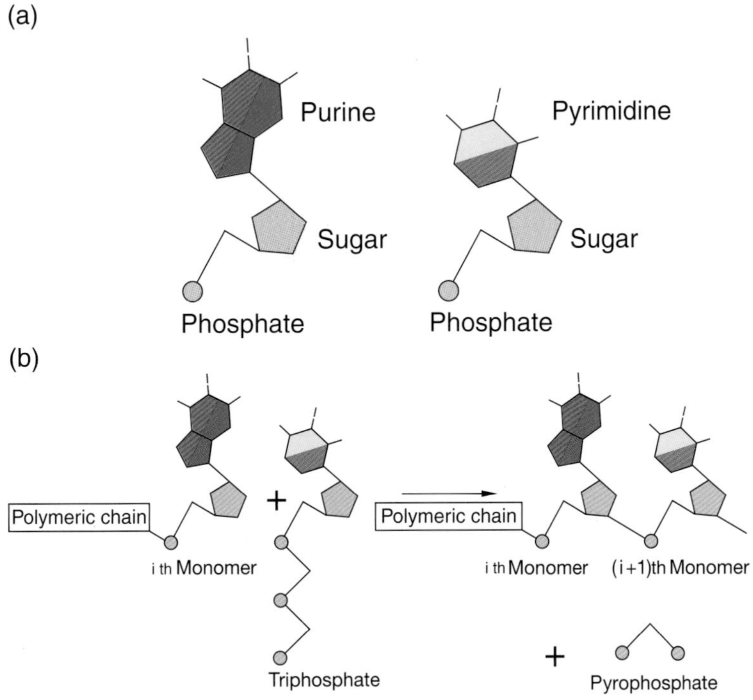

Figure 4.2.1

The formation of nucleic acid polymers

(a) The monomeric subunit of the nucleic acid is a defined chemical compound. It is composed of three molecular entities: a phosphate ion, a sugar molecule (ribose in RNA or deoxyribose in DNA) and a cyclic organic compound called a "base", which is either a purine (A or G) or a pyrimidine (C or U/T). (b) The synthesis of a polymeric chain starts from energy-rich monomers (e.g. triphosphates).

such problems. The reason I am presenting these numbers is to reveal the magnitude of the paradox that I have referred to above: the complexity we encounter in biology could come about – and we have to explain how it did so (reproducibly) – only through a drastic cut-back of the potential chemical complexity. Chemically speaking, this reduction of potency represents almost a castration. It destroys any form of "chemical libertinage". However, it does not thereby destroy complexity, but rather renders it controllable by the logically built-up principles of information.

The non-physicist may have difficulty in imagining the complexity I am talking about. In this section, as in the preceding one, I want to point out principles. The detail – and that's where the devil is to be found – will be discussed in later parts of the book, where I shall talk about biological reality. However, in order not to appear too mystical now I must explain some of the details of biological organisation, just enough to make clear what I mean in the present context.

As mentioned before, there are essentially two classes of biopolymers which are in full control of biological complexity. There is also a lot of other building material, in particular a rich chemistry that takes care of the supply of free energy. All this processing is "harnessed" by the two classes of molecules mentioned: the proteins and the nucleic acids. The proteins are the executives; they decide, by virtue of their (controlled) presence, which reaction has to proceed and when. They are extremely efficient catalysts, regulators, transport and defence systems – and in all these functions delicately tuneable. Sometimes they also just provide storage space or structural scaffolding. The catalysts, called enzymes and denoted by the ending "ase", master any type of chemical reaction needed to keep the life process going. We know many thousands of them: ones that shuffle electrons (oxidoreductases), or ones that break bonds by hydrolysis (introducing H_2O) as in the metabolic degradation of proteins, of nucleic acids (by proteases and nucleases) or of carbohydrates, esters and glycosides. Some enzymes transfer groups of atoms from one compound to another (transferases), others rearrange such groups within the same molecule (isomerases), others again link groups together (ligases or synthetases) or add small molecules such as water, ammonia or carbon dioxide (lyases).

The other class of biopolymers, the nucleic acids, are mainly legislators (some, especially RNA molecules, get involved in function too). The term "legislators" is to be understood in a broad sense. In an individual organism they don't "make" the law, they just "promulgate" it. They provide the instruction of what is to be done, both within the organism and between generations of organisms. Yet in evolution they also "made" the law that is constitutional for the whole species. And "law" is to be interpreted as legislative information (which I shall later explain more precisely).

The "chemical" complexity of these substances (Figure 4.2.1(b)) has been reduced for the sake of achieving an informational complexity that can then, in a controlled way, build up the realisable biological complexity, eventually up to organisms such as the human species. The nucleic acids are made up of four types of monomers, each having a molecular weight of about 300. All these monomers are built from three constituents: a phosphate group, a sugar and one of the four different heterocyclic compounds, simply called "bases" (because of their chemical property of reacting as bases, or proton-acceptors). The term nucleic "acid" derives from the acidic nature of the phosphate group. Phosphate and sugar are the backbone elements, common to all four monomers, joining them up into chains that ultimately comprise many millions of monomers, through a chemical linkage, an ester bond between the phosphate and the sugar of the subsequent monomer. The four bases belong to two classes: purines (adenine and guanine, referred to by their initials A and G) and pyrimidines (thymine and cytosine, or T and C). This holds for DNA, while in RNA (which actually preceded DNA in evolution) the methyl group at the C_5 position of the pyrimidine ring of thymine is replaced by a hydrogen atom (giving the base uracil, U). There is a lot more chemical detail involved that is outside the scope of the present section. The point I want to stress here is the pairwise complementarity between

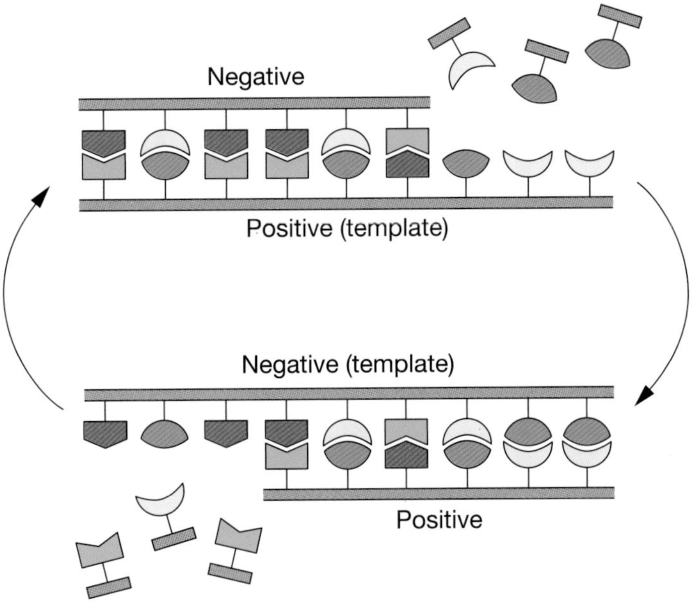

Figure 4.2.2

Purines and pyrimidines

These molecules are called “complementary”, which means that they can complement one another in forming base pairs: A with U in RNA or with T in DNA, and G with C in both RNA and DNA. The bases are, so to speak, the units of a four-letter alphabet. They are the symbols of the language in which genetic information is stored. A message in this chemical language can be thus represented as a sequence of four symbols. Reproduction (or copying) of information can be represented in the way shown here: as in photography, it proceeds by way of a “negative” of the template.

the purine and pyrimidine bases (G with C, and A with T or U), which is the basis of complementary reproduction (Figure 4.2.2). As such it can be interpreted as the ability to “read” and “write”, which is an *inherent* ability of nucleic acids, unique among biopolymers. At the same time it is an important factor in determining the double-helical structure of DNA (Figure 4.2.3) and the catalytic potential of single-stranded RNA molecules, based on their internal secondary and tertiary structures (Figure 4.2.4). All of these properties make use of the inherent pairwise complementarity among the four chemical entities G, C, A and U (or T). This is an important example of the fact that physical or chemical interaction is at the crossroads where mere interaction meets a logic associated with the abstract principles of information, with “life” as the result of this encounter.

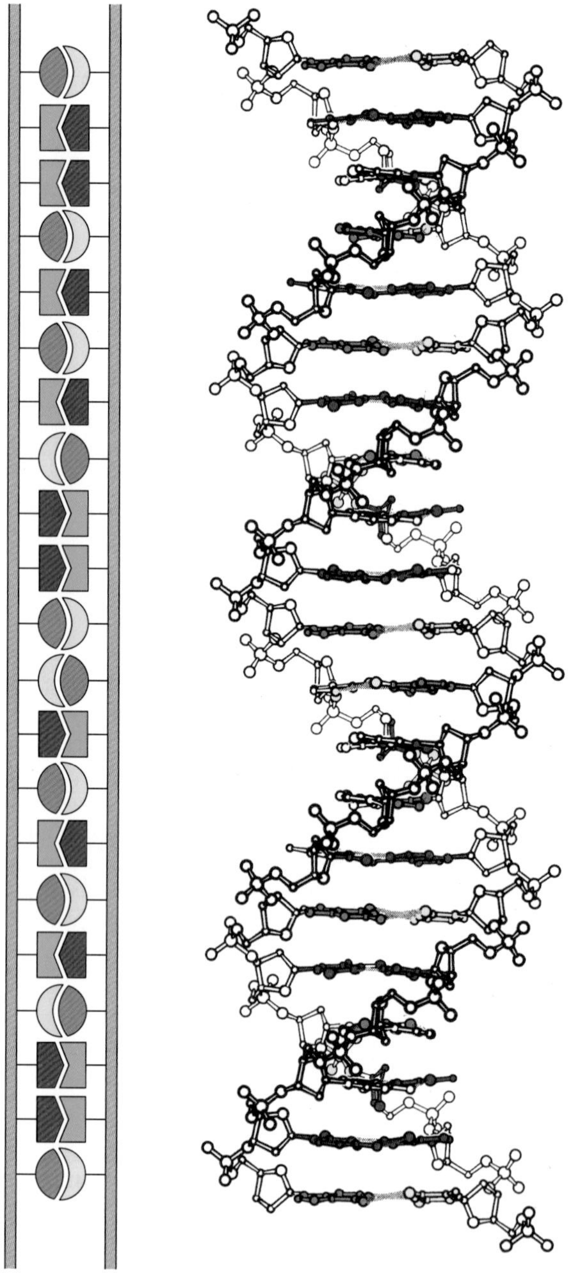

Figure 4.2.3

DNA exists in the form of double strands

The plus and minus strands are helically intertwined, enclosing the complementary base pairs.

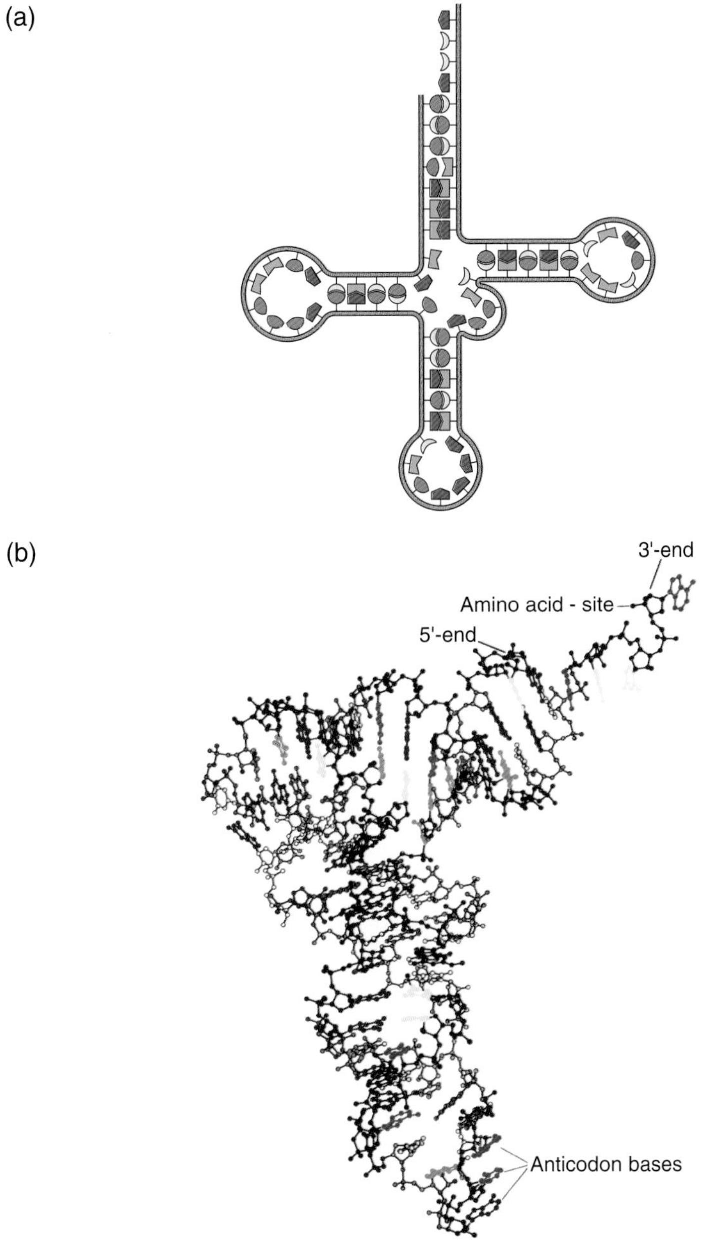

Figure 4.2.4

RNA usually occurs in the form of (spatially folded) single strands

---

By internal base pairing, RNA molecules fold into spatial structures, as proteins do. They are used as messengers in the processing of genetic information and as carriers of certain functions in the translation process. In recent years many other functions of RNA have been discovered, especially in eucaryotic cells. The figure shows a transfer-RNA, which has the rôle of an adapter that links a given amino acid (e.g. phenylalanine) to its three-letter codon in the genetic message (e.g. $\overrightarrow{\text{UUC}}$) by base-pairing with its exposed anticodon ($\overleftarrow{\text{AAG}}$). (a) Secondary (two-dimensional) base-pairing structure. (b) Tertiary (spatial) structure.

Why four such entities?

There are two ways to look at this question. First, we realise that there are many more possible candidates forming base pairs with sufficiently similar geometric configurations. If there had only been just those four bases fitting into this structure, then the reason would have been a purely chemical one. Moreover, we could modify the four bases by substitutions in many ways which, in fact, nature has done for some functional nucleic acids (e.g. for the transfer RNAs). So four bases constitute "reduced chemical complexity". But, given that fact, and given that present nucleic acids are the product of evolution that fulfils a certain purpose, why do we then have four bases and not only two? The answer is: we need two complementary pairs. The purpose of DNA, its present function, is quite clear: it encodes and carries the genetic information. It does something that our computers and telecommunication networks also do.

However, those networks use a binary symbol system, i.e. the two symbols 0 and 1, or "on" and "off". The numbers 0 and 1 are not complementary in the sense that they would naturally interact with one another. They can do so because messages do not have to copy one another. They are written and read by intelligent beings, and computers use them because they are programmed by logical rules. Intelligent beings are able to process binary messages, and I do not have to go into any detail of how they do it. Molecules, however, can only use their chemical interactions, such as I have listed in the preceding section, and for that it emerged that two interacting pairs instead of two non-interacting symbols were necessary and, as it turned out, sufficient.

We might further ask why nature uses this particular kind of backbone, why homogeneous chirality (see later), why then only one of the two possible choices and why distinctive types of sugar, i.e. riboses for RNA and deoxyriboses for DNA? Chemically, there would be a large repertoire at hand, and we can again only guess that it is a "purposeful" result of (an early phase of) evolution with the goal of "reduction in chemical complexity".

This is true in particular for the core constituents of nucleic acids, the ribose and deoxyribose phosphates. Needless to say, it is the sugar that confers chirality upon the nucleic acids, found throughout nature to be in the D-form. My friend and colleague at the Scripps Research Institute at La Jolla, emeritus member of the renowned chemistry department of the ETH (Eidgenössische Technische Hochschule) at Zürich, Albert Eschenmoser[20], spends a good portion of his time trying to find out why this particular kind of sugar is used. We know numerous simple forms of sugar, many of them having a formula that is a multiple of CH_2O. Because of this formula they were originally called "carbohydrates". However, this historical name has no chemical meaning because sugars are not "hydrates" of carbon. A sugar has always at least two hydroxyl and one aldehyde or ketone group (the latter two being carbonyl groups $>C=O$, the carbon atom of which in the first case is bound to a hydrogen and another carbon atom, and in the second case to two other carbon atoms). Simple sugars, for instance, containing three to seven carbon atoms are called trioses, tetroses, pentoses, hexoses and heptoses, respectively. Each of these, by spatial configuration, has different ways of arranging its atomic constituents.

Five- and six-carbon sugars frequently exist in ring forms, as is the case with glucose and also with ribose. They are classified according to their structural relationship to the simplest member in their group. Glucose is a hexose (i.e. having six carbons), but called a pyranose because it is structurally related to pyran (C_5H_6O), a six-membered ring, involving only five carbon atoms and one oxygen. Likewise, ribose is called a furanose because it is related to the five-membered ring furan (C_4H_4O), which contains four carbons and one oxygen. In deoxyribose ($C_5H_{10}O_4$) the hydroxyl in position 2 is replaced by hydrogen, causing a deviation from the $C_nH_{2n}O_n$ composition that provided the name "carbohydrate".

This is all, of course, textbook stuff. I remind the reader of it in order to demonstrate again the reduction of "chemical complexity" that took place in early evolution. Albert Eschenmoser, together with a group of highly able pre- and post-doctoral students (at both ETH Zürich and the Scripps Research Institute at La Jolla) has taken up the task in recent years not only of synthesising a large number of "artificial nucleic acids" based on alternative sugars (Figure 4.2.5), but also of measuring their physical and chemical properties and comparing them with those of the natural compounds.

A Sisyphean task?

Not quite, and therefore not at all!* Albert Eschenmoser was well aware of the stone that Sisyphus had to roll up the hill. He identified the problem as one of chemical evolution. His predecessor and fatherly friend at the ETH, the late Vladimir Prelog[21], once told me why an organic chemist prefers to study natural products: because, being selected by nature, they always unveil a secret, something surprising and usually interesting. Albert now complements this wisdom: if in addition one studies the possible alternatives, one may even understand why Nature made the particular choice that she did.

The case of proteins is another story with similar conclusions. Here we are dealing with the molecular executive of life, and we might have expected Nature to utilise whatever chemical complexity was available in order to serve this purpose. The result is astonishing. Twenty side chains of amino acids carrying different functional groups, arranged by the three-dimensional folding of a polypeptide chain comprising some hundred such units, achieve any fine tuning of catalytic function that may be required. The immune system provides wonderful examples of how quickly this works. Richard Lerner[22] of the Scripps Research Institute at La Jolla and Peter Schulz[23], originally of the chemistry department of the University of California at Berkeley, now also at the Scripps, showed how to do it. They prepared

* A logical inference borrowed from Thomas Mann.

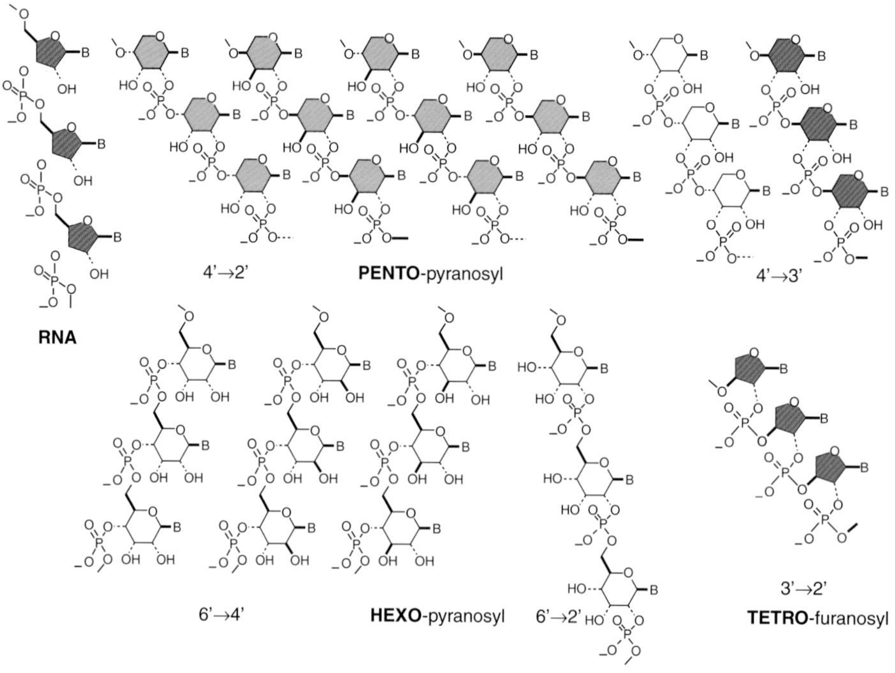

Figure 4.2.5

The various sugar ring structures

These structures were used by Albert Eschenmoser in his studies of alternatives for natural RNA and DNA. He also studied the nature of the sugars appearing in RNA and DNA (see text). Courtesy of Albert Eschenmoser.

catalytic antibodies which perform like enzymes. These antibodies were produced by immunising rabbits with a compound that mimics the substrate to be turned over in its reactive "transition state". Their work shows two important facts. First, enzyme function, in order to evolve, does not need millions of years; days to weeks suffice if the right conditions are fulfilled. Second, fine tuning (e.g. of the immune response) starts out from flexible, yet imprecise, multifunctionality and during a series of somatic mutations narrows down stepwise to precisely adapted folds having fine-tuned monofunctionality.

Proteins are chiral, too. They all use L-amino acids, although this doesn't mean that L-amino acids were the only ones around. There was probably no cumulative chemical symmetry breakage that preceded prebiotic evolution. An initial fluctuation may have sufficed if the subsequent process involved non-linear reproductive amplification (which it does). There might also have been some bias by minerals acting as surface catalysts. I am talking here about general principles, and not about the contingency involved in the historical process. It is certain that biochemical synthesis – however it started – adapted to the unique chirality, eventually not accepting the reverse enantiomer (i.e. the mirror image of the optically active group). Only the compounds made by the biosynthetic machinery of the cell show this unique homo-chirality. There are lots of short-chain polypeptides in living organisms, which operate as membrane carriers or hormones and which contain both L- and D-amino acids. Some are so-called depsipeptides, chains utilising amino and keto acids as, for instance, in gramicidin, an antibiotic against gram-positive bacteria. In keto acids the amino group is replaced by a ketone oxygen.

The pool of 20 natural amino acids used in proteins is the most elitist off-limits club I know. The proteins control everything that goes on in the world of biological chemistry. Sometimes they have to recruit co-factors, such as metal ions (as in the case of cytochrome c) or co-enzymes. This omnipotence is similar to that of the 26-odd letters in our language which allow us not only to describe everything happening in our experienceable world, i.e. the whole world of literature already written plus that to come, but also to express spontaneously and unequivocally our emotions.

In the preceding text I repeatedly used the term "purpose", which is foreign to physics and is dangerous because it can be misinterpreted in various ways. However, biological organisation is "purposeful", as is well expressed in a sentence I quote from Richard Dawkins:[24] "Natural selection, the blind, unconscious, automatic process which Darwin discovered, and which we now know is the explanation for the existence and apparently purposeful form of all life, has no purpose in mind." Purpose here is not something that is there beforehand – "purpose in mind" – but rather is something that comes out at the end and therefore is called "apparent". Moreover, true "purpose in mind" must also be seen as a result of some adaptive electrical and chemical cross-talk among nerve cells, admittedly on a time-scale quite different from that of chemical cross-talk (which sometimes is nothing but small talk) during evolutionary adaptation.

Perhaps it is this "apparent purposefulness" behind Darwin's principle of natural selection that made it seem suspect to many physicists. Here is the problem: proteins and nucleic acids and the viable organisations they build up are certainly structures in the sense of physics and chemistry, as they are determined by all those forces discussed in the preceding section. A particular sequence is a defined particular chemical structure, but it has a "meaning" that cannot be expressed in structural terms. Let me say it in a somewhat sloppy way: physical structure – up to the most complex chemicals we could imagine – is "structure for the purpose of structure". The biological structure I am referring to is "structure for the purpose of purpose" where the second "purpose" means function. I am saying this as a physicist because in this book I am going to try to find a physical principle for the formation of these kinds of structure. There is no reason for a proton to exist in the form it does other than its inherent quark structure. The DNA molecule that represents a gene owes its reproducible existence solely to its functional usefulness, which is not something that resides "inside" the molecule as do the forces that are necessary to secure its existence.

4.3. An Unsolved Mathematical Problem: P = NP?

The English mathematician Ian Stewart[25] has called the above question "one of the biggest unsolved mysteries in mathematics". Let me first explain the hieroglyphs P and NP before getting down to the problem to be discussed in this section, which in more general terms is combinatorial complexity. The evolution of life on earth is closely associated with solving problems of combinatorial complexity.

P is simply an abbreviation for "polynomial". A polynomial is an algebraic expression made up of terms that contain integral powers of a variable N, for example the quadratic form $a_0 + a_1N + a_2N^2$ where the a_i terms are constants. The variable N we shall talk about is a quantity that represents the complexity of the problem in question, for instance the length of a sentence expressed by the number of letters in it, or the number of bases in the DNA of a genetic programme. The genome of an organism as "simple" as the bacterium *E. coli* already comprises as many as $N \approx 4$ million base pairs. P then denotes a class of problems that can be solved by an algorithm causing a computer to run for a "polynomial time", i.e. a time proportional to N^k (or to a sum of those terms) where k is an integer such as 0, 1, 2 or larger.

Mathematicians call these problems "the easy problems" because they are relatively easy to solve, that is to say, easy compared with problems of "exponential complexity". Exponential complexity means that the complexity variable N appears in the exponent of some number, as, for example, in 2^N, e^N or 10^N. For larger values of N, exponential complexity always outruns polynomial complexity. Table 4.3.1 gives some impressive examples that need no further comment (a discussion of practical problems can be found in Refs. 31 to 33). In this table the variable N has been followed only up to N = 60. Remember that for the *E. coli* genome N is about 4 million, and the exponential complexity involved in the number of alternative sequences (i.e. 4^N) is larger than $10^{2.4\,\text{million}}$ (i.e. a 1 with about 2.4 million zeros). If the evolution of the *E. coli* genome had taken some "exponential time", micro-organisms would not exist on earth and neither would we be sitting here drawing such conclusions. Yet, we are here, and the problem I want to discuss right now is not our existence but rather the limits of computability.

I have not yet explained NP. What does it mean? It is also an abbreviation, however with a more mysterious content. NP stands for "non-deterministic polynomial". The word "non-deterministic" refers to a property that is quite difficult to explain because it is more related to a "hope" than to a "fact". The definition of the term does not refer at all to the process of finding the solution of the problem. Rather, it supposes that the correctness of a solution, however it was obtained (non-deterministically), can be checked within polynomial time. In more practical terms, it means that some number characteristic of the deviations from a correct solution can be minimised and shown within polynomial time to remain below a choosable limit. Take a jigsaw puzzle, consisting of a large number of small pieces which in isolation are hard to

Table 4.3.1 Comparison of polynomial and exponential complexity

Michael Garey and David Johnson in their book *Computers and Intractability* compare some straightforward polynomial and exponential functions of a complexity parameter n in respect of the required computer time. In this table corresponding values, adapted to the problems treated in this book, are reproduced. In our case n is, for instance, the length of a sequence. If identification of a defined sequence takes a certain time, the total time required for its identification – depending on what is called the "time complexity function" would be the value given in this table. The time unit for $n = 1$ is chosen to be 1 μs. The exponential functions chosen refer to obvious cases. The case 10^n has been omitted because it is trivial. (Note that 10^5 s are about 1 day, $\sim 4 \cdot 10^7$ s is 1 year and $\sim 4 \cdot 10^9$ s 1 century; this can be compared with 10^{54} s for $n = 60$.)

Time complexity function	Size n					
	10	20	30	40	50	60
n	0.00001 s	0.00002 s	0.00003 s	0.00004 s	0.00005 s	0.00006 s
n^2	0.0001 s	0.0004 s	0.0009 s	0.0016 s	0.0025 s	0.0036 s
n^3	0.001 s	0.008 s	0.027 s	0.064 s	0.125 s	0.216 s
2^n	0.001 s	1.0 s	17.9 min	12.7 days	35.7 years	366 centuries
3^n	0.059 s	58 min	6.5 years	3855 centuries	2×10^8 centuries	1.3×10^{13} centuries

identify, and suppose that the solution yields the face of a person we know very well. (Re)producing this face in the puzzle may take weeks, but once we have it, confirming its correctness may be a matter of seconds. Our eyes (or better: our brain) can do this at a glance. The difficult part, namely finding the solution, is not involved.

I have often heard quoted this very illustrative example, but I must confess that I have some qualms about it. The verification of the correctness involves an organ whose mechanism we do not really understand. Moreover, the checking is a physical process with limits of resolution that might not meet the mathematical requirements of a sufficiently fine-grained pattern. The ability to recognise almost instantaneously the face of one's grandmother is programmed from early childhood on. I am using this example because neurophysiologists once thought that there were specific cells in our brain destined for such a job. They called them "grandmother cells". Given the vast number, some 10^{10}, of neurons in the human brain, why shouldn't an appreciable part of them, linked by some intricate circuitry, be able to accomplish the task of identifying a rather complex pattern adaptively in a short time? Today, machines[26] are being constructed which simulate pattern recognition. Anyway, the concept of grandmother cells is not really *en vogue* any more.

In mathematics, NP designates a complexity class dealing with problems that can be solved by a non-deterministic Turing machine in polynomial time. In Section 3.6 we learned something about Turing machines, but what is a non-deterministic Turing machine? "Normal" Turing machines solve their problems by breaking them down to (possibly very long) sequences of single steps. Now think that each of N steps involves some tree-like branching process, producing two or more simultaneously possible computation paths. For a success it is required that at least one overall path (including all N "forks") exists that leads to the desired solution – all in all a typical evolutionary problem. NP problems may be easy or very difficult – in mathematicians' terminology, "hard".

Problems that are called NP-complete are among the hardest ones. They all seem to be on a par. If for one of them it could be shown that it were soluble in polynomial time, then its proof would hold for every problem in NP. However, such a proof does not exist (as yet). Mathematicians generally expect that $NP \neq P$ because the assumption $NP = P$ would lead to some rather far-reaching consequences. However, a non-existence proof can be very hard. One would have to consider all possible algorithms and show that they all fail. There is a dissymmetry between existence and non-existence proofs similar to the dissymmetry in the trapdoor functions discussed in Section 3.4.

I took up this problem because in my opinion it plays the key role in any theory of biological organisation. No one is able to "compute" the base sequence of the human genome, given its thousand millions of places or – if not all of them are functionally necessary – a "correct combination" of the 30,000-odd proteins with all their "functional genomics". This is a problem of exponential complexity and – given those numbers – totally impracticable. On the other hand, human beings came about within a time that is almost linearly related to their genome size (see Table 4.3.2).

Biological evolution is practically synonymous with the evolution of genetic programmes. This is not meant in the sense that the true problems of modern biology are limited to the genotypic level; it rather means that in evolution nothing endures

Table 4.3.2 Comparison of Complexity of Living Organisms and their Evolution Times

Apparently they are cases of NP.

	E. coli genome	Human genome
Number of base pairs (N)	4 million	3000 million
Number of genes (n)	~4000	~30,000
Time of evolution	~100 million years	~4000 million years
Combinatorial complexity 4^N	$\sim 10^{2,000,000}$	$\sim 10^{1,500,000,000}$
Combinatorial complexity 2^n	$\sim 10^{1000}$	$\sim 10^{10,000}$

unless it becomes fixed in genetic programmes. Whether or not fixation occurs depends on selective evaluation, which is based on phenotypes and their performance under environmental constraints. Genotypes, on the other hand, inherit a quantifiable combinatorial complexity. Figure 4.3.1 provides a simplified example. A sequence comprising six binary symbols, denoted "000000", is to be adapted to some task by changing its six symbols to "111111". As can be easily verified from the picture, there are altogether six factorial (6! = 720) possible (direct) routes from 000000 to 111111. This suggests some similarity to the travelling salesman problem, discussed below, in which a certain number of cities is to be visited, each only once, on the shortest possible overall route. N! is a number exhibiting exponential

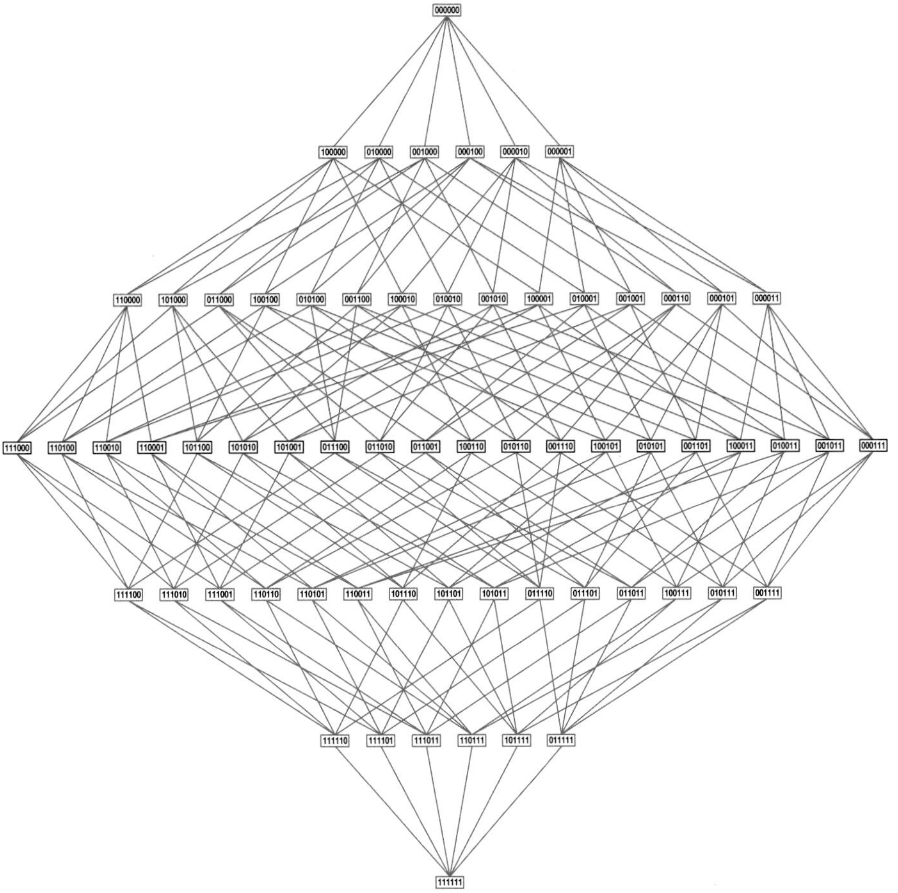

Figure 4.3.1

Space diagram of a sequence of six binary symbols

See text for a detailed description of the model.

complexity. For larger values of N it can be approximated by a formula, calculated in 1730 by James Stirling[27], which reads: $N! \approx (N/e)^N \sqrt{2\pi N}$, where e is the base of natural logarithms and π the well-known ratio between the circumference and the diameter of a circle. Already for $N = 10$ this approximation fits the true value of N! with better than 99% accuracy. For large values of N one may get good estimates by just considering the term $(N/e)^N$.

In the diagram in Figure 4.3.1, the point 000000 is taken to be the reference point. We could have started the procedure at any of the $2^6 = 64$ points of the diagram, finishing up at the respective antipodal point. In each case the diagram would have looked the same as the one shown here. This symmetry is due to the fact that we are actually dealing with a six-dimensional point space (which will be explained in more detail in Section 4.4). Whether or not detour-free routes exist in evolutionary optimisation, and which of the $6! = 720$ possible routes are to be taken, will also be considered later in this book. The example given here is in any case not very realistic. A single gene encoding one protein molecule, i.e. one of the functional molecules in living organisms, is a DNA sequence comprising something like 1000 code symbols. A virus, for instance the AIDS virus HIV, is an RNA sequence with about 10,000 such symbols.

Given these numbers, and appreciating all the tricks we are learning from molecular biology, we can make two statements regarding the NP/P problem. The vast numbers attendant upon the combinatorial complexity involved and the fact of the existence of optimally adapted structures together imply that the underlying problem of exponential complexity is in NP. The assumption that these structures came about by evolutionary optimisation suggests polynomial times, which would mean NP = P. However, this may not be due to any mathematical necessity, but rather to some specific and highly efficient approximative physical mechanism.

Let me illustrate this point by citing some more numbers. According to palaeontological findings, microorganisms, a present-day representative of which is the bacterium *E. coli*, existed already 3 to 4 thousand million years ago. The origin of the genetic code, according to some 1000 different sequences of code adaptors (tRNAs, of length 70–100 nucleotides), dates back almost 4 thousand million years. Given the age of our planet (some 4.5 thousand million years), and considering the time it needed to reach suitable physical and chemical conditions for the formation of tRNA-like structures, the origin of microbial life must have occurred within times of less than a few hundred million years and possibly even within a considerably shorter time span. The genomes of present-day microorganisms encode many thousand different proteins.

The length of the genomes of multicellular eucaryotic organisms is up to 1000 times larger, but the number of genes that encode functional phenotypes is only some 10 to 30 times as large as in procaryotes. On the other hand, multicellular organisms did not appear long before the early Cambrian, which means that their age is only about one-tenth that of early procaryotic microbes. A comparison of the evolutionary mechanisms of the two forms of organisms, their multiplicities and their population sizes certainly involves quite complex problems. However, there is no doubt that the times involved are clearly of polynomial (almost linear) rather than of exponential duration (see also Section 5.3). Exponential times, given that the differences in N amount to many millions, would be far outside anything realisable within geological times.

Again, I must emphasise that these numbers tell us very little about the possible validity of the mathematical problem: NP = P?. They simply suggest that there are physical conditions favouring mechanisms that achieve an optimal adaptation within polynomial times. A mathematical identity must hold for all possible cases, including pathological ones. Let me demonstrate what I mean by referring to the problem of the travelling salesman mentioned above and its "physical" solution using the method of simulated annealing.

The travelling-salesman problem is best characterised by the example[28] demonstrated in Figure 4.3.2. A merchant from the USA wants to visit customers in several European cities. He doesn't want to waste any money and, what is even more important to him, he wishes to save as much time as possible. To achieve this, he plans a route of the shortest possible length, in which each city is visited only once. Such a route, involving 100 cities, is shown in Figure 4.3.2. There are many similar situations, for instance in communal garbage collection, where careful planning could cut unnecessary costs and save time. Moreover, systematic planning should be extendable to an unlimited number of places, allowing a solution to be found within a short (computer) time – always supposing that many potential solutions exist.

A method ideally suited to the problem in question is that of "simulated annealing". It was proposed in 1953 in a seminal paper by N. Metropolis, A. Rosenbluth, M. Rosenbluth and Augusta and Edward Teller.[29] "Annealing" is a procedure used in metallurgy in order to perfect equilibration of a solidifying metal by allowing each atom to assume the position of lowest energy, corresponding to an ideal lattice. It involves heating and cooling the metal in the temperature range around its melting point, whereby the departures from the melting temperature decrease with progressing time. The speed and intervals of temperature changes then allow the atoms to find their correct positions. The Metropolis algorithm allows a simulation of this procedure. It is adapted to the Boltzmann distribution of energy states, which normally characterises an equilibrium distribution. The population of each energy state is weighted by an exponential of the ratio between the energy and the Boltzmann factor kT (where k is Boltzmann's constant and T is the absolute temperature; Chapter 2). The "simulated annealing" algorithm applies to an analogous procedure in order to obtain solutions for the combinatorial optimisation problem. It replaces the energy of a state by an "equivalent cost" function and temperature by a corresponding control parameter. Recent applications are based on independent papers[30] by S. Kirkpatrick, C.D. Gelatt jr. and M.P. Vecchi (1982 and 1983) and by V. Czerny (1985).

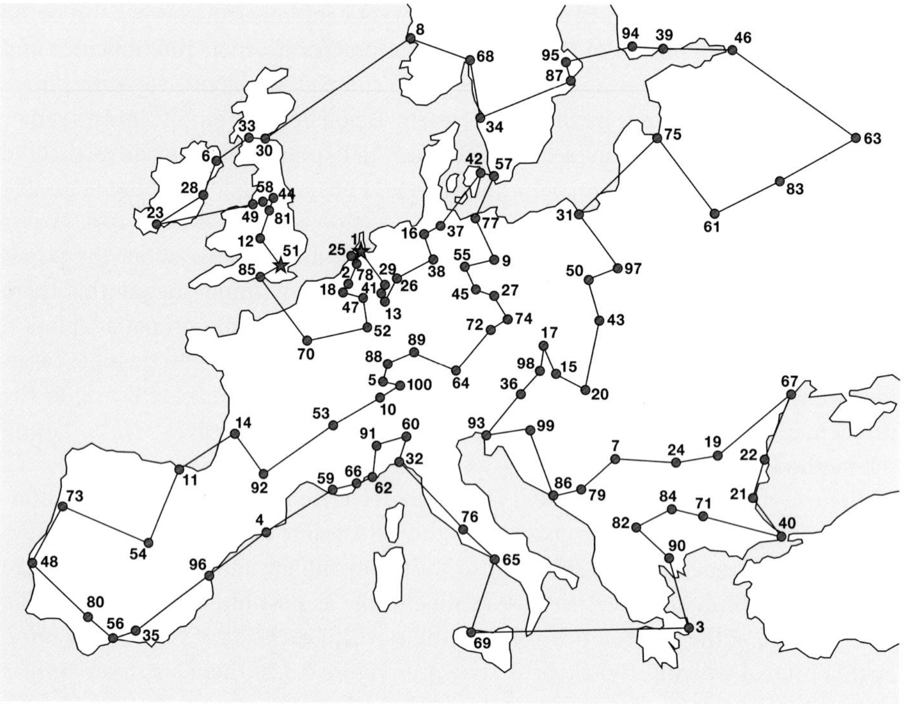

Figure 4.3.2

The Eur 100 travelling salesman problem

The picture shows a map of 100 major cities in Europe visited by a salesman. The solid line indicates the shortest possible tour. Adopted from the monograph *Simulated Annealing and Boltzmann Machines* by Emile Aarts and Jan Korst[31].

The travelling salesman problem requires us to find an optimum route including N cities for which the distance matrix is given. The length of the total route to be minimised plays the role of the total energy in the thermodynamic problem of annealing. Figure 4.3.3 provides a glimpse into the progressing evolution of the optimum solution.[31] It says more about how it actually happens than any verbal description could. Recently, an optimum solution for a tour that included 3038 cities was reported.[32] 3038! means more than 10^{9200} alternative routes.

I am not trying to give advice to travel agents or salesmen. Rather, what I wish to do is to analyse in more detail the method of simulated annealing and to see how it could apply to finding optimum routes of biological evolution. First of all, there seems to be only a very superficial correspondence between the two problems represented in Figures 4.3.1 and 4.3.2. In fact, simulated annealing seems to be almost ideally suited to solving the salesman's problem: the distance matrix is given, and hence

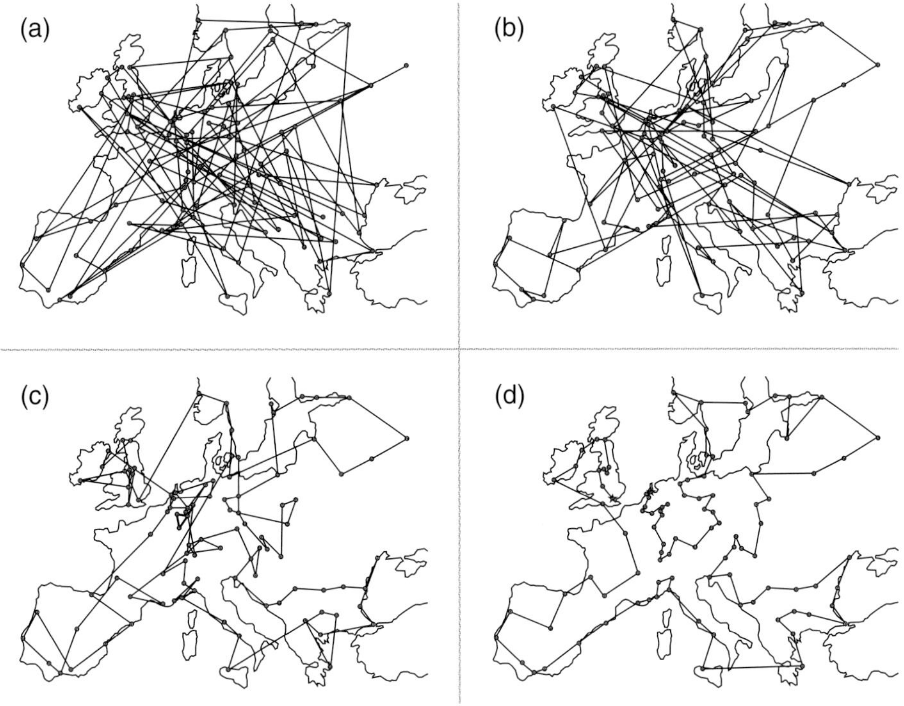

Figure 4.3.3
Solutions of the Eur 100 problem

Snapshots taken at four different stages of the simulation by the annealing algorithm[31]. Courtesy of E. Aarts and J. Korst. Reproduced with permission from John Wiley & Sons Ltd. © 1989.

each route has a defined total length. This is to be seen in analogy to metal crystals that involve various lattice defects. Each of the crystals has a defined total energy. An "optimal" crystal is one with a minimum of lattice defects, that is to say a crystal with a minimum of total free energy. Such a structure was shown in Chapter 2 to correspond to the completely established equilibrium that the metallurgist tries to achieve by annealing. This procedure is simulated in finding the shortest loop-free overall route for the travelling salesman. An evolutionary path, on the other hand, is not predetermined in any similar way. Nothing analogous to the distance matrix is known in advance, and an extremum principle such as $\delta F = 0$ (F = free energy) is not obvious – unless one were able to interpret "natural selection" in this way. The distances appearing in Figure 4.3.1 are Hamming distances of undefined "length", the whole evolutionary process being of notoriously non-equilibrial nature.

No doubt there is something that optimisation processes, such as those represented by Figures 4.3.2 and 4.3.3, have in common. In order to find out what it is, let us analyse both processes in some more detail. The complexity variable N in the first case is the number of positions of a binary sequence; in the example shown, $N = 6$. In the second case, N is the number of cities involved in the route of the travelling salesman: $N = 100$. The 2^N nodes in the first network are the true sequence states that are separated by Hamming distances. Any network of routes in the travelling-salesman problem would reasonably yield as nodes the N cities connected through $N(N-1)/2$ individual and usually quite different pair distances (Figure 4.3.3). In other words, we get two different kinds of network, and it makes no sense to compare them directly. Anyway, the total number of possible (non-redundant) routes in both cases is N!, as one can easily see.

In order to make a comparison meaningful, we should transform the salesman problem to a form more similar to the evolutionary problem shown in Figure 4.3.1. The nodes now represent different stages along given routes recording the cities (here numbered 1 to 6) that have already been visited once. The connecting lines among the nodes are no longer true geometrical distances. The new diagram, instead of representing a static spatial order, has now become a scheme of processional order, just as is true for the diagram of evolutionary pathways. The principal difference between the two similar-looking diagrams lies in the interpretation of how optimum routes come about.

Problems of exponential complexity are encountered repeatedly in molecular biology, a notorious example being protein folding, which has been shown to be NP-complete.[33] Protein folding is certainly a case where algorithms of annealing should find application. Simulated annealing describes an equilibration process; the force behind the natural process is due to the second law of thermodynamics. As was shown in Chapter 2, thermodynamic equilibrium – being a state of lowest free energy – is automatically restored after perturbation. The resulting finite affinities are "forces" behind chemical relaxation. The annealing process is helpful in overcoming barriers of frozen non-equilibrium situations. Evolution, on the other hand, is driven by "natural selection", which – as will be shown later – can also be characterised by gradients of "potentials". A "selected state", however, differs from an equilibrium state. Deviations are not characterised by Boltzmann factors, i.e. exponentials of some energy term divided by kT. The statistics of natural selection are different from Boltzmann statistics. Even for the travelling salesman problem there is no compelling reason why deviations from an optimum route should be Boltzmann distributed. It is merely the analogy of minimising the total length of an optimum route to minimising free energy in an annealing process that lies behind the success of the method of simulated annealing.

Many NP-complete problems are not so easily adaptable to simulation of thermodynamic equilibration, requiring multiple iteration. A typical example is the Tower of Hanoi game, shown in Figures 4.3.4 and 4.3.5. Each "tower" is a peg on which N rings of decreasing size are piled. The task is to reshuffle all rings from the first to the second peg, the third peg having an auxiliary function. The constraint is that each time only

Figure 4.3.4
Model of the Golden Tower of Hanoi.
Designed by Claus-Peter Adam.

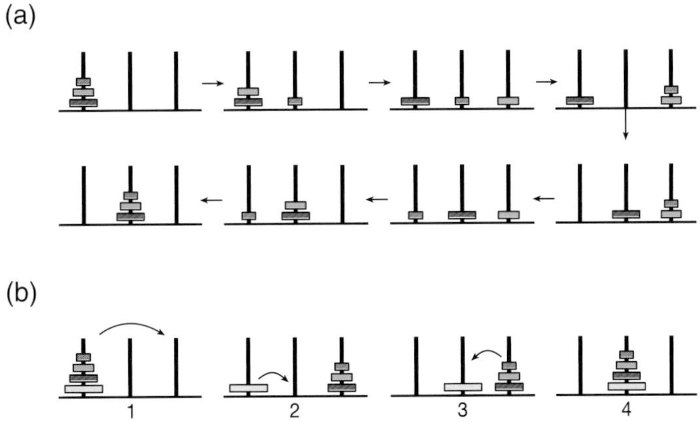

Figure 4.3.5
The Tower of Hanoi game with three and four rings

The rules are described in the text. (a) Three rings on the left-hand peg are replaced in seven steps according to the rules of the game. (b) The replacement of four (1 → 4) rings requires 15 moves, including the replacement (twice) of three rings (1 and 3) as described in (a) and the transfer of the largest ring in step 2.

one ring is to be moved and that a smaller ring is never allowed to be underneath one or more larger ones. In Figure 4.3.5 the procedure is demonstrated with three rings. If one is clever, one can do it in seven (but no fewer) moves, as is shown in Figure 4.3.5(a). Figure 4.3.5(b) shows how this procedure can be generalised. The general solution requires a minimum of 2^n-1 moves if n is the number of rings. The fairy tale, told by the inventor of the game, asserts that the end of the world will be close if a game involving 64 rings is successfully finished. The point is that one really has to carry out the mentioned minimum number of transfers, required by the iterative procedure. There is no chance of an accidental lucky strike if the game is played faithfully (which excludes anything like Alexander's solution of the Gordian knot problem, e.g. in our case by lifting all the rings together and placing them *en bloc* on the second peg).

Other types of optimisation problems have more of a selection-based evolutionary nature than an equilibration nature. The main point is that in an evolutionary process one usually does not know the final result until it has been achieved, in order to use it as a constraint for minimisation. Terrence Sejnowski[34] has applied the ideas of simulated annealing to the problem of adaptive learning in neural networks. And in this connection he has formulated the term "Boltzmann machine". I shall return to these highly innovative and exciting ideas in a second volume of this treatise. The question is whether one should not name "machines" to be assigned to learning problems "Darwinian" machines rather than Boltzmann ones because what will be required physically is selection rather than equilibration. Any further discussion at this point, however, would be premature. Needless to say, the ideas put forward in the later part of this section are physical rather than mathematical attempts to find an answer to the question: NP = P?

4.4. Are We Points in Hilbert Space?

Back in the 1970s, under the aegis of an organisation called the "Institut de la Vie", Maurice Marois of the University of Paris used to invite scientists from all over the world to an annual life sciences conference in the Trianon Palace in Versailles. The conferences were superbly organised, and we used to refer to them as meetings of the "Institut de la Bonne Vie". The scientists who came there were world experts in their respective fields of mathematics, physics, chemistry and biology; among them were many colleagues and personal friends. It was pleasant to listen to them, even if some of them did not bring along much expertise in the life sciences.

At one of these meetings a friendly old gentleman, the renowned Yale mathematician Henry Margenau, suggested considering a human being as a "point in Hilbert space". Hilbert space is a linear vector space that may comprise up to an unlimited number of dimensions. Its main importance for physics is that quantum mechanics can be given a complete axiomatic representation, where observables are described by operators and physical states by normalised vectors. The latter was shown in a classical paper by John von Neumann.[35]

Have you ever tried to transform yourself into Hilbert space? Don't worry, nothing will happen to you. All your 10^{28}-odd elementary atomic constituents will do what they always have done, i.e. change positional and momentum co-ordinates in their usual way, and even the "point", or vector, the only form in which you are allowed to exist in Hilbert space – given the many dimensions and Heisenberg's uncertainty relation – is not to be taken too literally.

I am now going to do something similar to what I have just described. I assume that our "alter ego" is represented by the blueprints we carry in our genes, being recombined DNA molecules we received from our parents. How is the space constructed in which each person's genetic "I" exists? Or, what kind of space allows the mapping and evolutionary development of our genetic information?

The main problem is the informational complexity of our genes. If we find a way of representing this information in a rational way, we may also find out how to study the processes that exert control over development. In Section 4.2 I gave examples of the huge informational complexity of our genes. Can we find a suitable space to represent this complexity?

We realise immediately that our three-dimensional Euclidean space cannot do the job. Take our universe and partition it into a cellular space. No matter how small one makes the single cells in this space, our universe would be much too small to accommodate what is needed. Let me illustrate this with a few numbers. If the cell size is that of a single gene or protein, say 10^{-24} m^3 (= 1 million cubic Ångstrøms), or that of one hydrogen atom (10^{-30} m^3 or 1 cubic Ångstrøm) or the smallest physically meaningful volume, i.e. the Planck volume ($\sim 10^{-105}$ m^3), the corresponding numbers of cells that might fit into the total volume of the visible universe would be 10^{102}, 10^{108} or 10^{183}, respectively. These are all certainly very respectable numbers, but they are much too small to allow a mapping of all the alternatives of even a single gene, which comprises about 1000 nucleotide monomers (with a length of about 3400 Å and a diameter of 20 Å). The number of alternatives is $4^{1000} \approx 10^{600}$. A number with 600 decimal places is inconceivably larger than one with "only" 183 decimal places.

Of course, this means that the space of genes is (and always has been) nearly empty. Nevertheless, we have to find the "fitting" positions within this space.

The construction of an information (or sequence) space has to satisfy two conditions:

(1) Each uniquely defined sequence must be represented by one and only one point or cell in the space. As we are dealing with genes (DNA or RNA molecules) made up of four classes of monomeric digits, the number of cells or points must be $2^{2\nu}$, where ν is the length of the sequence (in numbers of nucleotides).

(2) The cells or points must be arranged in such a way that all distances between them reflect the correct kinship relations among the sequences. This means that all sequences differing in only one position must be nearest neighbours. Those differing in k positions must be a distance k apart, and this condition must hold for any pair of sequences.

The idea of representing messages in this way goes back to Richard Hamming[36], who first applied it to binary sequences in coding theory. I used this concept in 1971[37] to describe the kinetics of self-organisation of replicating quaternary digit sequences, later developed into a general space concept together with Peter Schuster[38]. John Maynard-Smith[39] suggested a similar model for describing proteins, and Ingo Rechenberg[40] applied the idea to develop an evolutionary strategy for technical constructions.

Let me now systematically construct this space and – for the sake of simplicity – start with binary sequences. Instead of the digits 0 and 1 – common in informatics – I shall use the binary digits Y and R, which denote the pyrimidine (Y) and purine (R) bases of nucleic acids. The difficulty lies in our lack of ability to imagine geometries of more than three dimensions. However, with the help of an iterative procedure we may learn algorithms that allow us to replace imagination by logical inference.

The first diagram in Figure 4.4.1 represents the trivial case of a binary "sequence" of length 1, i.e. one digit, which can be either Y or R. The line that connects the two alternatives, as will all lines in subsequent diagrams, denotes a possible one-digit exchange (or one-error mutation). The next step is to add a second symbol (second diagram in Figure 4.4.1), so we are now dealing with two-digit sequences, and there are four such alternatives. Instead of simply alluding to the square which we obtain when drawing the mutational scheme, I would prefer to say: adding a second symbol requires doubling the first diagram and completing it with two more possible mutation lines among corresponding points. Now we proceed to the third symbol (third diagram), and I explain the cube obtained by repeating myself: Adding a further symbol requires doubling of the previous diagram and inserting the additional mutation lines. There is no reason why this instruction should not hold for any step from length n to length $n + 1$. So we have already learned the algorithm. We need it because we can no longer imagine the fourth diagram, the four-dimensional hypercube. However, just double the preceding diagram and add the mutation lines between corresponding points. This is a safe way to go on to any dimension: 5, 100, 1000, or as high as you wish.

Does this iterative trick mean that we can now also "imagine" the geometries in dimensions higher than three? Let's perform a test. If we look at the bodies with increasing dimensions (ν), we realise that they always contain 2ν faces (generally called hyperplanes) of the next lower dimension: a line has two ends, a square is limited by four lines, a cube has six planes at its surface, and in the same way the four-dimensional hypercube has eight cubes representing its "hyperface". It is as easy

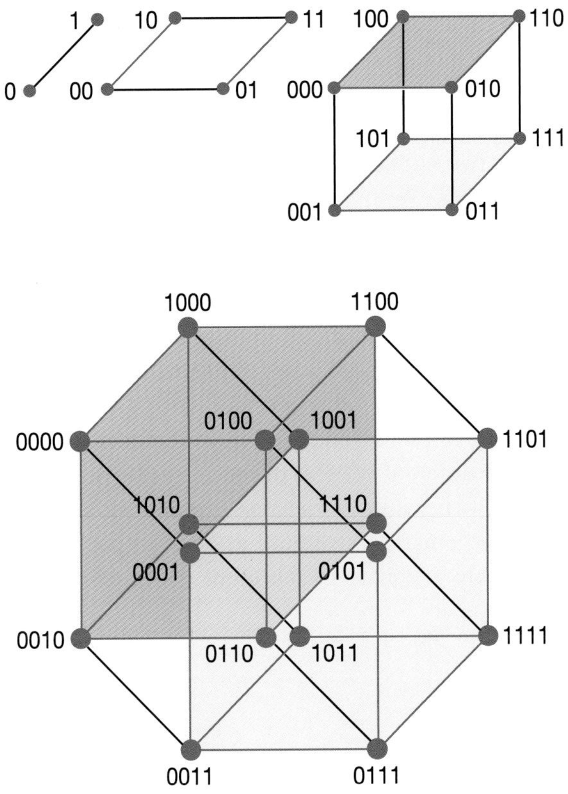

Figure 4.4.1

Iterative construction of sequence space

Starting with one position of a binary sequence, further positions are added iteratively, resulting at each step in doubling the previous diagram and drawing all lines that connect nearest neighbours. The binary sequence space is a ν-dimensional hypercube in which each vertex represents one of the 2^{ν} possible sequences. All lines connect nearest neighbours, i.e. mutants that differ in only one position.

as that: a ν-dimensional hypercube has a surface consisting of 2ν hyperplanes of the next lower dimension. Can you make out the eight cubes in the four-dimensional hypercube in Figure 4.4.1? If not, then go back to Figure 1.2.4, where they were identified individually.

There I showed the surface cubes of a four-dimensional hypercube in order to demonstrate an identity that is a central principle of algebraic topology: "The surface of a surface is zero." Take the plane. It is limited by four (vector) lines, of which two have always opposite orientation. By the same token, the six faces of a cube comprise three opposite pairs that cancel one another out. The principle, carried over to four-dimensional space–time, revealed important clues to Einstein's geometrodynamics, which I referred to in Chapter 1 (where the papers by Elie Cartan and John Archibald Wheeler are quoted). Can you now imagine what a hypercube "really" looks like? If so, try it with dimension five. I gave up at dimension four, which already requires two kinds of perspective in order to be represented on a two-dimensional sheet of paper. Nevertheless, in Figure 4.4.2 I present such a diagram for the transition from four to five dimensions, which at least gives some impression of a "compact" structure. However, in this representation on a two-dimensional sheet we are already dealing with three different but nevertheless equivalent perspectives. It becomes more and more difficult to imagine higher dimensions. For large dimensionalities we therefore choose a different manner of representation, which still reflects the connectivity among mutants. Figure 4.4.3 shows another means of representation, which can be extended to large dimensions, showing clearly the connectivity among the different vertices of space diagrams.

It is less important to be able to visualise n-dimensional spaces than it is to be able to use their mathematical structure as a device for mapping information. Hyperspace is completely adequate to represent the combinatorial complexity of

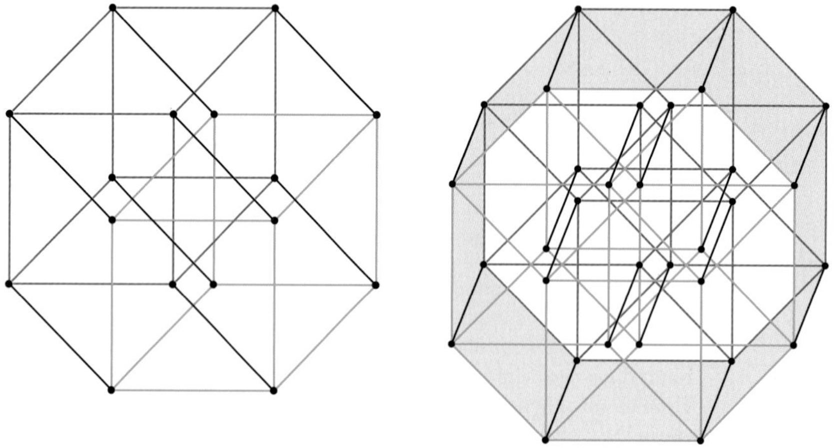

Figure 4.4.2

The five-dimensional hypercycle

As was shown in Figure 4.4.1, the four-dimensional hypercycle can be constructed from two three-dimensional cubes, which in the left-hand diagram are illustrated by the red and the green cubes. The two cubes are connected by eight (black) lines between corresponding points. The representation of the two (three-dimensional) cubes on a two-dimensional sheet introduces one perspective. The black connecting lines between the two cubes introduce a second perspective, since the five-dimensional hypercube is constructed form two four-dimensional hypercubes (red and green in the right-hand diagram) they introduce a third perspective by the 16 black lines connecting corresponding points.

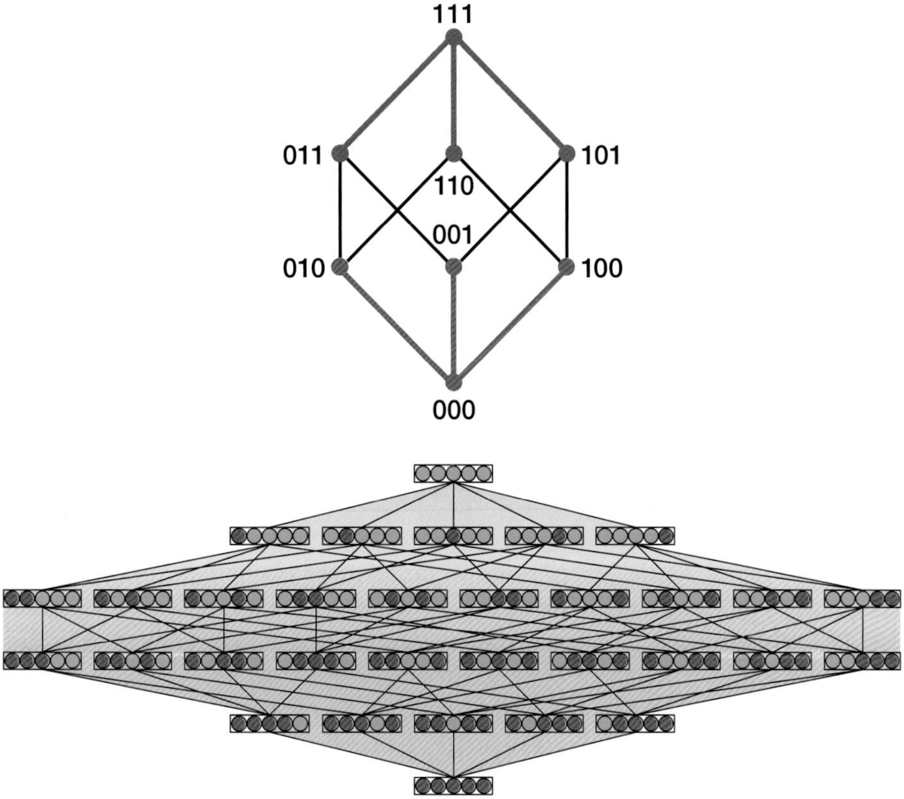

Figure 4.4.3

Simplified representation of higher-dimensional nets

The upper diagram shows a three-dimensional cube represented in a perspective in which the points 011, 110, 101 and 010, 001, 100 appear as two horizontal sequences. In the lower diagram this representation is generalised (for example, the five-dimensional hyperspace shown in Figure 4.4.2). In this arrangement any spatial imagination is lost; the black connecting lines inform us only about nearest neighbours. The advantage of this representation is that it can be generalised for any number of dimensions, while the spatial representation in Figure 4.4.2 would soon become impracticable. *Note*: Any true spatial representation must be compact, showing all points within a narrow circle, the radius of which does not exceed the diameter of the information space (e.g. sequence length ν).

binary sequences of length ν. The ν-dimensional hypercube has 2^ν vertices (corner points), which all lie on its surface. Each vertex corresponds to one of the 2^ν possible sequences. The vertices are connected by $\nu 2^{\nu-1}$ edges, each linking nearest neighbours. A "neighbourhood" may also include sequences that were obtained by successive mutations at several (k) positions. There are altogether $\sum_{x=1}^{k} \binom{\nu}{x}$ such sequences, having distances $\leq$k which belong to a neighbourhood thus defined. The distance is the smallest sum of edges that connect two points. It is called the Hamming distance in honour of Richard Hamming, who founded the concept in coding theory. Note that the metric does not coincide with the Euclidean metric, as expressed by Pythagoras' theorem. It is rather to be compared to the street block metric that characterises the city map of Manhattan. When you ask for the way from A to B the answer given to you will be in terms of numbers of street blocks rather than in distances.

Applying this concept to genetic information requires some further extension. Nucleic acids are made of four classes of constituents of their monomeric units. These constituents are denoted by the initials of their chemical names: A and T (or U), G and C. They belong to two types of organic molecules, the pyrimidines Y (T/U and C) and the purines R (A and G). (This is why I preferred Y and R as "binary symbols" instead of 0 and 1.) As we shall see later, mutations within one base class, called transitions, for various chemical reasons, occur more easily than mutations that change the base class, called transversions. For this reason it is not only useful, but also logically consistent to construct the sequence space in two consecutive binary steps.

As in Figure 4.4.1 one first assigns the base class, Y or R, obtaining a binary hypercube of dimension ν in which each of the 2^{ν} vertices denotes one of the sequences in RY notation. Each of these sequences then requires another binary decision for each of its symbols, producing at every vertex another binary hypercube of dimension ν as a subspace. This procedure is illustrated in Figures 4.4.4(a) and (b). For a sequence of four symbols it leads to the diagram represented schematically by Figure 4.4.5, involving 256 vertices, the construction of which is greatly aided nowadays by computer drawing. Another schematic representation is shown in Figure 4.4.6, which might have pleased the Dutch graphic artist Maurits Cornelius Escher. A three-dimensional model representing the genetic code has been presented in Section 3.4.

Note that according to Figure 4.4.4 all transitions occur within given subspaces and all transversions between different subspaces. Since they occur with differing probabilities, mutation distance is not as obvious from Figure 4.4.3(b) as is the Hamming distance in binary diagrams. Wherever distance analysis is performed (as in the construction of evolutionary trees) transitions and transversions should be considered separately.

There is one point that we should take notice of. Since each vertex in the transversion space (RY hypercube) has a transition subspace of equal dimension (ν) the total space now has an apparent dimension 2ν. Each of the 2^{ν} vertices of the RY hypercube contains a subspace having another 2^{ν}vertices, but not all vertices of two

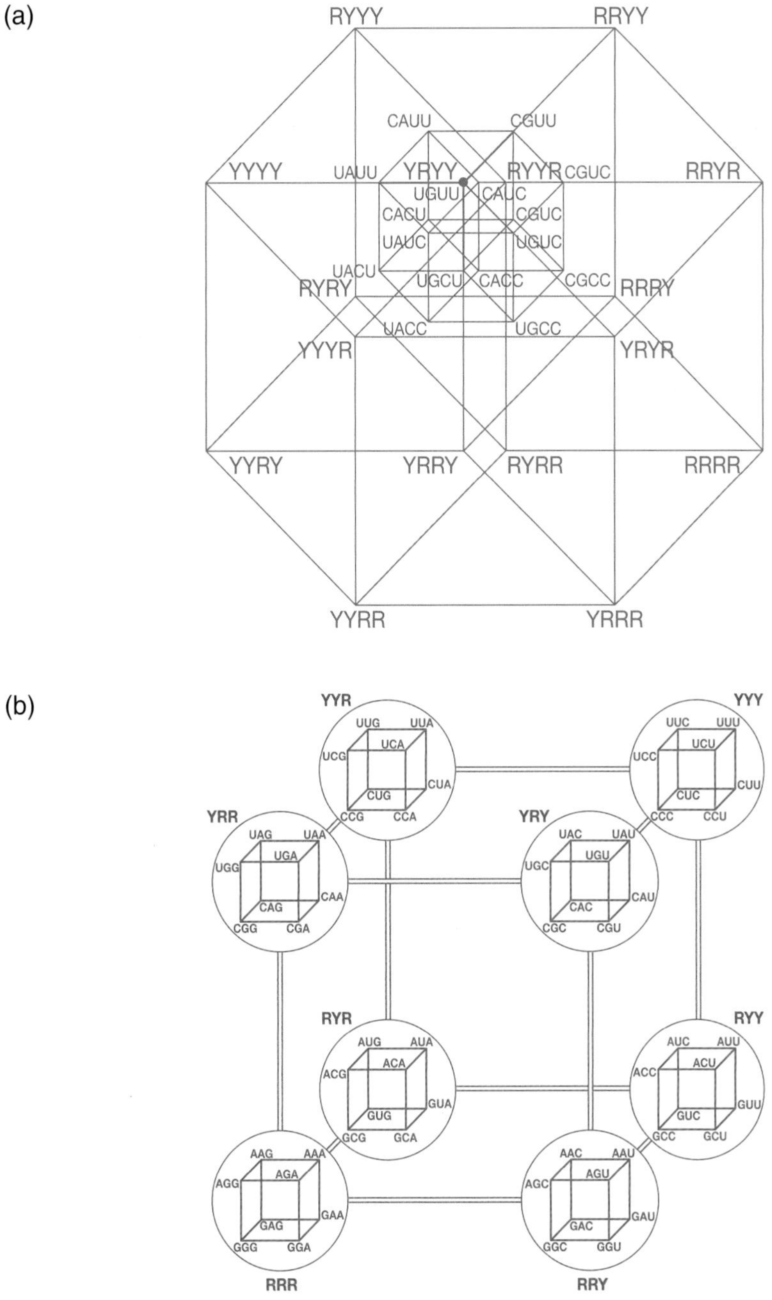

Figure 4.4.4

Quaternary sequences

(a) For nucleic acids, the quaternary sequence can be constructed in two consecutive steps, where in the first step a binary space in R and Y notation is drawn while in the second step each point of the binary RY space is supplemented by another binary space. In this subspace R is replaced by A or G and Y by C or U (this figure shows only the vertex YRYY). All lines in the first binary space refer to transversions, i.e. changes between the base classes R/Y (purine/pyrimidine) while all transitions (changes within base classes) appear in the binary subspaces at each vertex of the RY space. Both mutations occur with different probabilities. (b) Quaternary structure for a three-dimensional model.

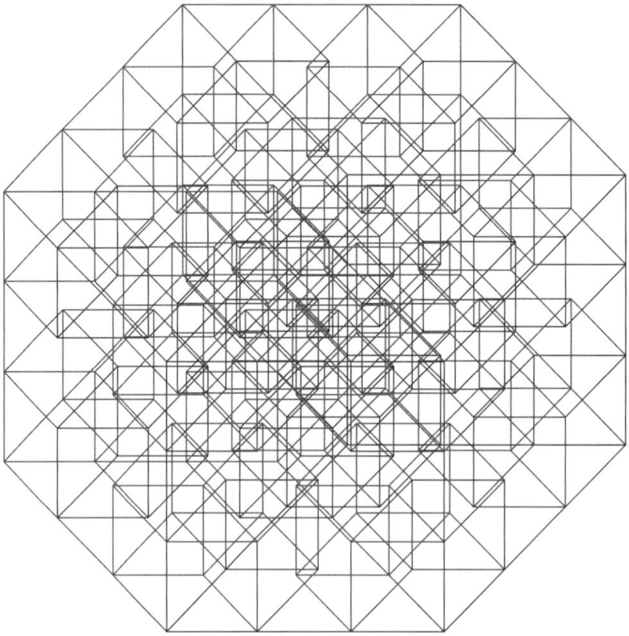

Figure 4.4.5

Schematic representation of quaternary space

Computer diagram of a quaternary network including 256 nodes. The diagram distinguishes between transitions and transversions.

neighbouring subspaces are connected with one another. Rather, each vertex of a subspace is connected to only two vertices of a neighbouring subspace, yielding a construction with $2^{2\nu}$ vertices and $\nu 2^{2\nu}$ edges, equivalent to a dimension of 2ν.

Let us discuss right away an application of sequence-space diagrams to genomics. RNA and DNA sequences of many organisms have become available, up to the entire human genome. These sequences, although referring to present-day organisms, also yield quantitative clues about their evolutionary past, if they are compared in an appropriate way. The "appropriate way" is prescribed by the multidimensional diagrams of sequence space. The evolutionary past, on the other hand, can be represented by a branching phylogenetic tree. Owing to accumulation of "noise" in present-day sequences the exact branching order may be obscured to a certain extent. Moreover, the "tree" need not adhere strictly to a topology of successive branching. Such non-adherence can be found especially in viruses, which by virtue of their high mutation rates try to evade the defence mechanisms of their hosts. The example that I am going to discuss addresses problems of exactly this kind, which will occupy us again in Volume 2 in connection with more biological questions.

Comparative sequence analysis rests on the correct alignment of related sequences. Suppose that the problem of alignment has been solved for a given set of

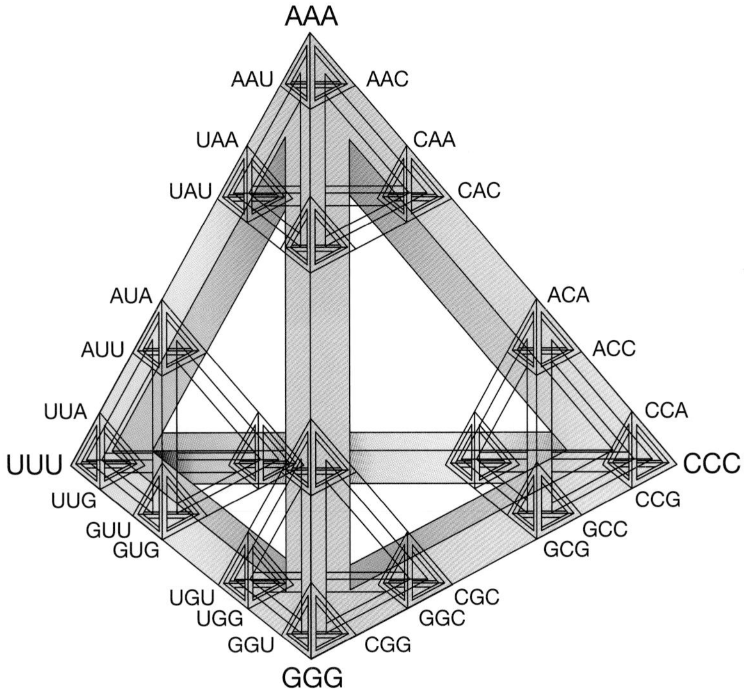

Figure 4.4.6

Alternative representation of a (four-dimensional) quaternary hyperspace

This picture is reminiscent of Maurits Cornelius Escher's artistic manipulations of space and perspective, but it does not distinguish between the different probabilities of transitions and transversions.

sequences. The data then can be represented in the form of a matrix of pair distances, similar to the distance tables that we find in road atlases. These pair-distance matrices are usually the basis for "tree construction". After applying certain corrections, one tries to find a tree that optimally fits all the data available. Various elaborate computer programs have been developed to do the job. However, a program that has been designed to generate a tree will do so, regardless of how well the "optimum" tree actually fits the data.

In using distance data, one loses quite a lot of information about positional fits. Hence, it is advantageous to use true sequence-space diagrams, rather than distances which sum up individual positions. We need a minimum of four sequences to carry out such an analysis. With four binary sequences we have for any position the following eight choices: (1) all four sequences agree, (2) three sequences agree and one differs, e.g. $A \neq B = C = D$ (the differing sequence can be any one of the four sequences), and (3) the four sequences agree pairwise, which can be realised in three possible ways: $A = B \neq C = D$, $A = C \neq B = D$, $A = D \neq B = C$. Other choices (e.g. $A \neq B \neq C \neq D$) do not exist for binary sequences. Hence there are seven

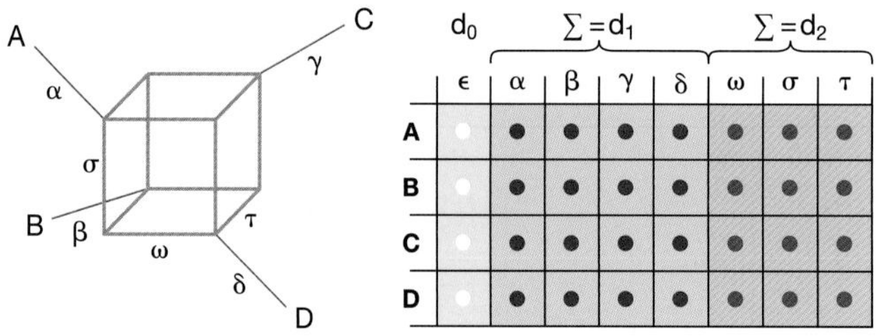

Figure 4.4.7

Schematic diagram representing distance correlations in binary space

For this analysis to be carried out, a group of (at least) four binary sequences (e.g. RY), supposedly of common origin, is required. Three classes of positions are to be distinguished:

- those at which all sequences show the same symbol, yielding the sum d_0
- those at which one of the four sequences differs from the three others, yielding the sum d_1
- those at which two of the four sequences differ from the other pair, yielding the sum d_2.

This covers all possibilities for binary sequences. From the model on the left we see that d_1 covers all protrusions α, β, γ and δ (dark points), while d_2 (dark points) is obtained from the sum of ω, σ and τ for all three box dimensions. Box distances come about through joint, parallel or reverse mutations. If all are close to zero, they indicate a bundle-like divergence. If only two of the box dimensions approach zero, while the third one has a finite value, they indicate ideal tree-like behaviour. In general, protrusions are clearly larger than the box dimensions for slightly noisy divergences of trees or bundles, while similarity of d_1 and d_2 indicates a high degree of randomisation.

different possible configurations for any non-congruent position in quadruples of binary sequences. The seven types of distance segments can be represented by the diagram shown in Figure 4.4.7. An analogous evolution for quaternary sequences is shown in Figure 4.4.8, where their limiting structures are also demonstrated. In order to obtain reliable averages for all seven types of segments (α, β, γ, δ, σ, τ, ω) one analyses a set of n sequences for which one can combine $\binom{n}{4}$ different quadruples (yielding, for $n = 100$ sequences, about 4 million diagrams), and determine the statistical averages extending to all positions in the sequences. We (which includes Ruthild Winkler-Oswatitsch, from the Göttingen laboratory, and Andreas Dress, at the mathematics department of the University of Bielefeld) have called this method "statistical geometry in sequence space".[41] Quadruples are a minimum requirement of this method because for triples of sequences the number of detectable segments does not match the number of unknowns in the equations. On the other hand,

4 Sequences : **A, B, C, D**
Positionally differentiated distance categories

0	Four of a kind	**A = B = C = D**	4 x 1
1	Three of a kind	**A = B = C ≠ D**	12 x 4
2	Two pair	**A = B ≠ C = D**	12 x 3
3	One pair	**A = B ≠ C ≠ D**	24 x 6
4	No pair	**A ≠ B ≠ C ≠ D**	24 x 1

256 Degeneracies

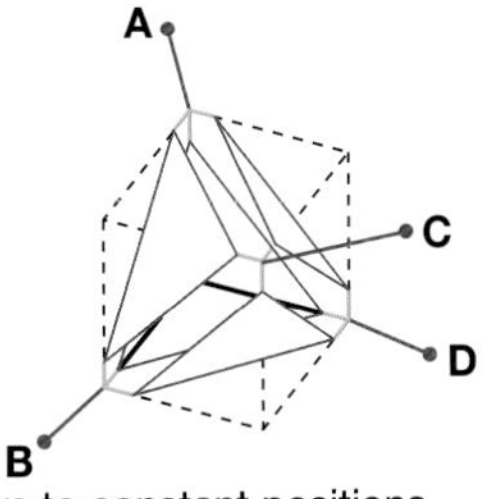

Category **0** refers to constant positions.
Category **1** refers to conservative positions with small variation.
Category **2** refers to positions of medium variability.
Category **3** and **4** refer to positions of large variability.
One large dimension in category **2** indicates treelikeness.

Figure 4.4.8

Four sequences in AUCG-space

All explanations are given in the figure.

one can prove that the divergence of a set of n sequences is exactly tree-like if all its $\binom{n}{4}$ quadruples have a tree-like structure (see the comparison of structures in Figure 4.4.9).

Figure 4.4.10 shows "statistical geometries" for the quadruples of sets of sequences of three different types of RNA viruses.[42] They refer to the binary changes for both transversions (Y ↔ R) and transitions (U ↔ C or A ↔ G). This figure shows how sequence-space diagrams – in comparison with tree diagrams – can greatly enhance our knowledge about the evolution of viruses. The analysis was carried out for the following viruses: influenza A (time-resolved tree), polio 1 and a group of HIV-1, HIV-2 and simian viruses (SIV). Each upper diagram refers to transversions, the lower to transitions. The three codon bases are considered separately.

The results speak for themselves. The time-resolved tree for a certain gene (the so-called non-structural gene) of the influenza A virus is confirmed by highly tree-like space diagrams. Transitions are accepted more frequently than transversions. Middle positions of codons are most stable, while terminal (i.e. highly synonymous) positions show lowest stability. The observed changes in a time-resolved tree are due mainly to highly variable positions. About two to three mutations per 1000 nucleotides are fixed during a period of 1 year. If all positions were to change at

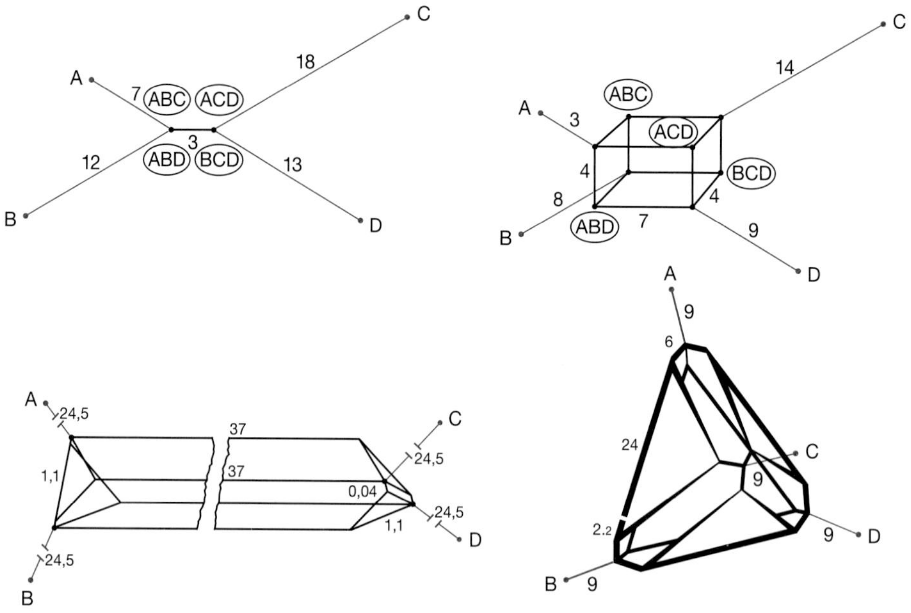

Figure 4.4.9

Comparison of diagrams in binary and quaternary space

The upper two diagrams show ideal trees (left) and noisy trees (right) as explained in preceding figures. In quaternary space, three distance classes of binary sequences are replaced by five classes of more complex definition, described above. The ideal limiting structures for bundles and trees come out to be the same as in binary space. The tetrahedral central body shrinks to a point, while only the protrusions are conserved, with correspondingly increased lengths. Nevertheless, one of the two lower diagrams, clearly indicating a tree-like structure, shows the intricate structure of the remaining diagram. The lower diagram on the left refers to the tree-like divergence of influenza A viruses, described in Figure 4.4.10.

such a high rate (which they do not), the virus would lose its identity within a few hundred years.

An entirely different pattern emerges for a 150-monomer-long region (called VPI) of polio-1 virus. While the first and second positions for both transitions and transversions are nearly invariant, yielding almost constant phenotypes, mutations accumulate in the third (highly synonymous) codon positions, which appear to be almost completely randomised, making a tree based on mere distances rather meaningless. The stability of the phenotype (which is due to the almost invariable first and second codon positions) of polio-1 virus is reflected in the susceptibility of this pathogen to vaccines. Influenza viruses, on the contrary, can only temporarily be suppressed by vaccines, as long as the immune system can cope with the slowly varying virus. This is radically different for the AIDS virus (HIV), where not only the immune system itself is the target of the virus, but also the large variability at

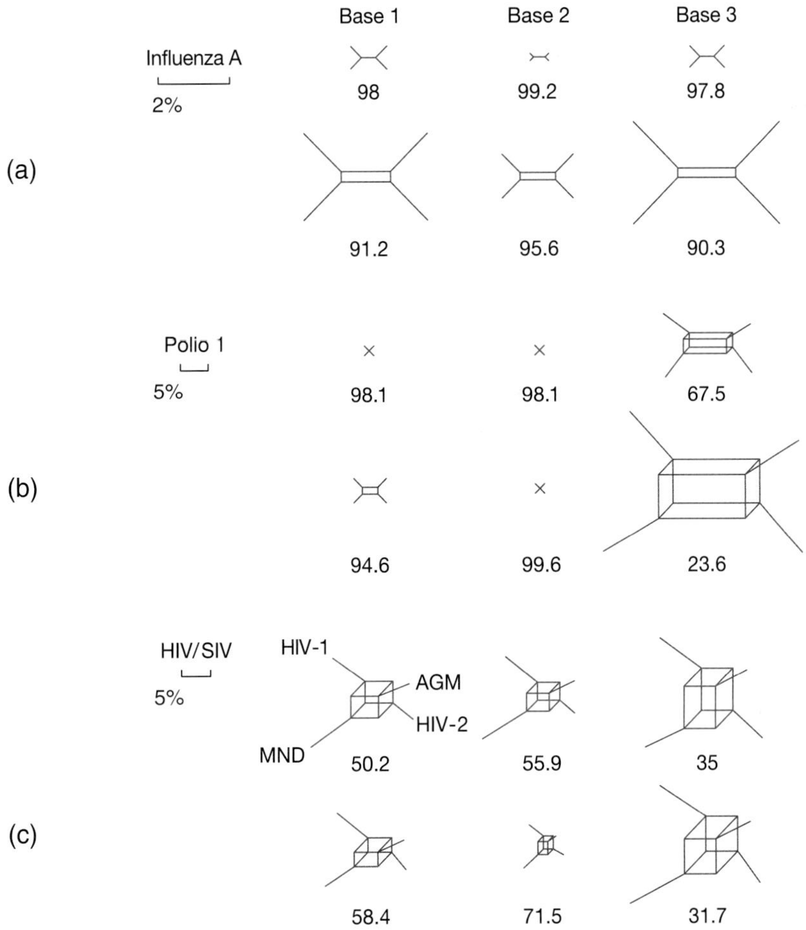

Figure 4.4.10

Sequence divergence for three different viruses

This figure presents three examples, of which the first one was chosen for its ideal tree-likeness. It refers to the time-resolved tree of the NS gene of the influenza A virus[42a], isolated from a given sample at intervals of 10 years. The gene shows fast changes that are probably due to mutations at highly variable positions. For all three cases the analysis was carried out separately for transversions (upper row) and transitions (lower row) at defined codon positions. As is seen for influenza A, the third (wobble) position of the codon is the least stable, the second (middle) position the most stable. This is different for the VPI region of the polio 1 virus[42b] where the first position shows very few changes, the middle position almost no change, while the third position (especially in the case of transitions) is almost completely randomised. We have discussed this case with respect to the better immune protection available for polio 1, compared with influenza A. Especially striking in this respect is the last case (the envelope gene of HIV and SIV), where mutational changes are large at all three codon positions. In the bottom diagram HIV-1 and HIV-2 refer to two types of human immunodeficiency viruses, while MND and AGM denote simian viruses (mandrill and African green monkey, respectively). The numbers in each row refer to the percentage of homologous positions. The figure shows clearly that entirely different conclusions could be drawn from representations in terms of averaged distances.

all genomic sites favours the establishment of escape mutants, with ultimately fatal results for the host organism.

Figure 4.4.10 shows the large variability of (human) HIV-1, HIV-2 and (simian) SIV. Randomisation of treelikeness shows up at all three codon positions and discriminates only slightly between transitions and transversion. In her PhD thesis[43], Katja Nieselt-Struwe used these sequence-space diagrams to obtain a history of divergence. She was able to establish three categories of positions for the envelope gene, ones that diverge in clearly distinguishable time intervals. While "constant" positions do not reveal any change over tens of thousands of years, variable and hypervariable regions have average lifetimes of about 1000 and 30 years, respectively. Such information cannot be extracted from distance-tree diagrams, where dating has led to much too short times because calibration was based on recent changes occurring mostly at hypervariable positions. According to these results, HIV is a descendant of an old family of retroviruses, and even the divergence of HIV-1 and HIV-2, or their simian precursors, dates back many hundreds of years.

The above discussion may have created the impression that information space is just a suitable way of representing kinship relations among sequences. I see it in a different way. We are dealing with a type of space that is as real as the physical space to which classical, quantum and relativity mechanics refer. Our access to information space is logical inference, as we have no sensory organs for perceiving it, as little as we have for phase space in quantum statistics. The concept turns out to be of the utmost importance for an understanding of basic problems of the origins and evolution of life and of thought. I therefore want to conclude this section with a survey of the issues that are introduced by the concept of information space.

The exponential complexity that underlies the problem of information requires an exponential space concept. There are three properties in particular in which the difference between information space and the more familiar Euclidean space expresses itself: volume, distance and connectivity.

(1) If a particular item of information is expressed in a string of N bits, the volume of information space is the total number of alternative bit strings of length ν, i.e. 2^{ν}. It is the volume of a ν-dimensional hypercube that has 2^{ν} vertices. To call the total number of vertices of a binary hypercube a "volume" is justified by assuming that each vertex is equivalent to a unit volume (which it actually would possess in "cellular" space). Adding another position to the bit string means doubling the space volume. Changing the number of positions, e.g. by making deletions, insertions or recombinations, does not influence existing configurations. As in Euclidean space we may define "spheres" having finite radii. A "sphere" of "radius" k includes all vertices up to a Hamming distance k from a reference point that defines the centre of the sphere. Its volume is $\sum_{k}^{x} = 1\binom{\nu}{x}$, where $\binom{\nu}{x}$ is the number of vertices in its surface x. As in Euclidean space, the surface of a sphere includes all points that have equal

distance from a centre point. Spheres might also be defined for subgroups of ν_i symbols (e.g. for single genes in a genome). For values of ν that are large compared with k the above sum representing the volume is given approximately by its last term $\binom{\nu}{k}$. Likewise, in continuous high-dimensional Euclidean space the volume lies almost completely localised in its surface layer.

(2) The second important clue from the hyperspace concept is seen in the fact that, despite the hugeness of its volume, distances remain relatively small. Wherever one wants to go, it's just around some corners. Yes, but one has to know which corners! The largest (loop-free) distance in this huge space, i.e. "from one to the other end" (corresponding to the two points 000…0 and 111…1 in Figure 4.3.1), is only ν (e.g. 6) in binary Hamming space or 2ν in quaternary gene space. The latter fact is very important for the effectiveness of evolutionary optimisation. The routes may not be detour-free. However, be assured that distance in sequence space is no problem. What we need is guidance to tell us around which corner to go, otherwise we get lost in the vast volume of sequence space. Random walk over large distances in physical space is already pretty ineffective. In high-dimensional space it is utterly hopeless. What one would need is guidance by some sort of potential gradient. Random walk along constant-potential surfaces by neutral mutations could be helpful in reaching new neighbourhoods.

(3) There is a third, and perhaps the most important, property of hypercubes of large dimension ν, which I call connectivity. What in landscapes on earth appears to be a mountain pass, a preferred starting point for a mountain tour, in information space becomes a huge multitude of crossroads. The number of edges exceeds the number of vertices by the factor $\nu/2$. Hence, each vertex offers many options for going on, and this repeats itself indefinitely, facilitating easy escape from traps.

In concluding this section I want to indicate how far the concept of information space will aid us in solving the problems of exponential complexity involved in generating information, both in life and in thought. The exponential space capacity together with moderate distances, the immense connectivity among all states and the presence of a fitness landscape offers a physical solution for bypassing the time-consuming procedure of checking all individual states. For this to work, a hierarchical principle of ordering is required.

My Belgian colleague and friend Christian de Duve in his book *Life Evolving*[44] – a marvel of biochemical common sense – proposes the idea of "modular combination", in order to cope with the problem of evolving exponential complexity of proteins or their genes. The (relatively small) polypeptide chain of say 100 amino acid residues is one among $20^{100} \approx 10^{130}$ possible alternative sequences, a multitude which cannot be accommodated in our universe, even with the closest packing possible. "Modular combination" supposes that the full-length chain has evolved by stepwise combination of smaller units or "modules". Christian de Duve refers to "undeniable traces of such a process in the structures of present day proteins, which are manifestly made of modules or motifs". Assume such a motif to consist of about 20 amino acid residues, these being the translation products of tRNA-like primordial genes. Iterative combination of such units – as products of natural selection – would yield any gene length, giving the impression of a "complete exploration of the available sequence space".

The idea as such, namely to apply geometrical rather than arithmetical progression to the problem of exponential complexity, is certainly a brilliant one, but the devil – once more – is in the detail. In fact, treating evolution as a process taking place in the exponentially complex structure of sequence space aims at the same goal. However, the process described above does not provide a complete exploration of the available sequence space. For instance, the units selected in the first step have explored a sequence space of, at the most, 10^{26} sequences, while the products of combination refer to a space of 10^{52} sequences, of which fewer than one in 10^{20} have been tested. In order to combine two modules the structures of which fit one another optimally and, in addition, are adapted to optimum functionality, one needs a considerable number of adaptive secondary mutations in both modules. In later phases of evolution such processes might be successful, and Nature indeed makes use of these in genetic recombination and in the maturation of the immune response.

Let me briefly describe how I think the idea of "modular combination" could be realised in sequence space by inclusion of secondary adaptive mutations. The model is illustrated in Figure 4.4.11. For simplicity of representation I have chosen an initial module of only four binary positions, described by a four-dimensional hypercube. The iterative nature of the procedure shows that I could have started with any configuration of higher complexity. The upper structure above the yellow circle shows the hypercube with its 16 possible sequential arrangements. Natural selection will favour one sequence as the centre of gravity of the quasispecies-like distribution which lends its reddish colour to the whole cycle. By adding four more positions one gets a new hypercube, now made up of $2^4 = 16$ reddish cycles. Each additional position requires a doubling of the former diagram (as was shown in Figure 4.4.1). Figure 4.4.11 shows an intermediate state with two positions added. Only if addition of a new position does not require any adaptive change in the initial hypercube will the reddish colour be conserved. Of course, natural selection will again favour one of the 16 cycles, which is indicated by a more intense red colour. The colour remains, since no change in the initial cycle was required. This, however, is more unlikely, on account of the arguments given above in the discussion of the "modular combination" model of Christian de Duve.

The quasispecies model, discussed in Section 4.7, covers a much larger portion of the sequence space of the final structure obtained by iteration of the above procedure, without giving up the idea of "modular combination". The essential point is that the

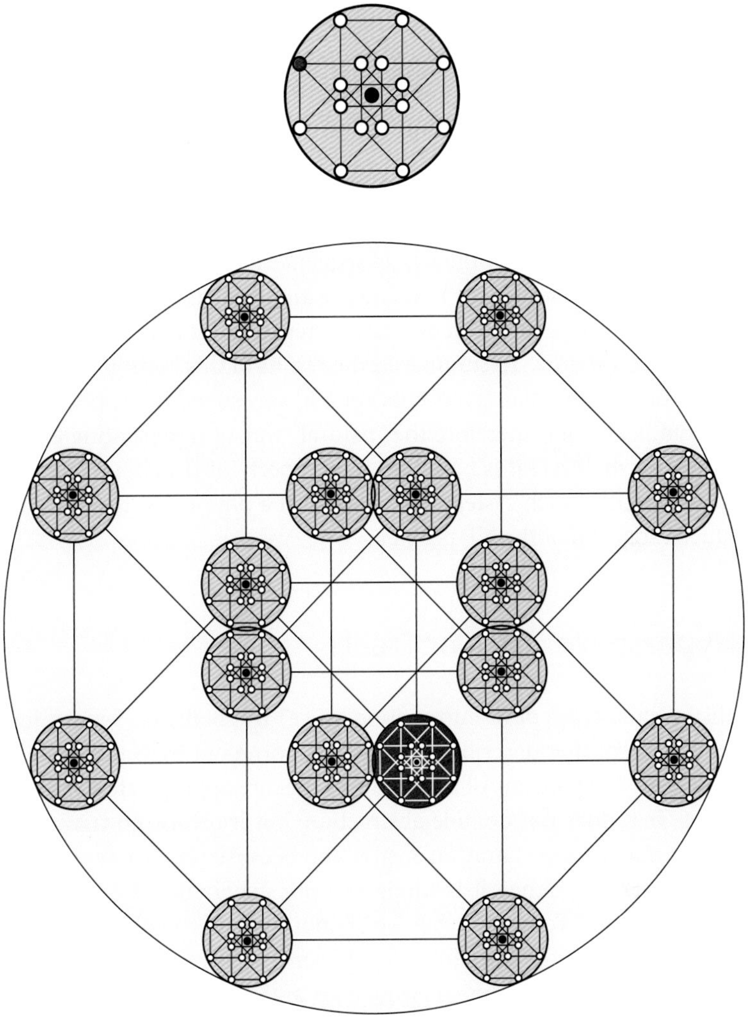

Figure 4.4.11

Schematic representation of the idea of "modular combination"

As an example, a binary sequence of four positions has been chosen because it can be illustrated by a four-dimensional hypercube. The general procedure is described in the text. In the example illustrated we start with a four-dimensional space centred by the black point. The red point represents the best-adapted mutant, which would become the centre in the next diagram (not shown). Instead, we go to the next-higher dimension, where another mutant establishes itself as the best-adapted one. In continuing this procedure one can scan a much larger space than by straightforward "modular combination" because one automatically screens the corresponding spaces of each selected sequence.

system remains a quasispecies during all steps of iteration. Another important point is that the iterative procedure is accompanied by an increase in fidelity of replication, in order not to violate the error threshold of the quasispecies (as will be explained in Section 5.5). Large populations greatly favour such hierarchical models.

The above example demonstrates the advantage of the concept of exponential space. Advantages gained in an early phase are conserved for later stages by keeping them omnipresent throughout the whole sequence space. The process is especially effective, since in a gene of length ν not all positions are of (equal) strategic importance, so that the complete hypercube can be replaced by hyperspheres with radii appreciably smaller than ν. To summarise the results of this section in one sentence, I would say: The "exponential space" concept is ideally suited to the problem of combinatorial complexity and therefore the "natural" way of representing evolutionary processes. As we shall see, it is based on the property of (inherently autocatalytic) self-reproduction. For such system it provides just a physical, rather than a general mathematical, solution of the NP problem.

4.5. Hyperspace: Trick or Treat?

How realistic is the concept of information space? By "realistic" I mean more than "suitable". A concept that describes phenomena borne out by Nature can be called realistic if it does so by means of assumptions that are appropriate to what is going on in reality. That does not exclude abstraction, but it means no trickery must be involved! Hyperspace is certainly an abstraction because we have no sensory organ to perceive it directly. Hence it is foreign to our experience. Can abstraction be adequate to reproduce correctly what we cannot directly follow up? That depends on whether we are able to start from a sufficient set of prerequisites. Mathematics could then tell us what the consequences are. So let us look in some more detail at the mathematical structure of information space (referred to for short as I-space).

In a hypercube of any dimensionality – as in the three-dimensional cube – all vertices lie at its surface. The distance between them has been introduced as the Hamming distance, which is the smallest sum of edges between any two vertices. It indeed fulfils the criteria of a true distance measure:

(1) The distance between identical points is zero.
(2) The distance between two points A and B is the same, regardless of whether it is measured from A to B or from B to A.
(3) The triangle inequality applies: the sum of two adjoining distances, say A to B and B to C, is at least as large as the direct distance from A to C.

Note that these criteria hold despite the fact that the metric differs from that of Euclidean space. Both cases of the third criterion, equality and inequality, can be easily verified by moving along the edges of a cube.

Nevertheless, visualisation of the point space should not distract us from the fact that we are dealing with the abstract concept of point spaces. Our lack of physical intuition is not to be blamed solely on our inability to envisage more than three dimensions. The four edges of a two-dimensional point space "surround" an empty plane, just as the 12 edges of a cube "enclose" an empty volume. Of course, we can aid our intuition by assigning a unit of finite extension to each point, which means replacing the point space by a "cellular space". The volume of a point space equals the total sum of all points that it includes (or the volume of all cells, if that helps). For a (binary) space of dimension ν the volume is then 2^{ν}. Table 4.5.1 describes what types of geometries are to be expected up to the general case of hypercubes of dimension ν and their hyperplanes of dimensions (ν–k). For example, the hyperface (k = 1) of a cube ($\nu = 3$) is its surface, consisting of six squares. The hyperplanes of next

Table 4.5.1 The structure of hypercubes

← Space →

← Subspace →

	Vertex	Edge	Square	3-cube	4-cube	5-cube	. . .	ν-cube
Point	1	2	4	8	16	32	. . .	2^{ν}
Edge	0	1	4	12	32	80	. . .	$\nu 2^{\nu-1}$
Square	0	0	1	6	24	80	. . .	$\binom{\nu}{2}2^{\nu-2}$
3-cube	0	0	0	1	8	40	. . .	$\binom{\nu}{3}2^{\nu-3}$
4-cube	0	0	0	0	1	10	. . .	$\binom{\nu}{4}2^{\nu-4}$
5-cube	0	0	0	0	0	1	. . .	$\binom{\nu}{5}2^{\nu-5}$
⋮								
k-cube	0	0	0	0	0	0	. . .	$\binom{\nu}{k}2^{\nu-k}$
⋮								
(ν-1)-cube	0	0	0	0	0	0	. . .	2ν
ν-cube	0	0	0	0	0	0	. . .	1

The horizontal rows indicate the numbers of subspaces that are contained in a given space denoted above each vertical column.

order down – the "one-dimensional surfaces" of a square – are its four edges, and the next are the two points at the end of each edge. Going up, a hypercube of dimension four has a "surface" (first hyperplane) consisting of eight cubes, as was shown in Chapter 1 (Figure 1.8.3). In Table 4.5.1 the columns refer to the (hyper-) space under consideration, while the rows tell us about their subspaces. For instance, a five-cube has as subspaces the following hyperplanes, of increasing order k: 10 hypercubes of dimension four ("surface"), 40 cubes, 80 squares, 80 edges and 32 vertices. In order to calculate these numbers one has to consider not only how many of the subspaces are involved, but also to what extent two different spaces are shared by the same subspace. (Each edge of a cube, for example, is shared by two planes, each vertex by three edges and hence also by three planes.)

We encounter similar abstract behaviour when we look at "spheres" in higher-dimensional hypercubes. At first, it looks awkward to talk of spheres when all vertices of the hypercube are supposed to belong to its surface. A sphere, in our imagination, is a space surrounded by a surface. How could a sphere fit into a point space in which all points lie at its surface? We should forget our prejudices and stick to the mathematical definition. Hamming introduced the idea into his theory of error-correcting codes by defining the surface of a sphere as being all points of a hypercube having equal distance from a reference point, which is thus analogous to the centre of the sphere. Any points with Hamming distances from the reference point that are smaller than the Hamming distance defining the surface of the sphere, together with the reference point, belong to the interior of the sphere, which accordingly may not be called an "empty" space.

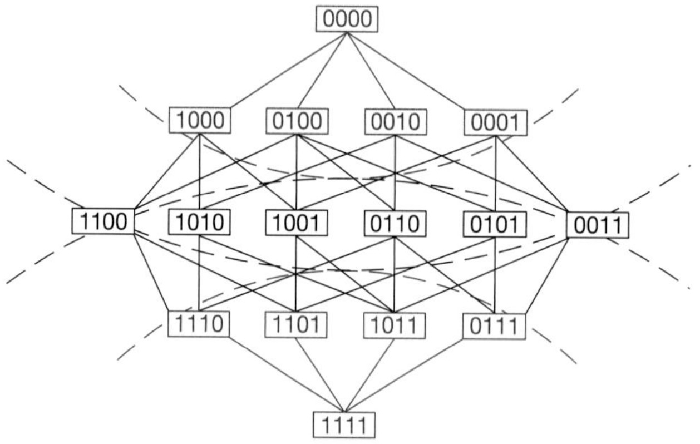

Figure 4.5.1
Spheres in four-dimensional hypercubes

The dotted lines indicate the surfaces of various spheres (one and two errors).

If we apply the above definition to points in the three-dimensional cube (see Figure 3.2.3), the resulting "spheres" do not really live up to their names. Figure 4.5.1 shows a corresponding diagram for dimension $\nu = 4$. Spheres of radius 1 do not overlap. In continuous Euclidean space two spheres can touch one another only at one point, called the "kissing point". Two touching spheres in I-space can "kiss" each other at several points, as do the two spheres with Hamming radii of 2 in Figure 4.5.1.

In Hamming's theory the spheres were supposed not to overlap because they were to isolate the message points from one another so that they can be individually checked in fulfilling certain parity relations with other points in the same sphere (see Section 3.4). While for error correction this condition requires non-overlapping spheres, evolutionary optimisation – as we shall see – requires the opposite of this: a maximum of overlap of spheres.

Anyway, the Hamming spheres fulfil their purpose only in spaces of higher dimension. Apart from the fact that we can hardly envisage such spaces, the pictorial analogy of a high-dimensional system that has clumped, and thereby gathered many points within a neighbourhood of small diameter, to form a sphere, is not a bad one. In the representation of six-dimensional hyperspace in Figure 4.3.1, the horizontal levels refer to states with equal Hamming distance from the two reference points 000000 or 111111. Hamming spheres of radius 1 have six points, those of radius 2 have 15 and those of radius 3 have 20 points on their surface. The corresponding volumes include seven, 22 or 42 points, respectively. Since such a diagram can be drawn for any reference point, a representation of this symmetry in three-dimensional space leads automatically to the analogy with a sphere, as was shown for the seven-dimensional case with 16 non-overlapping spheres of radius 1 in Figure 3.4.3. The spheres, having radii of 1 to 5 in (binary) hyperspace of dimension 20, are characterised by numbers, as shown in Figure 4.5.2.

Let us analyse this picture in more detail. The points in this "foamy" hyperspace, by symmetry, have to be arranged in concentric shells that are the surfaces of the corresponding Hamming spheres. We can extrapolate from Figure 3.4.3 that the number of points having the same Hamming distance "k" from a reference point in (binary) ν-dimensional space, which form the surface of the Hamming sphere, are given by $\binom{\nu}{k}$. The binomial coefficient $\binom{\nu}{k}$ for very large values of ν and $k \ll \nu$ can be approximated by the expression $\frac{\nu^k}{k!}$. The volume is the sum of binomial coefficients, where k varies from zero up to the value of k assigned to the surface under consideration. The approximation for the example presented in Figure 4.5.2 is not particularly good. Nevertheless, the true numbers of points in the five concentric shells with Hamming distances increasing from 0 to 5 are: 1, 20, 190, 1140, 4845 and 15504, yielding a total volume of 21,700 points. A sphere of radius 10 in a space of dimension 100 would have no fewer than nearly 30,000,000,000,000 (i.e. 30 million million, or 3×10^{13}) points at its surface, and these would constitute about 89% of

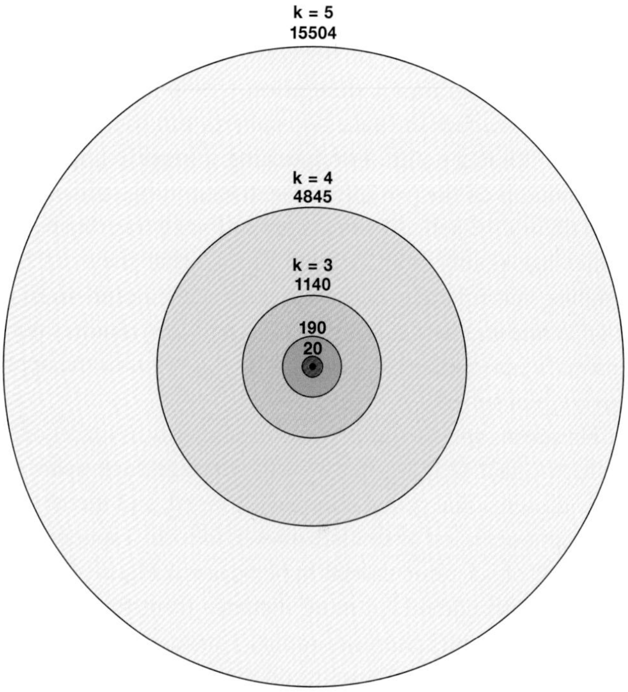

Figure 4.5.2

Concentric shells in 20-dimensional hyperspace

The radii chosen are 1, 2, 3, 4 and 5, starting with the deep red circle in the centre. The numbers attached to the circles give the respective numbers of points $\binom{\nu}{k}$ in the surface shell of the particular sphere.

its volume. If ν included 1000 positions, a sphere of radius 10 would comprise almost 3×10^{23} points, of which approximately 99% would be located in its surface shell.

The order of magnitude of 3×10^{23} reminds the chemist of Avogadro's number, which is just about twice this value. Our number is an estimate for a particular case and has nothing to do with Avogadro's number, except that it is of a quite realistic nature, if we think of molecular evolution. The dimension of 1000 corresponds to the length of a single gene, and a sphere radius of ten suggests a screening for mutants up to Hamming distances of 10, a fairly realistic situation in both Nature and biotechnology.

When Richard Hamming introduced his idea he had in mind a problem converse to the one I am exploring in this book. His problem was how to conserve information, e.g. in the transmission of a message, while I am asking how to change information and thereby how to generate variety, from which new information can be chosen by selection. In both cases we have to ask how to fill the space with spheres of radii that are smaller than – or even very small compared with – the largest Hamming distance in the space under consideration.

Sphere-packing in Euclidean space is a formidable mathematical problem, with many implications for coding theory and high-dimensional crystallography. These problems are described in an excellent book, *Sphere Packings, Lattices and Groups* by John Horton Conway and Neil James Alexander Sloane.[45] Did you ever try to fit solid spheres into a box? What is the most economical way of doing so? Two spheres can touch one another at only one point, which Conway and Sloane called the kissing point. The number of kissing points, or the "kissing number" plays an important role in the theory of sphere-packing. In two-dimensional space the hexagonal lattice solves the packing and kissing problem, but – as the authors write – "the miraculous enters" when those problems are studied in higher-dimensional spaces. Especially in eight and 24 dimensions, one encounters "unexpectedly good and very symmetrical packings that have a number of remarkable properties, not all of which are completely understood even today"[45].

If we turn back from Euclidean to discrete Hamming spaces, we find other similarly mysterious properties. The kissing numbers found here would make even Don Giovanni envious (*ma in Ispagna son già mille e tre*). The kissing number for each pair of neighbouring spheres in Figure 3.4.3 amounts to "only" 42 (i.e. seven contact points of one sphere with each of six nearest neighbours), but for a binary sequence of a length of 1000 and spheres having a "radius" of Hamming distance $d = 10$ it reaches vast numbers that ridicule Don Giovanni's modest *mille e tre*.

In information space sphere packing is greatly aided by large kissing numbers, although perfect fitting works only for very singular values of sphere radii and space dimensions. Radius-1 spheres are fairly easy to match certain space dimensions. The volume of the sphere times the number of spheres must equal the total space volume, or the ratio of space volume 2^{ν} and radius-1-sphere volume $\lambda = 1 + \nu$ must be a power of 2, i.e. 2^{μ}, where $\mu = \nu - \lambda$. This is true for ν equalling 3, 7, 15 …., i.e. for $2^{\lambda} - 1$, λ being an integer larger than 1. The example $\nu = 7$, yielding $\lambda = 3$, is shown in Figure 3.4.3. Spheres of radius 2 are more difficult to match because now the sphere volume $1 + \nu + \nu(\nu - 1)/2!$ must equal a power of 2. The lowest non-trivial case is $\nu = 90$, yielding $\lambda = 12$, i.e. $1 + 90 + 45 \times 89 = 4096 = 2^{12}$. (The space diagram in this case already is too complicated for graphical representation.)

Matching the total space with non-overlapping spheres requires centre-to-centre distances of $3, 5, 7, \ldots 2d + 1$ for spheres with radii corresponding to Hamming distances of $1, 2, 3, \ldots d$. However, perfect fit is not required, either for error detection and correction or for space screening in evolutionary optimisation. In the first case, all that is required is total isolation of all spheres from one another, which is fulfilled for sphere numbers $< 2^{\nu - \lambda}$. And for the second case we require overlap for evolutionary continuity, which also does not need to be complete. As has been mentioned before, the model becomes truly advantageous for large values of ν, with $\nu \gg d$. This, again, is true in both cases. The example shown in Figure 3.4.3 is therefore not really convincing. We have a sequence of seven positions, of which only four can be used for encoding the message, while three are used for checks that may show the presence of one error. That is not particularly economical. In a second example, we send 90 digits, 78 of which contain the message. The true advantage is seen only if ν reaches the order of magnitude of several powers of 10. A minor change in ν then allows each message point to be surrounded by a sphere that offers a sufficient number of parity relations to ensure that the message received is identical to the one originally sent.

In evolution, high dimensionality aids in a screening for advantageous mutants. Here an overlapping of spheres that are populated by mutant spectra is desirable. Any mutation may be deleterious, neutral or advantageous, the latter being the least abundant kind, its relative abundance depending on the level of adaptation. For binary sequences, each point mutation means a shift between the two subspaces that are the $(\nu-1)$-dimensional hyperplanes of the original ν-dimensional hyperspace. An advantageous mutation, once selected, is likely to shift the constellation among several of the other positions, allowing for a screening of the mutant space, as was indicated by Figure 4.4.8. Such a process, involving not only point mutations but also insertions and deletions of single digits and of clusters of them as well as recombinative reshuffling of segments, was one of the causes of the optimally combined genetic programmes which have formed the basis for the emergence of the wonders of life.

Information space, with its unlimited number of dimensions, hides a wealth of miraculous surprises. Remember the seemingly endless series of stories told by Scheherazade in her tales *A Thousand and One Nights*? The thousands of dimensions that characterise the information contained in each story allow for a wealth of possible clues.

Let me show, with some examples, how little our intuition based on experience can cope with problems occurring in connection with multidimensional spaces. In order not to overstress our imagination I start from continuous spaces of dimensions we are familiar with. The most symmetrical and therefore simple two-dimensional objects are the square and the circle, but their mutual relation is by no means trivial. One cannot square the circle! This was a problem that occupied generations of mathematicians until in 1882 Carl Louis Ferdinand von Lindemann proved the transcendence of π, while already in 1761 Johann Heinrich Lambert had proven π to be irrational.

π will introduce more mysteries at larger dimensions, as we shall see below. It is defined by the ratio of the circumference ($d\pi$) and the diameter (d) of the circle, which is the geometric locus of all points in a plane having the same distance from a given (central) reference point. This is a clear definition, but π itself, with its irrational uncertainty and its transcendental squaring claustrophobia, necessary to calculate precise values for the circumference and area of the circle, is less transparent.

Today we know the value of π to 206,158,430,000 decimal places (Yasumasa Kanada, Tokyo 1999), but this result may soon be outdone by another "Olympic" record. There is no straightforward relation as in the case of Euler's series for $e = \sum_{k=0}^{\infty} 1/k!$, the transcendence of which was proven by Charles Hermite 125 years after Euler's introduction of e. "Transcendent" is another word for "non-algebraic", which means that it cannot be represented by a solution of a polynomial equation involving rational numbers as coefficients. It was another relation, also found by Euler, in which e^{ix} is represented by $\cos(x) + i\sin(x)$ (i being the imaginary number $\sqrt{-1}$), that yields for $x = \pi$, $e^{i\pi} = -1$. It was this relation that also provided the key for a proof of the transcendence of π, as carried out by Lindemann, only nine years after the transcendence of e was proven by Hermite.

Let us now compare the circular with the square structures. A circle may touch a square at four points in two different ways, as is shown in Figure 4.5.3: the circle may be inscribed in the square or it may envelop the square. We ask what happens in both cases when the number of dimensions increases beyond any limit. The three-dimensional case is represented in Figures 4.5.4 and 4.5.5. The sphere inscribed into

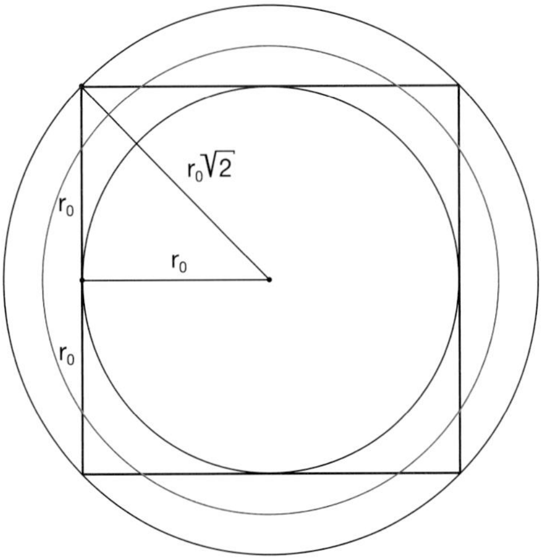

Figure 4.5.3
Enveloping and inscribed circles of the square

The edge of the square is $d_0 = 2r_0$, where r_0 is the radius of the inscribed circle. The radius of the enveloping circle is then $r_0\sqrt{2}$, i.e. $\sqrt{r_0^2 + r_0^2}$ according to Pythagoras' theorem. In n-dimensional Euclidean space, Pythagoras' theorem requires the summation over the squares of r_0 in all dimensions, i.e. $r_0\sqrt{n}$. The mysterious red circle in this diagram represents some average of both circles, which covers regions in- and outside the square. It will be of importance in connection with Hamming's paradox (discussed below).

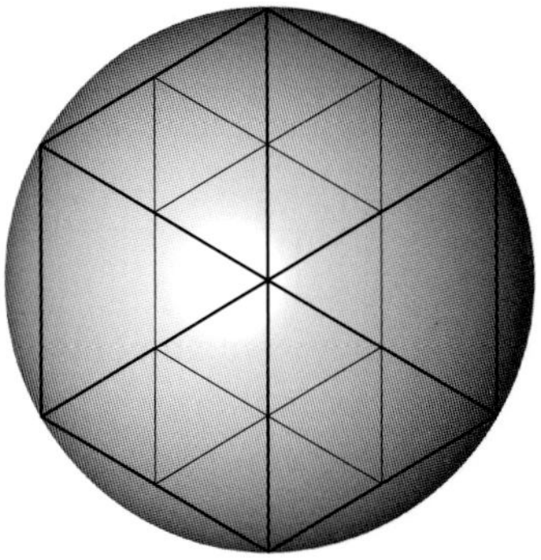

Figure 4.5.4
Enveloping sphere of a cube

See text.

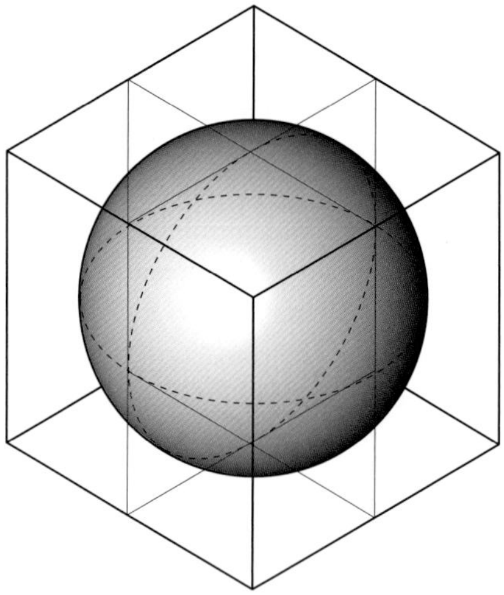

Figure 4.5.5
Inscribed sphere of a cube

See text.

Table 4.5.2 Cubes and spheres in n-dimensional Euclidean space

Volume of hypercube	$V_n = d^n$ (d, edge = distance between adjacent vertices; n = dimension)
Volume of hypersphere	Unit cube, d = 1; volume $V_n = 1$ for all values of n; Inscribed sphere, $V_n = C_n 2^{-n}$ (for $d_n = 1$)
Inscribed sphere	$d = 2r = 1; n = 2k$ $V_n = (\pi e)^k/(4k)^k\sqrt{2\pi k} = (\sqrt{\pi e/2n})^n \cdot (\pi n)^{-1/2} \to 0$ for $n \to \infty$
Enveloping sphere	Radius $r_n = \sqrt{n/2}$ $V_n = C_n, r_n^n = (e\pi/4)^{n/2}(\pi n)^{-1/2}$ Since $e\pi/4 = 2.135$, $V_n \to \infty$ for $n \to \infty$

The standard is the unit cube ($d = 1$). Relative to the standard, the volume of the inscribed sphere approaches zero, while the volume of the enveloping sphere tends to infinity for $n \to \infty$.

the cube touches the six walls, each at one kissing point, while a cube packed into the sphere has eight kissing points, the eight vertices of the cube. Let us assume that all edges have the length $d = 1$. Then, regardless of the dimensions, the volume will always be 1. We are dealing with unit squares (d^2), cubes (d^3) or hypercubes (d^k for $k \geq 4$). However, an analogous statement would not be true for spheres.

The area of a circle of radius r, with $2r = d = 1$ being its diameter, amounts to $\pi/4$. The volume of a sphere of diameter $d = 1$ is $\pi/6$. The general formula for a hypersphere of dimension k is $C_k \cdot 2^{-k} \cdot d^k$ (Table 4.5.2). Since d has been chosen to be 1, there is a steady decrease of the volume of the sphere inscribed in the hypercube,

Table 4.5.3 Stirling's approximation

$k! = (k/e)^k \sqrt{2\pi k}$

$n = 2k$	Stirling	n!	Accuracy
1	0.92214	1	0.92214
5	118.01916	120	0.98349
10	3,598,695.6	3,628,800	0.99170

Table 4.5.4 Volume of n-dimensional unit spheres $V_n(r) = C_n r^n$

n	C_n (expression)	C_n (numerical value)
1	2	2.00000...
2	π	3.14159...
3	$4\pi/3$	4.18879...
4	$\pi^2/2$	4.93480...
5	$8\pi^2/15$	5.26379...
6	$\pi^3/6$	5.16771...
7	$16\pi^3/105$	4.72477...
8	$\pi^4/4$	4.05871...
10	$\pi^5/20$	2.55016...
.	.	.
.	.	.
.	.	.
n = 2k	π^k/k	$\to 0$

Note that $C_n/C_{n-2} = 2\pi/n$. Radius $r = d/2$; d = diameter for unit sphere = 1.

the volume of which remains equal to 1 (unit cube). However, the hypersphere keeps touching the inner faces of the hypercube since its diameter remains constant (i.e. $d = 1$). What steadily shrinks is just its "volume", as is seen already in the transition from the circle to the sphere. The decrease from $\pi/4$ to $\pi/6$ is still quite moderate. The expression for the volume with $k = n/2$ contains for all even values of n the term $C_{2k} = \pi^k/k!$ (Table 4.5.3), which for larger values of k means an increasingly stronger diminution. The deeper reason for the complete vanishing of the sphere's volume (Table 4.5.4) – despite its constant radius of 1/2 – is the fact that the volume becomes more and more displaced into a surface layer, which gets thinner and thinner with increasing dimension until it approaches zero for $n \to \infty$.

The outer sphere, which envelops the constant volume of the hypercube, behaves in exactly the opposite way: its volume steadily increases relative to that of the unit cube. Knowing the expression for the volume of the envelope-hypersphere one can easily calculate that it increases relative to the hypercube, eventually approaching an infinite value for $n \to \infty$.

Richard Hamming, in his book *Coding and Information Theory*, explains the fact that the volume of convex multidimensional bodies is largely located in their surface. Specifically, this is true for the volume of the hypercube, which in the case of the unit cube remains constant at a value of 1. Take the difference between the volumes of two cubes, one of which is the unit cube with a volume $d^n = 1$ and the other is slightly smaller, with a volume $(d-\varepsilon)^n$ where ε is an arbitrarily small number, defining a thin surface layer of the hypercube. This difference, divided by the total volume, is the ratio of the volumes of the arbitrarily thin shell and the total sphere. Since $\varepsilon/d \ll 1$ we obtain for

$$V_{shell}/V_{sphere} \frac{d^n - (d-\varepsilon)^n}{d^n} = 1-(1-\varepsilon/d)^n = 1-e^{-n\varepsilon/d}$$

The exponential term disappears for arbitrarily large values of n, but maintaining the condition $n\varepsilon \gg d$, the shell can be arbitrarily thin, while the volume ratio of the shell and total sphere remains close to 1. The consequences of this result for the various spheres (relative to the unit cube) are contrasted in Table 4.5.2. The effect is a pure consequence of dimensionality, which in all expressions appears in an exponential way (cf. the difference with respect to polynomial growth in Table 3.3.1). We encountered this effect back in Chapter 2, in discussing statistical mechanics, where it was shown that nearly all of the energy of a system is localised in an extremely thin surface layer of 6N-dimensional phase space.

We are now well prepared to discuss an apparently quite mysterious example, which Richard Hamming described in his book and which he called "a paradox", in order to caution us about making intuitive guesses when treating problems in multidimensional space.

The problem is illustrated in Figure 4.5.6, where our model cases are two-dimensional squares and circles. An envelope square is subdivided into four areas of equal size, into which four unit circles touching one another are inscribed. The diameter of each circle is $d_0 = 2r_0 = 2$ (since $r_0 = 1$). The total box then has four edges of length $2d_0 = 4$ and hence an area of 4^2. The four inscribed circles fill more than three-quarters (exactly $\pi/4$) of the total space. They leave uncovered space not only in the four corners, but also at the centre between the four circles. Into this central space we fit another circle, which is tangential to the four larger unit circles. It has a radius of 0.414.

Now we start our procedure of progressively increasing the number of dimensions. In the first step we arrive at a cubic box having the volume 4^3 which houses eight unit spheres in its eight corners, and one sphere in the centre having now a radius of

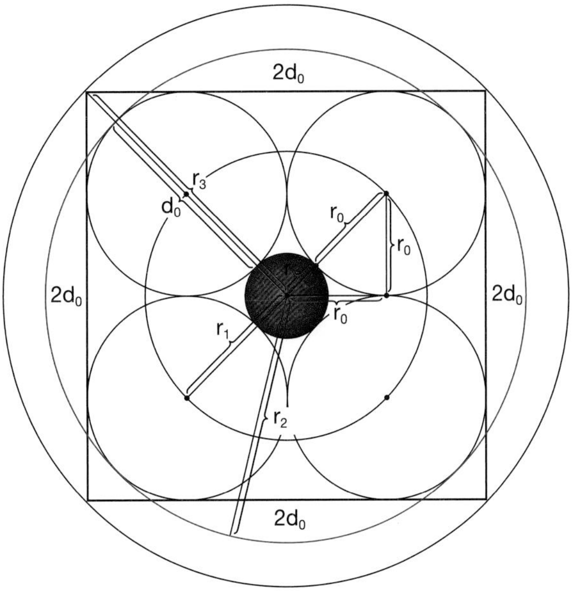

Figure 4.5.6
Hamming's paradox

The two- and three-dimensional cases are used for introducing the problem. A box (or square) with edges of $2d_0$ is compactly filled with corner spheres of radius $r_0 = d_0/2$: (a) four circles and (b) eight spheres. A sphere (red) is fitted into the free space at the centre. The radius of the inner sphere can be obtained using Pythagoras' theorem. The distance from the centre of the inner sphere to the centre of each corner sphere is $r_0 \cdot \sqrt{n}$, where n is the number of dimensions. This means that the radius of the inner sphere is $r_0(\sqrt{n}-1)$. For further explanation see the text. The comparison of (a) and (b) shows the increase in the size of the "inner sphere" with increasing dimension.

0.732 (see Figure 4.5.7). We note an increase in the latter value compared with the two-dimensional case, in accordance with our intuition.

Knowing the first two examples considered above, we may guess what will happen if we go on to a higher dimension (n). The hypercube engulfing all 2^n unit spheres will increase its volume to 4^n. The 2^n corner spheres of constant radii ($r_0 = 1$) will nevertheless shrink in volume according to C_n. Since the total volume blows up with increasing dimensionality, the unit spheres will recede more and more into the 2^n vertices of the hypercube. Correspondingly, the centre sphere will grow, and – hard to believe – at dimension 10 the inner sphere will already reach outside the total hypercube. The increase will go on, and the centre sphere will eventually reach an

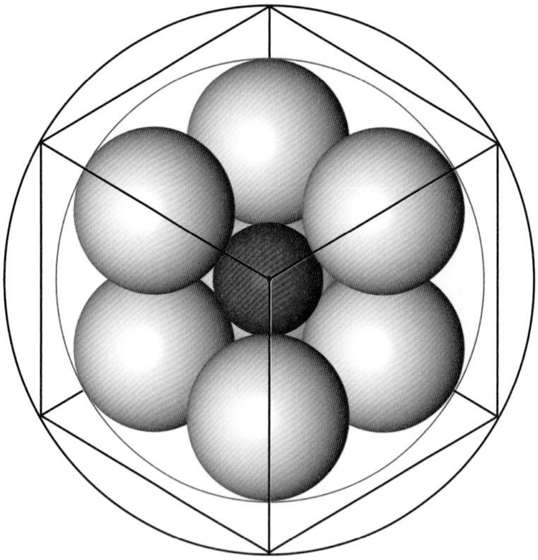

Figure 4.5.7
More on Hamming's paradox

A cubic box having the volume 4^3 which houses eight unit spheres in its eight corners, and one sphere in the centre having now a radius of 0.732.

Table 4.5.5 Hamming's paradox

Cube, $d = 4$; volume, 4^n
Inscribed corner spheres: volume C_n $(r = 1)$
Inner spheres: radius: $(\sqrt{n}-1)$ (see Figure 4.5.5)
Ratio of volumes: inner sphere/whole cube: $C_n(\sqrt{n}-1)^n/4^n$
Even n: $n = 2k$
Ratio: $\pi^k(\sqrt{2k})^{2k}(1-1/\sqrt{2k})^{2k}/4^{2k}k!$
Using Stirling approximation for k! and $\lim_{x\to\infty}(1-\frac{1}{x})^x = e^{-1}$
Ratio $= (\pi e/8)^k e^{-\sqrt{2k}}(2\pi k)^{-1/2}$
Since $\pi e/8 = 1.0675 (\pi e/8)^k$ increases more strongly than $e^{-\sqrt{2k}}$
$(2\pi k)^{-1/2}$ decreases for $k \to \infty$
The ratio approaches infinity for $k = n/2 \to \infty$

infinite volume – faster than the total hypercube which is supposed to engulf it. In other words, something which is meant to fit into a residual part of its host space eventually becomes infinitely larger than the host space itself.

This is the paradox described by Richard Hamming. He arrived at it by painstaking computations, which I summarise in Table 4.5.5. The conclusion in his plain and cool language was: "...the volume of the inner sphere becomes arbitrarily larger than the volume of the cube which contains all the 2^n unit spheres in its corners". No further comment, no explanation of the paradox! But I hear him saying: "I have done my part showing you the mathematical truth. This doesn't require any further explanation. It is now your task to find out what's wrong with your intuition."

So let's try it! The essential question we have to answer, I think, is: Why does the inner sphere grow faster towards an infinite value than its host cube, although it has to touch all corners filled with unit spheres from inside the host cube? I think we can give an intuitive answer after we have activated our intuition by the two examples discussed before. The inner sphere is not an inscribed sphere that has to touch the inner walls of the cube. It only has to touch the corners, leaving relatively little space for the inscribed unit spheres, but can easily exceed the cube at all other sites as the (mysterious) red cycle in Figure 4.5.3 does, which at lower dimensions cannot yet touch the corner spheres from inside the central hole. On the other hand, it cannot completely reach the envelope sphere because it must leave some space in the corners for the inscribed spheres, which at large dimensions do not require much of the available space. We know that the envelope sphere approaches infinity because apart from the vertices, which it only has to touch, its volume lies in a thin surface shell mostly outside the cube. The inner sphere may come close to it, but it must be somewhat smaller. How much depends on the relative size of its regions outside and inside the cube. The growth of the envelope sphere relative to the cube is given by a term $(\pi e/4)^k$, where $\pi e/4$ amounts to 2.135 and 2k equals n. For the inner sphere, the term which corresponds to $\pi e/4 = 2.135$ for the outer sphere must, in order to yield unlimited growth, also be greater than 1. This can only be shown by computation. It turns out that it happens to be just slightly larger than 1, i.e. $\pi e/8 = 1.0675$. So the conclusion is "paradox resolved"! Another lesson might be: "Intuition is good, mathematics is better!" Yes, but without intuition, there would be no mathematics.

What we have really learned from the examples above is how careful one has to be in extrapolating from daily experience to spaces of higher dimensions. The use of physical space has undergone some evolution from Newton's absolute frame as the "stage" on which all physical processes are performed to Einstein's space where it has an active role as a player, or to the multidimensional space of strings on which elementary properties of matter are based. In the following sections we shall need another form of high-dimensional continuous space, the n-dimensional simplex S_n, with its (bent) non-Euclidean geometry, representing population dynamics in information space, and last but not least the discrete non-Euclidean information space itself.

Discrete I-space differs in several respects from continuous physical space. The dichotomy of square and circle, expressed in the transcendence of π, does not occur. Volumes for cubes and spheres – the number of vertices in point space – are expressible by sums of binomials. But, again, the volume of a sphere is given mainly by its surface layer. The sum of binomials $\binom{\nu}{k}$ for $k < \nu/2$ is always dominated by its largest terms. For uneven values of ν the total binary space, involving 2^ν vertices, can be represented by two non-overlapping spheres. (For even values of ν the situation is a bit messier, but not different in principle.) The two spheres, which can be drawn for any sequences, are centred at two "antipodal" points representing sequences of binary digits, showing some kind of "supersymmetry". Every position in one sphere has an "antipodal" sequence in the other sphere at corresponding positions relative to two mirror axes. (By "antipodal" I mean a sequence in which all binary digits at given positions are exchanged: $0 \leftrightarrow 1$.)

Dual symmetries – for instance positive and negative electric charges, or the C-symmetry of matter and antimatter, or (the not yet identified) supersymmetry – are inherent to particle physics. Supersymmetry proposes a symmetry between bosons and fermions. Bosons (such as photons, mesons or nuclei consisting of an even number of nucleons) have integral spins resulting from symmetrical (quantum-mechanical) wave functions. Fermions (such as electrons or other leptons, quarks or their odd-numbered combinations, protons, neutrons etc.), on the other hand, possess half-numbered (odd multiples of 1/2) spins corresponding to antisymmetric solutions of their wave equations. Supersymmetry must be broken because so far no supersymmetric partners of known bosons and fermions have been identified. Supersymmetry in sequence space is also broken because of selective fixation of the sequences that carry genetic information.

I know that this analogy is of an almost allegorical nature, and may be too far-fetched to provide a basis for any physical model. If I failed to mention this I would be laying myself open to somebody as sarcastic as the late Wolfgang Pauli, who once responded to a model proposed by Heisenberg by drawing an empty frame on the blackboard and saying, "This is a beautiful Rembrandt – it's just the details that are still missing."

Now let us return to the question: "information space – trick or treat?". I-space is certainly a construct of our mind. If it existed in reality, one should be able to walk around in it, and there should be forces or causes to make this possible. So, isn't it after all just a nice metaphor? Well, as a construct of the mind it may be that, but "not quite and therefore not at all" (see footnote on p. 341). Why not?

First of all, information is older than we are. Information lies at the origin of life, and that event preceded us by nearly 4000 million years. The concept had to be implemented in reality before it could arise in our minds. Information is an invention of Nature. It is organised in a reproducible, mathematical way and has existed in reality from the time of its origin.

Secondly, it will be shown in Section 4.7 how moving through information space can be accounted for by very specific mechanisms that are of a chemical (and therefore also of a physical) nature. We shall see that this demands very specific chemical properties. Being of a chemical nature, these mechanisms are explained by quantum mechanics, but their property of being instrumental for life depends solely on their functional efficiency, which is only indirectly a consequence of their particular chemical structure, and hence of their quantum-mechanical basis.

Thirdly, there are "potentials" and "gradients" active in this space (see Section 4.9) which finally lead to a physical definition of what we might call the "quality", or "meaning", of information. Hence there is a causal relationship between the existence of information and the space that is spanned by it. This relation is not similar to the one between matter and physical space, but it establishes an analogy by its mere existence.

Hence I would not consider information space to be "just an allegory". Its mathematical structure is related to its physical reality. Hyperspace, in this respect, is more a "treat" than a "trick". As the tales of *A Thousand and One Nights* took shape only under the force of survival that caused Scheherazade to conceive them, the wonders of life only came about under forces that guaranteed "survival" in this space.

4.6. How Does Matter Move in Physical Space?

According to classical mechanics, a material body remains in its state of motion with constant velocity relative to any point of reference, as long as no forces act upon it, while the laws of dynamics describe how physical forces cause changes in the body's momentum. Those problems were discussed in Chapter 1. In daily life we move around by, for example, walking, riding in a car, taking a train, cruising the oceans, or flying aboard a plane. Obviously, by asking the above question I have in mind problems other than planning vacations or business trips. To move around is an inherent property of matter. We call it "thermal motion" and, according to quantum mechanics, it never ceases even if we approach arbitrarily closely to the absolute zero of temperature. The present section reviews our knowledge about such a spontaneous motility of matter in physical space. It is meant to lay the ground for the subsequent section, in which I introduce a new and abstract kind of motion that we encounter if we want to go "from here to there" in information space.

When I was a child, one day I happened to come across a greasy spot that had landed upon an absorbing surface. Playfully I took a pencil and encircled it. When I came back the next day the spot had expanded to about twice its former size. Now my curiosity was awakened. I encircled the extended spot again, but was disappointed that the growth seemed to slow down on the following days. I thought that the supply might perhaps have been exhausted. However, this did not prove true because within four days the diameter had doubled once again. Of course, at that time I had no idea of what was going on, until as a young student I heard a lecture in physical chemistry in which the laws of diffusion were explained; according to these it is not the distance, but rather its square that changes in proportion to time. The German mathematician and physiologist Adolf Fick formulated this law in exact mathematical terms around the middle of the 19th century.[46]

Fick's first equation correlates the diffusional flux density (the number of particles passing a unit area per unit time) with the gradient of concentration (which is the change in the number of particles per unit volume with spatial distances, given in differential terms, defining a "diffusion coefficient" (D) as the proportionality factor of this correlation. According to its definition, D has the dimensions of length squared over time. Fick's second equation then expresses the temporal change of concentration, i.e. the number of particles in a unit volume element by the divergence of the diffusional flow. The divergence of a vectorial quantity expresses the differential change of the flux vector along each spatial co-ordinate. In the case of diffusion it describes for a volume element the difference between in- and efflux along all space co-ordinates.

In simple physical terms, Fick's laws state: a macroscopic diffusional flux takes place between regions of higher and lower concentration. Continuity requires the temporal change of concentration in each volume element to be related to a finite difference between diffusional in- and efflux for that volume element. This is true only in the absence of any (non-vectorial) production or destruction of the substance under consideration, e.g. by chemical reaction.

Macroscopic diffusion flux requires (macroscopic) gradients of concentration. This does not mean that on a microscopic level there is no diffusional motion in the absence of a concentration gradient. A concentration gradient is, by definition, a macroscopic quantity. A single particle is not able to "sense" it. Zero macroscopic flux means a compensation of particle flows in all positive and negative directions. Single particles move according to a distance squared/time relation, independently of macroscopic gradients. In this case one speaks of "Brownian motion".

In a pamphlet appearing in 1828 under the title *A Brief Account of Microscopical Observation* the Scottish botanist Robert Brown[47] described a trembling motion of minute particles from living pollen grains of the primrose *Clarkia pulchella*. He first thought these motions to be characteristic of living material, but soon realised the same phenomena to be observable with dead pollen and also with powdered inorganic substances of comparable size. The insight that "Brownian motion" and diffusion down concentration gradients have a common origin, i.e. are caused by random walk of individual molecular particles, was first substantiated only half a century later by the Danish astronomer, actuary and mathematician Thorvald Nicolai Thiele.

I am going to discuss these phenomena, which even nowadays offer unexpected clues, from a present-day point of view, although without ignoring their historical context. Three aspects are of particular relevance with respect to problems discussed in this book:

(1) Diffusion as a random walk of individual particles.
(2) The self-similar nature and fractal dimension of diffusion.
(3) The "space-filling" nature of diffusion.

Let us look at these in turn.

Random walk

Diffusion is a consequence of the thermal motion of individual particles. The quantification of this insight goes back to Albert Einstein,[48] who studied the subject and gave his doctoral thesis the longish title: "On the Motion Required by the Molecular Kinetic Theory of Heat of Small Particles Suspended in Liquids in a State of Rest". His subsequently (1905) published paper[49] contained – as Abraham Pais calls it[50] – no less than a proof of the "reality of molecules", a view at that time strongly opposed by Einstein's prominent seniors. Among these were Ernst Mach and Friedrich Wilhelm Ostwald, who considered molecules and atoms to be mere mathematical fictions without any physical reality.

By the time Einstein wrote his paper, it was experimentally well established that the trembling frequency of the colloidal particles increases with decreasing size. It was in particular this effect for which Einstein's theory had a quantitative explanation. It could be extrapolated to the size of single molecules yielding another access to the value of Avogadro's (or Loschmidt's) constant, i.e. to the "reality of molecules". It was the French physicist Jean-Baptiste Perrin[51] whose carefully designed experiments brought Einstein's ideas on Brownian motion to fruition. It won him the 1926 Nobel prize in physics, which could well have been shared by Einstein.

Two other aspects of Einstein's work on Brownian motion – especially in the context of this book – are of enduring relevance. In Chapter 3 I asked whether we live in a "Markovian world". Einstein established that in a random walk every step is independent of the walk's history up to the time interval immediately preceding the step, thus ensuring the symmetry of probability with respect to the sign of displacement. He showed that this is independent of his original assumptions, which started from the ideal gas law or from Stokes' viscosity relations. Today we call this behaviour "Markovian". Andrei Andreievich Markov (cf. Section 3.5) had arrived at this view independently of Einstein. But his book, in which he formulated the general mathematical relations for sequences of "stochastic" events independent of particular physical realisations, appeared one year after Einstein's seminal paper.

The other enduring aspect of Einstein's paper concerns the distance squared / time relation, which he showed to be inherent to the mechanism of diffusion in a space of any dimension, regardless of whether or not macroscopic gradients of concentration are present. Consider first an ideal gas (as Einstein indeed did when he started from Boyle's ideal gas law). Such a gas is characterised by the absence (or neglect) of any internal interactions among the gas molecules. One considers only the interactions of the molecules with the walls of the vessel that encloses them. Such a state will always be reached at a sufficient degree of dilution. The molecules themselves, in their free motion, attain velocities at room temperature of some 100–1000 m s^{-1} (the latter being reached by the "lightest" molecules, such as hydrogen).

In a condensed phase, the situation is just the other way around: interactions among the molecules dominate over interactions with the periphery. The molecules inside a water droplet do not "care" what happens at their surface. They also move with correspondingly high velocities, but in doing so they hardly advance more than a fraction of 1 Å before they bump into a neighbouring molecule. Hence, although they move fast, they fail to make much progress. Einstein looked first at a space of dimension 1, i.e. a molecule moving along a line. One sees immediately that a particle cannot advance in one direction linearly with time (as freely moving particles in an ideal gas would do for a while). Since the probability of moving in the forward and backward direction – a linear space offers only these two options – is one half, the particle advances about equally well in both directions, and hence its distance from the starting (or any reference) point increases more slowly than time progresses. Einstein showed that on average it is the square of the distance that increases with time. The relation for linear diffusion can be tested in a long, thin capillary; this has three dimensions, but one of these extends much further than the two others.

Einstein's picture applies to Euclidean spaces of any dimension. The diagram in Figure 4.6.1 represents a diffusion process in two dimensions and this or a similar form can be found in various textbooks of physics. The nub of this figure is the fact that the number of line segments (36) is the square of the distance between the initial and final points (6), if measured in units that correspond to the average length of all single segments. Nothing is wrong with such a diagram, which indeed may represent a true diffusional path in two dimensions, but still something is misleading!

Here I do not mean the fact that the process could, on average, have terminated at any point of the red (or any black) circle I have drawn around the origin in Figure 4.6.1. What I mean is that the spatial distance between the initial and final points is *not* typically 6 if the true length of all distance segments is not identical with the average of all distance lengths. The mean square distance in Einstein's formula for the diffusion coefficient is the average of all squares, rather than the square of an average.

The reader may remember the story I told in Appendix A3 about a little mathematical genius. It was Carl Friedrich Gauss when he was a boy at elementary school. The teacher, tired of teaching and hoping for a quiet nap, gave his pupils the task of

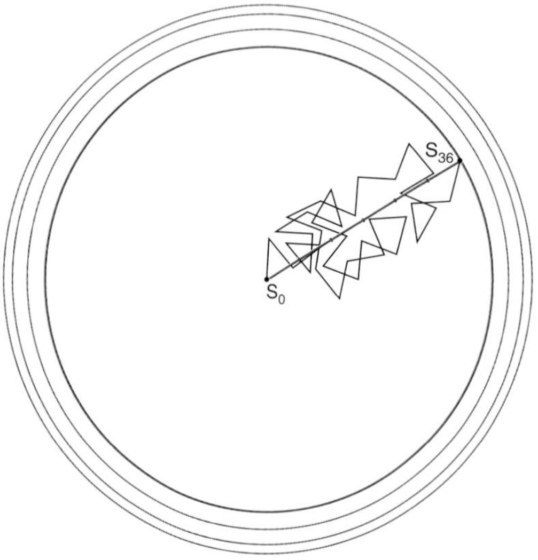

Figure 4.6.1

Schematic graphic representation of one of the possible diffusion paths in two dimensions, with the observations at 36 equidistant intervals

The total distance S_0–S_{36} (red line) in this diagram has a length exactly six times that of any individual distance segment. This, on average, is not precisely correct (see text). The three black circles in the figure refer to a Gaussian distribution.

adding up all numbers from 1 to 100. After a few minutes, little Carl had the result: 5050, and he was able to explain to the astonished teacher that all one has to do is to add 1 to 100, 2 to 99 and so on up to 50 + 51. That is simply $50 \times 101 = 5050$, or for any number n : $\frac{n(n+1)}{2}$

Now let us get the sum for all n squared: that is certainly more involved. Mathematicians have solved the problem for any sum of powers of natural numbers, and I have treated the problem from a different perspective in Appendix A3. For the special case of the power of 2, the sum can be obtained fairly easily by a trick similar to the one Gauss used (see Figure 4.6.2, reproduced from Appendix A3). It yields, for n equal to 100, $\sum_{k=1}^{100} k^2 = 338350$. The average of all squares is one hundredth of this number, i.e. 3383.50. It differs from the square of the average of all numbers 1 to 100, i.e. $50.5^2 = 2550.25$, by 833.25, which is an appreciable fraction of either of the two averages. The square root of this number is called the "variance" or "second moment" of the average. In our example it is nearly 29, which is rather more than half of 50.5, the average of the numbers 1 to 100.

Let's come back to diffusion in a plane. If one were to look at a larger number of diffusional paths all starting at the origin, and watch them for a given time, one would find some Gaussian distribution of end points, the maximum being a circle around the starting point the radius of which, however, is not usually identical with the square root of the length of all segments, as shown in Figure 4.6.1. If all (equidistant) time segments were to yield equal distance segments $r_i = \sqrt{x_i^2 + y_i^2}$, then (and only then) the average of squares would equal the square of averages and all end points would lie on the red circle. This, however, is usually not the case. Then $<r_i^2>$ is larger than $<r_i>^2$, not as much as in the numerical example given above, but rather more, as indicated by the red circle in Figure 4.6.1.

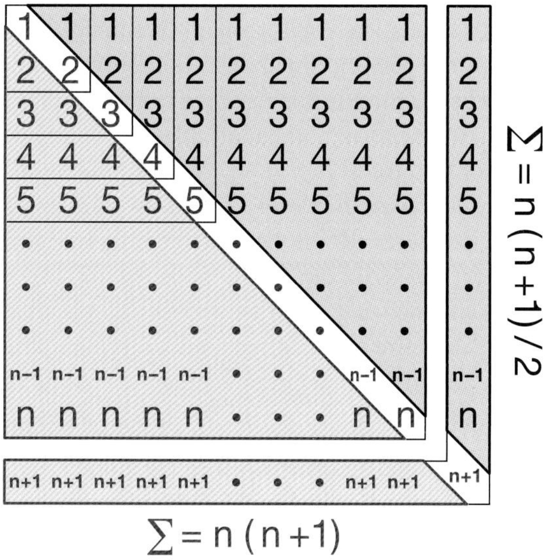

Figure 4.6.2

Reproduction of a table from Appendix A3 that derives the sum of squares by a two-dimensional geometric representation by applying tricks similar to those used by Carl Friedrich Gauss in his derivation of the expression for the sum of consecutive numbers. (The sum of squares is given by the sum of all the red numbers, i.e. the lower triangle plus the diagonal. Try to prove it yourself or look at Appendix A3.)

Why do I stress this point? Does it really matter? Well, it is not the (relatively small) numerical difference that matters – there is a more fundamental reason.

We are in Euclidean space with its Pythagorean metric. In a plane, r^2 is the sum of the squares of the co-ordinates, i.e. $r^2 = x^2 + y^2$. We see immediately that $<r^2>$ can be averaged to include all contributions from different co-ordinates, and this holds for any n-dimensional space where $r^2 = \sum_{i=1}^{n} x_i^2$, x_i being assigned to its co-ordinate i. One consequence is that Einstein's general expression $D_n = \frac{<r^2>}{2n\,\tau_r}$, where n is the number of dimensions and τ_r is the time required to walk over a distance defined by its mean square $<r_i^2>$, holds for an arbitrarily large number n of dimensions.

The fractal dimension of diffusion

We learned in the first part of this section that diffusion is characterised by a distance squared / time relationship, regardless of the dimensions of the space in which the process takes place. Diffusion has, so to speak, its own dimension. That's true – only we have a new word for it: we call it "fractal dimension". The great pioneer of the study of fractal behaviour and its impact on physics is Benoit Mandelbrot[52]. It was he who released it from the mathematical literature, where it had blossomed unnoticed since the work of Georg Cantor[53].

The fractal dimension characterises the scaling behaviour of self-similar curves. In Figure 4.6.3 I have again "constructed" a random-walk path (see also an artistic reflection in Figure 4.6.4). As we have learned from Einstein, the segments of such a path can be chosen arbitrarily. The points that are connected by straight lines and which make up the segments of the path simply refer to arbitrarily chosen moments of observation. If we look at any such segments of the upper path with higher time resolution, i.e. if we look – say 100 times – more often at a given segment we obtain the lower path in Figure 4.6.3, which is not identical with, but is of the same nature as, the upper path. We call such behaviour "statistically self-similar". Of course, there are always physical limitations that define a largest and smallest scale of observation. For diffusion this includes many orders of magnitude, for instance from observing diffusion in capillaries having the length of 1 m down to the molecular scale extending over less than a thousand millionth of a metre.

It was the German mathematician Felix Hausdorff[54] who – about 80 years ago – introduced a dimension that characterises such self-similar or fractal behaviour. Paths such as those in Figure 4.6.3, even if they are drawn with infinitely thin lines, cannot really be depicted in one dimension. They may eventually cover a whole area, just as landscapes on the surface of earth require three dimensions or topographic maps for their realistic representation. Mandelbrot referred to the west coast of Britain and showed that over the years its recorded length had steadily increased because of the increase in the resolution of cartography. If a simple curve, such as a straight line or a circle, is subdivided into smaller and smaller segments the product of the number of segments and their individual length remains constant. In

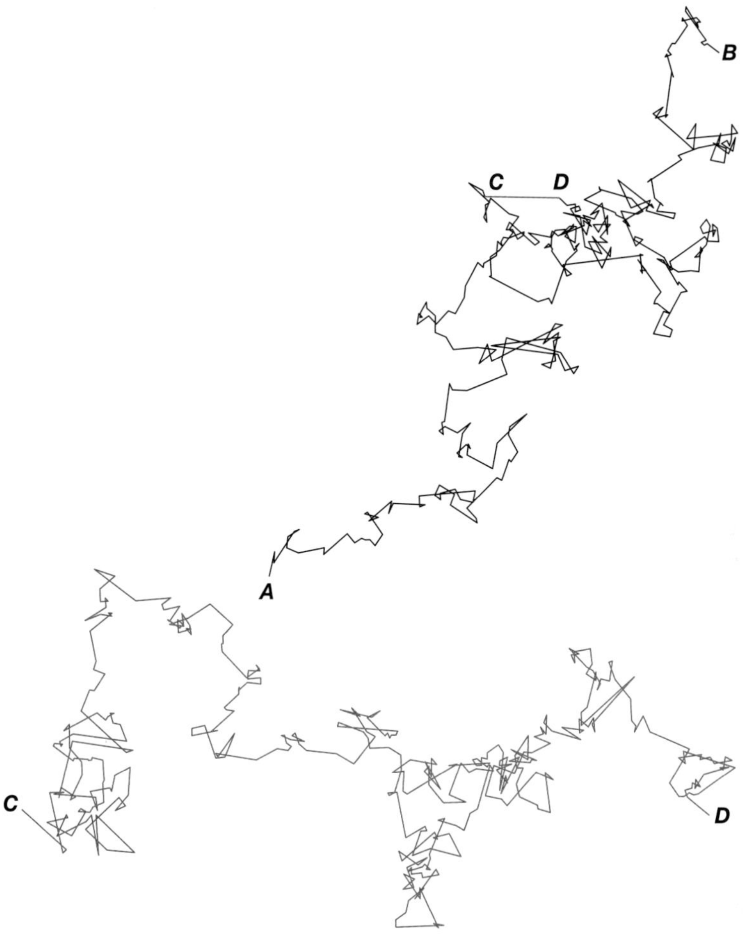

Figure 4.6.3
The self-similar nature of the diffusion process in two dimensions

The interval **C D** of the upper (black) trace is magnified in the lower (red) diagram. It is not the same as, but is in its nature similar to, the upper (black) diagram (see text).

fact, this was the way in which the number π was empirically determined, namely, by approximating a circle from above and below by polygons with increasing numbers of segments and extrapolating to the (coinciding) limiting values of both approximations. Such a method would not work for fractal curves. The smaller the segments, and the larger their number, the greater is their product. It was Hausdorff who realised that fractal behaviour may be describable by power laws. What remains constant is not the simple product (Nr) of the number (N) and length (r) of the segments; instead there is a limiting power >1 of r, i.e. r^{D_H}, for which

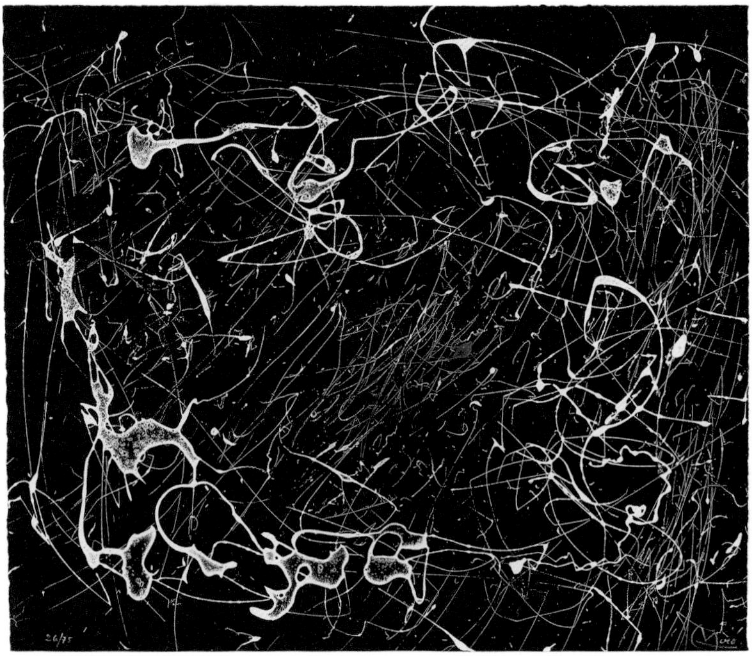

Figure 4.6.4
"Fissures" (lithograph by Joan Miro).
Author's collection.

Despite its title, I see this as a wonderful representation of a diffusion diagram such as the one shown in Figures 4.6.1 and 4.6.3, even though the random walk took place within the mind of the artist.

the product $N(r) \times r^{D_H}$ remains constant. D_H is called the Hausdorff dimension or fractal dimension, and it can be obtained if r tends to zero, i.e. as $D_H = \lim_{r \to 0} \frac{\ln N_{(r)}}{\ln 1/r}$, for diffusion – because of the distance squared / time relation – N(r) increases with increasing resolution (1/r), proportionally to $1/r^2$. Hence, the Hausdorff dimension of diffusion is 2. (A more exhaustive discussion on fractals and power laws can be found in an excellent monograph by Manfred Schroeder[55].) A consequence of the fractal dimension of 2 is that diffusion can be completely "space-filling" only in two dimensions and that only in such spaces is the probability of recurrence (i.e. revisiting the same "point") equal to 1 (cf. below).

The other headaches left to the physicists were caused by the fact that all the "curves" presented in Figures 4.6.1 and 4.6.3 (see also Miro's picture in Figure 4.6.4) are non-differentiable. Mathematicians since Karl Theodor Weierstrass have known how to deal with such problems, but to straighten them out for diffusion in space of any dimensionality was left to Norbert Wiener[56]. Theoretical physicists nowadays refer to these as Wiener processes. The strange mathematical concept of a non-differentiable "curve" thus manifests itself in the physics of Brownian motion. How far it remains an abstraction in physics as well could only be found out from a more detailed analysis of the physical process, extending down to the quantum mechanics of collisional processes. Moreover, diffusion in solids in physical detail is anything but a continuous process, requiring as it does the exchange of particles at lattice sites effected through discontinuous hopping into interstitial positions.

The importance of Wiener's work lies in a general clarification of the role of the space dimension in problems of space-filling and recurrence. In a paper in 1920 the American mathematician George Polya[57] showed that, because of the distance squared/time relation, a qualitatively different behaviour of diffusion is to be expected for dimensions larger than two. Norbert Wiener clarified this question in a most general way.

Norbert Wiener was the archetype of the absent-minded professor. The main building of the Massachusetts Institute of Technology, where he had his office, was a true labyrinth. In the early 1960s, when I gave a course of lectures there, I was amazed by the brightness of my students, but a colleague explained their remarkable ability by saying: "The others wouldn't have found your lecture room." Norbert used to walk through this labyrinth by a true Wiener process, thinking about a problem, in his left hand a manuscript into which he looked now and then, and his right hand keeping contact with the walls in order not to lose his way. He thus arrived at the door of my host, the physical chemist George Scatchard, who was just about to leave his room. George asked Wiener: "Norbert, what are you doing", but Norbert was so occupied with his problem that he immediately started talking about it to Scatchard. After he had gone through it he asked suddenly: "George, can you tell me which way I came?" George said: "No, but why don't you tell me where you want to go." Norbert said: "No, that's not the problem. If I was going in that direction I have already had lunch, but if I was going the other way, then I was going for lunch."

I cannot guarantee the truth of this story because it was told to me; but *se non è vero, è ben trovato*. In the second story I myself was involved, so I can guarantee its authenticity. Going to my lecture I used one of the elevators, where I met Norbert. He immediately started to talk to me about his new ideas, and for the next ten minutes we went up and down in the elevator, randomly stopping at floors where students came in and went out; again a true (one-dimensional) "Wiener process". After a while I got nervous and said: "Norbert, it is now five past two. At two I had to be at a lecture." Norbert (who was a generation older than me) looked at me and said somewhat indignantly: "What I am telling you here you will not hear in any lecture." I insisted: "Yes, Norbert, but I am the lecturer." He suddenly seemed to wake up: "What did you say, five past two? I have to give a lecture as well." and he left the elevator, which had just stopped, at a run.

The space-filling nature of diffusion

"How to find a target by random walk." The problem involves "space filling" and "recurrence" in relation to "dimensionality" (Figure 4.6.5). The early pioneer

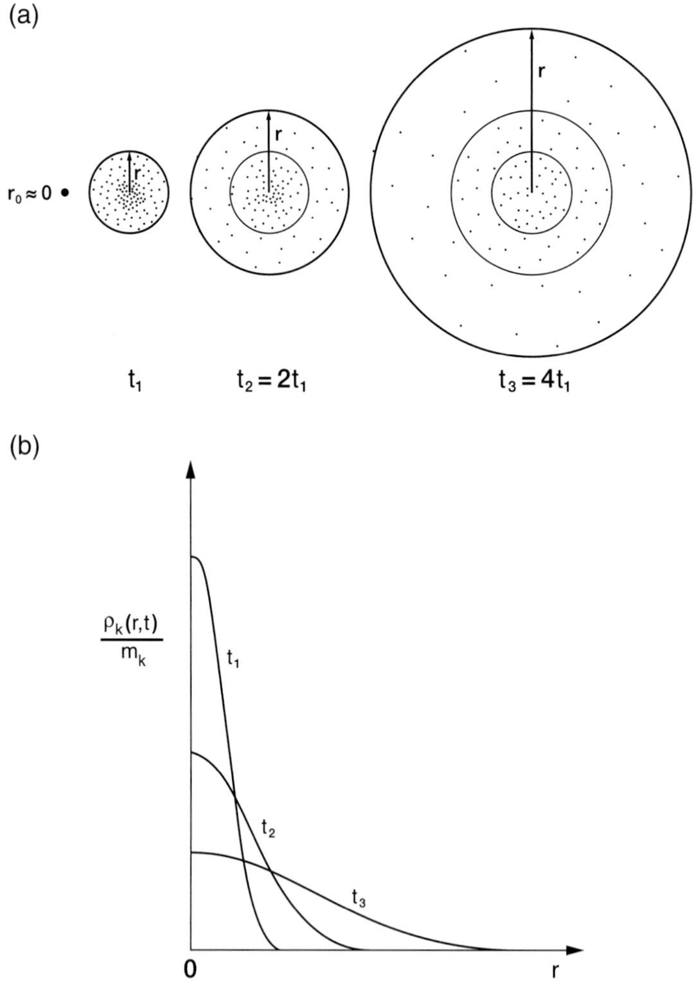

Figure 4.6.5
The "space-filling" nature of diffusion

(a) The diffusing particles could migrate in any direction in space, steadily expanding the territory they occupy (see text). (b) Calculated particle density as a function of time (t_1, t_2, t_3). The size of the space covered is reflected in the radius r.

was a Polish–Austrian physicist whose full name was Marian Ritter von Smolan-Smoluchowski. Almost simultaneously with Einstein, he started to think about Brownian motion, focusing more on practical problems in physical chemistry such as the diffusion-controlled aggregation of colloidal particles in solution. His first paper on this subject appeared in 1906,[58] and he fully acknowledged Einstein's seminal ideas, although most of his work was done independently before the appearance of Einstein's paper. When he died in 1917 Einstein praised von Smoluchowski in an obituary as an ingenious man of research and a noble and subtle human being (A. Pais 1982).[59]

How are we to hit a target if it is a particle of molecular, atomic or even more "elementary" size? The problem is as acute today as it was when von Smoluchowski raised the question of chemical aggregation, or when Rutherford started to probe the size of what then was thought to be the most "elementary" matter (cf. Section 1.5). In particle-scattering it is the "cross-section", i.e. the projected area of a three-dimensional object, which is the target of impact of accelerated particles, whatever final states are the product of such interactions. The cross-section here is defined as the ratio of scattering rate and incident flow (Figure 4.6.6) and measured in "barns", taken from the English expression "as broad as a barn door". One "barn (door)" in physics, written as 1 b, is defined as the minutely broad area of 10^{-28} m^2, typical of atomic nuclei.

Particle physicists nowadays use centre of mass energies of giga-electron volts (cf. Section 1.5). The large energies in this case are required in order to yield de Broglie wave lengths (cf. Section 1.3) that are short enough to resolve the "sizes" of elementary particles. The linear dimension in the latter case is of the order of magnitude of 10^{-20} m, the present limit of resolution for detecting electrons or quarks.

The cross-sections of atomic shells, as estimated from the Bohr radius of hydrogen, correspond to quite a huge "barn door". It is congruous with the range of cross-sections determined by atomic and molecular beams and as such representative for the chemical kinetics in dilute gases. However, the data above already show that the definition of the cross-section, as the ratio of scattering rate and incident flow, is not necessarily identical with the projected area of the three-dimensional target particle, but also reflects the efficiency of the interactions involved.

In the chemical kinetics of ideal or near-ideal gases we encounter quite similar problems. The frequencies of collisions among atoms or molecules when conditions allow sufficiently long mean free paths yield cross-sections in the (sub-)squareÅngstrøm range. As an example, consider the formation of a hydrogen molecule (H_2) from two hydrogen atoms (H + H). This is one of the fastest reactions in gas kinetics, where the collision represents the rate-limiting step of the chemical reaction. The thermal velocities of hydrogen atoms under normal temperature conditions are around several thousand metres per second. Rate constants are of the

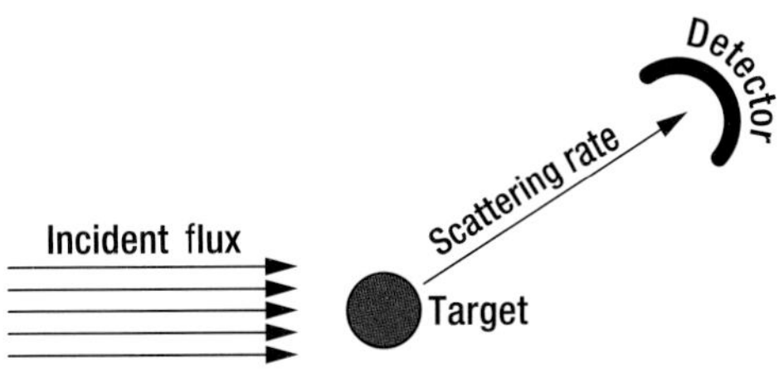

Figure 4.6.6

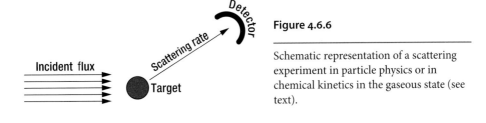

Schematic representation of a scattering experiment in particle physics or in chemical kinetics in the gaseous state (see text).

order of magnitude of 10^{-13} l (cubic decimetres) per particle and second, or on a molar basis (i.e. after multiplication by Avogadro's number) of $6 \times 10^{10}\,\mathrm{l\,mol^{-1}\,s^{-1}}$. The resulting cross-sections are in the range of a squareÅngstrøm and below. At a pressure of 1 bar (corresponding to about 3×10^{22} particles per litre) the collision frequencies are around 3×10^{9} per second. I use the litre as a volume unit because chemists (particularly in solution chemistry) are used to thinking in molar concentrations (1 molar = 1 mole per litre $\approx 6 \times 10^{23}$ molecules per litre).

For reactions in condensed phases we are dealing with physical conditions quite different from those in (dilute) gases. Frequencies of encounters among particles here are determined by the distance squared/time relationship of diffusion, and targets are not hit by molecules in free flight, but rather by "space-filling" motion in a certain "territory" (Figure 4.6.7). Accordingly I have replaced the word "collision" by "encounter". This is the legacy of von Smoluchowski, who was the first to derive an equation for the diffusion-controlled aggregation of colloidal particles. For this, he assumed uncharged reaction partners with finite particle volumes. At about the same time Paul Langevin[60] had derived a similar equation for the recombination of point-like oppositely charged particles, and it was Peter Debye[61] who in the 1940s combined the two concepts and derived an equation that applies to both charged and finite-sized particles in solution. Debye's theory showed that ionic reactions in aqueous media can be extremely fast processes, with rate constants of 10^{10} to $10^{11}\,\mathrm{l\,mol^{-1}\,s^{-1}}$. My own work in the early 1950s, involving the development of new relaxation methods to expand the experimentally accessible range from milli- to nanoseconds, was strongly influenced by the theoretical models of Peter Debye and Lars Onsager.

This gets me back to my theme: the space-filling nature of random walk. Isn't it surprising that the rate constants, for which the diffusion-controlled encounter represents the rate-limiting step, are of the same order of magnitude as those of

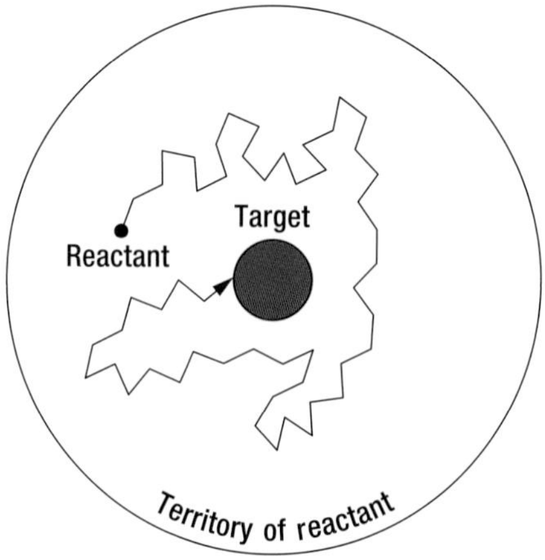

Figure 4.6.7

The territory of a particle A is a region of space of defined size, to be obtained by dividing the total volume by the total number of particles A present in that volume. In two-dimensional space the average time needed for a complete screening of the territory depends on the size of the territory. The number of encounters with other particles depends on their number and cross-section.

atoms in the gaseous state, which are determined by the high free-flight velocities of the reaction partners and their mutual cross-section?

I take, as an example for comparison, the known rates of formation of a hydrogen molecule from hydrogen atoms in the gas state[62] and compare it now with the diffusion-controlled neutralisation reaction in aqueous solution, i.e. the formation of a water molecule (H_2O) from a proton (H^+) and hydroxyl (OH^-) ion, both in a hydrated state. (The rate of the latter reaction was first measured in 1955 by my colleague Leo de Maeyer and myself, applying our (then) new electric-field relaxation technique to highly purified water.[63]) Both examples belong to the fastest reaction processes in their respective media, the neutralisation being the fastest chemical recombination process at room temperature that we know of so far. The value of its rate constant, $1.3 \pm 0.3 \times 10^{11}\,\mathrm{l\,mol^{-1}\,s^{-1}}$, means that – if it were possible to mix a 1-normal acid with a 1-normal base (1-normal meaning a concentration of $1\,\mathrm{mol\,l^{-1}}$ of acid or base groups) then the total second-order reaction $H^+ + HO^- \rightarrow H_2O$ would be complete within less than 10 pico seconds. However, one cannot mix two solutions within times shorter than a few tenths of a millisecond, even by using high-pressure flow devices. This is a problem where relaxation methods proved their usefulness. Such methods use ultra-short energy pulses for perturbing an equilibrium or a steady state where reaction partners are present in a perfect molecular mixture.

The true surprise – and that is the reason why I am quoting these examples – is the almost precise numerical coincidence of the rate constants of the two fastest reactions in gaseous and liquid media, respectively, although they occur in totally different physical states and are therefore based on quite different physical situations. The rate of the neutralisation reaction has been determined under almost ideal conditions, namely in highly purified water where the concentrations of both H^+ and OH^- are the lowest possible ones, i.e. $10^{-7}\,\mathrm{mol\,l^{-1}}$ in the presence of $55\,\mathrm{mol\,l^{-1}}$ of the reaction product H_2O. The diffusion coefficients of both H^+ and OH^- have values of around $10^{-4}\,\mathrm{cm^2\,s^{-1}}$, the largest found among ions in aqueous solution. The reason is the extremely high mobility of both excess protons (H_3O^+) and defect protons (HO^-) surrounded by hydrogen-bonded shells. The rate of migration is limited solely by the required rotation of water molecules. Otherwise the proton is almost freely mobile in the extended hydrogen-bond structures around the H_3O^+ and the HO^- ions (i.e. $H_9O_4^+$ and $H_7O_4^-$), yielding quite large reaction distances of nearly 10 Å. The rate constant for an individual encounter is determined essentially by the product of diffusion coefficients and reaction distance, and has the dimensions of (molar concentration $\times$ time)$^{-1}$.

The same dimensions are found for the rate constant of collision in the ideal gas state, but here on the basis of entirely different presuppositions. An equivalent low concentration of $10^{-7}\,\mathrm{mol\,l^{-1}}$ for hydrogen atoms is found at a pressure of about two millionths of a bar, ensuring "ideal gas" conditions. The mean free path length amounts to about 3000 Å (3×10^{-7} m). Free hydrogen atoms move at room temperature with a speed of about $3000\,\mathrm{m\,s^{-1}}$. The cross-section of atoms is near 1 Å^2. The product of cross-section and velocity (with the dimensions volume/time) determines the rate constant. The precise theoretical values contain some constants, such as 4π. Applying the rate constants to molar concentrations requires multiplication with Avogadro's constant. In this way we obtain for $H^+ + OH^- \rightarrow H_2O$ in water a rate constant of $1.3 \pm 0.3 \times 10^{11}\, l\, mol^{-1}\, s^{-1}$ and for the reaction $H + H \rightarrow H_2$ under ideal gas condition a value of $6 \times 10^{10}\, l\, mol^{-1}\, s^{-1}$, i.e., a value in the same order of magnitude.

The fact that the generally relatively "slow" diffusional motion (Figure 4.6.3) may even outrun the "fast" free-flight process seems, at first instant, most astonishing. However, I think that's just prejudice. There is certainly no over-riding reason why the two processes should yield such numerically concurrent results. In fact, neutralisation in an aqueous medium is a very special case because the mechanism of motion of (excess and defect) protons in the hydrogen-bonded structure of water is anomalous, as is the high velocity of the hydrogen atom in the gaseous state. The consequences for the neutralisation reaction are unusually large diffusion coefficients and reaction distances. This may account for as much as two orders of magnitude. No, what I mean by "prejudice" is the assumption that diffusion, compared with free flight, is a "slow" process. Is it really? Yes, it is increasingly slow for larger distances, as expressed by the distance squared / time relation. However, it is fast, and for the proton even ultra-fast, at small distances in the Ångstrøm range, where positions change within picoseconds. Free flight gets the particles within a short time from one end of a trajectory to the other, but it scans only a very thin line. Diffusion, on the other hand, rapidly scans its near neighbourhood. At their exceedingly low concentrations in pure water, all that single protons have to do is to screen a territory of about 3000 Å in diameter (Figure 4.6.7). And this is the conclusion that is most important for the problems discussed in this book: diffusion is a space-filling process that is best adapted to two dimensions, but is still pretty efficient in three dimensions.

After all, the two problems from a general point of view are not so different. Common to both is the task of finding a target that occupies a very small volume within a certain territory. Both processes work on a random basis and with no goal-directed guidance. Both mechanisms are constrained by physical prerequisites but otherwise carry out their task in a most efficient way.

The high efficiency of space-filling diffusion is even more dramatically demonstrated by an anomalously large value of the rate constant of recombination of a repressor, a protein molecule, with its binding site, the operator, located on an extended DNA chain (see Figure 4.6.8). The large DNA molecule has an almost negligible mobility, and the origin of the co-ordinate system can be placed at the operator site; but also the bulky protein molecule has a diffusion constant only in the range of $10^{-7}\, cm^2\, s^{-1}$ or less. The measured value of the recombination rate constant appears to be appreciably larger than expected for a diffusion-controlled reaction ($<10^{11}\, l\, mol^{-1} \cdot s^{-1}$) – unless we assume some unreasonably large electrostatic interaction or some 100-fold increase in size of the target. The latter turned

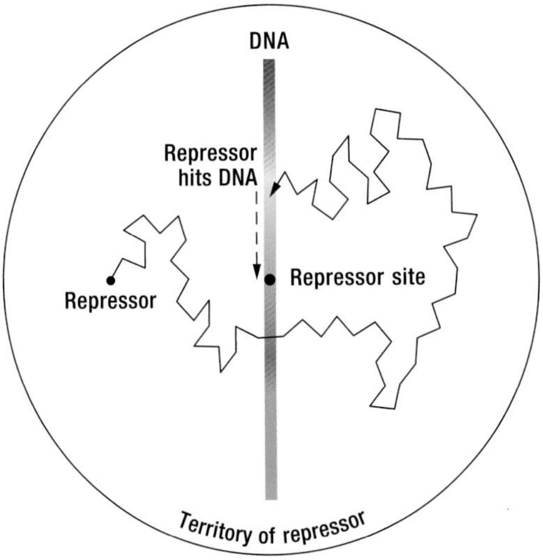

Figure 4.6.8

The encounter of a repressor molecule with the operator site of its target DNA is determined by a complex mechanism described in the text.

out to be the decisive factor. In co-operation with Peter Richter (who meanwhile had moved as professor of theoretical physics to the University of Bremen) we extended von Smoluchowski's work from spherical to spheroidal co-ordinates[64]. An extended DNA chain can be represented in these co-ordinates as a prolate rotational ellipsoid with one long axis 2a (a being the half-axis, e.g. 1000 Å) and two equal short axes of length 2b, with b corresponding to the radius of a DNA double helix (about 10 Å).

The surprising result of the theory was that the reaction distance is given largely by the long half-axis of the prolate ellipsoid, merely reduced by a factor $1 + \ln(2a/b)$. In other words, the almost thread-like ellipsoid of a length of $2a = 2000$ Å and an axis ratio $2a/b = 200$ acts like a sphere with a radius of about 200 Å. For comparison, the true volume ratio (sphere to ellipsoid) is almost 1 million, and the corresponding cross-section ratio almost 1000. Try to do the same experiment by shooting at two different targets of corresponding extensions and you have a direct demonstration of how efficient target-hitting by space-filling diffusion is, in comparison with free-flying projectiles.

The success story of repressor action on operator sites of DNA is quickly told. Positive charges on the repressor molecule keep it electrostatically bound in the potential valley around the negatively charged phosphate-diester chain of DNA. The electrostatic force acts perpendicularly to the chain, but not along its axis. Hence the repressor molecule can easily undergo one-dimensional diffusion as long as it is bound to the chain, defining a certain correlation length. So the repressor doesn't have to hit the DNA chain at the operator site, but rather at any position that lies within a distance from the operator equal to or less than the correlation length. This target-hitting efficiency, based on coupled three- and one-dimensional diffusion, is expressed in the anomalously high rate constants of recombination.

A coupling between three- and lower-dimensional diffusion is a trick used by Nature in various situations to compensate for the recurrence deficiency of three-dimensional as compared with one- and two-dimensional diffusion.

A prominent example of this is the sex attractant sent out by a female insect and detected by the male with incredible sensitivity (see Section 4.1). Gerold Adam and Max Delbrück[65], who presented a theoretical analysis of the mechanism of this process, opened their paper by quoting the mathematician George Polya, who first realised the discrepancy among diffusion processes in two and three dimensions, as expressed in more general terms in Wiener processes. Polya lets a drunkard (perhaps the one in Figure 4.6.9) ask fearfully "Will I ever, ever get home again?", and gives him the advice: "Just keep going and stay out of 3D" – which, indeed, he does.

Yes, this is exactly the answer for the pheromone problem explored theoretically by Adam and Delbrück. The female insect has to deliver her attractant first into three-dimensional space, where it can be carried away in any direction by air flow and is at the same time diluted to quite small concentrations. The male insect, in order to detect one or a few such molecules, has to remove them from three dimensions. For this purpose it uses its highly expanded net antennae. The pheromone molecule now stays in two dimensions and keeps moving until it safely reaches its tiny nerve-cell receptor, which it otherwise would never have found. The number of signals received even allows the male insect to detect a gradient that guides it to the female partner. There are many similar applications of coupling between three- and two-dimensional diffusion in Nature. The late Ephraim Katchalsky and his co-workers studied the kinetics of membrane-bound enzymes. Membrane-bound enzymes are the basis for particularly effective reaction chains, in which the product of one reaction is the substrate for a second reaction, and possibly more.

As Adam and Delbrück found, the mathematics of two-dimensional diffusion is more involved than that of the three-dimensional process. It becomes almost (but not completely) independent of particle size, and it is almost completely dominated by the size of the territory (the average space element available for each single particle). Only one-dimensional kinetics are entirely independent of size, and hold even for point-like particles, ultimately limited by quantum-mechanical indeterminacy. I spare the reader further detail; what I wanted to show in this section, and will be relevant in subsequent sections throughout, is the space-filling nature of diffusion. It is reflected in the rate constants, which apply in all concentration ranges for which the basic assumptions made by von Smoluchowski and by Debye are valid.

Rate constants – except in the case of first-order reactions – are not purely time constants. Reaction times depend on concentrations. For the neutralisation reaction in pure H_2O, where the concentrations of H^+ and OH^- at room temperature are around 10^{-7} mol l^{-1}, reaction times are between 10 and 100 μs. The larger the territory of the reactants, the longer is the reaction time. Think of single protein molecules in a territory of 1 l each. Their concentration would be the reciprocal of Avogadro's number (i.e. 1.7×10^{-24} mol l^{-1}). Assuming a diffusion coefficient

Figure 4.6.9

The drunkard apparently has no other chance than to "stay in two dimensions" unless he gets trapped in one dimension.
Drawing by Claus-Peter Adam.

of 10^{-7} cm^2 s^{-1} and correspondingly a rate constant of maximally 10^8 l mol^{-1} s^{-1} (as has been measured), their reaction time would then be longer than 100 million years. One particle per 30 litres would yield times comparable with the age of our planet. I mention this at the end of this section because there are models where the assumption is that life could have originated without reproductive self-amplification from single rare protein molecules. Yes, there are some 10^{20} l of water in the oceans, but that doesn't help. My advice to those authors is: sit down and think! It may be worth hundreds of millions (of years – sorry).

4.7. And How to Get from Here to There in Information Space?

The essential lesson of the preceding section was: target-finding by random walk in physical space is most efficient at low dimensionalities, preferably in one or two dimensions. Information space, by its very definition, includes as many as ν dimensions, where ν in the case of binary digits is the length of a sequence (i.e. its number of positions, which might be hundreds, thousands, millions or more). The advice given to the drunkard (see the preceding section) wouldn't make much sense. A major difference from physical space is the discrete structure and non-Euclidean metric of I-space, which in the language of the mathematician represents a "finite Galois field". Producing (semantic) information requires walking in this space. It is helpful to revisit a diagram already shown, namely in Figure 4.3.1. If the process starts at the point designated "000000", its Hamming distance with respect to this point initially does not increase with the square root of time, as would be typical of diffusion. Rather, there is a bias towards the "equator", which for large values of ν means, in the initial phase, an almost linear increase of distance with time. Only near the equator (Hamming distance close to $\nu/2$) will the probabilities of going forwards and backwards become equal. In Figure 4.3.1 this is expressed for any point by the number of lines going upwards and downwards.

Filling up the complete space by random searching nevertheless requires (on average) the testing of something like 2^ν points. And here we encounter an essential difference between continuous and discrete spaces. According to Einstein's relation, the diffusion coefficient in continuous space remains proportional to distance squared / time, but it changes with the inverse of the number of dimensions involved. In discrete (binary) space the difficulty of finding a target by a purely random search is correlated with the combinatorial complexity of the problem, in a more intricate way.

First of all, let us find out what the word "motion" with regard to this abstract space actually means. Localisation in this space seems quite different from what it is in physical space, and we do not know offhand any equivalent of force that could bring about a true biassed motion. "Information", regardless of how we ultimately define it, will originate or disappear in this space, whenever we "move" from one point to another. Hence, any form of motion in I-space is tantamount to a change of meaning, however we define it. It is true that information can travel through physical space, for instance as rumours or, in the popular children's game, whispered messages do, often undergoing a change in their content underway. However, this is not what I mean in the present context.

Localisation in I-space by definition is of a discrete nature. To be "present" at any of the 2^ν vertices of binary sequence space means to be present in the form of at least one – but preferably of many more – "physical" copies of that particular sequence of digits. Hence, we have to deal with populations of sequence copies, and for this purpose we have to augment our picture of sequence space by introducing a suitably defined population space. I introduced such a population space in Section 2.6 in order to represent problems in chemical kinetics. In the present case it is a linear vector space that assigns to each of the 2^ν vertices of binary sequence space an (almost) continuously variable population co-ordinate. A given distribution of sequences in I-space then would be represented by a "point" in population space, in a manner similar to the way in which, in statistical mechanics, a distribution of N particles with given spatial and momentum co-ordinates is represented by a phase point in (6N-dimensional) space (see Sections 2.2 to 2.5). In the present case we are interested mainly in relative population structures. Accordingly, we introduce relative population variables x_i with $0 \leq x_i \leq 1$ and $\sum_{k=1}^{n} x_k = 1$ where (for binary sequences) $n = 2^\nu$ includes all possible sequences. (At finite population size, however, most x_i values will be equal to zero.) The adequate population space then is the simplex S_n as introduced in Section 2.6 (for more details see Section 4.8). Note that the dimensionality of such a continuous non-Euclidean simplex space S_n increases exponentially (i.e. $2^\nu-1$) with the dimensionality ν of the "discrete" sequence space.

The question of "how to move in information space" has now become a problem of dynamics in population space. We see immediately that the production of sequences without some kind of instruction would be entirely useless. Owing to the enormous number of alternative sequences, any random mechanism will produce some (nearly) infinitely dilute "ideal information gas", expanded to any of the 2^ν positions of I-space. This would not make any sense for a plausible mechanism of generating (semantic) information.

The only meaningful way to populate a particular position in I-space is to reproduce the corresponding sequence, and the only way to "move" to any position at Hamming distance d_H is to make correspondingly many (i.e. at least d_H) errors in reproduction. Precise reproduction is a limiting case. First of all, it is physically hard to achieve since "noise is omnipresent" (Hamming). Second, if we want to populate any specified point, at least one copy has to get there in the first place by erroneous reproduction of some related mutants. Information carriers as a class have to inherit the faculty of error-prone reproductivity. In view of the tremendous complexity of Hamming space there is no way of defined motion through it except by occasional mutation in reproduction. Populations in I-space will therefore occur as "clouds" rather than (multiply populated) single points (as "survival of the fittest" might suggest). An important problem will be to find the optimum error rate that keeps the (moving) cloud stable, i.e. preventing its expansion to fill the whole space while at the same time keeping it from contracting into a single point. In addition, we need a (suitably defined) "fitness landscape", the gradients of which allow some guidance of movement. What we need sounds like "squaring the circle", namely, using the potentiality of a high-dimensional space geometry in order to construct a Wiener process (see Section 4.6) of lowest possible dimension.

I am now going to discuss a general population model of the system characterised above. It was originally derived in the late 1960s for the reproduction of information-carrying sequences, such as RNA and DNA. The theory, published in 1971[66], includes solutions for various models (including complementary reproduction, cyclic and hypercyclic networks) under different boundary conditions, such as constant fluxes or forces (i.e. chemical affinities).

In order to keep my promise to include as little mathematics as possible in the text I have to refer the reader who is interested in the details of how we approach this problem to the article by Peter Schuster in Appendix A4 or to Vignette 4.7.1 below. This refers to the rate equations that describe the temporal change of the relative population variables $x_k(t)$, with $0 \leq x_k \leq 1$ and $\sum_k x_k(t) = 1$. These equations hold for any individual information carrier I_k present in the system. The competition on which selection is based appears in the differences between the rates for excess production of individual information carriers I_k and the "average excess production rate" $\bar{E}(t)$ of the total population which under stationary conditions (e.g. in a flow reactor) may be balanced by an overall dilution flux. Only those individuals can reach stable steady-state values of $x_k(t)$ whose excess (i.e. formation minus decomposition) rate is able to match the average $\bar{E}(t)$. It is important to notice that the equations do not contain constant (i.e. x_k-independent) terms that would refer to any form of non-instructed *de novo* production of individual information carriers I_k, which is entirely negligible for any defined sequences in view of the huge number of possible alternatives of any given I_k that have a realistic length. On the other hand, since $\bar{E}(t) = \sum_k E_k x_k$ is a function of all variables $x_k(t)$, the term $\bar{E}(t)x_i(t)$ introduces – independently of the "reaction mechanism" involved – an inherent non-linearity to all rate equations, typical of the phenomenon of Darwinian natural selection.

Vignette 4.7.1. Natural Selection as a Physical Law

The problem

Perhaps the most important physical property with which we are confronted in biology is the extreme material complexity of any living state of matter. Theodosius Dobzhansky acknowledged this fact in his frequently quoted statement "Nothing in biology makes sense except in the light of evolution."

The physical states resulting from biological evolution, which Charles Darwin characterised as a never-ending chain of single events of "variation" and

Cont. ➲

➲ Cont.

"natural selection", are molecular organisations of a multifaceted form. Yes, they are regularities among processes in the sense used by Eugene Wigner (Section 4.9) and hence subject to physical theory. As a whole they cannot be understood as typical forms of physical structures; rather, they represent regularities in functional behaviour. They achieved their "exponential" combinatorial complexity within the "polynomially" short lifetime of our planet. The physical problem is discussed in greater detail in the article by Peter Schuster (see Appendix A4).

Some definitions

Natural selection is a dynamic state, far from any established thermodynamic equilibrium. The expressions to be derived therefore have to be of a kinetic form, although they nevertheless apply to a "steady state" of natural selection. What I had in mind in my 1971 paper[66], was the quantitative evaluation of results obtained by evolution experiments in molecular biological systems, such as nucleic acids, proteins, viruses and bacterial cells, some of which are described in Chapter 5

The general assumption is that all forms of life are under the control of genetic programmes, the information of which involves an unimaginable complexity (such as 10^{100} or 10^{1000} alternative states) as described in Sections 4.1 and 4.2. All these states have a limited stability and must therefore be reproducible (the reproduction can include small variations due to mutability). The kinetic equations have to deal with the following chain of events:

$$\text{reproduction} + \text{variation} \rightarrow \text{natural selection} \rightarrow \text{evolution}$$

Table V4.1 contains all the necessary definitions, and Tables V4.2 and V4.3 provide the kinetic equations, their solutions and a conclusion regarding physical law as the basis of Darwinian selection.

Table V4.1 Definitions

All population numbers n(t) or x(t) refer to the unit volume
$n_i(t)$ = number of compounds i *
$x_i(t) = n_i(t)/\sum n_k(t)$ = relative numbers i, $0 \leq x_i(t) \leq 1, \sum_k x_k(t) = 1$
$y_i(t)$ = normal modes of the vector system $\vec{x}(t)$

Cont. ➲

Cont.

Rate parameters of formation and destruction
Amplification due to reproduction and mutation: A_i**
Fraction of precise reproductions: $A_iQ_{ii}, 0 \leq Q_{ii} \leq 1^{\dagger}$
Fraction of mutations $i \rightarrow k : A_iQ_{ik}, Q_{ik}$ usually $<< Q_{ii}$
Any form of destruction: $D_i^{\dagger\dagger}$
Excess reproduction of $i = W_{ii} = A_iQ_{ii} - D_i$
Total excess production of the system $\bar{E}(t) = \sum_k E_kx_k^{\ddagger}$ with $E_k = A_k - D_k$.

*Since my original paper (*Naturwissenschaften* 1971[66]) was concerned with evolution experiments and evolutionary biotechnology the word "compound" includes biological units, of any form, that carry information.
**As explained below, A_i is the rate constant of inherent self- or complementary reproduction of a (necessarily) quasi-linear process.
†For a single-stranded RNA chain carrying ν_i symbols and having a copying fidelity q_i. Q_{ii} will be the product $\prod_{l=1}^{\nu_i} q_{il}$ or $<q_i>^{\nu_i}$, where $<q_i>$ is the geometric mean.
††In a bioreactor one can usually choose reaction conditions where $A_{ii} >> D_i$
‡In this case one has to remove the excess production, either under conditions of constant dilution flux $\Phi(t) = \bar{E}(t)$ or at constant force, as present in serial-dilution experiments (constant affinities).

Table V4.2 Rate equations

A dot on top of the concentration symbol generally means d/dt such as $\dot{n}_i(t) = dn_i(t)/dt$ or, for the general vector system $\vec{x}(t), \dot{\vec{x}}(t) = d\vec{x}(t)/dt$.

1) $\dot{n}_i(t) = W_{ii}n_i(t) + \sum_{k \neq i} W_{ik}n_k(t)$. In vectorial form: $\dot{\vec{n}}(t) = (W_{ik})\vec{n}(t)$.
   (W_{ik}) now is a matrix, W_{ii} its diagonal terms; $\vec{n}(t)$ represents a vector system. Such a system could be immediately be solved by main-axis transformation, yielding new normal coordinates $y_i(t)$ according to $\bar{y}_i(t) = \sum_i x_iy_i(t)$. Since all entries are positive it would yield a system of exponential growth that soon would run into difficulties. The normal modes $y_i(t)$ are linear combinations of the $x_i(t)$ variables. A way out of these difficulties is to choose x_i-coordinates.
2) $\dot{x}_i(t) = \{W_{ii} - \bar{E}(t)\}x_i(t) + \sum_{k \neq i} W_{ik}x_k(t)$, or in vectorial form:
   $\dot{\vec{x}}(t) = \{(W_{ik}) - \delta_{ik}\bar{E}(t)\}\vec{x}(t)$
   where (W_{ik}) is the above matrix and δ_{ik} the Kronecker delta, i.e. $\delta_{ik} = 0$ for $i \neq k$ and $\delta_{ik} = 1$ for $l = k$. This is the form derived in my 1971 paper[66].

An (apparent) difficulty is that the introduction of x makes the system of differential equations a non-linear one, but during my lecture in New York Marc Kac recognised immediately that a solution can be obtained as easily as for a system of linear equations. The most elegant way of solving this is given in a paper[68] quoted in the

Cont.

Cont.

text of Section 4.7; it is the substitution $= z_i(t) = x_i(t)\exp\int_0^t \bar{E}(t)dt$. It compensates for the nonlinear term $x_i(t)\cdot\bar{E}(t)$ yielding a linear system of differential equations of the form $\dot{\vec{z}}(t) = (W_{ik})\vec{z}(t)$. The trick is that $\exp\int_0^t \bar{E}(t)dt$ is a factor that holds for all components of the transformation and can therefore be easily converted back to reexpress everything in x-coordinates (i.e. $x_i(t) = z_i(t)\exp{-}\int_0^z E(t)dt$). The final result is the "strangely simple" form in which selection is expressed as a natural law in mathematical terms:

$$\dot{y}_i(t) = \{\lambda_i - \bar{\lambda}(t)\}y_i$$

To end with, let me add a few words. A genetic programme is represented by a spectrum of mutants because reproduction is not (and cannot and must not be) a copying process with 100% fidelity. In earlier treatments it was regarded – approximately – as almost such a process, an approach that might introduce serious errors due to the presence of neutral or nearly neutral mutants. Since the existence of quantum mechanics we have become accustomed to a treatment of such systems using vector spaces and matrix representations. The very fact that the system examined above leads to a formulation of a natural law is due mainly to certain peculiarities which are introduced by this treatment, and this in turn leads to three questions that demand an answer.

(1) Why use an exclusively linear ansatz?
At first sight, the answer seems trivial: otherwise we couldn't get complete solutions of the system of differential equations. However, this is not an adequate answer. We must say why we do not start from some more general function that is approximated by expansion into some power series. Here I have an answer: the first term, as a constant, would describe the formation *de novo* of an incredibly improbable state. Yes, "complexity" is the answer. It can only be produced by the copying of something obtained before – in a process that depends linearly on concentration and does not exclude copying errors, but which observes an error threshold.

(2) The next question would be: why do we not admit higher-order terms in the process of reproduction? Of course, we cannot exclude them, and they do occur in many situations. Yet the Darwinian mechanism – which requires variations – would exclude them because if progress comes from variation, then a single event must be able to compete with highly amplified states of

Cont.

Cont.

previous selection. A linear process – apart from stochastic fluctuations – is able to do this. A quadratic mechanism, for instance, is helpless in such a case. In fact, this Darwinian requirement even represents selection pressure for the development of homeostasis.

(3) The third question is closely connected with the two preceding ones. The system of linear differential equations would yield an (exponentially) complex set of eigenvalues. How can we deal with an arbitrarily large set of eigenvalues? If we look at our matrices, they can have only positive entries. Any negative terms would have to be non-linear. Under these conditions – in agreement with the first problem – the situation again becomes simple, allowing us to formulate a law. This is due to a principle, derived by the mathematicians Frobenius and Perron in the late 19th century, which says: in the whole spectrum of eigenvalues there is only one stable eigenvalue, and it is the largest one.

I will not go into further detail here. All these three points have been clearly decided by experimental results, which led to a new biotechnology based on the physical law explained above. Moreover, our interpretation of Darwinian behaviour has undergone drastic changes in our understanding of what natural selection means.

Natural selection as a physical law

The final result of Tables V4.2 and V4.3 reads simply:

$$\bar{E}(t) = \bar{\lambda}(t) \rightarrow \lambda_{max}$$

and therefore

$$y_i = \begin{cases} 1 \; \textit{for} \; \lambda_{max} \\ 0 \; \textit{for} \; \lambda_i \neq \max \end{cases}$$

Could there be a simpler formulation of a "natural law" (which I consider a physical law of the "regularity" of natural selection which – after all – remains a "complex familiarity")? It is familiar from all the facts and observations that led Charles Darwin to its formulation but this familiarity remains as complex as those of all natural laws. Its conclusions are: "fittest" and "wild-type" are not individuals, but rather "centres of gravity" in quasispecies distributions. Natural selection is a

Cont.

Cont.

phase transition in information space, fulfilling all the requirements of physical phase transitions. Neutral selection is an approximate description of the rôle of mutants, which was first recognised by Motoo Kimura. It prevents Darwinian evolution from getting stuck on any "minor foothills" of fitness as a consequence of macroscopic fluctuations that are the hallmark of critical phase transitions. It is expressed in the error-threshold relation:

$$(1 - < q_i >) \leq \frac{\ln \sigma_i}{\nu_i}$$

where $1 - < q_i >$ is the error threshold, ν_i is the information content and σ_i is a parameter of "superiority" in natural selection.

The non-linearity just mentioned is not the only one to be expected under natural conditions. RNA or DNA replication and decomposition usually involve enzymes that may be encoded by the sequence to be replicated. In nature, the various functions are highly controlled by "activators" and "repressors". Moreover, the assembly of polymeric sequences depends on the presence of the monomeric substrates that are linked up by the template-instructed enzymic process. All those terms expressing deviations from linearity can be summarised in a term, which appears in the equations in Vignette 4.7.1. In evolutionary experiments these individual non-linearities can be kept under control for the averages, as is usually also the case in biological environments. (More details below and in Section 5.6.) However, the non-linearity introduced by the simplex variables $x_k(t)$ and expressed in terms of $\bar{E}(t)x_k(t)$ must not be removed because it is at the heart of the process of "selection". Systems of non-linear differential equations are generally not solvable in closed form. The solution I presented in the 1971 paper[66] was obtained by second-order perturbation theory. For quasi-linear reaction mechanisms it reproduces Darwinian behaviour and shows, in particular the presence of an error threshold for mutation rates, whereby selection can be interpreted as a "phase transition" in I-space that occurs when the error threshold is violated (more details in Section 5.5).

When in the early 1970s I talked about this work at Rockefeller University in New York, the internationally renowned mathematician Marc Kac was present and suggested that my system of equations may be exactly solvable by a suitable transformation which removes (or "hides") the non-linearity, and reduces the solution to an eigenvalue problem. This suggestion was corroborated shortly afterwards in a paper by Colin J. Thompson and John L. McBride[67] (1974), and a year later by the Canadian scientists Billy L. Jones, Richard H. Enns and Sada S. Rangnekar[68], who independently used an alternative method. The solutions of these equations can be written in terms of the eigenvalues of the matrix (W_{ik}) representing the phenomenological coefficients of my original system of rate equations. David Rumschitzki of New York University studied the spectral properties of Eigen evolution matrices in more detail.[69]

With these equations I introduced a value concept that characterises semantic information by a properly defined "selective value". Let me tell a little story that is connected with this "value concept". In one of my lectures in the USA I was asked whether the term "eigenvalue" was named after me. My answer was that the "eigenvalue" or the "eigenfunction" is as little related to my name as the binomial coefficients (to which I often had to refer in my lecture) are named after some (fictitious) Italian mathematician with the name of "Binomi". The person who had asked the question obviously did not understand my attempt at a joke. He looked at me, shook his head and said: "But you have kept very well!". *Sic*!

One particular feature of the matrix (W_{ik}) representing the "excess production" is the fact that it has only real and positive entries and therefore refers directly to a theorem formulated in the early 20th century by the mathematicians Ferdinand Georg Frobenius[70] and Oskar Perron[71]. According to this theorem, the stationary solution of such a system of differential equations is governed solely by its maximum eigenvalue (λ_{max}), which is a simple, real and positive eigenvalue, associated with an eigenvector in R^n that has only positive co-ordinates. It can be retransformed into the x-co-ordinate system, yielding a "normal mode" $y_m(t)$ that is a linear combination of the $x_i(t)$ variables.

In terms of these normal modes the differential equation of "natural selection" describes in (strangely) simple terms some truly complex, but nevertheless familiar, facts:

$$\dot{y}_i(t) = (\lambda_i - \bar{\lambda}_i(t))\, y_i(t)$$

where $\dot{y}_i = dy_i(t)/dt$ is the time-derivative of the ith normal mode $y_i(t)$, which is associated with the eigenvalue λ_i, and $\bar{\lambda}(t)$ is the (time-dependent) average of all eigenvalues, representing the total excess production: $\sum_k \lambda_k y_k(t) = \sum_k E_k x_k(t)$, the latter being identical with the term $\bar{E}(t)$ introduced in Vignette 4.7.1). I have reproduced this equation in Vignette 4.7.1 because in this form it gives a clear idea of the simple mathematical representation of "natural selection" and hence of Darwinian behaviour. Look at the bracket term on the right of the equation above ($\lambda_i - \bar{\lambda}(t)$). The bracket is negative for all eigenvalues λ_i that are smaller, and it is positive for all eigenvalues λ_i that are greater, than the average $\bar{\lambda}(t)$. It means that the normal modes $y_i(t)$ for λ_i smaller than $\bar{\lambda}(t)$ decay, while those for λ_i that are greater than $\bar{\lambda}(t)$ grow exponentially. In this way the average $\bar{\lambda}(t)$ will become larger and larger until it represents the maximum eigenvalue λ_m. In other words, only the normal mode $y_m(t)$, associated with the maximum eigenvalue λ_m, will eventually survive, or in mathematical language:

$$\begin{matrix} y_m(t) \to 1 \\ y_{i \neq m}(t) \to 0 \end{matrix} \quad \text{for } t \to \infty$$

Let me "translate" this result into everyday language. Natural selection occurs in a self-reproducing system as a consequence of an extremum principle. It is a process of self-organisation under the guidance of a law of nature. As such it can be mathematically described in simple terms. The target of natural selection is not a single individual entity (or species), but rather a defined distribution of related individuals. We call it a "quasi-species" because it behaves almost like, or "quasi", a single individual species.

The law of natural selection is universal. It holds not only for biological species but just as well for any mixture of autocatalytically reproducing entities which, for instance, may be as "in"-organic as laser modes[72] (see Section 3.8). Perhaps it is the general principle that Darwin may have had in mind when he wrote his letter to George Charles Wallich, which I mentioned in Section 3.9.

Vignette 4.7.1 summarises the essential steps in deriving a physical law of natural selection in general terms. It considers the kinetics for each individual mutant that may (possibly) occur, the mechanism being of linear auto- or cross-catalytic form. This can be described by a matrix of rate coefficients. Growth limitation requires a removal of the excess production, introducing non-linearities, which can be removed by a suitable transformation. The result is a general solution in terms of "normal modes", which contain the contributions of all individual mutants. For millions or billions of possible mutants, solving the equations can yield a comparable number of eigenvalues, which contain the secrets about where the system would go in the evolutionary process, but which are still hopelessly complex. At this point one would usually abandon the calculation. However, in the case under consideration the situation is quite different, as shown in Vignette 4.7.1.

The system of differential equations for any non-linear reaction mechanism could not have been solved in any comparable general terms. Actually, many natural systems behave quasi-linearly – up to the evolution of man. Yet there are still many notoriously non-linear systems that do not show Darwinian behaviour. Non-linearities appearing in mechanisms, and our inability to deal with them in general terms, are not the reasons for their non-compliance with Darwinian behaviour, which combines selectivity with variability. A newly appearing mutant that is equipped with advantageous properties must be able to grow up from a single copy in the presence of many competing copies of previous "winners". This is most easily achieved by a linear mechanism that provides for exponential growth. The mechanism described in Vignette 4.7 is therefore the one best adapted to Darwinian behaviour. However, the interpretation of such behaviour introduces many new questions.

The interpretation of selection for a quasispecies deviates in several respects from that of selection of a classical biological species:

i) The *wild type* (or "fittest") is not a single type. Rather, it is defined by a consensus sequence which refers to some *centre of gravity* of the total distribution in I-space.

ii) The maximum eigenvalue, like any eigenvalue, is an *invariant* of the system. This means that it is defined by matrix coefficients of (W_{ik}) in the rate equations, irrespective of whether or not the stationary relative composition (for $t \to \infty$) has already been established.

iii) There exists an *error threshold* at which self-amplification abruptly turns into "snowballing" of errors, whereby the formerly stable consensus sequence, representing (semantic) information, disintegrates, either completely or in favour of a related, but better adapted, state.

iv) *Natural selection* appears to be a first-order phase transition in I-space if the target of selection identifies itself by some "selective advantage", while it appears to be a critical phase transition for neutral variants, involving large-scale fluctuations as is typical in the neighbourhood of critical points (observable as critical opalescence in liquid–gas transitions).

v) *Evolution* may then be considered as a succession of such phase transitions in sequence space, triggered by the advent of new mutants at the periphery of the distribution that destabilise the former wild type, shifting the centre of gravity to a new location in I-space. This may be interpreted as biassed motion towards regions of high fitness density in I-space.

vi) Because of the genotype–phenotype dichotomy the Darwinian phase cannot start by itself. It rather has to become organised in some non-linear initial phase. The target of selection is the genotype, while selective evaluation appears at the level of the phenotype. Hence there must be some feedback cycle involving the phenotype that is superimposed upon the reproduction cycle of the genotype (Section 5.3). We have called such systems "hypercycles"[73], and they may consist of quite complex reaction networks of nucleic acids and proteins. The biological detail is not the subject of this part of the book. However, one can easily see that a start could be greatly simplified if it were to involve only one class of molecules, i.e. if both genotype and phenotype were to be recruited from the same class of molecules, as in the example shown in Figure 4.7.1. This, in my opinion, is the strongest argument for the pre-existence of an RNA world. It is strongly backed up by the fact that even in

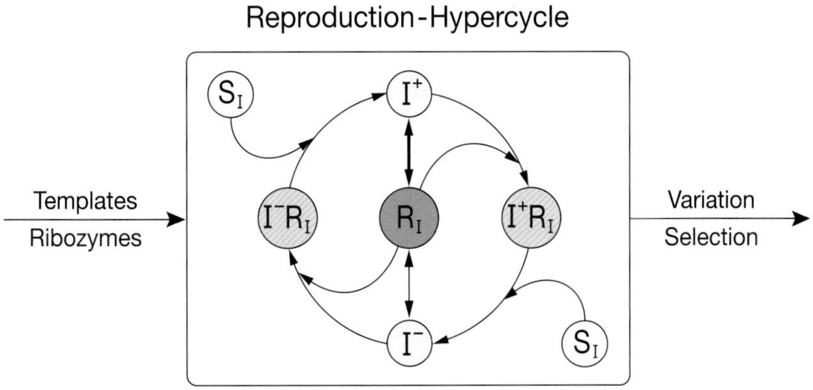

Figure 4.7.1

Catalytic reproduction of RNA by ribozymes?

If genotype and phenotype can be recruited from the same class of molecules, a selective start of RNA amplification would have been greatly aided. The mechanism of such a process is illustrated in this figure. I^+ and I^- are the two RNA templates (plus and minus strands) carrying the information (I) of the genotypes. S_i is the substrate for their formation and R_i is the ribozymic catalyst (i.e. the functional phenotype). Although in a simple case I^+ and I^- could be identical (and thus palindromic), and R_i could also be identical with one of these, this need not be the case. When I^+ and I^- are different, they have separate functions and both appear with their concentration terms in the rate equations; if they are identical, then their concentration appears as a square term. In other words, the hypercyclic mechanism (also in the case of buffered substrate) is non-linear, as mechanisms describing origins are. Unfortunately, this kind of ribozyme still is wishful thinking (see Section 5.3).

present-day organisms the translation machinery is mainly made up of RNA molecules. These include not only the messenger and transfer RNAs, but also the ribosomal RNAs, as well as shorter i-RNAs with a regulating function. According to recent findings[74], the RNAs are the main catalytically active constituents of the ribosome (the molecular machine that synthesises proteins according to the instructions given by the RNA messenger).

Replication of the RNA requires catalytic help, which in present-day organisms is provided by proteins, i.e. so-called RNA polymerases or replicases. In the RNA world, protein polymerases are replaced by RNA catalysts, so-called ribozymes (cf. the Nobel lectures by Sidney Altman[75] and Thomas Cech[76]). If the RNA template (genotype) and its ribozyme (phenotype) are identical, then there will be selective preference for this molecule. Nevertheless, the kinetics involve two such molecules, one being the template (genotype) and the other the catalyst (phenotype), requiring a rate equation containing the square of the concentration of the molecular species that embodies both functions.

Let me supplement the above list, emphasising another change which the matrix theory of Darwinian selection has brought about. Under realistic conditions the binomial (or Poissonian) error distribution is much too narrow to allow any effective exploration of sequence space, such as would be necessary for finding a promising evolutionary route. In fact, selection can be detrimental when it prevents a whole population from exploring new alternatives by fixing them to minor "foothills" in the fitness landscape of sequence space.

Now let us look to what the quasispecies model may offer instead. For comparison I take a sequence of length $\nu = 100$, which is relevant for the experiment described in Section 4.10. In classical genetic theory neutral mutants have been considered of being of minor importance. It was the Japanese geneticist Motoo Kimura[77] who came out with the – at that time heretical – idea that evolution, especially at advanced stages of life, might have depended as much on random drift among selectively neutral mutants at earlier stages as on natural selection in the Darwinian sense. In Section 5.4 I shall return in more detail to these ideas.

Life represents a combination of so many single traits that there must be many variations which for the whole organism may turn out to be neither advantageous nor injurious and which, as Darwin already noted in his Chapter 4 of the *Origin*, dealing with natural selection, "would be left either a fluctuating element or would ultimately become fixed". I don't see any difficulty in applying this to single genes. As modern studies of site-specific mutations suggest, there are changes, for instance at or near the active site of an enzyme, which are absolutely detrimental, but there are many other sites at which mutations are well tolerated. This is corroborated by the fact that different organisms encode equally efficient enzymes in quite different genes, showing appreciable divergences in their evolutionary trees. In fact, we all would look like identical twins if our genomes were nearly identical.

Quasispecies theory, which rates all mutants present according to their relative fitness, dispenses with the question of how to define "neutral". They all contribute to the maximum eigenvalue in which neutral or nearly neutral mutants dominate. Second-order perturbation theory, which I applied in the 1971 paper, yielded for the stationary population variables x_i expressions of the form:

$$x_i/\bar{x}_m = (1-q)^{d_i} \frac{W_{mm}}{W_{mm} - W_{ii}} f(d_i)$$

where $f(d_i)$ is a function of the Hamming distance between mutants i and m, $1-q$ is the average error rate per position, while W_{mm} and W_{ii} are the diagonal terms in the matrix of excess rate coefficients referring to wild type (m = maximum term) and mutant (i). The $f(d_i)$ are sum expressions taking care of all mutants that lie on the path from m to i (at Hamming distance d_i). For the first term, describing one-error mutants $d_i = 1$, $f(d_i)$ is equal to 1.

The detailed form of the above expression need not bother us in the present context. The expression – and that's why I quote it – was of much concern to me because of the difference (W_{mm}–W_{ii}) in the denominator (appearing also in all higher f(d) terms), which causes singularities in the case of neutral mutants, i.e. for $W_{ii} = W_{mm}$. I was well aware that those infinities had generally plagued second-order perturbation solutions, especially in the field of particle physics. They usually can be removed by renormalisation, and it happened that just at that time John McCaskill (who, coming from the Oxford physics department, bringing along experience in that field) joined our laboratory. His work on renormalisation[78] triggered our interest in phase transitions,[79] to which I shall return in Section 5.5.

The presence of neutral wild types causes large deviations from Poissonian types of distribution. In Figure 4.7.2 the example of a Poissonian distribution is illustrated. It requires an individual wild type that is clearly distinguished by some selective advantage. Appearing mutants are all very narrowly distributed around the wild type. This is depicted quantitatively by the sharp distribution in the centre of Figure 4.7.3, the area of which is shaded in dark red. The populated mutant spectrum reaches barely beyond the four-error mutants. Evolution under those conditions would be highly impeded.

Now let us look at some quasispecies models in which neutral mutants occur. We know nowadays that those neutral or nearly neutral mutants appear in almost all k-error classes for k values from 1 up to $\nu/2$. There are always positions that have no critical influence on a particular function. The triangle-shaped (light-red shaded) region in Figure 4.7.3 results if we assume that each k-error class contains just two mutants that are neutral with respect to the wild type. The wild type is now just the "centre of gravity" in the whole distribution, in which essentially only the well-adapted neutrals appear to be populated, showing under realistic conditions large fluctuations and not nearly the symmetry of distribution as represented by the curve in the centre of Figure 4.7.3. Neutrals can often be considered practically as one selected phase. Evolution under such conditions occurs in times that are many orders of magnitude shorter than in the classical examples, where it is impeded by the narrowness of the Poissonian distribution. The experiments described in Section 4.10 provide convincing examples.

The major difference is the replacement of the individual wild type by the complete ensemble, very much like the change of a dictatorial system into a democracy. The distributions are anything but the ones found in statistical physics. Motion in I-space includes all dimensions, although in the end it produces a quasi-one-dimensional path. How come? Brownian motion in physical space (see Section 4.6) was described by Manfred Schroeder as being "as close as we get in physics to a non-differentiable function". What about motion in I-space? In this respect it isn't very different. The trajectory goes into any of the many possible directions, and it does so via phase transitions that we call "natural selection". So, viewed superficially, the overall process might look similar in both cases, but there is an important difference! Although it is unlikely that any segment of random motion is exactly reversed, this is not true for single steps if only three dimensions are involved, which have just six possible choices

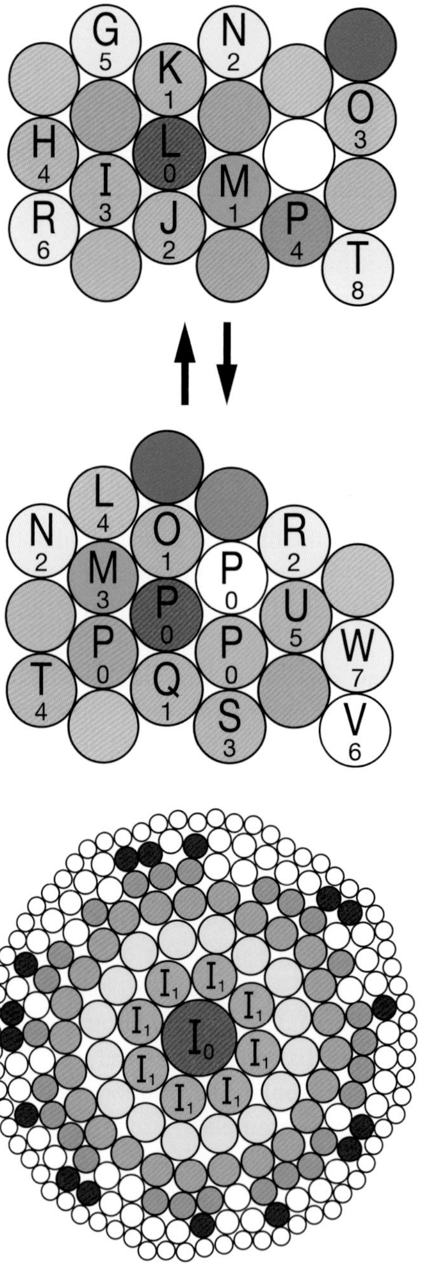

Figure 4.7.2

Mutant distributions

The classical picture of the target of selection was a "fittest" wild type surrounded by a Poissonian distribution of closely related mutants. Those distributions are quite narrow; their relative width decreases sharply with the length of the sequences involved, since the number of mutants depends exponentially on their length. This is schematically illustrated, where even for a relatively small gene a complete representation of all mutants of a given error class is limited to two or three positions. A completely different picture is obtained for neutral variants of the wild type, sketched below. Here, neutral or nearly neutral mutants appear with fluctuating population numbers in much wider ranges. Their mechanism of screening, required for evolution, is explained in the text. Quantitative estimates are illustrated in Figure 4.7.3.

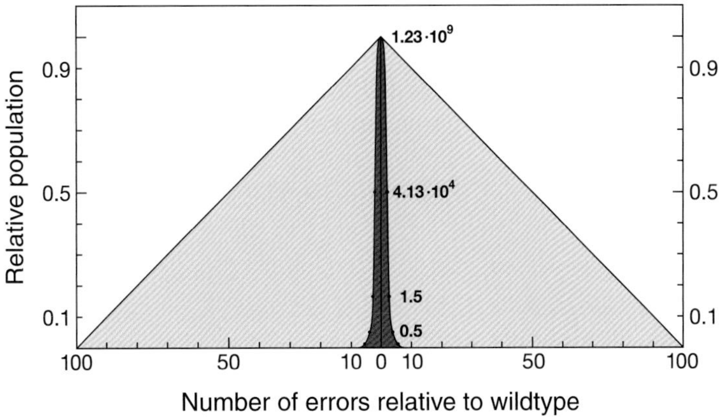

Figure 4.7.3

Quantitative comparisons of Poissonian with quasispecies distributions

The sharp distribution curve (dark red) in the centre of the diagram refers to case (a) in the preceding figure, i.e. the classical picture of a mutant distribution in the vicinity of an individual wild type being the target of selection. In this distribution the individual fitness of mutants is not weighted, and their probability of appearance depends simply on their Hamming distance to the wild type, which determines the number of alternative mutants in a given error class. The Poissonian distribution is a limiting case of binomial distribution for small probabilities and large numbers (cf. their "appearance" in Chapter 2). The (light red) triangular distribution owes its (formalised somewhat unrealistic) symmetrical appearance to the assumption of a completely regular distribution of (nearly) neutral mutants. The ones closest to the centre of the wild-type distribution show up most frequently. Since all mutants are individually "weighted" in the quasispecies theory, and since single neutral mutants (according to Kimura's theory) have a finite choice of growing up in the presence of highly populated competitors, all nearly neutral mutants in the region shown in light red may appear with time constants depending on their distance from the centre of the distribution. They will be the only ones reaching finite population numbers among the large number of possible mutants with various sequences. As Kimura recognised, these (quite numerous) neutrals are very important for the screening process necessary in evolution.

to change. As a consequence there is much overlap in single steps that is almost absent in high-dimensional space where the probability of reversing a dimension is low (i.e. $1/\kappa\nu$). The latter is also true for any multiple branching of routes. The consequence is a unique route that can be treated as if the process proceeded in only one dimension.

If I am asked for a physical explanation of this fantastic-sounding behaviour expounded by evolutionary motion in I-space, I like to refer to an example I found in Benoit Mandelbrot's[80] book *The Fractal Geometry of Nature.*

Mandelbrot asks the question of how a raindrop, falling on a fractal landscape, could find its way to the ocean. The general answer, of course is that gravity will always force the water to search for the lowest point in any landscape. But this requires slopes that avoid traps in the form of recesses or hollows. In Figure 4.7.4 the profile of a mountainous island is drawn according to Mandelbrot's ideas. For a real island, a vertical section may refer to different positions on its surface. Let us therefore start with the profile of a truly one-dimensional landscape. There is only one possible path that the raindrop can follow. Along that path it has to fill all the (one-dimensional) basins up to their brink in order to follow what is called a "devil's staircase" down to the ocean. In a (more real) two-dimensionally expanded landscape, given a fractal geometry, the water will always escape from the hollow at the lowest point of its (fractally shaped) brink (see Figure 4.7.5). In other words: if the vertical section in the figure represents an average profile the water does not have to fill the basins up to the average brink line, but only to the lowest point of the

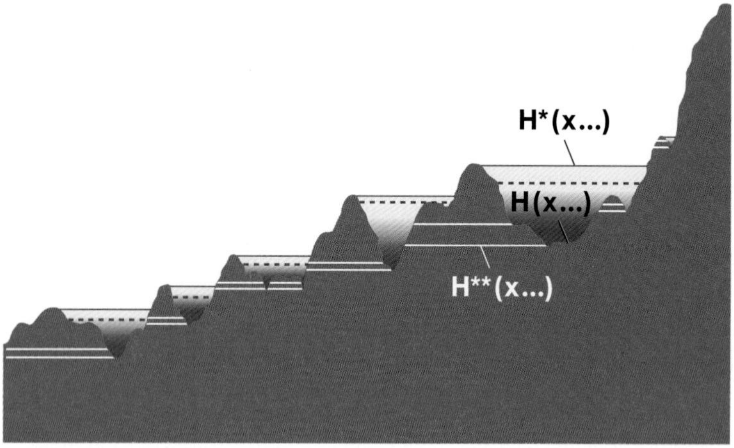

Figure 4.7.4

Benoit Mandelbrot's "raindrop problem"[80].

Benoit Mandelbrot has shown that the structuring of a mountain region can be represented by a fractal height distribution. In other words, the structures look similar on all scales. He considered a problem that one could regard as the reverse of an evolutionary process. In his case it reads: What chance does a raindrop have of reaching the ocean when it falls on an island that has a height distribution as illustrated here? If the island were actually one-dimensional, then all troughs would have to fill up before the raindrop could flow down to the ocean over a series of terraces (called a "devil's staircase"). An evolutionary process would proceed in an upward direction if the "gendarmes" (i.e. the block-like elevations) could be smoothed by "populating" them with mutants. For this to work, we need populations to move upwards, as in the quasispecies theory. A single raindrop falling onto the island never reaches the ocean.

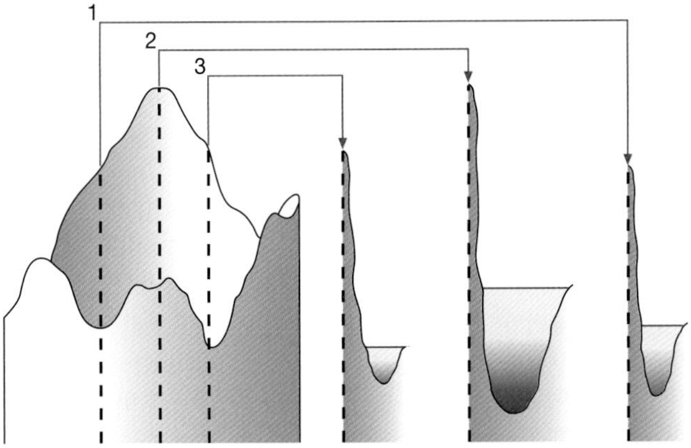

Figure 4.7.5

An optimum solution of Mandelbrot's "raindrop problem"

The problem with Figure 4.7.4 is that the surface of Earth is not one- but rather two-dimensional. In fact, that simplifies our problem. In a fractal landscape the edge of the real two-dimensional dip has all sorts of differing heights, and the water would run out if its level reached the lowest height of the fractal edge (case 3). Likewise, an evolutionary process does not work with an average, but rather with the highest possible selective value. Mandelbrot has shown that for each further dimension added to the picture the water level required would steadily go down until it reaches the lowest point. In mathematical language this reads: in a Hilbert space with an infinite number of dimensions, every raindrop would reach the ocean. This result had been obtained earlier by Mandelbrot's teacher, Pierre Lévy. Sequence space usually includes many dimensions, sufficient to approach an optimal solution. The take-home lesson is: evolution requires sufficiently large populations and some efficient screening procedure, as discussed in this section.

See also M. Eigen "Phase jumps – the physical basis of life" pp. 119–153 in "Between Science and Technology", A. Surlemijn and P. Kroes (Editors), Elsevier Science Publ. B.V. (North-Holland) 1990, and M. Eigen "The physics of molecular evolution", Chemica Scripta 1986 286 13–26.

two-dimensional brink. Now, Mandelbrot iterates this procedure by adding further dimensions. Following this principle the water level will steadily go down, until for a Hilbert space with an infinite number of dimensions every raindrop will reach the ocean, a result that was obtained in the 1930s by the French mathematician Pierre Lévy.[81]

The relevance of this example for processes of evolutionary optimisation is easily seen. One just has to turn the picture upside down. Then hollows become "gendarmes" in a rugged multidimensional, value landscape, in which climbing of fitness hills is enforced by natural selection. In sequence space this requires a population structure extending into all dimensions in order to find an optimum climbing path. In fact, the population around the tip of a gendarme now corresponds to the (bulk) water on the ground of a hollow. And it is important that there is a sufficiently large "population" of water molecules. Single individuals in this scenario cannot evolve, as little as single droplets can find their way to the ocean. There are several reasons why only populations can do this, among which connectivity and entanglement between individual mutant states in sequence space are of importance. Ernst Mayr arrives at a similar conclusion when looking at the problem from a quite different angle. He writes:[82] "The population concept adopted by most mathematical population ecologists was basically typological, in that it neglected the genetic variation among individuals of a population. Their "populations" were not populations in any genetic or evolutionary sense but were what mathematicians refer to as sets. The crucial aspect of the population concept to have emerged in evolutionary biology, by contrast, is the genetic uniqueness of the composing individuals." Yes, Mayr was correct in reaching this conclusion, but not when he concluded that mathematics (or physics) cannot deal correctly with those situations, and that therefore "a biological theory is different from a physical theory".

Mandelbrot's raindrop problem represents an instructive analogue of what happens physically in evolutionary optimisation. In Peter Schuster's contribution in Appendix A4 it is shown that the total average production has the character of a potential function. For the derivation of a force one has to take into account the non-Euclidean make-up of a simplex by applying Riemannian geometry. Moreover, there are other differences with respect to a mechanical force, apart from the fact that a "force" acting in information space is different in nature from a classical force. Nonetheless, the analogy does exist. It is interesting to note that, given the fractal structure of mountainous regions (almost) all rain water originating in the oceans will eventually find its way back to the oceans. The same is true – and greatly favoured by the high dimensionality of information space – for the inverse problem of evolutionary optimisation. Here as there, the paths are string-like, much as rivers on the surface of the Earth are.

In his *Tractatus Logicophilosophicus* Ludwig Wittgenstein says: "The Darwinian theory has no more to do with philosophy than has any hypothesis of natural science." Maybe Darwin's ideas in the 1920s were still considered to be purely hypothetical – at least from the point of view of logic – although Darwin himself had based them on a wealth of observational material. In fact, many physicists have expressed similar views. Darwinism – so-called, as if it were an ideology – did not seem to have any physical roots. What I have presented in this section is a mathematical formulation that takes Darwin's idea from the twilight of a hypothesis into the realm of provable science, in both theory and experiment. On an earlier occasion I expressed this using the slogan: "Darwin is dead, long live Darwin!"

Many very able young scientists, who today occupy chairs of biophysics in Europe and in the USA, have co-operated in this work in the past 40 years. Peter Schuster was involved from early on[83,84,85], and he has since built up a recognised school of theoretical biology from which several distinguished leaders in the field have emerged. He kindly agreed to write an appendix to this chapter (A4). John McCaskill joined our group in 1980 and made excellent and valuable contributions on stochastic theory,[86] renormalisation of selection theory[87] and theoretical and experimental work on space-expanded evolutionary processes. Later on he directed a most prolific group, exploring the technological future of these ideas, working with wet as well dry systems. David Rumschitzky[88] of New York University and Andreas Dress,[89] emeritus professor of mathematics at Bielefeld University in Germany, were also involved in developing a theory of molecular evolution. Last but not least, I should like to mention the experimental work of the late Christof Biebricher, who for many years co-operated closely with my group, showing that the ideas discussed here are, indeed, effective in Nature.[90,91]

What about my question in the title of this section?

Yes, information in physical space travels through communication lines, but that is not what I was asking about. "Getting from here to there" in information space is equivalent to generating (or decomposing) semantic information. What I have described in this section is a kind of motion that looks quite different from what we are accustomed to in physical space. Yet in both cases we "get from here to there" either by random drift or by purposeful motion, thereby occupying certain regions of space. Directed motion requires predetermined gradients, and this is even true if "potential gradients" occurred in our brains while we decided which is "the road (to be) taken".*

4.8. Can a Simplex be Complex?

From the point of view of language a complex simplex should be an oxymoron, a contradiction in itself. However, what the mathematician calls a "simplex" is well defined as a particular way of representation. So the answer to the above question will most probably be "yes", otherwise – as one may guess – I wouldn't have asked.

In Chapter 2 I introduced the term "simplex" for representing the reaction behaviour of systems of coupled chemical reactions. The simplex is a geometric construct derived from a multidimensional Euclidean phase space, spanned by its reaction co-ordinates. The word is derived from the Latin. Its first part refers to the word "simple", or even the "same". The second part, however, is not just a suffix. It rather derives from the verb "plecto", which means "twine", and hence indicates a network of interrelated parts, something that may get quite complex, indeed. The origin of the word, however, does not reflect any such sophistication. Its inventor, the

* "The Road Taken" is the title of a novel by Rona Jaffe.

Dutch mathematician Pieter H. Schoute, referred to the "simplest" regular geometric construct that is free of diagonals (see Section 2.6).

The simplex was shown to be a line segment in two-dimensional Euclidean space, an equilateral triangle in three-dimensional Euclidean space and a regular tetrahedron in four-dimensional Euclidean space (see Figure 2.6.7). That goes on in the same way for higher dimensions; the simplex S_n remains a "regular" hypertetrahedral space of dimension n in an $(n + 1)$-dimensional Euclidean hyperspace to which it refers – not really a very "complex" logical construct, supposing that you are not compelled to "imagine" a 100- or 1000-dimensional regular body. Moreover, in high-dimensional space of reaction co-ordinates it may represent very complex reaction behaviour.

Geometrically, "regularity" of the construct representing the n-dimensional simplex S_n means that each vertex has the same distance from every other one. Its iterative make-up was indicated in Section 2.6.

The simplex is an ideal map for representing systems in relative concentrations or population variables x_k for which $0 \le x_k \le 1$ and $\sum_{k=1}^{n+1} x_k \equiv 1$. Each vertex of such a simplex refers to a state where one of the k variables, i.e. x_i, becomes equal to 1 while all other variables approach 0. Complexity enters with evolutionary problems as treated in Sections 4.4, 4.5 and 4.7. Our imagination is hampered if n becomes of the order of magnitude of hundreds, thousands or larger and if we want to get a quantitative understanding of situations such as those described in Sections 4.5 and 4.7.

A simple case is the selective competition among three species in the absence of mutational coupling. The mutual coupling among the three species is due to the fact that they feed on a common source. Although time does not appear as an explicit variable, temporal changes in the relative composition of the reacting system express themselves in a trajectory which the system describes within the boundaries of the simplex. A stable fixed point can occur everywhere inside the simplex. Two examples are shown in Figure 4.8.1. The trajectory can start at any point inside the simplex according to the initial composition of the system. If we want to understand the course of the trajectory in physical terms, i.e. as an action of some "potential gradient" forcing the system to move in a certain direction, the example turns out not to be that "simple" because of the non-Euclidean nature of the metric valid in the constrained space of the simplex (see Appendix A4). This and some more "complex" examples are discussed in the appendix. In the absence of stable fixed points more complex trajectories result. In the case of stable fixed points Figure 4.8.2 shows the destination for various types of trajectories. Many simplices applying to evolution and the origin of self-reproduction can be found in the book *The Hypercycle. A Principle of Natural Self-Organisation*[85] as well as in the contribution by Peter Schuster in Appendix A4.

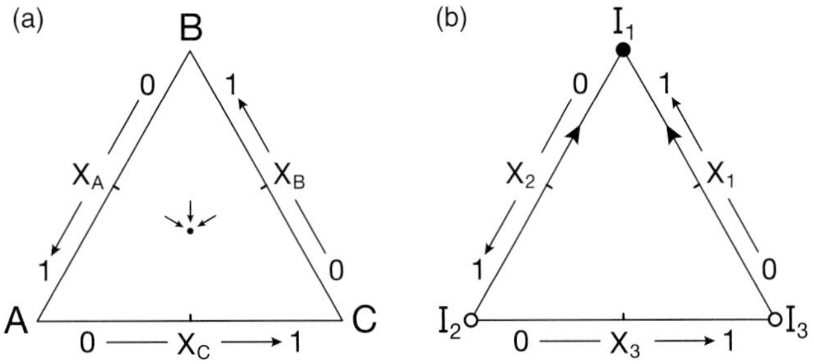

Figure 4.8.1

Stable fixed points in a triangular simplex

The triangular simplex is the (two-dimensional) representation of the ternary system of (three interacting) reactants under the constraint of conservation of the sum of their concentrations (see Section 2.6). Equilibrium among the three compounds yields a stable fixed point, called a "focus". (a) The focus approached by three equivalent partners (as referred to in Figure 2.6.5(b)). (b) The fixed point of selection, in this case among three independent (i.e. non-interacting) competitors (favouring I_1 in the upper corner of the triangle).

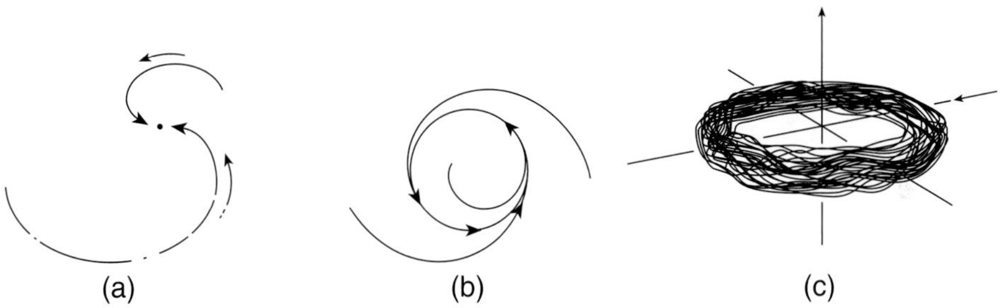

Figure 4.8.2

Examples of trajectories approaching stationary solutions

(a) Stable sink. (b) Stable harmonic oscillation. (c) Chaotic oscillation. Diagrams (a) and (b) describe the approach to stable solutions, such as a point-like sink (a) or a limit cycle representing an undamped oscillation among several reaction partners (b). Diagram (c) has no stable solution; it is characterised by a "strange attractor". In this system, any small perturbation will result in a trajectory which never reproduces itself, although it remains within a finitely limited region of phase space. It includes periodic orbits superimposed by periodic trajectories. Although this attractor is called "strange", it appears very frequently in nature. For examples of trajectories in evolutionary theory see the article by Peter Schuster in Appendix A4.

I became familiar with the term "simplex" in the late 1940s through the work of Karl Friedrich Bonhoeffer who, together with Paul Harteck[92] and simultaneously with (my teacher and thesis supervisor) Arnold Eucken[93], had discovered the two modifications of low-temperature hydrogen, i.e. *ortho-* and *para-*hydrogen. This work provided one of the early experimental corroborations of the validity of quantum mechanics and was mentioned as such in Heisenberg's Nobel laudatio.[94] Bonhoeffer, who introduced me to physical biology, was among the first physical chemists to use simplices for representing complex reaction behaviour based on non-linear differential equations (in his case, for mapping the oscillatory passivation of iron in nitric acid[95], which reproduces features appearing in Hodgkin and Huxley's model for the excitation of nerve cells[96]) In this work a new kind of complexity emerged which may be called "dynamical complexity", and on which I shall digress in this section.

Consider as an example a game of billiards. A typical trajectory of a billiard ball is drawn in Figure 4.8.3. It is physically entirely determined by the laws of elastic collision. A skilled player will therefore be able to guess pretty well the course of a trajectory that involves several reflections at the rectangular walls of the billiard table. Slight changes will result in equivalent slight parallel shifts of the trajectories, while small changes of the angle will cause deviations that increase linearly with the length of the trajectory. A general rule is that "small causes have small effects".

Now envisage a billiard table that contains, in addition, in its centre, a cyclic wall that acts as a reflector and thereby interferes with the trajectories of the billiard balls. This version of the game was introduced by the Russian mathematician Yakov Sinai,[97] who now teaches at Princeton University. Sinai has become famous through his work in statistical physics, which he did in co-operation with his mentor Andrey Nikolajevich Kolmogorov and which introduced a new concept of entropy[98] (called Kolmogorov–Sinai entropy).

As seen in Figure 4.8.4, reflection at the inner circular wall yields, for the incoming and outgoing ball, equal angles only with respect to the tangent of the circle which is perpendicular to the radial line at the point of deflection. The inclination of the

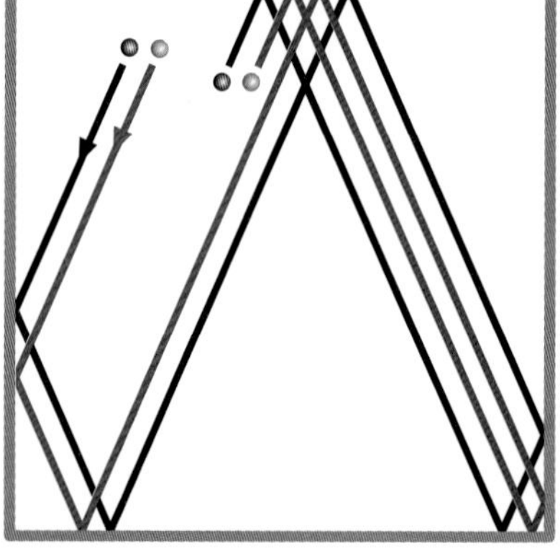

Figure 4.8.3
A typical trajectory of a billiard ball

As described in the text, small perturbations will cause only small parallel shifts in the trajectories of billiard balls.

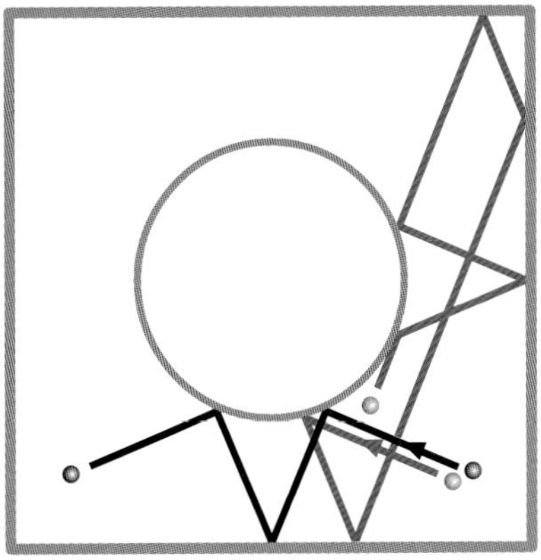

Figure 4.8.4
A circular wall at the centre of a billiard table

The circular wall will cause drastic changes in the trajectories. Initial deviations, although arbitrarily small, will grow exponentially and finally produce large and entirely unpredictable deviations (see text).

tangent changes very sensitively with the position along the circle. Here, small differences in the initial position as well as in the initial angle will cause strongly diverging trajectories. The deviations in this case grow exponentially, so that after relatively few reflections no similarity of the trajectories is recognisable any more, even if initial differences are "atomically" small. Here the small causes have large effects. The behaviour, although based on deterministic laws, becomes non-deterministic after the elapse of some finite time. We might call it (a form of) Russian roulette. In order to predict the behaviour over any protracted range of time we would have to know the initial conditions more precisely than is allowed by the laws of physics. This is what I mean by "dynamical complexity".

Today we know many examples of deterministic behaviour with entirely unpredictable outcomes. A century ago, Henry Poincaré noticed that the three-body interaction in astronomy is mathematically intractable[99] as was shown in Chapter 2 (cf. Figure 2.3.1). When we talk of unpredictability, the weather forecast immediately comes to our mind. Weather is definitely based on the laws of fluid- and thermo-dynamics, both of which are deterministic. Nevertheless, long-term prediction usually failed, until the MIT meteorologist Edward Lorenz[100] showed that the unpredictability of weather is of a fundamental nature, caused by the sensitivity of atmospheric dynamics to initial conditions. The story of the butterfly in South America causing a thunderstorm in Alaska can be found in most textbooks that deal with these phenomena. The term "(deterministic) chaos" has been coined in order to deal with the unpredictability caused by the sensitive dependence of deterministic processes on initial conditions. By virtue of its cause this kind of chaos is to be distinguished from truly undetermined random behaviour.

After mentioning two other early pioneers, David Ruelle, who in 1971 wrote a pivotal paper "On the Nature of Turbulence" together with Floris Takens[101], and Otto Rössler, who in the 1970s came up with a particularly simple mathematical model of a chaotic system[102], I am not going to rewrite here these stories about the discovery of chaos, all of which can be found in the still growing, fashionable chaos literature. Let me return to the examples that are more closely related to the theme of this book and which show the twofold aspect of complexity as we encounter it in information dynamics.

My story goes back to an observation made by the mathematically minded ecologist and former President of the Royal Society Lord Robert May. In a ground-breaking article, published in the British journal *Nature* in 1976[103], May analysed the daily fluctuations of the population number of a colony of blowflies, which were studied in classic experiments by A.J. Nicholson.[104] When I referred to May's article as being "ground-breaking" I mean just that: it was a major trigger of the flood of papers on chaotic dynamics which has built up during the last three decades.

Robert May's ingenious idea was to try to correlate population growth in an insect colony with an equation that has its roots in the early 19th century. It goes back to ideas coined by the British economist Thomas Robert Malthus and had much influence on Darwin's thinking about natural selection being the driving force of evolution. Malthus saw the limitations imposed by the rate of food supply, which usually occurs at an arithmetic (e.g. linearly increasing) rate, on the law of geometric progression in an (exponentially increasing) population. The rate equation that is representative of this situation is called the "logistic equation" and was explicitly formulated in the 19th century by the Belgian mathematician and sociologist Pierre-Francois Verhulst[105]. Its name emanated from the French "logistique" which in military science deals with the billeting and provision of troops. Verhulst's formula equates the growth rate with the product $x(1-x)$ describing a parabola, where x is a normalised relative population number (i.e. $0 \leq x \leq 1$). The logistic parabola is also called a "quadratic map". As long as x is small compared with 1, the rate is proportional to the population number, yielding geometric progression as considered in Section 4.7. If x, however, becomes so large as to approach 1, the term $(1-x)$ becomes very small and causes the rate to turn back to zero, thereby yielding saturation expressed by a constant population number. The logistic equation in differential form can be easily integrated and always yields for the population size a curve that starts exponentially from $x_0 \ll 1$ and approaches an eventually constant value $x = 1$.

The logistic equation in differential form applies well to molecular growth phenomena, if the supply of substrates is limited. Natural populations, however, often multiply in synchronised rhythms, e.g. diurnally, seasonally or annually. This was the case for the insect populations observed by May. Let us use a discrete logistic mechanism and repeat such an experiment on paper as follows. In order to get the population number in the time interval $t + 1$ we have to start with the population number at time t, multiply by an amplification factor and subtract a term that is the product of the amplification factor k and the square of the population number at time t. Expressed in the form of an equation it resembles the logistic form: $x_{t+1} = x_t + kx_t - kx_t^2$. Note that any x values are larger than 0, but smaller than 1. For $x_0 = 0$ no growth can occur and x_{t+1} will equal $x_t = 0$. However, the solution is unstable, as a small deviation from $x_0 = 0$ will trigger growth for any k value greater than 1. Examples are shown below, where I discuss the mathematical analysis, which includes the transition to chaotic solutions as observed by Robert May.

At this point let me interrupt the more serious expositions of "mathematical logistics" and tell a related story – just for fun.

Verhulst's equation, on which the logistic algorithm is based, is a phenomenological representation, meant to show the general behaviour of growth limitation due to shortage of supply or other kinds of exhaustion. Its explanatory power with regard to causes or mechanisms is limited and requires careful analysis of the "whereabouts" of every single case. The mathematician John Casti, in his book on complexity,[106] describes analyses of this kind made by Cesare Marchetti of the International Institute for Applied Systems Analysis in Laxenburg, Austria, which were applied to a diversity of technological, political, social and psychological phenomena, such as computer development in Japan, world airline traffic, the building of Gothic cathedrals in Europe, and – believe it or not – Mozart's musical compositions, all (seemingly) following the same simple dynamical pattern suggested by the logistic law.

In the case of Mozart's work, Marchetti proceeded as follows. The more than 600 compositions that Mozart wrote in the 35 years of his life were "typified" by some (obscure) statistics applied to (only!) 35 selected "major" (?) pieces. The cumulative number of these pieces, after normalisation, is a population number x that runs from 0 to 1. According to an integrated Verhulst equation, an exponential function is obtained which in a logarithmic plot is represented by a straight line. The annual sequence of cumulative numbers extracted from the analysis of Mozart's 35 selected "masterpieces" fits remarkably well to the logarithmic plot for the logistic curve. John Casti concludes from Marchetti's plots: "This [...] illustration is particularly intriguing since it suggests that each of us has some kind of "internal" program regulating our creative output until death. Moreover, as the saturation point is approached, people seem to die when they have exhausted 90 to 95 percent of their potential, as measured by the limiting values of their productive curve."

Surely you're joking, Mr. Casti!

(1) Mozart (trivially) could not continue to compose after he was dead, although he did so up to his last breath. Accordingly, the curve of his cumulative production, whatever its nature, reached a fixed value at the end of his life, although the approach need not have been asymptotic.

(2) It is true that different rate laws might be distinguished, if sufficient statistically weighted data were involved. Thirty-five selected works for a time span of 35 years is a very meagre selection, especially for a logarithmic plot, which is relatively insensitive to (quite well recognisable) fluctuations.

(3) A selection of 35 out of some 600 works is certainly biased, especially since the whole set is only "partially ordered", allowing for non-comparable alternatives.

(4) The musicological fact is that most of Mozart's important works bear high Köchel numbers (i.e. KV > 550), having been composed in his last years. Both the *Magic Flute* (KV 620) and the *Requiem* (KV 626) were composed in the very last year of his life, and the latter was not even finished. Other highlights are his late string quintets (KV 593 and 614) and his two clarinet works (the concert KV 622 and the quintet KV 581), as well as some wonderful piano works such as the *Adagio in B flat minor* (KV 549), or the *Gigue in G* (KV 574). These late compositions involve new stylistic elements, such as the provoking effects brought about by major sevenths and minor ninths, the intervals closest to an octave. The use of these intervals went far ahead of Mozart's time, eventually coming to fruition through the equalisation of all intervals by Arnold Schoenberg, Alban Berg and Anton Webern. On the other hand, Mozart does not fit easily into any scheme. He did important composition work before he was 10 years old. One of the early piano concerts, the *Jeunehomme Concerto* (KV 271), composed at the age of 20, is acknowledged to be as beautiful as his late concertos.

Mozart died from illness at the height of his artistic creativity. In our times, he would probably have survived and continued to produce further great works of music for many more years. At the same time, it is true that many people who live into old age become increasingly unproductive in their later days – and some even quite early in life, although Mozart certainly did not belong to this group.

May I repeat: I consider the above just an amusing story that demonstrates the pitfalls of applying mathematical analysis to the art works of geniuses and to their way of production, which does not follow simple laws. It by no means diminishes my respect for John Casti's excellent book, which contains lucid presentations of other scientific ideas, to which I shall now return.

The true revelation of non-linear dynamics came with the discovery of "deterministic chaos". Chaos theory as such has certainly several fathers, but who wants to be called a father of chaos? It is like calling James Watt the inventor of work.

In the 1970s Mitchell Feigenbaum, at a relatively young age, joined a research group on dynamical systems at Los Alamos. He came there without really knowing what he was supposed to do. While musing around and waiting for inspiration he played with his pocket calculator. All right! Isn't "musing" what scientists do for most of their time? And isn't a pocket calculator (or now a laptop) just the right tool for a present-day physicist to check new ideas quickly? Mitchell Feigenbaum most probably got as puzzled as anybody testing out the logistic algorithm, but, while May saw its biological relevance, Feigenbaum scented its fundamental mathematical implications. And then he had to get down to the nitty-gritty of mathematical proof in order to answer "nasty" questions set by his mathematical colleagues, like the one pithily pointed at him, after he had given a talk at Rockefeller, by Marc Kac: "Sir, do you mean to offer numerics or a proof?"[107]

The numerics were certainly those of the logistic equation, which May fathomed in the experiments with insect populations.

The appearance of bifurcations was not in itself earth-shaking news to physicists in the 1970s. One had to look a little more closely in order to find anything really new or surprising. In Figure 4.8.5 some x values obtained by the logistic algorithm for three discrete values of k, i.e. 3.20, 3.48 and 3.55 are recorded. A cursory glance reveals a similar up-and-down of the x values in the three series, indicating a rhythm of period two. A closer look, however, reveals that this is true only for the first series (referring to $k = 3.20$). If we define the period of oscillation as the time span (or number of discrete steps $t \rightarrow t + 1$) after which a former interval reproduces itself exactly (rather than just counting ups and downs) we realise that the three series have period lengths of two, four and eight. Figure 4.8.6 then shows that the period doublings involved occur at discrete values of k, the intervals $\Delta k = k_{n+1} - k_n$, which converge to 0 at the defined value of the growth parameter, $k = 3.569946\ldots$. In this series not only the Δk intervals themselves, but also their ratios shrink with increasing n. In other words, the ratios $\Delta k_n / \Delta k_{n+1}$ for two successive periods n and n + 1 are not constant but decrease towards a limiting value: $\lim_{n \to \infty} \frac{\Delta k_n}{\Delta k_{n+1}} = 4.6692016091029\ldots$[108] That's what Feigenbaum found out and the above number therefore is called Feigenbaum constant. It determines the limits of the growth parameter k where period doubling degenerates into "chaotic" behaviour.

A pretty good estimate of Feigenbaum's number had been obtained before by Siegfried Grossmann[109] from a bifurcation rate scaling law: $k_n = k_F - c_1 e^{-c_2}$, which for the logistic equation yielded the constants $k_F = 3.569946 \pm 1.3 \times 10^{-7}$, $c_1 = 2.628 \pm 0.13$ and $c_2 = 1.543 \pm 0.02$. This scaling law shows the strong convergence and allows one to calculate Feigenbaum's number to a limit expressed by the error rates. Both Feigenbaum and Grossmann did their work independently of one another at a time that apparently was ripe for the discovery of this new constant. A suitable symbol for the Feigenbaum–Grossmann constant is F. The letter F, as used by the Romans, derives from the "digamma", a letter in ancient Greek preceding the Attic alphabet. "Digamma", which means double gamma, is resembled by a superposition of two gammas (ΓΓ). The gamma in the Roman alphabet is the G, hence the letter F = double gamma recognises both <u>F</u>eigenbaum and <u>G</u>rossmann.

Feigenbaum showed that there is another scaling law for the decrease of the (properly defined) width of the pitchfork, which is related to the scaling law of the bifurcation rate.[110] In the limit of period doubling for $n \rightarrow \infty$, the initial parabola of the quadratic map becomes a transcendental function which, according to Feigenbaum, can be represented by an infinite power series. In particular, renormalisation revealed that this is a function that holds universally for all maps with quadratic maxima. (It should be noticed that period doubling as occurring in quadratic maps is not the only "route to chaos"). Given these facts, one can conjecture that F may be another (fundamental) number.

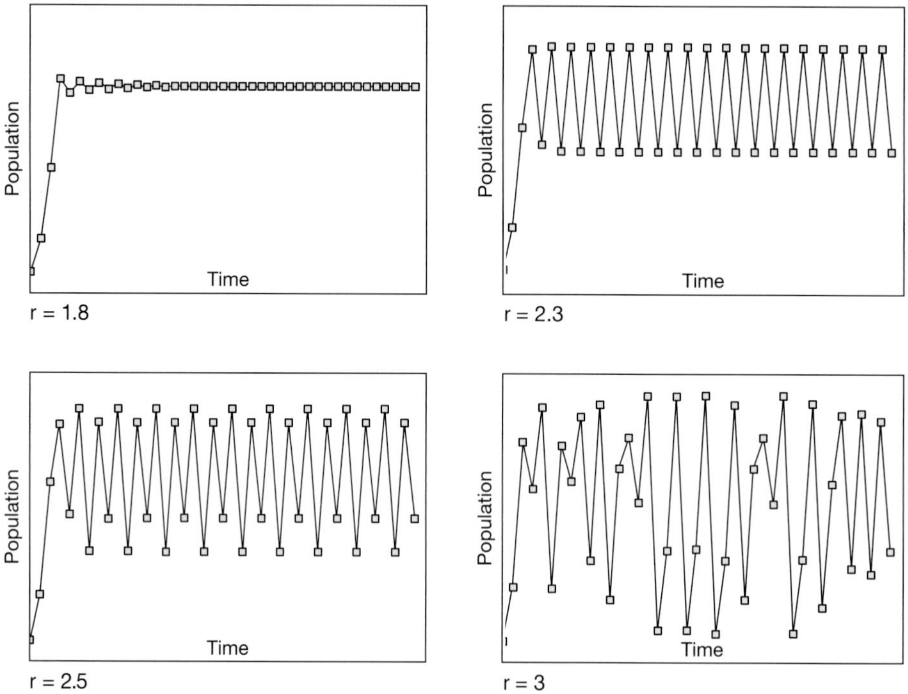

Figure 4.8.5
Chaotic dynamics in the growth of insect populations

This observation, analysed by Lord Robert May, was at the start of a new development in the natural sciences which may be epitomised by two catchwords: fractals and chaos. The four snapshots show what happens when population growth describable by a Verhulst equation proceeds in discrete steps. The dynamical law according to Peitgen and Richter takes the form:

$$x_{n+1} = f(x_n) = (1 + r)x_n - rx_n^2$$

where x_n describes the relative population size at the nth step and r is a constant $r = (x_{n+a} - x_n)/x_n$ describing the relative growth rate (e.g. increase per year). After n years the population size would be $(1 + r)^n x_0$. The four diagrams refer to the r values given beneath each diagram. At low values (e.g. $r = 1.8$) we obtain a constant stationary value for x, which for larger values (e.g. $r = 2.3$) becomes periodic. However, at $r = 2.570$ the process does not remain periodic at all: it jumps around to quite different values.
(see example for $r = 3.0$)

Another interesting aspect that I can only touch upon in passing is that of self-similarity and fractal behaviour. Benoit Mandelbrot is the great pioneer of these concepts.[111] Chaos, self-similarity and fractal structure are phenomena that are closely interrelated. The Mandelbrot set $Z_{n+1} = Z_n^2 + c$ also has a quadratic map, but it derives all its flavours of simplicity by being applied to complex variables. ("Complex" in this connection refers to the numbers involved, which are composed

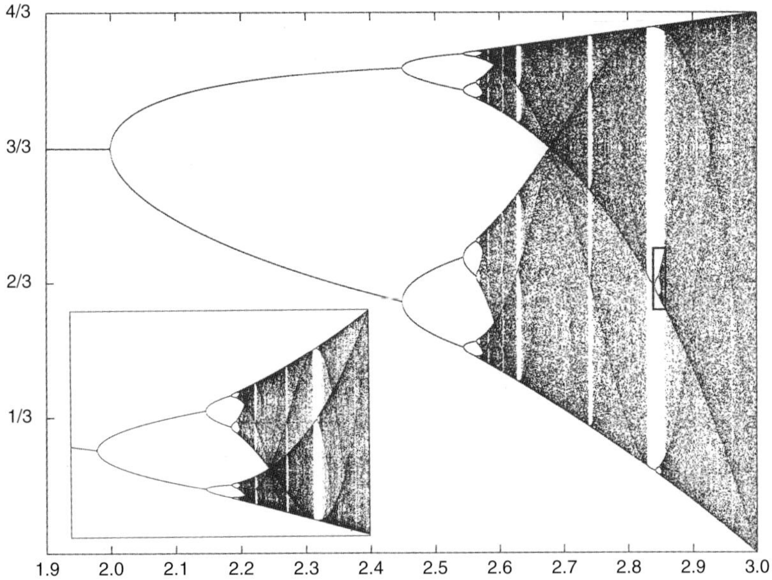

Figure 4.8.6

Bifurcation diagram of the Verhulst process

At larger r_n values we are dealing with a quite universal behaviour with period doubling, which occurs with bifurcations when the period 2^n becomes unstable in favour of period 2^{n+1}. The length of successive intervals $\delta_n = (r_n - r_{n-1} - 1)/(r_{n+1} - r_n)$ approaches a universal value of 4.669..., which also appears in a variety of other processes based on period doubling. This figure shows the r dependence of such asymptotic dynamics. More details can be found in the paper quoted above.[112]

of a real and an imaginary part: a + bi, a and b being real numbers and $i \equiv \sqrt{-1}$ the imaginary unit.) The "beauty of fractals" – this is the title of a book on complex dynamics by Heinz Otto Peitgen and Peter Richter,[112] magnificently illustrated with computer graphics – conveys direct sensory access to a magic new world in which science mixes with art (see Figure 4.8.7). My late friend the Hungarian composer György Ligeti was so excited about these graphs that he asked me, "Why hasn't a Nobel prize been given for this work?"

Can a simplex be complex? I think the answer to this question has been given by the above deliberations. "Deterministic chaos" represents the "chaotic" degeneration of a limit-cycle oscillation. The corresponding attractor is called "strange". Unlike the limit-cycle it does not reproduce itself, but rather fills part of phase space in a quasi-continuous way. In order to predict the exact path of the phase point, one would have to know the initial conditions more precisely than is ultimately permitted by physical law. While not all regions of phase space are accessible to the trajectory,

Figure 4.8.7

The Beauty of Fractals

This book by Hans Otto Peitgen and Peter Richter, which I have quoted in the text, is expressed fully in this picture, which was obtained from the expression $x \rightarrow x^2 + c$ with c being a complex parameter (i.e. containing a real and an imaginary part).
Reproduced with kind permission from Springer Science+Business Media.

there remains sufficient space that can be reached, causing dynamical complexity which may be interpreted as "unpredictable".

In dealing with complexity, an important question is that of the stability of attractors in the phase space of a simplex. Stability analysis – originally applied to the "stability of motion" – is associated with the Russian mathematician Aleksandr Mikhailovich Lyapunov.[113] The Lyapunov exponent is indicative of the divergence or convergence of two neighbouring trajectories in phase space. It can be positive, zero or negative and may be taken locally or as an average over the entire attractor. A positive coefficient indicates exponential divergence of the trajectories, i.e. instability or chaotic behaviour, while a negative value means convergence or stability. There are as many such numbers as there are dimensions of phase space. However, the individual numbers are not related to the co-ordinates, but rather refer to some "principal axes", the number of which equals that of the co-ordinates in phase space. Dynamical instability, expressed as a rate of growth of small deviations, was first defined mathematically by Sofya Kovalevskaya[114].

An analytically tractable algorithm that also applies to complex cases was obtained by Gregoire Nicolis, a member of the famous Brussels school of Ilya Prigogine. The rate function at the attractor $\mathbf{X}_s$ (the bold symbol indicates that it refers to a vector that has as many components as axes in phase space) is considered at perturbed positions $\mathbf{X}_s + \mathbf{x}_i$ (where $\mathbf{x}_i$ denotes small deviations in the co-ordinates of the vector) and then expanded as a power series of $\mathbf{x}$. The system of linear differential equations for the perturbed part (in which higher-order terms are omitted) can be solved. A decay of the perturbation, requiring all eigenvalues of the linear system to be negative, is a criterion of "asymptotic stability". An example is the relaxation behaviour of complex chemical systems near equilibrium.[115]

Linear analysis, however, is inadequate in cases of instability, including chaotic behaviour. A global treatment of the non-linear problem meets with unresolved difficulties and confines attention to local behaviour in the vicinity of bifurcations. A particular kind of bifurcation, the "pitchfork", has already been shown above to occur in the period doublings of the logistic parabola. There are other kinds of bifurcation that occur in non-linear chemical reaction systems (e.g. the Hopf bifurcation in Prigogine's "Brusselator")[116]. The transition from a fixed point to chaotic behaviour, as a rule, proceeds through a series of bifurcations describing an onset of oscillations of increasing complexity. Chaos may therefore be considered as a loss of stability of periodic solutions. In a way, it reminds us of the properties of real numbers: In the continuum characterised by an infinite number of decimal fractions the integers may be considered the analogue of stable fixed points, the rationals, showing periodic interactions in their fractional decimals, correspond to limit cycles where, again, quite complex periodicities may occur, and finally the irrationals, the majority in this continuum, uncover "chaos" in the sequential order of their decimals.

An elegant way of analysing chaotic dynamics was proposed by Henri Poincaré, who discovered this type of behaviour long before it became fashionable in physics. A more detailed description of his work is beyond the scope of this book. Instead, I refer to the excellent coverage of this field by Gregoire Nicolis[117]. Very instructive, especially to "newcomers", is also Garnett P. Williams' book[118] *Chaos Theory Tamed*, which gives a more popular account of the fundamentals of chaos theory, including an extensive discussion of information-theoretical concepts such as mutual information, redundancy and Kolmogorov–Sinai entropy.

At the end of this section I want to return to a problem left open in the preceding section, namely the representation of the phenomenon of natural selection by a simplex diagram. I take it from Appendix A4, written by my colleague and friend Peter Schuster (see also Figure 4.8.8). In this connection the question arose: Can we represent "natural selection" as being caused by some acting force (as suggested by the analogy to Mandelbrot's raindrop problem, discussed at the end of the preceding section? The answer is "yes", but we have to take into consideration some mathematical peculiarities, such as the non-Euclidean nature of the simplex space, as proposed by Siavash Shahshahani. In Appendix A4 Peter Schuster goes into more details of this sort of question.

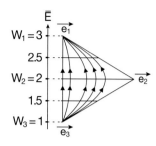

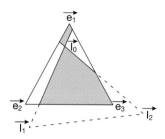

Figure 4.8.8

The simplex characterising Darwinian selection

Here I show a "complex simplex", which is explained in more detail in Peter Schuster's article in Appendix A4. This simplex may look simple, but it contains all the secrets of the complex familiarity of natural selection. The left-hand figure shows the curved trajectories which require application of Riemannian geometry as proposed by Siavash Shahshahani. The right-hand figure shows the ranges in which the fixed points of selection are to be found (see the contribution by Peter Schuster in Appendix A4).

4.9. What Does "Meaning" Mean?

In daily usage the word "information" is nearly exclusively associated with its semantic aspect. "Semantic", derived from the Greek (σεμαντικός), pertains to "meaning". Etymological identification does not yet provide a definition.

In his philosophical discourse *Die Einheit der Natur*[119] (*Unity of Nature*) the late Carl Friedrich von Weizsäcker formulated two theses in order to "objectify" the semantic aspect of information:

(1) Information is only that which is understood.
(2) Information is only that which produces information.

When in a private discussion I expressed some qualms about the first statement, arguing that the term "understand" defies my understanding more than the term "information" which, at least in molecular genetics, can now be well substantiated von Weizsäcker's immediate reaction was: "but every child knows whether or not it understands what it is told".

Yes, true! That's one of the "complex familiarities" which I addressed in the title of this book. Everyone knows what is meant by the word "understanding", yet nobody can tell what is really happening in our brains that creates the impression (or illusion?) of "understanding". Even to the neurobiologist much of it still is a book with seven seals.

In a text written some years later, von Weizsäcker[120] qualified his two statements as not intended to provide a definition, but rather being of some explanatory character. The second thesis, in particular, is supposed to illustrate what is meant by the first one, while both were used by him in support of his qualitative concept of "pragmatic information", which was discussed briefly in Section 3.6. An alternative interpretation of semantic information comes from the late Donald McKay, who said:[121] "The meaning of a message can be defined [...] as its selective function on the range of the recipient's state of conditional readiness for goal-directed activity." A comparative discussion of those qualitative concepts can be found in Bernd-Olaf Küppers' instructive monograph *Information and the Origin of Life*.[122]

In fact, McKay's formulation, although it refers to neural information, offers itself more directly to an interpretation of "meaning" associated with genetic information. What happens to a genetic message depends indeed on the "conditional readiness" of a molecular system to respond with "goal-directed activity", and this "selective function" is what characterises the "meaning of the message". In other words, the message must arrive in an environment that is "ready" (or able) to read and translate it, for instance, into some enzymic function which triggers "goal-directed activity". It also means – and this will turn out to be important later on – that the "meaning" of the genetic message had to evolve first in the context of an environment in which it could be "understood".

I should like to mention that I always cherished the clarity of Donald McKay's interpretations. In the 1960s we often shared a plane between London and Boston when attending Francis Otto Schmitt's Neurosciences Research Program meetings, which were hosted by the American Academy of Arts and Sciences at Newton, Massachusetts.

Semantic information or "meaning" is closely related to what Paul Davies[123] has called "specified complexity". A sentence comprising some 100-odd binary digits – or likewise a gene made up of a similar number of nucleotide monomers – represents complexity expressible in terms of Shannon entropy. What is involved is combinatorial complexity, i.e. the huge number of 2^{ν} (e.g. 10^{100}) alternative combinations of ν binary symbols that represent a typical sentence or gene. However, as Davies noted, the particular combination of the symbols in a sentence that "makes sense" has two aspects: "complexity", as expressed by contingency, and "specification", which cannot be explained by contingency, but rather must be due to some physical law acting in a given "environmental context". The paradox lies in the fact that both complexity and specification are inversely related to one another. The larger the complexity the smaller the chance of a specific outcome.

In a formal philosophical approach William Dembski[124] refers to Paul Davies' "specified complexity" and argues that no presently existing method is able to account for its occurrence. He, in particular, focuses on Darwinian theory and chooses as a primary target of his criticism Richard Dawkins' book *The Blind Watchmaker*.[125] In this book Dawkins tries to introduce Darwinian theory to a broader audience and I think he does it in an excellent way.

Dawkins' point of departure is an oddly shaped cloud, such as we sometimes see in the sky on a hot and sultry summer's day. Such a cloud often stimulates our fantasy to compare it to an animal, or a face, or what have you. This situation occurs in a scene of Shakespeare's *Hamlet*, where Polonius and Hamlet muse about the shape of a cloud and Polonius says: "Methinks 'tis like a weasel." As much as the example of an accidentally forming cloud bears relevance to the problem in question, Dawkins rather engages in the problem of how to bring about a meaningful sentence like the one above, i.e. a sequence of 28 characters (including the spacings between words). He contrasts an "all-or-none" procedure in which the whole sentence appears in one giant step with a stepwise accumulation of changes under some selection force.

For the first case he refers to the proverbial monkey who hammers at random at the keyboard of a typewriter, and asks for the probability that the monkey comes up with a given correct sequence. The chance of that happening he estimates to be one in 10,000 million million million million million million, or $\sim 10^{-40}$. Whatever we think of such "monkey business", it means that pure chance has no chance at all in "our business".

For the second case he starts from a random sequence which he duplicates, allowing some random inclusion of errors. At each step he chooses that sequence that matches best the previously known "target" sequence for further duplication, excluding any coinciding character from undergoing further changes. In three computer simulations Dawkins reached the target sentence after 43, 64 and 41 steps.

One may be surprised that, given 28 places, each having 26 alternative characters, the goal was achieved within such a short time. However, one must be aware that Shannon's theory reduces the number of symbols in binary coding to around 100, of which around 50% may agree accidentally with the target sequence. So what remains to be done on average are some 50-odd binary changes.

One may, of course, raise the objection that in this procedure the target sequence must be known in advance. No, it need not be known to the player who is informed. As in the game of Twenty Questions (see Section 1.10) it may be an external agent who decides what changes are to be accepted. Or there may be an environmental selection pressure which takes care of the job, as is the case for natural selection. What remains, then, is an objection against the entirely unrealistic nature of the selection procedure. Dawkins, in fact, anticipated such objections and stated clearly that in biological evolution the target structure is initially indeterminate and only takes shape during the evolutionary process. The situation is very similar to the particular version of the game of 20 questions described by Wheeler in Section 1.10. The essential point is the presence of some selection pressure.

Dembski[126] criticises Dawkins (and – as he says – other "fellow Darwinists") for using a procedure that does not *produce* specified complexity but rather *shuffles it around.* What does he mean? In order to arrive at any particular sequence of characters one certainly has to "shuffle around" symbols. The question is "how" one does it and whether it produces a specified sequence. In Dawkins' game it is a sequence that is unknown initially to the player and which eventually makes sense to him.

The objection that prior knowledge of the target sentence prevents a creation of "specified complexity" to me seems far-fetched. It was chosen in order to provide a simple selection procedure. However, it need not be done this way if other selection criteria can be defined. Some 20 years ago we proposed a similar game in order to demonstrate the role of the error rate in the evolutionary process, and we received similar complaints, which were in our view beside the point.[127]

For this game Ruthild Winkler-Oswatitsch had coined the sentence "take advantage of mistake" thereby expressing the aim of the game. Its German version reads "Lern aus den Fehlern". Both the English and the German versions start and end with the same sequence of letters (i.e. "take" or "lern"). This sequence is, so to speak, the promoter sequence for starting a reproduction (which was of special relevance in the hypercyclic version of the game, see Section 5.6). Otherwise the game proceeds like Dawkins' version. In our case, we varied the error rate systematically and demonstrated the existence of an error catastrophe, which appears abruptly when a certain threshold value of errors is crossed (see also Section 5.5). Vignette 4.9.1 reproduces the results obtained in two computer simulations, obtained in 1977, which were published in a book written together with Peter Schuster[128].

Vignette 4.9.1 The Hypercycle*
TAKE ADVANTAGE OF MISTAKE

Self-correction of sentences is the result of the evolution game, exemplified in this vignette.

TAKE ADVANTAGE OF MISTAKE

It has been chosen because it provides 'selectively advantageous' information with respect to the mechanism of evolution. Its special form permits a cyclic closure, whenever functional links among the single words are introduced. Using a code, in which each letter (and word spacing) is represented by a quintet of binary symbols, the information content amounts to $\nu_m = 125$ bits, allowing for about 4×10^{37} alternatives. This number excludes appearance of information by mere chance. The sequences of letters shown in these tables for given generations have been sampled as being representative for the total population of sequences in the computer store.

INITIAL SEQUENCE:	BAK GEVLNT GUPIF LESTKKM
DIGIT QUALITY $\bar{q}_m$: 0.995	
SELECTIVE ADVANTAGE PER BIT: 10	
1. GENERATION:	RAK GEVNNT GUPQF KESTKKM
5. GENERATION:	NAK AEZ,NS GEPOF MESTMKU
10. GENERATION:	VAKF ADV!NT.GE OF MISD!KE
16. GENERATION: (GOAL REACHED)	TAKE ADVANTAGE OF MISTAKE

The first example demonstrates that evolution is very efficient near the critical value $1 - \bar{q} \approx 1/\nu_m$, which with $\nu_m = 125$ amounts to $\bar{q}_m = 0.992$. Starting with a random sequence of letters the target sentence is usually reached within $20(\pm 6)$ generations for any value of $\bar{q}$ between 0.995 and 0.990. This efficiency near the threshold is even more evident if we compare the evolutionary progress at a given generation for various values of $\bar{q}_m$:

Cont. ➲

Cont.

INITIAL SEQUENCE: BAK GEVLNT GUPIF LESTKKM

SELECTIVE ADVANTAGE PER BIT: 10

$\bar{q}_m$	BEST SEQUENCE AFTER 11 GENERATIONS	NUMBER OF MISTAKES
0.999	LAKD AEV.NTAGU AF KISTQKM	9
0.995	TAKE ADV!NT GE OF MISTAKE	2
0.990	TAKEBADV!NTAGE OF MISXAKE	3
0.985	VATA ADBKMDI DHOD ?CSYBKE	16

Analogous behaviour is found for the disintegration of information for $\upsilon_m > \upsilon_{max}$. At an error rate $(1 - \bar{q}_m) = 1.5\%$, a selective advantage per bit of 2.5 corresponds to a σ_m value of about 5. Under these conditions the information is not stable any more. For small selective values (as in this example), however, disintegration (or accumulation) of information is a comparatively slow process.

INITIAL SENTENCE: TAKE ADVANTAGE OF MISTAKE

DIGIT QUALITY $\bar{q}_m$: 0.985

SELECTIVE ADVANTAGE PER BIT: 2,5

NUMBER OF GENERATIONS	BEST SEQUENCE	NUMBER OF MISTAKES
1	TAKE ADVANTAGE OF MISTAKE	0
5	TAKF !DVALTAGE OF MISTAKE	8
10	TALF ADVALTACE OF MISTAKI	5
20	DAKE ADUAVEAGE OF MJUTAKE	6
40	TAKE ADVONTQCU OF MFST!ME	7
71	TAKEB ?VALTAGI LV MIST!KE	8
71 GENERATION FOR $\bar{q}_m = 0.97$	?AMEBADTIMOACFHQEBA!STBMF	18

Comparison of the last two rows (referring to generation 71) shows that disintegration is much faster at an error rate of 3% ($\bar{q}_m = 0.97$).

For the other example (selective advantage of the bit = 10) the threshold is not yet passed at $\bar{q}_m = 0.985$; so that information is stable, as seen below, where the process starts with the correct sentence.

The fact that linear coupling – if it works at all – feeds all the advantage forward to the last member in the sequence provides a strong hint for a possible solution of the problem: The couplings should form a closed loop:

INITIAL SENTENCE: TAKE ADVANTAGE OF MISTAKE

DIGIT QUALITY $\bar{q}_m$: 0.985

SELECTIVE ADVANTAGE PER BIT: 10

71 GENERATIONS OUTPRINT OF 8 REPRESENTATIVE SENTENCES	NUMBER OF MISTAKES
TAKE ADVANTAGE OF MISTAKE	0
TAKE ADVANTAGIPOF MISTAKE	2
TAKE ADVANTAGE OF MISTAKE	0
TBKE !DVANTAGE OF MISTAKE	2
SAKE ADVANTAGE OF MGSTAME	3
TAOE ADVANVAGE OF MISTAKE	2
TAKE ADVAVTAGE OF MISTAKE	1
TAKE .DVANTAGE OF MISTAKE	1

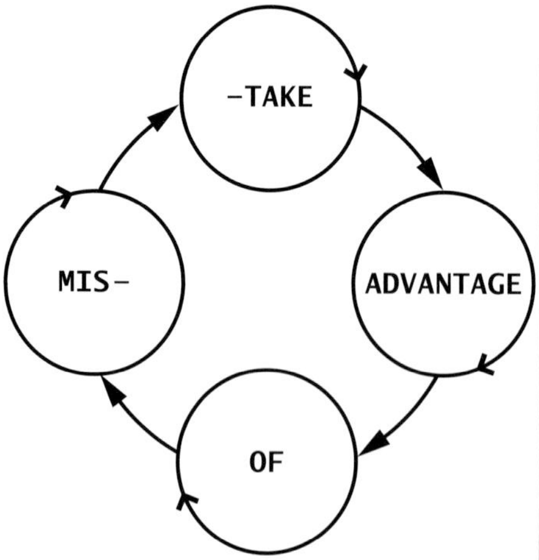

Cont.

Cont.

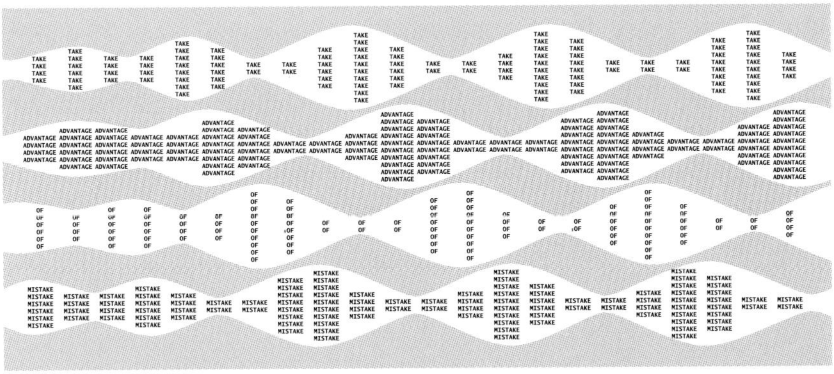

Then the enhancement due to the coupling will cyclically fluctuate through all words of the sequence. Our sentence actually was chosen as to provide automatically such a cycle overlap through the word 'mis*take*'. Since each word is a catalytic cycle (i.e. a self-replicative unit) the system represents a hypercycle of second degree according to our definitions. The result of the game is represented in the figure above. All four words show a stable steady-state representation with a periodic variation of their population numbers. The selective values of different words do not have to be the same – which would seem very improbable for any realistic system. Each word, furthermore, is represented by a stable distribution of mutants. Unless one of the words is wiped out by a fluctuation catastrophe (which becomes very improbable at a sufficiently large number of copies) the population numbers will continue to oscillate. In other words: The information of the whole sentence is stable.

Each word of the sentence TAKE ADVANTAGE OF MISTAKE represents a self-reproducing unit. The information of the sentence is stabilised by hypercyclic coupling among the words. In the graph, each printed word is representative of 10 copies in the computer store. All words in a vertical row refer to the same time, the intervals of which change discontinuously along the horizontal axis. The oscillation builds up from an initial equipartition of words and maintains a phase shift from word to word. The error distribution for each of the four words is stable.

*Text reproduced from Eigen, M. and Schuster, P. (1979). *The Hypercycle. A Principle of Natural Selforganization*. Springer–Verlag, Berlin, Heidelberg and New York.

Nowadays we are able to complement the computer simulations with test-tube experiments in order to study molecular evolution, both *in vitro* and *in vivo*. The first to do such experiments, in the 1960s, was the late Sol Spiegelman, who pioneered the "serial transfer" technique on which our present automated evolution machines are based.[129,130] The principle of serial transfer is explained in Figure 4.9.1, while Figures 5.9.3 to 5.9.7 in Chapter 5 show a modern technical realisation of the principle. Spiegelman isolated and purified the replication enzyme of the bacteriophage Qβ. His reaction mixture consisted of a cell-free solution providing optimum reaction conditions and containing as substrates the (energy-rich) monomers of RNA, as

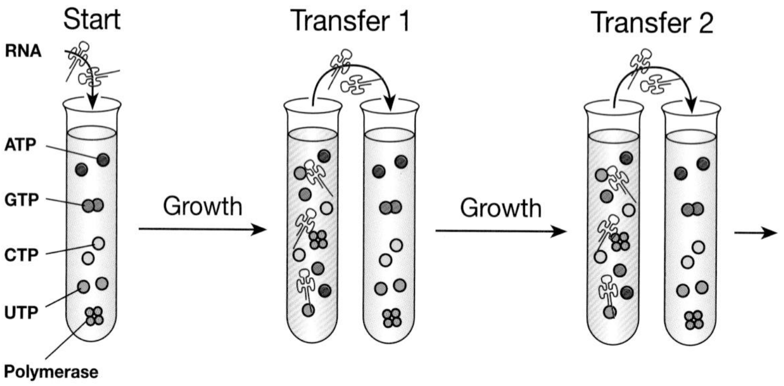

Figure 4.9.1

The principle of serial transfer

Serial transfer allows the observation of evolutionary phenomena under conditions of practically unlimited growth. The trick, to prevent an increase to an indefinitely large population, consists in alternating growth and removal in a stepwise procedure. The process starts in the far left test-tube, which holds the reaction mixture containing the four monomeric nucleotide substrates in energy-rich form (triphosphates: NTP), the specific polymerising enzyme (Qβ polymerase) and some required chemical factors (providing optimal ionic strength and pH). The template, a precisely calibrated amount of Qβ RNA, is added at time zero of the experiment. Now the replication process starts and is continued until a certain amount of RNA has been synthesised (e.g. a tenfold amplification of the initial amount of template). The growth process is then interrupted by a sudden temperature change, and an aliquot of the solution containing exactly the original amount of template is transferred to a new test-tube containing the original growth medium. This procedure can be repeated an indefinite number of times. It allows the evolution of optimally adapted mutants as in an unlimitedly growing population kept under constant environmental conditions during the whole procedure. In Section 5.9 it is shown how the procedure can be automated and developed into an evolution machinery with massively parallel production. It is interesting to note that the kinetics of amplification are observed only during the growth phases, which are not under the control of any flux that introduces non-linear terms (see Section 4.7). Selection occurs separately during the phases of serial transfer. This explains the fact that the solution depends only on the matrix describing the kinetics. The eigenvalues obtained refer to this very matrix in its diagonalised form.

a catalyst the purified Qβ replication enzyme and as the template a calibrated amount of wild-type Qβ RNA. In these first experiments he incubated his reaction mixture with the nearly full-length genomic RNA of phage Qβ, i.e. a molecule comprising some 3600 monomeric nucleotide units. (The full-length Qβ RNA contains about 4300 units.)

What would one expect to happen to the Qβ genome under these paradisiacal conditions? The RNA molecule finds in the test-tube environment everything that it needs for replication. The major part of the wild-type genome contains information on how to recognise and infect a bacterial (*E. coli*) cell, how to make use of the host's metabolism and how eventually to destroy the infected host. In the new environment this is all "ballast". The RNA molecule can replicate faster if it throws away the ballast, and that is what it does. After the fourth of the serial transfers, each of which initially involves an incubation time of 20 min, biological competence, such as infectiveness, is lost (Figure 4.9.2a). The acting selection pressure only implies: reproduce – as quickly as possible! To enforce it Spiegelman reduced the incubation time after the 13th, 29th, 38th and 52nd transfers successively to 15, 10, 7 and eventually 5 min (Figure 4.9.2b). The final, stable end product had reduced its length from 3600 to 550 nucleotide residues (but not further). The replication time had been shortened, not only in accordance with the reduction of length, but – over and above this – by an additional time reduction by, on average, a factor of 2.6 per inclusion of each monomer (Figure 4.9.2c).

Let us hear now how a critic, obviously resentful of "molecular" biology, commented on the above results. Brian Goodwin,[131] a renowned biologist, wrote: "This looks like Darwinian evolution in a test tube. But the interesting result was that this evolution went one way: toward greater simplicity. Actual evolution tends to go toward greater complexity, species becoming more elaborate in their structure and behaviour. But DNA on its own can go nowhere but toward greater simplicity. In order for the evolution of complexity to occur, DNA has to be within a cellular context; the whole system evolves as a reproducing unit."

Granting an equivalence of DNA and RNA (which actually was the evolving unit in Spiegelman's experiments) there are two points in Goodwin's interpretation which are hard to understand:

(1) Why should evolution only have "to go toward greater complexity", if no incentive for doing so is being offered? The correct Darwinian interpretation would exactly require the result observed, and that's what happens in nature too. What is true is that evolution in nature is adaptive.

(2) There are other results in the literature, which were available at the time the author published his criticism and which clearly demonstrated an evolution "toward greater complexity" under even simpler test-tube conditions. I am referring to serial transfer experiments by Spiegelman and his group in which

(a)

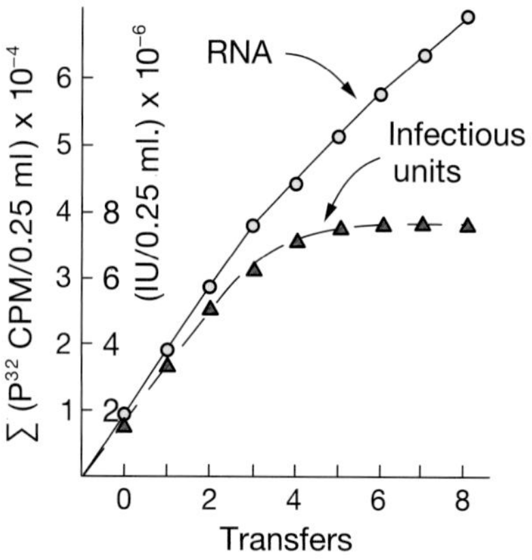

(b)

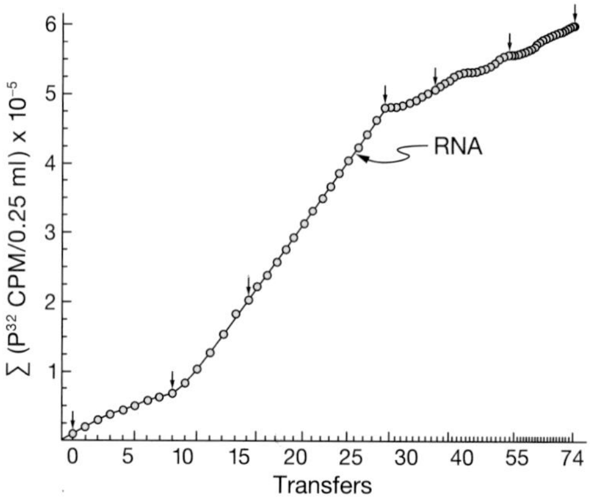

(c)

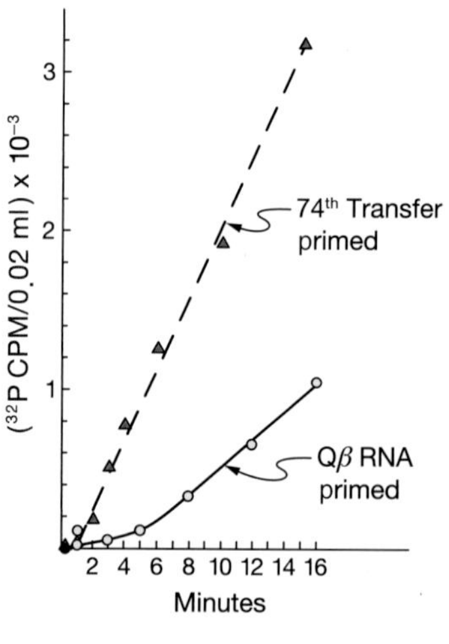

Figure 4.9.2(a–c)

Spiegelman's evolution experiment with phage Qβ

This is a quantitative representation of the rate behaviour in various phases of amplification (after Spiegelman[130]). Details are explained in the text. (a) The loss of infectious behaviour after four to six transfers. (b) The total growth curve. (c) Comparison of rates between the final transfer product and Qβ RNA.

> shorter variants of Qβ-RNA were adapted to the presence of substances that interfere with replication[132] or those favouring degradation of RNA.[133] Here RNA variants were obtained which developed a tolerance to an environment in which the original wild type would have succumbed. In other words, the resulting RNA molecules had not only retained the property of being a target for Qβ-replicase, but in addition retained the ability to withstand an interference by degrading enzymes or by preventing intercalation of substances that would have impeded replication. That is effectively an increase in complexity, although complexity need not increase if the species can adapt in some other way.

I admit that the effects obtained in Spiegelman's first test-tube experiments on evolution were still relatively small, simply because optimum conditions were not known at that time. When experiments were repeated under the conditions suggested by evolutionary theory, the results improved by orders of magnitude. An example of a quite sophisticated evolutionary optimisation will be described in Section 4.10. Meanwhile, a whole industry applying principles of molecular evolution has grown up and today enjoys a flourishing existence.

What is hard to understand is that Goodwin's argument was then used by a logician in order to "disprove" the Darwinian approach explaining biological evolution. Dembski even tried to generalise his arguments by referring to some obscure "no free lunch theorem". "No free lunch" was a jocular formulation used by Alan Guth[134] in his theory of "inflation" (see Section 1.9) to characterise the implications of any conservation law. In the absence of such a law – such as in the case in entropy-producing irreversible processes – it doesn't make much sense. The processes going on are driven by energy fluxes. So somebody else "pays" for the lunch.

What has all this to do with "meaning"? It may look as though I have lost my thread. Not at all! I think the experiments mentioned above demonstrate in an exemplary way how to generate meaning. Genetic information, apart from being "syntactically" organised, is uniquely of a semantic nature. A gene encodes the amino-acid sequence and thereby the precise spatial structure of a protein molecule which in a "suitable" chemical environment materialises a certain enzymic function. This function is the "meaning" of the genetic message, quite analogous to the meaning of a text communicated to us. The "suitable chemical environment" for a genetic program corresponds – in McKay's formulation – to the "conditional readiness" of the receiver of a message, the text of which is processed in his brain and responded to by "goal-directed activity".

The technical jargon of molecular genetics often borrows its terminology from the vocabulary of language processing. The latter, however, refers to processes that take place inside our brains, for which – so far – we bring along much less "understanding" than for the processes that deal with genetic information. Wouldn't we be better to do the opposite and use our knowledge from molecular genetics – in particular, that of how information is generated and processed – in order to try to comprehend terms such as "meaning" and "understanding" that constitute the "complex familiarity" of the phraseology of cogitation? Mind you, such an approach will be limited to the conceptualisation of semantic information. It will not tell us much about mechanisms.

A key question in this connection is: How does "meaning" come about in the first place? There is no meaning unless it has started to evolve in a certain environment. Semantic information is not subject to any conservation law, so it has to be produced. Moreover, it may deteriorate or become obsolete in a changing environment. In Spiegelman's serial transfer work using bacteriophage Qβ, the information for infecting bacterial cells became obsolete in the test-tube environment and therefore was discarded right away. In the automated evolution experiments discussed below (Section 4.10) the information that is generated allows RNA sequences to protect themselves against decomposition in an increasingly hostile environment, and they do this in a quite sophisticated manner.

"Specified information" is semantic information, and that is information representing meaning. The theory I presented in Section 4.7 is tailor-made for the problem of optimising genetic information. It tells us how meaning in this case comes about, and it shows the close analogy to physical processes that are driven by potential gradients. In our case it all happens in what I called information space.

The final goal of natural selection is existence. Shakespeare seems to have known this when he let Hamlet soliloquise: "To be or not to be – that is the question."

Yes, biological function, in order to come about, has to secure its existence in the form of a material carrier. What is selected is not the phenotypic function as such, whatever that may be; it is solely the value of this particular function in favour of the existence of its genotype, the material carrier that represents its meaning in the particular environment. Similarly, a certain firing pattern of nerve cells in our brain, if it represents meaning – endorsed by what is already stored in the memory – must somehow ensure its persistence. It most probably does so by selectively enforcing synaptic contacts in the brain's complex and (naturally) organised network of neurones (to be distinguished from what in computer technology is called an (artificial) "neural network").

Staying in the context of the present chapter, let me discuss the example of a sequence of binary symbols, with length $\nu = 100$. Such a sequence has $2^{100} \approx 10^{30}$ alternative symbol arrangements. Its simplex $S_{10^{30}}$ is correspondingly "complex". To picture such a construct is utterly impossible; the mere number of co-ordinates defies our imagination. Nevertheless, we can do calculations and compare numbers in order to find out whether such a problem of exponential complexity can be solved in polynomial time.

The scenario, in fact, is quite realistic. It is the scenario of the experiment I shall describe in Section 4.10. The RNA molecules that result in this experiment have a length of about 100 nucleotide units. RNA molecules are quaternary sequences, but if we consider only the base classes R and Y (see Section 4.4) we may use the binary picture. The experiments were carried out at nanomolar concentrations of RNA.

A few microlitres of such a solution contain a population of about 10,000 million (10^{10}) RNA sequences. The complexity of the mutant distribution produced suggests that in a population of 10^{10} molecules (nearly) all variants up to the five-error copies are expected to occur, while in the class of ten-error mutants still about 1000 different well-adapted individuals will be present. However, in order to find within a reasonable time a solution that requires a substitution at ten or more positions, we need those mutants to appear with a much higher probability than is predicted by the Poissonian distribution. The 1000 ten-error mutants appearing accidentally in this distribution are only about the 10^{-15}th part of the total number of ten-error mutants. Hence, to find accidentally a copy that exceeds by orders of magnitude the fitness of the wild type, required for survival under the conditions of the experiment, seems utterly impossible. It would rather require a quasi-species solution of the kind discussed in Section 4.7, and that's what was found in the experiment discussed in the next section. The sequences resulting from this experiment are able to respond to the changed conditions. They have "learned" how to cope with the new environment. Yes, they "understand" the new environment.

In the 1950s, Léon Brillouin, soon after the advent of Shannon's theory, wrote in his book *Science and Information Theory*[135]: "The present theory of information completely ignores the value of the information handled, transmitted or processed", and "It is only by ignoring the human value of the information that we have been able to construct a scientific theory of information by statistics, and this theory has already proved very useful."[136] Later on[137] he adds the almost prophetic sentence: "At any rate, one point is immediately obvious: any criterion for *value* will result in an *evaluation* of the information received. This is equivalent to *selecting* the information according to a certain *figure of merit*."

Yes, very true – but why did it take almost 50 years for semantics to become seen as a problem of evaluation and selection? Darwin's tenet of natural selection had been known by then for more than 100 years. Apparently physicists did not realise its physical nature, and before Watson and Crick's discovery (in 1953) biologists had not really recognised information to be a basic principle of life. In Section 3.9 I tried to visualise "how far it is from Shannon to Darwin". Genetic information is evaluated via natural selection.

The term "meaning" has different connotations in life and thought, since genetic and neural mechanisms are of quite different nature. There is no information without the condition that it must be readable. Reading is reproduction, in other words to (re)produce a copy that can be subjected to some form of processing. (Natural) selection is a direct consequence of reproduction, while "processing" is uncovering the meaning that guides selection.

Neuroscientists in the course of history have had quite diverse ideas about the structure and function of neural memory, beginning with complex clockworks and progressing to electric switchboards and molecular storage systems, similar to those found for genetic information.

Today we distinguish three levels of memory:

(1) Short-term memory, which is largely electrochemical in nature and is expressed in specific discharge patterns of linked nerve cells in the brain's neural network. It represents a first level of evaluation of received sensorial input and has to work with high temporal resolution, but has correspondingly a limited lifetime.

(2) Long-term memory, which provides an incorporation of the evaluated messages (which is a small fraction of all stimuli that have been received) into the long-term structure of the circuitry. It is more of a chemical nature and consists of changes in the synaptic connections among the nerve cells and their threshold of firing.

(3) The third level is a unique product of human evolution. It consists of a centre in the human brain in which the memory is reflected and transformed into a language that can be communicated in spoken or written form. This outlasts the lifetime of individuals and forms the basis of cultural evolution. In its written form it shows many similarities to genetic information, which was invented by nature almost 4000 million years earlier.

There are common physical prerequisites and principles which hold for the generation of meaningful information, regardless of their "technical" realisation, which differs greatly in genomics, proteomics, neurosciences and artificial intelligence. The origin of semantics requires a physical solution of problems of an NP-complete nature using a mechanism that responds to an extremum principle.

In concluding this section let me digress briefly on a question regarding the terminology I have used throughout this chapter. Some philosophers dislike the terminology, based on analogies, now current in genetics and molecular biology. They criticise in particular terms such as information, semantics, meaning, learning, memory, artificial intelligence etc. It is true that there is often quite a discrepancy in sophistication between the mechanisms that process information in the human brain and those that handle genetic information, rendering a comparison in many cases somewhat lame. Nevertheless, the principles for the generation of information are quite common, regardless of the mechanisms involved. The terminology used here certainly helps us to understand the beginnings – in our case the origin of (semantic) information, dating back 4000 million years. Furthermore, it may help to narrow the divide between the "two cultures" diagnosed by C.P. Snow[138].

So what does meaning mean?

In fact, there is no general answer to this question, i.e. there is no general physical theory of semantic information, as little as there is a general physical theory of processes. I am reminded of Wigner's words: "Physics doesn't deal with processes.

Physics deals with regularities, and *only* with regularities among processes." The amount of information is a common property that is enumerable, and therefore Shannon's theory is general. The meaning of information is a specific property. Language can be used to ask any question or to describe any thought. A general theory of semantic information would be a true "theory of everything". But wait, there is something that all portions of semantic information, all sorts of meaning, have in common. It is not the particular meaning itself, but the question "how it originates". The theory, presented in Sections 4.4, 4.5 and 4.7, answers the question "How can meaningful semantics originate?" This problem can at least be studied in analogy to the problem of genetic information, and we already possess quite a lot of information about possible mechanisms.

4.10. Pure Thought = Poor Thought?

The term "pure" has not much ambiguity. "Pure thinking" is meant to be a procedure based exclusively on constructs of our mind rather than on steady adaptation to empirical facts. The word "poor", however, has several connotations. If I interpret it as "deficient", I do not have in mind any lack of intelligence, excellence or elegance of thinking. What I mean is remoteness from reality and experience, and lack of adaptation to facts. The little digression on word games in Vignette 4.10.1 may illustrate the difficulties that so far have obscured the true problem.

There are only a few examples in physics where totally new insights have been gained by what may be called "pure" thinking. In fact, there are none, if we take the term literally.

What about the two theories of relativity? Didn't Einstein arrive at them by "pure thought", i.e. in the absence of any pressure from experimental results or of observation-based facts? Special relativity was the logical consequence of the Michelson–Morley experiment, but it is also true that at least general relativity went far ahead of its experimental verification. Nevertheless, there are few other examples for which this is true. Another one that comes to my mind is Dirac's theory of the electron (see Section 1.6), with its prediction of antimatter. But was it really "pure thought"? Einstein[139] expressed his opinion in the words: "...the creative principle resides in mathematics. In a certain sense, therefore, I hold it true that pure thought can grasp reality, as the ancients dreamed."

However, even mathematics is not the logically closed system of which Hilbert once dreamed. Given any consistent set of arithmetic axioms, there are "true" statements in the resulting system which, according to Gödel, cannot be derived from these axioms, thereby leaving some openness for consistency within the system. Consistency in mathematics is defined as "freedom from internal contradictions". In the case of Gödel's incompleteness theorem, it means that certain statements can be neither proved nor disproved within the system.

Vignette 4.10.1. Digression on word games

Word games are used in several places in this book to explain informational relations in terms of our language. The title of this section is an example. "Pure" and "poor" have similar sound patterns, but otherwise only little in common. The word "pure" generally means "clear", "uniform" or most frequently "uncontaminated", while "poor" is an expression for being "penniless", "meagre" or even "sub-standard". On the other hand, only scientists who use these words in connection with thought or theory could interpret "uncontaminated" as meaning "not using any empirical facts". What I want to make clear with this digression is that "poor thought" need by no means be "sub-standard" or "meagre". It may be logically entirely correct and even elegant, but nevertheless irrelevant for the problem in question. A good example is Wigner's use of S-matrix theory (see Section 1.4) in combination with stochastic matrices for reaching the conclusion that "replication is too improbable to apply to DNA reproduction in genetics". His conclusion is indeed true for many problems of "recurrence" in statistical physics (see Section 2.3). However, it does not apply to the replication of DNA sequences, where the chemical interactions among complementary base pairs outweigh the statistical effects due to thermal motion in the temperature range of life. In fact biology, because of its internal complexity, is prone to such effects, ones that do not fulfil Wigner's conditions for being "regularities among processes" (see Section 3.9).

Back to word games: I want to stress the differences of their appearance in different languages. They are good examples of the analogy with the molecular-biological interpretation of genetics in terms of language. The multitude of different word games became obvious to me when I tried to translate the above word game "pure" and "poor" into German. A first attempt to translate literally "pure theory = poor theory" gives the obvious result: "reine Theorie = schlechte Theorie", which destroys the play on words. However, inspiration struck, and I found a translation that reads "schlechthin Gedachtes = schlecht Hingedachtes", which in its meaning corresponds exactly to the original English version. *Schlechthin* means "plainly", while *schlecht Hingedachtes* is equivalent to "something not really applying". This translation is an even better word game than its English version, but it is different in its nature. "Pure" and "poor" are equivalent expressions at the sound level (approximately the same pronunciation), but they differ in spelling and meaning. My translation, by contrast, has exactly the same sequence of letters on both sides of the "equation", but differs in shifting the "blank" symbol (see Section 3.2) to a different position, leading to different accentuation and a different meaning. Such cases are also found in biology, for instance when an allosteric enzyme has to work in different environments.

When Einstein refers to the "creative principle" in mathematics, he is certainly correct, but in physics the assignment to a certain mechanism or structure comes first and requires a good deal of intuition before one can look for the most suitable mathematical formulation which is free from internal contradictions.

Our "thinking organ", according to all we know today, works primarily in an adaptive way. Each individual adapts through experience, through education and also through stimulated curiosity. During cultural evolution this organ has adapted to consistent rules of mathematical logic without being able to base its premises on its own rules. The difficulty enters with "physical reality". How are we to come to grips with it, i.e. to "grasp" it, other than by experience aided by experimentation? How could our brain "know" what "really *is*", other than by adapting to it under the constraint of observation? "Pure thought" would ultimately require reality to be pre-programmed into our brains.

Special relativity definitely started from a new experimental insight, namely, the fact that the velocity of light is finite (not so new) and that it is independent of the motion of its source or its observer (discovered through the precise measurements made by Michelson and Morley less than two decades earlier; see Section 1.2). This result revealed the velocity of light in a vacuum to be a natural constant, a property of physical space. Einstein combined this fact with the premise that the laws of mechanics must be the same for observers in different systems of unaccelerated motion. "General" relativity, then – as the name states – was a consistent generalisation of the "special" theory, introducing the presence of masses. Furthermore, it included a thoroughgoing realisation of the observable fact of gravitation, which Einstein interpreted as an influence of mass on the geometry of space. "Relativity" again meant the impossibility of distinguishing mere acceleration from the effect of gravitation. We thus cannot regard the spectacular new results brought about by both theories as the consequences of "pure" thought, although we have to admit that they preceded any experimental test or observation. Sophisticated feedback is involved in mental processes, and these are still largely uncharted. The – still incomplete – edifice of present-day physics is based as much on experiments as it is on logical abstraction.

The inadequacy of pure thinking becomes even more obvious in biology. Biological reality is complex. The objects of biology, living entities, are of a (combinatorial) complexity that vastly exceeds our imagination and always evades the deductive power of any theory. Does this therefore exclude any physical understanding of biology? Earlier in this book I quoted the geneticist Theodosius Dobzhansky, who said: "Nothing in biology makes sense except in the light of evolution". Darwin's theory is an abstraction of the experiences he collected during his voyage on the *Beagle*. Biologists (rightly) claim Darwin's principle to be an utterly "biological" tenet. But does it therefore evade "physical" understanding?

In this section I want to discuss a "Darwinian" experiment that I believe to be particularly relevant with respect to the question asked in the title of this section. It is a straightforward experiment involving replicating molecules, but the question it asks is by no means a trivial one. In fact, no colleague I consulted before the performance of the experiment was able to give me a satisfactory prediction of its outcome. The answer obtained turned out in retrospect to be reasonable, but it could in no way be readily anticipated, let alone subjected to quantitative computation. It is an instructive example of how to solve a problem of exponential complexity in polynomial time.

All we have to know in order to understand the experiment is that genetic messages are written in a chemical language using four letters that are arranged sequentially, as are letters (or codes) in human language. These molecular messages can be read and copied by lining up and linking the complementary letters A, U and G, C, supplied as monomers in energy-rich form. The linking of the monomeric letters into a sequence complementary to the original sequence is facilitated by an enzymic copying machine. The complementary strand obtained acts in the same way, as a template – like a photographic negative – to generate a copy of the original "positive". This kind of complementary reproduction amplifies the numbers of both the positive and negative strands (possibly at different rates).

Let me say a few words about the enzymic machinery with which our systems are equipped. The enzymes that do the job of molecular photocopiers are called *polymerases* or *replicases*. Some of them, like modern electronic machines, do not work unless they are shown some identification code. The identification code is usually a characteristic folding pattern, based on a defined sequence of the RNA strand. Since each sequence includes a lot of complementary characters, internal pairing interactions make the sequences fold up on themselves. Virus replicases usually respond only to the characteristic folding of the viral RNA that exposes a certain pattern, and hence reproduces it exclusively (as well as those mutants that have the same recognition pattern).

In addition to copying enzymes, we also need ones that steadily decompose the copies formed, otherwise the exponential growth of copies would soon clog the system. In order to keep evolution going indefinitely, we have to carefully balance growth and decomposition, steadily adding energy-rich material for growth and, correspondingly removing the energy-deficient waste. The destruction enzymes for RNA are called (ribo-)nucleases. They cleave RNA strands, usually at particular sites defined by the characters, and only where these are not protected by either base-pairing or spatial folds.

Now we are (almost) ready to perform the experiment. In such an evolution experiment one can exert selection pressure by adjusting the relative amount of copying and decomposing enzymes in such a way that only those sequences are amplified that escape decomposition by hiding the sites by which the decomposing enzymes could recognise them. We have built automated machines that do those experiments routinely and, by suitable variations, make them a basis for a new "evolutionary biotechnology". However, this is not the subject of this section.

Here we want to learn more about the principles used by nature to generate information. We started the series of experiments with a simple case, using a ribonuclease that cleaves strands only at positions following a G. It is called ribonuclease T_1. Application of an excess of this nuclease will favour the selection of those strands in which all Gs in non-base-paired positions (e.g. loops) are either hidden or replaced by A, U or C. However, the last statement is not quite correct. Any looped "finger" containing an unpaired C will appear in its complementary strand as an unpaired G. Thus the looped region will be all right only if one replaces G by A or U, but it will not be so if it is substituted by C because this would produce a G in the complementary strand. Even the replacement of G by U is problematic because the most frequent copying mistake is caused by a weak pairing between G and U.

Our experiments were done for strands of various lengths.[140] Artificial RNA strands of 100 nucleotides needed only hours to produce strands resistant to ribonuclease T_1, while single-stranded viruses whose RNA genomes had a length of about 4000 nucleotides took a few days.[141] In both cases the resulting sequences retained all the information for getting replicated and, in the latter case, for remaining infectious.

So far, the results have been straightforward, demonstrating a simple, not to say trivial, selection mechanism. No real intelligence was involved; it was just a matter of finding the strands that respond positively to the selection pressure. However, knowing what kind of substitution is required to make a successful selection possible, we can immediately think of a case that could not work in this simple way: replace ribonuclease T_1 by ribonuclease A.

Ribonuclease A has a specificity that distinguishes it from ribonuclease T_1: it cleaves after any of the two unpaired pyrimidine bases, i.e. after U or C, if it can gain access to them. Should we now simply replace unpaired U or C by A or G? Certainly not! Anything we do in order to save the "plus" strand will harm its complementary "minus" strand and *vice versa*, supposing that the two must adopt complementary folding structures, which indeed they do in this experiment.

The student who chose this experiment as the main problem for his doctoral theses, Günter Strunk[142], decided not to think through all the possible ifs and buts, he just went ahead and did the experiment. That sounds easy, but it isn't at all. One can't just take the substances, mix them together and see what comes out. One has to clone, to separate, to sequence. Furthermore the sample volumes are microlitres and the substance weights are nanograms. In short, one has to apply the whole arsenal of techniques of modern molecular biology, and – because of the high speed of the reactions – use automated equipment, i.e. serial-transfer machines monitoring all evolutionary changes by sensitive detection of physical signals reflecting the chemical changes that characterise evolutionary progress (see Section 5.9).

The catalyst employed was the replication enzyme isolated from a bacterial virus, called phage Qβ, our "pet". A similar one was also used in the evolutionary experiments of Sol Spiegelman, mentioned in Section 4.9. The enzyme and its kinetic characteristics have been thoroughly studied in our laboratory in recent years by Christof

Biebricher[143]. In particular, we know the folded structure of the RNA template that constitutes the recognition site for the enzyme (indicated in Figure 4.10.1). The template used was a microvariant of Qβ-genomic RNA, comprising only about 100 nucleotide monomers. It was discovered by Christof Biebricher[144] as a by-product of phage infection in *E. coli* cells which survives in the natural environment because of its comparatively large replication rate (with a doubling time of less than 1 min) and its high structural stability. The decomposing enzyme, ribonuclease A, is commercially available in highly purified form.

As I said, Günter Strunk decided to do the experiment first. The result was baffling: as in the transparent case of ribonuclease T_1 (mentioned above), after a few hours he obtained an RNA mutant that was resistant to a concentration of ribonuclease A several times greater than a concentration sufficient to completely wipe out the initial wild-type RNA. He then determined the sequences of the plus and the minus strand of the mutant RNA obtained. The results are shown in Figure 4.10.1 alongside the plus and minus sequences of the initial wild type.

Inspecting the sequences uncovers the following facts. The evolutionary process produces two shorter complementary sequences, showing exactly complementary folding structures. This was not the case for the initial sequences, for which folding algorithms yield differing secondary structures, since a (stable) GU pair in the plus strand would yield an (unstable) CA pair in the minus strand. Both strands conserve the recognition pattern for the replicase, including a ten-base-pair stem with the palindromic loop CUAG (the complement of which, read in the reverse direction, is also CUAG). Note that the evolutionary process must not change this (informational) feature, as otherwise replication is impaired. The binding to the replicase is much stronger than to the nuclease, and this means protection of the vulnerable palindromic loop. From the evolutionary point of view this all is very reasonable. However, what was puzzling was the fact that only one of the strands (the one we arbitrarily call the plus strand) was, apart from the palindromic loop, entirely free of cleavable pyrimidines. This did not make much sense because its complementary (we call the minus) strand is consequently strewn with cleavable sites. Why did this not result in a breakdown of the total system?

There must still be some hidden asymmetry between the plus and minus strands. Folding algorithms tell us that the two structures are precisely complementary. Hence, the difference must be rooted in different binding kinetics. Maybe, as a self-styled "laboratory of biochemical kinetics" we should have thought of this answer earlier.

Earlier, we had studied the kinetics of replication of complementary systems, such as RNA viruses. In such systems the total rate of amplification is the geometric mean of the individual amplification rates of the two strands. The geometric mean – unlike the arithmetic mean – contains the product of the two rates, and hence the total rate approaches zero if one of the rates becomes vanishingly small. As a consequence, we found that natural systems use certain symmetries to make the two rates (about)

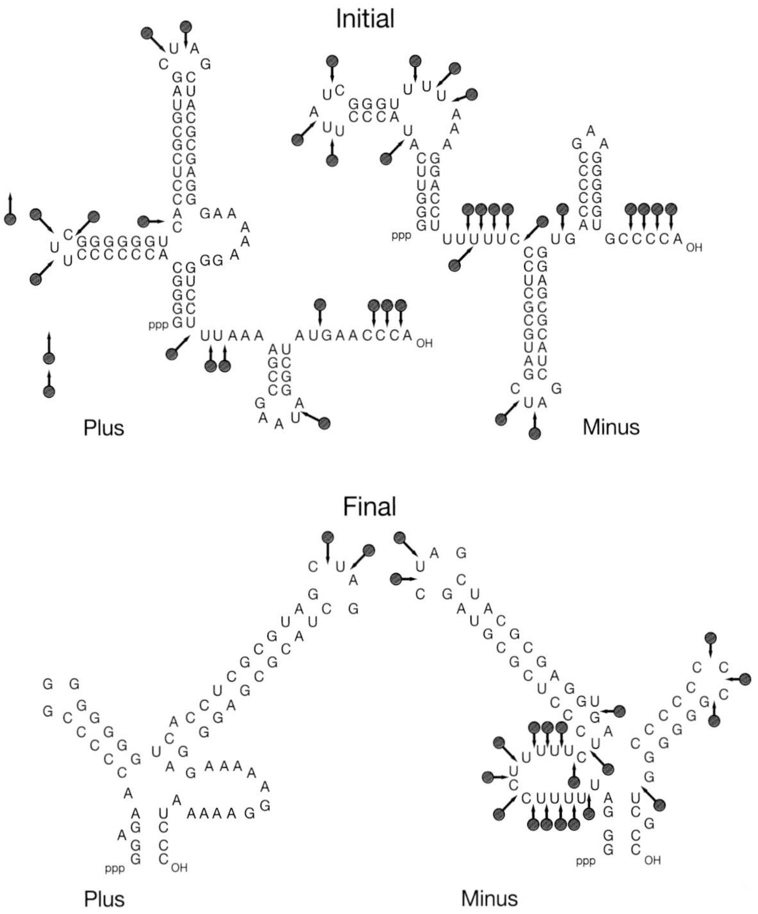

Figure 4.10.1

A Darwinian experiment applied to a mini-variant of Qβ RNA

The two structures in the upper row are the plus and minus strands of a mini-variant of Qβ RNA, isolated by Christof Biebricher from *E. coli* cell cultures infected with phage Qβ. They are, so to speak, the "wild type" of the mini-variant under normal conditions of phage infection. As is seen, this "wild type" has folding structures for the plus and minus strands that are not exactly complementary because of a GU pair in the plus strand which yields an (unstable) CA pair in the minus strand. (I should mention that in this case it is entirely arbitrary which of the two strands is called the plus and which is minus.) The lower row shows the strands resulting from the experiment, selected by evolution *in vitro*. In this case both strands show entirely complementary folding. All unpaired pyrimidines, susceptible to strand cleavage by nuclease A, are marked with a red dot. The only two free pyrimidines located in the recognition loop for Qβ replicase are protected by the bound enzyme. One notices a large asymmetry of both strands with respect to binding by the cleaving enzyme, nuclease A. For a detailed explanation, see the text.

equal and as large as possible. This is true for the initial compound that we call the wild type. Since the evolutionary products look even more symmetrical, we thought that they would work in the same way, especially as in our case it is one and the same enzyme that works on both the plus and minus strands.

Kinetic measurements performed subsequently showed that the mutant plus strands are produced from their minus strands as templates 100 times faster than *vice versa*, which as just stated was not true for the wild type. There are two competing enzymes: the replicase adapted to its recognition site and the nuclease, which is adapted to the cleavage site. Christof Biebricher[145] showed that for the cognate sequences the replicases are much stronger binding partners than the nucleases, which generally are not too sensitive to the sequence as a whole. On the other hand, the highly specific replicases are more sensitive to changes in their target sequences. The substitution of cleavage sites in the plus strand apparently goes hand in hand with a loss of specificity for replicase binding. The result is a dissymmetry in the production rates of plus and minus strands, resulting in a 100-fold excess of the plus over the minus strand. At the same time, the abundant plus strand becomes resistant to the nuclease, having all its cleavable pyrimidines replaced by purines, while the vulnerable minus strand remains bound to the replicase and thereby is protected against nuclease attacks. A formidable solution of Nature, indeed!

I am often asked how it is possible to do evolutionary experiments in the laboratory when evolution in Nature took millions or even billions of years. This is an especially good question if one remembers that the time spent in solving a problem must not exceed the lifetime of a PhD student. The answer is: evolution at the molecular level is not that slow – if the conditions provided are suitable.

Why, in our experiment, is the system able to head so directly towards an optimised solution? We do not need to propose some mystical property of living systems to explain this "goal-directed" property. Enzymes and templates in our experiment are present at concentrations between nano- and micromolar, while substrates, such as the energy-rich monomeric nucleotide units, are at least 1000-fold more abundant. "Nanomolar" in a sample as small as 1 μl (mm^3) means still about 1000 million particles, and "micromolar" 1000 times more. Hence "testing" at the molecular level is done in a massively parallel manner. The replication time for this relatively short-chained RNA molecule is less than 1 min. Growth by a factor of ten means slightly more than three generations, and that is repeated about 100 times in a typical experiment. So the numbers of copies to be tested are plentiful. Moreover, the only mutants likely to occur are close relatives of those populating the initial distribution, i.e. those which are already well adapted and probably not too far distant from those that match best the modified external environment. Figure 4.10.2 shows the initial and final sequences in an aligned form, revealing more clearly what drastic changes – including point mutations, insertions and deletions – took place. And that all happened within hours. Yes, evolution is indeed fast, if conditions are correct.

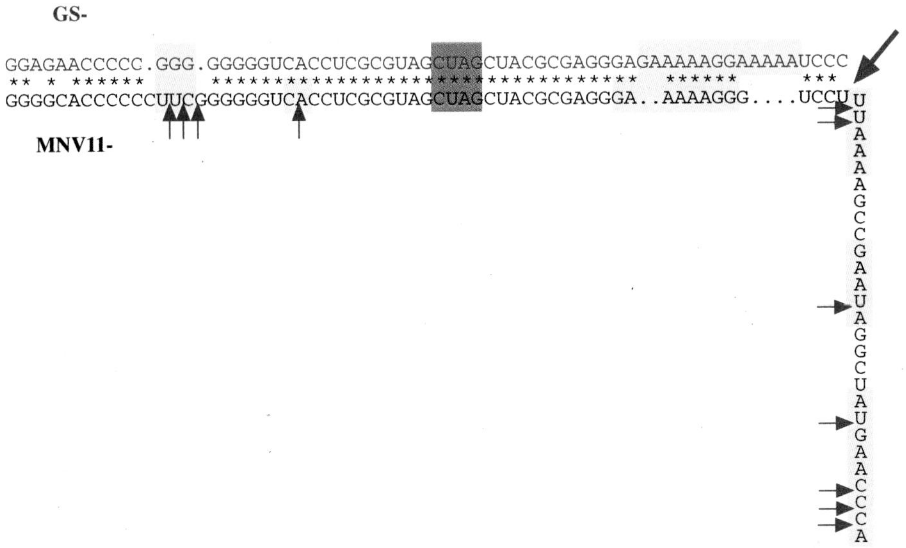

Figure 4.10.2

Aligned strands of initial and final sequences

The final sequence is shown with red symbols. One notices all sorts of mutations, including point mutations, insertions, deletions and cleavage of a strand segment. For further details see the text.

The truly surprising fact is that the result obtained could hardly have been guessed in advance. Given presently available concepts and computer times it could not have been derived from theoretical concepts. It would have required tools which precisely predict folding structures from sequences and which, given the folding structures, would allow one to predict the functions and calculate the rates. Perhaps this is another example of Orgel's second rule, which the late Leslie Orgel formulated as "evolution is cleverer than you are", and which Francis Crick[146] parodied as "evolution is cleverer than Leslie".

This experiment is also reminiscent of our previous discussion of exponential and polynomial complexity. Isn't the evolutionary solution, like folding of a protein, an example of an NP-type problem? After knowing the result of evolutionary adaptation we can show in polynomial time that it is indeed a clever, if not an optimal, solution (given the constraints that are operating).

The "solution" that resulted is certainly a particular structure in space, and the physical forces stabilising this structure are electromagnetic in nature. Can one then say that it is ultimately quantum mechanics that explains the origin of information? Well, there is a difference between these physical structures and those like that of a hydrogen atom. The hydrogen atom is a pure and unique result of physical forces governed by quantum mechanics – independent of environmental conditions. It is represented by an eigenvalue of the Schrödinger equation that does not allow for any alternative structure. RNA sequences, being chemical structures, are – albeit more complex – of a similar nature. Their existence, in the end, rests on quantum mechanics, too. However, the species we dealt with in our experiment have some 10^{50} alternatives of a more or less similar structural stability. Whether or not a particular sequence is selected depends solely on its functional response under the particular environmental constraints. It is "information", selected according to a dynamical extremum principle, that is generated and conserves itself. This "information" makes sense only in the context of the environmental constraints.

The complexity of the final product in this experiment results from an unlimited iteration of the replication algorithm. The experiment posed a complex problem with no trivial solution, but we could follow its solution through every intermediate step. Can we say that the algorithm *invented* kinetics in order to achieve its goal? No, we cannot say this. The algorithm is of a kind that makes use of any advantage being offered without "knowing" its nature or its cause. This makes it almost independent of the level of complexity involved, and only in this way does it generate information in an open and unlimited way.

Is pure thinking poor thinking?

I think that we must ask this question particularly in biology! In conclusion, let me rephrase this question in combination with the story of a most elegant theory in molecular biology, which was the result of pure thought and which turned out to be entirely wrong as soon as the first experimental results became available. The story was told in more detail by the late Francis Crick in his entertaining scientific autobiography *What Mad Pursuit*[147].

Soon after the discovery of the double helix, Crick received a letter from the Russian-born US physicist George Gamow, whose fame derived from important discoveries in atomic physics and cosmology. Gamow – correctly but on the basis of wrong "evidence" – had concluded that the genetic code had to account for 20 amino acids which, with the four bases of the Watson–Crick model, would require at least a triplet code, i.e. three bases encoding one amino acid. The difficulty of representing 20 amino acids by the 64 (4^3) possible triplet codons did not bother him too much because he thought of certain symmetries to account for the necessary redundancies. He attributed more importance to the discrepancy of spacing in the two aligned sequences. In the DNA template, three bases require 10 Å, compared with only 3.7 Å between neighbouring amino acids in the protein chain. So he came to the (wrong) conclusion that the code is read in an "overlapping" way, moving forward by just one base distance for each amino acid.

A more rigorous analysis of biochemical data, in which other molecular biologists, such as Sydney Brenner, Leslie Orgel, John Griffith and Alex Rich as well as

Francis Crick and Jim Watson participated, soon unearthed three facts, previously unnoticed:

(1) The number of "natural" amino acids to be considered by the code is indeed 20 (although quite different from Gamow's somewhat arbitrary list of compounds (cf. Section 3.8).

(2) The minimal code must be based on triplets. The spacing problem for a non-overlapping code (see (3)) then would require specific adaptors linking the amino acids to their respective codons (most probably utilising the specific patterns of hydrogen bonding of the various triplets).

(3) The code must be non-overlapping. Sydney Brenner calculated that the number of triplets in the case of a unique overlapping code would need to exceed 64, making all overlapping triplet codes impossible (supposing that the code was universal, which at that time was not yet established).

All three statements proved to be correct. However, they left unsettled the problem of how the correct codons for the 20 amino acids could be uniquely read out in a "comma-free" way. The adaptor hypothesis could account for an assignment of only 20 "sense" codons, of which all messages were to be made up. The obvious solution was that comma-free reading starts at some "assigned" place and simply proceeds three at a time in a gap-free manner.

Crick, Griffith and Orgel came up with a more elegant model. Suppose that all triplets with three identical bases, such as AAA, are prone to "out-of-phase" reading in sequences that have consecutive identical codons (AAAAAA...) and that such triplets must therefore be excluded. We are left with 60 possible codons XYZ, where X, Y and Z can be any of the four bases except for the four combinations in which all three bases are identical. By the same argument, as used for excluding codons with three redundant bases, all cyclic permutations of XYZ must be excluded too. In other words, if the codon $X_0Y_0Z_0$ is a sense codon, then $Y_0Z_0X_0$ and $Z_0X_0Y_0$ must be nonsense. Hence, only one-third of the 60 codons can make "sense", and that must be the ones that encode the 20 amino acids. This, indeed, looked like the correct interpretation!

The theory was published in 1957.[148] Only few years later, the first experimentally determined codon (Marshall Nirenberg and Heinrich Matthaei 1961)[149] became known. It was UUU, coding for phenylalanine, and the decisive experiment was done using a polymeric chain consisting only of Us. Thus, the very first actual evidence brought the beautiful theory crashing down.

Eventually the complete code was solved by experiment and not by theory. While it was coming out, attempts were made to guess the whole from the part, but these guesses were largely unsuccessful. It shows how easily one can be misled if one uses logic in a too straightforward way, in order to solve a biological problem.

So why was it "poor thinking"? It wasn't poor in the sense of being inelegant or unreasonable. No, it was poor in assuming *a priori* too much reason in nature. There was nobody around to set up such a rational construction. The only things "around" were molecular interactions and iterative selection. This is more than chance because it starts from physically favoured states and approaches "intelligent" solutions. It is intelligence *a posteriori*. An early code presumably started with fewer than 20 amino acids, that is, with those amino acids that from early on were present in sufficient abundance. If physical symmetries are to be incorporated and conserved, they must offer advantages. Only in this sense do present-day biological structures still reflect early "frozen accidents", possibly telling us something about their history.

"Pure thought" is devoid of one important prerequisite of thinking, namely empirical guidance, which in biology is of the utmost importance. Even in physics, with very few exceptions, the great theories have been founded on empirical experience. This section should remind us that understanding the phenomena of life in physical terms is hopeless if it is not based on experimental observation. Problems of complexity cannot be solved in the way suggested by the cartoon in Figure 4.10.3. The absence of complexity would not only make our life dull and flat. Complexity as such is one of the most important physical and chemical characteristics of life itself. The exciting years of the early history of molecular biology are described by Horace

Good
Evil
TYRANNY IS THE ABSENCE OF COMPLEXITY
-Andre Gide

Figure 4.10.3

Cartoon © Amedeo Henry 2003.

Judson in an elegant, profound, entertaining and witty way in a book entitled *The Eighth Day of Creation*[150].

At the end of this chapter, in which I introduce a theory that I think may be quite relevant for biology, I want to caution the reader. What I am expounding in this first volume is not yet biology: that is the subject of a second volume. The main theme of this volume is physics and my question, in particular, reads: Which kind of physics is relevant to biology? My answer is: it is the physics of information. If in this volume I frequently quote examples from biology, it is because biology contains the best examples of a physics of information. And if I start the following chapter with the question: "What is life?", then I do so not in order to give a full description of this mysterious wonder of matter, which would include thousands of answers, but rather in order to shed light on the necessary physical prerequisites that have to be fulfilled for matter to undergo the transition into a viable state.

LITERATURE AND NOTES

1. Quinkert, G., Engel, E. and Griesinger, Ch. (1996). "Aspects of Organic Chemistry: Structure". Verl. Helvetica Chimica Acta, Basel.
2. Ouellette, R. J. (1977). "Introduction to General, Organic and Biological Chemistry", 4th edn., Prentice Hall, Inc., Upper Saddle River, New Jersey.
3. Woodward, R. B. (1965). "Recent Advances in the Chemistry of Natural Products". Nobel Lectures, Chemistry 1963–1970. Elsevier, Amsterdam, 1972. Eschenmoser, A. and Wintner, C. E. (1977). "Natural Product Synthesis and Vitamin B12". *Science* 24: 1410–1420.
4. I am greatly indebted to Karl Ernst Kaissling for providing all the information about pheromones, including beautiful pictures of various butterflies.
5. Butenandt, A., Beckmann, R., Stamm, D. and Hecker, E. (1959). "Über den Sexuallockstoff des Seidenspinners Bombyx mori. Reindarstellung und Konstitution" ["On the Sex Attractant of the Silkworm Moth Bombyx Mori. Isolation and Structure"]. *Z. Naturforsch.* B 14: 283–284.
6. Nicolaou, K. C., Frederick, M. O. and Aversa, R. J. (2008). "The Continuing Saga of the Marine Polyether Biotoxins". *Angew. Chem. Int. Ed.* 47: 7182–7225.
7. Brenner, S. and Lerner, R. A. (1992). "Encoded Combinatorial Chemistry". *Proc. Natl. Acad. Sci. USA* 89: 5381–5383.
8. Leach, A. R. (1969). "Physical Principles and Techniques of Protein Chemistry: Pt. A". Academic Press, New York.
9. Born, M. and Oppenheimer, R. (1927). "Zur Quantentheorie der Molekeln". *Ann. Phys.* 20: 457–484.
10. Hartree, D. R. (1928). "The Wave Mechanics of an Atom with a Non-Coulomb Central Field". *Math. Proc. Cambridge Philos. Soc.* 24: 89–110; Fock, V. A.

(1930). "Näherungsmethode zur Lösung des quantenmechanischen Mehrkörperproblems". *Z. Phys.* 61: 126–148.
11. Hückel, E. (1931). "Quantentheoretische Beiträge zum Benzolproblem I. Die Elektronenkonfiguration des Benzols und verwandter Verbindungen". *Z. Phys.* 70 (3/4): 204–286; Hückel, E. (1931). "Quantentheoretische Beiträge zum Benzolproblem II. Quantentheorie der induzierten Polaritäten". *Z. Phys.* 72 (5/6): 310–337; Hückel, E. (1932). "Quantentheoretische Beiträge zum Problem der aromatischen und ungesättigten Verbindungen. III". *Z. Phys.* 76 (9/10): 628–648.
12. Heitler, W. H. and London, F. W. (1927). "Wechselwirkung neutraler Atome und homöopolare Bindung nach der Quantenmechanik". *Z. Phys.* 44: 455–472.
13. Woodward, R. B. and Hoffmann, R. (1965). "Stereochemistry of Electrocyclic Reactions". *J. Am. Chem. Soc.* 87(2): 395–397.
14. Bjerrum, N. (1920). "Der Aktivitätskoeffizient der Ionen". *Z. Anorg. Allg. Chem.* 109: 275–292.
15. Debye, P. and Hückel, E. (1923). "Zur Theorie der Elektrolyte. I. Gefrierpunktserniedrigung und verwandte Erscheinungen". [The Theory of Electrolytes. I. Lowering of Freezing Point and Related Phenomena]. *Phys. Z.* 24: 185–206.
16. Onsager, L. (1926). "Zur Theorie der Elektrolyte. I". *Z. Phys.* 27: 388–392; Onsager, L. (1927). "Zur Theorie der Elektrolyte. II". *Z. Phys.* 28: 277–298.
17. Eisenschitz, R. and London, F. (1930). "Über das Verhältnis der van der Waalsschen Kräfte zu den Homöopolaren Bindungskräften". *Z. Phys.* 60: 491–527; London, F. (1930). "Zur Theorie und Systematik der Molekularkräfte". *Z. Phys.* 63: 245–279; London, F. (1930). "Über einige Eigenschaften und Anwendungen der Molekularkräfte". *Z. Phys. Chem.* B11: 222–251.
18. Casimir, H. B. G. (1948). "On The Attraction of Two Perfectly Conducting Plates". *Proc. KNAW* 51: 793–795.
19. Sparnaay, M. J. (1957). "Attractive Forces Between Flat Plates". *Nature* 180: 334–335.
20. Eschenmoser, A. (2007). "The Search for the Chemistry of Life's Origin". *Tetrahedron* 63: 12821–12844.
21. Prelog, V. (1975). "Chirality in Chemistry". *Nobel Lectures, Chemistry* 1971–1980. Editor-in-Charge Tore Frängsmyr, Editor Sture Forsén. World Scientific Publishing Co., Singapore. 1993.
22. Tramontano, A., Janda, K. D. and Lerner, R. A. (1986). "Catalytic Antibodies". *Science* 234: 1566–1570.
23. Pollack, S. J., Jacobs, J. W. and Schultz, P. G. (1986). "Selective Chemical Catalysis by an Antibody". *Science* 234: 1570–1573.
24. Dawkins, R. (1986). "The Blind Watchmaker: Why the Evidence of Evolution Reveals a Universe Without Design". W. W. Norton, New York and London.
25. Steward, I. (1987). "From Here to Infinity. A Guide to Today's Mathematics". Oxford University Press, Oxford.

26. Haken, H. (1981). "Erfolgsgeheimnisse der Natur. Synergetik: die Lehre vom Zusammenwirken." Deutsche Verlagsanstalt (DVA), Stuttgart.
27. Stirling, J. (1730). "Methodus differentialis, sive tractatus de summation et interpolation serierum infinitarium". London. English translation: Holliday, J. (1749). "The Differential Method: A Treatise of the Summation and Interpolation of Infinite Series"; Tweddle, I. (2003). "James Stirling's Methodus Differentialis: An Annotated Translation of Stirling's Text". Springer, London. See also Hamming, R. W. (1980). "Coding and Information Theory". Prentice Hall, Englewood Cliffs, NJ, pp. 158–163.
28. Applegate, D. L., Bixby, R. M., Chvátal, V. and Cook, W. J. (2006). "The Traveling Salesman Problem: A Computational Study". Princeton University Press, Princeton. In 2005, Cook and others computed an optimal tour through a 33,810-city instance given by a microchip layout problem, currently the largest solved TSPLIB instance.
29. Metropolis, N., Rosenbluth, A. W., Rosenbluth, M. N., Teller, A. H. and Teller, E. (1953). "Equations of State Calculations by Fast Computing Machines". *J. Chem. Phys.* 21: 1087–1092.
30. Kirkpatrick, S., Gelatt Jr., C. D. and Vecchi, M. P. (1983). "Optimization by Simulated Annealing". *Science* 220: 671–680; Cerny V. (1985). "Thermodynamical Approach to the Traveling Salesman Problem: An Efficient Simulation Algorithm". *J. Optimiz. Theory Appl.* 45(1): 41–51.
31. Aarts, E. and Korst, J. (1989). "Simulated Annealing and Boltzmann Machines. A Stochastic Approach to Combinatorial Optimization and Neural Computing". Wiley, Chichester and New York.
32. See Ref. 28.
33. Berger, B. and Leighton, T. (1998). "Protein Folding in the Hydrophobic-Hydrophilic (HP) Model is NP-Complete". *J. Comput. Biol.* 5(1): 27–40; Unger, R. and Moult, J. (1993). "Finding the Lowest Free Energy Conformation of a Protein is an np-Hard Problem: Proof and Implications". *Bull. Math. Biol.* 55: 1183–1198; Fraenkel, A. (1993). "Complexity of Protein Folding". *Bull. Math. Biol.* 55: 1199–1210.
34. Churchland, P. S. and Sejnowski, T. J. (1992). "The Computational Brain: Models and Methods on the Frontiers of Computational Neuroscience". Bradford Books/MIT Press, Cambridge, MA.
35. MacRae, N. (1992). "John von Neumann", Chapter 6. Pantheon Books, New York.
36. Hamming, R. W. (1980). "Coding and Information Theory". Prentice Hall. Englewood Cliffs, NJ.
37. Eigen, M. (1971). "Selforganisation of Matter and the Evolution of Biological Macromolecules". *Naturwiss.* 58: 465–523.
38. Eigen, M., McCaskill, J. and Schuster, P. (1989). "The Molecular Quasi-Species". *Adv. Chem. Phys.* 75: 149–263.

39. Maynard Smith, J. (1970). "Natural Selection and the Concept of Protein Space". *Nature* (London) 225: 563–564.
40. Rechenberg, I. (1973). "Evolutionsstrategie: Optimierung technischer Systeme nach Prinzipien der biologischen Evolution". Problemata. Frommann–Holzboog, Stuttgart.
41. Eigen, M., Winkler-Oswatitsch, R. and Dress, A. (1988). "Statistical Geometry in Sequence Space: a Method of Quantitative Comparative Sequence Analysis". *Proc. Nat. Acad. Sci. U.S.A.* 85 (16): 5913–5917.
    Winkler-Oswatitsch, R., Dress, A. and Eigen, M. (1986). "Comparative Sequence-Analysis – Exemplified with Transfer-RNA and 5S Ribosomal RNA". *Chem. Scripta* 26B: 59–66.
    Eigen, M. and Winkler-Oswatitsch, R. (1990). "Statistical Geometry on Sequence Space." Doolittle, R. F. (Ed.). In: *Methods in Enzymology* 183. "Molecular Evolution: Computer Analysis of Protein and Nucleic Acid Sequences", Doolittle, R. F. Academic Press, San Diego and London, pp. 506–530.
42. Eigen, M. and Nieselt-Struwe, K. (1990). "How Old is the Immunodeficiency Virus". *AIDS*, 4 (Suppl. 1): 585–594.
    The data shown are based on:
    a) Influenza A (NS-gene): Buonagurio, D. A., Nakada, S., Parvin, J. D., Krystal, M., Palese, P. and Fitch, W. M. (1986). "Evolution of Human Influenza A Viruses over 50 Years: Rapid, Uniform Rate of Change in NS gene". *Science* 232: 980–982.
    b) Polio I: Rico-Hesse, R., Pallansch, M. A., Nottay, B. K. and Kew, O. M. (1987). "Geographic Distribution of Wild Poliovirus Type 1 Genotypes". *Virology* 160: 311–322.
    c) HIV/ SIV: All sequences available until 1990, see Ref. 42.
43. Nieselt-Struwe, K. (1992). "Konfigurationsanalysen kombinatorischer und biologischer Optimierungsprobleme". PhD Dissertation, Universität Bielefeld.
44. De Duve, Ch. (2002). "Life Evolving: Molecules, Mind and Meaning". Oxford University Press, Oxford and New York.
45. Conway, J. H. and Sloane, N. J. A. (1988). "Sphere Packings, Lattices and Groups". Springer Verlag, New York.
46. Fick, A. (1855). "Über Diffusion". *Poggendorf Ann. Phys.* 94: 59–86. English translation: Fick, A. (1855). "On Liquid Diffusion". *Philos. Mag.* 10: 30–39.
47. Brown, R. (1828). "A Brief Account of Microscopical Observations Made in the Months of June, July and August, 1827, on the Particles Contained in the Pollen of Plants; and on the General Existence of Active Molecules in Organic and Inorganic Bodies". *Phil. Mag.* 4: 161–173.
48. Einstein, A. (1905). "Eine neue Bestimmung der Moleküldimensionen". PhD Dissertation, University of Zürich.

49. Einstein, A. (1905). "Zur Elektrodynamik bewegter Körper". *Ann. Phys.* 17: 891–921.
50. Pais, A. (1982). "Subtle is the Lord: The Science and the Life of Albert Einstein". Oxford University Press, New York; Pais, A. (1979). "Einstein and the Quantum Theory". *Rev. Mod. Phys.* 51: 863–914; Pais, A. (1994). "Einstein Lived Here". Oxford University Press, New York.
51. Perrin, J. B. (1926). "Discontinuous Structure of Matter". Nobel Lectures, Physics 1922–1941, Elsevier Publishing Company, Amsterdam, 1965.
52. Mandelbrot, B. (1977). "The Fractal Geometry of Nature". W. H. Freeman & Co. Ltd, New York.
53. Cantor, G. (1883). "Über unendliche, lineare Punktmannigfaltigkeiten V". [On infinite, linear point-manifolds (sets)]. Math. Ann. 21: 545–591.
    See also Schroeder, M. (1991). "Fractals, Chaos, Power Laws: Minutes from an Infinite Paradise". W. H. Freeman, New York; Purkert, W. and Ilgauds, H. J. (1987). "Georg Cantor 1845–1918". Birkhäuser Verlag, Basel and Boston.
54. Hausdorff, F. (1919). See Schroeder, M. Ref. 55.
55. Schroeder, M. (1991). "Fractals, Chaos, Power Laws: Minutes from an Infinite Paradise". W. H. Freeman and Company, New York.
56. Wiener, N. (1923). "Differential Space". *J. Math. Phys.* 2: 131–174.
57. Polya, G. (1920). "Über den zentralen Grenzwertsatz der Wahrscheinlichkeitsrechnung und das Momentenproblem". *Math. Z.* 8: 171–181.
58. von Smoluchowski, M. (1916). "Drei Vorträge über Diffusion, Brownsche Molekularbewegung und Koagulation von Kolloidteilchen". *Phys. Z.* 17: 557–571, 585–599; von Smoluchowski, M. (1917). "Versuch einer mathematischen Theorie des Koagulationskinetik kolloider Lösungen". *Z. Phys. Chem.* 92: 129–168.
59. Pais, A. (1982). "Subtle is the Lord: The Science and the Life of Albert Einstein". Oxford University Press, New York.
60. Langevin, P. (1903). "L'Ionization de gaz". *Ann. Chim. Phys.* 28: 287–384; Langevin, P. (1903). "Sur la loi de recombination des ions". *Ann. Chim. Phys.* 28: 433–530.
61. Debye, P. (1942). "Reaction Rates in Ionic Solutions". *Trans. Electrochem. Soc.* 82: 265–272.
62. Bates, D. R. and Nicolet, M. (1950). "The Photochemistry of Atmospheric Water Vapour". *J. Geophys. Res.* 55: 301–327.
    $k = 10^9\text{–}10^{9,8}\ (L/Mol)^2 * sec^{-1}$. J. Phys. Chem. Ref. Data 34(3): 757–1397. (2005).
    A comparison with data for highly reactive partners in the liquid phase shows the surprising result that processes based on diffusion can reach higher encounter rates than those based on collisions among freely flying particles. It is true that the mechanisms in both cases are quite different. The rate constants

for the reaction $H + H \rightarrow H_2$ in literature are given in the dimensions $(l/mol)^2\, s^{-1}$, as compared to the more straightforward dimension $(l/mol)\, s^{-1}$, as they are directly measurable in relaxation times. One can, of course, calculate the rates under comparable conditions, obtaining, however, for diffusion-controlled rates in liquids by more than one order of magnitude, higher values than for collision-controlled processes in the gas phase, where often an additional partner must be available for transporting away any released energy. Refs. 63 to 65 show that the space-filling nature of random walk offers conceivable advantages for reaction partners at mutual distances that are not too great.

63. Eigen, M. and de Maeyer, L. (1955). "Die Geschwindigkeit der Neutralisationsreaktion" ["The speed of the neutralisation reaction"]. *Naturwiss.* 42: 413–414; Eigen, M. and de Mayer, L. (1955). "Kinetics of Neutralization". *Z. Elektrochem.* 59: 986–993.
64. Richter, P. H. and Eigen, M. (1974). "Diffusion Controlled Reaction Rates in Spheroidal Geometry. Application to Repressor Operator Association and Membrane Bound Enzymes". *Biophys. Chem.* 2: 255–263.
65. Adam, G. and Delbrück, M. (1968). "Reduction of dimensionality in biological diffusion processes". In: "Structural Chemistry and Molecular Biology", Rich, A. and Davidson, N. (eds). W. H. Freeman, San Francisco, pp. 198–215; Goldman, R. and Katchalsky, E. (1971). "Kinetic Behavior of a Two-Enzyme Membrane Carrying out a Consecutive Set of Reactions". *J. Theor. Biol.* 32(2): 243–57.
66. Eigen, M. (1971). "Selforganisation of Matter and the Evolution of Biological Macromolecules". *Naturwiss.* 58: 465–523.
67. Thomson, C. J. and McBride, J. L. (1974). "On Eigen's Theory of the Selforganisation of Matter and the Evolution of Biological Macromolecules". *Math. Biosci.* 21: 127–142.
68. Jones, B. L., Enns, R. H. and Rangnekar, S. S. (1976). "On the Theory of Selection of Coupled Macromolecular Systems". *Bull. Math. Biol.* 38: 15–28.
69. Rumschitzky, D. S. (1987). "Spectral Properties of Eigen Evolution Matrices". *J. Math. Biol.* 24(6): 667–680.
70. Frobenius, F. G. (1912). "Über Matrizen aus nicht-negativen Elementen". *S.-B. Preuss. Akad. Wiss.* 1912: 456–477.
71. Perron, O. (1907). "Zur Theorie der Matrices". *Math. Ann.* 64: 248–263.
72. Graham, R. and Haken, H. (1968). "Quantum Theory of Light Propagation in a Fluctuating Laser-Active Medium". *Z. Phys.* 213: 420–450.
73. Haken, H. (1981). "Laser Theory". Handbook of Physics, XXV/2C. Ed. S. Flügge. Springer, Berlin.
    Haken, H. (1964). "A Nonlinear Theory of Laser Noise and Coherence I and II". *Z. Phys.* 181: 96–124; Haken, H. (1965). *Z. Phys.* 182(4): 346–359; Graham, R.

and H. Haken (1968). "Quantum Theory of Light Propagation in a Fluctuating Laser-Active Medium". *Z. Phys.* 213(5): 420–450.

74. Yonath, A. E. (2009). "Hibernating Bears, Antibiotics and the Evolving Ribosome". Les Prix Nobel. The Nobel Prizes 2009. Grandin, K. (ed.). [Nobel Foundation], Stockholm, 2010.
75. Altman, S. (1989). "Enzymatic Cleavage of RNA by RNA". Nobel Lectures, Chemistry 1981–1990, Frängsmyr, T. and Malmström, B. G. (eds). World Scientific Publishing Co., Singapore, 1992.
76. Cech, Th. R. (1989). "Self-Splicing and Enzymatic Activity of an Intervening Sequence RNA from Tetrahymena". Nobel Lectures, Chemistry 1981–1990. Frängsmyr, T. and Malmström, B. G. (eds). World Scientific Publishing Co., Singapore, 1992.
77. Kimura, M. (1994). "Population Genetics, Molecular Evolution, and the Neutral Theory". University of Chicago Press, Chicago and London. (The book contains selected major papers, published between 1953 and 1987).
78. McCaskill, J. S. (1984). "A Localization Threshold for Macromolecular Quasispecies from Continuously Distributed Replication Rates". *J. Chem. Phys.* 80(10): 5194–5202.
79. Schroeder, M. (1991). See the Work quoted in Ref. 55.
80. Mandelbrot, B. (1977). "The Fractal Geometry of Nature". W. H. Freeman, New York. Reprinted 1982, 1983.
81. Lévy, P. (1932). See the more detailed discussion in Ref. 80.
82. Mayr, E. (1976). "Evolution and the Diversity of Life: Selected Essays". Belknap Press of Harvard University Press, Cambridge, Mass. and London.
    Mayr, E. (1959). "Typological Thinking versus Population Thinking". In B. J. Meggers (ed.). "Evolution and Anthropology: a Centennial Appraisal". Anthropological Society of Washington, Washington, D.C., pp. 409–412.
83. Swetina, J. and Schuster, P. (1982). "Selfreplication with Errors. A Model for Polynucleotide Replication". *Biophys.Chem.* 16: 329–345.
84. Schuster, P., Fontana, W., Stadler, P. F. and Hofacker, I. L. (1994). "From Sequences to Shapes and Back. A Case Study in RNA Secondary Structure". *Proc. R. Soc. London* B255: 279–284.
85. Eigen, M. and Schuster, P. (1979). "The Hypercycle. A Principle of Natural Self-organisation". Springer–Verlag, Berlin, Heidelberg and New York; Schuster, P. (1986). "The Physical Basis of Molecular Evolution". *Chem. Scripta* 26B: 27–41.
86. McCaskill, J. S. (1984). "A Stochastic Theory of Macromolecular Evolution". *Biol. Cybern.* 50: 63–73.
87. McCaskill, J. S. and Bauer, G. J. (1993). "Images of Evolution: Origin of Spontaneous RNA Replication Waves". *Proc. Natl. Acad. Sci. U.S.A.* 90: 4191–4195.
88. Rumschitzky, D. S. (1987). "Spectral Properties of Eigen Evolution Matrices". *J. Math. Biol.* 24(6): 667–680.

89. Dress, A. and Rumschitzky, D. (1988). "Evolution on Sequence Space and Tensor Products of Representation Spaces". *Acta Appl. Math.* 11(2): 103–115.
90. Biebricher, Ch. (1983). "Darwinian Selection of Self-Replicating RNA". In: "Evolutionary Biology", Vol. 16, Hecht, M. K., Wallace, B. and Prance, G. T. (eds.). Plenum Press, New York, pp. 1–52.
91. Biebricher, C. K., Eigen, M. and Gardiner, W. C. (1983). "Kinetics of RNA Replication". *Biochem.* 22: 2544–2559; Biebricher, C. K., Eigen, M. and Gardiner, W. C. (1984). "Kinetics of RNA Replication: Plus-Minus Asymmetry and Double-Strand Formation". *Biochem.* 23: 3186–3194; Biebricher, C. K., Eigen, M. and Gardiner, W. C. (1985). "Kinetics of RNA Replication: Competition and Selection among Self-Replicating RNA Species". *Biochem.* 23: 6550–6560.
92. Bonhoeffer, K. F. and Harteck, P. (1929). "Para- and Ortho Hydrogen". *Z. Phys. Chem.* B 4 (1–2): 113–141.
93. Eucken, A. Studies of the static (caloric) properties of ortho- and parahydrogen gas, described in more detail in Eucken's text book: "Lehrbuch der Chemischen Physik" II. Band. Akad. Verlagsges., Leipzig, 1943, particularly in Paragraphs 50 to 60.
94. Heisenberg, W. K. (1932). "The Development of Quantum Mechanics". Nobel Lectures, Physics 1922–1941. Elsevier Publishing Company, Amsterdam, 1965. The formulation reads: "for the creation of quantum mechanics, the application of which has, inter alia, led to the discovery of the allotropic forms of hydrogen".
95. Bonhoeffer, K. F. (1948). "Activation of Passive Iron as a Model for the Excitation of Nerve". *J. Gen. Physiol.* 32(1): 69–91.
96. Hodgkin, A. L. (1963). "The Ionic Basis of Nervous Conduction". Nobel Lectures, Physiology or Medicine 1963–1970, Elsevier Publishing Company, Amsterdam, 1972.
    Huxley, A. F. (1963). "The Quantitative Analysis of Excitation and Conduction in Nerve". Nobel Lectures, Physiology or Medicine 1963–1970, Elsevier Publishing Company, Amsterdam, 1972.
97. Sinai, Y. G. (1970). "Dynamical Systems with Elastic Reflections. Ergodic Properties of Dispersing Billiards". *Russ. Math. Surv.* 25: 137–191.
98. Kolmogorov, A. N. (1959). "Entropy per Unit Time as a Metric Invariant of Automorphism". *Proc. Acad. Sci. of USSR* 124(4): 754–755; Sinai, Y. G. (1959). "On the Concept of Entropy of a Dynamical System". *Proc. Acad. Sci. USSR* 124(4): 768–771.
99. Poincaré, H. (1884). "Sur certaines solutions particulières du probléme des trois corps". *Bull. Astronomique*, Serie I, 1: 65–74; Poincaré, H. (1891). "Sur le problème des trois corps". *Bull. Astronomique*, Serie I, 8: 12–24.
100. Lorenz, E. N. (1964). "The Problem of Deducing the Climate from the Governing Equations". *Tellus* XVI, I–II.

101. Ruelle, D. and Takens, F. (1971). "On the Nature of Turbulence". *Comm. Math. Phys.* 20: 167–192; Ruelle, D. (1990). "Deterministic Chaos: the Science and the Fiction". *Proc. R. Soc. London* A427: 241–248.
102. Rössler, O. E. (1976). "An Equation for Continuous Chaos". *Phys. Lett.* 57A(5): 397–398; Rössler, O. E. (1979). "An Equation for Hyperchaos". *Phys. Lett.* 71A(2,3): 155–157.
103. May, R. (1976). "Simple Mathematical Models with Very Complicated Dynamics", *Nature* 261: 459–467; May, R. (1974). "Stability and Complexity in Model Ecosystems". Princeton University Press, Princeton.
104. Nicholson, A. J. and Bailey, V. A. (1935). "The Balance of Animal Populations". *Proc. Zool. Soc. London* 1: 551–598.
105. Verhulst, P. F. (1845). "Recherches mathématiques sur la loi d'accroissement de la population". *Mem. Acad. Roy. Sci. Belles Lett. Brux.* 18: 1–42.
106. Casti, J. L. (1994). "Complexification: Explaining a Paradoxical World Through the Science of Surprise". Harper Collins, New York.
107. Gleick, J. (1987). "Chaos: Making a New Science". Viking Penguin, New York, p. 183.
108. How Feigenbaum arrived at the number, which now bears his name is not seen in the publications that finally appeared (e.g. Ref. 110), but it is vividly described in the book of Gleick quoted above (Ref. 107).
109. Grossmann, S. and Thomae, S. (1977). "Invariant Distributions and Stationary Correlation Functions of One-Dimensional Discrete Processes". *Z. Naturforsch.* 32a: 1353–1363.
110. Feigenbaum, M. J. (1978). "Quantitative Universality for a Class of Non-Linear Transformations". *J. Stat. Phys.* 19: 25–52.
111. Mandelbrot, B. B. (1982). "The Fractal Geometry of Nature". W.H. Freeman and Company, New York, and Mandelbrot, B. B. (1987). "Die fraktale Geometrie der Natur", Birkhäuser, Basel and Boston.
112. Peitgen, H. O. and Richter, P. H. (1986). "The Beauty of Fractals. Images of Complex Dynamical Systems". Springer-Verlag, Berlin, Heidelberg, New York and Tokyo.
113. Lyapunov, A. M. (1947). "Problème général de la stabilité du mouvement". Princeton University Press, Princeton.
114. Kovalevskaya, S. V. (1978). "A Russian Childhood" (translated and edited by B. Stillman with analysis of Kovalevskaya's mathematics by P. Y. Kochina). Springer, Berlin and New York.
115. Eigen, M. and de Maeyer, L. (1963). Review on chemical relaxation technology in "Technique of Organic Chemistry", Ed. A. Weissberger, Interscience, New York, Vol. VIII, part II, p. 89.
Eigen, M. (1967). "Immeasurably Fast Reactions". Nobel Lectures, Chemistry 1963–1970, Elsevier Publishing Company, Amsterdam, 1972.

116. Prigogine, I. and Lefever, R. (1968). "Symmetry-Breaking Instabilities in Dissipative Systems". *J. Chem. Phys.* 48: 1695–1700; See also Ref. 117.
117. Nicolis, G. (1995). "Introduction to Nonlinear Science". Cambridge University Press, Cambridge and New York.
118. Williams, G. P. (1997). "Chaos Theory Tamed". Joseph Henry Press/National Academy Press, Washington DC.
119. Weizsäcker, C. F. (1971). "Die Einheit der Natur". Carl Hanser Verlag, München, pp. 351–352.
120. Weizsäcker, C. F. The qualitative discussion of "pragmatic information" in so far is unsatisfactory as no quantitative definition of novelty and confirmation is given. Does it refer to symbol representation or its functional consequence? The theory of "natural selection" in this section deals with this problem.
121. McKay, D. M. (1980). Personal Communication; McKay, D. M. (1969). "Information, Mechanism and Meaning". MIT Press, Cambridge, MA.
122. Küppers, B. O. (1986). "Der Ursprung biologischer Information: zur Naturphilosophie der Lebensentstehung". R. Piper GmbH & Co. KG Munich, Germany.
Küppers, B. O. (1990). "Information and Origin of Life". The MIT Press, Cambridge, Massachusetts, London, England.
123. Davies, P. (1999). "The Fifth Miracle. The Search for the Origin and Meaning of Life". Simon and Schuster, New York.
124. Dembski, W. A. (2002). "Can Evolutionary Algorithms Generate Specified Complexity?" In: "From Complexity to Life. On the Emergence of Life and Meaning", Gregersen, N. H. (ed.). Oxford University Press, Oxford, pp. 93–113.
125. Dawkins, R. (1986). "The Blind Watchmaker". Norton, New York.
126. Dembski, W. A. (2002). "Can Evolutionary Algorithms Generate Specified Complexity?" In: "From Complexity to Life. On the Emergence of Life and Meaning", Gregersen, N. H. (ed.). Oxford University Press, Oxford, p. 107.
127. Winkler-Oswatitsch, R. and Eigen, M. Description of the glass bead game of R. W. O. in Ref. 128.
128. Eigen, M. and Schuster, P. (1979). "The Hypercycle. A Principle of Natural Selforganisation". Springer Verlag. Berlin, Heidelberg, New York.
129. Haruna, I. and Spiegelman, S. (1965). "Autocatalytic Synthesis of a Viral RNA in Vitro". *Science* 150: 884–886; Spiegelman, S., Haruna, I., Holland, I. B., Beaudreau, G. and Mills, D. (1965). "The Synthesis of a Self-Propagating and Infectious Nucleic Acid with a Purified Enzyme". *Proc. Natl. Acad. Sci. USA* 54(3): 919–927.
130. Mills, D. R., Peterson, R. L., Spiegelman, S. (1967). "An Extracellular Darwinian Experiment with a Self-Duplicating Nucleic Acid Molecule". *Proc. Natl. Acad. Sci. USA* 58: 217–224.

131. Goodwin, B. (1994). "How the Leopard Changed its Spots: The Evolution of Complexity". Scribner's, New York.
132. Saffhill, R., Schneider-Bernloehr, H., Orgel, L. E. and Spiegelman, S. (1970). "In Vitro Selection of Bacteriophage Q-beta Ribonucleic Acid Variants Resistant to Ethidium Bromide". *J. Mol. Biol.* 51(3): 531–539.
133. Strunk, G. and Ederhof, T. (1997). "Machines for Automated Evolution Experiments in Vitro Based on the Serial-Transfer Concept". *Biophys. Chem.* 66: 193–202.
134. Guth, A. H. (1997). "The Inflationary Universe. The Quest for a New Theory of Cosmic Origins". Helix Books, Addison Wesley, Reading, MA.
135. Brillouin, L. (1959). "La science et la théorie de l'information". Gabay, J. (ed.). Masson, Paris. English translation: Brillouin, L. (1956). "Science and Information Theory". Dover, Mineola, N. Y.
136. Brillouin, L. (1956). "Science and Information Theory". Dover, Mineola, N. Y. p. 294.
137. Brillouin, L. "Science and Information Theory". Dover, Mineola, N. Y. p. 297.
138. Snow, C. P. (1959). "The Two Cultures and the Scientific Revolution". Cambridge University Press, London.
139. Einstein, A. (1933). "On the Method of Theoretical Physics". The Herbert Spencer Lecture, delivered at Oxford, 10 June, 1933. Oxford University Press, New York.
140. Eigen, M. (1993). "The Origin of Genetic Information: Viruses as Models". *Gene* 135(1–2): 37–47.
141. Lindemann, B. Personal Communication.
142. Strunk, G. (1992). "Automatisierte Evolutionsexperimente in vitro und natürliche Selektion unter kontrollierten Bedingungen mit Hilfe der Serial Transfer Technik". PhD Thesis, Universities Göttingen and Braunschweig.
143. Biebricher, C. K., Eigen, M. and Luce, R. (1981). "Product Analysis of RNA Generated De Novo by QBeta Replicase". *J. Mol. Biol.* 148: 369–390.
144. Biebricher, C. K., Eigen, M. and Luce, R. (1981). "Kinetic Analysis of Template-Instructed and De Novo RNA Synthesis by QBeta Replicase". *J. Mol. Biol.* 148: 391–410; Biebricher, C. K., Diekmann, S. and Luce, R. (1982). "Structural Analysis of Self-Replicating RNA Synthesized by QBeta Replicase". *J. Mol. Biol.* 154: 629–648.
145. The following list provides a survey on Christof Biebricher's († 2009) work on Qβ-RNA.

Biebricher, C. K. (1983). "Darwinian Selection of Self-Replicating RNA". In: "Evolutionary Biology", Vol. 16, Hecht, M. K., Wallace, B. and Prance, G. T. (eds). Plenum Press, New York, pp. 1–52.

Biebricher, C. K., Eigen, M. and Gardiner, W. C. (1985). "Kinetics of RNA Replication: Competition and Selection among Self-Replicating RNA Species". *Biochem.* 23: 6550–6560.

Biebricher, C. K. (1986). "Darwinian Evolution of Self-Replicating RNA". *Chem. scripta* 26 B: 51–57.

Biebricher, C. K., Eigen, M. and Luce, R. (1986). "Template-Free RNA Synthesis by QBeta Replicase". *Nature* (London) 321: 89–91.

Biebricher, C. K., Eigen, M., Gardiner, W. C., Husimi, Y., Keweloh, H.-C. and Obst, A. (1987). "Modeling Studies of RNA Replication and Viral Infection". In: "Complex Chemical Reaction Systems; Mathematical Modelling and Simulation". Warnatz, J. and Jäger, W. (eds). Springer-Verlag, Berlin, pp. 17–38.

Biebricher, C. K. and Eigen, M. (1987). "Kinetics of RNA Replication by QBeta Replicase". In: "RNA Genetics", Vol. I "RNA-directed Virus Replication", Domingo, E., Ahlquist, P. and Holland, J. J. (eds). CRC Press, Boca Raton, FL, pp. 1–21.

Biebricher, C. K. (1987). "Reproduction and Evolution of RNA Species Synthesized by QBeta Replicase". In: "Proceedings of the First Latin American School of Biophysics", Fayad, R., Rodriguez-Vargas, A. M. and Violini, G. (eds). World Scientific Publishing, Singapore, pp. 5–84.

Biebricher, C. K. (1987). "Replication and Selection Kinetics of Short-Chained RNA Species". In: "Positive strand RNA viruses". UCLA Symposia New Ser. Brinton, M. A. and Rueckert, R. R. (eds). Alan R. Liss, New York, Vol. 54. pp. 9–23.

Biebricher, C. K. (1987). "Replication and Evolution of Short-Chained RNA Species Replicated by QBeta Replicase". *Cold Spring Harb. Symp.* 52: 299–306.

Biebricher, C. K., Eigen, M. and Gardiner, W. C. (1991). "Quantitative Analysis of Selection and Mutation in Self-Replicating RNA". In: "Biologically inspired Physics", Peliti, L. (ed.). NATO ASI Series B Vol. 263. Plenum Press, New York, pp. 317–337.

Biebricher, C. K. (1991). "Quantitation of Selection and Mutation in Self-Replicating RNA". *Biol. Chem. Hoppe-Seyler* 372(9): 635.

Biebricher, C. K. (1991). "Ribozymes: Biocatalysts Composed of Ribonucleic Acids". In: "Biotechnology Focus III", Vol. 3. Finn, R. K, Wagner, F. and Esser, K. (eds.). Carl Hanser Verlag, München, pp. 25–38.

Biebricher, C. K. (1991). "Evolutionary Research". In: "Interpreting the Universe as Creation", Brümmer, V. (ed.). Kok Pharos Publishing House, Kampen.

Lindner, A. J., Glaser, S. J., Biebricher, C. K. and Hartmann, G. R. (1991). "Self-catalysed affinity labeling of QBeta replicase", *Eur. J. Biochem.* 202: 249–254.

Eigen, M., Biebricher, C. K., Gebinoga, M. and Gardiner, W. C. (1991). "The Hypercycle. Coupling of RNA and Protein Biosynthesis in the Infection Cycle of an RNA Bacteriophage". *Biochem.* 30: 11005–11018.

Biebricher, C. K. (1992). "Quantitative Analysis of Mutation and Selection in Self-Replicating RNA". *Adv. Space Res.* 12: 191–197.

Biebricher, C. K. and Luce, R. (1992). "In vitro Recombination and Terminal Elongation of RNA by QBeta Replicase". *EMBO J.* 11: 5129–5135.

Biebricher, C. K. (1992). "Replication of Short-Chained RNA Species by QBeta Replicase". In: "Structural Tools for the Analysis of Nucleic Acid-Protein Interactions", Lilley, D. M. J., Heumann, H. and Suck, J. (eds). Birkhäusser, Basel, pp. 437–449.

Biebricher, C. K., Eigen, M. and McCaskill, J. S. (1993). "Template-Directed and Template-Free RNA Synthesis by QBeta Replicase". *J. Mol. Biol.* 231: 175–179.

Biebricher, C. K. and Luce, R. (1993). "Sequence Analysis of RNA Species Synthesized without Template". *Biochem.* 32: 4848–4854.

Biebricher, C. K. (1993). "Requirements for Template Activity of RNA in RNA Replication". In: "European Conference on Artificial Life 93", Goss, S. (ed.). MIT Press, Cambridge, MA. pp. 74–85.

Biebricher, C. K. (1994). "Amplifikation von RNA". In: "PCR im medizinischen und biologischen Labor – Ein Handbuch für Praktiker", Wink, M. and Wehrle, H. (eds). GIT Verlag, Darmstadt, S. 229–241.

Biebricher, C. K. (1994). "RNA Species that Multiply Indefinitely with RNA Polymerase". In: "Evolution of Self-Reproduction Systems", Fleischaker, G. R. and Colonna, S. (eds.). Kluwer Academic, Dordrecht, pp. 147–156.

Biebricher, C. K. (1994). "The Role of RNA Structure in RNA Replication". *Ber. Bunsenges.* 98: 1122–1126.

Zamora, H., Luce, R. and Biebricher, C. K. (1995). "Design of Artificial Short-Chained RNA Species that are Replicated by QBeta Replicase". *Biochem.* 34: 1261–1266.

Oberholzer, T., Wick, R., Luisi, P. L. and Biebricher, C. K. (1995). "Enzymatic RNA Replication in Self-Reproducing Vesicles – an Approach to a Minimal Cell". *Biochem. Biophys. Res. Comm.* 207: 250–257.

Rohde, N., Daum, H. and Biebricher, C. K. (1995). "The Mutant Distribution of an RNA Species Replicated by QBeta Replicase". *J. Mol. Biol.* 249: 754–762.

Domingo, E., Holland, J., Biebricher, C. K. and Eigen, M. (1995). "Quasispecies: The Concept and the Word". In: "Molecular Basis of Viral Evolution", Gibbs, A. and Calisher, C. H. (eds.). Cambridge University Press, Cambridge, pp. 181–191.

Biebricher, C. K., Nicolis, G. and Schuster, P. (1995). "Self-Organization in the Physico-chemical and Life Sciences". Office for Official Publications of the European Communities, Luxembourg.

Biebricher, C. K. (1996). "RNA Replication and Evolution". In: "From Simplicity to Complexity in Chemistry – and Beyond", Müller, A., Dress, A. and Vögtle, F. (eds). Vieweg, Wiesbaden, pp. 43–49.

Biebricher, C. K. and Luce, R. (1996). "Template-Free Synthesis of RNA Species Replicating with T7 RNA Polymerase". *EMBO J.* 15: 3458–3465.

Pop, M. P. and Biebricher, C. K. (1996). "Processivity and Fidelity of HIV-1 Reverse Transcriptase". *Biochem.* 35: 5054–5062.

146. Crick, F. quoted in Dennett, D. C. (1984). "Elbow Room: The Varieties of Free Will Worth Wanting". MIT Press, Cambridge, MA.
147. Crick, F. C. (1988). "What Mad Pursuit. A Personal View of Scientific Discovery". Basic Books, New York.
148. Crick, F. H. C., Griffith, J. S. and Orgel, L. E. (1957). "Codes without Commas". *Proc. Natl. Acad. Sci. USA* 43(5): 416–421.
149. Matthaei, J. H. and Nirenberg, M. W. (1961). "Characteristics and Stabilization of DNAase-Sensitive Protein Synthesis in E. coli Extracts". *Proc. Natl. Acad. Sci. USA* 47: 1580–1588; Nirenberg, M. W. and Matthaei, J. H. (1961). "The Dependence of Cell-Free Protein Synthesis in E. coli upon Naturally Occurring or Synthetic Polyribonucleotides". *Proc. Natl. Acad. Sci. USA* 47: 1588–1602.
150. Judson, H. F. (1979). "The Eighth Day of Creation. Makers of the Revolution in Biology". Simon and Schuster, New York.

5 Complexity and Self-Organisation

5.1.	What is Life – Now?	475
5.2.	Darwin for Molecules: Who Does the Selection?	497
5.3.	Does Natural Selection Require Linear Autocatalysis?	506
5.4.	Who Survives, the Fittest or the Luckiest?	523
5.5.	Natural Selection: A Phase Transition?	532
5.6.	Was the Watchmaker Really Blind?	545
5.7.	Where is the "Edge of Chaos"?	553
5.8.	Why Care What Other People Think?	563
5.9.	An Ultimate Machine?	575
5.10.	"It from Bit" or "Bit from It"?	590

5.1. What is Life – Now?

This question, put in this way, could easily be misunderstood. Obviously, it refers to Schrödinger's classic discussion of the physical aspects of life, based on lectures he delivered at Trinity College Dublin, Ireland, in 1943.[1] What may cause irritation in my title is the word "now". Suppose I had quoted Schrödinger's title as "What was Life – Then?" A reasonable answer – at least in most parts of the world at that time (1943–44) – would have been: "miserable!"

Schrödinger was certainly not the first to ask the question "What is life?". In his novel *The Magic Mountain* Thomas Mann tried to find an answer to his repeatedly reiterated question "What was life?". The chapter in question was written in 1920, and Mann's use of the past tense was probably meant to indicate that it was an old question, so far not satisfactorily answered: Life – what was it really? His answer – 23 years before Oswald Theodore Avery demonstrated the role of nucleic acids as the material of heredity in 1943 – is truly astonishing. Let me quote a few sentences from Mann's novel.[2]

In the chapter entitled "Research", Thomas Mann lets his protagonist Hans Castorp, in a snowy winter night in Davos, ponder about the very question of what life – "with its sacred, unclean secret" – is. The answers he found, although given in a highly sophisticated literary form, didn't really satisfy him. So he tried to unravel the secret of life by dissecting an organism into ever smaller parts – organs, single cells, their organelles – until he got down to the elementary units of life, which he called the "genes": "Those were the genes, the living germs, bioplasts, biophores ..." And then he went on to find a dialectical answer in a truly Hegelian way:

Thesis: "If they (the genes) were living they must be organic, since life depended on organisation"
Antithesis: "(But) if they were organised they could not be elementary, since an organism is not single but multiple"
Synthesis: "...however impossibly small they were, they must themselves be built up, organically built up, as a law of their existence.... Sooner or later, (further) subdivision must engender some kind between life and non-life, molecular groups, that embody the transition between viable organisation and mere chemistry."

Thomas Mann's diary reveals that he prepared himself carefully by reading Oscar Hertwig's *General Biology*, which appeared in 1916 in its third edition.[3] Hertwig, who already in 1875 had performed first studies of the fertilisation of the sea-urchin egg, was later the first to observe the reduction of the chromosome number in meiosis (in 1890). It also was he who recognised the nucleus as the carrier of heredity. Organic chemists at that time were not yet ready to identify the molecules of heredity with Friedrich Miescher's "Kernsäuren" (literally translated as "nucleic acids"), which he had discovered as early as in 1869. In his book, Hertwig presents a chemical analysis of the nucleus, according to which more than 50% appears to consist of nucleic acids. Correspondingly, although with some reservation, he expressed the view that they might be the source of inheritance, while the majority of biochemists until the 1940s favoured the proteins as candidates because of their enormous biochemical versatility.

If I am asked how Thomas Mann could find such a crisp formulation, which even today could still serve as a motto for a textbook on molecular biology, I have no satisfactory answer. I am reminded of my mentor, the physicist Peter Debye, who was famous not only for his great work, but also for presenting the most difficult problems in a very clear and understandable way. When asked how he managed to do this he answered (with a twinkle in his eye): "I just leave out the complicated bits." I don't think that Thomas Mann had any expertise in biology, but he informed himself by carefully reading Hertwig's text which, according to his diary, only took him a few months. He then processed the text with his masterful command of language and style, and "left out the difficult bits". Oscar Hertwig presented his own ideas with all the "ifs and buts" of critical analysis. Thomas Mann's novel is a classic of literature. Literary scholars whom I asked have praised his style and his gift of expression, but they did not seem to have noticed the intricacy of his deliberations on life.

Let me now return to Schrödinger's book. In 1994 a conference with the title "What is Life – 50 Years Later"? was held at Trinity College Dublin.[4] Stephen Jay Gould,[5] in his lecture, criticised Schrödinger's view, but not on the basis of its imperfections or its neglect of biological facts. No, he disputes the aim of physicists for "unification" as an "unquestioned dream and goal of science" in general. He said: "If Schrödinger's belief in reductive unification flowed from the 'unity of science' movement, then this movement, and its philosophical basis, also lay embedded within the even larger cultural force known as 'modernism', with its profound influence upon such fields as art, literature, and architecture. Modernism, above all, sought reduction, simplification, abstraction, and universalism." And – after applauding post-modernism's emphasis on playfulness and pluralism – he alleges: "A postmodernist could scarcely credit any unitary answer to such questions as 'what is life?' – particularly an answer like Schrödinger's, rooted in the modernist heartland of reduction to constituent basic particles."

Of course, the late palaeontologist had to keep his flag flying. We all embrace the results of palaeontology, as they can tell us *when*, *where* and *which* varieties of life came into existence during the historical course of evolution, and *how* they are all related to one another – quite a lot of questions for one branch of science to answer. Then, of course, the question "What is life?" has thousands of answers and "contingency appears to be a dominating factor". But I am afraid that that is not what Schrödinger had in mind when he asked his question – irrespective of whether or not we like the answers he gave. His question was not an expression of modernism, but rather the one of a "classical" scholar.

Physicists, such as Steven Weinberg[6] and Stephen Hawking, proudly call themselves reductionists, knowing well that reduction is only one – albeit an indispensable – part of physical analysis. Physics is not an ideology, and those who let it shrink to some "ism" don't understand what it is about. I even find the definition of physics given in Webster's dictionary ("the science that deals with matter, energy, motion, and force") too narrow. I think that all the chapters of this book also deal with physics, namely trying to *understand* the principles guiding the reality of our world. Life and thought are part of our real world. Often the physicist has to focus on a single variable, not because the problem is so simple as to depend only on that variable, but rather by finding out what happens if that particular variable is changed while the others are kept constant, and thereby sorting out its function. This holds true for theory as well as for experiment.

Stephen Gould was certainly justified in emphasising the importance of contingency when he considered the history of biological evolution. On the other hand, we may ask: Why is the phenomenon of life particularly prone to contingency? One answer is: In natural selection it is usually some singular event at the molecular level that becomes magnified to macroscopic appearance. In fact, Schrödinger himself, even before he engaged in quantum mechanics, was an apologist of chance. In his inaugural address to the natural science faculty of the University of Zürich in 1922[7] he said: "Research in physics has shown beyond a shadow of doubt that in the overwhelming majority of phenomena whose regularity and invariability have led to the formulation of the postulate of causality, the common element underlying the consistency observed *is chance*." What Schrödinger had in mind was the regularities of equilibrium thermodynamics, which at the macroscopic level have an entirely deterministic appearance.

The question "What is life?" may after all not be a good question unless one specifies its context. In this chapter I shall focus on the physical prerequisites of "life as such", and only in a second volume will I distinguish between "life" and "living beings". Gould clearly meant "living beings" and their historical contingency. The examples he is quoting appeared some 3000 million years after "life" originated. Schrödinger was referring – as he indicated in the subtitle of his book *The Physical Aspect of the Living Cell* – to a categorical property and its physical foundation. Historically, the transition from non-living to living matter certainly included contingencies as, for instance, expressed in the chirality of natural proteins and nucleic acids, or in the detailed structure of the genetic code and in other (possibly) "frozen" accidents. Contingency appears at all levels of life. It is physically inherent to the molecular scene of life and becomes visible at the macroscopic level through reproductive amplification. It thereby shows up not only at the molecular, but also at the cellular and organismic levels, and it ultimately may dominate societies and their cultures.

Let's go back to Schrödinger's book, *The Physical Aspect of the Living Cell*. Why did this book have such an impact, even though most of its ideas are now passé? James Crow,[8] a renowned population biologist of our time, in a paper published in 1992, echoed what impressed him most when he first read Schrödinger's book: "Perhaps it was his view of the chromosome as a message written in code. Perhaps it was his statement of 'life feeds on negative entropy'. Perhaps it was his notion that quantum indeterminacy at the gene level is converted by cell multiplication into molar indeterminacy. Perhaps it was his emphasis on the stability of the gene and its ability to perpetuate order. Perhaps it was his faith that the all too obvious difficulties of interpreting life by physical principles need not imply that some superphysical law is required, although some new physical laws might be." (Crow 1992, referred to by Gould (see Ref. 5).)

This is the best characterisation of Schrödinger's achievement that I have found so far, explaining why his book is a classic. "What is life?" is not just one, but rather a whole bundle of questions. And it was these questions that stimulated young physicists, such as Francis Crick, Leslie Orgel, Seymour Benzer, Walter Gilbert and many others, to think about problems in biology. The answers Schrödinger gave – as one may expect – fell victim to the facts when they emerged. Remember that (at least) in biology "pure" theory in the end turns out to be "poor" theory.

The main biological fact that Schrödinger neglected was Darwin's principle of natural selection. He mentioned it only in passing. Apparently it didn't make sense to him as a physicist. In fact, he simply did not like it. In another paper,[9] he referred to the "apparent gloom of Darwinism" and called the Lamarckian view "infinitely more attractive". And he added almost regretfully – because biologists might (or might not) have convinced him in the meantime – "unhappily, Lamarckism is untenable". This was 14 years after he had written "What is life?".

Darwin derived his principle from observations in nature. Only once, many years later, did he think of a more general theoretical foundation. In a letter[10] written in 1882 (during the last month of his life) to the zoologist and surgeon George C. Wallich,[11] Darwin confirmed that his work was concerned only with the manner of succession, and that he "left the question of the origin of life uncanvassed as being altogether *ultra vires* in the present state of our knowledge". But then he added that he thought he might somewhere have said "that the principle of continuity renders it probable that the principle of life will be shown to be a part, or consequence of, some general law". Such a general law, in fact, is a physical law, and I think we are now in possession of it. But let me first formulate three statements which reflect our present state of knowledge:

1) Life is not represented by any fundamental physical structure.

To exclude any misunderstanding: There are many physical structures involved in the life process. At the molecular level we find the globular structures of proteins, the double-helical arrays of DNA chains and the planar bilayer structures of lipid membranes. At higher levels, organs or whole organisms still appear to us in identifiable shapes. However, there is no uniquely defined physical structure of matter that is typical of life, in the way that there are universal structures in physics such as the quark, the electron, the proton and neutron, the atom, the large variety of chemical compounds, or macroscopic structures, such as crystal lattices. Yes, the double helix is a typical structure, but it is typical of life only in so far as it contains "readable" information, like a "tape" that encodes a protein molecule. In order for its symbol sequence to be readable it had to assume a stable linear shape, which is ensured by the double helix. A DNA sequence devoid of any such "information" is also double-helical; when it occurs in nature, this is often called "nonsense DNA" (which I think is a misnomer because it may have a purpose that we do not know).

Many attempts to find a universal structure of life have been as futile as Goethe's dream of a "primordial plant" that he called the *Urpflanze*. Erwin Schrödinger's "aperiodic crystal", which at the end of his book he called "the finest masterpiece ever made by the Lord's quantum mechanics" – and by this he really meant the structure of the chromosome – was an incredible prescience of the discovery made less than ten years later by Francis Crick and James Watson.[12] But it is the particular information being stored in DNA, and not the molecule's general physical structure, that is characteristic of life. The structure, as special as it appears to be, is determined by plain chemistry, which in turn is based on a physical theory for which Schrödinger must receive due credit as one of its authors.

2) Life is an overall organisation that is governed by functional rather than by structural principles.

A fundamental functional principle? To what end? The only answer I can give is "existence"! As we have seen (Section 4.9) Shakespeare seems to have known the answer as well.

Yes, existence is the guiding principle of natural selection and thereby of life. Isn't that another insult to man?

Copernicus removed us from the centre of the world, a world which at that time was the planetary system of our sun. But the horizon of the "world" has expanded together with our own "horizon". Today we can be glad *not* to be at the centre of our galaxy. It would be quite uncomfortable there because of a massive black hole that was discovered only recently.

It was Charles Darwin, with his principle of natural selection as the basis of evolution, who made us relatives of all animals, while Freud performed similar disenchantments on our mind. My (late) friend Sol Spiegelman once said: "Man is just a trick played by DNA so that it can reproduce even on the moon." To be sure, he meant it as a joke, but the joke contains a very important truth, encapsulated in the word "reproduce". No living being can come about except by reproduction of its chromosomes. "Be fruitful and multiply!" is one of the most important biblical edicts. In the early days of mankind reproduction was instrumental for survival – as it still is today, provided that we learn to keep it under control in an ethically acceptable way. We are all travelling on the same "spaceship" with quite limited resources.

The question "What is life?", of course, could have many answers in view of the large variety of living beings with their quite distinctive faculties. That was Steven Gould's main point. However, what I am asking for now is a principle that distinguishes life from non-life. Even the most primitive viable system is of a hyperastronomical complexity. So let me add a third statement:

3) In order for life to come about, there must exist some physical principle that controls complexity.

In fact, life at all levels appears to be organised. And – with a sidelong glance at ourselves – I have to add: the principle must hold at any level of development. Understanding this principle is, perhaps, the most important point in understanding the physics underlying biology. Yes, complexity is the main variable of biology, which requires "high throughput" methods of counting and screening.

The legacy of Schrödinger's "What is life?" is contained in his questions rather than in the answers, which were premature in his time. Many of the answers came later, from a generation that was stimulated by Schrödinger's questions. My Israeli friend Ephraim Katzir told me that when he was a child, his mother used to ask him when he came home from school: "And did you also ask a good question?"

From my late colleague Wilhelm Jost, I got the story of the discovery of anti-knocking agents, a field in which he himself did pioneering work. "Knocking" is pre-ignition of unburned fuel in the combustion chamber of an engine, ahead of the flame front that travels from the sparking plug towards the piston. It is triggered by compression and overheating of the gas before the flame front reaches it, and it may cause damage to the plugs. The prevention of knocking became an urgent problem in the development of engines with high compression, as used in aeroplanes. The solution came with asking the question: the answers first given were entirely wrong, but finally Thomas Midgley[13,14] found a highly efficient anti-knocking agent, lead tetraethyl, which retards the chain reactions of combustion. (Use of this agent is now discouraged as for many years it has contributed to environmental pollution. High octane numbers of petrol and the addition of aromatics are the present methods of choice.)

When Schrödinger at the end of his book referred to "the finest masterpiece made by the Lord's quantum mechanics" then he most certainly had in mind a very complex and sophisticated structure. Yes, there is a wealth of masterpieces in living organisms. But their secret doesn't lie in the principles of their structure; rather, it lies in the perfect adaptation of their structures to optimum function. The adaptation of structure to function requires algorithms for generating information, which are Darwinian, although not in any trivial sense.

Recently – half a century after Schrödinger – a new book with precisely the same title appeared which "rephrases the answer to Schrödinger's brilliant question" (Edward O. Wilson in a review on the book cover). The authors of this remarkable monograph are Lynn Margulis[15], Distinguished Professor in the Department of Geo-sciences of the University of Massachusetts at Amherst, Massachusetts, and famous for her discoveries connected with the origin of eucaryotic cells, and the science writer Dorion Sagan, a son of Lynn Margulis and the late astrophysicist Carl Sagan. In their third chapter, entitled "Once upon a planet", the authors list in particular three factors that are representative for the earliest microbial states of life: encapsulation, proteins and RNA (as a precursor of DNA). Under the heading "So, what is life?" (with which they conclude each chapter in a few sentences) they formulate with poetic flavour: "Life is the representation, the 'presencing' of past chemistries, a past environment of the early Earth, that, because of life, remains on the modern Earth.

It is watery, membrane bounded encapsulation of space–time ... Life is a nexus of increasing sensitivity and complexity in a universe of parent matter that seems stupid and unfeeling in comparison." Let us look somewhat more closely at the order they proposed, from our present point of view.

Figure 5.1.1 demonstrates "encapsulation" – today, as it is emphasised by Lynn Margulis with her slogan "cells first". The picture demonstrates (the details are unimportant here) that the capsule of a biological cell has a highly sophisticated structure allowing a lively chemical communication between the cell's interior and its environment.

Figure 5.1.2 represents a complex biochemical network based on the catalytic and regulatory properties of protein molecules – again today. All living organisms from unicellular bacteria up to multicellular plants and animals use chemical networks of this kind in order to meet their diverse metabolic needs.

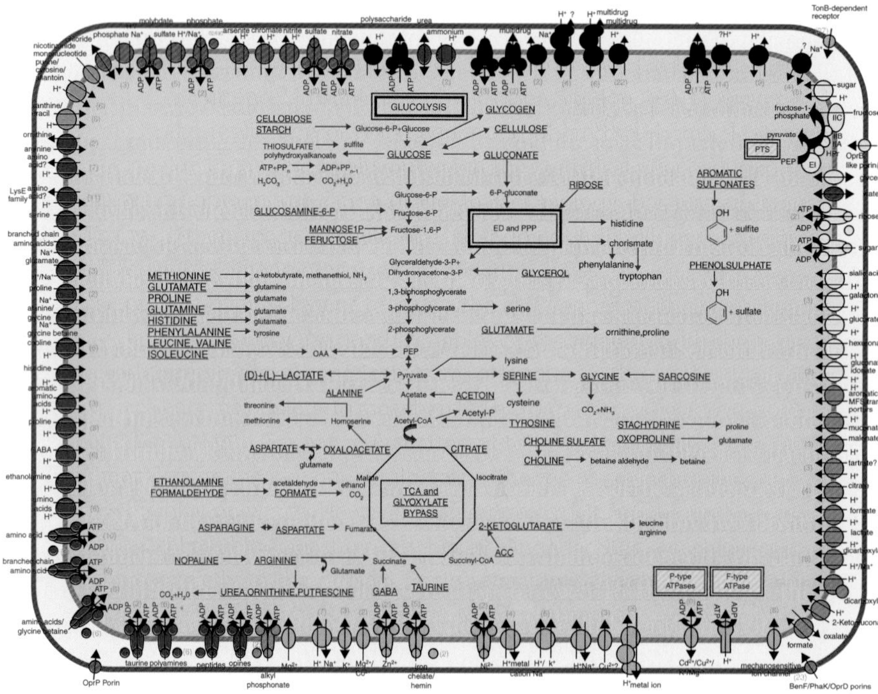

Figure 5.1.1
All living organisms are made up of cells

The picture clearly demonstrates the importance of "encapsulation" in holding the cell's contents together under fixed conditions and, at the same time, allowing a defined exchange of substances with the surrounding medium. This is more than plain encapsulation; it requires a highly sophisticated membrane structure, reflecting the complexity of the cell content. The cellular compartment shown in this picture belongs to the micro-organism *Pseudomonas putida* KT 2440 (cf. Figure 5.1.3). Courtesy of Burkhardt Tümmler.

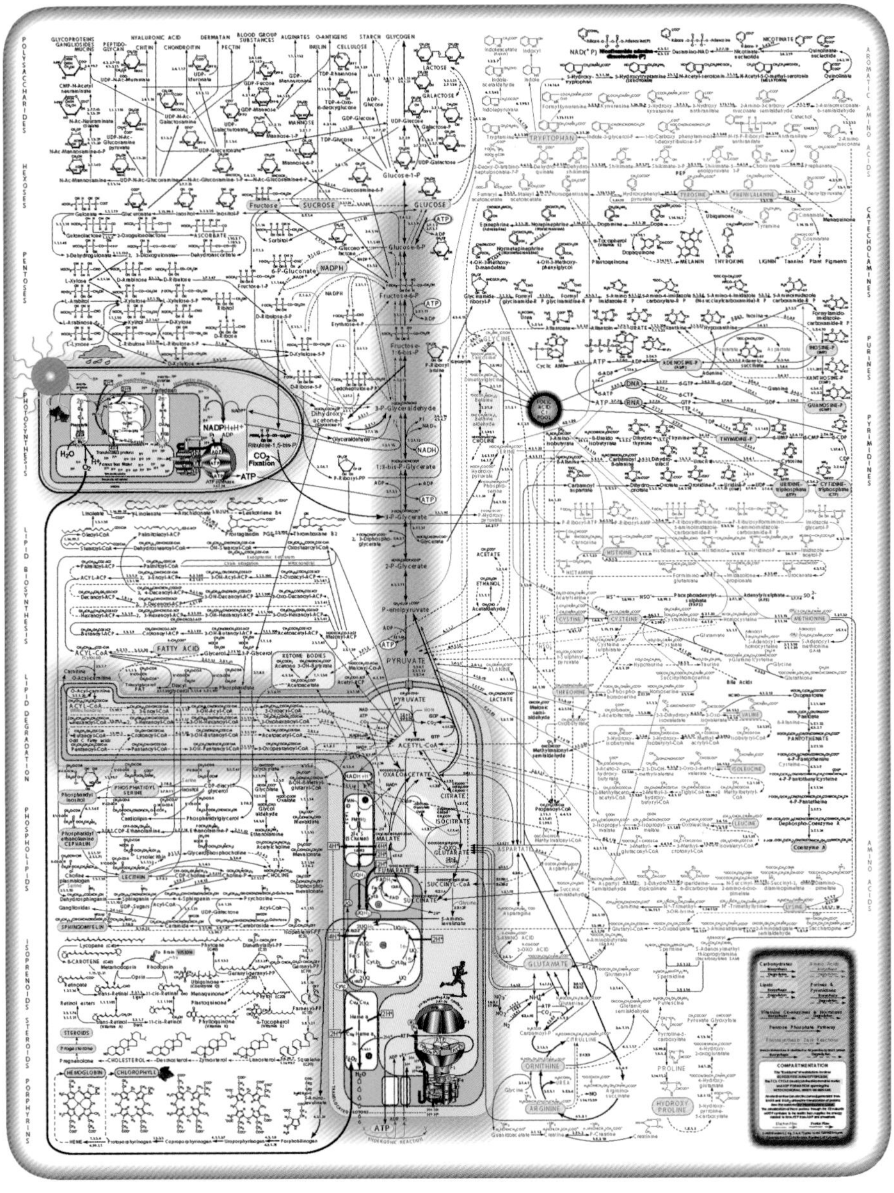

Figure 5.1.2

The complexity of the chemical network inside a living cell

The complexity of the chemical network inside a living cell is impressively documented in this picture. I have intentionally chosen a representation that leaves the detail difficult to resolve because that is not the focus of my discussion. The picture underlines all the more the complexity that is typical of all present-day cells, emphasising the importance of a highly regulated catalytic network of pre-programmed enzymes. However, it does not tell us how such a network could come about. Courtesy of Mary Osborn.

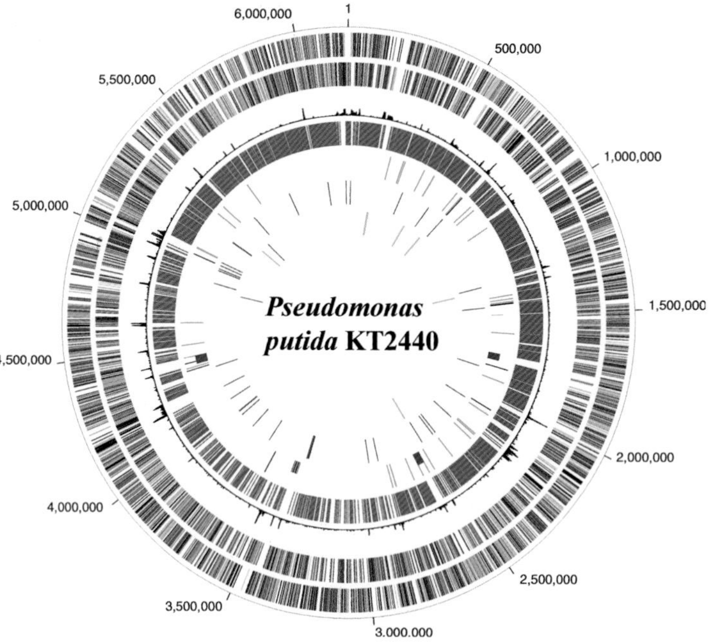

Figure 5.1.3

The genetic programme located inside each living cell

Even for the most primitive organisms, the genetic programme located inside each living cell has a complexity similar to that of the cell's complete enzymic network, including substrates and their reaction products, for which the cell membrane must provide a specific transport mechanism (see Figure 5.1.1). Each of the little bars indicates a gene, the unit for a genetic message. Courtesy of Burkhardt Tümmler.

Figure 5.1.3 shows a present form of the DNA genome, in this case that of the Gram-negative aerobic bacterium *Pseudomonas putida*, exhibiting the structural order of its hereditary "legislative", comprising thousands of individual genes.

The three pictures are representative of the above-mentioned main features of a living cell. In their combination, they constitute the cell as the smallest unit of life, and the biologist is well entitled to express it this way. There is no "life" below this level – now. What irritates me is that Lynn Margulis says: "cells first", not specifying whether "first" is meant temporally or causally.

Glancing at the three figures, the overwhelming impression one gets is: they are all extremely complex. No cell could have come about *de novo* in this perfect form. Does "cells first" mean that what is causally required is first the compartment and then – in the order discussed in Margulis' book – the proteins and finally the nucleic acids, i.e. the genes?

Any primordial medium in which life may have started must have presented some rich chemistry. Among the many compounds there will have been many amino acids, as Stanley Miller's[16] experiments suggest, and possibly also proteins or protein-like polymers, for instance proteinoids, as the late Sidney Fox,[17] who synthesised and named them, showed. Furthermore, there must also have been lipids or similar amphipolar substances, which may form all sorts of membrane vesicles, certainly precursors of membrane-enclosed biological cells. The study of those vesicular lipid systems in the 1970s was a flourishing field of activity in our Göttingen laboratory, as well as in other laboratories around the globe. My younger colleague, the physicist Hermann Träuble, after having discovered the quantisation of current in superconductivity, joined our group in 1972, together with the chemist Hansjörg Eibl. They founded an international centre of membrane research. It was a lively time with many discoveries and new ideas. Unfortunately, it met a premature end in 1976 with the tragic death of Hermann Träuble in a car accident.

That an empty vesicle in a chemically rich environment should be the precursor of a cell doesn't make much sense. It is rather the chemistry going on *inside* the vesicle that is of interest. Harold Morowitz[18] has suggested that, in the aqueous medium in which life arose, there had to be some barrier separating it from the vast expanse of the surrounding medium. The task of the cell membrane may either have been to protect what was produced inside the vesicle against degradation by external interference, or to keep the internal machinery together and prevent its deterioration by dilution.

Another possibility is just the formation of a multitude of different microenvironments of which only one, or a few, might become especially proliferative. Be that as it may; in all these cases there would have to have been simultaneous, selective transport of certain substances (possibly large molecules) into the vesicles, and of others out of them. The latter may be facilitated by active transport, by poration of the compartment (including electroporation if electric fields are present) or just by bursting when a certain growth level is surpassed. Whatever evolution takes place inside such a "cell", it must provide properties of the membrane that allow for some selective transport in and out of the compartment, as (according to Figure 5.1.1) would eventually be obtained at a final, optimised stage.

The two other main features of a living cell as highlighted in Figures 5.1.2 and 5.1.3 are regulated catalysis as effected by proteins, and inheritance as brought about by DNA. Which of the two came first?

"Proteins first" was the slogan of Sidney Fox, supported by many fellow chemists. Harold Morowitz, even more convincingly, stresses the fact that metabolism catalysed by enzymes follows the traces fixed by the chemistry of prebiotic times. In present-day organisms, the biosynthesis of proteins is instructed by nucleic acids, while nucleic acids require for their formation the catalytic support of proteins. It reminds one of the question brought up by scholastic philosophy: "Which came first, the chicken or the egg?" (see Figure 5.1.4).

"Damn you, Winkle, did you have to go and ask it which came first – the chicken or the egg?"

(Drawing by Dana Fradon; © 1971 The New Yorker Magazine, Inc.)

Figure 5.1.4

If "first", again, is meant in the strictly temporal sense, then the proteins might indeed have come first because on the primordial Earth the chemistry of the amino acids might have been accomplished sooner than that of activated nucleotides. But that would mean essentially random protein sequences. Sidney Fox always insisted that his proteinoids were not random. Of course, there were preferences for nearest neighbours and certain foldings in his sequences. What I mean is the absence of correlations among the way they form and the kind of catalytic function they eventually were able to fulfil. Their efficiency generally was quite weak anyway, and I do not see any *a priori* feedback between their function and their architecture that could enable them to achieve optimum performance. Even in present-day (optimised) enzymes, the relation between enzymic function, sequence and folding is a complex problem of NP-completeness. Let me tell a little story.

At one of our early Winter Seminars in the late 1960s, I talked about our results on the mechanism of allosteric enzymic catalysis. Kaspar Kirschner presented his results of relaxation studies[19] for glyceraldehyde phosphate dehydrogenase, an enzyme of the glycolytic cycle, and an example *par excellence* of evolutionary optimisation of enzymic regulation (see Section 2.1). In the discussion that followed, the question arose of why enzymes are so fantastically well adapted to their functions. I mentioned several examples from our 10-year-long studies of enzymic mechanisms using relaxation methods, which allowed us to "dissect" a reaction mechanism into its various elementary steps. In all the cases we studied, the result was a perfect fit of all steps in order to produce optimum efficacy of the overall reaction. Later experiments with site-directed mutations proved this picture: although some sites in the enzyme turned out to be extremely sensitive to any alteration, while others were less sensitive, there was no change that could improve the "overall performance", which turned out to be an optimum combination of rate and specificity, and not just some maximum rate.

Our biologists referred immediately to Darwin. My answer was: Darwin excluded (*expressis verbis*) any application of his theory to the origin of life, insisting that it dealt exclusively "with the manner of succession". His theory was focused on living beings rather than on molecular "individuals". Darwin knew as little of molecules as molecules know of Darwin.

Back in Göttingen, I thought more about the problem. The result was my 1971 paper in *Die Naturwissenschaften* on "Selforganisation of Matter and the Evolution of Biological Macromolecules".[20] The conclusion was: proteins are fabulous substances. They are accomplished experts for any problem of molecular catalysis and regulation. They are extremely skilful craftsmen or even "watchmakers", yet they are "blind watchmakers". They are programmed to do their job, but they can neither read nor write – they are "illiterate"!

As the pictures in Section 4.1 suggest, single-stranded polynucleotide chains are excellently suited for "reading" and "writing". Double-stranded DNA is otherwise rather unskilled. Its mother molecule, RNA, however, is also flexible enough to fold up, whereby it has been able to develop some technical skill, similar to – but not as perfect as – that of the proteins. Thus, RNA is capable of reading and writing as well as of doing some skilled work. Reading and writing are the basis of reproduction and, as such, they are the principal cause of natural selection and hence of evolution in the Darwinian sense. This is a major argument for a pre-existing "RNA world".

Let me return to the question of which property is the main determinant of the existence of life: cell-like vesicles, proteins or nucleic acids? What do I mean by "main determinant"? The word "main" is a substitute for the word "first", used by Lynn Margulis (cells first) or by Sidney Fox (proteins first). The word determinant, however, emphasises that I do not mean "first" in the temporal sense. "Cells first" thus stands for a "first" real unicellular organism, certainly not yet as sophisticated as we find it today, but consisting largely of both proteins and nucleic acids. And Fox's temporal "proteins first" does *not* mean present-day enzymes with defined functional properties, but rather some protein-like substances that might occasionally have possessed weak, accidental functions. After having elaborated what I am *not* aiming at, I should now explain what I do have in mind.

Which of the three properties – localisation by encapsulation, catalytic control of the ongoing chemical reactions, build-up and storage of genetic information – could have started a continuous evolutionary optimisation up to the present state of life? I think the answer is clearly "the origin" of genetic information. Why?

Take encapsulation first. As Lynn Margulis describes it, referring to the work of the physicist Erich Jantsch, it could lead to condensed "autopoietic" reaction systems with an inflow of free energy localised in chemical bonds of reaction partners and a corresponding outflow of entropy and reaction products, manifesting a simple form of metabolism. But what for? Oh yes, it can work if the autopoietic compartment contains certain proteins and nucleic acids and if it is able to solve the transport problems – in short, if it is a living cell.

Could it have started with certain proteins alone? No, it would have stopped as soon as the protein molecules initially present had decayed. Proteins could always form in a random manner, but such proteins did not have any inherent ability to reproduce themselves. Catalytic reaction cycles have been proposed that could reproduce themselves and thereby undergo natural selection. Such reaction cycles, apart from being extremely improbable, would have been able to "select" but not to improve themselves. In fact, no such cycle with an intrinsic power of evolution has been discovered so far. What is important: self-instruction is not an "inherent" property of proteins and hence does not allow their evolutionary optimisation. So, what about the nucleic acids?

Our candidate is RNA, or some precursor with similar properties. It could start on its own. Its sequence may represent the information for some phenotype, a ribozyme, that catalyses its own reproduction. In Section 4.10 an evolutionary experiment with a population of RNA molecules was described. We used normal enzymes as replicases and nucleases in order to let the reactions proceed with sufficiently high speed. Both types of reaction have been shown to be susceptible to ribozymes. At several places in this book I have stressed the necessity to present experimental data that allow a possible falsification of a theory – for biology an absolute necessity. The assumption of an early RNA world is based on hard experimental facts. Any origin requires at its start a non-linear phase of "nucleation". The great advantage offered by an RNA world is the fact that genotype and phenotype are represented by the same class of molecules, which is of great help in the nucleation process (see Section 5.3). Ribozymic phenotypes could have been replaced later by protein enzymes, which offer a much larger variety of functions. However, the start and the early phase of optimisation (see Section 4.7) are greatly enhanced if genotype and phenotype belong to the same class of molecules (see "hypercycles" in Section 5.3). In any case, the start as such requires some non-linear process of "nucleation", similar to the hypercyclic feedback phase in virus infection described in Section 5.3. Similar episodes might occur in later phases of evolutionary optimisation, whenever new mechanisms are to be introduced.

The "actors" on an RNA stage must have been quite efficient because their traces can still be seen in present-day living systems. Short RNA sequences of some ten monomeric units still fulfil many regulatory functions in present-day mechanisms of processing genetic information.[21] RNAs, larger but still relatively short-chained, with lengths closer to 100 units – the so-called transfer RNAs (tRNA) – are the universal adapters of the genetic code. They take care of the correct assignment of each of the 20 natural amino acids used in proteins to their cognate codons in messenger RNA (mRNA), as is necessary for error-free translation. Various kinds of ribozymes have been identified. Sidney Altman[22] and Thomas R. Cech[23] received the 1989 Nobel prize in chemistry "for their discovery of catalytic properties of RNA". And, last but not least, the ribosome, the protein factory of the cell, is a complex made up of both RNA and protein subunits of which, according to recent findings,[24] it is the RNA that is responsible for the catalytic function. Leslie Orgel called this fact the "smoking gun" that proves the complicity of RNA in the origin of life. The origin of the genetic code may well have required some encapsulation of the various constituents involved, but none of them has left witnesses as convincing as those found among RNA sequences. This field soon became a focus of research for younger groups of scientists in the USA, represented by names such as Jack Szostak and Gerald Joyce, who came up with exciting new ideas (see Section 5.3).

RNA, and its offspring DNA, introduced Darwinian behaviour, which is characteristic of life at all successive levels of cellular and organismic reproduction. As Theodosius Dobzhanski once said: "Nothing in biology makes sense except in the light of evolution." I might add: Nothing in evolution makes sense except in the light of natural selection. The latter is an inherent property of reproduction and variation, and hence of nucleic-acid-like, rather than of protein-like, structures.

In a recent book[25] with the title *Life Evolving* my senior colleague and friend Christian de Duve reaches similar conclusions. In his book the question "How did life arise?" is dealt with in three consecutive chapters: "The Way to RNA", "From RNA to Protein–DNA" and "The Birth of Cells". He starts with the RNA world, in which evolution on the basis of natural selection originated, and it is this principle that made subsequent steps possible. An RNA world as such was very limited, but through translation it could confer Darwinian properties upon the world of proteins, which is a world of almost unlimited chemical richness. Thanks to their newly won adaptive behaviour, enzymic catalysts could achieve complete control of everything going on at the chemical level. The door was opened for the build-up of an incredibly complex chemical network. This required the later replacement of RNA by DNA, in order to increase the storage capacity of information to a correspondingly large size. Moreover, encapsulation – which might have entered already in earlier phases – soon became an indispensable requirement: it was needed to keep everything together and to make the whole complex system a new optimisable unit, open to further and (almost) unlimited evolution.

When did all this happen?

Lynn Margulis in her book presents a "true-scale" timeline of the total history of our planet, which dates the first bacterial cells at 3900 million years ago. In 1989 we published in the journal *Science* a paper in which we dated the age of the genetic code at 3800 million years, with an error limit of some hundreds of millions of years.[26] Our method was based on a comparative sequence analysis of transfer RNAs, the adaptors of the genetic code, while Margulis' figure is obviously based on geological (i.e. sedimentary deposit) dating. The agreement between these two figures is interesting, since it seems to indicate that the first traceable bacterial cells and their DNA-based genetic apparatus do not have a very different age. Let me give a short description of our method before coming to any conclusion.

The method we used is an analysis of geometries of divergence in n-dimensional information space. We called it "statistical geometry" (see Section 4.4).[27] When I say "we", I include Ruthild Winkler-Oswatitsch of our Göttingen team and a mathematical colleague at Bielefeld University, Andreas Dress.

The transfer RNA molecule is specially suited to provide information about early stages of molecular evolution. First of all, thousands of sequences of these molecules are now known. Each of the large number of organisms studied contains about 40 different tRNAs, serving the 64 different codon combinations. Secondly, the length of tRNA is only 70–90 nucleotides, and the identity of some of them is constrained by base-pairing, so that the number of independently variable positions is reduced to about 30. Thirdly, tRNA is a quite conservative sequence and therefore particularly suitable for a study of early events in evolution. For instance, the sequences of phenylalanine tRNA of man and of the frog *Xenopus laevis* differ in only two positions.

The problem of deriving times from sequence changes depends on calibration. Extrapolation to the first node of bifurcation or even to the roots of the tree is associated with some uncertainly. All we need is two different kinds of time series which we can compare with one another in order to obtain some relative way of dating.

The first time series is that of phylogeny, schematically represented by Figure 5.1.5 and quite realistically modelled by the metal sculpture shown in Figure 5.1.6. It refers to the sequence of given tRNAs appearing in present-day organisms, including eubacteria, archaebacteria, eucaryotic protista, plants and animals as well as in cell organelles such as chloroplasts and mitochondria. There must exist about 40 different such trees, i.e. one for each of the 40-odd members of the family of tRNAs in a given organism. At the time of our study not all these data were available, yet a suffi-

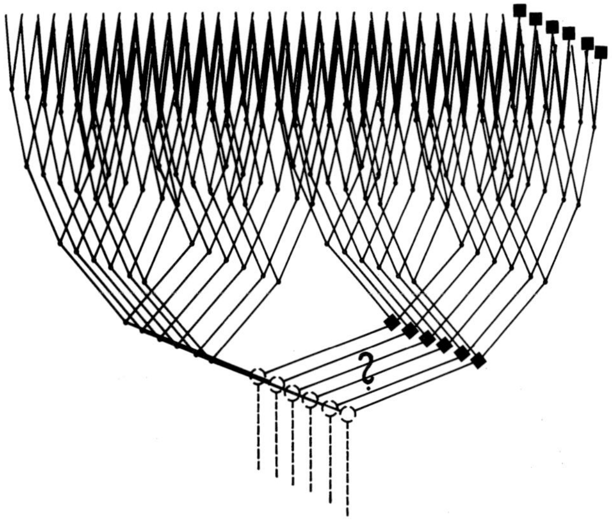

Figure 5.1.5

Superposition of phylogenetic trees of single tRNAs adapted to a specific anticodon (see text)

Figure 5.1.6

This tree by the artist Georg Meier-Henrich illustrates the complexity of evolutionary divergence in the real world, which, however, is more complex by far. From the author's collection.

cient number were. The successive phylogenetic changes produce typically tree-like diagrams yielding the earliest nodes for the tripods such as for eubacteria, mitochondria (chloroplasts) and archaebacteria (node I) or for eubacteria, archaebacteria and eucaryotes (node II). The dating of these nodes on the basis of geological assignments is most uncertain, introducing the wide error margins of the final data. The earliest bifurcation occurs between these two nodes, but it cannot be localised exactly, just as extrapolation of the distance down to the roots of the tree cannot be performed.

The second time series was obtained by statistical geometry. The 40-odd family members of tRNA for all given organisms were identical when the genetic code originated. Thereafter, each family as a unit "travelled" through all phylogenetic stages until the present-day organism evolved. During that time it accumulated mutations, which are reflected in the spread of sequence data for individual tRNAs in present-day families. Note that the divergence diagram of this spread is not tree-like, but rather bundle-like (see Figure 5.1.7). Hence it should project to a common node of origin that is identical for all families. A plot of these data (Figure 5.1.8) shows that this indeed is the case. The common point of convergence identifying one precursor sequence must lie below the plane at which all families become identical because the code originated from an assignment of different tRNAs to different amino acids. That is all I need to say about the principle. More details can be found in the papers quoted. Figure 5.1.9 shows the conclusions to be drawn from our data.

What the data mainly suggest is that the distance Δ_1 from the origin of the code to the first branching points of primordial cells is less than a third of the distance Δ_2 from the first branching points to the present time. This means a time difference of

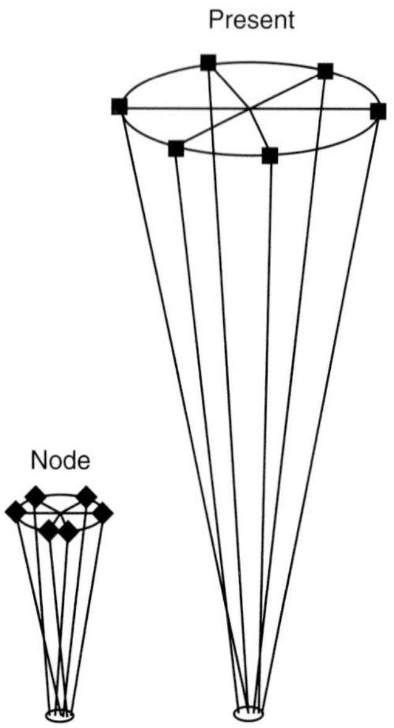

Figure 5.1.7

Alternative path of divergence found for the families of tRNA sequences contained in whole organisms (see text)

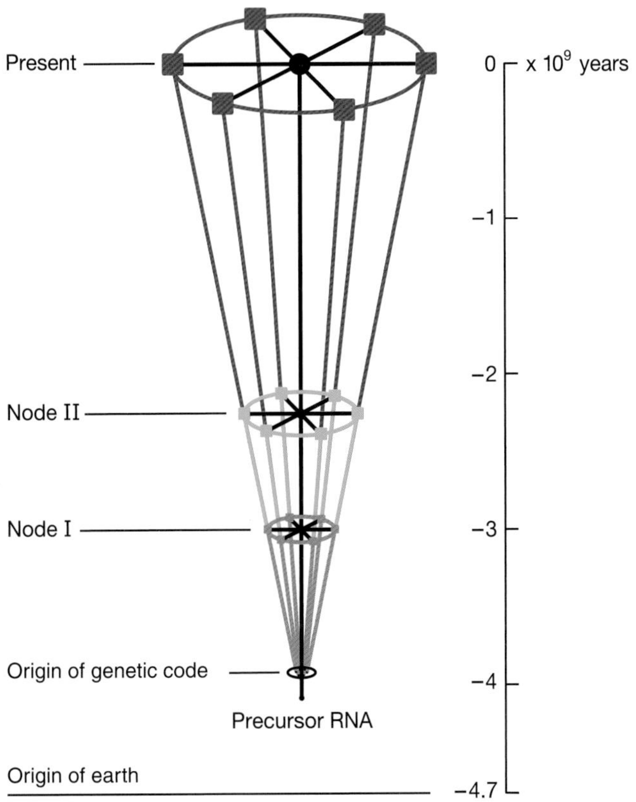

Figure 5.1.8

The divergence of consensus sequences of tRNA families from early nodes to present-day organisms allows an extrapolation back to the origin of the genetic code (see text)

less than 1000 million years for Δ_1 if the first nodes are dated at less than 3000 million years. Our more precise evaluation of the curves for distance segments of statistical geometry (see Section 4.4) yield an age of the genetic code of (3.8 ± 0.5) thousand million years. The large error limits are due mainly to uncertainties of geological dating of the earliest nodes.

I have never found these data quoted in the literature on early evolution. The paper in which we introduced the method of statistical geometry was hard to publish. I admit that the mathematics behind this method may be hard to understand for biologists. One referee wrote: "Don't these people know how to construct a tree? I shall be glad to send them my program." Yes, we knew how to construct a tree, but we also knew that tree programs will always come up with the best possible tree, even if the input data are random sets. Moreover, we know that there are divergence

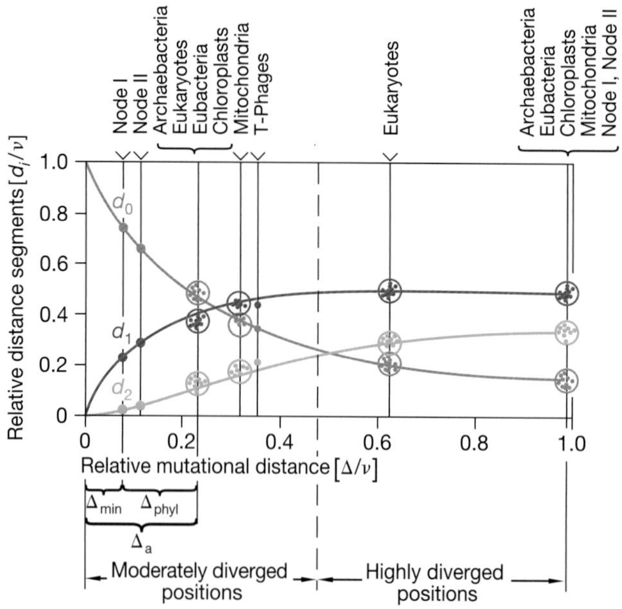

Figure 5.1.9

Evaluation of measured data in determining the age of the genetic code

diagrams that resemble topologies different from those of normal trees. It was the aim of the paper to derive mathematical criteria that tell us what kind of topology is involved and how to carry out a suitable analysis.

So, what is life now? Where is the demarcation line between the inanimate and the animate world?

The origin of life as I see it is the origin of information that is "inherently" reproducible and thus able to start an unlimited process of adaptation towards optimum function. This seems to me possible in an RNA world, which has left clear traces in present life, rather than in a protein world, for which we have not found any remnants. The problem still remains of how RNA came about, e.g. whether it had any precursors. We do not have any historical witnesses for such precursors in present-day life, as we have for RNA itself. Nor can we say whether the start of evolution had to await some favourable environment that allowed the synthesis of RNA. Here I could imagine some unspecific catalytic help provided by a proteinoid world. Whether or not one should call this "two origins of life", as Freeman Dyson[28] did for similar kinds of systems, depends on whether the state originally existing already had anything in common with "life" involving adaptation, a non-linear feedback mechanism of initiation, and finally the Darwinian process that optimises functionality. In this way one could define several "origins", each contributing just one additional improvement. I would rather call that "evolution", for which at later stages we have fossil witnesses. Evolution obviously did not proceed continuously, but in more or less large leaps, as was proposed by Eldredge and Gould[29] on the basis of fossil evidence. The present chapter reviews the physical mechanisms involved in the formation of biological information. A more coherent picture of how to envisage the origin and early evolution of life is reserved for Chapter 6 in the second volume of this book, "Self-organisation and Life".

Frankly, what to call "life" is a matter of definition. We know fairly precisely what the life that we experience today is. Transitory states between viable and non-viable matter don't exist any more. Perhaps we may one day be able to produce them again in the laboratory. Viruses are not autonomous living beings, yet inside a host cell they exhibit many properties of animate matter. They make use of the host's metabolism, they replicate, mutate and adapt, and they evolve, in some cases as quickly as the AIDS virus does. Similarly, we can reproduce certain properties, typical of life, in the test-tube, or in machines, causing inanimate matter to undergo an evolution usually encountered only in animate systems. This has led to a new "evolutionary" biotechnology, which at present is still *in statu nascendi.*

Life is a dynamic state of matter of nearly unlimited complexity which, despite this complexity, is able to conserve itself through reproduction. There is no viable matter in the world that allows self-reproduction to be dispensed with. However, the main feature of life is that it utilises this self-conservative ability for an evolution to unlimited complexity and sophistication. No end to this is apparent, unless everything is to terminate in a global catastrophe or self-made destruction. Life is synonymous with the generation of genetic information, which comes about in a series of phase transitions linked to symmetry breakages.

In the early 1970s, a group of Swiss chemists invited me to one of their very prestigious Bürgenstock Conferences to talk about the early stages of my theory of biological self-organisation. After my lecture, the grandseigneur of Swiss chemistry, Leopold Ružička, got up and said: "I have spent my life trying to find out what life is, and came to the conclusion that it is largely chemistry. Now you come here and tell us that it is all physics." Ružička – we called him Poldi – was known to me as a very jovial person, always ready for a joke. I had never seen him in such a distress as on that evening. It wasn't disapproval, it was simply deep concern (Figure 5.1.10) as expressed on a reprint of my 1971 paper found in his estate (Figure 5.1.11). Albert Eschenmoser[30] wrote about it in an obituary of Ružička, who was his mentor. In fact, it is not my opinion that "life is all physics". Most of it is chemistry, organised by physical principles, and it requires all the specific skills of a chemist: it is what these days we call molecular biology!

Let me conclude this section by reflecting my introductory remarks in recalling an episode that left a deep impression on me. In the autumn of 1975 Werner Heisenberg (whom I had known personally for years, having been his student in Göttingen) asked me to give a talk to a circle of physicists in Munich about my theory. However, the event could not take place, as Heisenberg became fatally ill. In a period of remission, he invited me together with Ruthild Winkler-Oswatitsch and some physicists from the Munich environment to his home, in order to discuss the theory with his colleagues. When, late in the evening, he accompanied us to the door he said:

Figure 5.1.10

Leopold Ružička listening to the author's lecture at the 7th Bürgenstock Conference in Switzerland. Courtesy of Albert Eschenmoser

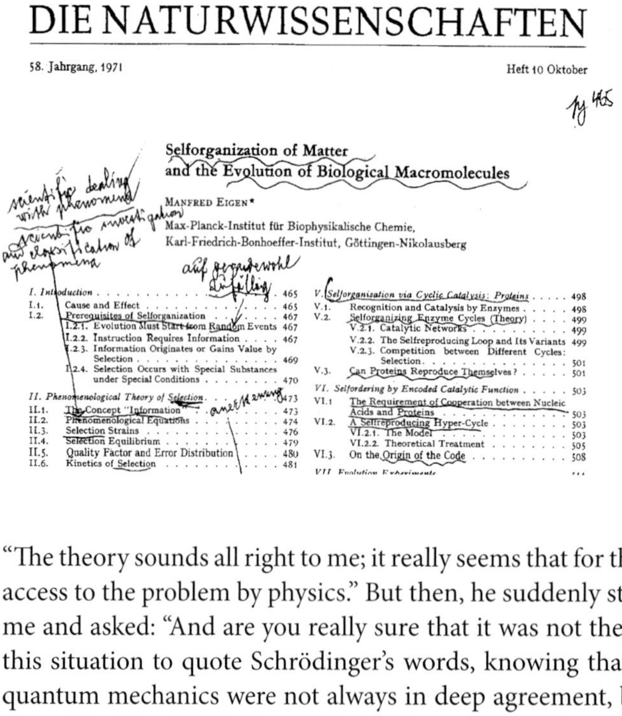

DIE NATURWISSENSCHAFTEN

58. Jahrgang, 1971 Heft 10 Oktober

Selforganization of Matter and the Evolution of Biological Macromolecules

MANFRED EIGEN*

Max-Planck-Institut für Biophysikalische Chemie, Karl-Friedrich-Bonhoeffer-Institut, Göttingen-Nikolausberg

I. Introduction 465
I.1. Cause and Effect 465
I.2. Prerequisites of Selforganization 467
I.2.1. Evolution Must Start from Random Events 467
I.2.2. Instruction Requires Information 467
I.2.3. Information Originates or Gains Value by Selection 469
I.2.4. Selection Occurs with Special Substances under Special Conditions 470

II. Phenomenological Theory of Selection 473
II.1. The Concept "Information" 473
II.2. Phenomenological Equations 474
II.3. Selection Strains 476
II.4. Selection Equilibrium 479
II.5. Quality Factor and Error Distribution 480
II.6. Kinetics of Selection 481

V. Selforganization via Cyclic Catalysis: Proteins 498
V.1. Recognition and Catalysis by Enzymes 498
V.2. Selforganizing Enzyme Cycles (Theory) 499
V.2.1. Catalytic Networks 499
V.2.2. The Selfreproducing Loop and Its Variants 499
V.2.3. Competition between Different Cycles: Selection 501
V.3. Can Proteins Reproduce Themselves? 501

VI. Selfordering by Encoded Catalytic Function 503
VI.1 The Requirement of Cooperation between Nucleic Acids and Proteins 503
VI.2. A Selfreproducing Hyper-Cycle 503
VI.2.1. The Model 503
VI.2.2. Theoretical Treatment 505
VI.3. On the Origin of the Code 508

VII. Evolution Experiments . . .

Figure 5.1.11

Remarks of Leopold Ružička on the author's original paper (*Naturwissenschaften* 1971). Courtesy of Albert Eschenmoser

"The theory sounds all right to me; it really seems that for the first time we have some access to the problem by physics." But then, he suddenly stopped walking, looked at me and asked: "And are you really sure that it was not the Lord?" I did not want in this situation to quote Schrödinger's words, knowing that the two great figures of quantum mechanics were not always in deep agreement, but I couldn't help saying something like: "Well, if it was the Lord, it must have been the same Lord who created your theory." Heisenberg agreed. It was the last time we saw him; he died about six months later.

5.2. Darwin for Molecules: Who Does the Selection?

In Section 5.1 I said that molecules knew as little about Darwin as Darwin knew (and I could have added: cared) about molecules. But now it is a molecule – DNA – which provides a sound physical basis for what long had been taken to be a strictly biological principle, not further reducible – so to speak an "axiom of biology". In order to underpin this statement, let me quote from the summary of Chapter 4 of *The Origin of Species*, in which Darwin introduces his central idea, the principle of natural selection:

> If under changing conditions of life, organic beings present individual differences in almost every part of their structure, and this cannot be disputed; if there be, owing to their geometrical rate of increase a severe struggle for life at some age, season or year, and this certainly cannot be disputed; then, considering the infinite complexity of the relations of all organic beings to each other and to their conditions of life, causing an infinite diversity in structure, constitution, and habits, to be advantageous to them, it would be a most extraordinary fact if no variations had ever occurred useful to each being's own welfare, in the same manner as so many variations have occurred useful to man. But if variations useful to any organic being ever do occur, assuredly individuals thus characterised will have the best chance of being preserved in the struggle for life; and from the strong principle of inheritance these will tend to produce offspring similarly characterised. This principle of preservation or the survival of the fittest, I have called Natural Selection.

Daniel Dennett has chosen this quotation of "two long sentences", the exact wording of which varies slightly in subsequent editions of Darwin's book, and commented that it is "fundamental", but "actually quite simple", adding that Darwin "himself could never formulate (it) with sufficient rigor and detail to prove (it)". Not by chance, I chose (part of) the same quotation in my 1971 paper, later adding Darwin's own assessment of the principle in his letter to George C. Wallich (28 March 1882), which I quoted above. I have shown that Darwin's principle can now be formulated and proven "with sufficient rigour", thereby uncovering much detail that until then had not been considered.

Darwin, although being somewhat envious of Lamarck's priority in postulating the concept of an "adaptive evolution" (cf. footnote below), did not completely reject his ideas. This only was done definitely in 1892 by the German biologist August Friedrich Leopold Weismann, who claimed the independence of the somatic cell line that makes up the body of an organism, and the germ line, the chain of cells that passes on the genetic information from one generation to the next. It is this crispening of Darwin's idea that received its substantiation by the discovery of DNA as the material basis of inheritance and which I referred to at the beginning of this section.

In many areas of biology there arose an almost endless discussion between "selectionists" and "instructionists". In our time, such debates have repeatedly heated up in fields as different as molecular biology, morphogenetics, immunology and neurobiology. I have already expressed clearly my aversion to any form of "ism" in science. Molecules obviously have no "intentions", they just "communicate" through physical (i.e. above all chemical) interactions. Should we call a straightforward binding process "selective" or "instructive"? If it is a hapten, which is recognised by an antibody, we usually call it "selective", regardless of whether it sets off a chain of events and is thereby the cause of an instruction. But what about a retrovirus, which first "instructs" the synthesis of a complementary DNA strand and then of a cDNA double strand that becomes integrated into the genome of the host? It provides "information" that wasn't there before, and from the host's point of view cannot be called selected – that term makes sense only for the infecting virus.

Is there a physical mechanism, or a class of mechanisms, that represents the basis of Darwinian selection? We have seen that the physical cause of an apple falling to earth, or water running downhill, is gravitation. In this connection I remember a rhyme written by the German writer Erich Kästner right after Hitler came to power. It may be translated as "Newton's Law of fall – applies on Earth to all. – To all!" (*Dem Gesetz des Falles gehorcht auf Erden alles – alles!*). Kästner's books were burned and he had to suffer persecution, but he survived the war.

Asking for a physical law behind Darwin's principle does not imply an attempt to make the whole process of evolution "calculable". Ernst Mayr,[31] who suspected physicists of such tendencies, is however right when he says: "Probably nothing in

---

Figure 5.2.1

Fundamental growth laws and selective behaviour

The time dependence of population growth for three different rate laws of product formation: (a) constant, i.e. independent of, (b) linearly dependent on and (c) quadratically dependent on population numbers of reaction products $n_i(t)$. The curves in column (i) represent the growth characteristics resulting from the above rate laws (a-i) and (c-i), having a singularity at $k_i t = 1/n_i$ $(t = 0)$. The curves in column (ii) reproduce the resulting selection behaviour, expressed in $x_i(t) = n_i(t)/\sum_k n_k(t)$, i.e. the "relative" population number with respect to all k reaction products present. What is shown by these curves is the entirely different selection behaviour that arises from the three rate laws:

(a-ii) Coexistence of all mutants (according to their individual k_i values), i.e. no real all-or none-selection takes place at all, but rather coexistence, similar to an "equilibrium".

(b-ii) Darwinian selection of the "fittest" entity, defined by a maximum k_i value. This allows advantageous newcomer copies, if their level has reached a stochastically defined threshold value (point of no return), to grow up regardless of the presence of formerly fixed "fittest" mutants.

(c-ii) "Once for all time" selection of a formerly fixed "fittest" mutant which resists the emergence of any newcomer mutants, even if these are characterised by larger growth constants. This entails a stop of further evolution, contrary to Darwinian behaviour.

(a) $\dot{n}_i = k_i$

$n_i(t) = n_i(t = 0) + k_i\, t$

(b) $\dot{n}_i = k_i\, n_i$

$n_i(t) = n_i(t = 0)\, e^{kit}$

(c) $\dot{n}_i = k_i\, n_i{}^2$

$n_i(t) = \frac{n_i(t=0)}{1-n_i(t=0)k_i t}$

$n_i(0)k_i t = 1$

$t = \frac{1}{k_i n_i(0)}$

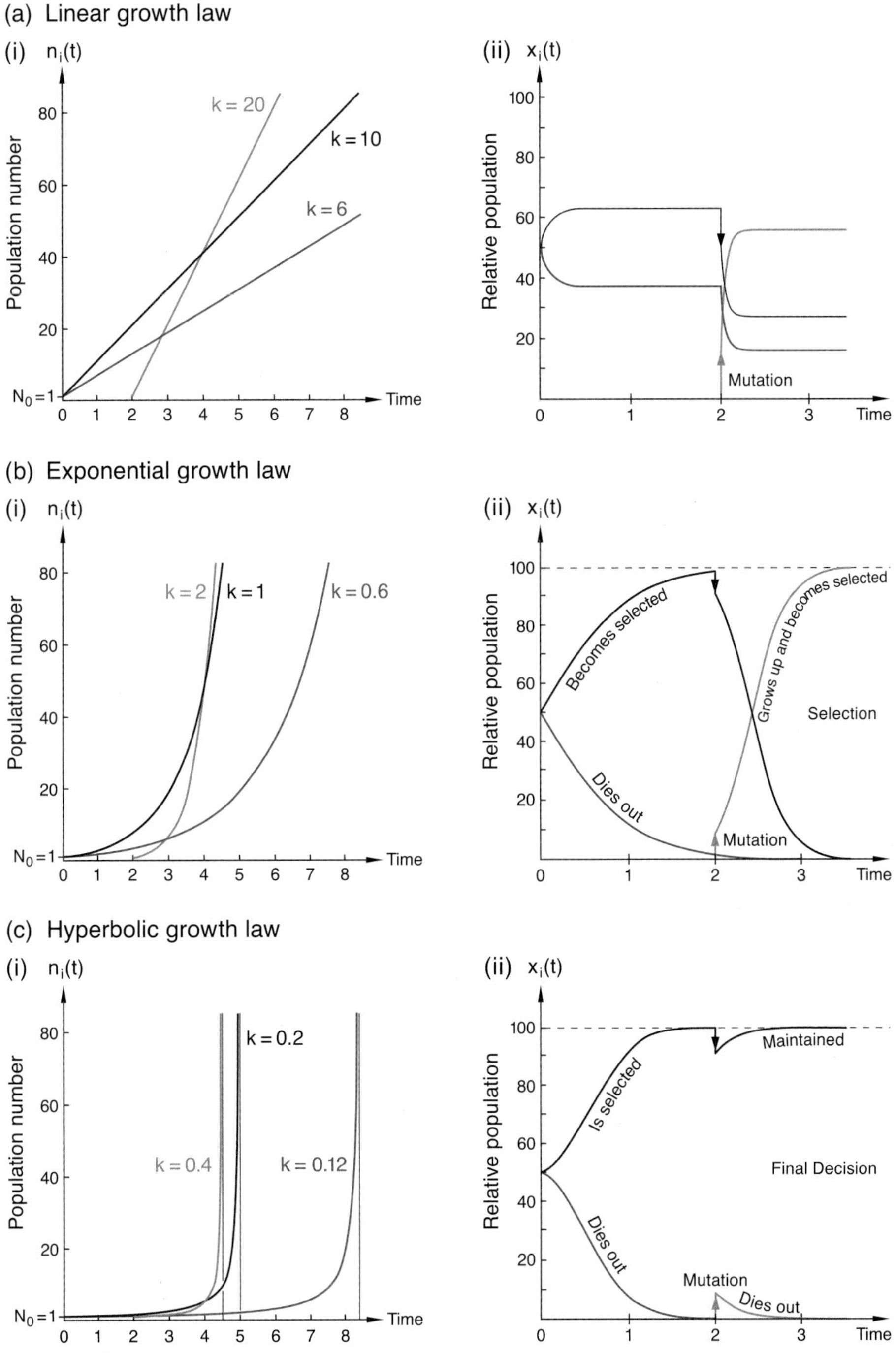

Figure 5.2.1 (Cont.)

biology is less predictable than the future course of evolution". The same author in his book lists the facts on which Darwin based his inferences: the finiteness or, at best, the linear increase of resources, which cannot keep pace with a potentially exponential increase in the population, results in some steady state which for the individual entails "struggle for existence". This inference Darwin adopted from Thomas Robert Malthus,[32] who had arrived at it in his *Essay on the Principle of Population* in 1798. Darwin combined it with the fact of the "uniqueness of the individual" and the "heritability of its traits" (cf. Darwin's own formulation, quoted above). It is noteworthy that Darwin, as early as 1842–44, refers in two essays (published only after his death) to the "enormous geometric power of increase" in population as a basic prerequisite of natural selection. Alfred Russel Wallace,[33] who had arrived later at the idea of "natural selection" independently, did not specify such a cause. For him the growth law played a secondary role. He argues that "the population is in any case constant, and even minor differences in succeeding generations should suffice to cause weaker and less well-organised forms to die out". This may be correctly observed, but at the same time it is devoid of any explanatory character. In a strictly logical sense it is even wrong: constancy of population is not sufficient for "natural selection". The decisive factor is the growth law.

Let me demonstrate this with the help of some fundamental growth laws in Figure 5.2.1.[34] The simplest form of growth is the linear increase of a population. The growth rate in this case is given by a constant term, independent of its (or any other) population number changing in time. Growth limitation in a system of several independently growing species leads to "coexistence". The various competitors exist with population numbers proportional to their growth constants. This case is represented in Figure 5.2.1 (a).

As a next example I consider exponential population growth. The rate law involves a positive linear dependence of the growth rate on population numbers. The solution of such a linear differential equation (or system of equations) is exponential. Hence, in this case the population grows exponentially (Figure 5.2.1(b-i)). The stationary state reached at growth limitation is one in which only one of the competitors survives, namely, the one that has the largest growth constant. All other population numbers decay exponentially to zero. This is clearly seen in Figure 5.2.1(b-ii), with one additional fact: After competition among all species initially present has singled out one survivor, which then dominates in the population, a single copy of a newly arriving competitor with a slightly larger growth constant is able to outgrow the entire population as formerly established. This is exactly true only for the deterministic case, while in the stochastic case there is a chance that a single copy, or a small number of copies, of the newly arriving competitor will die out before it reaches a "point of no return" in its amplification.

The third case refers to a hyperbolic growth law. The main difference between this and the exponential growth law is the sharper increase in the population number of the "survivor". This eventually leads to a singularity, which is to say that the growth curve reaches an infinite value after a finite time. An exponential function will reach infinity only for $t \to \infty$ (see Figure 5.2.1(c-i)). The rate law in this case is given by a term containing a power greater than 1 of the population number. The effect of limitation, again, is survival of only one competitor, but owing to the non-linear growth rate the system is now extremely selective. The important difference with respect to the exponential case is that "selected once" means "selected for ever". This kind of non-linear kinetics is well suited for any starting phase. However, if the population number of the survivor has grown to a sufficiently large value, no new competitor – even if it has a larger growth constant – will be able to outgrow the established population (see Figure 5.2.1(c-ii)). Such non-Darwinian behaviour would be injurious for any evolutionary optimisation, and therefore must reach a saturation phase where quasi-linear kinetics apply.

These examples, which represent the three fundamental growth laws "linear", "exponential" and "hyperbolic including a singularity", demonstrate also that "survival of the fittest" is not just a simple tautology. Darwinian behaviour, which includes both selective power and evolutionary potential, requires a mechanism that yields pretty nearly exponential growth. The rate law for such a system has to be – at least approximately – linear. The true surprise is how well all biological systems meet this requirement, as more realistic examples, treated in Section 5.3, will show. They provide the basis for the choice of the phenomenological rate equations in Vignette 4.7.1 explaining the phenomenon of "natural selection".

Growth in living systems is based on straightforward mechanisms of reproduction, an example *par excellence* being linear autocatalysis. Living beings usually show characteristic average lifetimes that express quasi-linear rate equations. Stronger deviations appear only under particular circumstances, where they are a consequence of some more or less strong non-linear interferences of a non-Darwinian nature. Such problems will be revisited in the following sections. In Figure 5.2.2 I show a method for an experimental study of these phenomena.

We may well ask what caused Darwin to emphasise so strongly the necessity of a "geometric ratio of increase" of the number of individuals that make up a population. I have always wondered how Darwin in this respect instinctively hit the nail on the head, because his book doesn't offer any mathematical deliberations on this point. In his Chapter 3 he gives numerical estimates, e.g. how many elephants – "being reckoned the slowest breeders of all animals" – would descend from a single pair, coming up with a number of nearly 19 million descendents, appearing within a period of only 750 years. He traces this back to the "doctrine of Malthus as applied to the whole animal and vegetable kingdom", and he argues that the large increase would require a correspondingly efficient sorting out in "the struggle for existence", thus explaining as a result the "survival of the fittest". I also think that in this point Darwin's deliberations went much deeper than those of Wallace.

Perhaps we get a little further if we adopt Dennett's[35] interpretation of the algorithmic nature of "Darwin's dangerous idea". Malthus' exponential growth law was simply a fact that applies at all levels of life from microbes to man and which even

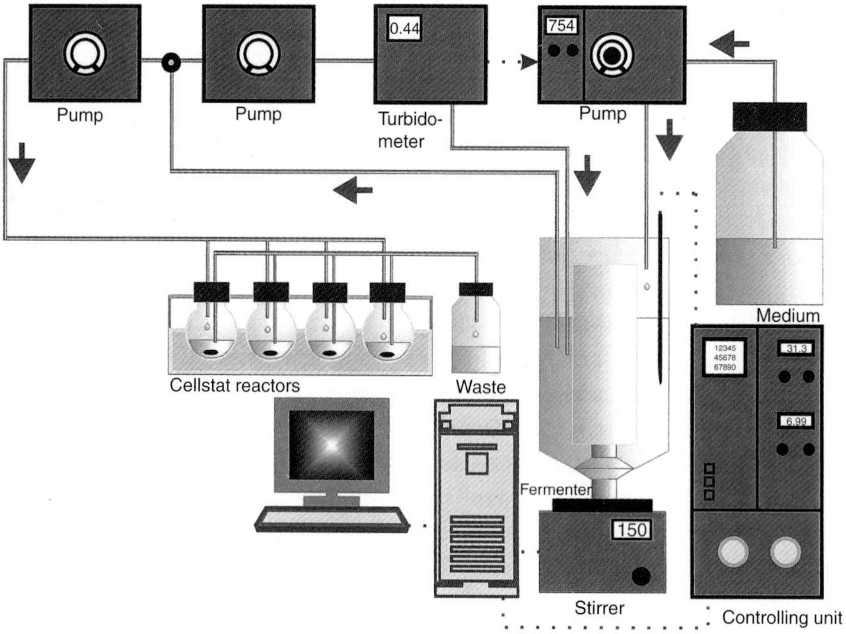

Figure 5.2.2
The cellstat

The cellstat is a bioreactor that was designed at our Göttingen laboratory for the purpose of studying the evolution of mechanisms of virus infection under defined host-cell conditions. A constant level of host cells of defined age is obtained by regulation of the turbidity of the liquid cell cultures (turbidostat). Virus infection under various conditions is carried out simultaneously using a battery of reactor cells, which are observed in parallel. The whole machine is under automated control and allows comparative evolutionary studies.

holds for certain subcellular composites that we usually do not term "viable". This offers the potency of an (almost) endlessly iterative application of the algorithm, which may explain the fine-tuned complexity of its eventual results. In order to keep the (mathematically required) adaptive mechanism functional, the target of selection had to "invent" all sorts of tricks, including compartmentation and individualisation, recombination, differentiation, etc. in order to keep up with the increase of complexity from virus-like RNA molecules to (sexual) combination, occurring first at a single-cell level and continuing throughout all higher kingdoms of living beings. The refinements in the mathematisation of population genetics, from Verhulst to Haldane, Fisher and Wright, did not deviate essentially from the linear growth kinetics imposed by the process of reproduction.

Darwin's "idea of the century", up to the present day anyway, has needed several refinements. Its main corroboration came from August Weismann's[36] clear distinction of the somatic line from the germ line, which proposes that the reproductive cells, ova and sperm, propagate directly from one generation to the next, unimpaired by somatic changes. The discovery of the two laws of inheritance by sexual reproduction, made in 1866 by the Austrian monk and botanist Gregor Mendel,[37] represent a major specification, if not modification, of Darwinian thinking; yet these laws remained unnoticed until 1900, when the Dutch botanist Hugo de Vries[38] rediscovered them in plant-breeding experiments. He interpreted his experiments correctly, only to find that Gregor Mendel had obtained the same results 34 years earlier. Mendel's laws were integrated into Darwinian theory only during the first half of the 20th century by the famous triumvirate of population geneticists Haldane, Fisher and Wright, and they have not lost their original flavour in Kimura's elegant theory of "neutral genetic drift". Finally, the rise of molecular biology in the second half of the 20th century, with all its exciting results, is an overwhelming affirmation of Darwin's ideas as being based on molecular behaviour. Now, the "fittest" wild type is no longer a single individual, but rather an extended distribution that includes many neutrals and is characterised by one maximum eigenvalue. This is valid up to the level of mankind. Otherwise we would all look alike – as genetically identical twins indeed do.

Coming back to the general problem of "natural selection" I want to emphasise the question of the physical conditions under which we may expect the principle to be effective. "Natural selection" is a special solution for the problem of "existence" far from equilibrium. There is no need for selection to be active under all circumstances. Physically it is contingent upon certain prerequisites, and it therefore works only when these are fulfilled. They are pretty closely associated with mutagenic reproduction. Mutagenicity refers to changes at the gene level such as erroneous copying (point mutations), deletions and insertions, and to various kinds of recombinative shuffling, while reproduction may be inherently auto- or cross-catalytic in nature, including replication and cell division. Since these processes are imperative for all stages of life, the Darwinian principle assumes its unlimited axiomatic applicability, as eloquently articulated by Dennett. On the other hand, there are many situations where the principle is put out of operation by additional constraints.

Recombinative shuffling of gene sections was later taken into account by the "synthetic theory" of the neo-Darwinian school which – as the name suggests – mainly provides a merger of Darwinian and Mendelian ideas.[39] Strictly speaking, genetic recombination as such doesn't introduce any new piece of DNA that was not in existence before. It changes combinations and thereby promotes the fast spread of advantageous changes among all individuals, which is of special importance in the case of larger genomes. The theories developed by the schools of John B.S. Haldane,[40] Ronald A. Fisher[41] and Sewall Wright,[42] and later by John Maynard-Smith[43] and Motoo Kimura,[44] have been applied successfully to many problems of population genetics. With the question "who selects?" I am asking, in particular, about the physical nature of the process of selection. Since it is a directed process, one would suspect the existence of a physical force that drives such a "non-vectorial flow". In the age of molecular biology one would guess that it is some kind of "chemical force". In fact, that's what it is – at least in principle.

The reactions on the basis of molecular, cellular or organismic evolution have to maintain states that are far removed from equilibrium. Phenomenologically, they involve two major terms: a positive one, representing growth, and a negative one, representing decay. I said "phenomenologically" because both these terms may include contributions from several sources. For molecules, the positive terms mean formation or synthesis, the negative ones decomposition or any other kind of removal. For living beings, they represent birth and death. We remember that Schrödinger recognised clearly (Section 5.1) that life cannot exist at or near equilibrium. Since equilibrium is a microscopically reversible state, any small deviation from equilibrium – regardless of non-linearities in the order of the reactions involved – can be expressed by the linear term of a power series that describes the deviations from equilibrium. These (chemical) relaxation processes were described in more detail in Chapter 2. Evolutionary optimisation could never occur under conditions near to equilibrium.

On the other hand, it has been possible in recent years to build up technical devices that realise evolutionary optimisation far from equilibrium. We have called them evolution machines (see Section 5.9). The technically most easily realisable case is that of a "constant flow reactor" (CFR) where both the input of reactants and the output of products occur at a constant rate. Such a machine, which we constructed in our laboratory in Göttingen[45] and used for a study of the "kinetics of phage infection", is shown in Figure 5.2.2. However, although technically easily realisable, this device is theoretically more difficult than a device in which the total amount of evolving species is kept constant. Why? The concentrations of the reactants and products have to be adjusted to the flow rates and thereby introduce non-linearities that would be absent if all concentrations were in a "buffered" state. This is the case if the "affinities" rather than the flow rates are kept constant. Hence, accordingly, flow regulation simplifies the problem. That is how we operate our "evolution machines". I will not go into further detail here, but will just indicate that with these considerations we are close to an answer to the question: "Who selects?" It must have something to do with chemical "forces", which are expressed by their affinities.

Affinity is the "force" driving a chemical reaction. In chemical thermodynamics affinity is a well-defined function, namely the partial derivative of Gibbs free energy with respect to a complementary reaction variable, called the "extent of reaction" (see Section 2.1). The latter reflects the relative stoichiometric concentration changes due to the particular reaction. The advantage of these two complementary functions, "affinity" and "extent of reaction", is that they are ascribed to individual reactions rather than to individual reactants. Regardless of how many compounds participate in a given reaction, they all change in defined stoichiometric ratios, allowing one to express their changes by one pair of variables. At equilibrium both variables, i.e. affinity and extent of reaction, become zero. The whole system consists of n coupled reactions, which near equilibrium can be described by a system of n linear differential equations, which has only negative eigenvalues representing a spectrum of relaxation times for the return to equilibrium. This is the basis of chemical relaxation spectrometry described in Chapter 2. The linear thermodynamic theory of such systems near equilibrium was first worked out in detail by Joseph Meixner,[46] a descendent of the famous Sommerfeld school. The early work on irreversible thermodynamics was dominated by two other pioneers: the Norwegian-born US scientist Lars Onsager[47] (Nobel prize-winner in 1968), who laid the foundations by his "reciprocity relations" and the Russian-born Belgian physical chemist Ilya Prigogine[48,49] (Nobel prize-winner in 1977), who made various important extensions to the theory of non-equilibrium steady states.

So let us return to the question "Who is doing the selection?" Being far from equilibrium, the driving force is ultimately chemical affinity. In fact, something similar will also hold for the generation of information in our central nervous system, which eventually became the basis of human culture. However, that's still a long way off. The equations derived in Section 4.7 contain a measure that essentially is represented by a chemical force. Just as the water droplet in Mandelbrot's problem (Section 4.4) is caused by gravitational forces to move towards the ocean, selection is the response to a force. The detail, of course, is quite a bit more complex. As Peter Schuster shows in Appendix A4.4, there is a potential function which owing to the bent structure of the simplex requires Riemannian geometry in order to yield a gradient perpendicular to the potential surface. The problem was first considered by Shashahani (loc. cit. in Appendix A4.4 by Peter Schuster) in connection with Fisher's fundamental equation.

The question "Who does the selection?" then is also related to the question "Which came first, the protein or the nucleic acid?" Natural selection is a consequence of self-organisation and, as such, is based on material properties of the system. The resulting "living" state is a material state with clearly defined properties. All living states we find today are subject to Crick's dogma: "DNA → RNA → protein → everything else", the first step of which is partly reversible. Neither proteins nor DNA include both abilities referred to above (direct copyability and the capacity to carry out functional tasks), although each of them is superior to RNA in one of the two properties required. When John von Neumann towards the end of his life came up with the idea of a "self-reproducing automaton"[50] he realised that the clue lay in the reproduction of the blueprint. The fulfilment of this requirement he could leave to a machine, constructed in an appropriate way. The "machine of life" had to organise itself, and such a self-organisation could start only with a substance that is in possession of both properties, which are ultimately chemical in nature. In other words, life is based on the laws of physics and, in particular, their expression in chemistry. I attach great importance to the way I phrase this. Life is based on physics, but its ultimate outcome is of a sophistication that transcends anything we can describe by any known physical law.

5.3. Does Natural Selection Require Linear Autocatalysis?

The three model cases treated in the foregoing section certainly prompt us to bring up this question. If only exponential growth, which is a result of linear autocatalysis, can lead to natural selection in the Darwinian sense, then an unduly narrow range of possible reaction mechanisms would be at our disposal in order to explain the origin of the extremely complex variety of living organisms by a Darwinian mechanism. What I am concerned with is the quasi-linearity of the catalytic reaction mechanisms which are the basis of natural selection, not the fact of natural selection itself, which is *per se* a non-linear phenomenon, as I showed in Section 4.7. In order to prevent any misunderstanding, let me add a few more words on this.

Natural selection is a consequence of the constraints that inhibit unlimited growth. In Section 5.5. I shall show how this introduces non-linearities into an otherwise linear autocatalytic mechanism. Another example of this sort was demonstrated in Section 4.8 with chaotic aberrations, a typical non-linear phenomenon occurring in certain insect populations, described by Robert May.[51] The description is based on the Verhulst equation (see Section 4.8) which owing to limited resources introduces saturation to an originally exponential growth of population. In the Verhulst equation the limitation is expressed by a negative quadratic term, reducing growth eventually to zero in an otherwise linear autocatalytic mechanism. The chaotic aberrations in addition require the reproduction of certain numbers of descendents within temporally discrete intervals.

When I use the term "linear autocatalysis" I refer solely to the mechanism of reproduction expressed by the matrix (W_{ik}), rather than to the selection mechanism, which turns out to be non-linear because of competitive growth. Non-linear autocatalytic mechanisms, on the other hand, as in the first and the third example discussed in the preceding section, do not lead to any properties that could be called "Darwinian". The question I want to discuss in this section is therefore concerned with the observed broader preponderance of quasi-linear processes, which constitute the basis of biological evolution. There is, of course, no absolute requirement for them in biology, where in several cases we need coexistence as well as once-for-all-time fixation, the latter occurring, for example, in the genetic code or in the uniform chiralities of biological macromolecules.

At first sight all processes in biology look quite complex, usually involving large varieties of different compounds and therefore anything but linear mechanisms. What immediately comes to my mind in this connection is the problem of "chemical relaxation", which is a scrupulously linear phenomenon, regardless of a possibly large molecularity or order of the reactions involved. In fact, we used relaxation spectrometry to elucidate complex biological reaction mechanisms, such as those of RNA and DNA replication, or of enzymic processes involved in allosteric regulation (see the examples discussed in Section 2.1). All these reaction processes are anything but linear. Nevertheless, they appear as linear phenomena in the recording of their relaxation–time spectra. The explanation is simple: linearity in all these cases is due to the fact that the reaction processes are recorded in situations very close to their equilibrium or to certain stationary conditions. Since the "affinities" and their complementary "extents of reaction" are zero at equilibrium they can be linearised in the vicinity of those states, regardless of their molecularity or reaction order.

For living systems, which are notoriously non-equilibrial, anything like this may seem to be entirely detrimental. Not quite! First of all, linear analysis may be applied to any neighbourhood of a reference state, and as such plays its role as a first approximation in the analysis of non-linear systems.[52] Moreover, living systems usually show the phenomenon of homeostasis. In the Oxford dictionary this is defined as "the tendency of a biological system to resist change and to maintain itself in a state of stable equilibrium". The latter, of course, is not identical with the physically well-defined state of thermodynamic equilibrium, but rather means some regulated stability of the internal state of the living system, protecting it from external perturbation of its functions.

In the following paragraphs I discuss two examples in more detail in order to illustrate the role of quasi-linear autocatalysis in the process of formation of genetic information. In Section 5.1 the importance of the origin of information for the origin of life was stressed. This was anticipated by John von Neumann in his discussion of the "self-reproducing automaton", a machine that optimises its function by applying the principle of iterative mutagenic reproduction to the blueprint for its construction.[50]

My first example is concerned with the molecular mechanism of replication of sequences of both RNA and DNA. This process is instrumental in generating and maintaining the information of any living entity, and it is based on the epoch-making discovery of Watson and Crick (Section 4.2). The studies I report here have been conducted at three levels: (i) kinetic measurements, (ii) computer simulations of the mechanisms involved and (iii) theoretical interpretations. All these studies have been done *in vitro* and hence are concerned exclusively with the corresponding chemical mechanisms. The question we really want to answer is: does the very simple ansatz of "linear autocatalysis" in Section 4.7 really apply to a system of such complexity? The system we consider entails the replication of the genome of a plus-strand RNA virus; this is a single RNA chain comprising a few thousand nucleotides. The experiments we did included only mini-variants of such viruses, which contained just some 100 monomeric units, but which sufficed for checking theory against practice in all kinds of ways.[53,54]

Nevertheless, the complexity of such a system is already enormous. Every individual sequence involves hundreds of different reaction steps: binding of enzyme and template, incorporation of monomeric nucleotides from energy-rich triphosphates, release of pyrophosphates, initiation, elongation and termination of chains of indefinite length, activation and inhibition, correct and incorrect inclusion of monomers –

yes, the problem on the molecular scale looks quite complex for any theoretical analysis. However, such an analysis in terms of averages can be carried out in a quite straightforward way, and the matrix in Section 4.7, which is representative of the linear system of equations, can be formulated explicitly.

The reader who is interested in more details of the mathematical treatment is invited to consult Appendix A5 because I intend to stick to my promise to keep the text largely free of mathematical excursions. How far the simplification by the linear ansatz in Section 4.7 goes is best seen from Figure 5.3.1, which represents the mechanism of RNA replication in all its detail. The indices in the rate constants refer to the mathematical treatment in the appendix and may be disregarded here, where I just want to give an impression of the complexity of the mechanism involved.

The surprising fact that emerges is the result of the mathematical treatment, which shows that the mechanism can indeed be well described by a quasi-linear ansatz, which has much similarity with the Michaelis–Menten form used in enzyme kinetics.

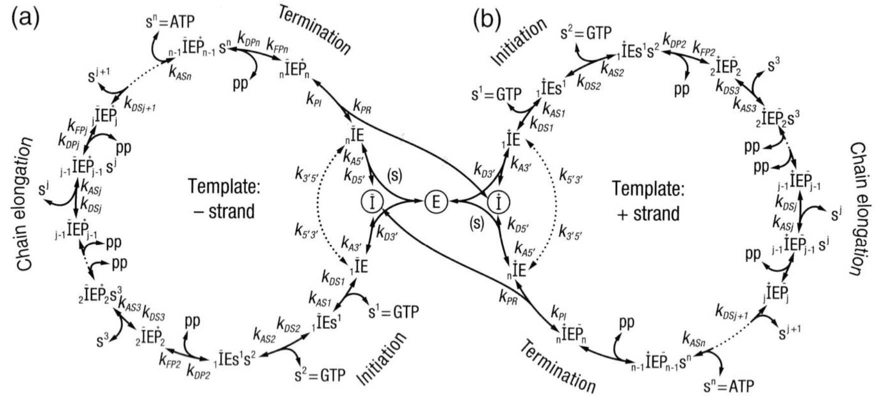

Figure 5.3.1

The mechanisms of RNA replication, reproducing all the steps involved

This picture was chosen not in order to deter the reader, but rather to convey some feeling for the true complexity that may be involved even in early evolution, which in the phenomenological presentations is satisfactorily described by the simple ansatz $dx/dt = kx$, k being a constant. In Appendix 5 the rate constants for different steps are explained in detail. Despite the repetitive nature of chain elongation, the individual rate constants may have quite different numerical values, depending on the spatial structure of the template, which has to unfold for replication. (Slow rates refer to so-called "pause" sites of the enzyme.) Two facts are of interest for the present discussion:

(a) Two cycles are involved in forming complementary replicas for both the plus and minus strands. For palindromic plus and minus strands, the two cycles involve identical steps and thus coalesce into one cycle. Otherwise, the total rate of replication results from the geometric mean of the rate contributions of the two cycles.

(b) Each cycle may be reduced to three principal steps (see Figure 5.3.2): 5'-template release and 3'-rebinding to enzyme; initiation of replica formation; replica chain elongation and release.

The Michaelis–Menten expression represents a simplified form that compresses a multistep mechanism into one equation, assuming stationary concentrations of intermediates. Nonetheless, it is successful in many cases and is therefore popular among biochemists. The analogy with the Michaelis–Menten case becomes more obvious if we simplify the mechanism of Figure 5.3.1 to the three-step form shown in Figure 5.3.2. The three major steps of the overall reaction are: (i) initiation, i.e. providing an active form of the enzyme-template-substrate complex from which a practically irreversible polymerisation reaction can start, (ii) chain elongation, providing the reaction product in form of a complementary template strand, and (iii) termination, which involves a dissociation of the enzyme from the 5' end of the template strand and its rebinding to the 3' end of either the plus or the minus strand. The assumption of a stationary state in this multistep mechanism leads to simplifications similar to those in the case of the Michaelis–Menten equation. However, we are now dealing with an autocatalytic reaction in which the catalytically active strand of the template represents the reaction product.

The analogy to straightforward enzyme kinetics is expressed clearly in the experimental data on the kinetics of RNA replication. Christof Biebricher and his co-workers have performed a large variety of such studies in our Göttingen laboratory. Some typical data are shown in Figure 5.3.3. Product formation in the

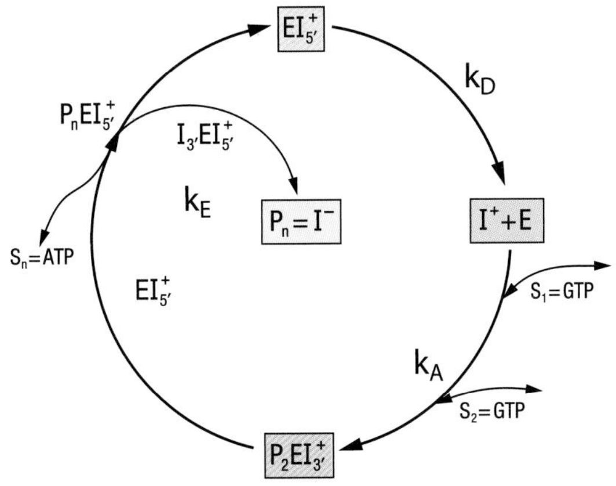

Figure 5.3.2
Simplified three-step mechanism of replica formation

The three principal steps of replica formation have been formulated in the legend of Figure 5.3.1: contracting template (E5'-I) release and 3'-rebinding, as well as replica-chain elongation and release, as suggested by the rate-limiting steps found in experiments. Of the two possibilities (replica and template release) the latter turned out to be the slower one, while a subsequent rebinding of template at its 3' end does not seem to cause any delay after its slow release from the 5' end. Replica release is usually fast compared with chain elongation. The same is true for chain initiation.

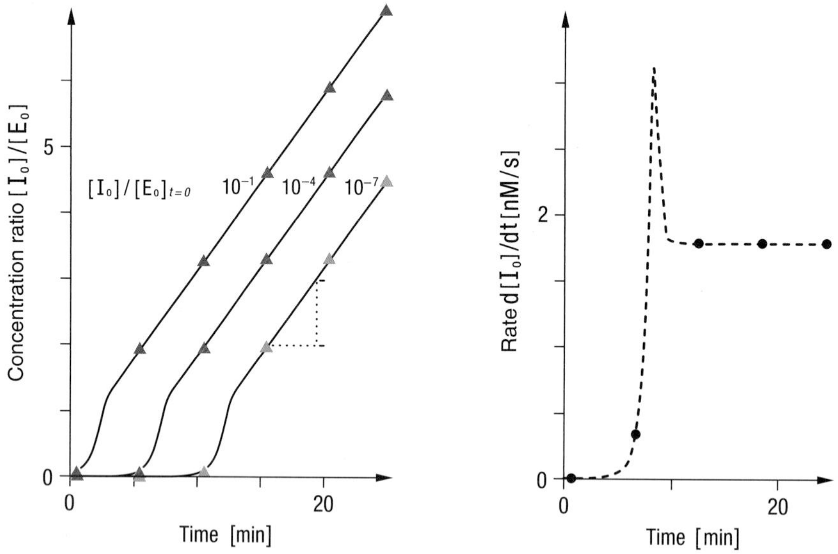

Figure 5.3.3

Calculated and computer-simulated growth profiles for the kinetics of RNA replication *in vitro*

Left: The exponential and linear phases of nucleotide incorporation as a function of time. Experiments were performed for various initial ratios of $[I_0]/[E_0]$, (i.e. template/enzyme), which exactly reproduce these profiles. The rates in the linear phase were taken from the slopes of the curves, those in the exponential curves from the ratios for the displacement of the curves, divided by the natural logarithm of $[I_0]$. Right: Simulated RNA synthesis rate profile for an initial ratio $[I_0]/[E_0] = 10^{-6}$. Exactly this behaviour (including the overshoot) was found in experiments. Note the difference in the two plots: on the left, concentration is plotted against time, and on the right, rate is plotted against time.

Michaelis–Menten mechanism usually (i.e. in the absence of co-operative effects) starts with linear growth of the product which, when the substrate concentration reaches the order of magnitude of the Michaelis constant, bends over to saturation. In our case of replication the curve starts with exponential growth due to uninhibited linear autocatalysis, while saturation in the Michaelis–Menten curve is now replaced by a linear growth phase (again caused by saturation of the enzyme, this time by the growing template).[54]

Of particular interest are the results for the exponential phase of replication. First of all, in this case one has to find a solution for the whole system of equations because assuming a stationary state for the intermediate steps of reaction does not make any sense in this case. A useful assumption, however, that does make sense is that of uniform rate constants for the various steps of chain elongation. At sufficiently large chain lengths this step will always become rate-limiting. We have tested by numerical simulation how well this assumption works for the averages in cases where inhomogeneities are present, e.g. in the form of pause sites. As shown in the appendix, with such assumptions the system of equations can be solved and eigenvalues can be given in the form of analytical expressions. What is of special interest is that the matrix we obtained turned out to be that corresponding to a Vandermonde determinant. This is the determinant of an n by n matrix, the ith row of which is given by: $1, x_i, \ldots, x_i^{n-1}$, with all the $x_i^k (k = 1, 2 \ldots n - 1)$ appearing as variables in a given polynomial equation that provides information about its roots. The determinant is named after the 18th-century French mathematician Alexandre-Théophile Vandermonde. A colleague of mine in mathematics who wrote a textbook on linear algebra[55] in which he referred to Vandermonde told me that he could not find any application of this interesting type of determinant in physics. He was very pleased when I showed him the above example.

Let me conclude this case by showing some diagrams that I find interesting because they demonstrate the physical build-up of exponential growth from defined uniform initial conditions for quasi-linear autocatalysis, which becomes exactly linear at large excess (or buffering) of the monomeric substrates, i.e. the nucleoside triphosphates of A, G, U and C (see Figure 5.3.4). What later on becomes a well-defined homogenous exponential function starts out from synchronised cycles in which the growing population of templates undergoes random mixing. This reminds one of the physical problem of measuring a finite lifetime of the proton, which is supposed to be greater than 10^{32} years, while the proton as such could not have come into existence at times earlier than 10^{10} years ago (the time of the "big bang").

Now let me move on to my second example, which in part is based on the first one, but refers to a higher level of complexity. My aim again is to show that for its evolutionary development the process can be well described by a quasi-linear mechanism of autocatalytic growth. The problem deals with the total process of viral infection *in vivo*, which can be analysed in all its individual steps. Our "pet" again is the bacterial virus, or bacteriophage Qβ, already used by Sol Spiegelman in his early pioneering experiments. The target of the virus is the bacterium *Escherichia coli.* The work was part of the doctoral thesis of my quondam student Michael Gebinoga. Christof Biebricher and William C. Gardiner were again partners in our team.[56]

My original idea was to find some experimental proof for the existence of catalytic hypercycles and their relevance in the wheel of life. The idea of the hypercycle had its origin in my 1971 paper in *Die Naturwissenschaften*, mentioned in Section 4.7. The hypercycle then became the focus of more extended theoretical studies together with Peter Schuster, culminating in a joint book on hypercycles with the subtitle *A Principle of Natural Self-Organisation.*[57] Why do we think of it so highly as to term it a "principle" of biological organisation?

The amount of information that could be accumulated by any linear autocatalytic process is severely restricted by the fidelity of copying, and is subject to an error threshold at which everything achieved previously would get lost. The phrase "error catastrophe" is a quite appropriate expression for such a situation. In early evolution this situation is aggravated by what I call the "genotype–phenotype dichotomy". Although important steps towards life might have occurred first in an RNA world,

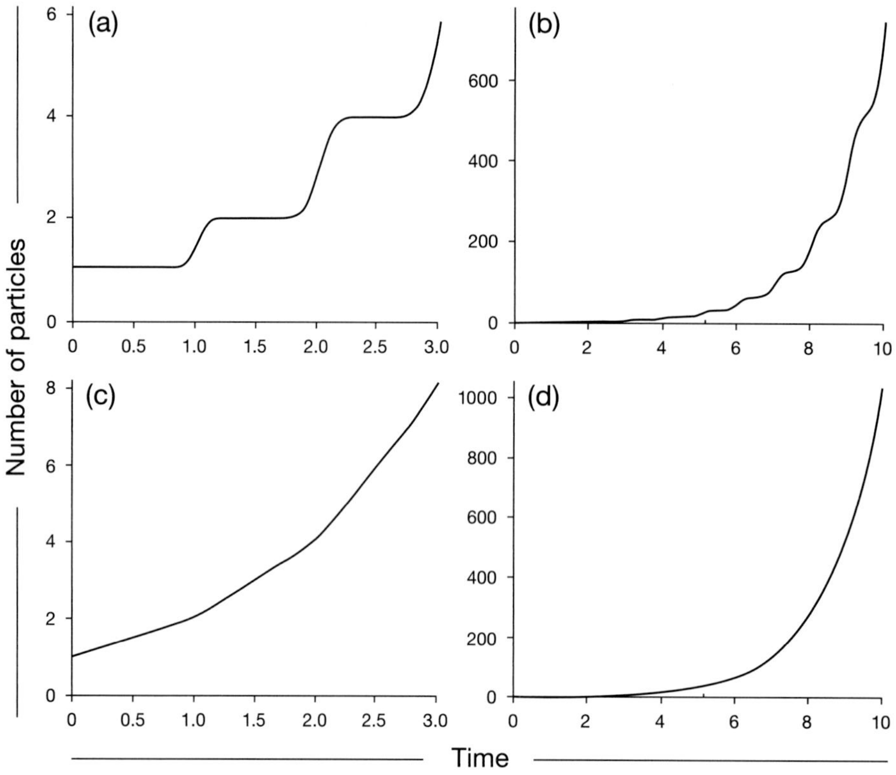

Figure 5.3.4

The start of exponential amplification

These examples refer to Appendix A5. They show that any exponential amplification of particles is based on the discrete process of doubling. This fact is only expressed clearly in the initial steps of the growth curve, starting from a single particle (The curve (c) has been according to curve (b) homogenously extrapolated). Apart from the exponential series (2^k for $k = 0, 1, 2, \ldots n$), the Fibonacci series is another good example. The series starts discretely but soon randomises, yielding a continuously increasing exponential function for the increase in the statistical expectation values. This is expressed differently in the expectation values for the exponential decay. We know that radioactive decay, despite a clearly defined half-time, can happen at any moment because it starts from a randomised initial state. The exponential increase from single events, by contrast, has a clearly defined initial state. The treatment presented in the appendix, leading to a Vandermonde determinant that can be solved exactly for special cases, showing clearly the discrete start from individual doublings, yielding for the starting phase quite pronounced deviations from a straightforward exponential increase. There remains the question of what happens in the initial phase of a radioactive decay when the half-time becomes larger – even reaching values that are large compared with the age of the universe. For proton decay the half-time according to the "standard model" of particle physics is estimated to be around 10^{32} years, but in our universe no material particle could be older than the time that has elapsed since the "big bang" (around 10^{10} years ago).

the enormous catalytic power of proteins turned out to be indispensable for life to reach any advanced level. Accordingly, proteins soon dominated the executive scene. However, we have seen that proteins can neither "read" nor "write". So they were in need of the nucleic acids to handle all "legislative" tasks in both selection and evolution. Any progress that appeared at the executive level had to be "written down" in the genomic "book of law". This requires a feedback loop from the executive level of function associated with proteins to the legislative level associated with nucleic acids. Functional advantages experienced by proteins can be made available to future generations only through fixation at the genetic level. Hypercyclic organisation in combination with compartmentation is a tool for bridging the dichotomy between phenotypic function and genotypic information. For the time being this was "pure theory", and we have learned that at least in biology this may be "poor theory". What we need is experimental proof.

The most direct contact between the phenotypic and genotypic levels may be mediated by an enzyme that catalyses the replication of its own genetic message. Something like this must have happened in early evolution and, although viruses might have appeared quite a bit later in evolution, it may still be an integral part of their infection cycle. So let us have a look at their mechanism in more detail. The best candidates for this are plus-strand RNA viruses such as phage Qβ. As quoted in our paper on the hypercycle,[57] as applied in biochemistry,[56] quite a number of authors have made important contributions to the subject. Here I particularly want to emphasise the work of Charles Weissmann and his school.[58]

Let me specify the problem clearly. What enters the host cell in our example is just a bare, single plus-strand of the virus carrying all the genetic information. "Bare" simply means that the virus does not bring along any of its proteins that it needs in order to carry out its eventually destructive functions. The bare viral RNA is entirely at the mercy of its host, which has nothing better to do than to let its 10,000 to 20,000 ribosomes translate the genetic message of the virus into those executive functions that will eventually destroy the host. That sounds like a far-fetched crime novel, but that's not the story I want to tell. The point at issue is the hypercycle.

One of the translation products in this initial phase of viral infection is an enzyme that specifically catalyses the replication of the virus' RNA. In fact, in the case of Qβ it is just one protein subunit, which joins up with three other protein units that the host itself uses for multiplication, the so-called elongation factors EF-Tu and EF-Ts and the ribosomal protein S1. Their combination with the virus protein yields a replicating enzyme that is highly specific for the phage genome, i.e. the plus strand of Qβ RNA, and is enormously successful in competition with any host replicase. Again, the host "graciously" provides everything needed for its own destruction. Now it really begins to sound like a crime novel, but I rather would draw the conclusion that viruses are late events in evolution, namely mechanisms that found a niche for survival by all the information they had gained in more highly sophisticated states of existence.

The above mechanism lives up to all its prerequisites. We have two differential equations describing two rates:

(1) The production of the polymerase that specifically catalyses the replication of infectious Qβ RNA (at a given concentration of the bacterial ribosomes) depends – in a good approximation – linearly on the number of RNA plus-strands present.

(2) The production of RNA plus-strands should be proportional to the product of the concentrations of the replication enzyme and the template. In the presence of a large excess of ribosomes the concentration of the replicase will come out to be proportional to the concentration of template strands present. Then the second equation just says that the number of RNA molecules produced (i.e. the rate of production of infectious strands) is proportional to the square of the template concentration.

The rigorous solution of the two equations yields a "classical" hypercycle. That's what theory says, but can we prove it? Can we find living examples showing the presence of hypercycles in nature?

Now I come to Michael Gebinoga's doctoral thesis. His first problem was how to get access *in vivo* to the kinetics of the production of proteins and of RNA. It is virtually impossible to measure synthesis rates accurately *in vivo* because the precursor concentrations cannot be measured, and precursors cannot be labelled by appropriate markers. Some tricks are needed.

Gebinoga decided to present the continuous process of sequential production of proteins and RNA as a series of discrete snapshots. For this he had to complete a full infection cycle *in vivo* by pulsed incubation. This means for the full cycle of infection, which takes about 40 min, something like 40 pulses of incubation, each lasting a little less than a minute. The samples taken during the pulsed incubation were analysed: details are given in our *Biochemistry* paper,[56] but the essential point is to extract the lipids from the cytoplasmic membrane by adding, at the end of each period of pulsed incubation, an organic solvent (toluene) to the incubation medium. This method was pioneered by R.E. Moses and C.C. Richardson[59] as well as by H.P. Vosberg and H. Hoffmann-Berling.[60] The semipermeabilised bacterial cell membrane allows rapid exchange of low-molecular-weight synthesis precursors with the surrounding medium, whereas the physical integrity of the bacterial cell and the locations and activities of the high-molecular-weight intracellular compounds are retained. While the infection process as such occurred under normal *in vivo* conditions, the analysis during the periods after interruption by addition of the organic solvent occurs under *in vitro* conditions. However, precise amounts of extraneously added low-molecular-weight synthesis precursors permit an accurate determination of relative incorporation rate profiles, which can be extrapolated to the *in vivo* conditions.

Let us have a look at the results obtained by Gebinoga in his dissertation thesis.[61] In Figure 5.3.5 the rates of RNA and protein synthesis *in vivo*, as well as the number of reproduced phage particles, are shown. They were analysed according to the method described above and plotted as functions of time. For better comparison they are all included in one and the same diagram with a common scaling for time. Note that what is plotted for RNA and the proteins is the rate of their synthesis (in number of particles per minute), rather than the amount synthesised (in terms of particle numbers or concentrations). The almost constant level of rate, reached after

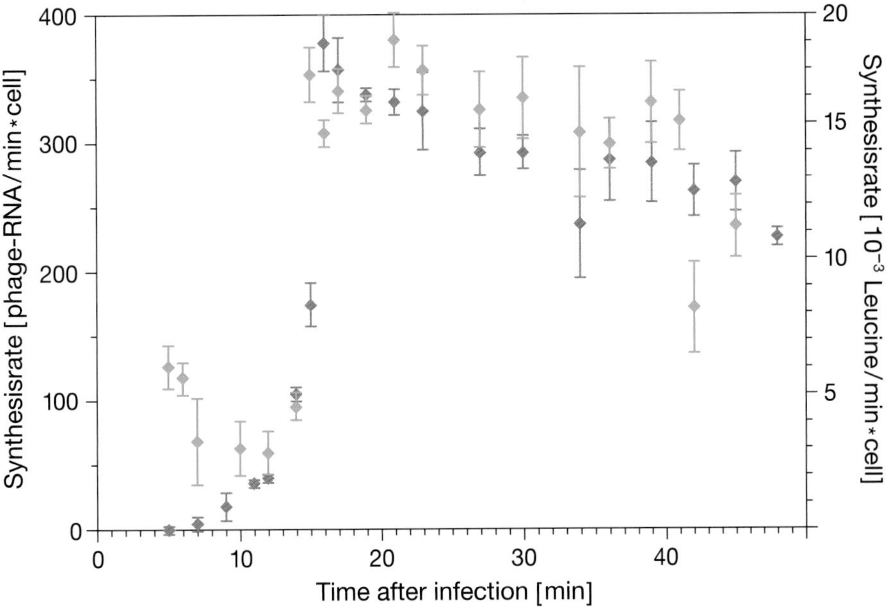

Figure 5.3.5

The sudden increase of virus template and its encoded proteins in early phases of phage infection

The formation of virus RNA and its protein translation products in early phases of infection has been measured experimentally *in vivo* by Michael Gebinoga (dissertation, Göttingen/Braunschweig 1990). The plot shows the formation rates for both viral RNA (green symbols, left-hand scale) and viral proteins (blue symbols, right-hand scale) as functions of time. What immediately catches one's eye is the fact that – after some initial protein accumulation – the formation rates of viral protein and viral RNA coincide, and after not quite 20 min a kind of "end point" is seen, as found in titration curves. The interpretation is indeed the presence of a "titration end point" at which the accumulating RNA template "saturates" the concomitantly accumulating RNA-encoded protein Qβ-replicase, i.e. the enzyme catalysing the replication of the viral RNA, while at the same time the more abundantly produced coat protein of the virus switches off the synthesis of Qβ-replicase by blocking the "Shine–Dalgarno box" of the replicase cistron. These curves are shown because they demonstrate quantitatively the presence of hyperbolic growth, which is typical of a hypercyclic mechanism.

about 20 min, means a linear increase in numbers. I have chosen this manner of representation in order to emphasise the simultaneity of the burst-like increase of both RNA and proteins, which immediately suggests a hypercyclic mechanism. The RNA synthesis rate includes both plus and minus strands, the ratios of which have to be determined separately. The same is true for the protein synthesis rate, which is dominated by the production of the coat protein, which is in great (and accurately known) excess over that of replicase units.

In order to understand the numbers in Figure 5.3.5 we have to know how the measurements were made. The rates of RNA synthesis were recorded by incorporation of radioactively labelled phosphate ([^{32}P]-CTP) and those of protein synthesis by incorporation of labelled leucine ([^{14}C]-Leu) of which 12 residues are contained in the (most abundant) coat protein. The numbers on the ordinate refer to the recorded count rates, rather than to the true rates of particle synthesis. The latter may be taken from Figure 5.3.6. Similar reservations apply to the numbers of the phage particles produced, the first of which appear about 20 min after infection, accumulating to a final "burst" size of 600 "infectious" particles, which, however, corresponds to more than 10,000 phage particles actually synthesised. Owing to a high accumulation of errors in later phases, only a fraction – less than 10% – of the phage particles turn out to be infectious. Furthermore, the fact that both RNA and protein synthesis rates reach a constant level after 20 min deserves an explanation. That is easy for the protein synthesis rate, which is finally limited to a constant level because of the constant number of ribosomes present. But why is there a constant level for the autocatalytic production rate of the RNA molecules? If the constant level of protein production leads to a linear increase in numbers, the number of replication enzymes should increase concomitantly, still yielding an exponential increase in the quantity of RNA. The answer was given beforehand and could be taken from the literature (cf. the references in the *Biochemistry* paper[56]).

The production of replicase after about 1000 such molecules per cell have been accumulated is down-regulated to zero, so that after 20 min a practically constant level of replicase (like the constant level of ribosomes) is maintained. This regulation is due to the increasing number of coat proteins which, apart from the function they are named for, are also engaged in regulation in the following manner. Coat protein binds to RNA at a site that overlaps with the Shine–Dalgarno sequence of the replicase cistron. (The Shine–Dalgarno sequence consists of five to nine nucleotides and precedes the start codon of prokaryotic messenger RNA; it is recognised by the ribosomal machinery for starting translation. It is named after its discoverers John Shine and Lynn Dalgarno, who proposed its function.[62]) The complex of the RNA with the coat protein suppresses the synthesis of the subunit of the replicase, but it can bind ribosomes for synthesising more coat protein or to replicate RNA. This complex can assemble with further coat protein (and other capsid proteins) to form mature viruses. Such regulation is a "wise" result of virus evolution, for two reasons:

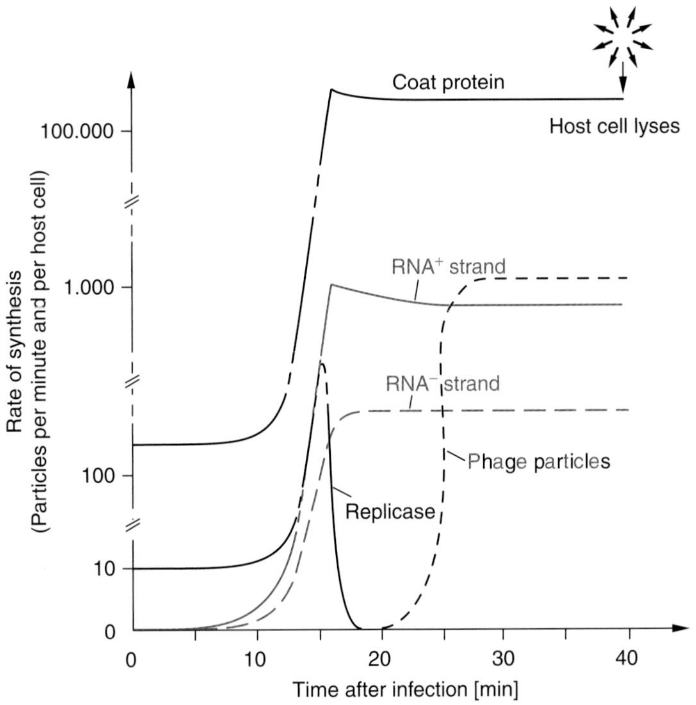

Figure 5.3.6

Schematic profile for individual production rates in the infection cycle of *E. coli* by phage Qβ

The ordinate refers to the rates (i.e. the numbers of particles formed per minute and per host cell) plotted as functions of time. The whole infection cycle lasts about 40 min. It starts with the penetration of one single RNA plus strand, injected by a phage particle that has docked at a pilus (i.e. a hair-like extension of the cell wall of the bacterium offering a tubus for the invasion of phage RNA). The single RNA^+ strand then has to wait until a few of the 10,000 to 20,000 ribosomes have accumulated a sufficient number of replicase molecules in order to compete with the ribosomes for replication of both plus and minus RNA strands. This leads to a hyperbolic increase both in viral proteins and in viral RNA, up to a titration end point corresponding to complete saturation of the replicase with template RNA. The much more abundantly accumulated coat protein of the virus then switches off further replicase production, so that from now on coat protein and RNA are produced at constant rates, thereby accumulating further in a linear fashion. Meanwhile, the production of complete phage particles has started. The host cell lyses about 40 min after infection, by which time about 10,000 or more complete phage particles have been produced, of which fewer than 10% (about 600 particles) turn out to be infectious. This diagram demonstrates a highly optimised mechanism. Not only is the number of the constituents forming complete phage particles with a balanced ratio of geno- and phenotypes (e.g. 180 coat proteins per RNA plus strand) regulated to fixed ratios (maintained in the later phases of infection), but down-regulation of RNA synthesis takes care of accumulating error rates in later phases of amplification, showing up in the surplus of non-infectious phage particles present when the host cell lyses. Nevertheless, some 600 infectious particles produced during the fixed period assigned to each replication cycle guarantee a wide spread of infectious particles resulting from a quasi-linear amplification mechanism, as is required for Darwinian selection.

it limits the formation of replicase at a stage where accumulation of errors in the template strands has already become quite large, and constant levels of ribosome and replicase can more easily match the production of RNA to that of the coat protein. At the end of the cycle a pool of coat protein comprising about 2×10^6 copies per infected cell has been synthesised.

Let us finally look again at Figure 5.3.6, in which a schematic profile of all rates in the infection cycle is presented. What strikes one immediately is the synchronous and quite sharp increase in production rates of both RNA and protein after about 10 min of infection, showing a cusp (a sharp peak) after about 16 min. This is an experimental proof of the presence of a hypercycle. A closer analysis clearly shows a hyperbolic rather than an exponential increase, so the cusp has the meaning of a sharp "end point" like that in a titration curve (see also Figure 5.3.3). What is being titrated here? The answer is ribosome with replicase! The faster-growing RNA exhausts the supply of ribosomes, which leads to constant rates of protein production and – owing to the down-regulated replicase production – also of RNA production, matching the two rates to one another. Why do we call this basic mechanism a "hypercycle"? Isn't it an obvious impression that one immediately gets by looking at Figure 5.3.7? That was my feeling when I coined the term "hypercycle". However, my partner Ruthild Winkler-Oswatitsch gave me a book with correspondence between the mathematicians Carl Friedrich Gauss and his Hungarian friend Wolfgang Bolyai in which Gauss, in one of his letters, also refers to the system he is discussing as a "hypercycle". Although my hypercycle in fact differs somewhat from what Gauss called a "hypercycle", I am quite pleased that I hit on an apparently useful name.

The non-linearity of the hypercycle in phage infection is nonetheless only an important peculiarity of the "internal" mechanism of phage infection, i.e. the mechanism occurring inside the host cell. It is responsible for yielding a well-defined time for the "burst" of the host cell at which a defined number of progeny phages are set free. The "external" overall mechanism of phage-particle production is then represented by a quasi-linear autocatalytic rate term resulting in exponential population growth. The take-home lesson is that internal homeostatic regulation quite regularly engenders the quasi-linear autocatalytic mechanism, resulting in exponential population growth. Typical cases are the doubling of the number of cells of *E. coli* within 20 min of normal growth, or defined generation times in higher organisms, which (at least over limited time intervals) all show exponential growth. Even human populations adhere to this rule, which indicates a quasi-linear type of autocatalysis of reproduction. However, for human populations this is not exactly true, as historical data show, because of "external" influences, such as improved availability of food, better hygiene and medical care, and not obvious non-linearities in the mechanism of human reproduction. A linearity of sexual reproduction may be surprising, since both sexes contribute in an equivalent (recombinative) way to the gene pool of their

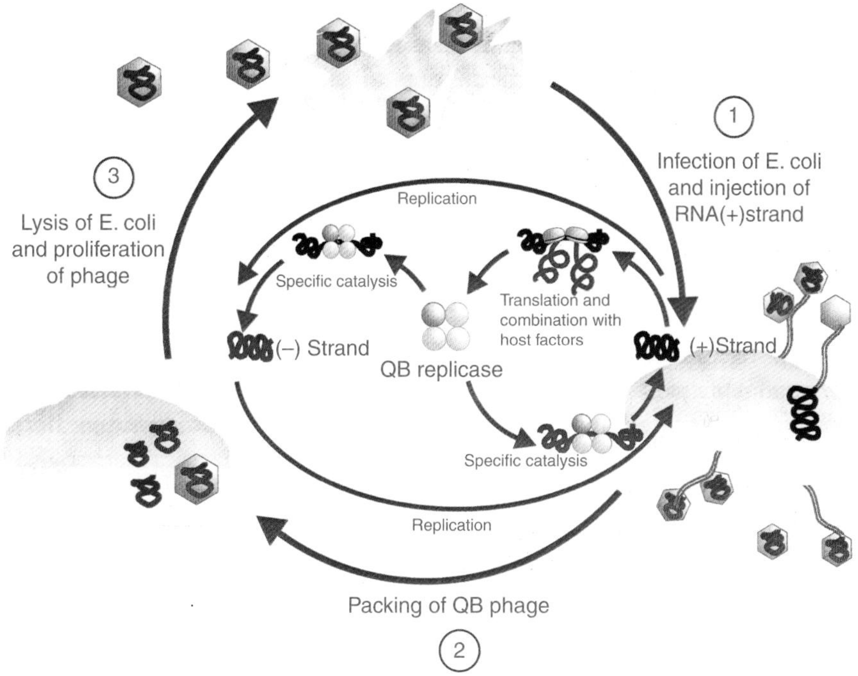

Figure 5.3.7

The hypercyclic nature of phage infection

A sketch to demonstrate *ad oculos* the mechanism of phage infection as described in the two preceding figures. It is obvious why we call such a hierarchy of cycles a "hypercycle". Less self-evident is the answer to another question: where in the mechanism is the "origin" of a new form of organisation? Doesn't the virus entering a host cell bring along all the information necessary for its own reproduction? That is true, but what is missing is the "phenotype", the machinery to start the process of reproduction. And this is the specific phage RNA replicase, which is not possessed by the host cell. The latter provides only the machinery of translation, which is required in all later phases and which we may therefore call a prerequisite of the environment. However, the replicase, which is necessary to start phage replication, cannot be provided by the host and therefore the mechanism is a good model for a new origin requiring the presence of both genotype and phenotype, the latter becoming amplifiable only with the help of the phage genotype (see also Figure 5.3.8).

progeny. However, while the contributions of both partners are well-balanced, the rates of reproduction on which population growth (or decline) is based are determined by the "rate-limiting" contributor of genetic material – in human populations generally the female – leading to quasi-linear autocatalytic reproduction rates. This is noteworthy because sex life is anything but a linear sequence of events involving two (or more) partners and therefore may occur in a highly non-linear fashion.

What the experiments clearly show is that quasi-linearity of the mechanisms resulting in exponential population growth is responsible for the Darwinian nature of evolutionary optimisation. It does, however, not hold for the "origin" of such behaviour. Wherever entirely new mechanisms originate, they require a non-linear start. The establishment of the feedback coupling between phenotype and genotype is an example *par excellence* of a hypercyclic mechanism. If genotype and phenotype can be recruited from the same class of macromolecules, as would be the case in an RNA world, then such a feedback mechanism could have started more easily than in a protein–RNA world requiring a complete translation "machinery" (Figure 5.3.7). In the simplest case, a replicase ribozyme would have to possess a specific palindromic recognition sequence for its plus- and minus-strand templates, which would initiate its own production by a quadratic rate law, since the reaction mechanism requires only two molecules of the same kind – one acting as a template and the other as a (ribozymic) replicase.

However, so far this is still wishful thinking. Reality is by no means that simple. First of all, the analogue of protein replicases makes use of enzymes that are able to reproduce both the plus and minus strands of RNA. For a ribozyme this would work best if the plus and minus strands to be amplified were palindromic, but efficient binding sites on the ribozyme for both selected strands would also offer certain advantages.

There are essentially two research groups who have looked into such problems in more detail: Jack Szostak and his co-workers at the Harvard Medical School and Gerald Joyce and his group at the Scripps Research Institute in La Jolla, California. They had to clarify many details because in most cases either the reactions faded away after one or two replication rounds or they didn't get going at all. Apparently the requirements for being at one and the same time a good template and a good catalyst are hard to fulfil. For single-stranded RNA, it may be much easier to act as a template than to possess the multifunctional activity of a polymerase. All the greater was my surprise when, quite recently, I found an article by Tracey Lincoln and Gerald Joyce with the title "Self-sustained replication of an RNA enzyme"[63] (see Vignette 5.3.1).

I knew Gerry Joyce before he became a professor at the Scripps Research Institute. He did his PhD work at the Salk Institute in the laboratory of the late Leslie Orgel, which was the focus for the study of early evolution and, in particular, the role of RNA. During his time at Scripps he repeatedly became a talking-point because of his brilliant work on RNA enzymes. Now he proposed an entirely new experimental approach to the problem of RNA multiplication. The method is based on two cross-replicating enzymes which amplify each other by linking oligonucleotide substrates. The method is sketched in Figure 5.3.8 with a short description in the legend. The amplification process could be shown to continue indefinitely, achieving an overall amplification of $>10^{25}$.

It is quite clear that this opens up a new way to approach the problem. On the other hand, in order to become effective the method must be generalised for linking together a large number of smaller oligonucleotides. The two large oligonucleotide

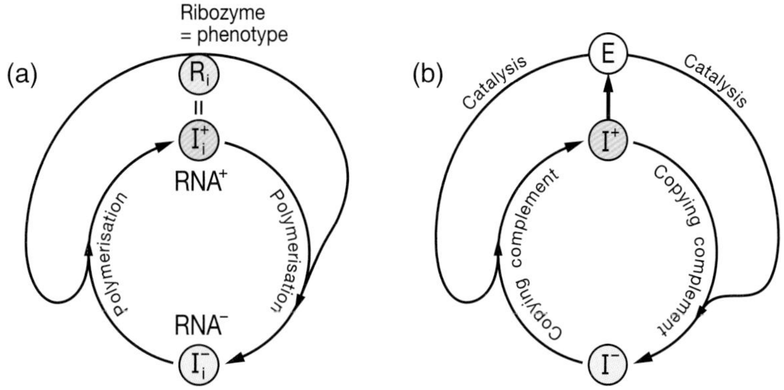

Figure 5.3.8

The role of the hypercycle in the origin of life

A system that is able to reproduce its genotype *and* to provide the phenotype necessary for the reproduction of the genotype is at the origin of an evolutionary start to life. A virus is not in itself an autonomous living being, but Figure 5.3.7 has taught us an important lesson. Yet at the beginning host cells had not come into existence. Can we find a substitute for them? Life at later stages would not have developed without the enormous catalytic power of proteins, but it is not clear how the complex machinery of translation could have been nucleated without reproducible catalytic help. As we know from systems in existence nowadays, such a machinery includes – besides the messenger RNA (mRNA) – some 40 different transfer RNAs (tRNA), which are the specific adapters of amino acids to their codons, i.e. base triplets, and a complex ribosomal machinery containing a number of ribosomal RNAs (rRNA). We might wonder why these (including their specific catalysts) are almost exclusively made of RNA. The magic word is "ribozymes"; these are catalytic RNA structures, and there are now many indications that an "RNA world" preceded the present-day "DNA–protein world". Among all the arguments in favour of the past existence of a primordial RNA world, one seems to me to carry the greatest weight: the origin of genetic information-processing requires both a genotype and a phenotype, and these must be linked by a hypercyclic mechanism. If this can be recruited from the same class of molecules, say RNA, then its origin can be imagined with a sufficiently high probability. The diagram shows how such a system could have worked. It includes a quadratic reaction mechanism for the phenotype favouring its own replication. The symbol "R" is assigned to this hypercyclic function.

pieces that were linked in these first experiments had to be synthesised beforehand. This approach is still in its infancy, although a number of problems involved have been solved. For instance, using tetramers of the form RY_3 was a wise choice because we have long known[64] that these have a binding strength that is ideal for experiments of this kind. We are very optimistic that the problem of RNA amplification, eminently important for the build-up of an "RNA world", will be solved before long.

Vignette 5.3.1 Self-sustained replication of an RNA-enzyme

(according to Tracey A. Lincoln and Gerald Joyce[63])

The description of a ribozyme as an RNA-enzyme that has the function of an RNA polymerase, such as Qβ replicase, was referred to earlier in this section as "wishful thinking". By this I meant that we have no experimental evidence that it has this function, which would involve the ability to move along a template and to include any of the four nucleotides into the growing chain of the replica complementary to the template. It may be that a ribozyme is not sufficiently flexible to do this. Hence the mechanisms of replication for protein-enzymes and for ribozymes might be based on different premisses, even if base-pairing between template and replica were to play a similar part in both.

Tracey A. Lincoln and Gerald Joyce have described an entirely new mechanism of "cross-catalytic replication" and have tested it experimentally. It works in the following way. There are two (complementary) enzymes E and E′ which reproduce one another from two complementary pairs of substrates (A, B) and (A′, B′). The two cross-catalytic reactions can be written as: $E'+A+B \rightarrow E'+E$ and $E+A'+B' \rightarrow E+E'$. Otherwise, the process is represented in its simplest form and Figure 5.3.8 is self-explanatory.

However, the mechanism involved is by no means that simple, as indicated by the right-hand picture in Figure 5.3.8. To understand it properly requires a detailed reading of the authors' original paper[63]. Here I only summarise some important points.

1. The two enzymes (ribozymes) E and E′ act as ligases in a cross-catalytic way, joining up the 3′ hydroxyl group of the substrate A or A′ with the 5′ phosphodiester, with simultaneous release of a pyrophosphate. The correct positioning of the substrate is effected by hydrogen bonds (base-pairing) with E′ and E respectively.

2. Evolution *in vitro* was used to optimise the rate of the tandem reaction. The ligase was configured first to self-replication (which turned out not to be very efficient) and then converted to the cross-catalytic format.

3. To obtain optimum results, the bases in both blocks in the two side arms were systematically replaced until the greatest catalytic effect was achieved.

4. (For these replacements, tetrameric sequences were used, with structures enclosing three A or U bases and one G or C base; these are ideal for fast base exchanges because of the moderate strength of their base pairs.)

Cont. ➲

➲ Cont.

The two G-U wobble pairs next to the linkage site (solid boxes) as well as the four base pairs next to the two ends are most important for enhanced catalytic activity. Those boxes in both enzymes E and E′ are active in each case at the 3′ end.

Another paper by Bianca Lam and Gerald Joyce[63a] describes how aptamers (which are double-stranded RNA structures that bind specific ligands and thereby change their structure) were used to produce active enzymes E and E′. I only mention that paper here, as a proper description of their work would certainly be beyond the scope of this vignette.

With regard to any practical application as an amplification method one may object that the substrates – which are sequences about half as long as the desired product – must be synthesized beforehand. The authors are well aware of this. What they want to demonstrate is the principle that catalysis might have come about in an RNA world, within which a protein/DNA world – and with it life as we know it – could later have arisen. At the end of their paper[63] the authors point to "using populations of newly formed enzymes to generate daughter populations of substrates. An important challenge for an artificial RNA-based genetic system is to support a broad range of encoded functions, well beyond replication itself". The fact that, even today, the important functions in the ribosomal translation system of all living beings are carried out by RNA rather than by proteins hints strongly at the original importance of an RNA world for the origin and early evolution of life.

5.4. Who Survives, the Fittest or the Luckiest?

One day my Japanese colleague and friend, Motoo Kimura, came to me and asked me the above question. In fact, he said, as I remember: "Manfred, shouldn't we reformulate the Darwinian principle as 'survival of the luckiest?'" My answer was: "Yes, Motoo, we may do so; but then we have to add that the 'luckiest' always has to be a member of the very élite club of the fittest." Motoo seemed to agree – although I had the feeling that his concurrence was not quite wholehearted. I knew that he was referring to the so-called "neutral mutants" and their importance in evolution. What I did not agree with was his characterisation of those processes as non-Darwinian. I shall return to this question later. Anyway, "neutral mutants" are today a cornerstone of modern population genetics, which they had not been before Kimura's discoveries.

The idea of "neutrals" is rooted in the work of the mathematically oriented schools of Haldane, Fisher and Wright in the first half of the 20th century. Haldane came from physiology, while Fisher and Wright were statisticians who actually converted population genetics to a branch of applied mathematics. It was the triumvirate of Haldane, Fisher and Wright who combined a mathematical approach to Darwinian theory with the Mendelian laws of genetics. The theory was called the "modern synthesis" or "neo-Darwinian theory". For the first time, it allowed rigorous enquiry about problems of population biology, such as "gene frequencies in finite populations", "genetic load" and "spread and migration" in relation to "speciation".

Modern theory in the second half of the 20th century is dominated by two names. One is Gustave Malécov, whose major work on stochastic theory is cherished by mathematically oriented scientists but has remained largely unnoticed by the majority of biologists, who have shown little interest in a concise description by mathematical theory. The other name is Motoo Kimura, whose theoretical work has been seminal for an understanding of modern biology, in particular with respect to the phase of evolution of higher life. Despite its rigorous mathematical foundation, Kimura's work eventually became quite popular among biologists, although some of them had initially reacted in an almost hostile way.

The late Motoo Kimura was one of the greatest Japanese scholars of the 20th century. I am not sure that this has been fully acknowledged in his own country. I would like here to express my admiration of the unique intellectual quality and originality of the work of this great thinker, who died some years ago. Motoo was the director of the renowned Japanese Institute of Genetics at Mishima, near Mount Fuji. We visited one another several times, so I became also familiar with his hobbies. He excelled in growing new variants of orchids – an artistic complement to his scientific work in genetics. And here he was as professional as in science, winning world-wide fame that was documented by various international awards. He once asked me whether he could name one of his beautiful prize-winning orchids after me. It is shown in Figure 5.4.1.

Let me now come back to the question of "non-Darwinian" evolution. In earlier parts of this book I have quoted Darwin, who in his *Origin of Species* addressed the problem of "neutrals" by saying: "Variations neither useful nor injurious would not be affected by natural selection and would be left either a fluctuating element as perhaps we see in certain polymorphic species, or would ultimately become fixed, owing to the nature of the organism and the nature of the conditions." Nothing is wrong with such a characterisation of "neutrals", the existence of which was also acknowledged in non-Darwinian theory, and in the late 1960s emphasised *expressis verbis* in two independent papers, one by Motoo Kimura (1968),[65] the other by Jack L. King and Thomas H. Jukes (1969).[66] The titles of these papers are interesting: King and Jukes refer for the first time in the title of their paper to the "random fixation of selectively neutral mutants" as "non-Darwinian evolution", an expression only later adopted by Kimura after he had already published a whole series of papers[67] (some together with Tomoko Ohta). According to Darwin's own words, a "random fixation" of "neutrals" is perfectly within the scope of his ideas, although he certainly

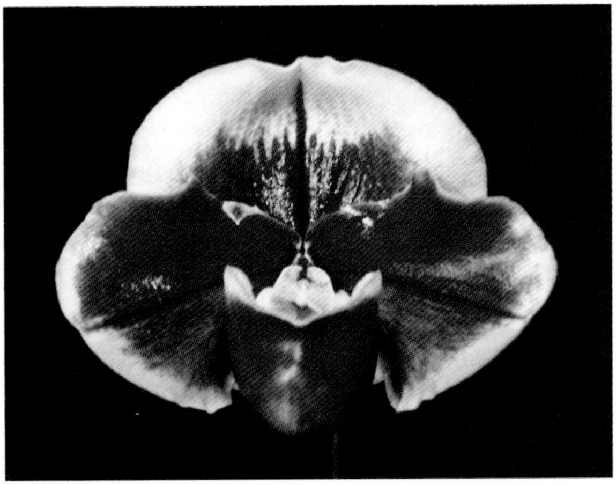

Figure 5.4.1

The clone "Eminent" of "Phaphiopedilum MANFRED EIGEN", registered in 1983 by Motoo Kimura

This magnificent flower won two awards, a silver medal from the Japan Orchid Growers' Association (JOGA) and the top prize from the Paphiopedilum Fanciers' Association.

did not realise its importance for the fact that "natural selection" still works at any advanced level of life. In our quasi-species model, neutral (and "nearly neutral") mutants are also imperative ingredients to guarantee a sufficiently large spread of the mutant distribution, necessary for evolution to proceed, especially in later phases of evolution at adequate rates (see Section 4.7). This, of course, could not be realised without a mathematical analysis of the processes involved – an analysis that Darwin himself never performed.

The first question we have to ask is how neutral is "neutral"? In other words, how do we define "neutral" or "almost" neutral? The definition used by Kimura refers to the population size: "'neutral' allows for deviations $\leq 1/N_e$ from a given value if N_e is the effective population size." This is a pretty arbitrary and therefore unsatisfactory definition. The population size, for instance in a human population, could be quite small, while for a bacterial colony it may easily include billions of individual cells. The doubling time of an *E. coli* cell is about 20 min. Maintaining conditions for uninhibited growth, one could produce within half a day more than ten times as many *E. coli* cells as the number of people on earth. In closest packing, all those cells would fill a space of no more than 10 mm^3.

In fact, a precise definition of "neutral" does not seem to me to be very important. The quasi-species model, which evaluates all mutants present, takes care of them anyway. However, an exact stochastic treatment of the quasi-species is still *ultra vires*, although useful approximations have been presented by Peter Schuster[68] and John McCaskill.[69]

The rigorous mathematical formulation of the concept of neutrals in a whole series of papers and the demonstration of its power in explaining biological facts, previously poorly understood, established the true importance of Kimura's work. Yet in the general literature on biology these concepts have not always met with an adequate appreciation. As in Darwin's time, the frequency of the occurrence of neutral (and nearly neutral) mutants in the close (and moderately distant) vicinity of the wild type is even today often greatly underestimated.

Ernst Mayr, whom Stephen Gould called "one of the greatest living evolutionary biologists", wrote as late as in 1997, in his book *This is Biology*[70] (called in a review by James D. Watson a "masterful overview"): "the study of allozymes by the method of electrophoresis in the 1960s revealed a far greater amount of genetic variability than anyone had previously suspected. Kimura, as well as King and Jukes, concluded from this fact and from other observations that much of the genetic variation must be neutral. This means that the newly mutated allele does not change the selective value of the phenotype. There is considerable argument whether the frequency of neutral mutations is really as great as claimed by Kimura (1983). What is far more controversial, however, is the evolutionary significance of neutral allele replacement." He concludes: "The replacement of neutral genes is considered by them [the naturalists] merely evolutionary "noise" and irrelevant for phenotypic evolution. If an individual is favoured by selection owing to the overall qualities of its genotype, it is irrelevant how many neutral genes it may carry along as hitchhikers."

Such arguments for me are hard to digest. Why should neutral genes be carried along only as "hitchhikers"? They represent – more or less – the information that is transmitted and thereby subsequently influences the rate of evolution, as specified in Section 4.7, where I used the title: "How to Get from Here to There in Information Space". Anyway, Kimura presents quantitative data. Moreover, the large amount of data obtained in more recent experiments on "site-specific mutagenesis" favour the view that many mutants are more or less "neutral", quite in contrast to earlier assumptions which, however, were not based on any concrete data. Site-directed or site-specific mutagenesis is a method of genetic engineering for modifying clearly identifiable sites of the amino-acid residues in the corresponding protein structure, in order to test their influence on enzymic activity. As intermediates to any progress in adaptation to a new situation, they are immensely important.

For me, however, more than anything else it is the mathematical rigour of Kimura's theories that really shows that neutral drift is more than just "evolutionary noise". Before I come back to this, let me provide some visual aids which may be of help in understanding Kimura's abstract models.

In Section 2.6 the glass bead games introduced by Ruthild Winkler-Oswatitsch were used to distinguish various mechanisms of natural selection and to characterise the types of fluctuations associated with them. I refer now to Figure 2.7.3, in which a random-walk mechanism was simulated in a game "played" by the computer. The situation was the following. As in the games described in Chapter 2, we use a "chessboard", on which 64 glass beads of two different colours can be placed. In this game we don't even need the co-ordinates that identify the 64 squares on the board. The game is totally dominated by chance and can be played by tossing a coin. "Heads" means that a sphere of colour "1" is to be replaced by a sphere of colour "2", while "tails" means the reverse of this.

Obviously this is a game of pure chance. The population structure, i.e. the relative frequencies of the colours, does not exert any influence on the procedure, as was the case in Ehrenfest's urn model or its converse that lies behind Darwinian selection. In other words: there is no "mass action" on decay or amplification. All that the population structure does is to maintain a constant occupation number. Figure 2.7.3 depicts the typical course of such a game. It was played on the computer many times. In fact, it is very boring. To make the game endless, coin-tossing has to go on after spheres of one colour have occupied all positions, so that the spheres of the other colour have a chance to re-enter the game.

With some recollection of earlier sections the reader will have realised that we are now just dealing with a plain diffusion process. The path recorded by Figure 2.7.3 differs sharply in appearance from the diagrams shown for the other game models that referred to equilibration or to "natural selection". Equilibration is a deterministic process in which the remaining fluctuations, on average, are confined to a range given by the square root of the number of particles involved. If these numbers are of an order of magnitude as large as Avogadro's number, then relative fluctuations are so small that we can call the chemical "law of mass action" a well-defined deterministic law. Yet Figure 2.7.3 seems to reveal entirely unpredictable behaviour . . . but perhaps not quite!

If we look at large numbers of such recordings we soon realise that they have two properties in common. They all reproduce repeatedly the whole range of possible population numbers between 0 and 64. Although they change in an entirely unpredictable way, they always – after the elapse of some time which again is not predictable for the individual period – reach the limiting numbers 0 and 64, and they do this on average after N^2 individual moves. If the number of particles involved were very large this recurrent behaviour would in fact become quite precise. This is the "law" of random walk, which for spatial diffusion shows up with the "distance squared over time" relationship, as was discussed in Section 4.6.

Kimura has always stressed the diffusion-like character of neutral selection. One of his very early papers has the title "Diffusion Models in Population Genetics".[71] Is the game described above applicable to his models? The ordinate value in Figure 2.7.3, for instance, could describe the sphere numbers, where one colour refers to a particular species and the other colour to the sum of all alternative (and in this respect undistinguishable) neutrals, so that one could look at the fate of a particular sphere. If we want to simulate natural processes we must take into consideration the fact that 64 moves represent one generation in which all spheres may "react in parallel". Unfortunately, the above model lacks one important property, namely, the dependence of colour change on the respective population number, which only for the balanced state of 32:32 yields equal chances for both colours to change. We would have to introduce more complex rules than simple coin-tossing.

Let us make some plausible attempts under simplified conditions. We first assume that for all neutrals (including the particular ones considered) the mutation rates are equal. Furthermore, the rate for any particular neutral mutation is quite small. We consider the competition between the one particular mutant that will turn out to be selected and the rest of the population, and we specify this mutant by the index 1. Category 2 now combines all neutral alternatives of 1. Then all neutral mutants of 1 contribute to category 2, while (nearly) all neutral mutants of members of category 2 will leave its sum unchanged because they will (almost) always produce another member of their own category. Hence under these conditions any of the extreme conditions of Figure 2.7.3 (e.g. $n_1 : n_2 = 64 : 0$) will almost never be reached. There will usually be a background of neutral mutants, the total sum of which may well equal or exceed the population level of mutant 1. One member of category 2 will eventually outgrow the rest and become an alternative of category 1. In other words, the result will most probably be a "neutral quasi-species", which as a whole is stable against any takeover by disadvantageous mutants, while its composition undergoes strong fluctuations. Such a compositional metastability involves "critical" phase transitions (see Section 5.5) and offers many advantages for exploring more extensive regions of sequence space, thereby lending wings to evolutionary progress.

A general stochastic treatment of the quasi-species model is still lacking. Only approximations with limiting assumptions have been successful so far. However, the growth of a single mutant out of a distribution of neutral variants – the problem I intended to discuss in this section – has been solved in a very satisfactory way by Motoo Kimura. That's easy to say, but what it means was best said by Sewall Wright, one of the pioneers of modern mathematical population genetics. After a talk given by Kimura, he rose to comment that "only those who tried to solve these problems could fully appreciate Kimura's great accomplishment".

Motoo's fundamental paper "Diffusion Models in Population Genetics", was published in 1964 and more recently reproduced in a book, with a foreword by J. F. Crow,[72] under the title *Population Genetics, Molecular Evolution and the Neutral Theory.*[73] This book contains a series of selected papers, which were stimulated by a "line of investigations of gene frequencies" by R.A. Fisher in 1922 and 1930,[74] continued by Sewall Wright in 1931[75] and by Gustave Malécot in 1948[76] as well as by other mathematicians. All these earlier papers were based on the Fokker–Planck equation of stochastic physics. This equation considers the probability density at a given time and its changes in a Markovian process. In its linear form this equation is identical with the diffusion equation treated in Section 4.6, but it contains an additional (negative) term describing some "slowing down", caused by friction or drift, which is proportional to the first derivative with respect to the spatial co-ordinates. As a partial differential equation in space and time the Fokker–Planck equation can be solved exactly only under special conditions. The equation is named after the Dutch theoretical physicist Adriaan Daniël Fokker and Max Planck. Fokker used the steady-state form in 1914[77] for treating an optical problem, while Planck extended it in 1917[78] to the general form now used in physics.

A rigorous mathematical formulation of the problem was given in 1931 by Andrei Kolmogorov,[79] who derived the two equations called Kolmogorov's backward equation and Kolmogorov's forward equation or, in short, the "master equation", a term by which the physicists now honour Kolmogorov's fundamental contribution. The equations deal with Markovian systems (see Section 2.5) and "backward" and "forward" refer to recursions to the initial and final state.

Kimura, following in the footprints of Fisher and Wright in his various papers, eventually preferred Kolmogorov's approach. However, unlike his mentors, who studied individual problems "with great ingenuity",[80] Kimura attacked the whole process of change of gene frequencies. What this meant to population genetics is expressed clearly in Sewall Wright's comment after Kimura's 1955 lecture at Cold Spring Harbor quoted above. I shall not now go any further into the details of his theory, but rather use the random-walk game and compare it with the alternative model that describes more correctly the situation of "neutral selection", in order to demonstrate more lucidly its physical background. We can then take a look at some results of Kimura's theory.

Coin-tossing is obviously not well suited for simulating mutations that result from copying processes because it is independent of the population structure. In other words, even if one of the two colours (e.g. blue) represents one individual neutral, while the other colour (e.g. yellow) stands for the sum of all neutral alternatives, and if neutral selection has happened (i.e. yellow:blue = 0:64), then coin-tossing would always bring in yellow again with a probability of 50%. In the reality of neutral selection, however, yellow could enter again only through a mutation of blue, i.e. with a quite small probability. We therefore have to change the rules of the game accordingly. New spheres are allowed to enter only through reproduction or through occasional mutation of spheres present on the board. The following rules would do justice to this situation:

(1) We again use two octahedral dice to identify the co-ordinates of the squares according to the occupation of the board by blue and yellow spheres.

(2) The dice are thrown alternately for removal and insertion of the spheres. In the first throw the sphere on the square with the co-ordinates shown by the dice

will be removed, while the second throw determines the colour (blue or yellow) by which the empty square is refilled. Strict alternation of the two throws of the dice takes care of perpetual full occupation of the board.

(3) In addition, occasional mutation of the selected form (i.e. blue) according to some key may be introduced by corresponding additional throws of the dice.

In order to demonstrate the changes relative to the random-walk game, shown in Figure 2.7.3, let us look at the extreme situation of complete neutral selection of one individual (i.e. blue). In the random-walk game each throw (i.e. each toss of a coin) would bring in the other colour (yellow) with 50% probability, and that would go on for any further throw. By contrast, in the neutral selection game the chance of the other colour (yellow) for any re-entry is much lower (according to the key chosen). Since in the extreme situation of neutral selection any throw for removal or for insertion would hit a square occupied by a blue sphere, the chance for a yellow sphere with a mutation probability of 1 in 64 would be 1/64, or one per generation. (We remember that 64 throws give each sphere on the board, on average, one chance to change. In reality this would occur in parallel and therefore comprise one generation.) Hence while in the random-walk game "yellow" could increase its population number by about 32 within one generation, it is on average only one sphere in the case of the neutral-selection game. The next generation has a 50% chance for each doubling or reduction to zero, and so on. What I want to show is that only in the second case can we speak of true selection, which, however, is not stable over long periods. Whenever the sum of neutral competitors (yellow) has grown up there will again be a chance for one of them to become dominant. The random-walk game, on the other hand, although it occasionally reaches the extreme situation of 0:64, never does so for any extended period. Rather, it reaches, over a time-average, any possible occupation ratio.

Now let us have a look at Kimura's schematic representations of the results obtained by stochastic theory. Figures 5.4.2 and 5.4.3 express what I tried to demonstrate qualitatively with the game models considered above. Neutral selection is a true process of selection, although of a metastable kind. Figure 5.4.3 certainly shows an extreme situation of complete selection. In reality there will be many intermediate situations in which selection is not complete and the emergence of new mutants is not as straightforward as shown in the picture chosen. However, what is shown is quite realistic for (nearly) complete selection. Once reached it tends to persist because in the extreme situation appearance and emergence of new neutral mutants is quite slow. Such cases are discussed in more detail in Kimura's paper (1976),[81] from which Figure 5.4.3 was reproduced. Figure 5.4.4 was taken from an earlier paper by Kimura and Ohta.[82] It shows a "typical" pattern of extinction and multiplication of neutral mutants. In order to understand what is meant to be demonstrated one should know that most of the mutants appearing in each generation don't survive

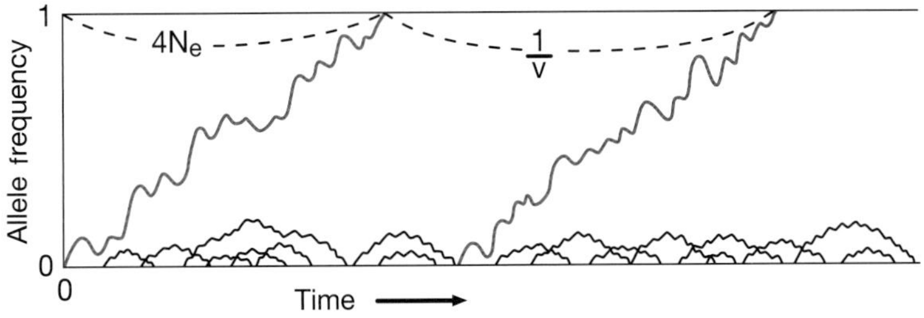

Figure 5.4.2

Neutral mutant alleles growing up in a finite population

Schematic curves representing "neutral selection", redrawn from M. Kimura.[73] These alleles may become new centres of selection.
Courtesy of Tomoka Ohta.

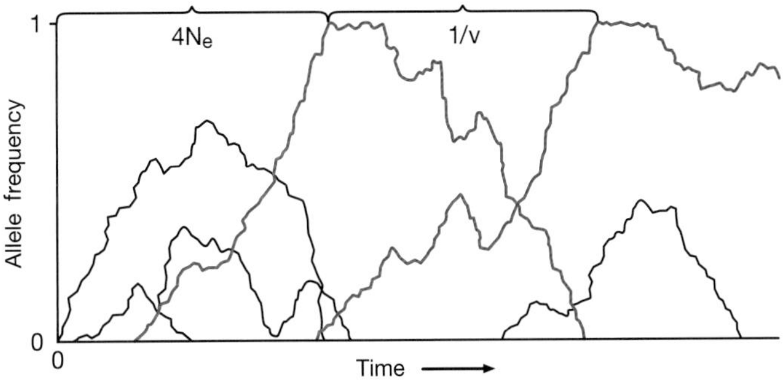

Figure 5.4.3

Typical pattern for the multiplication and extinction of selectively neutral mutants

The neutrals occur as one mutation every ten generations ($4N_e v = 0.2$).

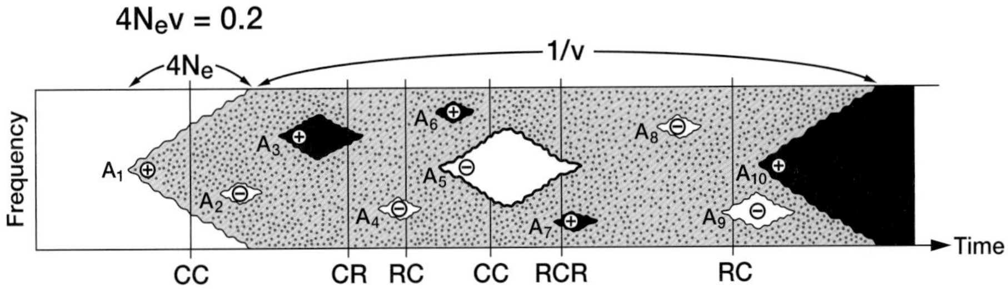

Figure 5.4.4

Quasi-species-like occurrence of neutral mutants in a finite population

Parallel appearance of neutral mutants in a quasi-species distribution favouring the largest eigenvalue (schematic representation).
Redrawn from M. Kimura and T. Ohta[82]. Courtesy of Tomoka Ohta.

for more than a few generations. What is shown in this picture are those mutants that have attained frequencies higher than about 10%, and since they cannot be recorded continuously they might become "visible" when they have reached higher frequencies (otherwise all patterns would start at the same frequency of 10%). The patterns shown indicate schematically that only a small fraction of them will eventually become selected, while most will die out even after they have reached appreciable frequencies.

Let me now reach a conclusion. In this volume I am not presenting a systematic approach to biology. What I want to uncover are the physical roots that become apparent when we try to understand the logic of life. The question in the title of this section has no simple answer with regard to either of the two alternatives. The earlier interpretation of Darwin, stressing "selection of the fittest", presents us with evolutionary changes that may soon reach a deadlock in a complex, rugged fitness landscape. In the next section it will be shown that evolution through natural selection can be represented by a series of phase transitions, including both those of the first kind, such as the freezing and melting of information, and those of a "critical" nature (see Appendix A2), which include macroscopic fluctuations by neutral drift. It was Motoo Kimura who opened our eyes to the latter. They may be mandatory for explaining our existence.

In his early career Motoo was not always treated very well by some of the leaders in the field. He once came to me in great distress complaining that it must be his East Asian origin that was the cause of the hostility he experienced in Western countries. I tried to convince him that it had nothing to do with his origin. Such rejection is normal for anybody doing something that is entirely new, and in most cases it is simply lack of understanding which, for that very reason, engenders hostility. Perhaps he accepted my explanation. At least his trustful behaviour toward me nourished this hope.

What is my answer to the question in the title of this section? The survivor is the quasi-species with the largest eigenvalue. The wild type is then a fluctuating centre in this quasi-species, i.e. at any instant the luckiest within the "club" of fitness of this quasi-species.

5.5. Natural Selection: A Phase Transition?

One of the properties most characteristic of the quasi-species is the existence of an error threshold. Consider a macromolecular sequence consisting of a defined (and limited) number of different classes of monomer. Its reproduction in nature is always associated with a finite error rate. The higher the error rate, the more mistakes will occur and need to be removed with the help of some error-correction mechanism, but why is there an error threshold, the violation of which would completely destroy the information contained in the text? I have often been asked this question. A boy at school is expected to make certain mistakes. The larger the number of mistakes, the worse his marks will be. Why should there be a singularity where his total knowledge breaks down?

This was also a question asked of me by Richard Feynman one evening after I had given a lecture. We didn't reach agreement that evening. Next morning, when we met at breakfast, he said that he had done some calculations and he could not see that for the replication of nucleic acids such a threshold should in general exist. I gave him a piece of paper on which I had written down the derivation of my threshold relation in a few lines. He glanced briefly at it and said immediately: "Oh yes, if you make such a strange assumption". My "strange" assumption was that I described the production of new copies just by one single linear term, rather than by a more general expression resulting from an expansion of the rate equation into a power series. He was certainly right, but in biology we know that even the smallest genes are sequences that have some 10^{hundreds} possible alternatives, and the only way of producing any given sequence is by copying it from one that is already present. This excludes absolutely constant terms representing any sort of *de novo* synthesis. It is certainly true that higher-order terms may occur under special reaction conditions as, for instance, in hypercyclic systems (see Section 5.3). Such systems, however, have to be considered separately. They do not interfere with the error threshold. Of course, it was my mistake not to have told this to Richard beforehand.

Let us first look at some computer simulations carried out by Peter Schuster and his coworker Jörg Swetina.[83] Figure 5.5.1 shows the relative fractions (x_i) of mutants, occurring stationarily in the reproduction of a binary sequence of length $\nu = 50$, as a function of the error rate $(1 - q)$ per symbol. The letter q represents the fidelity of reproduction, so $(1 - q)$ is the complementary error rate per symbol.

Before discussing this figure, let me make a few remarks on the model. Fifty binary positions yield a total of $2^{50} \approx 10^{15}$ possible alternative sequences, which would mean for any unbiassed appearance of each individual sequence a probability of only 10^{-15}. This model, however, refers to a "wild type" (fraction x_0, the index "0" meaning zero errors) which has an (effective) replication rate ten times higher ($\sigma = 10$) than that of any of its mutants, which are assumed to be "degenerate" in respect of their replication rates. Owing to the symmetry of this model we can introduce error classes occurring with fractional population numbers x_1 to x_{50} (representing 1 to 50 errors) in which the individuals are indistinguishable. The number of possible individuals in each class follows a binomial distribution $\binom{\nu}{k}$, where ν is the total number of positions (i.e. $\nu = 50$) and k is the number of erroneously occupied positions (k = 1 to 50). This distribution is symmetrical with respect to $k = \nu/2$, i.e. error class "k" corresponds to error class "ν–k". We could call the 50-error sequence "complementary" to the master (i.e. 0-error) sequence, or each

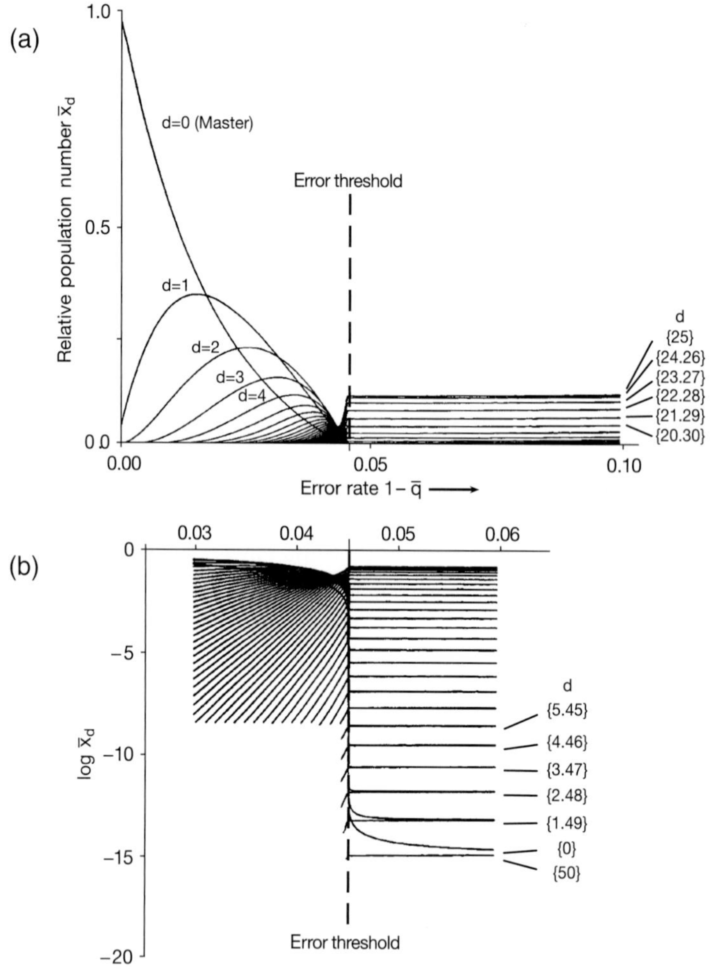

Figure 5.5.1

Computer simulation of quasi-species selection by Peter Schuster and Jörg Swetina

The model assumes binary sequences of length 50 with an error rate per digit of $(1 - q)$, given at the abscissa (q being the fidelity of copying a single digit). All error copies I_i are assumed to be degenerate in respect of their replication rates (A_i), i.e. it is assumed that they have identical rates, except for the master copy (I_o), which has a tenfold higher replication rate (A_o), i.e. $A_o/A_d = \sigma = 10$. d is the Hamming distance with respect to the master copy I_0. (a) Plots of relative population numbers of the master copy I_o. x_{d_i} for the sum of all d-error copies, marked by their Hamming distance with respect to the master: d_i. (b) Logarithmic plot of the curves shown in (a). Above the error threshold all species appear according to their binomial frequency $\binom{\nu}{k}$, i.e. the information represented by the consensus sequence has been completely lost.

k-error-"complementary" to each (ν–k)-error sequence, both equal in their numbers of individuals, which is $\binom{\nu}{k} = 50$ for k $= 1$ as well as for k $= 49$. The maximum of this distribution is at k $= \nu/2 = 25$, which comprises almost 10^{14} individual sequences (i.e. nearly 10% of the total sum). All x_i, each having a value between 0 and 1, form a sum that is equal to 1. These comments may be helpful in understanding what is seen in Figure 5.5.1, which I shall discuss now.

Looking at the figure as a whole, we get the overwhelming impression that we are indeed dealing with some sort of "phase transition". However, we should not be misled into interpreting "phase transition" as something that occurs instantaneously on the time axis. The abscissa is the "error rate per symbol" and the ordinate values stand for relative population numbers, which assume their values only after quite some time. As a result of competition it is a process that may, and in fact often does, proceed quite slowly. Therefore these phase changes are not easily observable in real time.

At zero error rate (1 – q), the "master sequence" I_o is the only finitely populated species and may therefore be taken as representative of Darwin's "survival of the fittest". However, an error rate of zero is an unrealistic idealisation. In fact, evolution would not occur under such circumstances. With increasing error rates we see the build-up of a molecular quasi-species. The relative population number of the "master sequence" decreases while that of the mutants rises. First it is the class of one-error sequences that gets populated, followed by the two-error, three-error etc. classes, until a point is reached that we have termed the error threshold. In Figure 5.5.1 it occurs at a value of (1 – q) equal to $\ln(\sigma/\nu)$, as suggested by "ideal" theory. In our case the value of $(1-q)_{thr}$ amounts to 0.046. We see that close to this value the population of master sequences has decayed to quite a low value, while mutants up to an appreciable distance from I_o are becoming populated. Close to the error threshold we find ideal conditions for evolutionary changes. However, just across the error threshold all individual population variables assume their unbiased *a priori* probabilities of $2^{-\nu}$ for each individual, which means a maximum representation for the class of 25-error mutants and a minimum for the single "master sequence", whose expectation value changes at the threshold abruptly by many orders of magnitude – as to be expected for a first-order phase transition.

What clearly changes at the error threshold is the "information" represented by the consensus sequence of the quasi-species. It is the sequence that one would find by sequence analysis using conventional methods. Figure 5.5.2 shows the consensus sequence determined from the population variables simulated in a computer experiment by the Spanish physicist Pedro Tarazona,[84] who spent some time as a guest scientist in Peter Schuster's laboratory at Vienna University. The green curve in Figure 5.5.2 refers to the consensus sequence of the corresponding quasi-species, which I calculated from Tarazona's data. What really changes at the error threshold

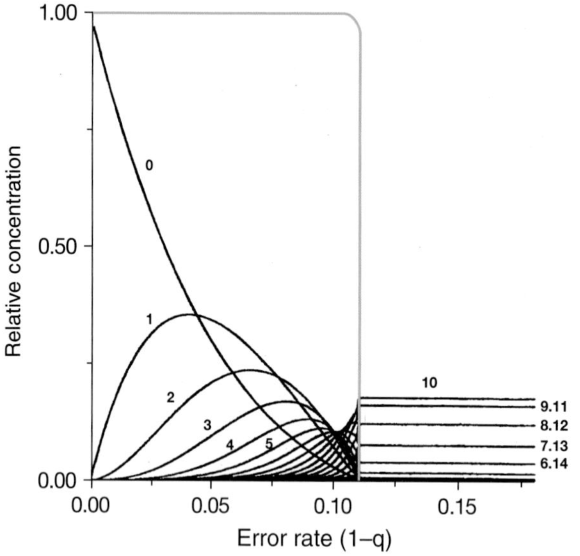

Figure 5.5.2

The "melting" of quasi-species information

The simulations in Figure 5.5.1 were done by Pedro Tarazona during his visit as a guest professor at Peter Schuster's laboratory in Vienna. In these examples the sequence length was only 20. The green curve represents the consensus sequence of the quasi-species and therefore the information contained in it. Even for this relatively short length it already shows an almost ideal first-order phase transition.

is the "information" contained in the quasi-species, rather than that of any of the individual sequences, which all change simultaneously (in terms of their stationary population values, not necessarily in time). Admittedly, Schuster's model is an idealised one, which as such may not be found anywhere *in vivo*, but it nevertheless is quite representative of the principles involved.

The disintegration of the quasi-species at the error threshold appears as a dissolution of the "information" that it represents, rather than as any degradation of its material properties. As such it can be compared to the melting or vaporisation of some ensemble of material particles, in which only their manifestation rather than their material composition changes, usually in the form of a first-order phase transition, which is characterised by a discontinuous change in the energy content of the material ensemble. A "true" discontinuity at any melting or boiling point requires an infinitely expanded system. Only then does the first derivative of the energy content (e.g. with respect to temperature) really approach infinity. One calls this the "thermodynamic limit". In the thermodynamic limit, first-order phase transitions show infinite first derivatives (e.g. specific heats) at the transition points, i.e. true singularities.

When I call selection a "phase transition" I mean exactly this type of behaviour, except that we are now dealing with information rather than with energy. It can be regarded as a jump of information at the error threshold, i.e. a jump in the number of sequences that can be characterised as a clearly defined order expressed by their consensus sequence – that is, by a variety of $2^{50} \approx 10^{16}$ different sequences rather than the one binary sequence that happens to be identical with the consensus sequence. In other words: it refers really to the total of 10^{16} sequences that represent the

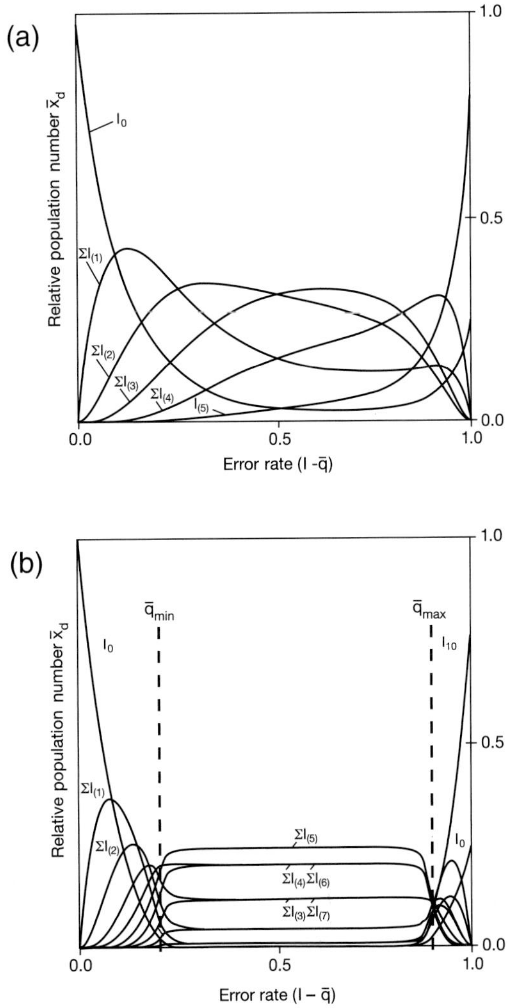

Figure 5.5.3

The build-up of the quasi-species phase transition from small sequence lengths ν, according to Peter Schuster and Jörg Swetina[83]

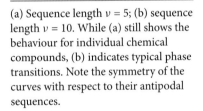

(a) Sequence length $\nu = 5$; (b) sequence length $\nu = 10$. While (a) still shows the behaviour for individual chemical compounds, (b) indicates typical phase transitions. Note the symmetry of the curves with respect to their antipodal sequences.

quasi-species. Without any reservation I would call this a first-order phase transition (see Figure 5.5.1). Nevertheless, this order has to build up. Peter Schuster and Jörg Swetina[83,85] did some computer experiments to look into this question. They studied the build-up of co-operativity with sequences of smaller length. Examples for lengths 5 and 10 are shown in Figure 5.5.3. The curves for length 5 still have the appearance of a superposition of individual chemical reactions of five different pentamers, while the curves for length 10 allow the build-up of a phase transition to be recognised; for a length of 20 this is already pretty complete, as Figure 5.5.2 shows.

Let us return to the simulations for length 50. The following example[86] has turned out to be quite instructive with respect to the phase transition-like nature of selection.

We consider now a model slightly different from that discussed in Figure 5.5.1. There we had only one dominant "wild type", which replicated ten times more efficiently than all its (degenerate) mutants. Now we consider, in addition, an almost equally efficient "antipode" I_{50}. This has an efficiency that is 90% of that of the wild type I_o, but its 50 single-error mutants (I_{49} of the antipode I_{50}) are assumed to have an efficiency of 50% of that of the wild type, i.e. five times of that of all other (degenerate) mutants of I_o. This is illustrated in Figure 5.5.4 (a). The simulated curves in Figure 5.5.4 (b) now show two phase transitions. At the beginning (i.e. $(1-q) = 0$)I_o again is the dominant wild type and its survival curve starts exactly as in Figure 5.5.1. Note that we now use a logarithmic scale for the ordinate and show what happens only for a part of the abscissa, i.e. the range of error rates (1 – q) between 4% and 5%. In this way we can establish more precisely the sharpness of the second phase transition at an error rate of about 0.045. While at lower error rates the wild type I_o is the clear "winner" of the competition, it is superseded by I_{50} at the error rate of about 4.5%. Here the now (partly) populated one-error mutants of the antipode I_{50}, i.e. the 50 I_{49}-sequences, have meanwhile built up a quasi-species-like structure around I_{50}, which as a whole is superior to the quasi-species around I_o with its less potent one-error mutants I_1. The picture clearly shows that it is the quasi-species, rather than the individual wild type, that is rated in natural selection and that the process – despite the involvement of many different individuals – is extremely sharp over many orders of magnitude, as is to be expected for a phase transition of the first order. It does not reflect in any way a process of gradual accumulation of mutants.

The models discussed so far tell us a lot about the nature of selection as a process of self-organisation and of generation of specifiable information. We may, however, be somewhat dissatisfied with this kind of model. The internal symmetry of the mechanisms considered in the above figures is far from being realistic. The corresponding fitness landscapes consist of isolated peaks surrounded by entirely homogeneous mutant regions. In fact, we wouldn't be sitting here to reflect on our origin if such landscapes had any actual relevance. In the neighbourhood of any real peaks in the fitness landscape there must be many points of similar "height", as in a mountainous countryside. Therefore let us look at another model,[87] one that consists of two nearly neutral peaks at a distance that can be bridged. Figure 5.5.5 depicts such a model. The reason why I have chosen it is that it is typical for fitness landscapes of mutant distributions in the quasi-species model, where isolated peaks are quite unlikely. In this case, evolutionary progress is achieved along mountain ridges of the fitness landscape. Curves a and b in Figure 5.5.6 show how a gap between two peaks can be bridged by populating intermediate mutant positions. These two curves refer to two extreme situations. More realistic distributions would yield values between these curves. If the two peaks were surrounded by mutants with fitness values as assumed in Figure 5.5.1 the gap could not be bridged (curve d in Figure 5.5.6). Neutral and nearly neutral mutants are of an utmost importance for evolutionary "tinkering".

(a)

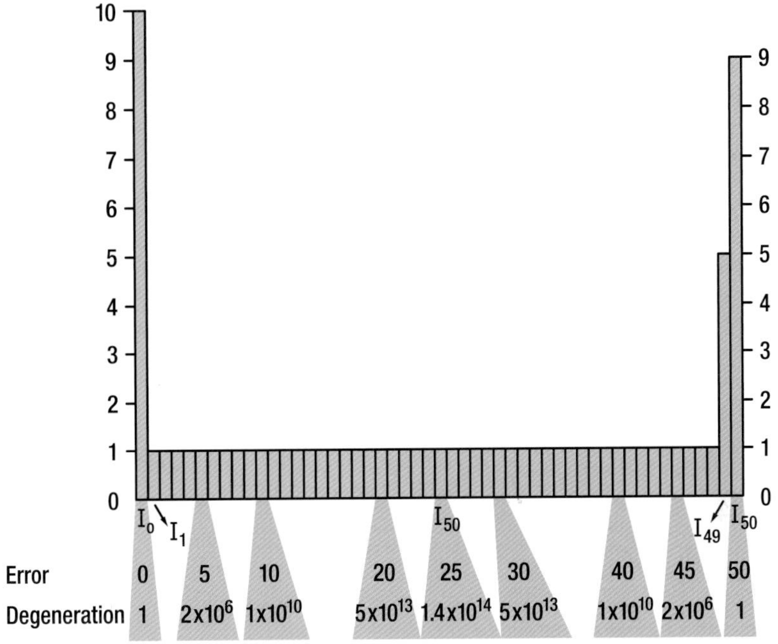

(b)

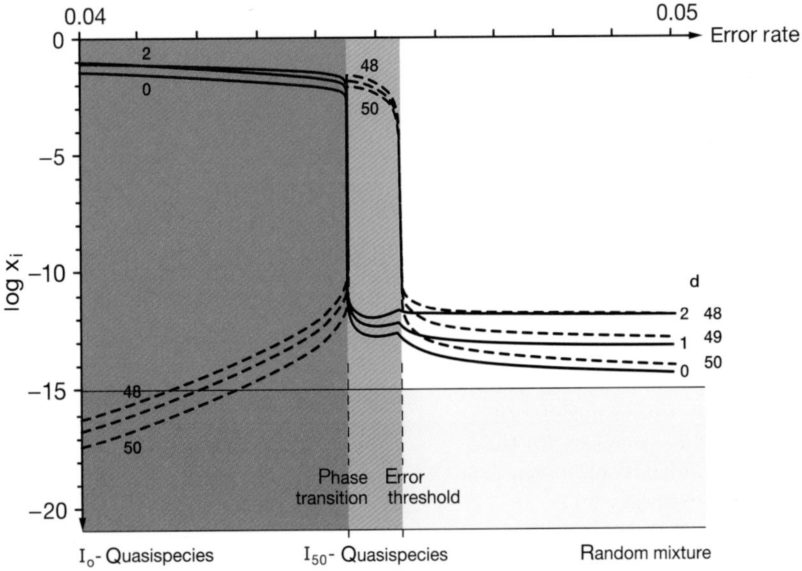

Figure 5.5.4
(Cont.)

Figure 5.5.4

A typical quasi-species competition

(a) The graph of replication rates for all error copies. The model assumes one master copy with a replication rate ten times greater than those of all error copies with the exception of the antipode of the master copy, which has a rate nine times that of the others, and its one-error mutants, which have rates five times that of the rest of the mutants (which form the majority). (b) The relative population numbers show that at an error rate near zero, the master copy (with a tenfold higher rate) is the clear winner of the competition. However, with increasing error rates the quasi-species of the antipode copy (the "second master copy", with a ninefold higher rate) builds up more efficiently than that of the first master copy because of the fivefold excess rate of its one-error mutants, compared with that of the one-error mutants of the first master copy. Finally, a second phase transition occurs, in which the second master copy – with the help of its quasi-species – becomes the final winner. This experiment is a wonderful demonstration of the fact that it is the whole quasi-species and not just its singular master copy that is evaluated for selection.

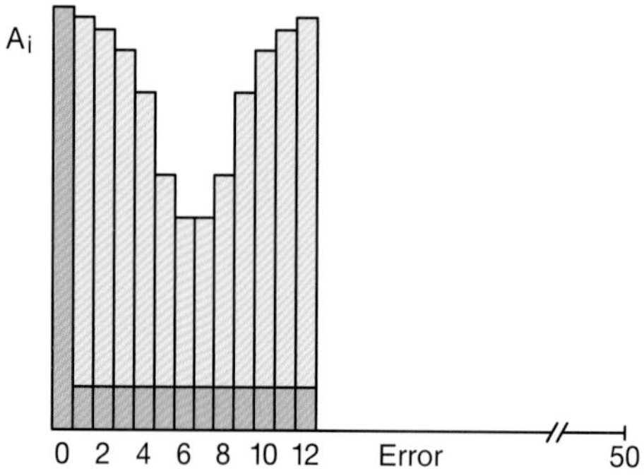

Figure 5.5.5

Phase transitions in more continuous mutant distributions

The scheme of replication rates for all mutants. It is seen that a practically continuous transition to a second maximum value takes place.

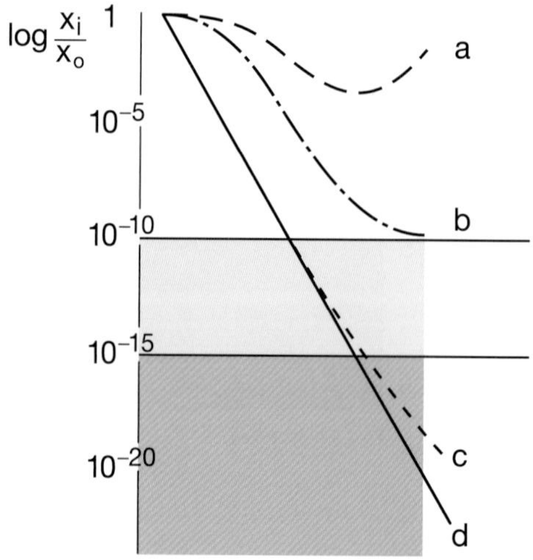

Figure 5.5.6

Population curves for the scheme in Figure 5.5.5

(a) The above scheme holds for all members of the error class. (b) The above scheme holds only for one path linking the two peaks. (c) For comparison, the connection linking the two peaks consists of mutants with tenfold lower replication rates. (d) The curve obtained if all mutations between the two peaks are deleterious.

Rather than continuing with a discussion of specific, more realistic fitness landscapes – a subject for the more biological chapters in a second volume – let me come back to the physical nature of the error threshold and its role in phase transitions in information space.

First of all, the presence of the error threshold in all biological organisms and its dependence on the size of the genome is a well-established experimental fact. Table 5.5.1 shows some examples for RNA and DNA genomes, illustrating the universal nature of this principle. The striking differences between RNA and DNA systems are due to the differences in RNA and DNA replication, which are explained in more detail in Section 5.7.

The physical nature of the phase transition is based on three prerequisites: reproduction, co-operativity of this process and "snowballing" of error production at the threshold, resulting in unlimited correlation lengths. The physicist will realise immediately the correlation with the prerequisites for phase transitions in material systems. The first condition is trivial. What is in existence must have had a way to come about, otherwise no phase could build up. The second condition requires that what comes about through the first condition must be positively influenced by what is already there. The third condition is important for making co-operativity strong enough to produce a true phase transition. In other words: the co-operativity must have a sufficiently large correlation length to allow it to spread through the whole phase and not remain limited to the close vicinity of the reproductive centre. As Lars Onsager[88] had found out, co-operativity for that purpose has to act at least in two dimensions. Physical processes of this kind are considered in more detail in Appendix A2. True phase transitions in the mathematical sense always refer to the so-called "thermodynamic limit", in which the correlation length can approach infinity. Strictly speaking, the correlation length must not be smaller than the extension of the system. How far are these conditions fulfilled for our systems, which do not refer to physical space, but rather to what we call "information space"?

In fact the first condition, which I called "trivial", already presents a major obstacle. Nothing as complex as is required in order to be "alive" can come about *de novo* with any adequate probability. Hence we have to assume that what we call "the origin of life" must after "nucleation" be associated with some process of adaptation. This is still a problem of prebiotic chemistry, far from being finally solved. It requires some initial state from which biological evolution involving reproduction and adaptation in a Darwinian sense could have started. Replication certainly fulfils the second condition for phase transitions. It produces something that is positively influenced by what is already there. This co-operativity acts in two directions. Sequences as such are co-operative units. In addition, their temporal sequence, because of replication,

Table 5.5.1 Error rates and genome sizes of RNA viruses as compared with autonomous organisms

Virus[a]	Genome size: υ (number of nucleotides or base pairs)	Error rate: $(1 - q)$ (per replication round and per nucleotide)[b]	Error rate: $\upsilon\,(1 - q)$ (per replication round and per genome)[b]
RNA			
Bacteriophage Qβ	4200	3×10^{-4}	1.3
Polio-1 virus	7400	3×10^{-5}	0.2
Vesicular stomatitis virus	11000	1×10^{-4}	1.1
Foot and mouse disease virus	8400	1×10^{-4}	0.8
Influenza-A virus	14000	6×10^{-5}	0.8
Sendai virus	15000	3×10^{-5}	0.5
HIV-1 (AIDS virus)	10000	1×10^{-4}	1.0
Avian myeloblastosis virus	7000	5×10^{-5}	0.4
DNA			
Bacteriophage M13	6400	7×10^{-7}	4.6×10^{-3}
Bacteriophage λ	48500	8×10^{-8}	3.8×10^{-3}
Bacteriophage T4	16600	2×10^{-8}	3.3×10^{-3}
E. coli	4.7 million	7×10^{-10}	3.3×10^{-3}
Yeast *(Saccharomyces cerevisiae)*	13.8 million	3×10^{-10}	3.8×10^{-3}
Neurospora crassa	41.9 million	1×10^{-10}	4.2×10^{-3}
Human	3000 million	$\sim 10^{-12}$	$\sim 3 \times 10^{-3}$

[a] Even today reliable data are still scarce. Examples are taken from Eigen (1993)

[b] q is the fidelity of replication per nucleotide (nt), i.e. the (normalised) probability for inclusion of the correct (complementary) nt. $(1 - q)$ then is the probability for including a non-complementary nt.

[c] There is an error threshold relation (Eigen, 1971)[20], according to which the product $\upsilon\,(1 - q)$ has to remain below a threshold value $\ln \sigma$, where σ rates the fitness of the best adapted mutant relative to its competitors in the quasi-species. For RNA viruses σ is sufficiently larger than 1 and $\ln \sigma$ of the order of magnitude of 1, so that the critical error rate is inversely proportional to υ, the number of nt in the genome. For DNA genomes σ is found very close to one yielding $\ln \sigma$ between 10^{-2} and 10^{-3}.

reflects co-operativity, as shown in analogy to a spin system in Figure 5.5.7). In other words, the replicated sequences are similar, if not identical, to their templates. As Figure 5.5.7 shows, it is true for replicative processes in space and time. The latter fact opens the way for unlimited correlation lengths. This has to be imagined in the following way.

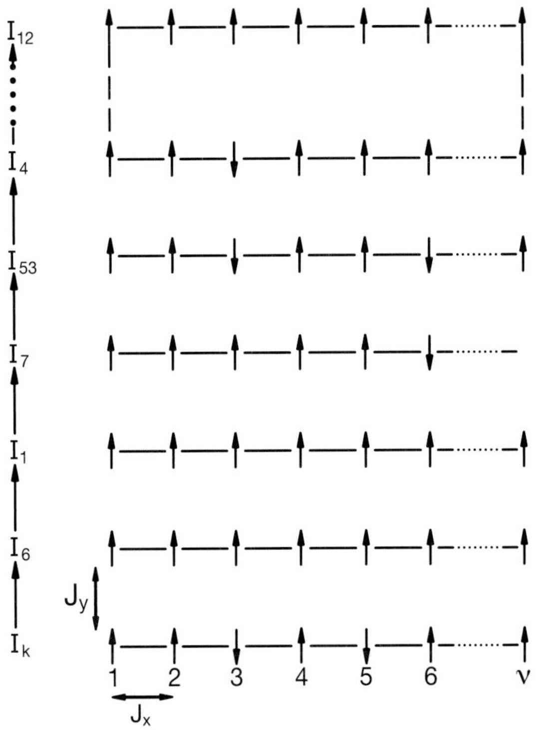

Figure 5.5.7

The two-dimensional co-operativity present in natural selection

In this picture a two-dimensional model genealogy is compared to a two-dimensional spin lattice. Individual spins exist in two states (spin-up or spin-down). Every row can be represented by a binary sequence, as in the binary model of a polynucleotide sequence. "In-row" interaction in the spin lattice, as described by the spin–spin coupling constant J_x, is co-operative in nature. It is to be seen in analogy to the co-operative interaction among all monomeric units within a sequence representing, as a whole, genetic information for the functional properties of a certain individual protein molecule. On the other hand, co-operative interaction in the lattice of spin states is also introduced by the "vertical" coupling constant J_y, which in a model genealogy depends on the individual mutation-prone replication kinetics. The resulting effect is two-dimensional co-operativity in sequence space, similar to that found for spin glasses as temporally changing co-operativity in physical space.

Whenever the error threshold is reached, the "wild type" (however we define it) disintegrates faster by mutation than it can restore itself by precise reproduction. This means it becomes unstable with respect to its numerous mutant neighbours. They themselves again become unstable with respect to their even more frequent neighbours, and so on – altogether yielding an unlimited "avalanche", which destroys all information that had been collected before through positive selection.

This behaviour is based on the inherently non-linear term in the selection equation (see Table 5.5.1). There is a far-reaching analogy between systems showing natural selection and thermodynamic systems at equilibrium; it includes co-operativity with unlimited correlation lengths, becoming infinite in the "thermodynamic limit". These analogies have been addressed in several papers.

We are still left with one unanswered question: what about selection among neutrals? Let us go back to selection in physics. The liquid–gas transition is a good example. In the phenomenological van der Waals description, first-order behaviour of the transition goes on until a "critical point" is reached at which the density of the liquid equals the density of the resulting gas (Figure 2.8.3). This problem has only quite recently found a satisfactory solution, namely by Kenneth Wilson's treatment using the renormalisation group, as described in his Nobel lecture.[89] I suppose that in selection among states of equal fitness value we have a situation similar to the one found in liquid–gas transitions near the critical point where the two states reach equal densities, showing large macroscopic fluctuations that result in "critical opalescence". Kimura's stochastic theory of the emergence of single species also suggests large macroscopic fluctuations because "selection" is a phenomenon that appears at the macroscopic level. Such a result would have quite important consequences.

If neutral selection causing neutral drift occurred in an unpredictable, so to speak "random", way any process of evolution would be an interplay of deterministic selection of favourable mutants with indeterministic selection of neutrals, and the overall process would thus be indeterminate. This would explain Eldredge and Gould's "punctuated equilibrium" – perhaps not in the same way as Gould had expected (see Section 5.1). Be that as it may, it would explain and unify older views which as yet I do not want to analyse any further – except with respect to the more general conclusion, that evolution of life is indeed a sequence of phase transitions towards optimal adaptation of a "fittest state", which is not identical with a "fittest individual" and which can act on a much wider range of mutants than "classical selection" of individuals.

The ideas I have discussed in this section are pretty well independent of the detailed mechanisms, which will be considered in a second volume, where we shall look for their reality in biology. Here I have discussed general prerequisites. Semantic information can only come about by selection, which is the opposite of equilibration. Selection is physically based on replication. Replication hence fulfils two tasks: the conservation of information, without which complexity could never come under control, and providing the basis for phase transitions, such as the formation of new information in an unlimited way. Why did Nature choose to accomplish this by a phase transition? A possible answer may be that this is the best way to stabilise one (i.e. the selected) microstate against the macrostate of "the rest of the world".

Similar prerequisites will turn out to be necessary (although not sufficient) for the mental processing of information. The mechanisms that realise selective evaluation in material form may differ, but the fundamental constraints ought to be of a similar nature. The replication of a neural state requires both memory and autocatalytic feedback. Time and space here acquire new dimensions.

5.6. Was the Watchmaker Really Blind?

On the jacket of Richard Dawkins' excellent and influential book *The Blind Watchmaker*[90] we find appraisals by two prominent contemporary scientists in the field of evolution who themselves have made great contributions to our understanding of biology in the widest sense of the word. Their assessment of the book, as being the "best general account of evolution" in recent years, is one that I can subscribe, and therefore I have to specify in more detail what may sound like a critical undertone in the title of this section: the word "blind" could easily be misunderstood and so it happened. A sentence that appears on the jacket of Dawkins' book, right next to the two appraisals, reads: "Natural selection – the unconscious, automatic, blind, yet essentially non-random process that Darwin discovered – has no purpose in mind." Yes, a nice analogy, and almost correct – but not quite.

The title of Dawkins' book refers to William Paley, a 19th-century theologian, who compared life with the complex construction of a watch which, if you find it "upon the ground", would after inspection immediately give the impression that it was constructed according to purposeful design and did not come about by some natural cause – as, for instance, did the stone, lying on the ground next to the watch. Thus far, the "blind" watchmaker is a wonderful allegory.

What, in my opinion, could be misunderstood in the jacket text referred to is the association of the words "blind" and "unconscious". Sure enough, the eye is part of the brain, but blindness doesn't exclude consciousness. That's trivial and may not have been meant either. My qualms are concerned with the use of the word "unconscious", a word that everybody believes they understand, but which nobody can explain. Dawkins tells a wonderful story about the self-organised nature of life resulting from the continuously reiterated interplay between "chance and necessity", a term powerfully promulgated in Jacques Monod's book,[91] where he says: "chance, and nothing but chance" is the cause of variation, while necessity, i.e. something enforced by physical law, is superimposed as "natural selection". However, the interplay of chance and law is not that simple. Chance, an important element of biological evolution, is not "totally blind". If it were, Nature would have had to test far too many alternative combinations to find the very few that are suitable for selection. And "consciousness" is something else, a "complex familiarity" that we so far do not understand "consciously". However, the "consciousness" required for selective self-organisation is different. It represents the deviation of mutational chance from total blindness. On the other hand, mental consciousness also is a complex phenomenon that may start from random firing of nerve cells, which only by selective interaction with synaptically linked neurons, through excitation and inhibition, achieve the orderly behaviour we express as "consciousness". But this theme will be taken up in the second volume.

Perhaps these qualms of mine are too subtle for a popular representation, as Dawkins' book was intended to be, which necessarily has to leave out some of the complex detail. Anyway, let me say what I have in mind. The result of mutation depends not only on the nature of the mutational process as such, but at least as much on the composition of the state that undergoes mutation. The process of mutation as such can certainly occur at any position of a DNA or RNA sequence because it is usually triggered by thermal or otherwise random fluctuations. However, the result, i.e. the kind and frequency of the resulting mutant, depends just as much on the composition and abundance of the mutating precursor, which need not be the average wild-type sequence itself, but could equally well be any abundantly populated mutant in the wild-type distribution. Hence the resulting mutant is not an entirely random choice; rather it is a close relative of an abundantly populated mutant and so has a high probability of being well adapted too. There are similar relations that cause a bias towards fitness.

Clearly, all mutant sequences appear within a certain region of sequence space, with their abundances decreasing in proportion to the increasing Hamming distance relative to the centre. However, this distribution is anything but random, as has been considered in classical models, e.g. represented by some Poissonian or polynomial distribution. Each mutant has a precursor, which in a quasi-species is part of an evaluated distribution of mutants. What is entirely blind and occurs entirely "unconsciously" is the mutational process as such, not the emergence of the mutants. In the quasi-species model each individual mutant is evaluated in this respect. Correspondingly, we have to deal with a large system of coupled differential equations, involving a matrix with positive entries for all non-diagonal terms, but mathematics tells us that this system of equations contains only one eigenvalue describing stable selection, regardless of how large the number of equations happens to be (see Section 4.7). Here the presence of partly active and, in particular, of neutral mutants causes large deviations from the relatively narrow mutant distributions of a Poissonian type.

There are three influences that we can distinguish. First and foremost, the quite irregularly distributed neutral mutants of the wild type have the largest number of progeny and hence represent peaks in population space around the "centre" of the quasi-species.

Secondly, there will be some clustering of all sorts of peaks in the vicinity of those neutrals due to certain similarity relations, just as a clustering of peaks occurs frequently in mountainous landscapes.

Thirdly, deleterious or poorly adapted mutants usually then appear in more distant and possibly quite extended regions with flat-land character, where advantageous mutants can hardly be expected.

As a consequence of these effects the population is biased, consisting almost solely of (more or less) well-adapted mutants. In other words, only a relatively small fraction of all mutants around neutral peaks are the ones that are screened; the optimal wild type eventually selected appears within a self-organising distribution of limited size. This is also what we found in our experiments. Adaptation to externally imposed environmental changes occurs at unexpectedly high rates, which cannot be explained on the basis of entirely "blind" mutation.

What is typical for a quasi-species distribution is the absence of simple symmetries. Therefore, such distributions reach much farther into space and consequently they are able to screen a larger variety of well-adapted mutants than simple polynomial types of random distributions could. Such behaviour is greatly aided by "neutral drift". The much narrower Poissonian types of distributions around individual wild types are devoid of those properties, ultimately needed for efficient evolutionary progress. Our distributions look quite like peak distributions in mountainous landscapes, albeit in a space of large dimensionality.

Darwinian processes, as we understand them today, occur "automatically". This is not adequately expressed by the word "unconsciously", I wouldn't even say "with no purpose in mind". That may be correct in a trivial sense, yet we do not know sufficiently well what happens in our mind when we devise or execute some purposeful plan. The "purpose" of selection is "to be", and the process of selection is physically adapted to this purpose, in the present case utilising the non-trivial structure of the mutant distribution. Altogether, this can still be called a process of "tamed" chance, connected with the necessity of selection. Richard Dawkins, by not going into such detail, does excellently by just portraying the basic nature of "Darwinian processes".

Daniel Dennett sees the main success of Darwin's idea in its unlimited application. Each advancement, as small as it may be, if taken as the new platform of further advancement, will eventually exceed any expectation, in a way similar to that in which the sum of all natural numbers, adding only one in each step, eventually tends to infinity. There is no doubt that this aspect of Darwin's idea is a very important one, fully justified against the background of 19th-century thinking, and losing not a bit of relevance by undergoing its two great amendments: recombinative inheritance, due to Gregor Mendel, and the clear distinction of soma and germ line by August Weismann. The greatness of these ideas is appreciated nowadays, more than ever before, as a consequence of our insights into the mechanisms of inheritance provided by molecular biology. Nevertheless, we live in this modern age and our questions have shifted to more quantitative aspects; here, changes may involve large numbers of different leaps as postulated by "punctuated equilibrium".

In Appendix A2, where I discussed physical phase transitions, we realised that a relatively small quantitative discrepancy between an observed critical coefficient of about one-third and its formerly theoretically predicted value of one-half has brought about a fundamental revision of the way in which continuous phase transitions are described, yet without making nonsense of earlier ways of thinking. A similar situation may now hold in molecular biology, where quantitative experiments have become possible.

The idea of an evolutionary "design from within" is such a development, triggered by quantitative analysis. Owing to the fact that biological evolution involves problems of "exponential complexity", its quantitative consequences can be tremendous. A Poissonian (i.e. random) distribution around a wild-type gene of length 1000 in a population of about 1000 million individuals can hardly be checked for all its four-error mutants, the number of which for binary sequences amounts to the binomial coefficient of $\binom{1000}{4}$ or just over 10,000 million (10^{10}) different individuals. However, a quasi-species distribution – as indicated above – might be able to find the evolutionarily most important 10- to 20-error copies, whereas random checking would have to involve some 10^{24} to 10^{42} copies. The unexpectedly high efficiency of experiments in molecular evolution has borne out the relevance of these estimates. A quantitative effect here may become a new quality. Molecular engineering with 10- to 20-error mutants opens new roads along which evolutionary biotechnology is now proceeding. At the same time, these estimates show that our picture is still far from complete. A binary sequence of 1000 digits has $2^{1000} \approx 10^{300}$ different alternatives. Being able to check these even in steps of 10^{50} at a time reduces the complexity of the problem just from 10^{300} to 10^{250}, a big leap, but not a solution of the problem. There must be other issues involved, such as the fractal nature of fitness landscapes and other effects discussed in this section. Note that I am not talking about particular biological mechanisms, which are not a subject of this volume. I am still talking about physical problems of complexity associated with the generation of information.

Daniel Dennett, in his book *Darwin's Dangerous Idea*[35] has appreciated the relevance of those thoughts, but rejected their "revolutionary" nature. I agree with him regarding the term "revolutionary", which I never have used myself. How could I, being a strong adherent of evolutionary thinking – quite apart from the fact that I hate revolutions, which are often "overthrows" in which everything gets out of control? The scientific literature is full of "revolutionary" ideas. But should they not rather be described as "evolutionary" leaps? The prototype of a non-political revolution is the "Copernican revolution", which can lay claim to this epithet if any scientific development can. However, Nicolaus Copernicus himself never made such a claim. His title *De Revolutionibus Orbium Coelestium* does not refer to what was later called the Copernican Revolution. The word "revolutionibus" in the title of his treatise is the ablative plural of "revolution", meaning the rotations (or orbits) of heavenly bodies ("orbium coelestium"). The "Copernican revolution", an expression coined centuries later, refers to the fact that Copernicus upset the Ptolemaic system and thereby removed humans from the centre of the world. This indeed could be called a revolution, not just in the scientific sense. At the same time it was a radical change in how humans saw themselves in this world.

What type of self-organisation is necessary in order to come to grips with the problem of complexity? It certainly requires a more sophisticated interplay between variation and selection than simply trial and error. Both selection and neutral drift are based on replication. In the vicinity of the error threshold we find the most flexible responses: large-scale fluctuations that diffuse along plateaus of the fitness landscape and sharp, but labile selection in a rugged "countryside". We have seen how multidimensionality provides escape routes more easily. Moreover, Werner Ebeling[92] and his co-workers at the Humboldt University of Berlin have demonstrated how gaps in a mountain ridge can be crossed in a way analogous to that in which potential barriers can be penetrated by matter in quantum-mechanical tunnelling, although the mechanism of this penetration is not specifically quantum-mechanical. If I said that evolution is based on selection, this means that it can occur only in replicating populations (whatever the particular mechanism of replication may be). Single entities cannot evolve!

How can we imagine the self-organised generation of information? Peter Schuster[93] and his colleagues at Vienna University have studied an instructive example: the formation of RNA sequences with maximum stability of secondary structures, involving a maximum number of base pairs. The most striking result is the presence of a large number of equivalent structures at various levels of base-pairing. For the sake of simplicity, let us look again at binary sequences with only one kind of base pair. Starting at an arbitrary point in sequence space characterising a random sequence, the Hamming distance to the optimal structure turns out on average to be much smaller than half of the number of binary positions in the sequence. (With respect to any final sequence about half of the initial positions should be occupied "correctly" anyway.) It turned out to be more closely proportional to the square root of the total number of positions (see Figure 5.6.1). "Selective advantage" was measured as an increase in the number of base pairs formed. The actual time curves of approach to the optimum state were typically of the form shown in Figure 5.6.2. Evolutionary progress takes place in steps that are followed by more or less extended plateaus. However, the latter do not mean that evolution comes to a standstill. These plateaus rather involve neutral structure reorganisations, i.e. refoldings without an increase in the number of base pairs.

The conclusion from these experiments, carried out on computer as a "theory of neutral networks", is: large parts of sequence space are covered by "neutral networks". Admittedly, the model chosen is highly favourable for the formation of neutral

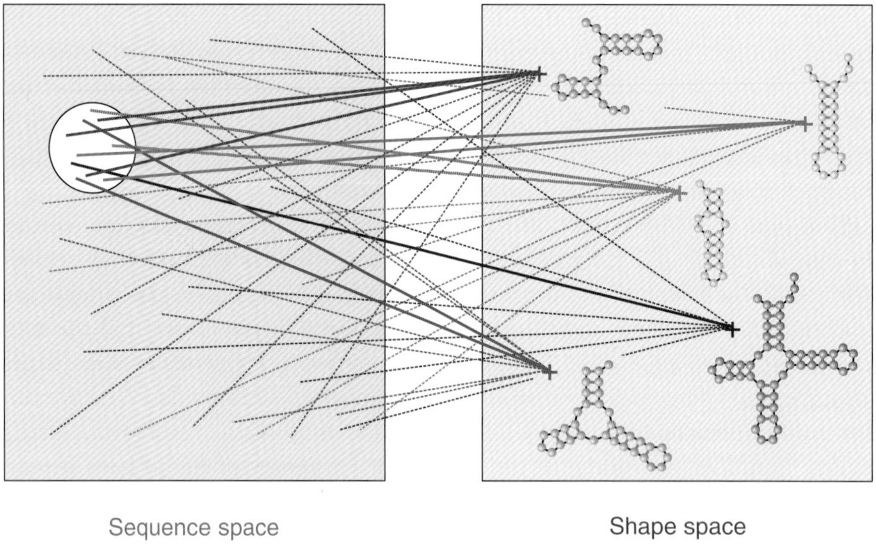

Figure 5.6.1
Shape space covering

RNA secondary structure formation is sketched as a mapping from sequence space into shape space: One particular structure is assigned to every sequence; the inverse, however, is not true since many sequences may form the same secondary structure. Accordingly, many points of sequence space are mapped onto a single point in shape space representing one particular structure (indicated by a cross in the sketch). RNA sequence-to-structure mappings are characterised by relatively few *common structures*, which are readily found by evolution or by screening random sequences, and many *rare structures*, which occur at such low probabilities that they play no role – either in nature or in laboratory experiments. The sequences forming a given common structure are distributed almost randomly in sequence space. Consequently, one or more sequences folding into any of these common structures can be found in an arbitrary spherical environment with a radius much smaller than the dimension of sequence space. Its value is closer to the square root of the total number of dimensions. More details are available in the three papers listed below.

Schuster, P. (2003). Molecular Insights into the Evolution of Phenotypes. In: J. P. Crutchfield and P. Schuster, editors, *Evolutionary Dynamics – Exploring the Interplay of Accident, Selection, Neutrality, and Function*, pp. 163–215. Oxford University Press, New York.

Schuster, P., W. Fontana, P. F. Stadler, and I. L. Hofacker (1994). From Sequences to Shapes and Back: A Case Study in RNA Secondary Structures. *Proc. Roy. Soc. London* B 255: 279–284.

Hofacker, I. L., P. Schuster, P. F. Stadler (1998). Combinatorics of RNA Secondary Structures. *Discr. Appl. Math.*, 89:177–207.

networks. Schuster and his colleagues have shown how the model is modified by the introduction of two kinds of base pairs. The folding of protein structures with particular catalytic sites is certainly a problem of much higher complexity. However, the ease with which such structures form in catalytic antibodies[94,95] indicates that this problem is by no means beyond experimental grasp.

The question in the title of this section was stimulated by a metaphor in the title of the book by Richard Dawkins which refers to some cursory objections against Darwin's ideas that were easily refuted and should thereby have been delegated to

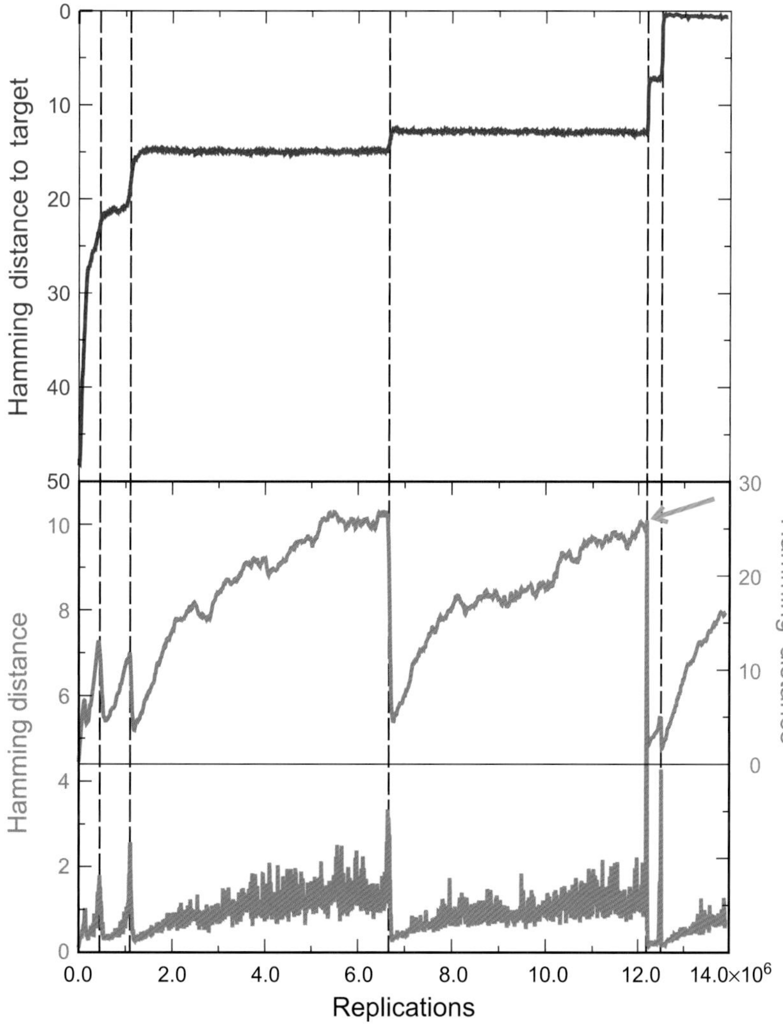

Figure 5.6.2 (Cont.)

the "garbage of history". But, as Dawkins shows, this has not happened. The slogan of "intelligent design", which has arisen in our time, is a good example of this. As far as I can see, "intelligent design" is a product of the ideas of people who themselves do not fulfil one very important requirement of intelligence, namely, the willingness to compare an idea with its alternatives. Of course, their notions are based entirely on their own religious beliefs, and we should respect this because it may mean much for their life. However, we should make it clear to them that the results of science are not an alternative "hypothesis" and are even not in contradiction to their religious belief; rather, they are the best outcome of a process of logical inference and experimental evidence acquired in an endeavour by the whole of human society, in other words the best possible explanation that can be offered by human society.

The term "intelligent design" is, of course, a human construct. According to Webster's definition "intelligence" is the "manifestation of a high mental capacity", an activity that includes "learning, reasoning and understanding", of purely human origin. Applied to any divine faculty, I think any religion must consider anything like this pure blasphemy. All religions of Mosaic origin forbid in their commandments *expressis verbis* any kind of imagination of God. If it was He who created "Nature",

---

Figure 5.6.2

Evolutionary optimisation of RNA structure

The graphs show a typical trajectory of a simulation of RNA optimisation with the secondary structure of $tRNA^{phe}$ as target (S_τ). The population size was N = 3000 molecules, a fitness function of the form $f_k = (\alpha + d_s(S_k, S_\tau)/n)^{-1}$ with $\alpha = 0.01$ and n = 76 was chosen, and the mutation rate amounted p = 0.001 per site and replication. The graphs show as functions of time: (i) the distance to the target, averaged over the whole population, $E\{d_s(S_i,S_\tau)\}(t)$ (red curve), (ii) the mean Hamming distance within the population $E\{d_p\}(t)$, as a measure of the width of the population in sequence space (blue curve, right ordinate), and (iii) the mean Hamming distance between the populations at time t and $t + \Delta t, E\{d_c\}(t,\Delta t)$ (green curve, left ordinate) with a time increment of $\Delta t = 8000$ replications. This quantity is a measure of the migration velocity of the consensus sequence of the population. The ends of plateaus (vertical broken lines) are characterised by a collapse in the width of the population and a peak in the migration velocity corresponding to a bottleneck in diversity and a jump in sequence space. The green arrow indicates a remarkably sharp peak at the end of the second long plateau ($t \approx 12.2 \times 10^6$ replications) corresponding to a jump in the consensus sequence of Hamming distance 10. On the plateaus the centre of the spreading cloud stays practically constant (the speed of migration is Hamming distance 0.125 per 1000 replications), corresponding to a constant consensus sequence. Each new adaptive phase is preceded by a drastic reduction in genetic diversity, $E\{d_p(t)\}$; the then diversity increases during the quasi-stationary epochs and reaches a width of Hamming distance more than 25 on long plateaus. The spreading of populations on plateaus is, apparently, only a function of time (or, more precisely, of the number of replications) whereas the jump size seems to be independent of the length of the preceding plateau.

then this includes the laws of Nature, which science – as a cooperative endeavour of humans – tries to unravel. Religious people should respect this in their personal beliefs.

Scientists, on the other hand, should bear in mind that their objective findings never answer the question of the existence of a God. This is a pure matter of belief. A large proportion of non-scientists – and this includes many individuals of the highest capability – will be unable to understand all the results of scientific endeavour. For such people, these results must in any case remain a matter of belief. Modern state constitutions guarantee the freedom of personal belief, and this is to be respected for all individuals. In this connection I would like to point out the recent comments made on this subject by two renowned scientists, who in my view address the problem in a most appropriate way.

The first comments occur in a book entitled *The Creation* by Edward O. Wilson,[96] who expresses his opinion in a series of "letters to a southern Baptist pastor". The other comments appear in a collection of essays by the late Stephen Gould,[97] in particular "Darwin and the Munchkins of Kansas", published in 1999 in a note to *Time Magazine*. These two publications express excellently all I could have said on the subject.

5.7. Where is the "Edge of Chaos"?

During my annual stays at the Scripps Research Institute at La Jolla, I repeatedly had the chance to visit Font's Point in the Anza Borrego desert of Southern California (see Figure 5.7.1). Font's Point is located at an edge that separates the monotonously formed desert from the badlands with their nearly ideal fractal structures, preserved in stone and documenting one of Nature's examples of the *Beauty of Fractals* (so the title of a book by Heinz-Otto Peitgen and Peter Richter[98] from which an "image of a complex dynamical system" was represented in Figure 4.8.7). There is a profound relationship between fractals and chaos, which I alluded to when discussing "strange attractors" in Section 4.8.

"Life at the edge of chaos" is a catchy title that we find in the later works of Stuart Kauffman. Let us unpack it. Deterministic chaos in modern terminology means the loss of regularity in non-linear periodic behaviour. This assignment of the word "chaos" to an inherently deterministic mechanism has become so popular that its indeterministic counterpart, the entirely disordered tohubohu (from the Hebrew *tohu wa-bohu*), characterising the initial state of the world, has almost come out of perspective and is often mixed up with its deterministic counterpart. The term "chaos" certainly involves something mysterious, and "the edge of chaos" commends itself as a backdrop for some mysterious origin of life. Nothing wrong with such a terminology in popular literature! Since information somehow must have originated

Figure 5.7.1
The fractal structure of the Bad Lands of Anza Borrego at Font's Point

Photograph by courtesy of David Usher.

from non-information, and since chaos of all sorts abounded right at the original tohu wa-bohu, I see no deeper meaning in an origin of life "at the edge of chaos" than would be expressed by saying that: all rationals are "at the edge of the chaos of irrationality", which certainly is a correct statement.

Deterministic chaos may occur in connection with non-linear oscillations showing limit-cycle behaviour. A special case of transition from periodic order to deterministic chaos in population growth has been found by Robert May at Oxford University, as discussed in Section 4.8. It appears in the discrete and multiple population growth of certain insects. However, I can't see in those special cases a general phenomenon of life, which usually needs to escape chaotic behaviour. The development of homeostasis is a typical example of the general tendency of living matter to maintain internal stability and thereby counteract chaos in any form. The same can be said of the nervous system, in which the chaotic firing of nerve cells may indeed occur, but must generally be seen as a pathological phenomenon. Let us look at some typical dynamical systems as encountered in biology.

The system farthest away from the "edge of chaos" is a "closed" system that establishes internal equilibrium. The term "closed" means the absence of fluxes of energy and matter across the boundaries of the system. It is homeostatic in a trivial sense because it possesses an internal mechanism that counterbalances fluctuations, with a strength that increases with the deviation from equilibrium. In the vicinity of chemical equilibrium all reaction processes, regardless of their order, are linearly related to their "extent of reaction" (see Section 2.1). This behaviour has its roots in the closedness of the system, which ensures microscopic reversibility, detailed balance and the conservation of mass and energy. Whatever is in excess in one component must be deficient in one or several other components, causing fluctuations in any of them to decay exponentially towards their "asymptotically stable" equilibrium values. Equilibration is equivalent to the death of any form of life, as was stated clearly by Schrödinger (see Section 5.1). So let's not waste any further time: Equilibrium does not allow the generation of any new information.

However, we can find steady-state systems that behave in a very similar way to an equilibrium, even though a steady flux of matter and/or energy crosses their boundaries, causing a positive rate of entropy production. Such a system is portrayed in Figure 5.2.1(a). The substances in this system are produced at a constant rate k, independently of their own presence (diagram a-i). If unperturbed they would be produced at a constant rate and would grow linearly with time. If growth is restricted by flow, or by removal through decomposition, then we obtain the steady-state diagram (a-ii). In both cases the number of particles removed per unit time for the various substances is proportional to their amount present. The result is a situation similar to that of a system at equilibrium, namely, constant proportions of the various substances according to their rates of production and removal. Statistical fluctuations have no effect on the average representation. The difference with respect to equilibrium is the absence of microscopic reversibility and internal conservation. Fluctuation of A has no effect on B and *vice versa*. The result is just some stationarily weighted coexistence, and nothing that can be called either "equilibrium" or "natural selection".

First of all, there is a finite rate of entropy production, which distinguishes this state from equilibrium. The entirely different nature of this state of stationary coexistence, as compared with equilibrial coexistence, becomes obvious when we consider reactions of higher order, for instance those described in Figures 5.2.1 (b) and (c), yielding various forms of "natural selection" – something that never could happen in systems at equilibrium. In the laboratory one can produce such situations with the help of a "constant-flow reactor" (often abbreviated as CFR), a device popular in industrial chemistry as well as in microbiology. Figure 5.2.2 shows a special design of CFR, the "cellstat", which was constructed in our laboratory and used for *in vivo* studies of bacterial viruses and their evolutionary behaviour.

My subject in this section is the "edge of chaos". Deterministic chaos cannot be observed very often in the devices described above, unless one deliberately uses systems that are prone to it, such as the growth processes studied by Robert May and mentioned in Section 4.8. They are based on "Verhulst dynamics" as expressed by the so-called "logistic equation", which contains a linear production term, causing exponential growth, in combination with a negative non-linear (quadratic) term that ultimately constricts the growth process to a constant stationary population size. In its differential form the equation does not provide a chaotic solution, but for a synchronously stepwise growth (e.g. in annual steps) it may show chaotic instability. For this to occur, the growth parameter has to exceed a certain threshold value, above which a whole series of bifurcations occurs along with an increase in the growth parameter.

Populations of mutants can be considered as networks, and the normal-mode character of the quasi-species reflects this property. John Ross at Stanford University has studied this property and compared it to John Hopfield's neural network models,[99] with which it has much in common.[100] In fact, Stuart Kauffman's early model of a Boolean network shows interesting analogies when applied to models of neural networks. His later applications to larger networks of protein catalysts[101] – probably in unawareness of the quantitative results that were published – do not live up to early expectations by the author in 1971.[20] The trouble with deterministic chaos interfering with the evolution of life is its primary target, which is the dynamical process itself rather than population numbers of individual mutants. This, in other words, may mean the collapse of the function of a total network, rather than of its single constituents. It is best seen if we compare it with the "error threshold" in natural selection. The chaos produced by violation of the error threshold is of a quite different nature. It does not destroy the mechanism but, it results under certain boundary conditions. In evolution one is looking for alternative variants with advantageous properties, and not for changes in the mechanism of selection, i.e. the mechanism of selective reproduction, which acts on all variants in a similar manner.

The slogan the "edge of chaos" could of course easily be applied to the mechanism of natural selection. In this case there is real pressure on the experimenter to work as closely as possible to the error threshold, which represents a true "edge", i.e. a borderline that must not be crossed. In fact, the closer a wild type approaches this limit, the larger will be the variety and extent of its mutant spectrum, which itself is a determining factor for further evolutionary progress. Even an occasional violation of the error threshold, which may be detrimental for a particular mutant, does not necessarily lead to a complete breakdown of the distribution, which can easily become dominated by other mutants. This lends some robustness to the process of evolution. So far, life has not been overthrown by any global chaos in the time since its origins in the tohu wa-bohu more than 4000 million years ago.

How uniquely the quasi-linear rate law of replication fits the needs of selection as a true category of material behaviour, like equilibration or chaos, is seen if we study other types of rate laws. A case in between a constant rate and a linear rate is the square-root law (or any other root law). The square-root law was first proposed as an alternative relation for template instruction by Eörs Szathmáry[102] and was substantiated both experimentally and theoretically for enzyme-free amplification of oligonucleotide templates by Günther v. Kiedrowski.[103] The resulting organisation still involves coexistence of competitors, but also to some degree selective behaviour, different from the cases described before (see Figures 5.7.2 and 5.7.3).

On the other hand, rate laws that involve dependences stronger than linear ($x; \sim kx^n$ with $n > 1$) usually lead to singularities in the growth curves. If the rate law can be assumed to be quadratic ($x; \sim kx^2$), the result is hyperbolic growth. The rate constant k here is a second-order rate constant (dimension: reciprocal concentration times reciprocal time). A characteristic rate, measured in terms of a reciprocal time (obtainable from expansion into a power series), is then the product of the second-order rate constant and a concentration. This means that any species that has grown up to a macroscopic population is in such a large excess over a single mutant copy that this lonely copy could never compete for growth with a populated ensemble, even if its growth constant k in itself had a superior value. Selection under such conditions is "once for all time". Non-linear "nucleation processes" of this kind are

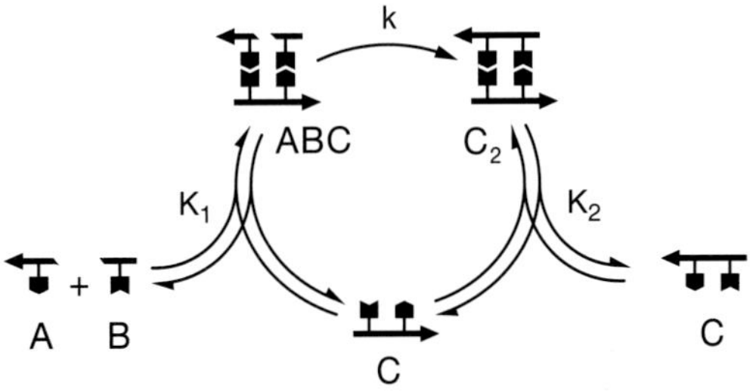

Figure 5.7.2

Minimal reaction model of the autocatalytic reaction cycle in synthetic self-replicating systems

K_1 and K_2 define the overall formation constants of the complexes **ABC** and C_2, respectively.
Courtesy of Günter von Kiedrowski.

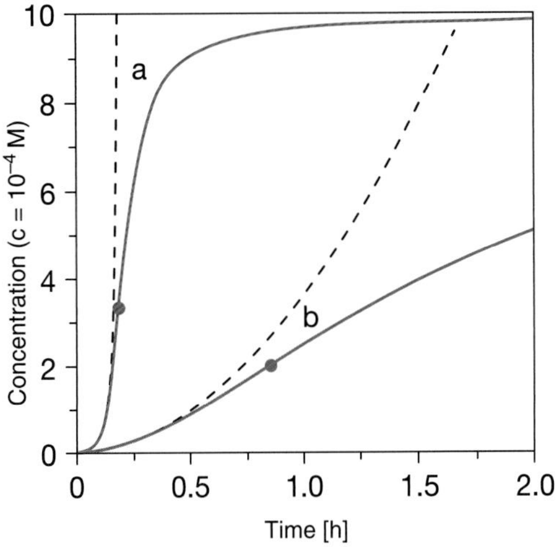

Figure 5.7.3

Growth curves according to Figure 5.7.2

Exponential (a: $c = c_0 e^{kt}$) and parabolic (b: $c=[\sqrt{c_0} + kt]^2$) growth. The dashed curves represent the ideal case of unlimited growth. Each dot indicates a point of inflection.

necessary at the very early beginning of selective evaluation and may be reflected by the universal genetic code and its biosynthetic machinery as well as by the universal chiralities of the macromolecular components of present-day biological cells. What is even more important is that – unless one defines this already fairly sophisticated phase as the "origin of life" – these universalities do not require one single primary event. In Nature there might have been very many early attempts from which one was singled out "for all time" in the non-linear "nucleation" phase.

Deterministic chaos is a domain of non-linear dynamics. The usually quite complex mechanisms of biological reactions ought therefore to provide many examples of deterministic chaotic behaviour. However, in general this is not true, perhaps because of the need for the involvement of robust mechanisms. For evolutionary processes this may be the reason for a preference for quasi-linear mechanisms. Let me conclude this section by discussing two typical examples of reproduction mechanisms.

A simple and straightforward case is the replication of a single-stranded RNA virus, as discussed earlier. I start with this example in order to compare it later with the more complex case of semi-conservative reproduction of double-stranded DNA in pro- and eucaryotes. We have seen in detail that even single-strand replication involves a multitude of single steps, and that this may lead to quite complicated expressions for the rate constants that describe the overall process. The replication of single-stranded RNA is a fairly error-prone process, which limits the length of strands to a threshold value for stable reproduction.

If the probability of the precise copying of a single position "i" is called q_i then the probability Q_ν of obtaining a precise copy of a strand including ν positions is given by the product of all positional probabilities q_i; for the whole strand this may be expressed by the geometric mean $\bar{q}$ to the power of ν, i.e. $Q_\nu \approx \bar{q}^\nu$. The complete truth is a bit more complicated still. The primary copy of a plus strand in this process is its complementary minus strand, like a "negative" in photography. It also has to be copied before a true copy, i.e. a "positive", of the information-carrying plus strand is obtained. So the kinetics for the total system of plus and minus strands involves two coupled differential equations, yielding two eigenvalues. Being the roots of a quadratic equation, one of the eigenvalues is positive and the other is negative. The positive one describes the growth (or selective behaviour) of the "plus and minus strand" couple, while the negative one causes both plus and minus strands to approach a fixed ratio under steady-state conditions. Again, in this "simple" case the common values of both time constants are the geometric means of the individual values of the plus and minus strands.

This is a lucid and meaningful result, showing that for the replication of single-stranded RNA – despite the fact that the information is usually confined to one of the strands – both plus and minus strands are equally important. If one of the fidelities (and thereby its associated rate) becomes zero, then the product of the two values, which enters the geometric mean, becomes zero too. (This is not the case for an arithmetic mean, which contains the sum of the two values.) More detail can be found in the 1971 paper.[20]

If I were to conclude that Nature has adapted itself to this mathematical logic, then I would have mixed up cause and effect. It is science that has adapted to natural reality by using the logic of mathematics in order to understand the logic of evolution. The upper part of the data in Table 5.5.1 refers to single-stranded RNA viruses. We are working under conditions where the geometric mean of the single-digit fidelity $\bar{q}$ is close to one, so that the probability of obtaining a correct copy of the whole strand $Q = \bar{q}^{\nu}$ becomes equal to $e^{-\nu(1-\bar{q})}$, where $\nu(1-\bar{q})$ could be called the expectation value for errors in the total strand. This relation has purely mathematical reasons based on the fact that $(1-\bar{q})$ is small compared with unity.

For an experimentalist it is always rewarding to see one's data backed up by a lucid and straightforward mathematical relation. However, if we look at the lower part of Table 5.5.1, which refers to data for DNA replication, we may at first be puzzled by the fact that the relation $(1-\bar{q}) \approx 1/\nu$ is apparently violated – indeed, by several orders of magnitude. Nevertheless, at least the average single-digit errors $(1-\bar{q})$ change in proportion to $1/\nu$. Moreover, $\nu(1-\bar{q})$ is homogeneously shifted – by two to three orders of magnitude – to smaller values, for all pro- and eucaryotes from *E. coli* to humans, even including DNA viruses. This again hints at a common cause, having perhaps some clear-cut explanation. For this purpose we must analyse more closely the mechanism of DNA replication, which accords only superficially with that of RNA replication in facilitating the exponential growth of individual sequences.

The doubling of a given message stored in a DNA sequence, which in the case of RNA required two separate steps (i.e. plus strand → minus strand and minus strand → plus strand) occurs for DNA in one overall process, called semi-conservative reproduction. In this process the original double strand disappears in favour of formation of two new double strands, one instructed by the plus strand of the parent helix and the other by its minus strand. So what's different with respect to RNA if both plus and minus strands are exactly complementary to one another? In the case of RNA (after two rounds) we get two plus strands, the old one and a new one, formed with the help of an intermediate minus-strand template that is later discarded. While in the case of DNA we obtain two new double strands at the expense of the former double strand, which *de facto* disappears. So far, there should not be any difference between RNA and DNA – if we could disregard the errors occurring in the process of copying. But we cannot. In order to see why, we have to look at Figures 5.7.4 and 5.7.5, where the mechanism of DNA replication is schematically illustrated (and explained in an extensive legend).

From the quite different mechanism of DNA replication represented in Figures 5.7.4 and 5.7.5 we can deduce three facts, all relevant for our discussion:

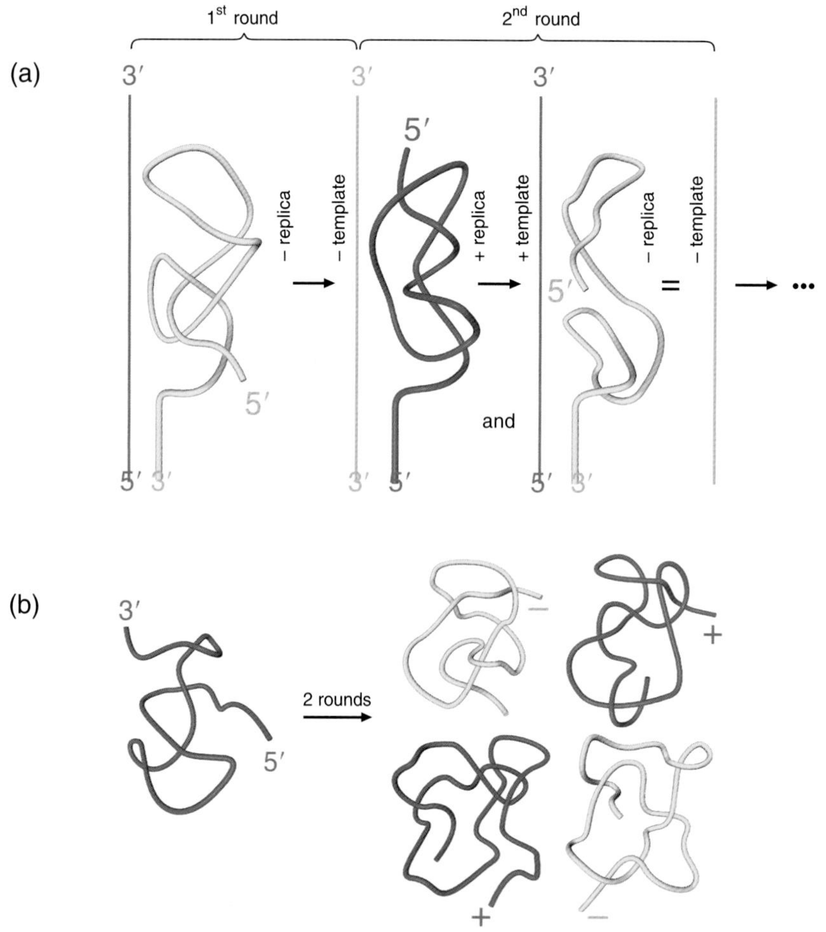

Figure 5.7.4

Mechanism of RNA replication

(a) Two rounds of single-strand RNA replication are required in order to double the plus strand (e.g. of a virus) containing the genetic information. The process consists of straightforward chain elongation along the template strand in the 3' to 5' direction, building up the replica from its 5' to its 3' end. During the full round of replication the template remains essentially intact, while mispairing of bases may occasionally introduce errors in the replica strand. (The fidelity of RNA replicase tolerates about one error in 10^4 nucleotide monomers.) Under natural conditions this type of copying error appears much more frequently than mutations due to radiation damage or chemical interference. Hence it is mainly the reading error occurring in the replica which limits the length (according to the error-threshold relation), but which is also responsible for evolutionary changes.

(b) Two rounds of replication lead to four separate copies, two positive and two negative.

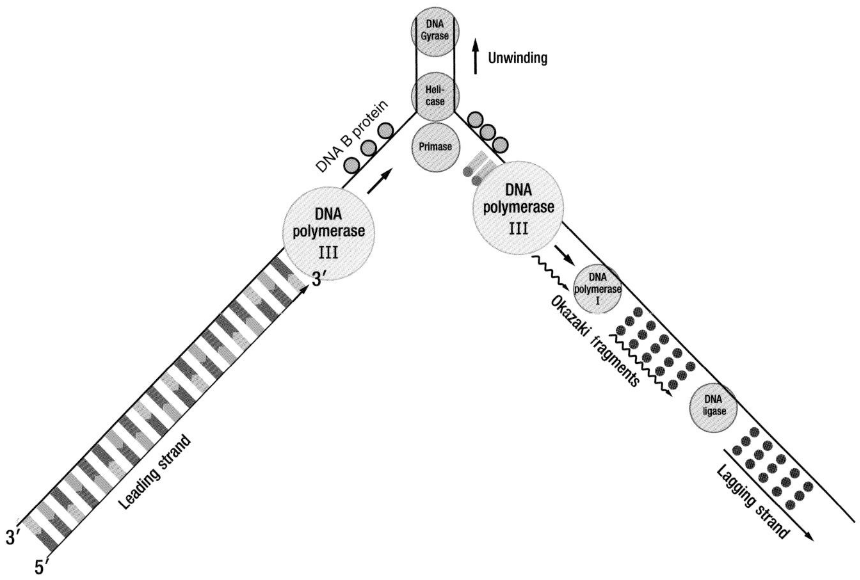

Figure 5.7.5
Mechanism of semiconservative DNA replication

One round of semiconservative DNA replication suffices in order to double the information contained in template DNA. There was an evolutionary demand for a faithful copying of long DNA sequences as, for instance, the total genome of a single prokaryotic cell, comprising as many as millions of nucleotide monomers in a single chain. Concomitantly, the fidelity of replication had to increase by many orders of magnitude in order to observe the restrictions implied by the error threshold relation. Since the polymerising enzyme always reads a template strand from its 3' to its 5' end, the mechanism of replication necessarily greatly differs for the two strands. Since the "lagging" strand has to form from smaller pieces, the so-called Okazaki fragments, which have to be ligated to the full-length replica, more steps are involved, making the process more prone to errors and requiring additional error correction. At the same time, this allows the leading strand to reproduce with much higher fidelity, as is reflected clearly in the error threshold of the combined process of duplication of DNA double strands (see Table 5.5.1). We therefore have a situation similar to that in single-strand RNA replication. Although the original double-strand template disappears, to be replaced by two replica double strands, one of them (the one containing the "leading" strand) resembles almost precisely the original template, while the other (the "lagging" strand) adapts to the error threshold relation of the overall process.

(1) The synthesis of leading and lagging strands involves different error rates. The leading strand builds up continuously in a relatively uniform manner which, according to the mechanistic needs of the enzyme, has to proceed always in the $5' \rightarrow 3'$ direction of the newly synthesised strand. It has to do the same in the synthesis and ligation of the Okazaki fragments in the lagging strand (named after their discoverers, the Japanese biochemists Tsuneko and Reiji Okazaki). The latter process, where the synthesis of all Okazaki fragments

must be initiated by RNA primers that have to be removed before ligation, involves a lot more errors than the straightforward synthesis of the leading strand.

(2) Both leading and lagging strands undergo proof-reading, which is effected in several steps by an enzyme machinery. The DNA-polymerase complex includes an automatic proof-reading mechanism for every nucleotide included at the 3' end. It doesn't proceed unless the newly included nucleotide is correctly paired with its template partner. This is a property that had to be "learned" during evolution and which helps enormously in reducing the number of errors. If the 3' nucleotide does not fit, it is cleaved off and replaced by the correct partner, otherwise chain elongation does not proceed. Note that this need for error correction may explain why DNA-strand synthesis always proceeds in the direction 5' to 3'. However, even this does not yet suffice to prevent violation of the error threshold relation and must therefore be supplemented by further proof-reading by an enzyme that recognises and removes mispaired nucleotides from the (specially marked) newly synthesised strand.

(3) Since the lagging strand is the template for the leading strand in a subsequent round of replication, the fidelity of its synthesis must fulfil all requirements for not violating the error threshold $\nu(1 - \bar{q}) \leq 1$. Since the leading strand is subjected to the same error-correcting procedures as the lagging strand, it will be synthesised with a fidelity $\bar{q}$ very close to 1. This suggests that the double strand resulting from leading-strand synthesis can be assumed, to a good approximation, to be identical with the former double strand, which in semi-conservative DNA replication is supposed to disappear. The overall process of semi-conservative DNA replication can thus be assumed to be formally identical with RNA replication, in which the former template is conserved while only the newly synthesised strand contains errors that have to fulfil the simple and lucid threshold relation.

The data in Table 5.5.1 suggest such an explanation, supposing that the values $\nu(1 - \bar{q}) \ll 1$ refer to leading-strand replication while the error threshold $\nu(1-\bar{q}) \approx 1$ is fulfilled by the synthesis of the lagging strand. Since many of the data refer to eucaryotic genomes we first thought of possible influences caused by recombinative processes and by the fact that introns might require less precise copying than exons, involving regulatory functions tolerating a larger number of neutral mutants than exons, for instance by mechanisms as proposed by Klaus Scherrer[104] in recent papers. There are also recent data from a statistical analysis of the language-like characteristics of exons and introns which point in that direction. Much of this detail is still awaiting further clarification. However, the error-threshold problem discussed above does not seem to be influenced by such detail, for two reasons.

First, any error-correction mechanism, effective in DNA replication, will act on both the leading and lagging strands in a similar fashion. Secondly, if the copying of exons and introns involves differences in fidelity the error-threshold relation applies to the error rates observed in exons. The data on DNA replication in Table 5.5.1 do not indicate any fundamental differences between pro- and eucaryotes, while the differences with respect to RNA replication seem to result from the different mechanisms of RNA and DNA replication.

The question in the title of this section is concerned with the "edge of chaos", and I don't think we care too much of "where" this edge occurs, but rather what it means for the evolution of life. We have seen that two kinds of chaos can be distinguished: "deterministic" chaos, as it has been called, and truly "random" chaos, as we may call it here, just for the purpose of distinction. Deterministic chaos may occur anywhere in the solutions of equations of non-linear dynamics, where the "edges" may be defined by the appearance of bifurcations as shown in the example discussed in Section 4.8. They appear in mechanisms which in simplex representation become "deterministically" confined to certain regions but otherwise "destabilise" a system in an unpredictable way. This may be the reason why these mechanisms are rarely found in life processes. They are entirely absent from linear systems, the most pronounced example being the vicinity of chemical equilibrium, representing as it does an "asymptotically stable" state. It is characterised by zero affinities and zero extents of reaction. Any deviation from equilibrium expresses itself in linearly growing values of both affinity and extent of reaction, causing the system to respond with increasing resistance to such a perturbation, resulting in the linear response of "relaxation".

Such self-regulating power is missing in systems far from equilibrium. Talking about the "edge of chaos" in such systems makes sense to me only in one particular way, namely when the "edge" is not just a limit, but that being close to it rather offers an advantage or even represents some *sine qua non*, a condition indispensable for the process in question. This is the case for the error threshold in evolutionary optimisation. Crossing the error threshold means disintegration into *The Jungles of Randomness* (the title of a book by Ivars Peterson[105]). However, not being close to it would mean a standstill of evolution. This is why in this section I went into more details of the errors involved in RNA and DNA replication. Referring to an "edge of chaos" in this connection makes sense. But here it is random chaos, close proximity to which does not endanger the robustness of the mechanism of natural selection under the conditions of a stable quasi-species.

5.8. Why Care What Other People Think?

Again a title from Richard Feynman[106]! But unlike *Surely You're Joking* this title derives from a rather sad story. It was meant in a very personal way, written by his fatally ill young wife Arlene, reminding him that some personal secrets are so important that it doesn't matter "what other people think" if they know of them. The story shows the most brilliant physicist of the post-war generation – as the *New York Times* wrote – being "a man, always earnest and sometimes troubled, who thought more deeply than most others about the power and limitations of science".

Here I am asking the question before the neutral and emotionless background of science. Science is at once the most feudal and the most democratic system. Ideas are purely autocratic, but no idea endures unless it can be reproduced and is accepted by the world-wide collective of scientists. Nevertheless, a lot of emotion is involved too. Emotion is a basic ingredient of our ability to "understand". The mathematicians Jacob Schwartz and Mark Kac told us: "Shocking as it may be to a conservative logician, the day will come when currently vague concepts such as motivation and purpose will be made formal and accepted as constituents of a revamped logic, where they will at last be allotted the equal status they deserve, side by side with axioms and theorems." Patients whose brains were damaged by stroke or physical accidents, impairing their emotional behaviour but leaving intact their skills of logical reasoning (proven by high scores of IQ) have been found to act highly irrationally, often detrimentally to their own well-being.

It's mainly emotion "that cares what other people think" – in both directions. It is (at the very least) highly exasperating to see one's own ideas exploited without receiving any credit. Conversely, if one has an idea, one is not too keen to find out whether somebody else has already had the same idea or a similar one. In this section, let me tell some stories that will serve as examples of the subject matter in previous sections.

How do we proceed if we have to explain some new finding? In order to explain a given situation one can proceed in two fundamentally different ways: start at the simplest possible level and approach reality by building up complexity, or start from the most complex possibility and reduce complexity until a clear picture emerges. My piano teachers Rudolf Hindemith (Paul's brother) and his wife Maria Landes-Hindemith told me that before a performance one should practice everything in several artificially complicated variations. Having been played this way first, the real piece becomes a lot easier. There is good reason to assume that life had to start out from complexity, the complexity of chaos, or tohu wa-bohu. In the beginning of life there was just chemistry. That's it! We cannot call this chemistry part of biology. In order to become biology this chemistry had to invent something that transcended it, as information does.

Prebiotic chemistry was by no means compelled to stay as chaotic as it originally may have been. Reaction–diffusion coupling, surface catalysis and "all that" may have created forms and structures that departed quite far from total randomness. I especially like an approach by Leo Buss and Walter Fontana,[107] who applied λ-calculus to an arbitrarily complex system of chemical reactants in the absence of replication, showing that these systems organise themselves in particular ways that may have been relevant in prebiotic chemistry. λ-calculus is an algorithm for universal computation that was developed in the 1930s by Alonzo Church,[108] paralleling the work of Alan Turing and Emil Post (cf. Section 3.6). An experimental contribution to this problem came from Sidney Fox,[109] who used solid-state chemistry to synthesise – in the absence of any externally applied instruction – protein-like substances from mixtures of natural amino acids, calling the resulting products "proteinoids". Their structures were not completely random, probably because of interactions among nearest neighbours, etc. They also showed a multiplicity of relatively weak catalytic activities. These results are certainly of importance with regard to the role of prebiotic chemistry. Fox interpreted these results as a strong indication of the validity of his "proteins first" hypothesis, which states that the proteins, with their nearly unlimited functional capacity, preceded the nucleic acids, which in all present-day living systems provide the information according to which the protein molecules can form. If correct, it would have been a molecular answer to the old question "Which came first, the chicken or the egg?".

If "proteins first" is taken as meaning proteins in the form of their early precursors, such as proteinoids, and "first" as a purely temporal assignment, then I have no difficulty with this thesis. What I do not see is how those early proteinoids could have had more in common with present-day forms than their chemical nature, e.g. than being polypeptides. Their non-randomness is based on purely structural causes, such as preferred nearest-neighbour interactions or preferred folding, or on requirements for chemical synthesis. These constraints, imposing a particular spatial structure on the polypeptide chain, could at best casually – and then not in an optimum way – coincide with the requirements of functional performance. The complexity involved in adapting structural constraints to optimal functional performance is so immense that I do not see any other way than evolutionary optimisation by replication, mutation and the resulting selection – all properties that early proteins were as devoid of as most present-day ones still are. This does not exclude the possibility of an important role of protein-like structures as environmental factors in early prebiotic chemistry.

I am asking the question of what is necessary in order to engender the inheritable information of life. Whether or not this should be called the "origin of life" is a matter of definition; I prefer a state that opens the possibility of unlimited evolutionary adaptation. We find this at the molecular level whenever it became possible to generate and store information representing biological function. I cannot see how this could have materialised without an involvement of nucleic acids.

Now I read in two books by Stuart Kauffman,[110] which appeared in close succession, that he thinks he has found a quantifiable solution to the question of how proteins could have become self-reproducible.[111] In fact the arguments in his books are based on a paper of his with the title "Autocatalytic Sets of Proteins"[112] which uses elements of Boolean algebra that he had introduced in two earlier papers and applied to "randomly constructed genetic nets". Let me quote from one of his books:

"Since Darwin's theory of evolution, Mendel's discovery of the 'atoms' of heredity, and Weismann's theory of the germ plasm, biologists have argued that evolution requires a genome. False, I claim." And on the same page, he supplements his harsh (and brave) statement: "a collection of molecules has the property that the last step in the formation of each molecule is catalysed by some molecule in the system". He calls it "self-organisation at the edge of chaos" or "crystallisation of connected metabolism as a percolation problem".

What Kauffman calls "connected metabolism" is a general chemical network far from equilibrium in which all reactions that go spontaneously are "metabolic", which essentially means "driven by differences in free energy". My system was related to catalytic cycles of enzymes, which by closure become autocatalytic, just in the sense of his "crystallisation of connected metabolism". Moreover, we both meant "early" non-adapted systems at the threshold of life, and we thought, at first independently, about this problem some 40 years ago, both papers having been published in 1971. However, we came to entirely different conclusions. Stuart's treatment was a general application of the theory of Boolean networks without referring to any particular molecular model, which can be found only in his later papers and in the two books mentioned. I think that his general reference to Boolean networks is a brilliant idea. I could see fruitful general applications of his light-bulb model with its binary (yes or no) decisions to neural networks because they are much closer to those ideas. I could also see special applications in cell differentiation. However, I am afraid that the way he uses Boolean nets in his books on molecular self-organisation simply doesn't work.

In fact, this was my conclusion 1971, i.e. at a time when I didn't yet know Stuart Kauffman and when he had not yet specified quantitatively his ideas about protein nets. My treatment dealt with the kinetics of such particular catalytic networks of proteins (cf. Figure 5.8.1, as opposed to Kauffman's network in Figure 5.8.2), using a normal-mode treatment which showed that these networks, on cyclic closure, do have a positive eigenvalue and hence may possess the potential for Darwinian selection (cf. Section 5.6). However, they are not able to evolve and thereby to adapt to optimal performance. Do they need to? Yes, they do. And why aren't they able to do so? The answer in short is: proteins ain't light bulbs! This is not just another of the many well-known light-bulb jokes. It means that light bulbs are manufactured, while proteins have to evolve. Or, better, light bulbs are made to be "on" or "off", while proteins are substances of combinatorial complexity. Most of them will always be "off", and those which are "on" are most probably at first quite dim. They need a self-organising, mutagenic mechanism in order to become bright. That mechanism is not only missing in Kauffman's theory, it is automatically excluded. In order to see this, let's go step by step through his assumptions.

Consider a set of polypeptides of maximum length M containing all possible linear combinations of monomers. We have already seen that such a set with 20 different monomers (amino acids), even for moderate length M, will be of a gigantic

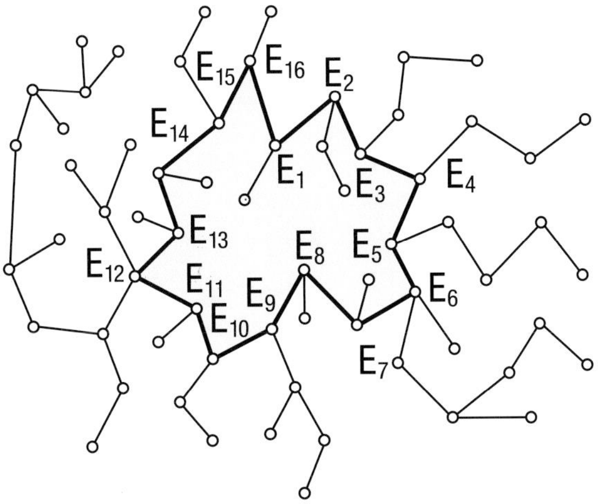

Figure 5.8.1

A self-replicating catalytic cycle of enzymes, as proposed by the author (1971)[20]

Owing to the cyclic closure of a chain of reactions, in which each step consists of the enzyme-catalysed formation of a protein molecule which in the subsequent step represents the enzymic catalyst, the system is self-reproductive for all the members of such a cycle. Such a system may come about in a sufficiently large ensemble of polypeptide chains, including those sequences that can fold into catalytically active structures, containing with finite probability all enzymes required for such a process. An indefinitely large set should be able to achieve this property. While such a system is self-reproductive and hence subject to natural selection, large cycles of this sort become entirely devoid of any evolutionary power. For this purpose let E_1' be a mutant offering some selective advantage in producing E_2. Then the cyclic chain from E_1 to E_n would end up with E_1 rather than with E_1'. In order to use its selective advantage in favour of the whole cycle it would have to produce a whole chain $E_1' \rightarrow E_2' \ldots \rightarrow E_n'$ that favours E_1' over E_1. The 1971 paper contains a mathematical treatment of the kinetics involved, showing that the enormous functional capacity of proteins cannot compensate for their missing power of "reading" or "writing", necessary for sequential reproduction.

size (e.g. 10^{130} for M = 100). The combinatorial problem involved could be as well demonstrated with only two classes of monomers. Length M = 100 then would allow about 10^{30} states. The important point in the theory is the assumption that any reaction, for instance the formation of any of the members in the set, can be catalysed. This assumption seems to me to be justified on the basis of what we know about natural enzymes and catalytic antibodies. If for a given set these functions are to be fulfilled by the members of the set, we must assume the set to be correspondingly large. In such a set there must be cycles of catalysed reactions which then are of an autocatalytic nature. These, if all prerequisites for their existence are fulfilled, must grow up and eventually dominate the network. This is demonstrated in more detail in Figures 5.8.3(a) and (b).

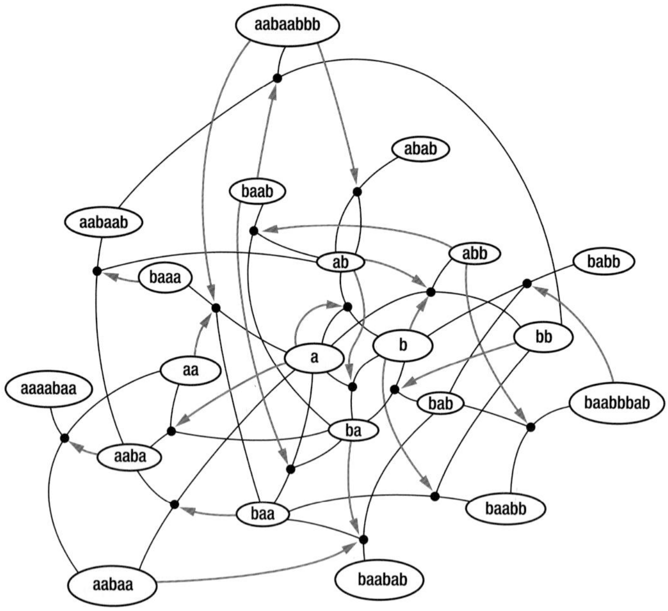

Figure 5.8.2

Stuart Kauffman's "autocatalytic set" model[111]

This model has much in common with the catalytic cycle shown in Figure 5.8.1. In order to be "autocatalytic" it has to involve cyclic closure, which takes care of its need to have reproductive properties.

While it is acceptable to simplify the model to binary classes of monomers, it is not acceptable to assume also a binary value function, one which rates catalytic efficiency by "on" or "off". If "on" means optimum efficiency, then such cycles would (almost) never form (the "almost" based only on my precaution: "never say never"). As I see it, the catalysts, formed at random, would be more likely to have the efficiency of what Sidney Fox found in his proteinoids. This would require Kauffman's cycles to evolve towards optimum efficiency. How can they do so, when they are devoid of replication? In a cycle of catalysts A to Z, where A catalyses the formation of B, B of C and so on until Z feeds back to A, any mutation from A to A′ must trigger a series of mutations down to Z′ which ensures the preferential reproduction of A′. An "on–off" model may disregard all cycles in which "negative" catalytic effects occur. By that, I mean proteolytic enzymes, i.e. ones that break peptide bonds. According to the model such "proteases" should occur as frequently as peptide-bond-making enzymes, as indeed is the case in real biological systems. All this means that the system, despite being selective, is devoid of evolutionary power. The model is correct as far as the combinatorial problem is concerned: If the system is sufficiently large,

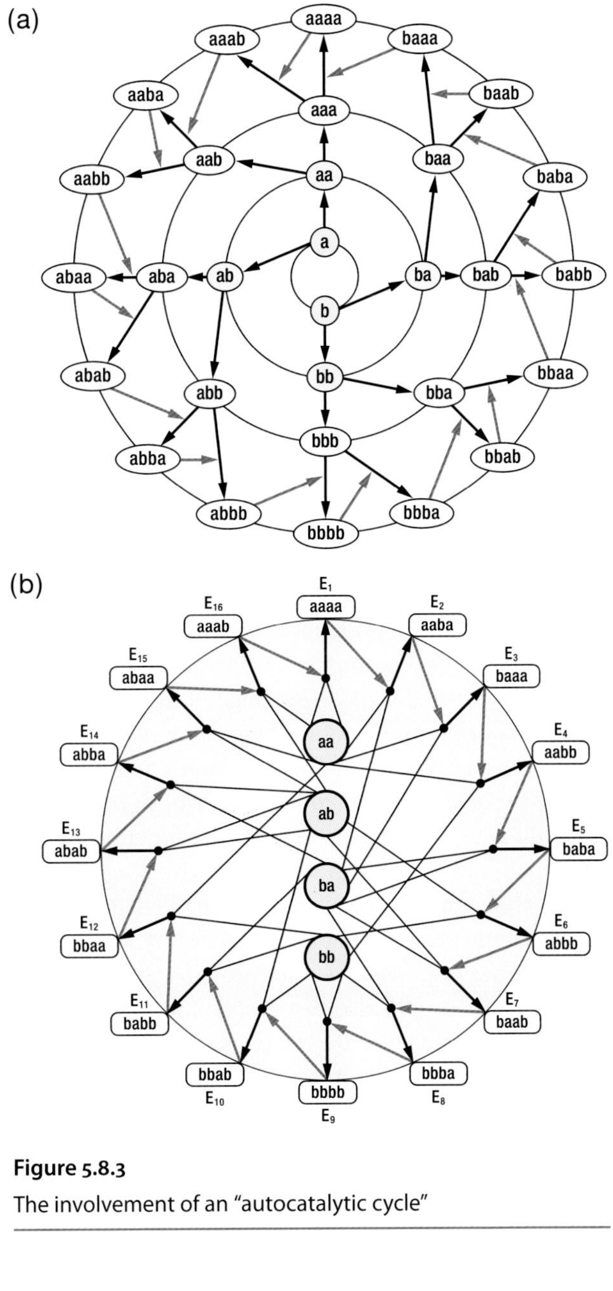

Figure 5.8.3

The involvement of an "autocatalytic cycle"

it must contain all sorts of catalytic activity. Proteins with a length of more than 100 amino-acid residues might do this in present-day systems.

The next question is: how large do they have to be, in order to develop their amazing catalytic properties? In addition, realistic systems have to obey the laws of molecular kinetics. Both questions are accessible to experimental testing, and we have looked at both. Taking a metabolic enzyme with a length of about 100, we were able to reduce its length to about 30 before most of its catalytic faculties were drastically impaired or disappeared. Furthermore, our kinetic studies with enzymes allowed us to "dissect" their mechanisms and to find out in detail which requirements would have to be fulfilled. The main point is not combinatorial, but rather that systems would have to obey the laws of molecular kinetics.

Let us assume that the "planetary laboratory" of earth allowed evolution to "experiment" in "test-tubes" that contain as much as 10^{20} (100 million million million) litres of an aqueous medium. (That's about the volume of all the oceans on earth combined.) Suppose a protein concentration of 1 g per litre – that is 10^{-4} molar in proteins that have a molecular weight of 10,000, which corresponds to about 100 amino-acid residues. Such a "thick" soup has never existed during the 10^{17} s of existence of our planet, but it serves as a safe upper limit for my estimate. Such a scenario means the presence of, at most, about 10^{40} protein molecules at any instant. One may say: how wonderful – among such a large number of molecules I shall find anything I need. It is true that 10^{40} is a huge number, and I make no attempt to express it in products of thousands of millions. However, I cannot imagine what this number really means, i.e. to say whether it is large enough, too small or in huge excess. The total number of alternative sequences of length 100 is 10^{130}. Who can tell whether the few useful ones that contain the catalytic functions required are present in the vanishingly small fraction (one in 10^{90}) of the total set?

This is a question that could be resolved by suitable experiments. With nucleic acids one can demonstrate such principal properties by looking at the important steps. Nothing like that has ever been tried for proteins. Such experiments, allowing an estimate of overall rates, should also be possible for protein catalysts. Moreover, the RNA world has left remnants that are still of use in present-day systems, while in the case of proteins no witnesses exist in today's world.

While the above consideration calls into question the relevance of the model under realistic conditions, Shneior Lifson,[113] one of the fathers of chemical dynamics, pointed out some intrinsic and fundamental mistakes. They are concerned with the assumption of a constant probability P for any of the polymers to catalyse any of the reactions. A polymer of length M can be formed in many ways, e.g. by condensation of shorter segments. Given a constant probability P for any polymeric chain to be formed catalytically, Kauffman calculates that for large enough M there is a chance that the formation of a given polymer of any desired length can be catalysed at least in one way. Lifson challenges this result, pointing out that the underlying assumption that P is independent of the length M cannot be right. When M increases, the number of sequences in the set increases exponentially, and so does the number of bonds that have to be formed catalytically. Lifson finds that the complexity increasing with length M modifies the probability P in such a way that the chance of any polymer being formed catalytically remains far too small.

Let me make some quantitative estimates in order to substantiate Lifson's complaint. Let's assume that at least 15 positions within a longer polymeric chain have to be occupied correctly, while the rest may be assigned arbitrarily – certainly an extreme, and therefore quite dubious, assumption. In that case, one in 20^{15} ($\approx 10^{20}$) polypeptide sequences would be a compound showing enzymic activity. A second-order rate constant of binding to a substrate, even in today's perfectly adapted enzyme systems, cannot have values larger than those for diffusion-controlled reactions, and for enzymes usually has a value in the order of magnitude of 10^7 (mol/l). The substrate for the next step must be another polymer, namely the one catalysing the subsequent step in the autocatalytic cycle, i.e. again only one among some 10^{20} sequences present. The reaction time for each step in product formation would then reach the orders of magnitude beyond the age of our planet or even our universe. Sydney Brenner, who is always quick with his comments, said, after I told him about the model, "the earth might be a little too small to allow it".

Let me say in conclusion, the idea behind Stuart Kauffman's model has a certain appeal: to overcome complexity by assuming a sufficiently large population, which allows checking of all possible solutions. After all, antibodies have managed to cope with all sorts of binding problems. However, molecular self-organisation results from mechanisms based on physical and chemical interactions that should not be in conflict with molecular kinetics.

The general lesson of our work in enzyme kinetics is that those proteins which have undergone evolutionary optimisation, as all natural enzymes found today have done, show most remarkable performances as catalysts, often accelerating a reaction by factors of a million to a billion. Equally impressive is the binding specificity of proteins, expressed particularly in antibodies. It is the consequence of the minutely detailed evolutionary design of active sites, which in turn requires the flexibility offered by sufficiently long polypeptide chains involving 20 different functionally active groups. For this to work, the length of the chain has to exceed by far the linear dimensions of the active site, which – through precise folding – comprises several groups (possibly distant in the chain) in close contact. Examples have been given in Section 4.2. We do not know any single enzymes that catalyse specifically (and precisely) the formation of longer sequences of polypeptides or – in the absence of templates – of polynucleotides. Our finding of synthesis *de novo* of reproducible chains of polynucleotides by RNA polymerases (which was initially received with great suspicion among biochemists) is fully explicable within the limits of the "dogma of molecular biology", which states that the vectorial character of protein synthesis (RNA $\rightarrow$ protein) is not reversible. The puzzling result of obtaining reproducibly defined uniform sequences is not a curiosity, but rather a prime example of the Darwinian behaviour of self-reproductive molecules.

The point I am stressing is complexity, which in the world of the living is always good for a surprise. Only few years ago – with the outbreak of "mad cow disease" in England – a new infectious particle was discovered and related to a number of previously known neurodegenerative diseases (such as Creutzfeldt–Jakob disease in man or scrapie in sheep). The infectious particles found were called "prions" by Stanley Prusiner, who for his pioneering work in characterising the nature of this infectious agent was awarded the 1996 Nobel prize for physiology and medicine. (For the more detailed story see Prusiner.[114]) Prusiner not only named the infectious particles, he introduced the abbreviation PrP (I am not sure whether here Pr stands for protein, prion or Prusiner). He demonstrated convincingly that this particle which, after infecting an organism, seemed to be able to multiply like a virus or pathogenic microbe, was a protein molecule. A self-reproducing protein? Molecular biologists throughout the world could not believe that. However, Detlev Riesner[115] was able to show convincingly that the purified, yet infectious, substance contained *less than one* molecule of nucleic acid larger than 100 nucleotides per infectious unit. Moreover, Stanley Prusiner had already shown that the infectious prion protein (he called it PrP^{Sc}, Sc from the sheep disease scrapie, caused by prions) has a non-infectious pendant (called PrP^{c}, c for cellular) in every host that in its polypeptide sequence is identical with that of the pathogen, while showing traits that point to a quite different folding structure. So the formation of PrP^{sc} is not based on self-reproduction but rather on the catalysis of a conformational transformation by the infectious agent.

The identification of the host protein turned out to be vital in understanding the nature of the prion puzzle. Instrumental for establishing this was an observation made by Charles Weissmann, another pioneer in the field, and his co-workers.[116] They found that knock-out mice do not become ill after infection with the pathogenic agent. (Knock-out mice are not specially bred boxing champions but rather animals in which a gene – here the gene encoding the host protein PrP^{c} – has been made inactive (or "knocked out").) Weissmann's experiments showed unambiguously that the host protein is required for the spread of the disease in the infected individual. Prions (PrP^{Sc}) are like ideologies that convert only people who are susceptible to certain ideas.

Why do I tell this story? Because it is exciting? Yes, but as such it does not really belong here – except that it is yet another example of autocatalysis that is to be clearly distinguished from the inherent autocatalysis of replication. This autocatalysis doesn't originate information, it just spreads information that existed before. In this case it is something that puts existing molecules "into shape". For this it needs more than just a single contact: it needs three-dimensional molecules that co-operatively shape the "plastic" conformation of the protein molecule. This behaviour is well-known and common among proteins.

Can we conceive of proteins or protein-like structures that are self- or cross-reproductive? I consider the attempts that have been undertaken in this direction to be as instructive as corresponding studies with analogues of nucleic acids have been. A particularly interesting case of cross-reproduction was reported recently by Reza Ghadiri.[117] It's especially these very specific successful attempts that teach us about chemistry's resources. In particular, they tell us why Nature chose nucleic acids for the purpose of processing existing – and generating new – information.

Strong adherents of a "chicken (i.e. protein) first" view are found in particular among many of my colleagues in physics. Perhaps the Darwinian principle seems too cloudy to them, or perhaps some biologists have told them that it is a typical "biological" and not a "physical" principle, whatever that may mean. We have already heard that Schrödinger clearly expressed his discomfort with Darwin, finding Lamarck's ideas more elegant and intriguing, although he admitted with resignation that "unhappily Lamarckism is untenable" so that "we appear to be thrown back on the gloomy aspect of Darwinism". The use of the ending "-ism" indicates that he – still in 1956 – considers those biological "tenets" to be philosophy rather than physics.

Shneior Lifson[113] refers in his paper to another such case, again an eminent representative of theoretical physics, one of the architects of quantum electrodynamics: Freeman Dyson. Let me quote from Dyson's book *Origins of Life*:[28] "I have been trying to imagine a framework for the origin of life to be homeostasis rather than replication, diversity rather than uniformity, the flexibility of the genome rather than the tyranny of the gene, the error tolerance of the whole rather than the precision of the parts."

Lifson comments on these statements, stressing that what is required is "homeostasis *and* replication, diversity *and* uniformity, tolerance *and* precision". I would add: whatever it is that in any situation is most important, we should let Nature tell us how it is and not try to tell Nature how it has to be. This was also the maxim of another theoretical physicist, Max Delbrück, who thereby became a great biologist. Dyson's style is so precise and lucid that Lifson, as he remarks, could easily spot the point where he needs selection too. In the third of his ten assumptions Dyson states: "There is no Darwinian selection. Evolution of the population of molecules within a cell proceeds by random drift." However, he restricts random drift in assumption seven: "The active monomers are in active sites where they contribute to the ability of a polymer to act as an enzyme. To act as an enzyme means to catalyse the mutation of other polymers in a selective manner so that the correct species of monomer is chosen preferentially to move into an active site." Lifson comments: "An enzyme cannot choose the 'correct' species and the active site because they are undefined before the effect of the mutant on the cell has been tested." I just refer to my earlier saying: in biology the logic follows from interactions and not *vice versa*.

Nowadays many physicists are concerned with the theoretical foundations of biology. A particularly proliferative think-tank has been created in New Mexico near to (and inspired by) the Los Alamos National Laboratories: the Santa Fe Institute. Their programme is the study of complex systems, and for this they have recruited a number of first-class physicists, among them Murray Gell-Mann and Phil Anderson. I know that at least some of them agree with me that there is not just one theory of complexity, and I hope that they pay sufficient attention to the fact that biology is not as dry a science as is (at least good part of) physics. If in this section I have criticised some approaches of physicists to biological themes, I do not mean that I regard their attempts as futile. I am deeply convinced that physics provides a basis for biology, which does not mean that biology can be reduced to physics. As important as Stuart Kauffman's Boolean networks, Freeman Dyson's non-linear mechanisms or Per Bak's sand-piles may be in the context of biology, before applying them to concrete biological problems we have to take notice of the experimentally testable facts. Having now mentioned Per Bak's sand-piles let me conclude this section by discussing briefly this further example of a physical approach.

One doesn't have to be born in Denmark (with 1.33 m of coastline per inhabitant) in order to play as a child at the beach with sand-piles. So we all have experience of what happens when one tries to pile up dry sand on a given area. The pile reaches only a certain height before an avalanche is triggered and the sand slides down. Sometimes the avalanche is small and leads only to a small correction of the height, but occasionally it is large, leading to more drastic reduction. The late Danish theoretical physicist Per Bak, who spent some time at Brookhaven as well as making some visits to the Santa Fe Institute, studied the phenomenon in detail. His conclusion was what he called "self-organised criticality". The model is indeed instructive, or even explanatory, of several phenomena in biology, economy and cultural life, where Bak was certainly right.[118]

Looking at the fluctuations in the sizes of avalanches, one realises an analogy to the "critical phenomena" that I discussed in Section 5.5. I suppose that Per Bak, being a theoretical physicist, analysed his sand-piles by applying renormalisation in a way similar to that done for other critical phenomena associated with phase transitions. If so, he must have realised the similarity with the "critical" error threshold that we observe in molecular selection. As was shown by Kimura's stochastic models, large-scale fluctuations, similar to those observed by opalescence near the critical point of the gas–liquid phase transition, are to be expected near the error threshold of mutation only among (almost) neutral replicators. Otherwise the fluctuations become small, as is the case for first-order phase transitions far below the critical point (cf. the experimental examples given above).

Sand-piles, and their "criticality at the edge of chaos", are certainly instructive models. Applying them to biological phenomena means a careful analysis of the biochemical processes involved. I cannot see that they tell us beforehand whether they are more realistic for Stuart Kauffman's Boolean networks than for the experimentally testable error-threshold phenomena in molecular evolution. The latter involve as much freedom from "the tyranny of the gene", "diversity" and "homeostasis" as

Freeman Dyson wished for his system and, in addition, guarantee survival of the animate state of matter.

The physics required for the understanding of "animate matter" has been most aptly described in our Winter Seminars by Shneior Lifson, whose balanced remarks I have often compared with those of the wise man in Gotthold Ephraim Lessing's play *Nathan der Weise*:

"Evolution of galaxies and solar systems means the long-time course of a random drift of macroscopic systems toward ever-increasing entropy, or ever-decreasing free energy. Darwinian evolution of autocatalytic systems is no exception in this respect. However, its random drift is directed by natural selection towards all those specific properties that characterise animate matter, namely fitness, adaptability, organisation and the like. Such properties are neither related directly to thermodynamics nor measurable by its methods. As a consequence, animate organisation and inanimate order are totally different and unrelated concepts. Unfortunately, their superficial similarity is a source of much confusion in the context of life and its origin."

5.9. An Ultimate Machine?

A machine is a man-made device. Today, we also use the word in a figurative sense. The replicative or metabolic machinery of a cell is not man-made. Cause and effect are here inverted. Webster distinguishes "simple" from "complex" machines, and lists among simple or elementary machines the lever, the wheel, the axle, the pulley, the screw, the wedge and the inclined plane, admitting that there may be more than these items. Which was the first machine that our forefathers invented? I must admit I don't know, and I don't know whether anybody knows, and neither do we know who were the inventors of those elementary machines. In early times, the lever was certainly among the most important inventions, allowing large masses to be moved. It was the Greek mathematician and physicist Archimedes who demonstrated mathematically how the lever works, although he did not discover it, as by then it had long been in use: "The ratio of the effort applied to the load raised is equal to the inverse ratio of the distances of the efforts and load from the pivot, or fulcrum, of the lever" (Random House / Webster 1997). So he – perhaps – constructed the first machine that reveals a principle, the subject of this section. His insight led him to the proud statement: Give me but one firm place to stand, and I will move the earth. To cut the story short, many complex machines have been invented since. One of the most influential was the printing machine of Johann Gensfleisch Gutenberg, which using movable type revolutionised book-printing and thereby triggered an avalanche in the distribution of texts – a major source of human culture.

It is not my intention to give an exhaustive account of the development of machines. What I want to get to is a special kind of machine, idealised machines that represent certain principles. Gutenberg's machine did not embody any new mechanical or metallurgical principles; it was constructed with real material and its purpose was to ease the important procedure of serial duplication.

A prototype of the machines I have in mind is Carnot's machine, which I mentioned in the Section 1.1 of this book. Carnot's machine is an ideal reversible heat engine based on the cyclic procedure, named after the French engineer Nicolas Léonard-Sadi Carnot. The Carnot cycle consists of four consecutive reversible steps, shown in Figure 5.9.1(a).

The most economical representation of the cycle is in a temperature–entropy diagram (see Figure 5.9.1(b)). This yields a straightforward rectangular shape, since temperature does not change in the two isothermal processes, while entropy remains constant in the two adiabatic changes. Assuming ideal conditions (e.g. neglecting friction and heat conduction), we find that the ratio W/Q for the work (W) obtained on the expenditure of heat (Q) is (independent of the nature of the gas): $1 - T_2/T_1$. Since T_2 is smaller than T_1 the Carnot efficiency is always between 0 and 1, approaching 1 for $T_2 \ll T_1$.

Sadi Carnot, although an engineer, was not to earn the fruits of his deep insights. His work was long ignored, until Emile Clapeyron, also an engineer with deep scientific insight, referred to it in 1834. Carnot had died in the Parisian cholera epidemic in 1832.

I am now making a big leap from energy to information, which also has been characterised by "ideal" machines named after Alan Mathison Turing and John von Neumann.

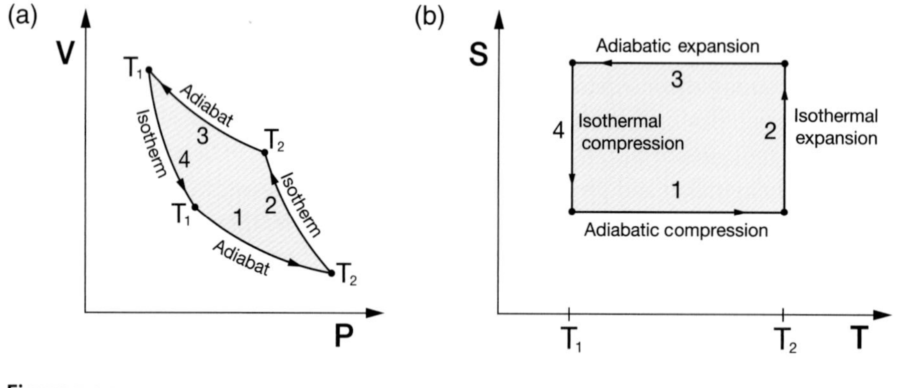

Figure 5.9.1
The Carnot cycle

(a) The usual representation of the Carnot cycle as a pressure–volume diagram. (b) The Carnot cycle as a temperature–entropy diagram, representing the two isothermal processes (i.e. T = constant) and the two adiabatic processes (i.e. S = constant) as two vertical and two horizontal lines forming a rectangle. The Carnot cycle is used to show that the circular integrals $\oint pdV$ or $\oint TdS$ have finite values, establishing the second law of thermodynamics.

Turing's device is purely hypothetical. It is supposed to compute whatever is computable according to the rules of mathematics. In Section 3.6 I described how Turing envisaged the idealised version of such an abstract machine. He was able to show its relation to Gödel's incompleteness theorem: his machine will never stop if it tries to solve a problem that in Gödel's sense is an undecidable one. The concept of this machine dates back to the year 1936. In 1950 Turing went as far as to suggest constructing a machine that is able to learn and to carry out intelligent operations. He proposed a procedure which now is called the Turing test,[119] in which a human subject in one room is separated from the machine in another room while an interviewer in a third room asks questions that are to be answered by both the machine and the subject. If the interrogator cannot distinguish between the two with the help of his questions, the machine is assumed to possess human-like intelligence. Roger Penrose[120] challenges the view that it is possible to simulate human intelligence with the help of a computer. I agree that neither a Turing machine nor any present-day computer, including the highly efficient connection machines with their enormous degree of parallel processing, resembles a human brain. Another question is that of how many functions of a brain can be simulated by a computer, and I do not agree at all with the opinion that anything in the hardware would have to be based on a new "fundamental" physical principle, such as quantum gravity (which still has not yet been realised in material form). Even a new physical principle, once it becomes known, can be simulated. For me, the more interesting question is that of how hardware has to be constructed so that in principle it resembles our brain. But this is not a question I shall discuss in this section.

Alan Turing was one of the leaders of a team in World War II that cracked the German Enigma cipher code and thus rendered outstanding services to his country. After the war he was prosecuted for naturally rooted behaviour that is nowadays entirely accepted. He committed suicide at the age of 42.

Modern computers are based on Turing's concept, yet despite their enormous capacity they do not live up to Turing's intentions. When he coined his idea in 1936 there was nothing like it around, although as early as in 1822 Charles Babbage[121] had devised a programmable general-purpose machine on a mechanical basis using input programs that were entered on punched cards. Even more amazingly, Ada Byron Lovelace,[122] a daughter of Lord Byron, in her short life had worked out a complete method of programming that would have allowed her to tackle problems, for instance in astronomy, that up to her time were not only unsolved but thought to be *ultra vires*.

Returning to the 20th century, it was the Hungarian-born mathematician John von Neumann who more than anybody else was responsible for the realisation of the idea and hence is to be called the father of the modern computer, which some day may find its perfection in the form of the "quantum computer". At the Institute for Advanced Study in Princeton he supervised the construction of the first computer that used a flexible program, named MANIAC-I. von Neumann was one of the most brilliant minds of 20th century, and he made important contributions to the mathematical foundation of quantum mechanics. In 1944, together with Oskar Morgenstern,[123] he laid the foundations of modern game theory with its many applications in economics. His fellow countryman Eugene Wigner called his mind "a true miracle" and Hans Bethe wondered whether his brain "did not indicate a species superior to that of man". The reason that I am focusing here on von Neumann is that it was he who proposed another abstract machine that indeed is closely related to the origin of information, treated in this chapter. The von Neumann machine is an abstract, self-reproducing automaton, a computer that reproduces not only the machine, but also – what is more important – the program for building it. The theory of self-reproducing automata was coined at about the same time as – but in ignorance of – the discovery of the molecular form of such a machine: DNA. It is a tragedy that von Neumann died at a young age. The book *Theory of Self-Reproducing Automata* was reconstructed posthumously by Arthur Burks[124] in 1965 from von Neumann's unfinished manuscript.

The original model of the von Neumann machine (1950) was conceived in (unnecessarily) realistic terms. The machine was supposed to move around in a warehouse in order to gather all the parts needed for its reconstruction. As mentioned before, one of the most important steps involved is the reproduction of its own blueprint in order to ensure its survival through successive generations. That is exactly what living systems do. Reproduction of the program is the decisive step. The translation of a program into reality must suppose the existence of some physical environment that is to be exploited for the execution of the program. Much of the refinement of von Neumann's idea is due to his colleague and friend Stanislaw Ulam,[125] who worked on the idea of cellular spaces. (I have most pleasant recollections of meeting this very witty, jovial and warm colleague on several occasions.)

We may think of a cellular space as being a board subdivided into squares, each of which may assume a finite number of states, depending on the states of neighbouring squares and on certain correspondingly defined transition rules. von Neumann was able to prove that a configuration of 200,000 cells, each assuming one of 29 possible states, would meet all the requirements of a self-reproducing automaton and at the same time be able to carry out all sorts of mathematical operations, like a Turing machine. The English mathematician John Horton Conway,[126] at the Institute for Advanced Study (at Princeton), devised games – he called them "life games" – that can be played on suitable boards. The "beasts" coming to "life" in Conway's (albeit deterministic) game can be pretty complicated. von Neumann's idea of a self-reproducible automaton certainly belongs to a class I am calling Darwinian machines. Its closest realisation is represented by what is called "DNA computing", which utilises the possibility of replicating programmed DNA molecules in a massively parallel fashion as a purely computational device.[127]

George B. Dyson[128] wrote a book with the title *Darwin among the Machines* – a title he borrowed from Samuel Butler (1863), a vehement critic of Charles Darwin.

Dyson's book might as well have been called "von Neumann among the machines". It is von Neumann who has by far the largest number of entries (some 108) in Dyson's subject index and who is the true hero of this story. George B. Dyson is the son of Freeman Dyson. I know that sons don't like to be introduced as the sons of great men, but if he himself has a son he could name him Freeman and then, when asked, respond: "No, I am his father" (as indeed the son of Max Born did). George Dyson appears to have inherited from his father a dislike of Charles Darwin. Like Samuel Butler, he rather favours Charles Darwin's grandfather Erasmus, and remarks "Charles had inherited – not invented – the evolutionary faith." That's what Darwin, in fact, said of himself in a footnote of the third edition of his *Origin of Species*. Perhaps he was jealous of Lamarck and therefore credited the idea to a general evolutionary trend coming up in Europe, mentioning his grandfather Erasmus in England, Etienne Geoffroy Saint-Hilaire in France and Johann Wolfgang von Goethe in Germany (see the footnote in Section 5.2). Anyway, it is the clear formulation of the biological principle of natural selection that we credit to Charles Darwin. Would we know it if we just had read Erasmus' *Zoonomia*? We would perhaps know it today, but then we would have to credit it to somebody else. So I can't agree at all with Butler and George Dyson, except that the name Erasmus has some appeal to educated people.

In the Vienna of the late 19th century there was a famous critic: Eduard Hanslick, a friend of Johannes Brahms and of the renowned surgeon Theodor Billroth. He disliked the music of Richard Wagner, Anton Bruckner and Peter Tchaikovsky. When after the première of Wagner's *Rienzi* he was asked for his opinion, he answered: "There was much I really liked, except for one thing that was just awful – the music." If I were to say that George Dyson's book is "charmingly written", one might too easily put my remarks above into the same category as Hanslick's critique of Wagner. That would be unjust! Dyson's book is not only charmingly written, it is a jewel of a rare and special kind, a true treasure trove of historical detail. And not only that: it says something about the subject of this section, the "ultimate machine". Nowhere else have I found it so clearly said: the golem of an "ultimate machine" can't be a single entity like a human being, it must be an "entirety", such as human culture. And whether such an "entirety" can have a "supermind" is still an open question.

Let me try to say this more clearly in order to be understood not only by my fellow scientists. I quote from Dyson's book: "Von Neumann was as reticent as Turing was outspoken on the question of whether machines could think." Edmund C. Berkeley, in his otherwise factual and informative 1949 survey *Giant Brains*, captured the mood of the time with his declaration that "a machine can handle information; it can calculate, conclude and choose; it can perform reasonable operations with information. A machine therefore can think." von Neumann never subscribed to this mistake. He saw digital computers as mathematical tools. The fact that they were members of a more general class of automata that included nervous systems and brain-like structures did not imply that they could think. Our modern computers have not yet achieved this goal. von Neumann, having built one computer, became less interested in the question of whether such machines could learn to think but rather asked the question of whether such machines could learn to "live", i.e. to reproduce. To say this in von Neumann's own words: "Complication' on its lower levels is probably degenerative, that is, that every automaton that can produce other automata will only be able to produce less complicated ones. There is, however, a certain minimum level where this degenerative characteristic ceases to be universal. At this point automata which can reproduce themselves, or even construct higher entities, become possible".

There will be no true artificial intelligence without artificial life, i.e. without "Darwin among the machines". Once we are able to define life, we may use the definition as the basis of a computer program. Having the program, we can start some speedy evolution. How far we get depends on how intelligently we constrain the environment in which the evolutionary process goes on. To call this "artificial life" is as much (or as little) justified as one may call certain parasitic programs "computer viruses". "Artificial life" is not really a new idea, since the machines are among us! When Peter Schuster came to my laboratory in the late 1960s, in order to study fast reactions, he offered his help in simulating my kinetic models of molecular evolution in the computer. In the acknowledgements of my 1971 paper I thanked him saying: "We had much fun in betting about the outcome of evolutionary competitions – the computer always won."

The term "artificial life" has now become popular for a "dry" form of life mainly due to the work of Thomas Ray[129] and Christopher Langton.[130] We decided to experiment with artificial life on a wet basis, i.e. studying evolution in the test-tube (as, for instance, described in Section 4.10). These studies have now evolved into a new type of "evolutionary biotechnology", handled by machines we constructed in Göttingen. Darwin certainly is "among those machines".

In Figures 5.9.2 to 5.9.7 I show some examples of the technological realisation of the principle of evolution by "critical selection", following the experiment described in Section 4.10. Figure 5.9.2 shows the principle behind this technology, and in Figures 5.9.3 to 5.9.7 the various steps of perfecting this technology are described. The physical variable of biology is complexity. Hence what the machines must be able to do is "fast screening". The device described in Section 4.10 uses just one sample, but does processing on a broad, parallel, molecular basis. The next machine (Figure 5.9.4 and 5.9.5) was already able to handle 1000 samples in one run. The machines becamequite complex because they must at the same time also handle all the chemical and biological procedures (Figure 5.9.6). Finally, with modern equipment built in a company founded for such purposes, we are now able to process about a million samples a day (Figure 5.9.7(a) and (b)).

An "ultimate machine" will certainly have to involve human intelligence. This is not the place to discuss human intelligence in any detail, or whatever we know about

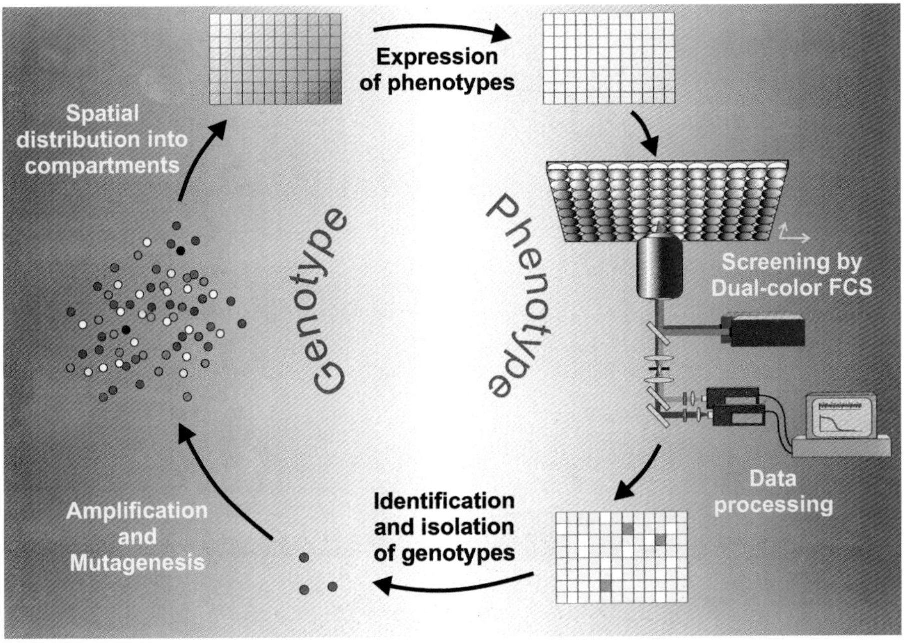

Figure 5.9.2

The principle of the evolution machine (schematic representation)

The cycle starts in the upper corner by physical identification (in this case using dual-colour fluorescence correlation spectroscopy, FCS) of single copies of phenotypes that are isolated, the genotypes of which were subjected in subsequent steps to amplification and mutagenesis. The distribution obtained is isolated in single compartments, where individual phenotypes of mutants are expressed and subjected to fitness tests. A selected set then starts a new cycle, which will be iteratively repeated. The specific physical test was used since it applies to single molecules. The complexity of the total scheme is the reason for the automation of the whole procedure.

it, which will have to be the subject of parts of the second volume. Our present interest lies in the implications of an "ultimate machine" for life and thought. If I said that the evolution of life is equivalent to a continuous and unceasing generation and accumulation of information, then thought is the continuation of this process at an advanced level. In both cases, general principles and requirements are of a similar nature, so that Gerald Edelman speaks of "neural [similar to "evolutionary"] Darwinism". What is common to them is the search for optimum solutions of problems, subject to certain constraints. A general flavour of the "problems" is their complexity, characterised by a huge number of alternative solutions. Isn't an "ultimate machine" one that can cope with any particular task – as do the various present-day machines – but finds itself an optimum way to do it, as if it were constructed for its particular job by human intelligence? In any case, the many problems involved will cause an

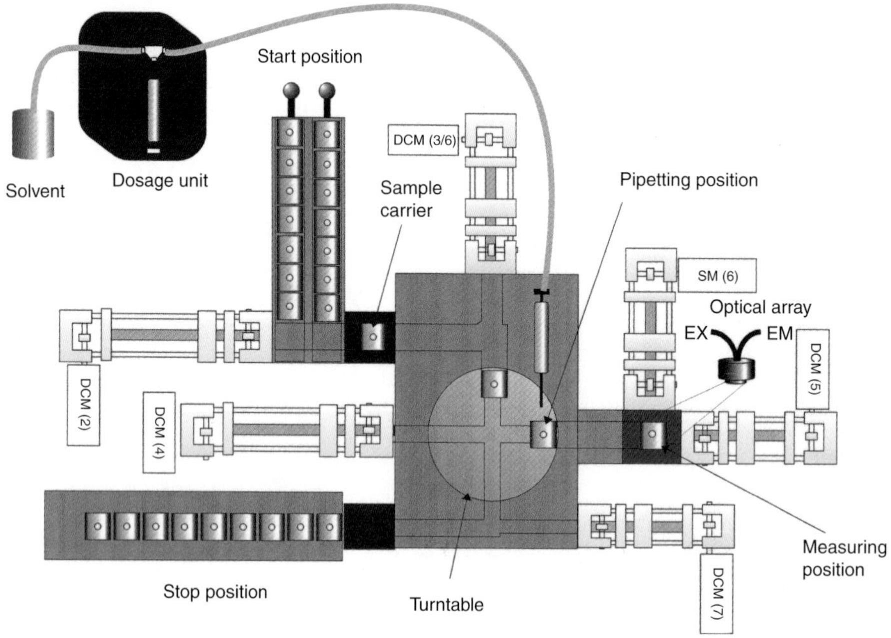

Figure 5.9.3

Schematic realisation of an automated serial transfer machine

Spiegelman's principle of "serial transfer" (see Figure 4.9.1) has been materialised in this construction of a machine. Automation was commensurate since replication of single copies, in order to build up whole populations, took times of only seconds to minutes. The transfer occurs between little plates that can be easily shifted around. The plates contain cavities that can take up a few microlitres of sample solution. An automatic pipette in the centre carries out the serial dilution by transferring aliquots of amplified material to sample holders filled with fresh reaction mixtures. The transfer is controlled by an optical device, which when RNA or DNA is the amplifiable material records either fluorescence or absorption in the near UV and triggers plate motion and pipetting by electrical signals that control a fast start and stop of amplification in the optically controlled sample holders.

"ultimate machine" to be an ensemble of many machines, each being adapted to an optimum performance of its specific problem.

In their book *The Computational Brain* Patricia Churchland and Terrence Sejnowski[131] discuss prototypes of mechanisms that, applied to neural networks, could resemble the control functions that allow optimum problem solution. They favour a device they call the "Boltzmann machine". It utilises the strategy of "simulated annealing", which was discussed in Section 4.3 in connection with finding solutions to problems of exponential complexity within polynomial times. Kirkpatrick *et al.*[132] (see Section 4.3) invented a strategy that is based on the well-known metallurgical procedure of annealing. A piece of metal solidifying from a

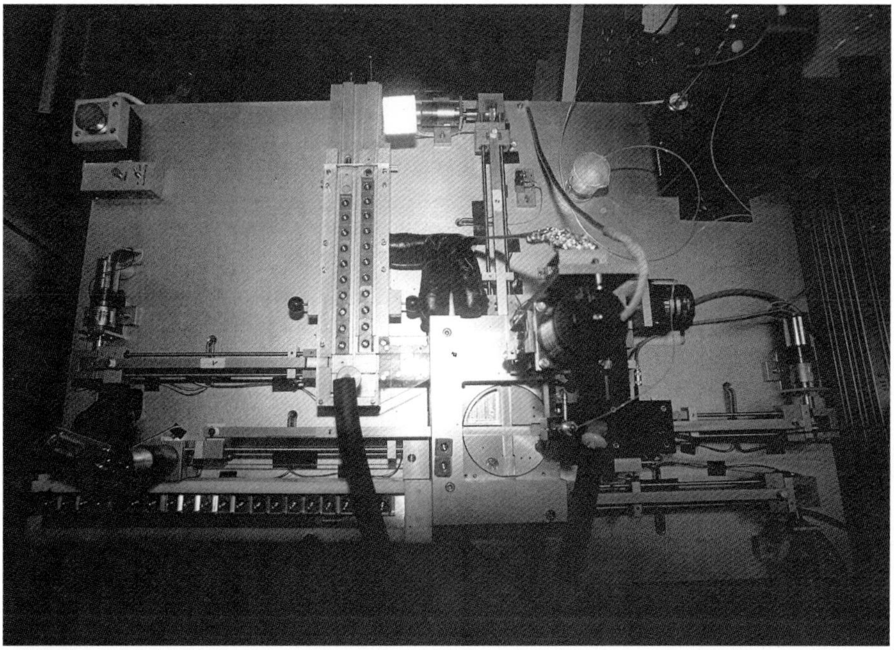

Figure 5.9.4
Technical realisation of the serial-transfer machine

The technical realisation follows almost exactly the schematic representation given in Figure 5.9.3. However, the functioning of the device involves a few experimental tricks. A fast start and stop of amplification was effected by fast temperature changes within the sample solution. For this purpose, the thin platelets holding the sample solution were made from pure aluminium, having a temperature response time of less than 1 s. The amplifying enzyme did not work in the aluminium cavities, which therefore had to be lined by coating them with a thin layer of gold. Otherwise the machine worked like a train set. It contained – depending on the number of evolutionary steps – about 100 platelets (the rolling stock) that were shunted from one track to another at the turntable near the pipette.

melt usually does not form an ideal crystal lattice structure, but rather includes many atoms at interstitial positions as well as empty lattice sites. Annealing means control of lattice formation in crystallisation by slow cooling and reheating around the melting-point, thereby allowing a gradual transition from the liquid to the solid phase where each atom can find its position of minimal potential energy. The authors therefore emphasise that this method would enable one to find a "global" minimum. That is correct in principle, but rarely in practice. At any finite temperature the system would not establish the singular microstate of minimal energy, but rather establish an equilibrium among all states according to their Boltzmann distribution. Hence, Sejnowski therefore called the machine based on this method a "Boltzmann machine". The difficulty for real material systems in working at very low temperatures

(a)

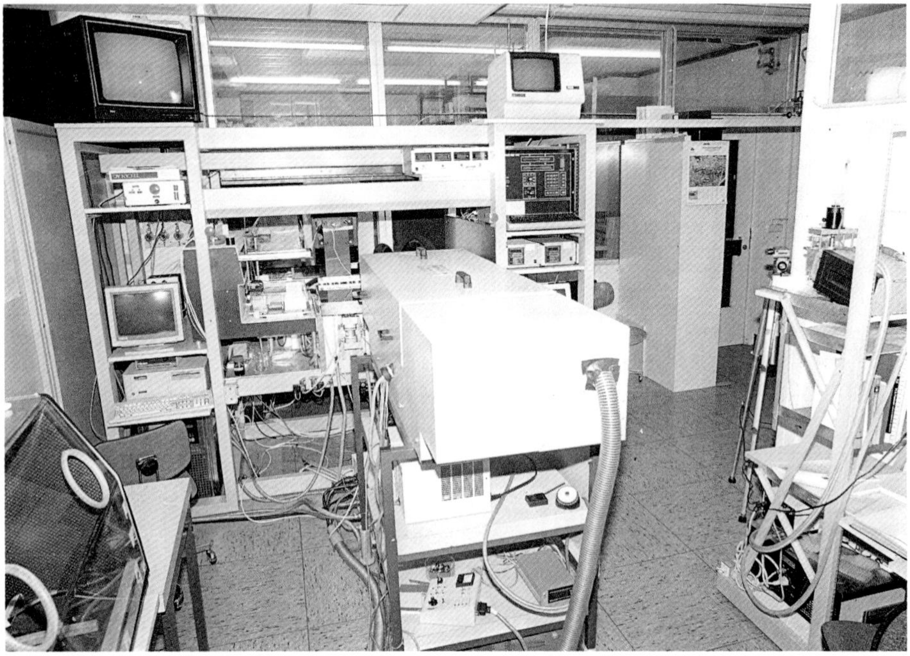

(b)

Figure 5.9.5 (Cont.)

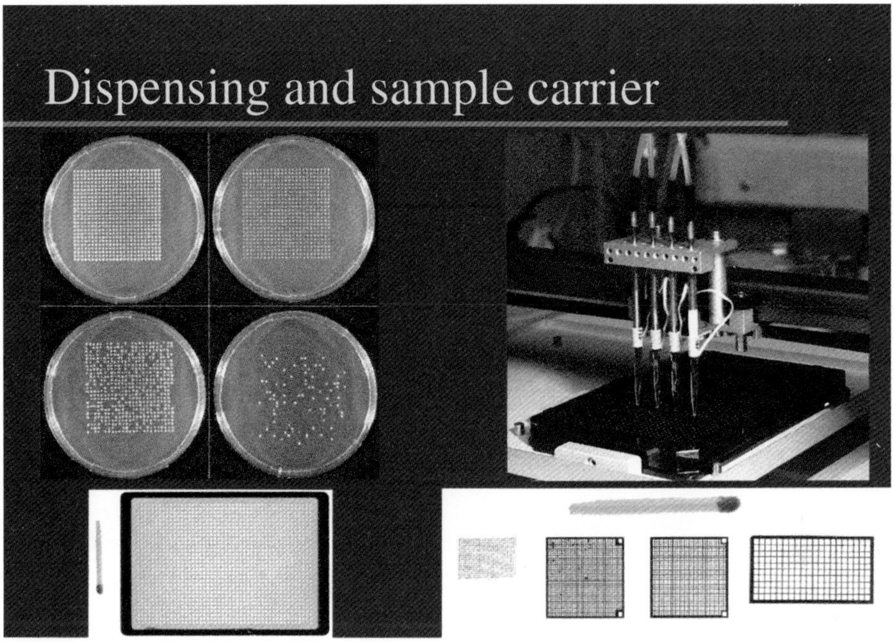

Figure 5.9.6

Multiple pipetting of reactants and products in molecular evolution

This is typical for automatically controlled machines and requires the development of specifically adapted robotic techniques.

Figure 5.9.5

Generalisation of the serial transfer principle

Evolution is a property of populations, which have to be analysed for their complexity (i.e. the spectrum of alternative sequences produced by mutagenic reproduction). The efficiency of the single-channel machine of Figure 5.9.4 was greatly increased by the "thousand-channel" machine shown here. The (nearly) thousand "channels" are represented by a corresponding number of microcavities carved into a plate of pure aluminium (of large heat conductivity), which is thermoregulated very rapidly by a heating and cooling solution from a thermostat, a device applicable to polymerisation chain reaction, a technique used in molecular biology for amplifying single chains of RNA or DNA in short times. In fact we use it for that purpose, here in a massively parallel way. Otherwise the "thousand-channel" machine works in a way similar to the one described in Figure 5.9.4, involving multichannel fibreglass optics for single-molecule recording and a sample plate moving back and forth between different levels of temperature for amplifying mutant products under different environmental constraints.

(a) General view. The white box is a laser-powered fluorescence detector.

(b) View showing the multichannel sample plate (right, foreground) and the optical detection unit (black, in the middle).

Figure 5.9.7

Industrial realisation of "evolutionary biotechnology". Two views of an advanced machine

These photographs show a device first constructed and brought to the level of technical performance by Andre Koltermann and Ulrich Kettling. Both worked originally for their PhD theses in the Göttingen Max Planck Institute and subsequently founded a company called "Direvo" (for "directed evolution"). Their arrangement included robotics for all the chemical (synthetic and analytical) and the typically biological operations involved.
Courtesy of Andre Koltermann and Ulrich Kettling.

results from the small mobility of atoms under those conditions, which makes the relaxation process extremely slow. This difficulty does not exist if the annealing principle is used in designing an abstract program that simulates the natural process. However, the difficulty that remains is that discussed in Section 4.3. If the problem involves exponential complexity, the capacity of the program would soon be exhausted, as it doesn't reduce exponential to polynomial complexity. Any application to information-generating networks requires the identification of individual "microstates". Hence a global minimum may be found only in systems of relatively restricted size. "Simulated annealing" – unlike its thermodynamic analogue – is not really an equilibrium method, since in any application it involves selective interference. Despite its limitations it shows impressive results, as was demonstrated with the results quoted in Section 4.3. Sejnowski's Boltzmann machine applies to neural networks. These are so-called "associative networks", which "associate" a (possibly very large) number of inputs with a related number of outputs via some non-linear internal coupling scheme. The simplest coupling devices may be resembled by the McCulloch–Pitts[133] neuron or, if learning facilities are incorporated, by the Rosenblatt perceptron.[134] More realistic networks make use of changeable elements such as Hebbian synapses (cf. Carpenter and Grossberg[135]). I shall not go into more detail here, since the subject of neurobiology will be revisited in the second volume, while I am dealing here with principles only.

Churchland and Sejnowski contrast the Boltzmann machine with another feedback network, which in its original form restricts the optimisation process to local minima rather than to a global minimum. A sufficient richness in states characterised by local minima could serve to represent computational states of a network. John Hopfield[136] published a paper in 1982 in which – as its title says – he saw such a physical system as a unit "with emergent collective computational abilities". The physical analogue Hopfield uses is the spin-glass. A spin-glass differs from a magnetic metal in that only some of its constituents are atoms with large magnetic moments, randomly distributed throughout the crystal lattice of the non-magnetic material. The spins of the magnetic atoms are assumed to be either "up" or "down", resembling a two-state model of a neural network, the cells of which are either "on" (i.e. firing) or "off" (i.e. silent). The spin system at elevated temperatures represents a random mixture of both spin orientations, involving attractions and repulsions. On sudden lowering of the temperature, the system will seek the nearest local minimum of potential energy. This includes so-called "frustrated" states in which composites with different orientations face one another and hence don't allow the collective to behave uniformly like a ferromagnet. The treatment of Hopfield's model benefits from the enormous amount of theoretical work done on the physical system. Hopfield's decision problem is based on the fact that the spin-glass equations demonstrate the existence of a network of stable configurations, a necessary attribute of computational networks.

In Section 4.7 I outlined how selection at the molecular level due to mutational coupling behaves in a manner analogous to spin-glasses. John Ross has shown the mathematical equivalence of Hopfield's network with our quasi-species model. Peter Schuster, John McCaskill and I have repeatedly stressed these analogies on the basis of the equations for the two-dimensional Ising model involved.[87] The potential surface of the spin-glass has its parallel in the fitness landscape of the quasi-species.[137] In the case of the latter, one of the dimensions is time. In other words, apart from the consonant behaviour of the total sequence, characterised by its gross (context-dependent) dynamical parameter, there is the sequence of generations which owing to (erroneous) reproduction behaves "co-operatively in time". In this way the fixation to "local" optima is surmounted and a "global" optimum can be approached – with similar restrictions as in the case of simulated annealing. In fact, simulations (Katja Wang and Kay Nieselt-Struwe[138]) have shown that optimisations on the basis of the quasi-species model can be done most efficiently. The seminal point is to use a sufficiently broad spectrum of mutants with parallel processing. Nature, of course, can be much more efficient in this respect.[139]

I would call machines of this type "Darwinian machines" and Hopfield's model, given a dynamic interpretation, would certainly fall into this category. I think that an "ultimate machine" will be (at least partly) a "Darwinian machine", a machine built according to criteria that guided the coming-about of the human species. A serious general argument is: how can one execute evaluation other than by intrinsic selection? I think this is one of the main reasons why Gerald Edelman[140] speaks of "neural Darwinism", while the series of his experimental gadgets he calls Darwin 1, 2, 3 ...; I think he has now arrived at least at Darwin 10. If information in a neural network is defined by the firing pattern of neurons we could use the concept of information space, introducing neighbourhoods as defined by synaptic connections. And we could treat the dynamics in such a space, including reproductive feedback, which may cause phase transitions such as were found in sequence space and apply to molecular evolution.

Anyway, there are many alternatives, and a lot of sophistication is still required in order to construct an "ultimate machine", which – I am sure – will be of a Darwinian nature. So far I have talked only about some basic principles. The second volume will enlarge on these in more detail, but it will also show how far apart we still are from such an "ultimate" goal. If it can be reached at all it may take us to an "ultimate" limit of time.

Nevertheless, talking about an "ultimate machine" I do not want to be misunderstood. A universal machine makes as little sense as does a universal "theory of everything". Imagine that there were a machine that could resemble human intelligence. Something approaching or even surpassing it is behind the idea of an "ultimate machine". But at the same time we realise that most machines were constructed in order to surpass our capabilities, although only in a very special and restricted way. One of the oldest machines, the lever, allowed man to move masses like the gigantic stone blocks of the pyramids. Our present-day computers do not possess human intelligence and cannot do anything their program doesn't tell them to do, but they can compute at an awesome speed and thereby solve previously unsolvable mathematical problems. Many branches of science flourishing now were unthinkable only a few years ago. And, as George Dyson has convinced us, Darwin is also among those machines, which he calls "an entirety, such as human culture".

This gets me to the question: where will all this end up? Even nowadays people are sometimes afraid of machines, not necessarily for the same reason as the Luddites, who in early 19th century England destroyed manufacturing machines because they expected to be robbed of their jobs. No, there is a general uneasiness, which is hard to define, but which is expressed most clearly in the arts. Perhaps there is the fear that the "ultimate machine" may turn out to be a "useless machine" (cf. Figure 5.9.8).

Figure 5.9.8
Jean Tinguely's representation of a "Useless Machine"

This machine is thought to do everything and therefore to be of no use for anything, or – as the artist comments – an emblem of a totalitarian occidental mercantilism.
From the author's collection.

5.10. "It from Bit" or "Bit from It"?

Some 10 years ago, at the 30th anniversary of our Winter Seminar at Klosters in the Swiss Alps, John Archibald Wheeler presented a talk with the title "It from Bit". It is published as one of the – both poetically and prophetically fascinating – essays in his book *At Home in the Universe* in the series *Masters of Modern Physics*[141] from the American Institute of Physics. I took the opportunity to give my talk, dealing with the origin of "viable matter", under the title "Bit from It". Neither title is to be seen as contradicting the other; they are rather two independent melodies in polyphonic music that complement one another. In this section I shall make them a contrapuntal conclusion to the theme of this book.

The "bit", a contraction of the words "binary digit", certainly has to be seen as a human invention. Binary numbering is generally credited to Gottfried Wilhelm Leibniz, who in 1679 introduced it as the primary choice of positional numbering. For him the digit 0 meant "the void", while the digit 1 was assigned to "God". The principle of binary numbering is recorded on a medal (Figure 5.10.1), the coinage

Figure 5.10.1

Leibniz's medal representing the principle of binary numbering

Binary numbering was proposed to Duke Rudolf August of Wolfenbüttel as the basis for coinage around 1700.
Courtesy of the Gottfried Wilhelm Leibniz Bibliothek.

of which Leibniz proposed to his patron, the Duke Rudolf August of Wolfenbüttel. However, Leibniz was not aware that almost 100 years before him the English mathematician Thomas Harriot had devised – although regrettably not published – a general treatment of positional numbering.

A binary response, "yes" or "no", is easily materialisable in an electric circuit by switching it "on" or "off". This turned out to be instrumental for its use in our cyber age, both in communications networks and in electronic data-processing. The application of the binary response in telegraphy, in fact, predated this by more than 100 years. I have mentioned the electromagnetic telegraph, invented independently by Samuel Findlay Morse in the USA and by Carl Friedrich Gauss and Wilhelm Weber in Germany, in Chapter 3. In Figure 5.10.2 a copy of an original page from the diary of Gauss, dated 1833, is shown, which is kept at the University Library of Göttingen. It is interesting to note that Gauss had already anticipated the two major coding schemes which nowadays are in use: the "block code" and the "variable-length code" (see Section 3.4).

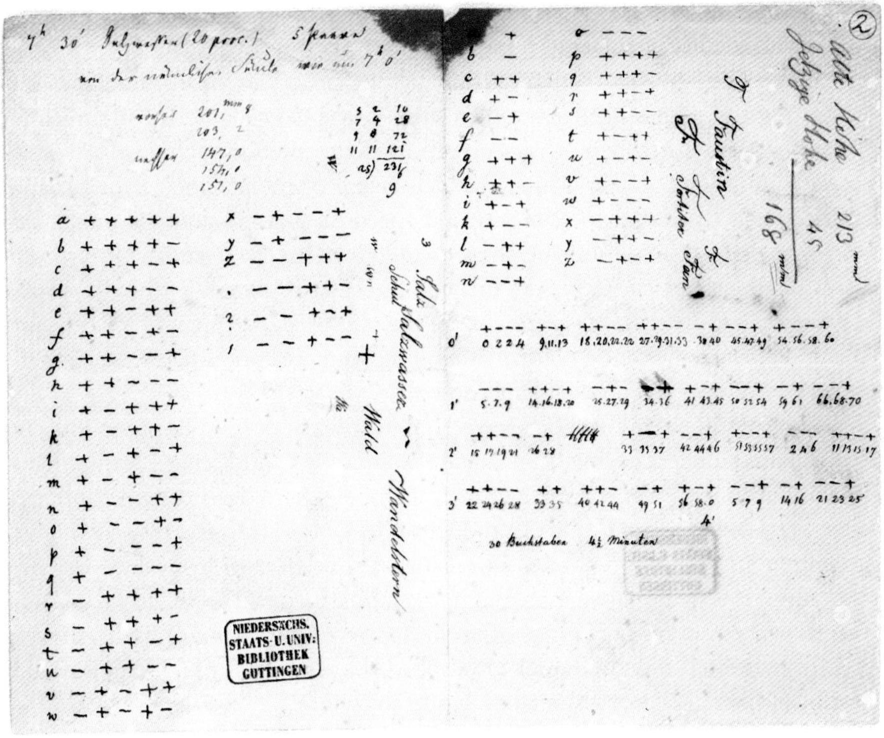

Figure 5.10.2

Copy of a page in the diary of Carl Friedrich Gauss

Courtesy of the University Library of Göttingen.

Above I have called the bit a human invention. This is not quite correct, if we have in mind the use of binary coding. This is a human re-invention of something that happened in Nature long ago and of which we did not know until quite recently. The genetic code (Figure 3.4.1) is a block code of 64 triplets of four different kinds of code symbols, and it is about 4000 million years old. Yes, it consists of four symbols, while 4 is $2 + 2$, 2×2 and 2^2, and this means that the binary logic had to be applied twice in order to produce a code that is able to organise itself (see Section 4.5). The binary machine code as well as the machines that handle this code today have been invented by intelligent beings. The genetic code had to come about by evolutionary self-organisation. Its logic is based on physical interactions and they come in twice. First, the four symbols belong to two different classes of chemicals (purines and pyrimidines), which show equally shaped "complementary" pair structures. Secondly, two pairs with discernible binding stabilities are required, in order to produce a code of unlimited combinatorial complexity, at the same time allowing both encoding and the build-up of a processing machinery. This is explained in more detail in Section 4.2. In the present context it suffices to note that it all happened at a quite early stage of chemical evolution, when our planet was still fairly young (see Section 5.1). Neither Gauss nor Morse could have known of it.

A block code obviously offers advantages in processes of evolutionary self-organisation and, as discussed in Chapter 3, this is true as well for telecommunication and data-processing involving error-correction and adaptation to channel capacities. According to Figure 5.10.2, Gauss was well aware of the advantages of variable-length codes with respect to economy and speed of transmission. He might have been afraid that a fast-running messenger would beat his electromagnetic telegraph for any lengthier message to be sent through the 1200-m-long line from the roof of the astronomical observatory to the roof of the physics department of his colleague Wilhelm Weber in the centre of town. This conjecture was presumably the basis of a number of (more or less credible) anecdotes.

Modern coding and information theory has its roots in the middle of the 20th century, its main pioneers having been Claude Shannon[142] and Richard Hamming.[143] This work deals with the "representation, transmission and transformation of information", as Hamming said in his introduction to a book with the title *Coding and Information Theory*. John Wheeler, in raising the question "It from Bit?", certainly had in mind this quantitative aspect of information theory. The "It", as he explains, is "every particle, every field force, even the space-time continuum itself", in other words, "everything" existing in our physical world. As such it must refer as well to all forms of "animated matter", including ourselves.

All life we encounter is instructed by genetic programmes which have a material basis. As will be shown in more detail in the second volume, their origin is tantamount to the origins of life. If we identify "It" with the quality of certain matter of being "viable", then this "It" indeed rests on "Bits". If, on the other hand, "It" – as in Wheeler's interpretation – refers to matter as such, then the "Bits of life" have their origin in "It", since the evolving code rests on material carriers. This prompted me to choose for my talk the title "Bit from It", and not just because of the alliteration in both titles. I wanted to demonstrate the problem associated with the word "origin", regardless of whether we speak of life, of the universe or of any "it" that is a product of our thoughts. "Origins" involve a feedback between "It" and "Bit". Unless they are linked to one another right from the outset, the one cannot originate without the other.

Wheeler, if I am correct, had in mind two aspects of his theme. The first runs under the title "participatory universe" and ends up in the loop: "Physics gives rise to observer participancy; observer participancy gives rise to information, and information gives rise to physics." No objection! After all, "matter is a state of mind". On the other hand, his second aspect is based on the principle that "mind is a state of matter", matter that was around before we started our observations. Wheeler quotes as an example the entropy of a black hole. For him it is "a symbol, also, in a broader sense, of the theme that every physical entity, every it, derives from bits". Let us look more closely at this example.

A black hole, the ultimate state of matter, crunched under the influence of its own gravitational forces, is characterised uniquely by just three properties: its total mass, its electric charge and its angular momentum. The simplest case of an uncharged and non-rotating black hole was considered right after the arrival of Einstein's theory of general relativity in 1915 by the German astronomer Karl Schwarzschild.[144] He showed that the radius of the black hole (R_{BH}) is directly proportional to its total mass (M_{BH}), the proportionality constant being $2G/c^2$ (G = gravitational constant, c = light velocity). The radius R_{BH} is defined as the distance between the centre and the so-called event horizon of the black hole, the latter demarcating the extension of a space from which, because of the gravitational pull, even light quanta cannot escape any more.

Outside the event horizon, the familiar laws of physics retain their validity, while we are pretty ignorant about whatever may go on inside the space demarcated by the event horizon. General relativity – as Stephen Hawking and Roger Penrose have shown[145] – predicts a space–time singularity at which the mass density becomes infinitely large and the laws of physics must break down. Such a singularity, of course, is hard to digest by a quantum model. Apart from the singularity, the tidal forces active inside the black hole will have disrupted any individual particle, causing a true inferno reminiscent of Dante's *Divina Commedia*: "Lasciate ogne speranza, voi ch'intrate" ("All hope abandon, ye who enter here").

What we do know, nonetheless, is that total mass, charge and angular momentum remain conserved. Let's look again at the Schwarzschild model. Its remarkable result, reported above, is the fact that the total mass (or energy) is reflected by the radius R_{BH} of the event horizon. Jacob Bekenstein[146] and Stephen Hawking (independently) realised that the surface area A_{BH} of the black hole at its event horizon ($\sim R_{BH}^2$) can be interpreted as its entropy S_{BH}, for which they derived the relation (named after them) $S_{BH} = kc^3 A_{BH}/4G\hbar$ (k is the Boltzmann constant, G is the gravitational constant and $\hbar$ is Planck's number, $h/2\pi$; c is the velocity of light; while the index BH of A may be interpreted as either "Bekenstein–Hawking" or "black hole"[147]).

Roger Penrose in his book[148] notes that the formula for S_{BH} is the first expression in which all fundamental constants (light velocity, gravitational constant, Planck's number and Boltzmann's constant) appear simultaneously. (As shown in Section 2.1, Boltzmann's constant can be set to 1, which yields an entropy in absolute units.) Now we are back to thermodynamics, which as a general phenomenological theory should also be applicable to such strange cases. Knowing the energy $E_{BH} \sim R_{BH}$ and the entropy $S_{BH} \sim R_{BH}^2$, Hawking derived a temperature T_{BH} from dS/dE = 1/T, yielding $T_{BH} \sim R_{BH}^{-1}$.

In Section 2.1 we learned that in conventional thermodynamics entropy is suitably written as a function of internal energy U and volume V: (S(UV) (see Table 2.1.3). Then, with $dS(U,V) = (\partial S/\partial U)_V dU + (\partial S/\partial V)_U dV$, the temperature T is given as $(\partial S/\partial U)_V = 1/T$. If we want to adapt this procedure to the black hole, with its strange relations between energy, entropy and size, we realise that any change of entropy with energy automatically involves a change in volume. The only conserved parameters of the black hole are its mass energy ($U = E_{BH}$), its charge Q and its angular momentum J. The two latter parameters in the Schwarzschild model are assumed to be 0. Hence any change is solely reflected in the Schwarzschild radius R, and we have for the definition of a temperature T: $(\partial S/\partial U)_{Q,J} \equiv 1/T$. Since $U \sim R$ and $S \sim R^2$ we obtain $1/T = 2S/U$ and $\int_0^R TdS = \int_0^R dU$. The interpretation is that the internal energy U is entirely of entropic nature, as we would formally expect for an ultimate state of matter, in accordance with the laws of thermodynamics. This has been expounded, in much greater detail, by Stephen Hawking in his paper "Black Holes and Thermodynamics".[149]

It might be more impressive to present some orders of magnitude, which is done in Table 5.10.1 for a black hole of a mass equivalent to 10^{80} protons, representing the order of magnitude of the visible mass of our universe. The true mass of the universe may be larger by orders of magnitude, owing to the presence of invisible matter. The numbers demonstrate vividly the oddity of this state of matter.

The fact that the entropy of a black hole is given essentially by its surface at the event horizon is the most surprising fact that emerges from Schwarzschild's calculations. Susskind and t'Hoff offer as an explanation a holographic model,[151] which is an exciting idea, but it demonstrates clearly the strangeness of the experimental facts.

Of course, in physics numbers make sense only if they are seen in a frame of reference. 10^{80} is a huge number; but we expect the universe to be huge. So we just take notice of it. But what about the enormously large entropy of the black hole? It is many orders of magnitude larger than any entropy we would estimate for our "real" universe – and here we meet with a true surprise. Apart from the fact that it is the surface area of the black hole that is correlated with its entropy, the Bekenstein–Hawking formula also uncovers a direct relation with the Planck scale.

Table 5.10.1 Black Hole Physics

Planck dimensions	Relations	Values
Mass m_p	$\sqrt{\hbar c/G}$	2×10^{-5}g
time t_p	$\sqrt{G\hbar/c^5}$	5.3×10^{-44}sec
length l_p	$\sqrt{G\hbar/c^3}$	1.6×10^{-33}cm
area $l_p^2 = a$	$G\hbar/c^3$	$2.61 \times 10^{-66}\text{cm}^2$
Black hole properties	Relations	Values for $M_{bh}\,10^{80}$ protons
Surface area A_{bh}	$8\pi\, M_{bh}^2\; G^2/c^4$	
Entropy S_{bh}	$\frac{A_{bh}}{4}(c^3/G\hbar)$	10^{123} e.U.
	$2\pi(M_{bh}/mp)^2 = A/4a$	bits
Hawking–Bekenstein number N	2.77a	

In Section 1.6 we made the acquaintance of Planck units, which are derived from the three fundamental constants c, G and $\hbar$. If we look at Bekenstein–Hawking entropy we realise that it is essentially the ratio of the surface area A_{BH} and the square of the Planck length ($l_p^2 = G\hbar/c^3$). In other words, apart from the factor 1/4, the entropy represents the number of Planck squares that cover the surface area of a black hole at its event horizon (the Boltzmann constant k is set to 1 in this estimate).

The statistical interpretation links entropy with the logarithm of the phase–space volume (see Section 2.5). John Wheeler refers to a paper of Kip Thorne and Wojciech Zureck, who represent the phase volume by 2^N possible, but indistinguishable, configurations at the Planck lattice, the logarithm of which reproduces the Bekenstein–Hawking expression. A string-theoretical approach was provided by Andrew Strominger and Cumrun Vafa,[151] who were able to count the degrees of freedom lost in the degradation of matter in the black hole. Their results support the interpretation given by Bekenstein and Hawking.

It may seem surprising that a theory that is concerned with the most elementary properties of matter is claimed to explain a macroscopic property of matter. It is the extreme nature of the black hole, as expressed in its huge entropy, indicating a degradation of mass into its finest possible grains (Planck units), which demands an explanation. String theory deals with matter at the Planck level and asks what properties of matter could emerge at this level. If matter is successively partitioned into smaller and smaller units the extreme would be a point-like unit, i.e. a unit in which all dimensions tend to 0. In our imagination, for instance, the electron and possibly also the quarks as the “most elementary” particles known seem to be point-like. String theory proposes that different dimensions in this respect behave differently.

If in three-dimensional space only two dimensions are allowed to converge to 0 (whatever 0 on the Planck scale means) we obtain a string-like or looped shape. States of a particle then may be represented by standing waves along the string. If two of its dimensions remain finite, we get membrane-like structures. String theory or – if combined with supersymmetry – superstring theory is consistent as a quantum theory only in spaces of higher dimension. Its viable candidate, called M-theory, requires 11 dimensions. Most of the additional dimensions in four-dimensional space–time are assumed to be curled up on the Planck scale. These small distances of 10^{-35} m are associated with huge energies of about 10^{19} GeV, which far exceed the range presently accessible to experimentation. On the other hand, the large energy range they cover makes these theories good candidates for unifying the four fundamental interactions.

If in a black hole matter is degraded to its finest level, then string theory may show us a way to estimate the degrees of freedom lost by such a fine partition. This allows a calculation of the phase volume ϕ, the logarithm of which determines its entropy. At the same time it clearly confirms that entropy represents the amount of "true" information irretrievably lost.

The question in the title of this section – and hence the subject of my discussion – is "It from Bit?", and I want here to remind the reader that entropy is not what we usually call "information". Entropy rather is the amount of lost, or missing, information. The black hole is to be seen as an "ultimate state" of matter – like a book that has been dissolved into its individual letters. In a similar way, all its information about structure has become lost – except that of its total mass, charge and angular momentum, and of some echo reverberating from the Planck world. The latter is expressed by the size of the Planck square, as it occurs in the Bekenstein–Hawking formula. And this leads us to the true problem behind Wheeler's question: "It from Bit?".

The three fundamental constants of physics G, c and $\hbar$ (or their combinations) may be seen as the primordial information that guides the formation of "It" in a nascent universe. There are three choices:

(1) The natural constants with their present values are inherent to any physical world.

(2) The constants are more or less random choices for any universe, but became fixed at an early phase of the big bang. We happen to live in a sustainable universe where the constants accidentally assumed values that allow our existence (cf. the weak anthropic principle; Section 1.10).

(3) The constants evolved during an early growth period to values that warrant most efficiently a sustainable universe.

It was John Wheeler[152] who in 1973 took the third case into consideration, proposing a universe that goes through cycles of expansion and contraction. Through repeated changes, occurring in the singularity of contraction, the universe achieved a fine tuning of the natural constants to their present values.

A more concrete model of this kind was quite recently proposed by Lee Smolin[153] in his book *The Life of the Cosmos*. Lee Smolin is a theoretical physicist who has made important contributions to string theory. Leonard Susskind,[154] the early pioneer of (physical) string theory, pays homage to Smolin by calling his idea "a valiant effort that deserves serious thought".

Smolin shapes his theory in close analogy to the Darwinian principle: "evolution through natural selection". This indeed is an intriguing modification of Wheeler's idea" as Roger Penrose[155] calls it. I should add that – considering all the differences with respect to the biological situation (which is strongly backed by observation as well as by experimentation) – the idea must be called quite courageous. Smolin knows this when he unpretentiously concedes: "What I am presenting in this book is a frank speculation, if you will a fantasy." Let us look a bit more closely at his fantasy, which he calls *expressis verbis* a Darwinian model.

It rests on three assumptions which, indeed, are entirely speculative:

(1) The black hole is not a final state of matter. Rather, its singularity provides the source for the birth of a new universe.

(2) The resulting new universe still has some memory of its ancestry. It involves fundamental constants not identical to, but similar to, those of its "parental" universes.

(3) "Fitness" of each individual universe is based on the number of black holes and hence of "viable" offspring universes produced during its lifetime.

These theses sound Darwinian, but they are Darwinian only in a quite superficial way. Evolution in biology can occur only in populations (see the arguments of Ernst Mayr in Section 4.7). A single isolated gene as such would never "evolve". The competitive character of "internal selection" is expressed in the population numbers of mutants. In his *Origin of Species* Darwin stresses the Malthusian point of view as the basis of competition. He contrasts the geometrical ratio of increase of a population with the (at best) linear increase of the resources. This has two consequences: On the one hand there is a "non-hostile" competition among individuals, producing a steady selection pressure within the total population of a given species and, on the other hand, a (possibly) peaceful coexistence with populations of other species living on different resources.

Here I see a cleft between Smolin's model for the universe and true Darwinian behaviour, a cleft I cannot see how to bridge. Smolin's universes favour those which produce a maximum number of black holes. In this respect our universe is quite proliferative. (Astrophysicists suppose that a black hole may be found at the centre of nearly every galaxy.) What will happen if each of them produces a new universe? Will they all appear within the space dimensions of the present universes, which would yield a pretty chaotic situation, or will they make up their own spaces, having no contact with one another and thereby behaving like biological species in "niches"? What are the resources on which they "live" and why should such a process favour universes with natural constants hospitable to life?

Being a physical chemist by training (starting out from physics) I have strong qualms about reversing a fundamental law of thermodynamics without the slightest hint from experiment or observation. A large black hole is a state of defined mass-energy having the highest possible entropy and the lowest possible temperature (near absolute zero) and, according to Smolin, is supposed to change spontaneously into a state of exceedingly large temperature (such as 10^{32} K) and minute entropy, as is assumed for the origin of the universe in the "big bang". Roger Penrose,[155] in addition, particularly stresses the point "that some presently unknown physics cannot convert the space–time singularity of collapse into a 'bounce', but also readjusts the fundamental physical constants when this happens." The phenomenon of "mutation during replication", as convincing as it has turned out in biology, has no counterpart in cosmology or particle physics.

Leonard Susskind, in his recent, excellent book,[156] mentions two facts which he sees as being in striking disagreement with Smolin's prerequisites: New results of string theory (not available when Smolin wrote his book) make "small" variations (of the natural constants) in successive generations unlikely. Moreover, he doubts that the thesis of maximum production of black holes has any physical backing.

Leonard Susskind regrets the harshness of his reaction to Smolin's book, but stresses that his harshness "is directed against particular technical points, not against Smolin's overall philosophy". He thinks that "Smolin deserves great credit for getting the most important things right". The most important point, as I see it, is the proposal of an alternative to the so far prevailing opinion that we have to assume a vast number of accidentally formed universes in order to make our existence plausible. The appearance of a sustainable universe is more likely to have been the consequence of some self-organisation than a matter of pure chance – just as the appearance of human beings was the result of an optimising evolution rather than of pure chance.

I think that an appropriate model for the origin of a sustainable universe does not need to follow too closely any biological pattern. The situations really differ too much when we come to any detail. The natural constants or certain combinations of them are – so to speak – an initial information content of a fundamental physics. It may be that other properties also have to be explained. The standard model of physics still leaves us with quite a collection of unexplained numbers (see Section 1.6).

Phenomenologically, what we would need is a theory for the origin of information, and this may answer the question "It from Bit". For the molecular evolution of life, such a theory was described in Chapter 4 and in this chapter. What I will do now is to show how this kind of thinking may also apply to the "information" expressed by matter. Note that I am not going to develop any specific theory for the origin of the universe. This cannot be my task. Such a theory would require a solution of the problem of quantum gravity, which is well beyond my expertise.

My question is concerned with the origin of information on which a sustainable world is based. And here we should focus on a possible initial state. As a general scheme let us consider the following.

If a black hole is an ultimate state of mass-energy we must assume a primordial state of (almost) vanishing entropy and high temperature (on the Planck scale of some 10^{32} K). Exceedingly low entropy entails that only one or very few quantum states are involved, but the estimated temperature of about 10^{32} K suggests an enormous energy density. A closed dynamic universe would then require a zero-sum game between mass-energy and potential gravitational energy (Wheeler[158]), while a finite and positive cosmological constant would allow an uptake of energy from the quantum vacuum, which apparently happened to quite some extent during the inflation period around 10^{-35} to 10^{-33} s after the big bang (see Susskind[156]). We have no idea about the physics of these early states of the universe. However, it is unlikely that they are described by Smolin's model. I have no idea how the big bang came about, but it must have been a non-linear event that only could have been followed by Darwinian phases, in which the natural constants became fixed to optimal values. There were some 10^{43} units of Planck time within the first second.

Although the physics of the early universe largely remain a closed book, there is some consensus among particle physicists and cosmologists about the evolution of the universe as indicated in Figure 1.8.6. Starting with a Planck world at about 10^{-43} s after the big bang, three major phase transitions took place in the early universe as a consequence of symmetry breakages separating the four principal forces:

(1) The "freezing out" of gravity within the first 10^{-40} s.

(2) The "split-off" of strong forces between 10^{-38} and 10^{-32} s, most probably connected with an exponential inflation period of around 10^{-35} s.

(3) The segregation of the weak and electromagnetic forces after around 10^{-12} s.

Phase transitions, because of their inherently autocatalytic nature, yield ideal conditions for evolutionary optimisation (see Sections 4.7 and 5.5). Moreover, there are sufficiently many orders of magnitude – more than for biological evolution – that may take care of a perfect fit for a sustainable physical world.

The attractive aspect is that (in the best case) we might end up with three equations for three unknown fundamental physical constants or eigenvalues that contain the three fundamental natural constants or combinations of them. This could be a first approximation of a more detailed theory yielding all the parameters involved in the standard model. However, if the prerequisites of the Frobenius–Perron theorem (see Chapter 4) are fulfilled there is only one, namely the largest, eigenvalue, which is determinant for the whole system. A multiverse with many "bubbles" having non-overlapping co-ordinates during its phase of optimisation would not be necessary in such a model.

In order to describe the physical problem involved better, and to show why we do not have to populate an indefinitely large number of states in order to find the most suitable outcome, I shall now focus on the targets of optimisation. And here I find a true analogue to the problem of evolution of life within our universe. What is it that has to be selected?

The structures in particle physics – leptons, quarks, nuclei and atomic scaffoldings – are more prone to symmetry operations than the macromolecular structures in biology, which have the appearance of very complex machines. Read what John Wheeler has said about symmetry in physics: "Law derives from symmetry; but symmetry hides the mechanism of mutability. . . . In order to understand structure we have to understand the mechanism that produces it."[141] Let me quote a (perhaps more relevant) example from laser physics, presented by Hermann Haken,[157] which I discussed in Section 3.9. In a laser the autocatalytic process consists of stimulated emission, occurring in a population of excited states produced by "optical pumping". The "phase transition" in this case appears as a population "inversion" in which the phases of all released photons become "enslaved", as Haken called it, which means they assume one highly synchronised common phase. (Note that the word "phase" is used here in two different ways, namely, as an "optical phase of oscillation" and a "state of aggregation" that undergoes a phase transition.)

An analogous example in particle physics might be the conversion of electromagnetic radiation energy into matter, as can nowadays still be seen in pair production, i.e. a photon with a sufficiently large energy may produce an electron–positron pair by interaction with an atomic nucleus. I mention this case because it might prove to be an example of autocatalysis in the formation of material particles from radiation. It is the charge structure of the atomic nucleus that may provide autocatalytic help in pair production. The cross-section of this process can be calculated according to an expression derived by Hans Bethe and Walter Heitler.[159] The cross-section is a measure of the efficiency of the energy transformation that occurs in this process. It shows a very sharp dependence on the magnitude of the electric charges involved, as illustrated in Figure 5.10.3. What is seen from this plot is that, within its error limits, there is a sharp maximum in the rate of energy transformation for a charge equalling that of the electron. In other words, the value of the elementary charge is at an optimum with respect to efficient autocatalytic energy transfer.

This case was related to me personally by the solid-state physicist Hans Queisser of the Stuttgart Max Planck Institute as a possible example from physics that falls in with our theory of natural selection. I shall not dwell here on Queisser's further conclusions regarding the involvement of masses showing a preference for the muon relative to the electron, nor will I discuss consequences regarding the preference of matter over antimatter. What I want to show is the interdependence of natural

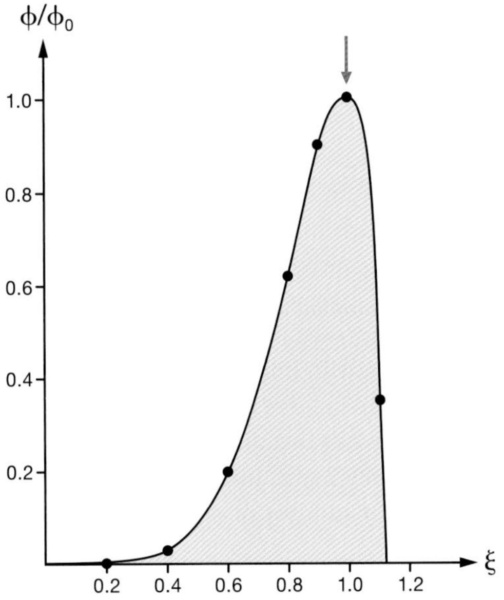

Figure 5.10.3
Relative cross-section of pair production: $h\nu \rightarrow e^{+} + e^{-}$

This is according to an expression of Bethe and Heitler[158]. The relative cross-section ϕ/ϕ_o is plotted as a function of ξ, where ξ refers to the fractional charge ξe, e being the elementary unit of electric charge. The cross section ϕ_o refers to $\xi = 1$, i.e. to the elementary unit of charge. The Bethe–Heitler relation has the form $\phi_{(\xi)} = \xi^3\,(3.11 \cdot \ln(2.6/\xi^2) - C)$, where ξ is a parameter related to the electric charge e involved in the scattering process and C is a constant having a value around $C \approx 2$. A plot of $\phi_{(\xi)}$ as a function of ξ yields the sharp maximum for e at the value of the electronic charge e_o, as shown in the figure. ($\xi = 1$ refers to $e = e_o$.) This at least establishes a clear internal relationship between h, c and e_o. The reaction is catalytically favoured by the presence of matter which has the charge structure based on e_o. This means that the catalyst maps its charge structure to the products of transformation of radiation energy. In the early evolution of the universe, where this reaction occurred far from equilibrium, autocatalytic transformation of radiation energy may have been the cause for selecting the elementary charge e_o, which we know today is the most efficient catalytic entity. In this case e_o was an early product of the evolution of our universe, which nowadays is uniquely represented. An experimental proof does not seem to me to be out of reach. It will require a careful study of the kinetics of this process occurring under slightly perturbed conditions for selection.

parameters, based possibly on adaptive kinetic processes as they may have appeared during any of the three phase transitions consequent on symmetry breakages among the fundamental forces within the earliest phases of our universe. At that time, unlike today, there was no "equilibrium" in this case, i.e. no balance between pair production and annihilation; such imbalance is an important prerequisite for evolutionary change.

We may consider a dynamic universe in which a kind of "natural selection" occurs in combination with phase transitions associated with the separation and specification of gravitational, strong, weak and electromagnetic interactions as a consequence of symmetry breakages. It may be that the present values of the natural constants evolved and became fixed during these events as optimum choices for a stable universe with prolonged existence. This would be in close analogy to the evolution of life in our universe. However, such an analogy would have to be the result of the (still largely unknown) physics of the early universe and should not simply be postulated with reference to the biological case, where the prerequisites are quite different.

Information in biology is organised in the form of long sequences of symbols and therefore has much in common with linguistic information. Information expressed in a physical process is organised in quite different ways, but it may have in common with biological information the fact that only a minute fraction of a huge number of possible combinatorial events represents information about existence and hence is a target for selection (cf. the huge number of possible combinations, as estimated by Leonard Susskind[156]).

The question in the title of this section is of a similar nature to the one raised by the Scholastic philosophers: "Which came first, the chicken or the egg?" The answer suggested by the results of modern biology is: neither of the two came "first". They both had to originate in a (non-linear) feedback cycle and evolve to some optimal joint performance, which includes reproductive multiplication, adaptive variation and proliferation of well-functioning descendents. Altogether this is an open-ended process of "complexification". Reflecting on Wheeler's[141] "It from Bit" I would conclude that neither the "Bit" nor the "It" had any form of existence *a priori*; they both originated in a feedback loop with subsequent adaptation to a universe of stable existence. Table 5.10.2 compares informational with material physics.

Table 5.10.2 Material versus Informational Physics

Information = individuality	No individuality
A posteriori symmetries (if advantageous)	Fundamental *a priori* symmetries
No structural principle; instead, informed functionality	Structure from symmetry
Optimisation according to an extremum principle of dynamics	Minimisation of energy
Combinatorial complexity	Strange simplicity
Averages useless	Averages valid
Unlimited action	Defined sinks
Open systems	Closed systems
Complex	Simple
Familiar	Strange

LITERATURE AND NOTES

1. Schrödinger, E. (1944). "What is Life?". Cambridge University Press, Cambridge.
2. Mann, Th. (1924). "Der Zauberberg". S. Fischer, Frankfurt. English translation: "Magic Mountain".
3. Hertwig, O. (1918). "Das Werden der Organismen". (The Development of Organisms). 2nd edn. Verlag von Gustav Fischer, Jena.
4. Murphy, M. P. and O'Neill, L. A. J. (eds.) (1995). "What is Life? – The Next Fifty Years. Speculations on the Future of Biology". Cambridge University Press, Cambridge.
5. Gould, S. J. (1995). "What is Life? as a Problem in History". In: "What is Life? – The Next Fifty Years. Speculations on the Future of Biology", M.P. Murphy and L. A. J. O'Neill (eds). Cambridge University Press, Cambridge, pp. 25–39.
6. Weinberg, S. (1992). "Dreams of a Final Theory. The Scientist's Search for Ultimate Laws of Nature". Chapter III. Pantheon, New York.
7. Schrödinger, E. (1929). "Was ist ein Naturgesetz?" *Naturwiss.* 17, (1929), 9–11. English translation: Schrödinger, E. and Murphy, J. (ed. and translation). (1935) "What is a Law of Nature?" In: "Science and the Human Temperament". pp. 107–118. Allen and Unwin, London.
8. Crow, J. F. (1992). "Erwin Schrödinger and the Hornless Cattle Problem". *Genetics* 130(2): 237–239.
9. Schrödinger, E. (1958). "Mind and Matter". Cambridge University Press, Cambridge.
10. Darwin, Ch. (1882) in a letter to George Wallich, London, written March 28, 1882. (Darwin died on April 19, the same year.)
11. Letter 13747—Darwin, C. R. to Wallich, G. C., 28 Mar 1882. Darwin Correspondence Project. www.darwinproject.ac.uk.
12. Watson, J. D. and Crick, F. H. (1953). "Molecular Structure of Nucleic Acids; a Structure for Deoxyribose Nucleic Acid". *Nature* 171 (4356): 737–738.
13. Midgley, Th. and Boyd, T. A. (1922). "The Chemical Control of Gaseous Detonation with Particular Reference to the Internal-Combustion Engine". *Ind. Eng. Chem.* 14: 894–898.
14. Midgley, T. A. and Boyd, T. A. (1922). "Detonation Characteristics of Some Blended Motor Fuels". *SAE J.* 451.
15. Margulis, L. and Sagan, D. (1995). "What is Life?: The Eternal Enigma". University of California Press, Berkeley and Los Angeles.
16. Miller, S. L. and Urey, H. C. (1959). "Organic Compound Synthesis on the Primitive Earth: Several Questions about the Origin of Life have been Answered, but much Remains to be Studied". *Science* 130: 245–251.
17. Fox, S. W. and Dose, K. (1972). "Molecular Evolution and the Origin of Life". W. H. Freeman, San Francisco.

18. Morowitz, H. J. (1992). "Beginnings of Cellular Life: Metabolism Recapitulates Biogenesis". Yale University Press, New Haven and London, p. 8.
19. Kirschner, K., Eigen, M., Bittman, R. and Voigt, B. (1966). "The Binding of Nicotinamide Adenine Dinucleotide to Yeast d-Glyceraldehyde-3-phosphate Dehydrogenase: Temperature Jump Relaxation Studies on the Mechanism of an Allosteric Enzyme". *Proc. Natl. Acad. Sci. USA* 56: 1661–1667. See also Kirschner, K. (1968). "Allosteric Regulation of Enzyme Activity". *Curr. Top. Microbiol. Immunol.* 44: 123–146.
20. Eigen, M. (1971). "Selforganisation of Matter and the Evolution of Biological Macromolecules". *Naturwiss.* 58: 465–523.
Elbashir, S. M., Harborth, J., Lendeckel, W., Yalcin, A., Weber, K. and Tuschl, Th. (1999). "Duplexes of 21-Nucleotide RNAs Mediate RNA Interference in Cultured Mammalian Cells". *Nature* 411: 494–498.
21. Hamilton, A. J. and Baulcombe, D. C. (1999). "A Species of Small Antisense RNA in Postranscriptional Gene Silencing in Plants". *Science* 286: 950–952.
22. Altmann, S. (1989). "Enzymatic Cleavage of RNA by RNA". Nobel Lectures, Chemistry 1981–1990. Frängsmyr, T., Malmström, B. G. (eds.). World Scientific Publishing Co., Singapore, 1992.
23. Cech, Th. R. (1989). "Selfsplicing and Enzymatic Activity of an Intervening Sequence RNA from Tetrahymena". Nobel Lectures, Chemistry 1981–1990. Frängsmyr, T., Malmström, B. G. (eds.). World Scientific Publishing Co., Singapore, 1992.
24. The story of an early involvement of RNA in evolution is excellently described by Monroe W. Strichberger in his "Evolution", 2nd edn, 1996. Jones and Bartlett Publishers, Inc., Sudbury, Massachusetts. Boston, London, Singapore. In particular, one can find a critical review of the evidence for an "RNA-world", which extends from Tom Cech's work on "self-splicing RNA" to Harry Noller's exciting finding about the catalytic role of ribosomal RNA. Some early literature (until 1993): Cech, T. R. (1987). "The Chemistry of Self-Splicing RNA and RNA Enzymes". *Science* 236: 1532–1539; Sassanfar, M. and Szostak, J. W. (1993). "An RNA Motif that Binds ATP". *Nature* 364: 550–533; Joyce, G. E. and Orgel, L. E. (1993). "Prospects for Understanding the Origin of the RNA-World". In "The RNA-World", R. F. Gesteland and J. F. Atkins (eds). Cold Spring Harbor Laboratory Press, Cold Spring Harbor NY; Noller, H. F., Hoffarth, V. and Zimniak, L. (1992). "Unusual Resistance of Peptidyl Transferase to Protein Extraction Procedures". *Science* 256: 1416–1419. The whole field is now one of the "hot" areas in molecular biology.
25. De Duve, Ch. (2002). "Life Evolving: Molecules, Mind, and Meaning". Oxford University Press, Oxford, New York.
26. Eigen, M., Lindemann, B. F., Tietze, M., Winkler-Oswatitsch, R., Dress, A. and von Haeseler, A. (1989). "How Old is the Genetic Code? Statistical Geometry of tRNA Provides an Answer". *Science* 244(4905): 673–679.

27. Eigen, M., Winkler-Oswatitsch, R., Dress, A. (1988). "Statistical Geometry in Sequence Space: A Method of Quantitative Comparative Sequence Analysis". *Proc. Natl. Acad. Sci. USA* 85: 5913–5917.
28. Dyson, F. (1985). "Origins of Life". Cambridge University Press, Cambridge and New York.
29. Eldredge, N. and Gould, S. J. (1972). "Punctuated Equilibria: an Alternative to Phyletic Gradualism". In: "Mode in Paleobiology", Schopf, T. J. M. (ed.). Freeman, Cooper, San Francisco, pp. 82–115.
30. Eschenmoser, A. (1990). "Leopold Ružička – From the Isoprene Rule to the Question of Life's Origin". *Chimia* 44: 1–21.
31. Mayr, E. (1988). "Toward a New Philosophy of Biology", Part I. Belkamp Press of Harvard University Press. Cambridge MA and London, pp. 16–36.
32. Malthus, Th. R. (1798). "An Essay on the Principle of Population, as it Affects the Future Improvement of Society". London. Printed for J. Johnson, in St. Paul's Church-Yard.
33. Wallace, A. R. (1858). "On the Tendency of Varieties to Depart Indefinitely From the Original Type". *Proc. Linn. Soc. Lond.* 3: 53–62. (Manuscript sent to Darwin on 9 March 1858 and read together with Darwin's paper to the Linnean Society of London on 1 July 1858.)
34. Eigen, M. (1981). "Darwin and Molecular Biology". *Angew. Chem. Int. Ed. Engl.* 20(3): 233–241.
35. Dennett, D. C. (1995). "Darwin's Dangerous Idea. Evolution and the Meaning of Life". Simon Schuster, New York.
36. Weismann, A. (1892). "Aufsätze über Vererbung und verwandte biologische Fragen". Verlag von Gustav Fischer, Jena.
37. Mendel, J. G. (1866). "Versuche über Pflanzen-Hybriden". *Verhandlungen des naturforschenden Vereines in Brünn*, Bd. IV: 3–47.
38. De Vries, H. in the year 1900 rediscovered Mendel's law (found 20 years earlier, see Ref. 37).
De Vries, H. (1901). "Die Mutationstheorie. Versuche und Beobachtungen über die Entstehung von Arten im Pflanzenreich, von Hugo de Vries". Leipzig, Veit & comp., 1901–03.
De Vries, H. (1909). "The Mutation Theory. Experiments and Observations on the Origin of Species in the Vegetable Kingdom". Translated by J. B. Farmer and A. D. Darbishire. Volume II. "The Origin of Varieties by Mutation with Numerous Illustrations and Six Colored Plates". Chicago, The Open Court Publishing Company.
39. See "Synthetic Theory", as quoted in Refs. 40 to 44.
40. Haldane, J. B. S. (1932). "The Causes of Evolution". Longmans, Green & Co., London.
41. Fisher, R. A. (1930). "The Genetical Theory of Natural Selection". Clarendon Press, Oxford.

42. Wright, S. (1931). "Evolution in Mendelian Populations". *Genetics* 16(2): 97–159.
43. Maynard-Smith, J. (1986). "The Problems of Biology". Oxford University Press, Oxford; Maynard-Smith, J. (1989). "Evolutionary Genetics". Oxford University Press, Oxford.
44. Kimura, M. and Ohta, T. (1971). "Theoretical Aspects of Population Genetics". Princeton University Press, Princeton.
45. Biebricher, C. K., Eigen, M., Gardiner, W. C., Husimi, Y., Keweloh, H.-C. and Obst, A. (1987). "Modeling Studies of RNA Replication and Viral Infection". In: Warnatz, J. (1987). "Complex Chemical Reaction Systems: Mathematical Modelling and Simulation: Proceedings (Springer Series in Chemical Physics)". Jäger, W. (ed). Springer-Verlag, Berlin, pp. 17–38.
46. Meixner, J. (1943). "Absorption und Dispersion des Schalles in Gasen mit chemisch reagierenden und anregbaren Komponenten". *Ann. Phys.* 43: 470–487; Meixner, J. (1953). "Die thermodynamische Theorie der Relaxationserscheinungen und ihr Zusammenhang mit der Nachwirkungstheorie". *Colloid and Polymer Science* 134(1): 3–20.
47. Onsager, L. (1931). "Reciprocal Relations in Irreversible Processes. I". *Phys. Rev.* 37: 405–426.
48. Prigogine, I. (1977). "Time, Structure and Fluctuations". Nobel Lectures, Chemistry 1971–1980. Frängsmyr, T., Forsén, S. (eds.). World Scientific Publishing Co., Singapore, 1993.
49. Glansdorff, P. and Prigogine, I. (1971). "Thermodynamic Theory of Structure, Stability and Fluctuations". Wiley-Interscience, New York.
50. von Neumann, J. (1966). "Theory of Selfreproducing Automata". Burks, A. W. (ed.). University of Illinois Press, Urbana, IL.
51. See Section 4.8.
52. Nicholis, G. (1995). "Introduction to Nonlinear Science". Cambridge University Press, Cambridge and New York.
53. Biebricher, C. K. (1983). "Darwinian Selection of Self-Replicating RNA". In: "Evolutionary Biology", Vol. 16, Hecht, M.K., Wallace, B. and Prance, G.T. (eds). Plenum Press, New York, pp. 1–52.
54. Biebricher, C. K., Eigen, M. and Gardiner, W. C. Jr. (1983). "Kinetics of RNA Replication". *Biochem.* 22: 2544–2559; Biebricher, C. K., Eigen, M. and Gardiner, W. C. Jr. (1984). "Plus-minus Asymmetry and Double-Strand Formation". *Biochem.* 23: 3186–3194; Biebricher, C. K., Eigen, M. and Gardiner, W. C. Jr. (1985). "Competition and Selection Among Self-Replicating RNA Species". *Biochem.* 23, 6550–6560.
55. Kowalsky, H. J. (1979). "Lineare Algebra", 9th edn. Walter de Gruyter, Berlin and New York.
56. Eigen, M., Biebricher, C. K., Gebinoga, M. and Gardiner, W. C. Jr. (1991). "The Hypercycle. Coupling of RNA and Protein Biosynthesis in the Infection Cycle of an RNA Bacteriophage". *Biochem.* 30: 11005–11018.

57. Eigen, M. and Schuster, P. (1979). "The Hypercycle. A Principle of Natural Self-Organization". Springer-Verlag, Berlin, Heidelberg and New York.
58. Weissmann, C., Feix, G., Slor, H. and Pollet, R. (1967). "Replication of Viral RNA XIV. Single-Stranded Minus Strands as Template for the Synthesis of Viral Plus Strands in Vitro". *Proc. Natl. Acad. Sci. USA* 57(6): 1870–1877; Weissmann, Ch., Feix, G. and Slor, H. (1968). "In Vitro Synthesis of Phage RNA. The Nature of the Intermediates". *Symp. on Quant. Biology*, Cold Spring Harbor, XXXIII, 83–100; Weissmann, C., Billeter, M. A., Goodman, H. M., Hindley, J. and Weber, H. (1973). "Structure and Function of Phage RNA". *Annu. Rev. Biochem.* 42: 303–328; Weissmann, C., Borst, P., Burdon, R. H., Billeter, M. A. and Ochoa, S. (1964). "Replication of Viral RNA. IV. Properties of RNA Synthetase and Enzymatic Synthesis of MS2 Phage RNA". *Proc. Natl. Acad. Sci. USA* 51: 890–897.
59. Moses, R. E. and Richardson, Ch. C. (1970). "Replication and Repair of DNA in Cells of Escherichia coli Treated with Toluene". *Proc. Natl. Acad. Sci. USA* 67(2): 674–681.
60. Vosberg, H.-P. and Hoffmann-Berling, H. (1971). "DNA Synthesis in Nucleotide-Permeable Escherichia Coli Cells". *J. Mol. Biol.* 58: 739–753.
61. Gebinoga, M. (1990). "Kinetik der intrazellulären Entwicklung des Phagen Qß in Escherichia Coli". PhD Thesis, Braunschweig (Technical University Carolo-Wilhelmina) and Göttingen (MPI for Biophysical Chemistry).
62. Shine, J. and Dalgarno, L. (1974). "The 3'-terminal Sequence of Escherichia Coli 16S Ribosomal RNA: Complementarity to Nonsense Triplets and Ribosome Binding Sites". *Proc. Natl. Acad. Sci. USA* 71(4): 1342–1346.
63. Lincoln, T. A. and Joyce, G. F. (2009). "Self-Sustained Replication of an RNA Enzyme". *Science* 323(5918): 1229–1232.
63a. Lam, B. J. and Joyce, G. F. (2009). "Autocatalytic aptazymes enable ligand-dependent exponential amplification of RNA". Nat. Biotechnol. 27(3): 288–292.
64. Pörschke, D. (1968). PhD Dissertation, Technical University Braunschweig, Braunschweig. See also Eigen, M., Maass, G. and Pörschke, D. (1967). "Dynamische Untersuchungen an Oligo- und Polynucleotiden". *Angew. Chem.* 79(9): 417.
65. Kimura, M. (1968). "Genetic Variability Maintained in a Finite Population Due to Mutational Production of Neutral and Nearly Neutral Isoalleles". *Genet. Res.* 11(3): 247–269.
66. King, J. L. and Jukes, Th. H. (1969). "Non-Darwinian Evolution. Most Evolutionary Change in Proteins May be due to Neutral Mutations and Genetic Drift". *Science* 164(3881): 788–798.
67. Kimura, M. and Ohta, T. (1973). "Mutation and Evolution at the Molecular Level". *Genetics* 73(Suppl.): 19–35.
68. Schuster, P. and Sigmund, K. (1984). "Random Selection–a Simple Model Based on Linear Birth and Death Processes". *Bull. Math. Biol.* 46: 11–17.
69. McCaskill, J. S. (1984). "A Stochastic-Theory of Macromolecular Evolution". *Biol. Cybern.* 50(1): 63–73.

70. Mayr, E. (1997). "This is Biology: The Science of the Living World". The Belknap Press of Harvard University Press, Cambridge and London.
71. Kimura, M. (1964). "Diffusion Models in Population Genetics". *J. Appl. Probab.* 1: 177–232.
72. Crow, J. F. (1994). Foreword. In: "Population Genetics, Molecular Evolution, and the Neutral Theory", Takahata, N. (ed.). The University of Chicago Press, Chicago and London.
73. Kimura, M. (1994). Selected Papers. In: "Population Genetics, Molecular Evolution, and the Neutral Theory", Takahata, N. (ed.). The University of Chicago Press, Chicago and London.
74. Fisher, R. A. (1930). "The Genetical Theory of Natural Selection". Clarendon Press, Oxford. See also Fisher, R. A. (1922). "On the Dominance Ratio". *Proc. Roy. Soc. Edinb.* 42: 321–341.
75. Wright, S. (1931). "Evolution in Mendelian Populations". *Genetics* 16(2): 97–159.
76. Malécot, G. (1948). "Les mathématiques de l'hérédité". Masson et Cies, Paris.
77. Fokker, A. D. (1914). "Die mittlere Energie rotierender elektrischer Dipole im Strahlungsfeld". *Ann. Phys.* 43: 810–820.
78. Planck, M. (1917). "Über einen Satz der statistischen Dynamik und eine Erweiterung in der Quantumtheorie". *S.-B. Preuss. Akad. Wiss.* 5: 324–341.
79. Kolmogorov, A. N. (1931). "Über die analytischen Methoden in der Wahrscheinlichkeitsrechnung". *Math. Ann.* 104: 415–458.
80. Feller, W. (1951). "Diffusion processes in genetics". In: Proceedings of the 2nd Berkeley Symposium on Mathematical Statistics and Probability, Neyman, J. (ed.). University of California Press, Berkeley, CA, pp. 227–246.
81. Kimura, M. (1976). "How Genes Evolve: a Population Geneticists View". *Ann. Genet.* 19(3): 153–168.
82. Kimura, M. and Ohta, T. (1973). "Mutation and Evolution at the Molecular Level". *Genetics* 73 (Suppl.): 19–35.
83. Swetina, J. and Schuster, P. (1982). "Self Replication with Errors, a Model for Polynucleotide Replication". *Biophys. Chem.* 16: 329–345.
84. Tarazona, P. (1992). "Error Thresholds for Molecular Quasispecies as Phase Transitions: from Simple Landscapes to Spin-Glass Models". *Phys Rev.* A 45: 6038–6050.
85. Schuster, P. and Swetina, J. (1988). "Stationary Mutant Distributions and Evolutionary Optimization". *Bull. Math. Biol.* 50: 635–660.
86. Model proposed for showing the quasispecies nature. The population chooses the lower peak at larger concentrations as consequence of its more favourably populated neighbourhood. Schuster, P. (1999). "Mendel-Vorlesung: Evolution der Moleküle. Einblicke in die Natur mit Perspektiven für die Biotechnologie". TBI Preprint No. 99-pks-010. Vienna University, Vienna.

87. Eigen, M. (1985). "Macromolecular Evolution: Dynamical Ordering in Sequence Space". *Ber. Bunsenges Phys. Chem.* 89: 658–667; Eigen, M. (1986). "The Physics of Molecular Evolution". *Chem. Script.* 26B: 13–26.
88. Onsager, L. (1944). "Crystal Statistics. I. A Two-Dimensional Model with a Order-Disorder Transition". *Phys. Rev.* 65: 117–149.
89. Wilson, K. (1982). "The Renormalization Group and Critical Phenomena". Nobel Lectures, Physics 1981–1990. Frängsmyr, T. and Ekspång, G. (eds.). World Scientific Publishing Co. Singapore, 1993.
90. Dawkins, R. (1986). "The Blind Watchmaker. Why the Evidence of Evolution Reveals a Universe without Design". W. W. Norton, New York.
91. Monod, J. (1970). "Le Hasard et la Nécessité. Essai sur la philosophie naturelle de la biologie moderne". Edition du Seuil, Paris. English translation: "Chance and Necessity: An Essay on the Natural Philosophy of Modern Biology". (transl. A. Wainhouse). Knopf, New York.
92. Ebeling, W., Engel, A., Esser, B. and Feistel, R. (1984). "Diffusion and Reaction in Random Media and Models of Evolution Processes". *J. Stat. Phys.* 37(3–4): 369–384.
93. Schuster, P. (2006). "Prediction of RNA Secondary Structures: From Theory to Models and Real Molecules". *Rep. Prog. Phys.* 69: 1419–1477.
94. Tramontano, A., Janda, K. D. and Lerner, R. A. (1986). "Catalytic Antibodies". *Science* 234(4783): 1566–1570.
95. Pollack, S. J., Jacobs, J. W. and Schultz, P. G. (1986). "Selective Chemical Catalysis by an Antibody". *Science* 234: 1570–1573.
96. Wilson, E. O. (2006). "The Creation: An Appeal to Save Life on Earth". W. W. Norton, New York.
97. Gould, S. (1999). In: Wilson, E. O. (1999). "The Creation: An Appeal to Save Life on Earth". W. W. Norton, New York. See also: Gould, S. (2006). "Darwin and the Munchkins of Kansas". In: "The Richness of Life", McGarr, P. and Rose, S. (eds). W. W. Norton, New York and London, pp. 616–618.
98. Peitgen, H. O. and Richter, P. (1986). "The Beauty of Fractals: Images of Complex Dynamical Systems". Springer-Verlag, Berlin, Heidelberg and New York.
99. Hopfield, J. J. (1982). "Neural Networks and Physical Systems with Emergent Collective Computational Properties". *Proc. Nat. Acad. Sci. U.S.A.* 79: 2554–2558. See also: Hopfield, J. J. and Tank, D. W. (1986). "Computing with Neural Circuits: A Model". *Science* 233: 625–633.
100. Pichler, E. E. (1990). "Selforganisation in Biology and Chemical Model Systems". PhD Dissertation, University of Vienna, Vienna. The dissertation contains a comparison of the Hopfield–Little neural network and the chemical reaction network of an evolutionary process.

101. Kauffman, S. A. (1986). "Autocatalytic Sets of Proteins". *J. Theor. Biol.* 119(1): 1–24.
102. Szathmáry, E. (1991). "Simple Growth Laws and Selection Consequences". *Trends Ecol. Evol.* 6: 366–370.
103. von Kiedrowski, G. (1993). "Minimal Replicator Theory I: Parabolic Versus Exponential Growth". *Bioorg. Chem. Front* 3: 113–146.
104. Scherrer, K. (1989). "A Unified Matrix Hypothesis of DNA-directed Morphogenesis, Protodynamism and Growth Control". *Biosci. Rep.* 9: 157–188; Scherrer, K. and Jost, J. (2007). "Gene and Genon Concept: Coding Versus Regulation. A Conceptual and Information-Theoretic Analysis of Genetic Storage and Expression in the Light of Modern Molecular Biology". *Theory Biosci.* 126: 65–113.
105. Peterson, I. (1998). "The Jungles of Randomness: A Mathematical Safari". John Wiley, New York.
106. Feynman, R. (1988). "What Do You Care What Other People Think?" W. W. Norton, New York and London.
107. Fontana, W. and Buss, L. W. (1994). "The Arrival of the Fittest: Toward a Theory of Biological Organization". *Bull. Math. Biol.* 56: 1–64; Fontana, W. and Buss, L. W. (1994). "What would be conserved if the tape were played twice?". *Proc. Natl. Acad. Sci. USA* 91: 757–761.
108. Church, A. (1941). "The Calculi of Lambda Conversion". Annals of Mathematics Studies No. 6. Princeton University Press, Princeton NY.
109. Fox, S. W. and Dose, K. (1972). See Ref. 17.
110. Kauffman, S. (1993). "The Origins of Order: Self-Organization and Selection in Evolution". Oxford University Press, New York; Kauffman, S. (1995). "At Home in the Universe: The Search for the Laws of Self-Organization and Complexity". Oxford University Press, Oxford and New York; Kauffman, S. (2000). "Investigations". Oxford University Press, New York.
111. Kauffman, S. (1986). See Ref. 101.
112. Kauffman, S. A. (1969). "Metabolic Stability and Epigenesis in Randomly Constructed Genetic Nets". *J. Theor. Biol.* 22(3): 437–467.
113. Lifson, S. (1997). "On the Crucial Stages in the Origin of Animate Matter". *J. Mol. Evol.* 44(1): 1–8; Lifson, S. and Lifson, H. (1999). "A Model of Prebiotic Replication: Survival of the Fittest versus Extinction of the Unfittest". *J. Theor. Biol.* 199(4): 425–433.
114. Prusiner, S. B. (1997). "Prions". Nobel Lectures, Physiology or Medicine 1996–2000. Jörnvall, H. (ed.). World Scientific Publishing Co., Singapore. 2003.
115. Kellings, K., Prusiner, S. B. and Riesner, D. (1994). "Nucleic Acids in Prion Preparations: Unspecific Background or Essential Component?". *Philos. Trans. R. Soc. Lond. B Biol. Sci.* 343(1306): 425–430. See also: Riesner, D. (1991). "The Search for a Nucleic Acid Component to Scrapie Infectivity". *Semin. Virol.* 2: 215–226.

116. Büeler, H., Aguzzi, A., Sailer, A., Greiner, R. A., Autenried, P., Aguet, M. and Weissmann, C. (1993). "Mice Devoid of PrP are Resistant to Scrapie". *Cell* 73(7): 1339–1347; Sailer, A., Büeler, H., Fischer, M., Aguzzi, A. and Weissmann, C. (1994). "No Propagation of Prions in Mice Devoid of PrP". *Cell* 77(7): 967–968.
117. Lee, D. H., Granja, J. R., Martinez, J. A., Severin, K. and Ghadiri, M. R. (1996). "A Self-Replicating Peptide". *Nature* 382(6591): 525–528.
118. Bak, P. (1997). "How Nature Works: The Science of Self-Organised Criticality". Oxford University Press, Oxford and New York.
119. Turing, A. M. (1950). "Computing Machinery and Intelligence". *Mind* 59: 433–460. Reprinted in Hofstadter, D. R. and Dennett, D. C. (1981). "The Mind's I. Fantasies and Reflections on Self and Soul". Basic Books, Harmondsworth.
120. Penrose, R. (2005). "The Road to Reality. A Complete Guide to the Laws of the Universe", 2nd Edition 2005, Alfred A. Knopf, New York.
121. Babbage, Ch. (1835). "On the Economy of Machinery and Manufactures". Charles Knight, London.
122. Toole, B. A. (1992). "Ada, the Enchantress of Numbers: A Selection from the Letter of Lord Byron's Daughter and Her Description of the First Computer". Strawberry Press, Hill Valley, CA.
123. von Neumann, J. and Morgenstern, O. (1944). "Theory of Games and Economic Behavior". Princeton University Press, Princeton, NJ.
124. von Neumann, J. (1966). "Theory of Self-Reproducing Automata". Edited and completed by A. W. Burks, University of Illinois Press, Urbana and London.
125. Ulam, S. M. (1970). "Cellular Space". In: "Essays on Cellular Automata", Burks, A. W. (ed.). University of Illinois Press, Urbana and London.
126. Conway, J. H. (1970). See Gardner, M. (1970). "Mathematical Games: The Fantastic Combinations of John Conway's New Solitaire Game 'Life' ". *Sci. Am.* 223: 120–123. Also: Gardner, M. (2001). "The Colossal Book of Mathematics". W. W. Norton, New York.
127. Adleman, L. M. (1994). "Molecular Computation of Solutions to Combinatorial Problems". *Science* 266(5187): 1021–1024 (early paper on DNA computing).
128. Dyson, G. B. (1997). "Darwin Among The Machines: The Evolution Of Global Intelligence". Helix Books, Addison-Wesley, Reading, MA, Menlo Park, CA and New York.
129. Ray, T. S. (1996). "An Approach to the Synthesis of Life". In: "The Philosophy of Artificial Life". Oxford Readings in Philosophy. Boden, M. A. (ed.). Oxford University Press, Oxford, pp. 111–145. Ibid. other related papers.
130. Langton, Ch. G. (1995). "Artificial Life: An Overview". MIT Press, Cambridge, MA.
131. Churchland, P. and Sejnowski, T. J. (1992). "The Computational Brain: Models and Methods on the Frontiers of Computational Neuroscience". Bradford Books, MIT Press, Cambridge, MA.

132. Kirkpatrick, S. (1984). "Optimization by Simulated Annealing–Quantitative Studies". *J. Stat. Phys.* 34: 975–986.
133. McCulloch, W. and Pitts, W. (1943). "A Logical Calculus of Ideas Immanent in Nervous Activity". *Bull. Math. Biophys.* 5: 115–133.
134. Rosenblatt, F. (1962). "Principles of Neurodynamics: Perceptrons and the Theory of Brain Mechanisms". Spartan Books, Washington DC; Minsky, M. and Papert, S. (1969). "Perceptrons". MIT Press, Cambridge, MA.
135. Grossberg, S. (ed.) (1988). "Neural Networks and Natural Intelligence". MIT Press, Cambridge, MA; Grossberg, S. (1988). "Nonlinear Neural Networks Principles, Mechanisms and Architectures". *Neural Networks* 1: 17–66.
136. Hopfield, J. J. (1982). "Neural Networks and Physical Systems with Emergent Collective Computational Abilities". *Proc. Nat. Acad. Sci. U.S.A.* 79: 2554–2558.
137. Leuthäusser, I. (1986). PhD Dissertation, University Göttingen, Göttingen. See also: Leuthäusser, l. (1986). "An Exact Correspondence between Eigen's Evolution Model and a Two-Dimensional Ising System". *Chem. Phys.* 84: 1884–1885.
138. Nieselt-Struwe, K. (1992). "Konfigurationsanalysen kombinatorischer und biologischer Optimierungsprobleme". PhD Dissertation, Universities Göttingen und Bielefeld. See also: Kreutz, R., Dietrich, U., Kühnel, H., Nieselt-Struwe, K., Eigen, M. and Rübsamen-Waigmann, H. (1992). "Analysis of the Envelope Region of the Highly Divergent HIV-2ALT Isolate Extends the Known Range of Variability within the Primate Immunodeficiency Viruses". *AIDS Res. Hum. Retroviruses* 8(9): 1619–1629.
139. Eigen, M. and Nieselt-Struwe, K. (1990). "How Old is the Immunodeficiency Virus?" *AIDS* 4 (Suppl. 1): S85–S93.
140. Reeke, G. N., Finkel, L. H., Sporns, O. and Edelman, G. M. (1990). "Synthetic Neural Modeling: A Multilevel Approach to the Analysis of Brain Complexity". In: "Signal and Sense: Local and Global Order in Perceptual Maps". Edelman, G. M., Gall, W. E. and Cowan, W. M. (eds). Wiley-Liss, New York, pp. 607–707. Edelman's machines Darwin I, II, III ... are described on pp. 647–699.
141. Wheeler, J. A. (1996). "At Home in the Universe" in the series of "Masters of Modern Physics", Chapter "Beyond the Black Hole", pp. 271–294; Chapter "It from Bit", pp. 295–311. Springer-Verlag New York, Inc.
142. Shannon, C. E. A. (1948). "Mathematical Theory of Communication". *Bell Syst. Tech. J.* 27: 379–423, 623–656. Shannon-Weaver-Model. For Shannon's "Main Theorem" see Chapter 3.
143. Hamming, R. W. (1980). "Coding and Information Theory". Prentice Hall, Englewood Cliffs, NJ.
144. Schwarzschild, K. (1916). "Über das Gravitationsfeld eines Massenpunktes nach der Einstein'schen Theorie". *S.-B. Preuss. Akad. Wiss., Kl. Math. – Phys.-Tech.* 189–196.
145. Hawking, S. W. and Penrose, R. (1970). "The Singularities of Gravitational Collapse and Cosmology". *Proc. R. Soc. London* A. 314(1519): 529–548.

146. Bekenstein, J. D. (1973). "Black Holes and Entropy". *Phys. Rev.* D 7: 2333–2346.
147. Hawking, S. W. (1975). "Particle Creation by Black Holes". *Commun. Math. Phys.* 43: 199–220.
148. Penrose, R. See Ref. 121.
149. Hawking, S. W. (1976). "Black Holes and Thermodynamics". *Phys. Rev.* D 13: 191–197.
150. 't Hooft, Gerard (1993). "Dimensional Reduction in Quantum Gravity". Essay dedicated to Abdus Salam. In: "Salam Festschrift". World Scientific Series in 20th Century Physics, Vol. 4. Ali, A., Ellis, J. and Randjbar-Daemi, S. (eds.). Utrecht preprint THU-93/26. gr-qc/9310026.
151. Strominger, A. and Vafa, C. (1996). "Microscopic Origin of the Bekenstein-Hawking Entropy." *Phys. Lett. B* 379(1–4): 99–104.
152. Misner, C. W., Thorne, K. S. and Wheeler, J. A. (1973). "Gravitation". W. H. Freeman, San Francisco.
153. Smolin, L. (1977). "The Life of the Cosmos". Oxford University Press, Oxford and New York.
154. Susskind, L. Susskind's approach to string theory is described in his book, quoted in Ref. 156.
155. Penrose, R. (2004). See Ref. 120.
156. Susskind, L. (2005). "The Cosmic Landscape: String Theory and the Illusion of Intelligent Design". Little, Brown & Co, New York and Boston.
157. Haken, H. (1964). "Statistische Nicht-Lineare Theorie des Laserlichts", ["Non-linear theory of laser noise and coherence. I"]. *Z. Phys.* 181: 96–124. See also: Graham, R. and Haken, H. (1968). "Quantum Theory of Light Propagation in a Fluctuating Laser-Active Medium". *Z. Phys.* 213: 420–450; Graham, R. and Haken, H. (1970). "Laser Light—First Example of a Second-Order Phase Transition Far Away from Thermal Equilibrium". *Z. Phys.* 237: 31–46.
158. Bethe, H. and Heitler, W. (1934). "On the Stopping of Fast Particles and on the Creation of Positive Electrons." *Proc. R. Soc. A* 146: 83–112; Heitler, W. (1957). "The Quantum Theory of Radiation", 3rd edn. Clarendon Press, Oxford, Chapters 25 and 26.

Conclusion

The contents of this book are in many ways reflected by its title. The main focus is on the physical nature of information and its role in life processes. Physicists today are familiar with the quantitative aspect of information as introduced by Claude Shannon in his "theory of communication". A communication channel must work for any kind of information, independent of its content. What counts in Shannon's theory is: how much information can be transmitted in the shortest possible time and with the highest possible fidelity. In biology we are interested in a different issue: finding out how the information of life has come about. For this, we have to occupy ourselves with the question of how the semantics of biological information became fixed in our genes. The path of the evolution of life, starting from some complex random chemical background, is guided by a logical scheme shown in Vignette C1.

Vignette C1: The "strange simplicity" of the evolution of life.

Our theoretical approach to life is based on three statements quoted repeatedly in this text:

1) Theodosius Dobzhansky: "Nothing makes sense in biology except in the light of evolution".

2) Charles Darwin, whose greatest achievement was establishing natural selection as the chief cause of biological evolution, said toward the end of his life: "I believe [...] that the principle of continuity renders it probable that the principle of life will hereafter be shown to be a part, or consequence, of some general law".

3) Eugene Wigner: "Physics doesn't deal with processes. Physics deals with regularities, and only with regularities among processes".

Cont. ➲

➲ Cont.

Life, whatever else it may be, is certainly a regularity among material processes. No living being comes about by a means other than reproduction. Hence physics should – at least – deal with this regularity of matter.

Figure C1 of this vignette presents the scheme that any theory of the origin of life should deal with. A constitutive aspect appears to be the feedback cycle sketched into the centre of the scheme, to which the theory (expressed in vectorial form on the blackboard in Figure C2) refers. It is an abstract formulation of John A. Wheeler's thesis "It from Bit" (Figure C3), here applied to the biochemical problem of the genotype–phenotype dichotomy (Figure C4).

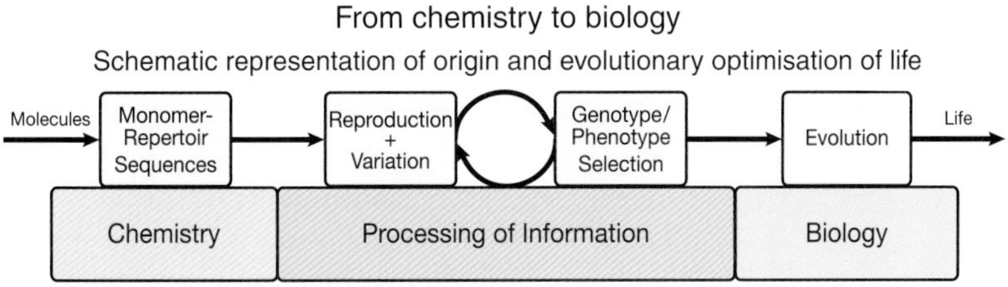

Figure C1

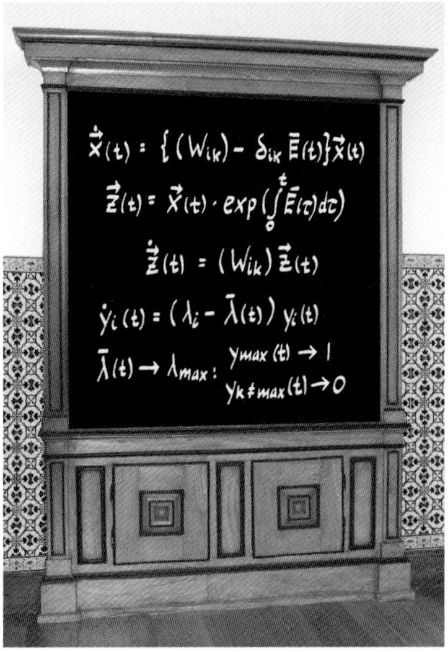

Figure C2

Cont. ➲

Cont.

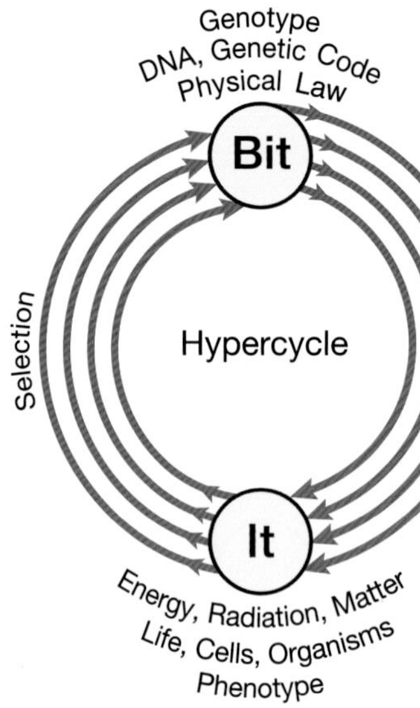

Figure C3

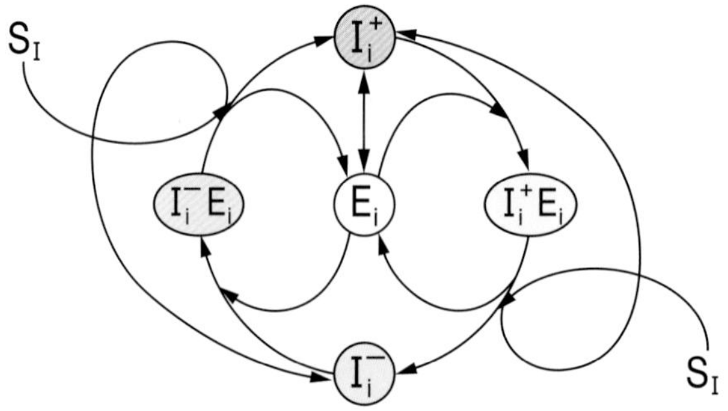

Figure C4

The whole process started in a chemically rich environment. With the transition of molecular diversification into informational complexity the door was opened for an unlimited process of evolution, from chemical reaction cycles to biological cells and networks – leading eventually to the complex structure of the human brain. The work described in this book provides a formal theoretical proof of Darwin's results that were culminating in formulating the principle of "Natural Selection". However, the interpretation of this principle, that has developed since the days of Darwin and Wallace, does not reflect our modern understanding of an evolution starting in a chemical environment and developing via self-organisation of its typical autonomy consisting in optimal biological functionality. Theory as such can only tell us which properties are necessary and where additional functions (for instance, as provided by enzymatic catalysis or by cellular compartmentation) are required. It cannot specify which of the functions, now being active in life, were available right at the beginning. It is the "if-then" logic on which processes are based, which today can be subjected to experimental proof. The complexity of any present biological system is so large that I do not see any other than an evolutionary way for it to come about in a reproducible form. I could not find any alternative theoretical approach to the problem of life and its complexity.

The most important question that a conclusion should answer, reads: What is new in this book? One of the most important introductions into the 20th century physical theories was the use of matrix algebra as was used in quantum mechanics. That is also true for the "biological" theory of the origin and evolution of life. It is required for a correct representation of multiple couplings among the various mutants present in any distribution of genes. The theory answers the question: "What is natural selection?" and comes up with the solution: a true "phase transition" according to all physical criteria being applied, however, not in physical space, but rather in information space. Another unique particularity, not observable in inanimate systems, is the "genotype–phenotype dichotomy"; its most important consequence is the basic non-linearity of the chemical reaction mechanism, which comes to expression in the origin, but has to be suppressed in the further evolution of life. Darwinian selection requires a quasi-linear chemical mechanism, because advantageous single mutants have to be able to compete with established populations. Apart from fulfilling this requirement by quasi-linearity of the chemical mechanism, the non-linearity of the selection process is due to a superimposed term which becomes effective under steady-state conditions.

In the following, I am going to stress some important points:

(1) The population of a certain genotype consists of a mutant distribution in which coupling exists (in principle) among all individual members. The original theoretical treatment in population genetics focused only on single individuals as "winners" in a "competition for existence". Among these, there would have been quite a large number of (nearly) neutral mutants, which should have had some possible influence on their respective "wild types". It turned out to be difficult even to define the term "neutral". The question "How neutral is 'neutral'?" is on a par with "How equal is 'equal'?". Anyway, the mathematically adequate approach is to consider each individual type of mutant and write down the kinetic (differential) equation for its formation or disappearance. Taking all mutants into account leads to a huge set of coupled

equations. Here, “huge” means a (small) gene that includes (only) 300 nucleotides as code units allows for $4^{300} \approx 10^{180}$ different mutant types, and the number of possible solutions for such a set of equations is comparably large. The size of this number is almost impossible to grasp: compare the age of the universe in seconds (not even 10^{18}) or its diameter in centimetres (a mere 10^{29}). Owing to the mutational coupling between individual genotypes, the solutions for such a set do not refer any more to individual types, but rather to combinations of terms for many (possibly all) mutants. The solutions are the eigenvalues of a matrix $((W_{ik}))$ determining the kinetics of the coupled reaction system. The set of those equations was first published in my 1971 paper in *Naturwissenschaften*, quoted in the text (see Chapter 4 and Appendix A4). The solutions usually represent a huge spectrum of eigenvalues.

However, there was a wonderful surprise: The properties of this special set of equations are such that the normal modes belonging to the eigenvalues except the largest one die out. This is described by the mathematical theorem of Perron and Frobenius (quoted in Section 4.7), according to which the largest eigenvalue of such a matrix (which has only non-negative entries) is “single, positive and real”.

This was a great relief to us. Dealing with many billions of mutants would have rendered our attempts at a solution of the problem completely futile. Now, however, we can say that the term “fittest survivor”, as chosen by Darwin, is formally justified, although nobody in Darwin's time could explain what it really meant, other than “something that is able to survive”. It is important to note that the eigenvalue of a matrix is an invariant that does not change during the whole process. The term “fittest” here does not refer to any single individual rather it is a property of the whole distribution of mutants, which we call a “quasispecies”. The maximum eigenvalue contains contributions from most of the members of the quasispecies. In other words, the wild type, in the sense of a fittest individual, does not in general exist. “Fitness” is a property of the whole quasispecies.

The fact that selection is not the business of isolated individuals, but rather a matter for the population of a complete quasispecies, has much in common with the comparison of a dictatorship with a democracy. The comparison is only incomplete insofar as the voter in a democracy does not have to pass an ability test, as each individual in the quasispecies has to do. The result of the test in the quasispecies gives a weight to the vote of each individual that is absent in political voting.

(2) One of the most interesting aspects of the quasispecies is that the process in which it emerges has to be described as a phase transition. This is correct according to all the criteria used in physics to define phase transitions. However, our process does not take place in physical space and therefore cannot be observed in the usual sense of the word. Rather, it occurs in information space, which we do not perceive with our senses. The most direct access to it is provided by sequence analysis, with the help of which we are able to “screen” the sequence space, looking for suitable regions of fitness.

In physics, phase transitions are classified by their order. First-order phase transitions, such as the transition from solid to liquid or from liquid to gas, are generally characterised by a latent heat, i.e. by an abrupt or jump-like change of a state parameter such as enthalpy. Phase transitions of second (or higher) order show singularities in their higher derivatives, e.g. in heat capacities etc. Typical second-order transitions are observed in ferromagnetism, in antiferromagnetism, in low-temperature physics (superconductivity, superfluidity) and – last, but not least – in the neighbourhood of critical points, such as the liquid–gas transition, where the densities of liquid and gas approach the same value. Owing to their common behaviour, we today generally call these second-order phase transitions "critical" phenomena (see N. Goldenfeld, *loc. cit.*, Appendix A2). Many analogues of this can be found in the kinetic phenomena of natural selection. The presence of isolated regions of sequence space with clear advantages in fitness leads to first-order phase transitions, as illustrated in the two examples presented by Peter Schuster and Jörg Swetina (see Figures 5.5.1 and 5.5.4). The second example is particularly instructive, since it clearly demonstrates the quasispecies nature of what would earlier have been called an "individual wild type". Selection processes that occur in analogy to "critical phase transitions", with macroscopic fluctuations among neutral regions near the critical error threshold, will be highlighted as a separate item because of their special importance with respect to the speed of evolutionary adaptation.

Consider a liquid–gas transition at the critical point, where the densities of liquid and gas become equal. We have three types of region: liquid, gas and mixtures of these. We observe "critical opalescence", which tells us that the fluctuating regions have sizes up to macroscopic dimensions. In a mutant distribution close to the critical error threshold we also should find macroscopic fluctuations, as we know from the work of Motoo Kimura. Using stochastic theory, based on the Fokker–Planck equation as well as on Kolmogorov's *Ansatz*, Kimura was able to show that single neutral mutants of RNA or DNA that are selected under critical conditions occur with large fluctuations and thereby mimic evolutionary changes despite being optimally adapted. This, of course, is a very important discovery, and in my view it has not received the attention among biologists that it deserves. It explains the observed fact that genetic sequences do not stop changing, in this case among macroscopic neutral regions, although they have become optimally adapted. Especially in higher organisms, they prepare the system to remain flexible in case of unexpected changes in their environment. Nigel Calder, generally praised for his exceptional clarity and style as a writer, strongly defended the – as he calls them – "Japanese heretics" in his recent *Magic Universe: The Oxford Guide to Modern Science*, in an essay entitled "Molecules Evolving". Calder mentions not only Kimura, but also his early co-worker Tomoko Ohta, who included "nearly neutral" genes in Kimura's theory, and his contemporary Susumu Ohno, who advocated the hypothesis of gene duplication, an idea that found strong support in the studies of Hans-Werner Mewes with yeast genes at the Max Planck Institute for Biochemistry in Munich. All this work has shown that neutral mutants do a lot more than just travel along as "hitch-hikers of molecular evolution".

In considering natural selection to be a "phase transition" of a total population, we interpret "neutral selection" as a "critical phenomenon" that includes densely populated and highly extended regions, covering large mutant distances. Unfortunately, the process is much more complex than a critical phase transition in physics between two states in media of homogeneous composition such as a liquid and a gas near their critical point, or two domains in magnetic material at the Curie temperature (where transitions between para- and ferromagnetism occur). The latter is true for the Néel temperature in the analogous case of anti-ferromagnetism. Using the renormalisation group, Kenneth G. Wilson found a general solution for these kinds of systems showing critical phase transitions. The greater complexity of critical phase transitions in information space results from the much higher complexity that is consequent upon the many dimensions involved in information space. Nucleation and growth of cooperative couplings depend on the structure of the fitness landscape, in particular on the presence of "bridges" among (nearly) neutral fitness regions.

After coming up with a theory that offers such a straightforward and articulate solution, one may ask – in view of the expected complexity of the mechanisms involved – how much one can learn about the general nature of biological evolution. In the text I have mentioned the example of quantum-mechanical theory, which in chemistry is required for a full understanding of the covalent bond, but with regard to the formation of any (possibly quite complex) organic compound may also require information about more details. The same is true for any specific application of evolutionary theory.

Molecular evolutionary theory tells us that for the storage of information in macromolecular structures *de novo* processes (to be expressed by constant terms) can be completely neglected owing to their exceedingly small probability of appearance. The same is true for terms of higher order than linear in the case of optimal adaptation. All non-diagonal terms in the matrix of rate coefficients, then, are represented exclusively by positive entries, causing the matrix to have only one stable eigenvalue.

One may still ask how it comes about that so many different reactions that do appear in the molecular storing of information adhere to the restricting condition of being linear. All living beings come about by reproduction. In his novel *The Magic Mountain* (*loc. cit.*, Section 5.1), Thomas Mann expresses it with the words: "... no life form could be proved that did not owe its existence to propagation by a parent". For populations of living beings one quite generally observes growth laws that are close to exponential. An exponential function represents the solution of a linear differential equation, as is required for selection according to the Darwinian model. In Section 5.3 we have described in great detail an experimental and theoretical study of the infection of a colony of *E. coli* cells by the bacterial virus Qβ. The mechanism consists of many steps, including an initial one resembling an ideal hypercycle. However, the overall mechanism can be expressed in the form of a linear differential equation with a universally found time constant of 40 minutes, after which infected cells lyse and proliferate new viruses. Likewise, a tremendously more complex human population also grows, according to an exponential or nearly exponential law, as proven by frequent observations. Some of the possible reasons for such a behaviour have been discussed in the text.

I mention this problem in the Conclusion because its solution is of great importance for understanding the speed of evolution, given the huge complexity of information space. Single isolated points of high fitness, even if they are surrounded by Poissonian distributions of mutants, would not allow efficient screening of large portions of their total information content within sufficiently short times because of the extreme narrowness of Poissonian distributions. You would not be sitting here reading these lines if evolution had not used tricks of the kind described above. On the other hand, those tricks are mainly responsible for what the biologist calls homeostasis. It provides some constant internal conditions which may be responsible for the observed quasi-linearity of many evolutionary mechanisms.

(3) Do the new insights gained in this book have any practical use or benefit? All living beings, despite – but ultimately because of – their tremendous complexity are subject to the "physical" law of natural selection. Darwin lived 150 years ago. So how can a physical understanding of his theories provide us with new practical ideas? In large parts of this book I chose a relatively thorough representation of the physics of inanimate matter in order to emphasise the particular characteristics of animate matter which ultimately add up to "semantic information" and its material origin. In view of the tremendous complexity that we encounter in viable matter, I have always tried to bear in mind my own warning that "pure theory [at least in biology] is poor theory". Therefore, much effort in our work has been dedicated to experiments involving possible systems of early (molecular) evolution. In this volume I could only describe individual examples, mostly in Chapters 4 and 5. They show that Darwinian theory, in order to describe "origins", has to be supplemented by non-linear terms. We call those systems which control the first steps in the origin of life "hypercyclic". Hypercycles are feedback cycles that establish the dichotomy of genotype and phenotype.

Needless to say, the main spade-work – including the development of a new technology – was of an experimental nature, and took about 80% of our time. So far, no serious disagreements with the results of theory have emerged. Meanwhile, evolutionary biotechnology has found its way into the industrial world: once more, Nature has acted as a mentor for the development of refined products in chemistry and especially in pharmacology.

(4) Finally, let me emphasise Nature's methods as an optimal way of solving problems that involve the origin and evolution of information in a more general sense. This has been clearly demonstrated for the life process. Proteins, which dominate the molecular executive in present-day organisms, took over this role because they are

unmatched by any other class of molecules; however, in order to play it out in a reproductive way they had to know how to "read and write", a property inherent to nucleic acids. This is what we have learnt from theory and experiment, and it is still reflected by present-day organisms which in their genetic apparatus up to the level of translation make use of RNA, in both its legislative (information processing) and executive (ribozymic) functions. Our studies of the age of the genetic code suggest that this has been the case for the past 4000 million (4×10^9) years. Why is this so? Because only RNA, as a class, fulfils all the necessary prerequisites for a molecular origin and adaptive evolution of semantic information.

As a physicist I immediately ask whether an analogous process could be responsible for other phenomena in inanimate matter in which information has to be generated and adapted to some optimised performance. An obvious example of such a process, discussed in this book, is the production of coherent light by lasers (Section 3.9). In particular, I asked the above question after it was stimulated by an essay of the late John Archibald Wheeler with the title "It from Bit". His essay was concerned with nothing less than the origin of our universe. "Bit" stands for "information", here in the form of the physical laws, while "It" – in Wheeler's words – means "every particle, every field of force, even the space–time continuum itself..." which may have an immaterial source and explanation. Yes, that's what I meant by semantic interpretation; but is it meaningful to ask which came first? Isn't this a kind of "the chicken or the egg" question? Didn't both have to come about at the same time in a feedback cycle, e.g. a hypercycle as suggested in Figure C4? This would lead us to think about the origin of the universe in the way that is suggested by our abstract phenomenological theory of origins.

In cosmology, theories are entertained that start from a huge set of parallel universes which – appearing in different coordinates of multidimensional space – do not interact with one another. Their huge multiplicity results from the precision which the values of natural constants have to achieve in order to produce a universe like ours, in which life and thought could develop. According to a theory that our universe would be just one among a huge number of alternatives in which anything like this could have happened. Similar theories also were discussed (and developed ad absurdum) in earlier phases of molecular biology.

I must say that models of this kind, i.e. ones that produce sufficiently large numbers of individual solutions in order to find one that fortuitously fulfils the constraints of low probability, are fairly primitive in comparison with any of the evolutionary processes. The theory proposed in this book is a phenomenological one, based on general physical prerequisites. The process of optimal adaptation requires (positive) linear terms. Constant terms are excluded because of the vastly low probability of *de novo* formation, while terms of higher than linear order introduce difficulties in an up-growth of new solutions. Such systems can occur only during growth phases.

Present-day particle physics suggests a possible existence of those phases during the "freezing out" of interactions, starting from one universal interaction at small distances (on the Planck scale) with the "freezing out" of gravitation, followed by strong nuclear, then weak and finally electromagnetic interactions. An example, for instance, is the "inflation" period put forward by Alan Guth.

A procedure in which one would systematically test the physical details that determine the overall process has a great advantage over simple *ad hoc* assumptions. I am not a particle physicist, but the theoreticians among them might be interested in such a procedure. Against the background of my own work, evolutionary models such as the one described by Lee Smolin are interesting because of their "evolutionary" nature, but they should avoid implausible *ad hoc* assumptions.

Anyway, these are questions which, for the time being, I will leave as questions, ones that will require a lot more theory and observation before they can be solved.

Almost every year we are confronted with new discoveries brought about by experiment and observation in particle physics and cosmology. However, some of these appear to me so far to be not much more than an introduction of "new words". Let me use as an example one of the latest new ideas: "dark energy".

What is dark energy? Up to now it is just a new term. At the turn of the 18th/19th centuries, the German poet Johann Wolfgang von Goethe in his masterpiece *Faust* made some unflattering comments about the way in which science sometimes purports to introduce new ideas by simply creating new words. In *Faust* it is the Devil, dressed in the gown of the scholar Faust, who tells a young student keen to understand how science is done (in my translation):

> "Wherever an idea is lacking,
> A suitable word will soon be found."

So what does the new term "dark energy" actually mean? As we do not even know what energy really is (Section 1.1), we are entirely helpless when it comes to "dark energy". One might think it an analogue of "dark matter", i.e. matter inside a black hole. However, this is highly misleading. Although we do not know the exact fate of matter within the black hole's event horizon (Section 1.7), at least the phenomenon of black holes – huge pieces of definable mass that have collapsed under their own gravitational force – is well established in present-day physics.

This is not the case for "dark energy". Nobody is sure what it really means. Its existence was proposed on the basis of measurements of the red shift of the light from supernovae in distant galaxies in relation to their brightness. This observation told us that the expansion of the universe was once slower than it is today. Therefore the universe is expanding at an ever-increasing rate, and additional energy might be a possible explanation. It is called "dark energy" because we cannot "see" it, but sometimes I wonder whether the real reason is that we are completely in the dark about its nature and origin. In fact there are other possible explanations for the expansion of the universe, for instance that we live in a region of space that is emptier than average.

Thus, the jury is still out on "dark energy", as it is also for the origin of information of the physical laws that guided the start of our universe. However, the question may one day be clarified by precise measurements. As we have seen repeatedly in biology: pure theory can be poor theory. In other words, our thinking has to be guided by empirical experience, obtained by observation and experiment.

In this book I have tried to pave a consistent path from the "strange simplicity" to the "complex familiarity" of our world. The subtitle of this work says that it is a "treatise on matter, information, life and thought". So far, I have not really talked systematically about life and thought. Where I occasionally mentioned examples, I used them in order to demonstrate their physical background, rather than their biological role.

In the past 50 years I have spent my time doing both theoretical and (even more so) experimental work in molecular biology. The present volume has outlined some of the principles that my work and that of many others have brought to light. A second volume is planned, in which the main emphasis will be on "life, thought and human culture" – keeping in mind their physical background. Its themes and topics will continue along the path set out in the present volume:

- Self-organisation and life
- Life and living beings
- Living beings and thought
- Thought and culture
- Culture and the future.

The second volume will be a joint scientific (ad)venture together with several colleagues who have occupied themselves with these matters for many years. At my age I cannot be sure that I shall be able to complete this task. However, I am confident that someone will take over the baton before it falls out of my hand, in order to continue the never-ending relay race of scientific endeavour and progress.

Appendix A1.1

Manifestations of Energy in the Physical Universe

1. Dimensions and abbreviations

Quality	Standard dimension	Explanation
Length	metre [m]	1 Ångström [Å] = 10^{-10} [m]
		1 light-year [ly] = 9.46×10^{15} [m]
		1 parsec [pc] = 3.0857×10^{16} [m]
Time	second [s]	1 day [d] = 86400 [s]
		1 year (of 365 [d]) = 3.1536×10^{7} [s]
Mass	kilogramme [kg]	= 10^3 [g]
Force	Newton [N]	= [kg m s^{-2}]
Energy	Joule [J]	= [kg m^2 s^{-2}] = 6.25×10^{18} [eV] = 1 Volt-Ampère-second [V A s]
	calorie [cal]	= 4.184 [J]
	electron volt [eV]	= 1.602×10^{-19} [J]
Power	Watt [W]	= [J/s] = 1 Volt-Ampère [V A]
Electric charge	Coulomb [C]	= Ampère-second [A s]
	1 Faraday [F]	= 9.65×10^4 [C]
		charge of electron e = 1.602×10^{-19} [C]
Temperature	centigrade [°C]/Kelvin [K]	T difference of 1 [K] equals T difference of 1 [°C]

Prefixes combined with the standard measures are: c = centi-(10^{-2}), m = milli-(10^{-3}), μ = micro-(10^{-6}), n = nano-(10^{-9}), p = pico-(10^{-12}), f = femto-(10^{-15}), a = atto-(10^{-18}), k = kilo-(10^3), M = mega-(10^6), G = giga-(10^9), T = tera-(10^{12})

2. Ranges of energy

The assignment of dimensions of energy started independently in different areas of physics. The mechanical standard of energy, the Joule, expresses the energy required to accelerate a mass of 1 kilogramme from the resting state to a speed of 1 metre per second. It agrees with the definition of the unit of electrical power of 1 watt [W], dissipated in a conductor carrying 1 ampère [A] of current between two points of 1 volt [V] potential difference within 1 second, i.e. 1 [J] = 1 [W s] = 1 [V A s]. Energy in the form of heat was originally assigned the unit "calorie", which is the amount of energy necessary to heat 1 gram of water from 15 to 16 [°C]. The modern unit of 1 kilocalorie, still used by many chemists in connection with reaction enthalpies and entropies, and popular in the nutritional evaluation of food, is redefined to be precisely 1 [kcal] = 4.184 [kJ]. Dealing with thermal energies of $\sim$0.6 kcal mole^{-1} near to room temperature (T = 300 [K]) the kilocalorie is the unit closest to our experience.

The physics of our time favours the electron volt as a comparative unit of energy. Here we have to take notice of the fact that the electron volt is a unit that refers to the properties of single particles, its definition being the energy that an electron takes up when it is accelerated by a potential difference of 1 volt. The joule, on the other hand, refers to the kilogramme, which is a macroscopic unit. The two worlds – macroscopic experience and the microcosm – are bridged by Avogadro's (or Loschmidt's) constant: $N_A = 6.022 \times 10^{23}$. It tells us how many molecules are contained in 1 mole, defined as M grams of a substance whose molecular weight is M, where M = 12 refers to the atomic form of carbon (C). Thus 1 mole of water weighs 18 grams, or 1 [kg] of water contains 55.55...moles. Hence, one electron volt is "only" 1.602×10^{-19} [J], but it characterises a reaction enthalpy of nearly a hundred thousand (97081) joules or 23.2 [kcal] per mole, i.e. a respectable chemical reaction (or activation) enthalpy. A Boltzmann factor of $e^{-1[\text{eV}]/k_B T}$ at T = 300 [K] reduces a reaction rate by nearly 17 orders of magnitude.

The following energy ranges are typical for the various areas in physics:

Thermal energy $k_B T$
(k_B = Boltzmann's constant: 8.62×10^{-5} [eV · K^{-1}])

Temperature	Range	Energy
10^{-9} to 10^{-8} [K]*	Bose–Einstein condensation	10^{-12} to 10^{-13} [eV]
2.17 [K]	Superfluid helium	(1 to 4) $\times 10^{-4}$ [eV]
300 [K]	Room temperature	0.026 [eV]
$\sim 10^{8}$ [K]	Temperature inside fixed stars	$\sim$10 [keV]
$\sim 10^{32}$ [K]**	Planck range (see Section 1.6)	$\sim 10^{19}$ [GeV]

*Lowest temperature reached so far in experiments.
**Beyond present-day physics.

3. Chemical interactions

According to Coulomb's law: $e^2 = 1.44$ [eV · nm],
where 1 [eV] = $1.602 \cdot 10^{-19}$ [J] and 1 [J] = 1 [$kg^{1/2} \cdot m^2 \cdot s^{-2}$].
Hence $e = \sqrt{1.44 \text{ [eV} \cdot \text{nm]}}$ in equivalent mechanical dimensions
becomes $e = 1.52 \cdot 10^{-14}$ [$kg^{1/2} \cdot m^{3/2} \cdot s^{-1}$].

Compound	Interaction	Energy
H atom	proton and electron at distance a_o^*	27.22 [eV]
Na^+, Cl^-	in water at a distance of 10 [Å]**	0.018 [eV]
$H^+ + OH^- \rightarrow H_2O$	in liquid water at T = 300 [K]***	0.93 [eV]
$ATP^{4-} \rightarrow ADP^{3-} + P^-$	in liquid water at T = 300 [K]***	0.315 [eV]
H_2 molecule	chemical bond	4.5 [eV]
N_2 molecule	chemical bond	9.7 [eV]
outer electrons	ionisation energies	[eV]
inner electrons	in heavy atoms	[keV]

*a_o = Bohr radius = 0.0529 [nm].
**Hydrated ions in ion cloud.
***Dissociation equilibrium.

4. Electromagnetic radiation

(with h = Planck's constant: 4.14×10^{-15} [eV s])

Frequency ν[Hz]	Wavelength λ	Range	Energy
3×10^9 to 3×10^5	10^{-1} to 10^3 [m]	Radio	4×10^{-10} to 1.24×10^{-5} [eV]
3×10^{12} to 3×10^9	1 [mm] to 10^{-1} [m]	Microwave	1.25×10^{-5} to 0.0125 [eV]
4×10^{14} to 3×10^{12}	700 [nm] to 1 [mm]	Infrared	0.0125 to 1.66 [eV]
7.5×10^{14} to 4×10^{14}	400 [nm] to 700 [nm]	Visible light	1.66 to 3.1 [eV]
3×10^{17} to 7.5×10^{14}	1 [nm] to 400 [nm]	Ultraviolet	3.1 to 1.25 [keV]
3×10^{20} to 3×10^{17}	1 [pm] to 1 [nm]	X-rays	1.25 [keV] to 1.25 [MeV]
$>3 \times 10^{20}$	<1 [pm]	γ-rays	>1.25 [MeV]

5. Matter: mass energies

(with c = velocity of light in a vacuum = 2.99792458×10^8 [m/s])

Particle	Existence	Interaction	Mass energy
electron: e^-	free, atomic shell, metals	electro weak	0.511 [MeV]
muon: μ^-	free (τ)	electro weak	105.7 [MeV]
tauon: τ^-	free (τ)	electro weak	1.777 [GeV]
electron neutrino: ν_e	free, atomic nuclei	weak	<7 [eV]
quarks: u, d	neutron, proton, meson	strong	$\left.\begin{matrix} u{\sim}5 \\ d{\sim}10 \end{matrix}\right\}$ [MeV]
gluons:	atomic nuclei	strong	<10 [MeV]
W^+, W^-, Z^0	free, atomic nuclei	weak	$\left.\begin{matrix} 80 \\ 91.2 \end{matrix}\right\}$ [GeV]
proton: p^+	free, atomic nuclei	strong electromagnetic	938.27 [MeV]
neutron: n	atomic nuclei, free (unstable)	strong	939.57 [MeV]
atomic nuclei	atoms	strong + electroweak	[GeV] range

6. Cosmos: gravitational energy

(with G = Newton's constant = 6.6726×10^{-11} [$m^3\ kg^{-1}\ s^{-2}$])

Celestial body	Mass [kg]	Radius [km]	Acceleration* g [$m \cdot s^{-2}$]	Potential energy* [eV] Proton (p)	Electron (e)
Earth	5.97×10^{24}	6380	9.78–9.81	0.65	0.00035
Sun	1.99×10^{30}	696000	274	1990	1.08
White dwarf	$\sim 1.4 > 3 \times M_{sun}$	$\sim 10^4$	$\sim 2 \times 10^6$	220000	120
Neutron star	$\sim 1.4 \times M_{sun}$	~ 10	$\sim 2 \times 10^{12}$	220 [MeV]**	120000**
Black hole	$>3 \times M_{sun}$	~ 10***	singularity, no individual existence		

*Values refer to the surface $g = \frac{GM}{r^2}$. Potential energy $= \frac{GM}{r} \times \begin{cases} m_p \\ m_e \end{cases}$

**The values approach the order of magnitude of mass energies. Relativistic treatment!

***For masses > $3\,M_{sun}$ there is no stable existence of individual particles in the black hole. As "radius" the event radius, which for spherical symmetry is identical with the Schwarzschild radius $r_s = 2\,GM/c^2$, has been listed.

Appendix A1.2

Mathematical Concepts in Physics

Peter Richter*

1. Scalars, vectors, tensors

Physics and mathematics have somewhat different notions of scalars, vectors, and tensors. The difference is that in mathematics, these objects are defined in an abstract way, without reference to a space in which they "exist", whereas in physics, they are conceived as objects in space and/or time, so-called "fields". In this book, we usually imply the physical meaning, but for an explanation it is convenient to start with some mathematics.

Scalars derive their name from the property that they can be read off a scale, hence mathematically they are numbers. Of the different types of numbers such as natural numbers, integers, rationals, real numbers, complex numbers or even quaternions, the most common scalars are either real numbers (in classical physics) or complex numbers (in quantum mechanics).

Vectors are objects with several components, such as $\boldsymbol{v} = (v_1, v_2, \ldots, v_n)$ where the number n of independent components (which may be infinite) is called the dimension of the vector space of possible n-tuples $\boldsymbol{v}$. The components v_i are usually real or complex numbers, but sometimes they are more involved objects like differential operators, for example. For objects $\boldsymbol{u}, \boldsymbol{v}, \boldsymbol{w} \ldots$ to qualify as elements of a vector space, an addition $\boldsymbol{u} + \boldsymbol{v}$ must be defined, and it must be possible to multiply them by certain scalars, with the usual laws for addition and multiplication. For example, the space of points $\boldsymbol{r} = (x, y, z)$ in a three-dimensional Cartesian system (the 3D Euclidean space) is a vector space over the real numbers because given two points $\boldsymbol{r}_1 = (x_1, y_1, z_1)$ and $\boldsymbol{r}_2 = (x_2, y_2, z_2)$, the sum $\boldsymbol{r}_1 + \boldsymbol{r}_2 = (x_1 + x_2, y_1 + y_2, z_1 + z_2)$ is again a point, and so is any $\lambda\boldsymbol{r} = (\lambda x, \lambda y, \lambda z)$ if λ is a real number. Vectors are the objects of linear algebra. Most common vector spaces are equipped with

*Address: Institut für Theoretische Physik, Universität Bremen, Otto-Hahn-Allee, 28359 Bremen, Germany, email: prichter@uni-bremen.de

a structure called *scalar product*: given two vectors $\boldsymbol{u}$ and $\boldsymbol{v}$, their product $\boldsymbol{u} \cdot \boldsymbol{v}$ is defined to be a scalar, i.e., a number. For example, a natural scalar product in the 3D Euclidean space is given by

$$r_1 \cdot r_2 = x_1 x_2 + y_1 y_2 + z_1 z_2. \tag{A1.2.1}$$

With this scalar product, it is possible to express the length r of a vector $\boldsymbol{r}$ as $r^2 = \boldsymbol{r} \cdot \boldsymbol{r} = x^2 + y^2 + z^2$ (Pythagoras).

Tensors are objects with two or more indices, like T_{ij} where both i and j run from 1 to n so that there are altogether n^2 components. Tensors of rank 2 have 2 indices, tensors of rank 3 have 3 indices, and so on. They generalize the notion of vectors which may be viewed as tensors of rank 1. Again, it must be possible to add tensors (of identical rank), and to multiply them by scalars. A way to generate tensors of rank 2 from vectors $\boldsymbol{u}$, $\boldsymbol{v}$ is to consider the dyadic product $T_{ij} = u_i v_j$.

Given a tensor T_{ij} , it is possible to obtain a scalar T by the operation called contraction: $T = \sum_i T_{ii}$. If T_{ij} is the dyadic product of two vectors $\boldsymbol{u}$ and $\boldsymbol{v}$, then the contraction $T = \sum_i u_i v_i$ is their scalar product.

A common use of tensors is to "apply" them to vectors. The result of this application is another vector: $\boldsymbol{v} \mapsto \boldsymbol{v}'$ where $v'_i = \sum_j T_{ij} v_j$. If $T_{ij} = u_i w_j$ is the dyadic product of vectors $\boldsymbol{u}$ and $\boldsymbol{w}$, then the action of T_{ij} on any vector $\boldsymbol{v}$ gives a vector in the direction of $\boldsymbol{u}$:

$$\sum_j T_{ij} v_j = u_i \; \boldsymbol{w} \cdot \boldsymbol{v}. \tag{A1.2.2}$$

So much for the mathematical background. Let us now discuss the more specific terminology of physics. We begin with classical Newtonian physics where space is conceived as the 3D Euclidean space, and time as a real scalar. The two are completely independent of each other. The notion of vectors and tensors is then closely tied to the particular nature of Cartesian vectors in 3-space. But what *is* the nature of these Cartesian vectors $\boldsymbol{r} = (x, y, z)$? If we view them as points in physical space, then the values (x, y, z) depend on how we choose the orientation of the Cartesian coordinate system. The same point $\boldsymbol{r}$ is represented by different sets of coordinates (x, y, z) depending on how the axes are chosen. For example, if K is the original coordinate system and K′ obtained from K by a rotation about the z-axis, with angle α, then we have

$$\begin{pmatrix} x' \\ y' \\ z' \end{pmatrix} = \begin{pmatrix} \cos\alpha & \sin\alpha & 0 \\ -\sin\alpha & \cos\alpha & 0 \\ 0 & 0 & 1 \end{pmatrix} \begin{pmatrix} x \\ y \\ z \end{pmatrix} \tag{A1.2.3}$$

It turns out that the behavior of (x, y, z), for the same physical point, under general *rotations* of the coordinate system, is their most important characteristic. If $\mathbf{D}$ is used as a symbol for rotations, we write the relation between the coordinates in the two systems $\boldsymbol{r}' = \mathbf{D}\boldsymbol{r}$, or in components

$$r'_i = \sum_j D_{ij} r_j. \tag{A1.2.4}$$

The essential property of rotation matrices is $\mathbf{D}\mathbf{D}^t = 1$, or in components

$$\sum_j D_{ij} D_{kj} = \delta_{ik}, \tag{A1.2.5}$$

which implies that scalar products are invariant under rotations,

$$\boldsymbol{r}'_1 \cdot \boldsymbol{r}'_2 = \boldsymbol{r}_1 \cdot \boldsymbol{r}_2. \tag{A1.2.6}$$

This expresses the isotropy of space: it does not matter how the Cartesian frame of reference is oriented.

A 3-vector $\boldsymbol{v}$ is now defined to be any object with three components which under rotations transforms as $\boldsymbol{v}' = \mathbf{D}\boldsymbol{v}$. Velocities $\dot{\boldsymbol{r}}$ and accelerations $\ddot{\boldsymbol{r}}$ are 3-vectors in this sense. Masses do not change under rotations, hence they are 3-scalars. Forces are products of scalars (masses) and 3-vectors (accelerations), hence they are 3-vectors. Energies are scalar products of 3-vectors (forces and displacements), hence they are 3-scalars.

Generalizing this point of view, second rank 3-tensors in the sense of physics are defined as objects with two indices T_{ij} which transform like 3-vectors with respect to each index:

$$T'_{ij} = \sum_{kl} D_{ik} D_{jl} T_{kl}. \tag{A1.2.7}$$

Examples in classical mechanics are the tensor of inertia (which is a symmetric tensor) or the angular momentum (an anti-symmetric tensor).

The important difference between the general mathematical definition and the specific use in physics is that physicists incorporate the structure of space, in particular its isotropy. 3-scalars are defined to be invariant under rotations $\mathbf{D}$, 3-vectors and 3-tensors are defined by their transformation laws under rotations. In Newtonian mechanics, physical laws must not change under rotations of the coordinate system. This is guaranteed if the laws are formulated in terms of 3-scalars, 3-vectors, and 3-tensors. This principle is called *Newtonian covariance.*

Einstein discovered that space and time are not independent. They are to be combined as a four-dimensional Minkowski space-time, with points $\boldsymbol{x} = (ct, \boldsymbol{r}) \equiv (x_0, x_1, x_2, x_3)$ and basic scalar product

$$\boldsymbol{x} \cdot \boldsymbol{x} = c^2t^2 - (x^2 + y^2 + z^2) = \sum_{i,j=0}^{3} g_{ij} x_i x_j. \tag{A1.2.8}$$

g_{ij} is called the *metric tensor* (of special relativity[1]); its only nonvanishing components are $g_{00} = 1$ and $g_{11} = g_{22} = g_{33} = -1$. The transformations of space-time which leave this product invariant, are the famous Lorentz transformations

$$x' = \mathbf{L}\,x, \tag{A1.2.9}$$

where **L** must fulfill the relation

$$\mathbf{L}^t\mathbf{g}\mathbf{L} = \mathbf{g} \quad \text{or} \quad \sum_{j,k} L_{ji}g_{jk}L_{kl} = g_{il}. \tag{A1.2.10}$$

In relativity theory, the role of spatial rotations **D** is taken over by the more general Lorentz transformations **L**, but then everything is very similar. Instead of 3-vectors we have 4-vectors $\boldsymbol{v} = (v_0, v_1, v_2, v_3)$, defined by their transformation behaviour

$$v_i' = \sum_j L_{ij}v_j, \tag{A1.2.11}$$

and instead of 3-tensors of rank 2 we have 4-tensors of rank 2,

$$T_{ij}' = \sum_{kl} L_{ik}L_{jl}T_{kl}. \tag{A1.2.12}$$

Examples of 4-vectors are

- the 4-velocity $\boldsymbol{u} = \gamma(c,\ \dot{\boldsymbol{r}})$, where $\gamma^2 = 1/(1-\dot{r}^2/c^2)$, with invariant 4-scalar $\boldsymbol{u}\cdot\boldsymbol{u} = c^2$;
- the energy-momentum $\boldsymbol{p} = (E/c,\ m_0\gamma\dot{\boldsymbol{r}})$, with invariant 4-scalar $\boldsymbol{p}\cdot\boldsymbol{p} = m_0^2c^2$;
- the 4-force $\gamma(\boldsymbol{F}\cdot\dot{\boldsymbol{r}}/c,\ \boldsymbol{F})$, where $\boldsymbol{F}$ is the 3-force acting on a particle;
- the 4-current density in electrodynamics, $(\rho c, \boldsymbol{j})$.

The electromagnetic fields $\boldsymbol{E}$ and $\boldsymbol{B}$ form an antisymmetric 4-tensor

$$F_{ij} = \begin{pmatrix} 0 & -E_x & -E_y & -E_z \\ E_x & 0 & B_z & -B_y \\ E_y & -B_z & 0 & B_x \\ E_z & B_y & -B_x & 0 \end{pmatrix} \tag{A1.2.13}$$

When physical laws are formulated in terms of 4-scalars, 4-vectors, and 4-tensors, it is guaranteed that they preserve their form under Lorentz transformations. This is called *Einstein covariance*.

[1] In general relativity the metric tensor is no longer of this simple form. It is still symmetric, $g_{ij} = g_{ji}$, but tends to depend on space-time $\boldsymbol{x}$. For purposes of convenient notation, it has become customary to distinguish between *contravariant* vectors with upper indices, $(x^0, x^1, x^2, x^3) \equiv (ct, x, y, z)$, and *covariant* vectors with lower indices, $x_i = \sum_j g_{ij}x^j$, but that need not concern us here.

2. Schrödinger's and Heisenberg's Quantum Mechanics

The two basic concepts of quantum mechanics are wave functions, or *states* $|\psi\rangle$, and linear operators **A**, or *observables*, acting on these states. The measurement of a quantity **A** in a state $|\psi\rangle$ means to determine the average $\langle\psi|\mathbf{A}|\psi\rangle$. Quantum theory predicts the time development of these values. The following is a summary of how this is done, and how it compares to its classical counterpart.

Recall how the **state** of an elementary particle (a point mass) is defined in classical mechanics. At any instant of time, the particle has a position $\boldsymbol{q} = (q_1, q_2, q_3)$ and a momentum $\boldsymbol{p} = (p_1, p_2, p_3)$. Together they form a "point" $(\boldsymbol{q}, \boldsymbol{p})$ in six-dimensional phase space. These points are the classical states of a particle.

In quantum mechanics, the states are conceptually very different. In one of the standard representations, they are complex functions $\psi(\boldsymbol{q})$ on the space of positional variables. These functions are called wave functions, or probability amplitudes. The latter name derives from the interpretation of $|\psi(\boldsymbol{q})|^2 d^3\boldsymbol{q}$ as the probability to find the particle within the volume element $d^3\boldsymbol{q}$ at the point $\boldsymbol{q}$. This interpretation puts a restriction on the functions $\psi(\boldsymbol{q})$, namely, that they be normalizable,

$$\int |\psi(\boldsymbol{q})|^2 d^3\boldsymbol{q} = 1. \tag{A1.2.14}$$

The set of all functions that can be so normalized is called the *Hilbert space* of squareintegrable complex functions over the given manifold of positions $\boldsymbol{q}$. The Hilbert space is a vector space (of infinitely many dimensions) because functions may be added and multiplied by complex numbers.

Instead of functions $\psi(\boldsymbol{q})$, one may consider functions $\widetilde{\psi}(\boldsymbol{p})$ on the space of momentum variables, as states of a particle. The two kinds of functions are not independent of each other but related by Fourier transformation:

$$\widetilde{\psi}(\boldsymbol{p}) = \int \mathrm{e}^{-\mathrm{i}\boldsymbol{q}\cdot\boldsymbol{p}/\hbar}\psi(\boldsymbol{q})d^3\boldsymbol{q}, \quad \psi(\boldsymbol{q}) = \int \mathrm{e}^{\mathrm{i}\boldsymbol{q}\cdot\boldsymbol{p}/\hbar}\widetilde{\psi}(\boldsymbol{p})\frac{d^3\boldsymbol{p}}{(2\pi\hbar)^3}. \tag{A1.2.15}$$

The normalization condition for $\widetilde{\psi}(\boldsymbol{p})$ reads

$$\int |\widetilde{\psi}(\boldsymbol{p})|^2 \frac{d^3\boldsymbol{p}}{(2\pi\hbar)^3} = 1, \tag{A1.2.16}$$

which shows that $|\widetilde{\psi}(\boldsymbol{p})|^2 d^3\boldsymbol{p}/(2\pi\hbar)^3$ is the probability that the momentum has a value in the range $d^3\boldsymbol{p}$ around $\boldsymbol{p}$.

In fact, the functions $\psi(\boldsymbol{q})$ and $\widetilde{\psi}(\boldsymbol{p})$ are just different representations of the same vector $|\psi\rangle$ in the particle's *Hilbert space*. $\psi(\boldsymbol{q})$ is called the wave function in positional representation, $\widetilde{\psi}(\boldsymbol{p})$ is called the wave function in momentum representation.

The transition from one to the other corresponds to choosing different bases of states in Hilbert space.

The notation $|\psi\rangle$ for vectors in Hilbert space is due to Dirac. It specifies a state without reference to a special representation. The same state can also be characterized by $\langle\psi|$, the adjoint vector of $|\psi\rangle$. The difference is the same as between column ($|\psi\rangle$) or row ($\langle\psi|$, with implied complex conjugation) for a vector in Cartesian space. The main difference between Hilbert spaces and Cartesian spaces is that the latter are real and finite-dimensional whereas Hilbert spaces are (usually) complex and infinite-dimensional. Both spaces possess a scalar product. If $|\psi\rangle$ and $|\phi\rangle$ are two states from a particle's Hilbert space, then $\langle\phi|\psi\rangle = \langle\psi|\phi\rangle^*$ is their scalar product, usually a complex number. (The * means complex conjugation.) It is interpreted as the *probability amplitude* that the particle in state $|\psi\rangle$ behaves as if it were in state $|\phi\rangle$.

If $|\boldsymbol{q}\rangle$ is the state where the particle is exactly at position $\boldsymbol{q}$, then $\langle\boldsymbol{q}|\psi\rangle$ is the probability amplitude that a particle in state $|\psi\rangle$ behaves as if it were at $\boldsymbol{q}$. Hence we have $\psi(\boldsymbol{q}) \equiv \langle\boldsymbol{q}|\psi\rangle$, and similarly $\tilde{\psi}(\boldsymbol{p}) \equiv \langle\boldsymbol{p}|\psi\rangle$, if $|\boldsymbol{p}\rangle$ is the state where the particle has a definite momentum $\boldsymbol{p}$. The question may then be asked how a particle at a definite position $\boldsymbol{q}$ is represented in momentum space, and vice versa. The Fourier transformation formulas above may be interpreted as

$$\langle\boldsymbol{p}|\boldsymbol{q}\rangle = \mathrm{e}^{-\mathrm{i}\boldsymbol{p}\cdot\boldsymbol{q}/\hbar} \text{ and } \langle\boldsymbol{q}|\boldsymbol{p}\rangle = \mathrm{e}^{\mathrm{i}\boldsymbol{p}\cdot\boldsymbol{q}/\hbar}. \tag{A1.2.17}$$

This shows that a particle with a definite position is evenly spread out over the entire momentum space, and a particle with definite momentum is evenly spread out over the entire position space. In other words: it is impossible to have the particle in a quantum mechanical state where both position and momentum are sharply defined.[2]

So much on states. Let us now discuss **observables**, or physical quantities that an experimenter may want to determine in a measurement. In classical mechanics, any function $f(\boldsymbol{q},\boldsymbol{p})$ on phase space is such an observable. As examples we may consider the components of the angular momentum $\boldsymbol{l} = \boldsymbol{q} \times \boldsymbol{p}$, or the energy $H = \boldsymbol{p}^2/2m + V(\boldsymbol{q})$ if m is the particle's mass and $V(\boldsymbol{q})$ the potential of the forces acting on the particle. The most elementary observables are the components q_i of position $\boldsymbol{q}$, and the components p_i of momentum $\boldsymbol{p}$. Remember that the q_i and p_i are said to be "conjugate" to each other. This means they are not completely independent: the space of observables possesses a structure, called the *symplectic structure*, by which they are related:

[2] Never mind that states like $|\boldsymbol{q}\rangle$ or $|\boldsymbol{p}\rangle$ cannot be normalized and hence, in a strict sense, do not belong to a particle's Hilbert space. This problem has been solved mathematically in terms of the theory of distributions.

$$
\begin{aligned}
\{q_i,\ p_i\} &= 1 \\
\{q_i,\ p_j\} &= 0 \quad \text{if } i \neq j \\
\{q_i,\ q_j\} &= 0 \\
\{p_i,\ p_i\} &= 0.
\end{aligned}
\tag{A1.2.18}
$$

Here the brackets $\{\cdot,\cdot\}$ are the so-called Poisson brackets; for any two functions $f(\boldsymbol{q},\boldsymbol{p})$ and $g(\boldsymbol{q},\boldsymbol{p})$, they are defined by

$$
\{f,\ g\} = \sum_i \frac{\partial f}{\partial q_i}\frac{\partial g}{\partial p_i} - \frac{\partial f}{\partial p_i}\frac{\partial g}{\partial q_i}. \tag{A1.2.19}
$$

In classical mechanics, the relation $\{q_i,\ p_i\} = 1$ does not exclude the possibility for a particle to have definite values of $\boldsymbol{q}$ and $\boldsymbol{p}$ at the same time, but as we will see, the corresponding relation in quantum mechanics is at the heart of the uncertainty relation.

Again, quantum mechanics has a very different notion of observables. They are linear *operators* **A** acting on the states $|\psi\rangle$. As such, they have a *spectrum of eigenvalues*, and these eigenvalues are what can be measured in a single experiment. To the extent that measurements must produce real values (as opposed to complex numbers), the operators must have real spectra, i. e., they must be Hermitean, or self-adjoint: $\mathbf{A}^* = \mathbf{A}$; if **A** is represented as a matrix, this means the matrix must not change if it is transposed and subject to complex conjugation.

To be specific, let us discuss the energy operator **H**, also called the *Hamiltonian* of the system. Its spectrum of real values E is determined by the eigenvalue problem

$$
\mathbf{H}|\psi\rangle = E|\psi\rangle, \tag{A1.2.20}
$$

which is to be read as follows. Given the Hilbert space of a particle and its Hamiltonian **H**, look for vectors $|\psi\rangle$ of the Hilbert space such that the action of **H** reproduces $|\psi\rangle$ up to a constant. The constant is then called an eigenvalue, and the vector $|\psi\rangle$ an eigenstate. Depending on the nature of the problem, the spectrum $\{E\}$ is a discrete or a continuous set of values, or a combination thereof. For example, a harmonic oscillator of frequency ω has a discrete spectrum of energies $E_n = \hbar\omega(n + \frac{1}{2}), n = 0, 1, 2, \ldots$; a free particle has a continuous spectrum, with eigenstates for any $E > 0$; the hydrogen atom has a discrete energy spectrum for negative E, and a continuous spectrum at $E > 0$. Whatever the case, we use an index i (discrete or continuous) to label the eigenenergies E_i and corresponding eigenstates $|\psi_i\rangle$.

When energy is measured in an arbitrary state $|\phi\rangle$ which is not an eigenstate of **H**, then each single measurement produces a value E_i from the spectrum of **H**, but the values differ from experiment to experiment. What can be said, however, is that in a large number of identical experiments, $|\langle\psi_i|\phi\rangle|^2$ is the probability to measure E_i because $\langle\psi_i|\phi\rangle$ is the probability amplitude that $|\phi\rangle$ behaves as if it were $|\psi_i\rangle$.

In other words: the average value of energy measured in a state $|\phi\rangle$, also called the *expectation value* of H in $|\phi\rangle$, is

$$\langle\phi|\mathrm{H}|\phi\rangle = \sum_i E_i |\langle\phi|\psi_i\rangle|^2. \tag{A1.2.21}$$

Quantum mechanics cannot do better than predict the values of such averages. Of course, if $|\phi\rangle$ happens to be one of the eigenstates $|\psi_i\rangle$, then each single measurement gives E_i. But in general, the outcome of a single experiment is unpredictable.

So far the discussion of operators has been somewhat abstract. In most applications, states are considered in positional representation $\langle \boldsymbol{q}|\psi\rangle$, and then it is possible to give explicit representations of the various observables of interest. For example, the operator $\mathbf{q}_i$ corresponding to the i-th component of position acts on states $|\psi\rangle$ in such a way that

$$\langle \boldsymbol{q}|\mathbf{q}_i|\psi\rangle = q_i \langle \boldsymbol{q}|\psi\rangle, \tag{A1.2.22}$$

i. e., at each position $\boldsymbol{q}$ it produces the coordinate q_i. The operator $\mathbf{p}_i$ corresponding to the i-th component of momentum is represented as a differential operator:

$$\langle \boldsymbol{q}|\mathbf{p}_i|\psi\rangle = -\mathrm{i}\hbar \frac{\partial}{\partial q_i} \langle \boldsymbol{q}|\psi\rangle. \tag{A1.2.23}$$

States $\langle \boldsymbol{q}|\psi\rangle = \langle \boldsymbol{q}|\boldsymbol{p}\rangle = \exp\{\mathrm{i}\boldsymbol{p}\cdot\boldsymbol{q}/\hbar\}$ with definite momentum $\boldsymbol{p}$ are eigenstates of $\mathbf{p}_i$ because the differentiation in the above formula produces the value p_i. To obtain explicit expressions for the energy operator H, we may now use the classical energy function $H(\boldsymbol{q}, \boldsymbol{p})$ and replace $\boldsymbol{q}$ and $\boldsymbol{p}$ by their operators $\mathbf{q}$, $\mathbf{p}$. In many cases the energy function has the form $H = \boldsymbol{p}^2/2m + V(\boldsymbol{q})$, and then the Hamiltonian operator reads, in abstract form,

$$\mathrm{H} = \frac{\mathbf{p}^2}{2m} + V(\mathbf{q}), \tag{A1.2.24}$$

and in positional representation, i. e., when acting on functions $\langle \boldsymbol{q}|\psi\rangle$,

$$\mathrm{H} = -\frac{\hbar^2}{2m}\frac{\partial^2}{\partial \boldsymbol{q}^2} + V(\boldsymbol{q}). \tag{A1.2.25}$$

The Poisson structure (19) in the space of classical phase space functions has an important correspondence in quantum mechanics. If A and B are two operators acting on Hilbert space, then

$$[\mathrm{A}, \mathrm{B}] = \mathrm{AB} - \mathrm{BA} \tag{A1.2.26}$$

is called their *commutator*. With (22) and (23) it is straightforward to check that (18) is replaced by the fundamental commutation relations

$$\begin{aligned}[\mathbf{q}_i, \mathbf{p}_i] &= \mathrm{i}\hbar \\ [\mathbf{q}_i, \mathbf{p}_j] &= 0 \quad \text{if } i \neq j \\ [\mathbf{q}_i, \mathbf{q}_j] &= 0 \\ [\mathbf{p}_i, \mathbf{p}_j] &= 0. \end{aligned} \tag{A1.2.27}$$

From there, it is possible to derive commutation relations involving the energy operator $\mathbf{H}$ or the angular momentum operator $\mathbf{L} = \mathbf{q} \times \mathbf{p}$. It is amazing that in many cases these algebraic relations are sufficient to determine the entire spectrum of a given operator.

The commutation relations are a central part of the uncertainty principle. Given a state $|\psi\rangle$ and an operator $\mathbf{A}$, the uncertainty in the quantity $\mathbf{A}$ is given by the square root of the mean square deviation from the average value over many measurements,

$$\Delta A = \sqrt{\langle\psi|(\mathbf{A} - \langle\psi|\mathbf{A}|\psi\rangle)^2|\psi\rangle}. \tag{A1.2.28}$$

If and only if $|\psi\rangle$ is an eigenstate of $\mathbf{A}$ can this uncertainty be zero. Now consider two operators $\mathbf{A}$ and $\mathbf{B}$. It is a matter of elementary analysis to show that the product of the two uncertainties ΔA and ΔB can be estimated from below as follows:

$$\Delta A \cdot \Delta B \geq \left| \frac{\mathrm{i}}{2}[\mathbf{A}, \mathbf{B}] \right|. \tag{A1.2.29}$$

This means the two quantities $\mathbf{A}$ and $\mathbf{B}$ cannot both assume sharp values unless they commute. If they are conjugate like position and momentum, the relation $[\mathbf{A}, \mathbf{B}] = \mathrm{i}\hbar$ implies that $\Delta A \cdot \Delta B \geq \hbar/2$.

Let us finally discuss the **time development**, first in classical and then in quantum mechanics. The phase space point (q, p) of a classical particle moves according to Hamilton's canonical equations,

$$\dot{q} = \frac{\partial H}{\partial p} \quad \text{and} \quad \dot{p} = -\frac{\partial H}{\partial q}. \tag{A1.2.30}$$

From there the chain rule tells us that any function $f(q, p)$ changes according to

$$\frac{d}{dt} f(q, p) = \{f, H\}, \tag{A1.2.31}$$

where $\{\cdot, \cdot\}$ is the Poisson bracket defined in (19). (We assume that the Hamilton function does not explicitly depend on time.)

How does this translate into quantum mechanics? To make contact with measurements, the time development of the expectation values $\langle\psi|\mathbf{A}|\psi\rangle$ must be determined. Now there is a choice. Schrödinger chose to consider the operators $\mathbf{A}$ as fixed and the states $|\psi\rangle = |\psi(t)\rangle$ as time dependent. Heisenberg did it the other way round; he considered the state of a system to be given once and for all, and the operators $\mathrm{A} = \mathrm{A}(t)$ to be time dependent. Of course, both versions of quantum theory must give the same results for $\langle\psi|\mathrm{A}|\psi\rangle$.

In Schrödinger's theory, the wave function $|\psi\rangle$ changes according to the famous equation

$$\mathrm{i}\hbar\frac{d}{dt}|\psi\rangle = \mathrm{H}|\psi\rangle, \tag{A1.2.32}$$

which has the formal solution $|\psi(t)\rangle = \exp\{-\mathrm{i}\mathrm{H}t/\hbar\}|\psi(0)\rangle$. For the expectation value this implies

$$\langle\psi(t)|\mathrm{A}|\psi(t)\rangle = \langle\psi(0)|e^{\mathrm{i}\mathrm{H}t/\hbar}\mathrm{A}e^{-\mathrm{i}\mathrm{H}t/\hbar}|\psi(0)\rangle. \tag{A1.2.33}$$

If we now interpret $e^{\mathrm{i}\mathrm{H}t/\hbar}\mathrm{A}e^{-\mathrm{i}\mathrm{H}t/\hbar}$ as a time dependent operator $\mathrm{A}(t)$, we find that it obeys the equation of motion

$$i\hbar\frac{d}{dt}\mathrm{A} = [\mathrm{A}, \mathrm{H}] \tag{A1.2.34}$$

which is the "quantized" version of (31). This is the basic equation in Heisenberg's version of quantum theory. To the extent that classical mechanics is formulated in terms of Poisson brackets, and quantum mechanics in terms of commutation relations, the correspondence is indeed very intimate:

$$\{\cdot,\cdot\} \leftrightarrow \frac{1}{\mathrm{i}\hbar}[\cdot,\cdot]. \tag{A1.2.35}$$

Appendix A2

The Nature of Physical Phase Transitions*

In Section 5.5 natural selection is shown to be equivalent to a physical phase transition. According to Charles Darwin, natural selection represents the basis of evolutionary optimisation, which biologists have tended to regard as an irreducible biological phenomenon beyond any physical interpretation.[1] Physicists, on the other hand, were not too impressed by the undefined nature of selection, as it was overstressed by many biologists and often misinterpreted as a case of tautological behaviour. Among the physicists there was one exception: Ludwig Boltzmann,[2] who said that the 19th century "will one day be remembered as the century of Darwin, rather than that of iron, or of steam, or of electricity".

Be that as it may, my conclusion is based squarely on the properties described in Section 5.5. The only difference I see with respect to physical phase transitions is the fact that natural selection is a phase transition in high-dimensional information space rather than in any physical environment represented by space–time co-ordinates. At first sight it may look awkward to call selection a physical phase transition, since information is a non-material property, even though information requires material carriers. What then is the process that produces something that transcends matter without changing its physical nature? We might think of a particular order, such as that of different phases in a given ensemble of matter. However, a physical phase change comes about by breakage of symmetry, causing the acting forces to produce different states of spatial order. Therefore, in this appendix I want to focus on the differences by characterising in some more detail a physical phase and its changes.

The physical phase is a state of order in homogenously distributed matter. H_2O, for instance, may form ice, exhibiting crystalline long-range order, or it may, as in liquid water, present a condensed phase with some order limited to shorter distances;

* This is an attempt at a purely verbal characterisation of the physics of phase transitions, leaving aside their well-developed mathematical treatment, which can be found in the literature cited.

or it may exist as water vapour, where in the limit of low pressure no intermolecular interactions are present. These three fundamental phases – solid, liquid and gas – by no means exhaust the potential of matter to assume particular forms of order. The term "phase" is also used for amorphous, glassy or superfluid states, or for states exhibiting certain magnetic and electric behaviour – such as ferro- and anti-ferromagnetism, or superconductivity, occurring in certain crystal lattices at low temperatures – or for plasma, a state of all matter existing at high temperatures that shows electrical conduction due to thermal ionisation.

A phase transition is simply the passage of matter from one state of order into an alternative one. However, the word "simply" should not create the impression that we are dealing with anything that can be understood as "simply" as it is defined. In a similar way, selection may be defined by a transition between two states, one representing a certain meaning, the other having no, or another, meaning. In order to characterise the physical nature of phase transitions we have to dig deeper. Let me quote first Nigel Goldenfeld,[3] who right at the beginning of his *Lectures on Phase Transitions and the Renormalization Group* raises the question I consider most important in my context: "Why do phase transitions occur at all?"

In Section 2.3, I discussed statistical thermodynamics, the essential result of which is the correlation of Gibbs free energy with the logarithm of the partition function of statistical mechanics. The latter is a sum of the exponentials of Hamiltonians $\hat{H}$, divided by k_BT(k_B being Boltzmann's constant and T absolute temperature), the sum taken over all degrees of freedom contributing to $\hat{H}$. This sum usually involves the exponentials of analytical functions. How can it give rise to the non-analytical behaviour we are accustomed to associate with the discontinuity of a phase transition? If a macroscopic piece of ice melts, enthalpy is absorbed as "heat of melting", without showing up in any detectable change of temperature. The heat capacity (which is the temperature derivative of enthalpy) accordingly becomes infinite at the melting point. Correctly speaking, this is true only in the so-called thermodynamic limit, supposing an infinitely extended system.

Paul Ehrenfest proposed a classification of phase transitions, calling them n-th order if any of the n-th derivatives of Gibbs free energy with respect to an intensive variable (e.g. $-(\partial G/\partial T)_P$ = entropy, or $-(\partial^2 G/\partial T^2)_P$ = heat capacity) shows a discontinuity. Critical phenomena, such as a gas–liquid transition at the "critical temperature", the onset of magnetisation below the Curie temperature in a ferromagnet or the transition from normal to superfluid helium4 (^4He), which is distinguished by a λ-shaped temperature dependence of heat capacity, have been considered to be second-order phase transitions according to Ehrenfest's classification. Recent insights, however, have proven such a classification to be incorrect. These phase transitions show divergences, but still turn out to be "continuous". The λ-shaped heat capacity of ^4He involves a "cusp" rather than the true discontinuity previously assumed. Three examples of transition curves of first and second derivatives may be mentioned:

a) The liquid–gas transition near a critical point is connected with density changes that observe the relation $|T - T_c|^\beta$, where β is universally close (but not equal) to 0.33 (e.g. 0.327 ± 0.006 for sulphur hexafluoride[4]) and where T_c is called a "critical" temperature. At temperatures above T_c liquid and gas are indistinguishable, while below T_c liquid and gas, separated by phase boundaries, coexist.

b) The magnetisation M in the limit of zero external magnetic field density for a ferromagnet at $T \leq T_c$ (T_c being the Curie temperature) follows the same scaling law, i.e. $M \sim (T_c - T)^\beta$. The exponent β agrees with the above value (within experimental error) and is found to be valid for both ferro- and antiferro magnets.

c) The heat capacity of ^4He as a function of temperature above and below the λ-point follows a scaling law $\sim |T_c - T|^{-\lambda}$ with $\lambda = 0.013 \pm 0.003$.

How do phase transitions come about? What is the reason for their (apparent) "all-or-none" behaviour, as found in some of the variables?

Consider a single molecule that can exist in two different configurations and let the two structural isomers differ in their enthalpy content (i.e. $\Delta H \neq 0$). The equilibrium constant relating the two forms is determined by a Boltzmann factor ($e^{-\Delta G/RT}$, cf. Section 2.8) involving the difference of free energies (ΔG) between the two states. If the two isomers are present at equal concentration, $\Delta G(= \Delta H - T\Delta S)$ is then zero, and ΔH equals $T\Delta S$, where T is the temperature and S is the entropy (see the main text). As is seen in the figure, the transition from one form into the other as a function of temperature is quite smooth, the steepness depending on the magnitude of ΔH referring to the temperature range in which reaction enthalpy (ΔH) equals reaction entropy, multiplied by temperature T ($T\Delta S$). The situation is adequately described by thermodynamics in the above way.

Now think of a molecule like haemoglobin, the protein that constitutes the red colour of our blood. The task of this molecule is to transport oxygen from the lung to those sites in the body where it is required for metabolism, i.e. for "burning" food, thus producing energy-rich molecules that promote chemical work. Haemoglobin includes four (interacting) subunits, each providing a site for binding one oxygen molecule. Evolution has brought about a mechanism that allows for sophisticated regulation of the binding and release of oxygen. In the lung, where oxygen is taken up and therefore is present in large excess, all subunits assume a structure that provides high affinity for oxygen binding. In other words, haemoglobin acts like a sponge, sucking up as many oxygen molecules as possible. On the other hand, at metabolic sites where free oxygen is rare, all subunits co-operatively undergo a change in conformation that lowers their affinity for oxygen considerably. As a consequence, oxygen is released most efficiently at the sites where it is required.

Chemists have called this regulatory mechanism "allosteric" because the presence of oxygen (and its binding or release in one subunit) triggers the affinity of other ("allo") subunits. Different models have been proposed in order to describe this prototype of a "co-operative" regulation mechanism in terms of the structural properties typical of proteins (see the example in Section 2.1). Haemoglobin is just one particular example of an allosteric protein – a prominent one, because it was the first structure to be elucidated in the early 1950s by X-ray crystallography in a heroic effort by Max Perutz,[5] for which he shared the 1961 Nobel prize in chemistry with John Kendrew,[6] whose work was focused on the structure of myoglobin, a (one-subunit) relative of haemoglobin.

With co-operative transformations, as we find them in allosteric enzymes (Table 2.1.4 in the main text, Section 2.1), we are on our way to phase transitions, but so far we just have co-operative behaviour. The higher the degree of co-operativity, the larger the number of entities that simultaneously undergo a change. If temperature is the external variable, transitions become confined to an increasingly narrow range, since enthalpies become larger and larger. In Figure 2.8.2 (text) I have drawn transition curves for several enthalpy values, all centred at the midpoint of transition where $\Delta H = T\Delta S$ (ΔH and ΔS being the enthalpies and entropies of reaction and T the absolute temperature; cf. Section 2.8). For all the (calculated) curves in Figure 2.8.2 I could immediately quote examples from biochemistry.

The blue curve in Figure 2.8.2 corresponds to a chemical transformation such as the closing or opening of relatively weak chemical bonds, e.g. forming a base pair in nucleic acids. The larger values for ΔH and ΔS in the red curve are often encountered in protein reactions, including co-operative changes of structure that involve large numbers of "weak bonds". The red curve refers to a tenfold increase in reaction enthalpy (and entropy). Such larger values are typical of covalent bonds, but when they are associated with large entropy values they also indicate strongly co-operative interactions. The aggregation of deoxyhaemoglobin S, a process that is associated with sickle-cell anaemia and known as "gelation", is a typical example. Kinetic studies have shown that the primary step involves the co-operative formation of a complex of 30 protein molecules. Such a change occurs in a very narrow range of temperature and concentration.[7] Another example is the infectious nature of prion proteins,[8,9] causing an autocatalytic progression of the structural transformation of a related host protein. It manifests itself in "mad cow disease" in cattle, in scrapie in sheep and in Creutzfeldt–Jakob disease in humans, for which various models have been developed.[10,11] A similar example of medical importance may be the cause of plaque formation in Alzheimer's disease.

Relevant as these co-operative changes of structure in biology and medicine are, what do they tell us about phase transitions? Certainly, they show that one may obtain quite sharp transitions (such as the black curve in Figure 2.8.2) because of the co-operative involvement of many partners, even though at the same time the individual nuclear interactions involved are of quite modest strength. Yet all such curves are still continuous and analytical, while true phase transitions typically show non-analytical behaviour. We might argue that such behaviour occurs only in the thermodynamic limit, i.e. in an infinitely extended system, which ultimately will be the case; however, something is still missing. Let us consider, as an example, the condensation of matter in the transition from the gaseous to the liquid state.

Johannes Diderik van der Waals devised an empirical equation that describes the relations between pressure, volume and temperature in fluid states of matter where (co-operative) interactions among the particles cannot be neglected. Already in the 17th century the Anglo-Irish chemist Robert Boyle and the French physicist Edme Mariotte had (independently) discovered a law for ideal (i.e. highly dilute) gases, according to which – at constant temperature – the pressure and volume of the gas are strictly inversely proportional to one another. Van der Waals realised that this law could be modified by two influences that become effective with increasing density:

i) The attractive forces that act among the particles and that eventually cause the gas to condense to the liquid phase may be considered as "internal pressure" to be added to the external pressure term. The distance relationship of these van der Waals' forces (experimentally detectable in "real" gases) justifies a term for the internal pressure that is inversely proportional to the square of the volume. The correct quantum-mechanical description of such (relatively weak) forces among electrically neutral particles is due to Fritz Wolfgang London, yielding a force term inversely proportional to the sixth power of the mutual distance between the two interacting particles – like an induced dipole–dipole interaction – coming out with changes at close distances. In the text of Chapter 2.8 a more superficial description was given, assuming a negative free energy term $\sim 1/V$, yielding for the pressure term with $p = -(\partial F/\partial V)_T$, a term inversely proportional to V^2.

ii) The volume likewise has to be corrected by a term representing a "hard sphere" volume of the particles, which has to be subtracted from the volume available for accommodating all particles. The "hard sphere" volume refers to a repulsive interaction among the particles which changes so steeply with distance (tenth power) that a constant term is a good approximation.

Van der Waals' equation was a stroke of genius that won him the 1910 Nobel prize for physics. Figure 2.8.3 depicts the pressure–volume relationship for various values of the temperatures as parameter. This condensed description is based on a discourse by R. Stephen Berry, Stuart A. Rice and John Ross.[12] These three authors are personally well known to me and I consider their treatment to be the most profound given in any textbook of physical chemistry that I have seen. The curves in Figure 2.8.3 show two ranges separated by a "critical" curve passing through a point of inflection. This curve refers to the critical parameter T_c (374°C for H_2O) in the density–temperature phase diagram shown in Figure 2.8.4. All curves above the critical curve reproduce the hyperbolic form we know from the ideal gas law, although the singularity $p \to \infty$ appears at a finite volume and the curve refers to a fluid, quite different from an ideal gas, in which one cannot distinguish between gas and liquid. In passing through this regime it is always possible to switch between pure gas and liquid without touching the region in which gas and liquid are in coexistence, separated by phase boundaries. This is the case for all curves below the critical curve. Here, van der Waals' equation exhibits two extrema, a maximum and a minimum. The dashed (parabolic) line marks two points on each isotherm, i.e. two volumes that refer to the same pressure, the left one referring to the pure liquid and the right one to the pure gaseous state. Both points are stable, requiring $(\partial V/\partial p)_T < 0$, thereby indicating the stable coexistence of two phases. However, between these two points there is a region – indicated by the (parabolic) dotted line – where the change of volume with pressure is positive, i.e. $(\partial V/\partial p)_T > 0$. States within this region are unstable, and phase transitions – in this semi-empirical mode of description, and in the absence of supersaturation – are supposed to occur along the line separating the two shaded areas of equal size.

In this semi-empirical interpretation, a phase transition is characterised by an instability of one phase relative to the other. But what about the critical point? As this is a point of inflection (for H_2O at 218 atmospheres pressure), both the first and the second derivatives of pressure with respect to volume, i.e. $(\partial p/\partial V)_T$ and $(\partial^2 p/\partial V^2)_T$ are zero. The isothermal compressibility at this point becomes infinitely large, in other words vanishingly little work is required in order to accomplish either compression or expansion. This suggests large fluctuations in density near the critical point, and these are indeed observed as corresponding non-uniformities in the refractive index, causing the fluid to become opaque (critical opalescence). The approach to the critical point is reflected in the convergence of the densities of liquid and gas, i.e. with a (non-rational) critical coefficient β close to, but definitely not equal to, one-third. Furthermore, the temperature dependence of the heat capacity shows the λ-shaped form of transition involving latent heat, as is associated with the coexistence range below the critical point. Infinite heat capacity results from the constancy of the temperature despite heat release during phase transition. But what does it mean when we say "at" the critical point, where liquid and vapour become indistinguishable? Isn't there something missing in the above deliberations on stability which is of fundamental importance? The answer to these questions is given below.

As mentioned at the beginning of this appendix there are other equally puzzling examples of critical behaviour, such as ferromagnetism or antiferromagnetism. The critical temperature T_c here is named after Pierre Curie. Raising the temperature above the Curie point results in a sharp change from ferro- to paramagnetic behaviour. In the absence of an external magnetic field the magnetic dipoles, through co-operative interaction, can align in a reversible way to form macroscopic magnetic moments that disappear at the Curie temperature, where the orientation of the dipoles becomes randomised.

In a similar way, antiferromagnets align their atomic dipoles in opposite directions, so that their fields exactly cancel one another out. Again, this behaviour changes abruptly to paramagnetism with random orientation of the atomic dipoles at a certain temperature, named after the French physicist Louis Eugène Félix Néel. It was he who pioneered research on antiferromagnetic substances, which today are of great practical importance as memory devices in computer technology. The first theoretical treatment of the critical behaviour of such magnetic materials is due to another French physicist, Pierre-Ernest Weiss. He identified the critical behaviour, including the magneto-caloric phenomena, associated with the phase transition. Both van der Waals' and Weiss' theories explain correctly the co-operative and non-analytical nature of critical phenomena, but – being "mean field" approaches – do not match quantitatively the data established by accurate measurements.

What is meant by a "mean field" approach? In order to explain this we have to address the problem from a molecular-statistical point of view. In 1925 the German physicist Ernst Ising proposed in his doctoral thesis a simple model in which only interactions with nearest neighbours were taken into account. The model was meant to apply to ferromagnetism, but was carried through by Ising for linear chains in which just two possible orientations, describable by spins of either +1 or −1, are allowed. Nearest neighbours can thus align in either a parallel or an antiparallel fashion. Other simplifications, e.g. replacing the Hamiltonian operator by the energy, made the model easily tractable. The model showed that there are no phase transitions for a one-dimensional model, and Ising concluded that neither would they exist for any nearest-neighbour model, regardless of dimension. The first conclusion turned out to be correct, the second did not.

More detailed elaborations of the one-dimensional model confirmed Ising's result. Curiously enough, his unrealistic one-dimensional magnetic model found applications for linear macromolecular chains, such as nucleic acids or polypeptides. The transitions between structures helically ordered by hydrogen bonding (as in protein α helices) or by base pairing (as in DNA) on the one hand and random coils on the other can be described satisfactorily by the model. Bruno Zimm[13] and Shneior Lifson[14] proposed such models which, after Donald Crothers[15] joined our group from Zimm's laboratory, were demonstrated experimentally by Gerhard Schwarz[16] (summarised in my Harvard Lectures 1965). We – besides Don Crothers, Gerhard Schwarz, Rufus Lumry, Norio Ise and others – had much pleasure in studying such models in the early 1960s using our methods of relaxation spectrometry. The gist of one-dimensional theory is that a kind of correlation length exists that is proportional to the square root of the co-operativity parameter σ, which reveals the co-operative increase in binding strength for any nearest neighbour after nucleation of the first bond. For nucleic acids we find that a base pair next to an existing one can form with an about 10,000-fold higher probability than the first single base pair. The square root dependence then suggests that the correlation length includes about 100 base pairs. In other words, the co-operative "melting" of a helix behaves as though it were occurring in pieces of about 100 base pairs. This yields quite steep "melting" curves, but these curves never get infinitely steep for arbitrarily long sequences. Infinite steepness would require infinitely strong nearest-neighbour interactions, which is physically not meaningful.

It is straightforward to extend the idea to propose a nearest-neighbour model for two or three dimensions. However, as Nigel Goldenfeld remarks "What may be simple to state may not be simple to solve!" A special case, the two-dimensional Ising model for a magnet at zero external field, was solved exactly by Lars Onsager[17] in 1944. Mathematically, his solution was a *tour de force*, unprecedented in the history of solid-state physics. (Yet it wasn't this paper, but his basic ideas of non-equilibrium thermodynamics, that won him the 1968 Nobel prize in chemistry.) Let me add a few personal remarks on this greatest among the theoretical chemists of the last century.

As brilliant as his thinking was, Onsager had much trouble communicating his ideas. His students called his lectures "Norwegian I" and "Norwegian II". He used to smile and laugh when he got close to the solution, which he didn't explain further, assuming that everybody could guess it for themselves and not realising that everyone had got lost long before. Almost 30 years before I met him personally, Lars introduced himself to Peter Debye with the words: "Are you Professor Debye? Do you know that your theory is wrong?" Debye's answer, as he later told me, was: "Come, let's sit down and have a cigar."

I met Lars and his charming wife Gretl at many occasions, in the warm atmosphere of their home at New Haven, where we once got caught for several days by a snowstorm, and at his farm in New Hampshire, and in Göttingen, where Lars spent a year as a Gauss Professor. I often asked myself why his presentations in science were understood only by a few of his colleagues. The reason was probably that Lars himself did not realise how far ahead of everybody else he was. In daily life, during country walks, with his expertise in wines and cuisine, there was nothing Lars could not explain in every detail, causing my partner once to remark: Lars really knows everything!

Let me get back to phase transitions. The two-dimensional Ising model for the ferromagnet in the absence of external fields was to remain a unique example of an exact solution. As I said, it was a *tour de force*, requiring highly specialised and sophisticated mathematical techniques that did not readily lend themselves to generalisation, and so far it has not been possible to extend them to more than two dimensions or to the presence of external fields. Nevertheless, a model that can be solved exactly appears to be invaluable as a benchmark for testing approximations.

A very fruitful approximation consisted of replacing the individual interactions with neighbours (and their fluctuations) by a "mean field". This "classical" approach, which is also implicit in the more empirical approaches of van der Waals and Weiss, found its most "succinct encapsulation" in Landau's theory of phase transitions. Landau[18] assumed a function similar to, but not identical with, free energy, in which only configurations with a certain magnetisation density M are considered and which is analytic in M. In the absence of external fields this function cannot depend on the sign of magnetisation, and therefore it contains only terms with even powers of M (i.e. $\sim M^2$, M^4 etc.). The true free energy refers to the minimum of Landau's function with respect to all possible values of M. Assuming the analyticity of M with temperature, Landau showed by minimising his function that M can be finite at temperatures T below the Curie temperature T_c, but disappears towards T_c in the form $M \sim (T_c - T)^{1/2}$. The exponent ½ disagrees with the empirically found values (according to Figure 2.8.3), which are close (but not equal) to ⅓, the difference being clearly outside experimental error.

Nevertheless, Landau's theory should get credit for the fact that most of its predictions were correct, and this theory, more precisely than any of its predecessors, accounted for the fact that in the thermodynamic limit we are dealing with a true phase transition. At the Curie temperature the correlation length for co-operative alignment becomes infinite. In this theory, the loss of analyticity of M at the point of phase transition (in the thermodynamic limit) results from the averaging procedure connected with minimising free energy with respect to magnetisation. The implicit assumption is made that all space-dependent fluctuations in M average out. Landau's theory even predicts where this assumption is justified. The fact that this is not the case near a critical point reveals that Landau's theory – as Nigel Goldenfeld phrases it quite bluntly – "contains within it the seeds of its own destruction". Something was still missing.

Since Maxwell and Boltzmann, physicists have become well accustomed to the presence of fluctuations at the microscopic level. In many (but not all) cases the fluctuation builds up a (statistical) "force" for its own reversal, as illustrated most lucidly by Ehrenfest's model in Chapter 2. At the same time, I have shown that there are situations where just the opposite occurs, i.e. microscopic fluctuations (e.g. base substitutions in nucleic acids) that build up and eventually become manifest (e.g. by selective fixation) at the macroscopic level. Likewise, macroscopic states may appear to be indifferent towards fluctuations, which thus show up on all scales of length. This is the case near a critical point. Take as an example the liquid/vapour system. If the densities of both liquid and vapour become the same, there is nothing to distinguish them any more other than minute differences, e.g. regions that are a tiny bit more liquid-like than gas-like or *vice versa*. Hence there is not a "force" restricting their expansion, and these fluctuations will appear at all levels of magnitude. Similar situations are met near the critical points of magnetic transitions. Regions of all sizes of magnetisation will occur which eventually align for $T \to 0$. The fact that near the critical point we find power laws with identical critical coefficients suggests the universal nature of these phenomena.

The obvious feature common to all these physically diverse situations is the occurence of fluctuations on all length scales, from microscopic to macroscopic. The consequence is that certain measurable quantities depend upon each other in the form of power laws (see Section 4.8). The scaling of these laws, typical of critical phenomena, is decisively determined by these fluctuations, which do not cancel out. The non-negligibility of fluctuations – next to co-operativity and infinite divergence of correlation length – is the third important ingredient of the present-day theory of continuous phase transitions (see the work by Fisher, Widom[19] and Kadanoff[20] described by Kenneth Wilson[21]).

The next important step after Landau's theory was Ben Widom's empirical notion of the scaling law near the critical point, which as an equation of state was able to satisfy experimental data for quite different materials. Moreover, for the case of magnetisation Kadanoff proposed an intuitive explanation of scaling by grouping spins successively into blocks of increasing size that look the same on all length scales, interacting with their corresponding "effective" magnetic moments. The breakthrough came in the early 1970s with a series of papers by Kenneth G. Wilson, who introduced his method of the "renormalisation group" to the physics of condensed matter.

In his Nobel lecture (1982) Ken Wilson[21] tells us the exciting story of the renormalisation group, which started in particle physics. Renormalisation, by perturbation procedures, was first applied by the architects of quantum field theory in the 1940s to remove divergences that appear in higher terms. It is a kind of reparametrisation, applied in quantum electrodynamics (cf. Section 1.6) to bring the electron's charge and mass into agreement with measurable parameters. Elementary particles in quantum field theory have a composite internal structure on all size scales, and experimentally determined scattering cross-sections can be explained only by taking account of this fact. This provides a physical need for renormalisation to allow problems at all length scales to be dealt with. Any acceptable field theory today is supposed to be a gauge theory. "Renormalisation group" is now a formalised procedure, so to speak, a "theory about theories". The term appeared first in the work of Murray Gell-Mann,[22] where it was based on perturbation theory, while Ken Wilson later gave it a new, non-perturbative geometrical interpretation.

The amalgamation of these ideas in particle physics with the problems of large-scale fluctuations in phase transitions, as expounded in Ben Widom's and Michael Fisher's chemistry seminars at Cornell in the early 1970s, turned out to be the key to Ken Wilson's success in solving a puzzling outstanding problem of physical chemistry. All this happened in the last 40 years, during which a "new look" evolved that has its value in both condensed matter as well as high-energy physics. I had the privilege of spending some time at Cornell during the early 1960s as Andrew D. White Professor-at-Large, after being introduced there in 1954 by Peter Debye. I remember many lively discussions with Ben Widom and Michael Fisher – apart from those with my hosts Frank Long, Simon Bauer and Harold Scheraga. Ken Wilson, whose father I knew well from my Harvard lectures, was just about to settle down in the physics department.

The preceding discussion is not meant as a substitute for a scientific text on phase transitions. However, an answer to the questions about phase transitions with regard to natural selection in Chapters 4 and 5 requires a thorough knowledge of what is meant by the term "phase transition" in condensed matter physics, in particular by its three features: co-operativity, infinite divergence of correlation and multiscale fluctuations. This is important because in this book I am applying those criteria to an entirely different, non-material situation. It is true that information in our world could not have come about without material carriers that were subject to processing mechanisms. Yet the substrate is not of material nature. Genetic information is not identical with DNA, much as literature is not identical to printer ink or sound waves.

If I say that information "comes about by selection", does this process fulfil the criteria that we have derived for a physical phase transition? What in this case is the order parameter we have to focus upon? How does co-operativity come about, and is it of an "all-or-none" nature? These problems are addressed in Chapters 4 and 5.

We shall see that, although of non-material nature, the "phase transition" associated with selection is analogous to physical (i.e. material) phase transitions, and thus is dependent on the three principles: (1) co-operativity, (2) correlation lengths that surpass the borders of extension of the material phase (reaching infinity in the thermodynamic limit) and (3) fluctuations in a critical range, in our case represented by the simultaneous selection of substantial numbers of neutral mutants.

References

1. Mayr, E. (1988). "Toward a New Philosophy of Biology". The Belknap Press of Harvard University Press. Cambridge, MA and London. There are quotations about earlier work on this question.
2. Boltzmann, L. (1886). "Der zweite Hauptsatz der mechanischen Wärmetheorie". Almanach der *K. Akad. Wiss., Wien* 36: 225–59. See J. A. Barth (1905). "Populäre Schriften". Leipzig, p. 28.
3. Goldenfeld, N. (1992). "Lectures On Phase Transitions And The Renormalisation Group". Addison Wesley, New York.
4. Holmes, L. M., Van Uitert, L. G. and Hull, G. W. (1971). "Magnetoelectric Effect and Critical Behavior in the Ising-Like Antiferromagnet, DyAlO3". *Solid State Commun.* 9(16): 1373–1376.
5. Perutz, M. F. (1962). "X-ray Analysis of Haemoglobin". Les Prix Nobel en 1962, Stockholm, Imprimerie Royale, P. A. Norstedt and Söner.
6. Kendrew, J. C. (1962). "Myoglobin and the Structure of Proteins". Nobel Lectures, Chemistry 1942–1962. Elsevier Publishing Company, Amsterdam. 1964.
7. Hofrichter, J., Ross, P. D. and Eaton, W. A. (1976). "Supersaturation in Sickle Cell Hemoglobin Solutions". *Proc. Natl. Acad. Sci. U. S. A.* 73(9): 3035–3039.
8. Prusiner, S. B. (1982). "Novel Proteinaceous Infectious Particles Cause Scrapie". *Science* 216(4542): 136–144; Prusiner, S. B. (1997). "Prions". Les Prix Nobel, Almquist and Wiksell International, Stockholm.

9. Weissmann, Ch. (1991). "A 'Unified Theory' of Prion Propagation". *Nature* 352: 679–683; Sailer, A., Büeler, H., Fischer, M., Aguzzi, A. and Weissmann, C. (1994). "No Propagation of Prions in Mice Devoid of PrP". *Cell* 77(7): 967–968. See also *Cell* 73: 1339–1347 (1993), and *Nature* 389: 795–798 (1997).
10. Bessen, R. A., Kocisko, D. A., Raymond, G. J., Nandan, S., Lansbury P. T., Caughey, B. (1995). "Nongenetic Propagation of Strain-specific Properties of Scrapie Prion Protein". *Nature* 375: 698–700. Further work of P. T. Lansbury Jr. quoted in ref. 11.
11. Eigen, M. (1996). "Prionics or the Kinetic Basis of Prion Diseases". *Biophys. Chem.* 63(1): A1–A18.
12. Berry, R. S., Rice, S. A. and Ross, J. (2000). "Physical Chemistry", 2nd edn. Oxford University Press, New York and Toronto.
13. Zimm, B. H. and Bragg, J. K. (1959). "Theory of the Phase Transition between Helix and Random Coil in Polypeptide Chains". *J. Chem. Phys.* 31: 526–531.
14. Lifson, S. and Roig, A. (1961). "On the Theory of Helix-Coil Transition in Polypeptides". *J. Chem. Phys.* 34: 1963–1974.
15. Crothers, D. M. (1964). "The Kinetics of DNA Denaturation". *J. Mol. Biol.* 9: 712–733.
16. Schwarz, G. (1964). "Zur Kinetik der Helix–Coil-Umwandlung von Polypeptiden in Lösung". *Ber. Bunsenges. Physik. Chem.* 68: 843–850; Schwarz, G. (1965). "On the Kinetics of the Helix-Coil Transition of Polypeptides in Solution". *J. Mol. Biol.* 11: 64–77.
17. Onsager, L. (1944). "Crystal Statistics. I. A Two-Dimensional Model with an Order-Disorder Transition". *Phys. Rev.* 65: 117–149.
18. Landau, L. D. and Lifschitz, E. M. (1980). "Statistical Physics", Part I, 3rd edn. Pergamon, New York. See also Prix Nobel 1982.
19. Widom, B. (1965). "Equation of State in the Neighborhood of the Critical Point". *J. Chem. Phys.* 43: 3898–3905. See 1975 in "Fundamental Problems in Statistical Mechanics", Vol III (E.C.G. Cohen, ed.). North Holland, Amsterdam; Fisher, M. E. (1964). "Correlation Functions and the Critical Region of Simple Fluids". *J. Math. Phys.* 5: 944–962; Essam, J. W. and Fisher, M. E. "Padé Approximant Studies of the Lattice Gas and Ising Ferromagnet below the Critical Point". *J. Chem. Phys.* 38: 802–812.
20. Kadanoff, L. P. (1966). "Scaling Laws for Ising Models near Tc". *Physics* 2: 263–272. In this paper the idea of "blocks of magnetic moments" near the critical point was developed. The scaling laws of Widom and Fisher (see ref. 19) were derived in this paper.
21. Wilson, K. G. (1982). "The Renormalization Group and Critical Phenomena". Nobel Lecture 1982, Almquist and Wiksell International, Stockholm.
22. Gell-Mann, M. and Low, F. E. (1954). "Quantum Electrodynamics at Small Distances". *Phys. Rev.* 95: 1300–1312. Ken Wilson called this paper "the principal inspiration for his own work".

Appendix A3

On the Nature of Mathematical Proof

1. Mathematical induction

The two main methods of mathematical proof are called "constructive" and "indirect". Let me start with the constructive method of "mathematical induction" or, as it is also called, "complete induction". It is the prototype of a constructive proof and the main principle for investigating the qualities of natural numbers: if a certain property holds for $n = 1$, the smallest of all natural numbers, and if from its validity for some n a similar validity for $(n + 1)$ follows, then this assertion is true for all natural numbers. The proof may alternatively be based on a hypothesis, as is stated in Richard Courant and Herbert Robbins' classic book *What is Mathematics?*[1] (newly revised by Ian Stewart), to which I shall refer repeatedly in this appendix. I quote: "The fact that the proof of a theorem consists in the application of certain simple rules of logic does not dispose of the creative element in mathematics, which lies in the choice of the possibilities to be examined. The question of the origin of the hypothesis belongs to a domain in which no very general rules can be given; experiment, analogy and constructive intuition play their part here. But once the correct hypothesis is formulated, the principle of mathematical induction is often sufficient to provide the proof."

In the examples I am going to discuss, the results (some of which are also quoted in the above-mentioned book) are all well known. Moreover, to prove their correctness by mathematical induction is a relatively easy job, so that the authors (in certain cases) left this to their readers. I felt sufficiently challenged by one of the problems considered, namely the sums of arbitrary powers of natural numbers. One rainy weekend in my Californian refuge – such days occur anyway too rarely in Southern California – I recalled Courant's words quoted above. My ambition did not admit to contenting myself with a mere verification of the known solutions. Rather I wanted to show how certain surprising results can be visualised by geometrisation. I think that some of the problems treated can be recommended for advanced courses in high-school mathematics. The problem I chose was the general solution for the sum of the q-th power of the first n natural numbers ${}^{q}\sum_{n} = \sum_{k=1}^{n} k^{q}$ for any power q of k. The classical solution of this problem is well known. It requires a knowledge of certain finite sums such as those defining the Euler and Bernoulli numbers[2] (see below). My real motivation was to learn more about n-dimensional point spaces, which play a part in my thoughts on information as described in Chapters 4 and 5.

1.1 The sum of integers ${}^{1}\sum_{n}$

I start with the trivial case of ${}^{q}\sum_{n}$ for $q = 1$. There is an amusing story of a boy at elementary school. The teacher, tired from lecturing all morning, gave his charges the problem of adding up all the numbers from one to a hundred, hoping to get some time for a little nap. But after a few moments one of the youngsters arrived at the solution, 5050, and he had to enlighten his startled teacher as to how he had got the result that quickly. The boy's name was Carl Friedrich Gauss, who did in his head what is depicted in Figure A3.1. This graphical depiction of the trick used by Gauss is one-dimensional in nature. I say this because the method I am going to describe is based on n-dimensional geometry. The fact that two one-dimensional lines are involved is solely due to the arithmetical trick that Gauss used, which in its generalisation is based on mathematical induction.

1.2 The sum of squares ${}^{2}\sum_{n}$

The problem for $q = 2$ is only slightly more complex if we use a suitable geometrical illustration, which in this case has to be two-dimensional (see Figure A3.2).

Closer inspection of Figure A3.2 reveals:

1) The sum of each vertical column, and hence as well of the diagonal, is given by ${}^{1}\sum_{n} = n(n + 1)/2$.
2) All horizontal rows are n-fold repeats of their respective positional number k in the vertical column. Their sum for (any) position k is $n \cdot k$.

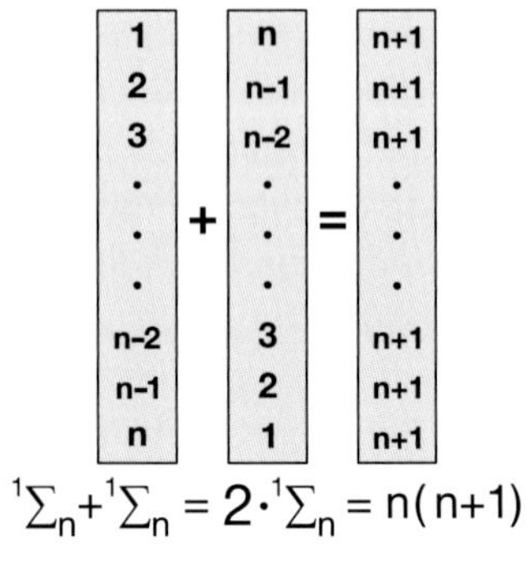

Figure A3.1

Graphical representation of what happened in the mind of the elementary-school boy whose name was Carl Friedrich Gauss

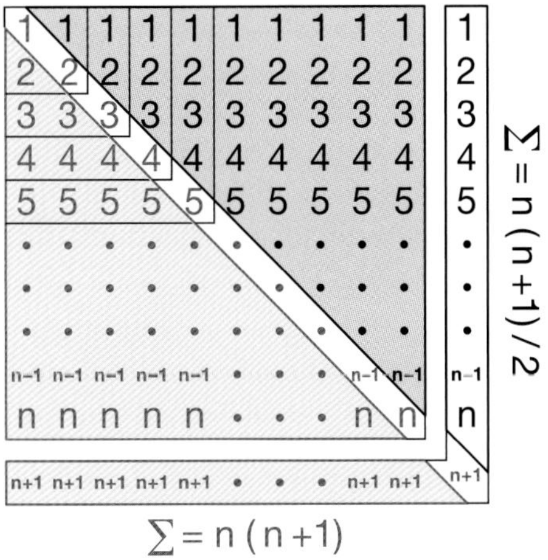

Figure A3.2

Two-dimensional illustration for the calculation ${}^{2}\sum_{n}$, including a geometrical visualisation of complete induction

3) The total sum of all numbers in the table is $n\,{}^{1}\sum_{n} = n^2(n+1)/2$.
4) The sum of all squares, i.e. $\sum_{k=1}^{n} k^2$, is then represented by all red colored numbers of the lower-left triangle plus the diagonal.

These are the conditions on which the construction of the diagram is based. In order to calculate the sum of squares ${}^{2}\sum_{n} = \sum_{k=1}^{n} k^2$, we need a fifth non-trivial statement that can easily be proven by mathematical induction. This statement reads:

5) The sum of all red numbers in the lower-left triangle (plus diagonal) for any n is twice the sum of all black numbers in the upper-right (shaded) triangle.

The proof by complete induction is greatly aided by the geometrical construction in Figure A3.2. First we compare the sums of each horizontal row of the lower-left triangle with their complementary vertical sum in the upper-right triangle. The ratio of horizontal to vertical sum starts with 2:1, followed by 6:3, 12:6 etc., the value of each ratio being two. The same holds for the ratios of the sums in corresponding triangles. The proof can now be completed by induction – in our diagram by transition from n to $n+1$. This means for the lower-left triangle the addition of one horizontal row, the sum of which is $(n+1)n$. (Note that the diagonal does not belong to either of the two triangles.) For the upper-right triangle the transition means the addition of one vertical column: $\sum_{k=1}^{n} k = n(n+1)/2$, which holds for any number n. It proves that the corresponding horizontal sums in the lower-left triangle (plus diagonal) of all red numbers are twice as large as those in the upper-right triangle. Thus, the "geometrical" construction in this case provides a direct visualisation of "complete induction", yielding $(2n+1)/3 \cdot {}^{1}\sum_{n}$ or $(n^2+n)(2n+1)/6 \equiv {}^{2}\sum_{n}$.

However, we see clearly from this simplest case of a higher (i.e. two) dimensional representation that we are not dealing with straightforward squares (i.e. n^2) but rather with "weighted" ones. The "value" of each number in the k-th horizontal line is "k". This means a kind of "stretching" of one dimension, which will also occur in all higher-dimensional cases.

1.3 The sum of cubes ${}^{3}\sum_{n}$

The geometrical representation introduced with ${}^{2}\sum_{n}$ can be generalised by any higher dimension q for calculating ${}^{q}\sum_{n}$. It appears to be more involved because we now require some spatial imagination. For the sum of cubes ${}^{3}\sum_{n}$ the procedure is illustrated in Figures A3.3a and b.

The cube consists of n planes, each containing n^2 equal numbers k, where k runs from 1 to n. From an inspection of the geometrical model we can make the following statements:

1) The sum of the cubes of integers ${}^{3}\sum_{n}$ is given by the sum of all numbers in the (red) shaded planes, which include the "diagonal plane" from the front upper-left point to the lower-right edge (Figure A3.3a).
2) The total sum of all (red and black) numbers in the cube is ${}^{3}\sum_{tot} = n^2 \cdot {}^{1}\sum_{n} = n^3(n+1)/2$.
3) The diagonal, the sum of which again equals ${}^{1}\sum_{n}$, but which does not need to be considered separately, divides the cube into two parts, i.e. the numbers in the red shaded area, the sum of which is ${}^{3}\sum_{n}$, and the non-shaded rest of the numbers, the sum of which we call the "rest sum" ${}^{3}\sum_{rest}$.
4) There is no fixed proportion of ${}^{3}\sum_{rest}$ to ${}^{3}\sum_{n}$, as was the case for the sum of squares. It rather depends on n. We therefore proceed to calculate ${}^{3}\sum_{rest}$ using mathematical induction, which is represented geometrically in Figure A3.3b.

Enlarging the cube from $(k-1)^3$ to k^3 positions requires an addition of three faces, which for convenience I have called the xy, xz and yz faces. Adding the bottom face (xy) is equivalent to adding k^3, i.e. k^2 (shaded) positions each having the value k. For the xz and yz faces we add $(2k-1)$ sums of integers running from 1 to $(k-1)$. Since the bottom face belongs to the shaded part, the sums run only from 1 to $(k-1)$. Furthermore, since the two faces (xz and yz) have one edge in common, there are only $(2k-1)$ sums running from 1 to $(k-1)$. Hence ${}^{3}\sum_{rest}$ in each single step changes by:

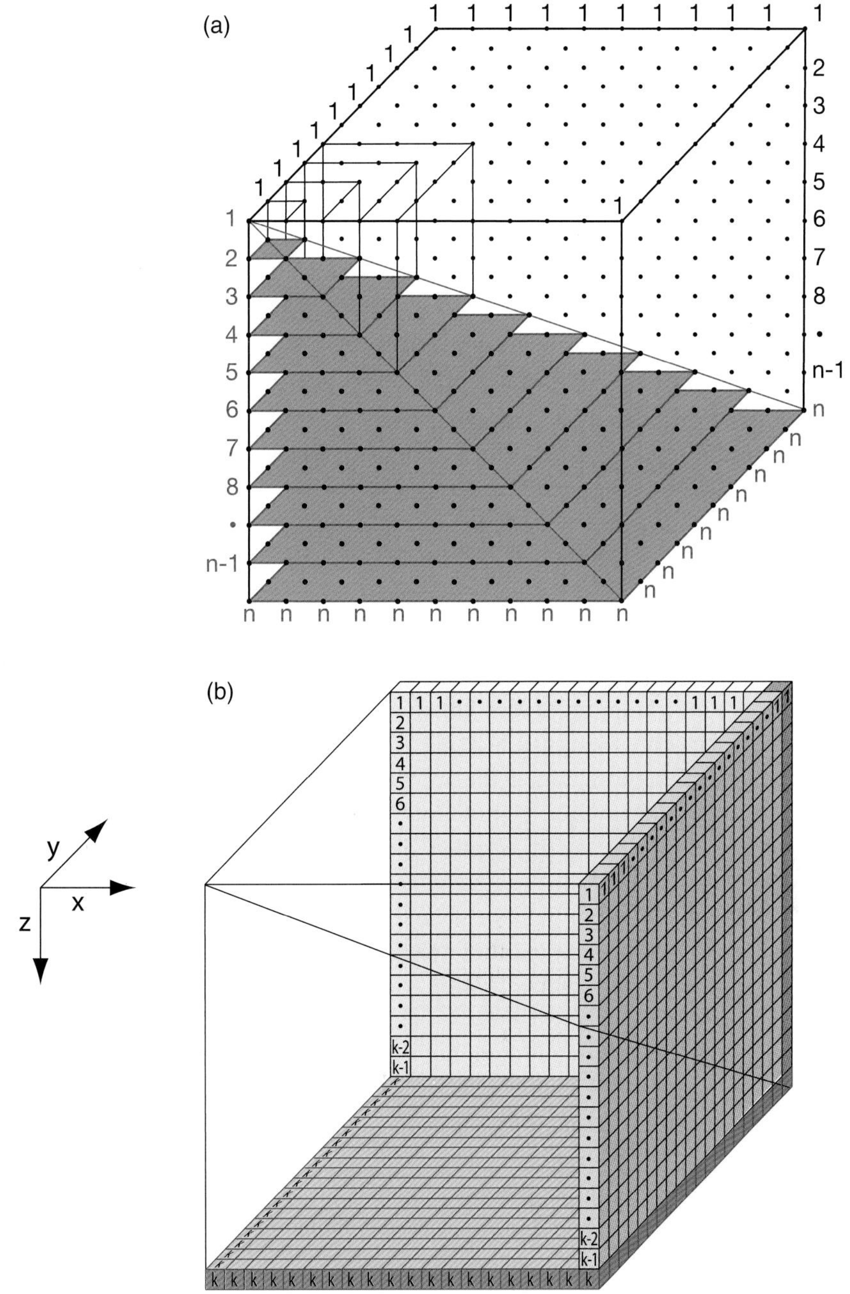

Figure A3.3

The three-dimensional representation shows the way to treat the n-dimensional problem (see text). (a) The points in the red-shaded area yield the sum of cubes: ${}^{3}\sum_{n}$
(b) How to proceed by mathematical induction by enlarging the cube (see text).

$$\delta\,{}^{3}\!\sum_{rest} = (2k-1)\sum_{l=1}^{k-1} l = k\,(k-1)\,(2k-1)\,/2 = k^3 - 3k^2/2 + k/2$$

Now we can add up for the rest sum of the whole cube as defined above:

$${}^{3}\!\sum_{rest} = \sum_{k=1}^{n} k^3 - \tfrac{3}{2}\sum_{k=1}^{n} k^2 + \tfrac{1}{2}\sum_{k=1}^{n} k$$

Knowing the sum of squares ${}^{2}\!\sum_n = \frac{(2n+1)}{3} \cdot {}^{1}\!\sum_n$ we arrive at:

$${}^{3}\!\sum_{rest} = {}^{3}\!\sum_{n} - \tfrac{3}{2}\,{}^{2}\!\sum_{n} + \tfrac{1}{2}\,{}^{1}\!\sum_{n} = {}^{3}\!\sum_{n} - n\,{}^{1}\!\sum_{n}$$

This indicates to us already a symmetry, resulting from the particular geometry. Since the sum ${}^{3}\!\sum_n + {}^{3}\!\sum_{rest}$ yields the "volume" of the whole space ${}^{3}\!\sum_{tot} = n^2\,{}^{1}\!\sum_n$ the sum ${}^{3}\!\sum_n$ comes out to be $\frac{(n^2+n)}{2}\,{}^{1}\!\sum_n = ({}^{1}\!\sum_n)^2$.

For $n \to \infty$ the sums ${}^{3}\!\sum_n$ and ${}^{3}\!\sum_{rest}$ converge to the same value $\frac{n^2}{2}\,{}^{1}\!\sum_n$ yielding ${}^{3}\!\sum_{tot} = n^2\,{}^{1}\!\sum_n$.

Let me quote once more from the book of Courant *et al.*[1], who refer in particular to this formula: "It should be remarked that although the principle of mathematical induction suffices to *prove* the formula once this formula has been written down, the proof gives no indication of how this formula was arrived at in the first place; why precisely the expression $[n(n+1)/2]^2$ should be guessed as an expression for the sum of the first n cubes, rather than any of the infinitely many expressions of a similar type that could have been considered."

My first reaction was, too: "Wonders will never cease". However, geometry offers a suggestive explanation. The particular space I have called ${}^{3}S_{tot}$ (for which ${}^{3}\!\sum_{tot} = n^2\,{}^{1}\!\sum_n = {}^{3}\!\sum_n + {}^{3}\!\sum_{rest}$) yields a special symmetry among the two sums ${}^{3}\!\sum_n$ and ${}^{3}\!\sum_{rest}$ that is not present for any other dimension. In order to explain this let us now consider first the generalisation to any arbitrary dimension q.

1.4 The general case ${}^{q}\!\sum_n$

So far I have applied mathematical induction more in a "physical sense", namely by deriving results simply from quantitative inspection of geometrical structures. Let me conclude now by using mathematical induction to generalise the rules for an application to structures of higher dimensionality that are harder to visualise because of our lack of imagination. Nevertheless, the geometrical interpretation with the help of the three sums ${}^{q}\!\sum_{tot}$, ${}^{q}\!\sum_n$ and ${}^{q}\!\sum_{rest}$ will remain the basis of our more abstract deliberations.

The four-dimensional hypercube in Figure A3.4a will be the model that applies in a formally analogous way to any dimension q. We define, again, one co-ordinate where k runs from 1 to n and call it the "vertical" axis. For each value k of the

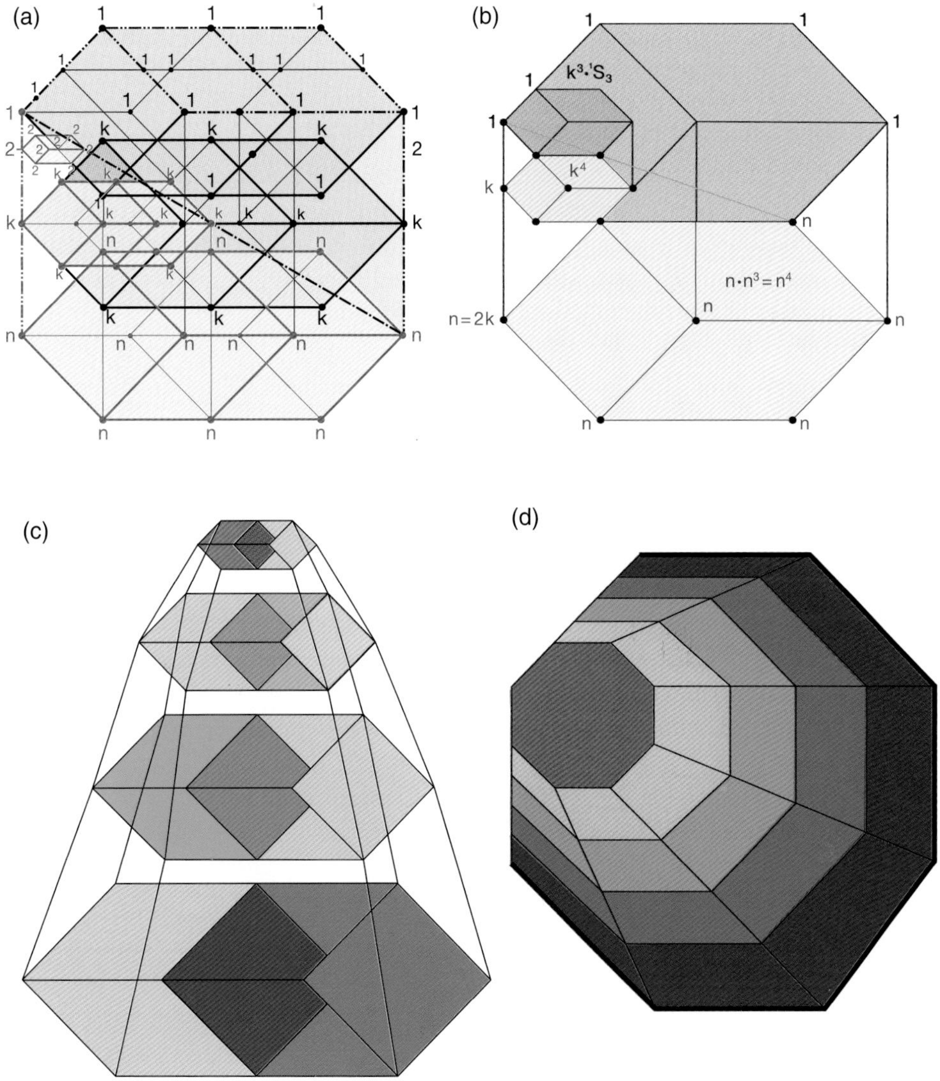

Figure A3.4

In this figure the four-dimensional case (q = 4), which we still have some ability to imagine, is used to introduce the generalisation. The proof cannot gain much further advantage from a geometric visualisation. The abstraction takes us close to the classical proof, which can be found in textbooks. Nevertheless, the geometrical model still provides some general insight in the case of large dimensions. (a) The complete state model for q = 4. (b) The enlargement of the subspace for increasing n as described in the text. (c) and (d) Artistic models of the enlargement.

"vertical" axis there is a (q – 1)-dimensional hyperface in which all positions have the same value k, quite analogous to the two-dimensional planes in Figure A3.3b. Each of these hyperfaces contains n^{q-1} positions. In Figure A3.4b the three-dimensional hyperface, referring to position k on the "vertical" axis, is highlighted in red, in order to distinguish it clearly from hyperfaces referring to other positions and to stress the analogy to the planes in Figure A3.3a. For dimensions $q > 4$ the situation is similar, except that the hyperplanes are more complex, higher-dimensional structures with various overlappings of common faces in their lower-dimensional hyperplanes, as shown for the four-dimensional example in Figure 1.2.5 in the text. We again compare the total volume ${}^q\sum_{tot}$ with the two volumes ${}^q\sum_n$ and ${}^q\sum_{rest}$. Hence, as in the three-dimensional case, the following statements hold:

1) The total sum of numbers in the hypercube ${}^q\sum_{tot}$ is given by $n^{q-1} \cdot {}^1\sum_n = (n^{q+1} + n^q)/2$.
2) The total sum consist of two parts: ${}^q\sum_n = \sum_{k=1}^{n} k^q$, i.e. the sum of powers q of integers 1 to n (the "unknown" we want to determine) and the rest sum ${}^q\sum_{rest} = {}^q\sum_{tot} - {}^q\sum_n = n^{q-1} \cdot {}^1\sum_n - \sum_{k=1}^{n} k^q$.
3) In Figure A3.4b ${}^q\sum_n$ is given by all red-coloured symbols, while ${}^q\sum_{rest}$ is represented by all black symbols. Note that in the figure the colours red and black distinguish two different classes of number arrangement.
4) As introduced with the three-dimensional case we use the above relation between ${}^q\sum_{tot}$, ${}^q\sum_n$ and ${}^q\sum_{rest}$ in its variational form in order to obtain an explicit expression for ${}^q\sum_{rest}$. Its geometrical interpretation, as given for ${}^3\sum_{rest}$, is still obvious for ${}^4\sum_{rest}$, but becomes increasingly cumbersome for larger dimensions because of our inability to imagine high-dimensional structures and their sharing of common faces with lower hyperplanes.

The variational increase of the total sum when going from (k − 1) to k is:

$$\delta\,{}^q\sum_{tot} = k^q\,(k+1)/2 - (k-1)^q \cdot k/2 = k^q\left\{(k+1) - k\left(1-k^{-1}\right)^q\right\}/2$$

where $(1-k^{-1})^q$ can be expressed by its (finite) binomial power series yielding:

$$\delta\,{}^q\sum_{tot} = \frac{k^q}{2}\left\{1 + q - \frac{q(q-1)}{2!}k^{-1} + \frac{q(q-1)(q-2)}{3!}k^{-2} - \ldots \right.$$
$$\left. \pm \frac{q(q-1)(q-2)\ldots(q-p+1)}{p!}k^{-q+1}\right\} \text{ (last term } \pm\, p = q)$$

In order to obtain the variational expression for the rest sum we have to subtract the hyperplane k^q from the above expression:

$$\delta^{q}\sum_{rest} = \frac{q-1}{2}k^{q} - \frac{q(q-1)}{2\cdot 2!}k^{q-1} + \frac{q(q-1)(q-2)}{2\cdot 3!}k^{q-2} - + \ldots \text{ (last term } \sim k)$$

yielding the total rest sum by adding up all δ terms:

$$^{q}\sum_{rest} = \frac{q-1}{2}\sum_{k=1}^{n}k^{q} - \frac{q(q-1)}{2\cdot 2!}\sum_{k=1}^{n}k^{q-1} + \frac{q(q-1)(q-2)}{2\cdot 3!}\sum_{k=1}^{n}k^{q-2} - + \ldots \text{(last term} \sim \sum_{k=1}^{n}k)$$

The equation $^{q}\sum_{n} = {}^{q}\sum_{tot} - {}^{q}\sum_{rest}$ then contains as the only unknown $^{q}\sum_{n}$, all other terms referring to sums with powers $< q$:

$$^{q}\sum_{n} = \sum_{k=1}^{n}k^{q} = \frac{n^{q+1}}{q+1} + \frac{1}{q+1}\left\{n^{q} + \frac{q(q-1)}{2!}\sum_{k=1}^{n}k^{q-1} - \frac{q(q-1)(q-2)}{3!}\sum_{k=1}^{n}k^{q-2} + \ldots\right\} \text{ (last term} \sim \pm\sum_{k=1}^{n}k)$$

Now let us look at hidden symmetries among the various sums.

$^{q}\sum_{tot}$, which is the sum of $^{q}\sum_{n}$ and $^{q}\sum_{rest}$, has the simple form of $(n^{q+1} + n^{q})/2$. The sum $^{q}\sum_{n}$, according to the expression given above, is dominated by its "leading term" $n^{(q+1)}/(q+1)$, which represents its limiting value for large values of n. For $n = 100$ the leading term is already about hundred times larger than all of the rest. We therefore write:

$$^{q}\sum_{n} = \frac{n^{q+1} + n^{q}}{q+1} + \frac{^{q}\Delta_{n}}{q+1}$$

where $^{q}\Delta_{n}$ includes all other terms listed above. Then $^{q}\sum_{rest}$ can be written as

$$^{q}\sum_{rest} = \frac{q-1}{2}\;\frac{n^{q+1} + n^{q}}{q+1} - \frac{^{q}\Delta_{n}}{q+1}$$

Here we see immediately the symmetry of $^{q}\sum_{n}$ and $^{q}\sum_{rest}$, the leading terms of which become identical for $q = 3$, where the sum of both equals the leading term of $^{q}\sum_{tot}$, which is given by $(n^{(q+1)} + n^{q})/2$. This symmetry is unique for $q = 3$ because it does not exist for any other arbitrary q value. Moreover, for $q = 3$, the first term in the above expression for $^{q}\sum_{n}$ becomes equal to $n^{2}\cdot {}^{1}\sum_{n}/2$ while $^{q}\Delta_{n}/(q+1)$ is about $n\cdot {}^{1}\sum_{n}/2$, so that the whole expression $^{3}\sum_{n}$ equals $({}^{1}\sum_{n})^{2}$. For analogous reasons the term $n \cdot {}^{1}\sum_{n}/2$ has to be subtracted from the leading term, so that ${}^{q}\sum_{rest}$ becomes equal to ${}^{1}\sum_{n} \cdot {}^{1}\sum_{n-1}$, again a unique symmetry for $q = 3$.

Now assume that any symmetry exists that allows us to write:

$${}^{q}\sum_{n} = {}^{i}\sum_{n} \cdot {}^{k}\sum_{n}$$

Representation of the sums by their "leading terms" would yield

$$\frac{n^{q+1}}{q+1} = \frac{n^{i+1}}{(i+1)} \cdot \frac{n^{k+1}}{(k+1)}$$

which means that

$$(i+1) + (k+1) = q+1 = (i+1)(k+1)$$

having the only solution:

$$i = k = 1; \quad q+1 = 2 \times 2 = 2+2 = 4 \quad \text{and} \quad q = 0+1+2 = 3$$

With this, the uniqueness of $q = 3$ is reduced to the trivial facts of the theory of natural numbers, namely 3 being the only number that is the sum of all its precursors and 4 being the only number given by $x + x = x \cdot x = x^{x}$, for $x = 2$.

The above rules hold also for any higher-order products ${}^{q}\sum = {}^{i}\sum \cdot {}^{k}\sum \cdot {}^{l}\sum \ldots$ that can always be reduced to a product of two sums. The general validity depends essentially on the fact that the leading terms for any $q > 3$ are devoid of the above symmetry, since $\frac{(q-1)}{2}$ is always (increasingly) larger than 1 for increasing $q > 3$.

In our case the solution for ${}^{q}\sum_{n}$ builds up iteratively from the solutions for smaller q values. Hence it is expressed in terms of powers of n^{q}, n^{q-1}, n^{q-2} etc. The same result was achieved in the classic solution in terms of Bernoulli numbers B_n defined by the generating function[2]

$$\frac{te^{xt}}{e^{t-1}} = \sum_{n=0}^{\infty} B_n(x)\frac{t^n}{n!} \quad \text{for} \quad |t| < 2\pi$$

with $B_0 = 1, B_1 = -1/2, B_2 = 1/6, B_4 = -1/30, \ldots$

It is interesting to notice certain symmetries, although none of them reaches that of ${}^{3}\sum_{n} = ({}^{1}\sum_{n})^{2}$ with its much simpler geometry.

My main incentive for treating the above problem is not the reproduction of expressions already known for the sums of powers of integers, but rather stressing the geometrical visualisation of the problem. Higher-dimensional spaces are of interest in several parts of this book, for example in Sections 1.2, 1.8, 1.9, 2.9, 3.6, 4.4, 4.5 and 4.8. They are a major tool in treating the problems involved in the "generation of information".

The final part of the proof, in which we go beyond dimension $q = 4$ and where our spatial imagination begins to fail, shows how the geometrical depiction converges with the known general proof in which the Bernoulli numbers appear. My mathematician colleague and friend Andreas Dress tried to find out whether the proof given here was known in the literature, so that I could quote some reference. However, he could not find any reference, although the general opinion of his colleagues was that "geometrisation" of the problem might be familiar and considered to be "folklore". To be serious again, I did not expect to present new mathematics in this book and I would be truly surprised if those deliberations have not been made before. Let me come to some conclusion.

1.5 Summary and assessment

The method introduced above involves geometrical visualisation. It uses the word "space" in the mathematical sense as an ordered set with a mathematically well-defined structure. It is a geometry of sequences of natural numbers dominated by the sequence $\sum_{k=1}^{n} k = (n^2 + n)/2$, calculated by Carl Friedrich Gauss when he – as the story goes – was a youngster at elementary school (see Figure A3.1). Dealing with discrete numbers the geometrical representation has to use point spaces, which in general are hyperspaces of arbitrary dimensionality q, denoted ${}^{q}S_{tot}$. The index "tot" = total refers to the fact that this space consists of two subspaces: ${}^{q}S_{n}$ and ${}^{q}S_{rest}$, with ${}^{q}S_{n}$ comprising the set of the first n natural numbers of power q, i.e. $\sum_{k=1}^{n} k^q$ and ${}^{q}S_{rest}$ representing the set of all points of ${}^{q}\sum_{tot}$ that do not belong to ${}^{q}\sum_{n}$. These spaces have the "seemingly simple" form of hypercubes, e.g. ${}^{q}\sum_{tot} = q^{(n-1)}\,{}^{1}\sum_{n}$, with q^{n-1} being the ((n – 1)-dimensional) hyperplane of the hypercube n^q.

Why then do I call $q^{(n-1)}\,{}^{1}\sum_{n}$ a "seemingly simple" form of a hypercube? A "true" hypercube should be an analogue of a cube in three-dimensional space. The quantity q^{n-1} is indeed a "true" hyperplane of the "true" hyperspace q^n, but $q^{(n-1)}\,{}^{1}\sum_{n}$ is not because ${}^{1}\sum_{n} = (n^2 + n)/2$ yields a cube with numbers valued differently in each of the n dimensions. It certainly is a hyperspace with peculiar properties and the total space ${}^{q}S_{tot}$ could be interpreted as a "hypercube" that "changes" its size in some complicated manner. "How" is seen by the calculations presented above, which involve "mathematical induction" in building up the stretched form. The calculated sums are derived from the three hyperspaces ${}^{q}S_{tot}$, ${}^{q}S_{n}$ and ${}^{q}S_{rest}$, and therefore still represent a "geometrisation" of the problem.

This can be shown in a simple way if we look at the largely dominating leading terms of the sum expressions obtained by the detailed calculations. The leading term is the one with the highest power of k that represents a good approximation at large values of n (such as 100 or more). The maximum error then may be around 1%. The leading terms for all values of q have the general form for ${}^{q}\sum_{n} : n^{q+1}/(q + 1)$, for ${}^{q}\sum_{tot} : n^{q+1}/2$ and for ${}^{q}\sum_{rest} : ((q - 1)/2)\cdot(n^{q+1}/(q + 1))$. If we compare the expression for ${}^{q}\sum_{n}$ with that for a "true" hypercube, such as q^n, we see that with the increase in n the hyperspace ${}^{q}S_n$ undergoes inflation of its size (or volume) in the "vertical" direction in proportion to n/2. This is pictured for a four-dimensional space ($q = 4$) in Figure A3.4b, c and d, where the latter two figures are more an artistic reflection on this effect. Compared with a normal hypercube n^q the "volume" of which is the sum of n of its hyperplanes, i.e. $n \times n^{q-1} = n^q$, the hyperspace ${}^{q}S_{tot}$ consists of n of its inflating hyperplanes $n(n^{q-1}n/2) = n^{q+1}/2$. Going back to ${}^{q}\sum_n$ and ${}^{q}\sum_{rest}$ we see that for $q > 3$ the sum ${}^{q}\sum_{rest}$ inflates faster than ${}^{q}\sum_n$ by the factor $(q - 1)/2$. This is the gross behaviour, neglecting all the details of the sums which in the geometrical picture are caused by overlapping of hyperplanes (of all dimensions) planes and edges. If we want to calculate these details precisely we have to apply "mathematical induction" in each step of the iteration because the changes occur by going from k to $k + 1$ for all k values from 1 to n. A geometrical visualisation of this procedure (for $q = 2$, 3 and 4) was shown in Figures A3.2, A3.3 and A3.4. The symmetries in ${}^{q}\sum_n$ found for those spaces can be traced back to the peculiarities that distinguish the numbers 2, 3 and 4 from larger numbers (see Figure A3.2). I am sure that the geometrisation will uncover further secrets at higher dimensions, but this is not the place to go into more detail.

In physics we do not yet have an answer to the questions "why strings" and "why do we live in a three-dimensional world for which our brain developed some intuitive access"? Fundamental theories of physics (such as string theory) are based on higher-dimensional spaces. Maybe the symmetries observed in our geometrical visualisation are of some relevance in this connection. At the "big bang" they started from a singularity that could "materialise" only in the form of a one-dimensional string inflating further to higher dimensions.

In conclusion let me return to Richard Courant[1], who has reminded us that "constructive intuition" as much as the "rules of logic [such as mathematical induction] play their part" in the extraordinary efficiency of mathematics for the problems of physics, which left such a deep impression on Eugene Wigner. I met Richard Courant on several visits to the USA. He always was curious to hear about Göttingen, where he lived until 1933. After the war, the city of Göttingen made him an honorary citizen. Since the same honour was granted on me, many years later, our pictures now hang in the city hall opposite one another, allowing us to indulge forever in "virtual communication".

2. Indirect proof*

Cantor's diagonal argument is a famous example of an indirect method of mathematical proof. Reasoning in this case is based on assuming tentatively the contrary

* I think that this section should be included in view of the problems treated in this book, although it is a short summary of something treated in detail in the literature.[3,4]

of the statement to be proven and then demonstrating some absurdity or antinomy in the resulting conclusions. Hence the truth of the original statement is proven as a consequence of the logical principle of the "excluded middle" (*tertium non datur*). Cantor's diagonal argument[3], which I mentioned repeatedly in other sections, has found numerous applications. I choose it as an example in this appendix because of its relevance for an appreciation of biological complexity and its astonishing consequences, discussed in Chapters 4 and 5. The following presentation is based on both an early text by Cantor's contemporary Felix Klein[4], one of the great representatives of the Göttingen mathematical school (see Section 3.6), and the book by Courant and Robbins quoted in the first part of this appendix.

Cantor's argument, at first sight, might sound simple. However, it is anything but simple or trivial, and accordingly it found several opponents among his contemporary colleagues, many of them highly distinguished mathematicians. I would rather call the concept ingenious. What is the difference? Something trivial must be both obvious and true, while an ingenious idea cannot be obvious – otherwise anyone could have seen it. Yes, it may even be strange but, nevertheless, it must be true.

Let me therefore go back somewhat further. Cantor[3] is one of the founders of set theory and thereby provided a mathematical analysis of infinity. The most prominent example – even for non-mathematicians – is that of the infinite set of the sequence of positive integers, 1, 2, 3, ... referred to in Appendix 1. It has "no end"; that is what the Latin adjective "infinitus" means. Adding 1 to any number n in this sequence defines the next larger integer. And this never reaches an end; but it establishes one of its most important characteristics. The infinite set of all positive (as well as of all negative) integers is "denumerable".

Infinity is usually expressed by the symbol ∞, but this does not mean that ∞ can be thought of being something like an ordinary number. This leads us immediately to some important consequences regarding the power or potency (German: *Mächtigkeit*) of a set, which is called its "cardinality". If a set were finite, e.g. if the sequence of positive integers were to stop at any finite natural number n, then the total set of n natural numbers could not be equivalent to any of its proper subsets. But for an infinite set, Cantor showed that this is possible. For instance, there is a one-to-one correspondence between the set of all natural numbers and its subset of even numbers:

$$\begin{array}{ccccc} 1 & 2 & 3 & 4 & 5\ldots \\ \updownarrow & \updownarrow & \updownarrow & \updownarrow & \updownarrow \\ 2 & 4 & 6 & 8 & 10\ldots \end{array}$$

The two are equivalent denumerable sets.

Cantor denoted the cardinality of a set by the first symbol of the Hebrew alphabet, called aleph (ℵ). The "smallest" infinite set is that of natural numbers, being assigned the cardinality $\aleph_0$. Many other, more complicated, sets have the same cardinality because one can establish a one-to-one correspondence with the set of integers, and it is the fact of countability or denumerability that is expressed in the cardinality of a set. Accordingly, $\aleph_0$ has all sorts of odd properties, such as

$$\aleph_0 + 1 = \aleph_0 \quad or \quad n \times \aleph_0 = \aleph_0 \quad or \quad \aleph_0^n = \aleph_0$$

where n is a finite number. The most surprising fact is that the set of all rational numbers also has the same cardinality $\aleph_0$. This is surprising because we are used to the fact that rationals distribute themselves (fairly) densely between any two subsequent integers. Yes, there are infinitely more rationals than integers, but they are denumerable because each is made up of a pair of integers (p, q), i.e. p/q, yielding the cardinality of $\aleph_0^2 = \aleph_0$. The denumerability of the set is explained in the scheme shown in Figure A3.5a and b (following Courant *et al.*[1]).

The sequence obtained (right-hand figure) contains every rational number, showing that the total set is denumerable. The important point for the discussion below is that all points can be connected by a line which, however, covers a two-dimensional plane – according to the unlimited variation of both p and q.

Does the countability of rational numbers mean that any infinite set is denumerable? The answer to this question is Cantor's legacy: There are infinities that are uncountable and therefore must be "bigger" than the infinity of natural numbers. And here is Cantor's proof in the words of his contemporary Felix Klein[4]: "Let us write the set C_1, containing the real numbers $0 < x < 1$, as decimal fractions, all being brought into a denumerable series where the a, b, c, . . . are numerals 0, 1, . . . , 9 of any possible choice and sequence" (Figure A3.5 is redrawn from Klein's monograph). Klein continues with a warning against the choice of the numerals 0 and 9

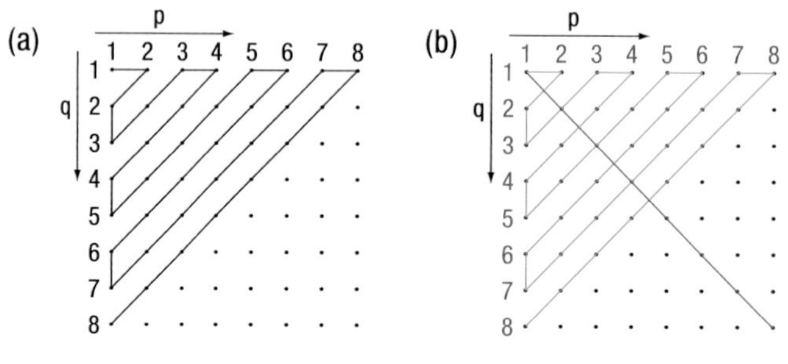

Figure A3.5

(a) If p and q are all natural numbers, each point in the diagram represents one rational number. The diagram shows clearly that all rationals can be connected by one continuous line. The total set of rationals is therefore denumerable and has the same cardinality $\aleph_0$ as all natural numbers. (b) The red diagonal connects all points p/q = 1. Above this line, p/q is larger than 1, below the line it is smaller than 1. Both sequences of natural numbers p and q are unlimited.

because 0.999... and 1.000... (both continued infinitely) are defined as being equal to one another. He then also stresses the necessity of always using an infinite number of decimal places, which for all real numbers excludes those with a finite number of decimal places, and replacing them by infinite decimals that close with an unlimited number of nines. He then continues: "In order to obtain a decimal fraction x' which differs from all sequences x_i contained in the table we give prominence to the numbers $a_1, b_2, c_3, \ldots$ which in the above representation occupy the diagonal of the table (red colour) and replace them by $a'_1, b'_2, c'_3, \ldots$, each a'_1, b'_2, c'_3 and so on being different from a_1, b_2, c_3, etc". In order to prevent x' from terminating at any finite position one should avoid 0 and 9 in any replacement. Felix Klein then remarks that the emphasis upon the diagonal determines the name of the method (see graph in Figure A3.6).

Cantor's proof is of an indirect nature. It is shown that by assuming C_1 to be denumerable contradictions are introduced. In order to demonstrate this, Cantor considered the "diagonal" consisting of the numerals $0, a_1, b_2, c_3, \ldots$, keeping in mind that he was dealing with an infinite number of those numerals. He replaced all the numerals appearing in the diagonal of x by different symbols $x' = 0, a'_1, b'_2, c'_3, \ldots$, again continued for the whole infinite set. Two non-terminating decimal fractions can only equal one another if all numerals agree. This is not the case for x_1 and x'_1, and neither is it the case for any x_i in the infinite set. Hence the decimal fraction x' differs from all numbers $(x_1, x_2, x_3, \ldots)$ of the system which we had assumed to be denumerable, proving the non-denumerability of the continuous set C_1.

Klein also explains that the x' obtained is a true decimal fraction, which excludes $0.999\ldots = 1$ or any decimal with finite number of places. He emphasises that x' must differ from all decimal sequences x_i contained in the original scheme. I do not want to go into further detail of his exposition, except to mention that he also generalised the sequence of numerals a, b, c, ... by complexes of numerals that are surrounded by zeros. Following a suggestion by J. König[4,5] he called those complexes "molecules", perhaps showing a premonition of applications to macromolecular sequences as discussed in Chapters 4 and 5.

Figure A3.6

Cantor's "diagonal method" for disproving the denumerability of all *real* numbers. C denotes the entirety of all real numbers in the form of decimal fractions in the interval $0 < x < 1$; a, b, c being the numerals 0, 1, ..., 9 occuring in all possible combinations and sequences. The diagonal is highlighted in red. It is shown in the text that the numerals in all non-terminating decimal fractions cannot be made to agree, and therefore the system is proven to be non-denumerable.

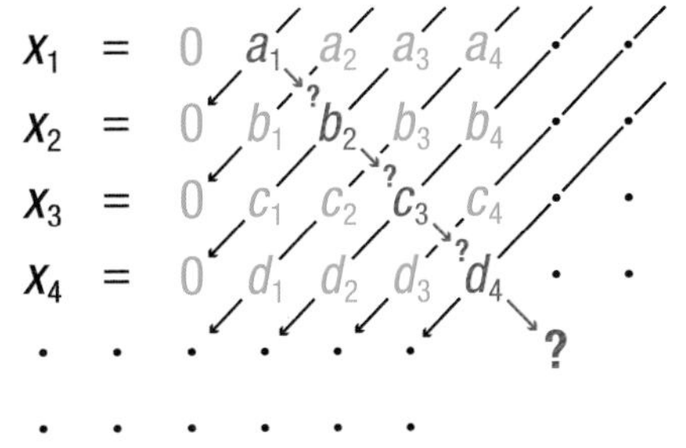

As I mentioned earlier, Cantor's proof sounds simple but is not simple at all. Yes, the method can be called "diagonal", but the proof is that of the "non-existence of that diagonal". This becomes apparent if we try to construct a regular scheme by starting with decimal fractions of finite length. Of course, we are then doing something that is not allowed for irrationals, which by definition have an unlimited number of places, producing a table containing an infinite number of decimal sequences. I use this scheme only in order to demonstrate that with every additional decimal the number of possible sequences increases by a factor of 10, ending up with a factor of 10^n for n additional places. In the case of irrationals, n ultimately becomes infinite. The proof then shows that such a total set has no diagonal in the sense of that in quadratic matrices. In other words, there is no way to draw a line, as for the case of a quadratic matrix. This can easily be shown for binary sequences, which at each step double the number of equations. Since the number of steps is infinitely large, this exponential complexity causes the difficulty of demonstrating non-denumerable cases.

In my introduction to this appendix I said that it would not provide any new results to the mathematician. That is not quite true. If he is going to apply his thinking to biology, he should be aware of what we consider to be the true mathematical problems in biology.

References

1. Courant, R. and Robbins, H. (1996). "What is Mathematics? An Elementary Approach to Ideas and Methods". Revised by Ian Stewart. Oxford University Press, New York.
2. Gradshteyn, I. S. and Ryzhik, I. M. (1999). "Tables of Integrals, Series and Products". Jeffrey, A. (ed.). Academic Press, London.
3. Cantor, G. (1874). "Über eine Eigenschaft des Inbegriffes aller reellen algebraischen Zahlen" ("On a Characteristic Property of All Real Algebraic Numbers"). *Crelles J. f. Mathematik* 77: 258–262. Cantor's main work, in particular the papers on set theory between 1870 and 1900, are discussed in most major textbooks of mathematics.
4. Klein, F. (1924). "Elementarmathematik I" in Band XIV der "Grundlehren der Mathematischen Wissenschaften". pp. 271–288.
5. König, J. Quoted in ref. 4, p. 279.

Appendix A4

The Mathematics of Darwinian Systems

Peter Schuster*

Abstract: Optimization is studied as the interplay of selection, recombination and mutation. The underlying model is based on ordinary differential equations (ODEs), which are derived from chemical kinetics of reproduction, and hence it applies to sufficiently large – in principle infinite – populations. A flowreactor (*continuously stirred tank reactor*, CSTR) is used as an example of a realistic open system that is suitable for kinetic studies. The mathematical analysis is illustrated for the simple case of selection in the CSTR. In the following sections the kinetic equations are solved exactly for idealized conditions. A brief account on the influences of finite population size is given in the last section.

1. Replication in the flowreactor

Replication and degradation of molecular species $\mathbf{I}_i$ ($i = 1, 2, \ldots, n$) in the flowreactor (Figure A4.1) follows the mechanism

$$\begin{aligned} * &\xrightarrow{a_0\ r} \mathbf{A} \\ \mathbf{A} + \mathbf{I}_i &\xrightarrow{k_i} 2\mathbf{I}_i \\ \mathbf{I}_i &\xrightarrow{d_i} \mathbf{B} \\ \mathbf{A}, \mathbf{B}, \mathbf{I}_i &\xrightarrow{r} \varnothing\ , \end{aligned} \tag{A4.1}$$

*Address: Institut für Theoretische Chemie der Universität Wien, Währingerstraße 17, 3. Stock, 1090 Wien, Austria, email: pks@tbi.univie.ac.at

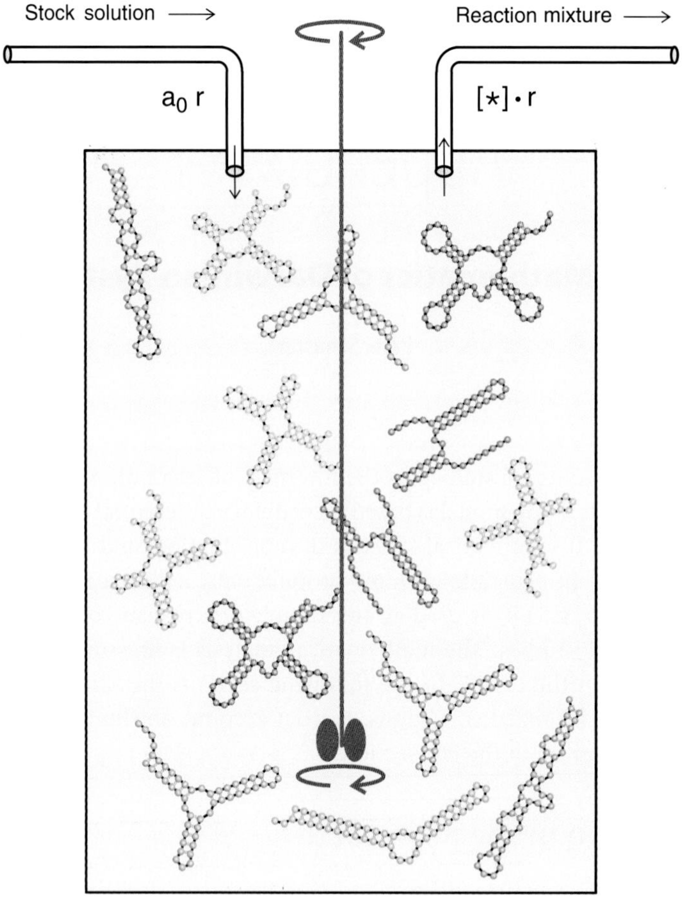

Figure A4.1
The flowreactor for the evolution of RNA molecules.

A stock solution containing all materials for RNA replication ($[\mathbf{A}] = a_0$), including an RNA polymerase, flows continuously at a flow rate r into a well-stirred tank reactor (CSTR) and an equal volume containing a fraction of the reaction mixture ($[*] = \{a, b, c_i\}$) leaves the reactor (for different experimental setups see Watts).[1] The population of RNA molecules ($I_1, I_2, \ldots, I_n$ present in the numbers $N_1, N_2, \ldots, N_n$ with $N = \sum_{i=1}^{n} N_i$) in the reactor fluctuates around a mean value, $N \pm \sqrt{N}$. RNA molecules replicate and mutate in the reactor, and the fastest replicators are selected. The RNA flowreactor has been used also as an appropriate model for computer simulations. [2, 3, 4] There, other criteria for selection than fast replication can be applied. For example, fitness functions are defined that measure the distance to a predefined target structure and mean fitness increases during the approach towards the target.[4]

and is described by the following $(n+2)$ kinetic differential equations

$$\dot{a} = -\left(\sum_{j=1}^{n} k_i c_i\right) a + r\,(a_0 - a),$$

$$\dot{c}_i = (k_i\, a - (d_i + r))\, c_i,\ i = 1, 2, \ldots, n \text{ and} \tag{A4.2}$$

$$\dot{b} = \sum_{j=1}^{n} d_j\, c_j - rb.$$

The variables $a(t)$, $b(t)$, and $c_i(t)$ are concentrations, $[\mathbf{A}] = a$, $[\mathbf{B}] = b$ and $[\mathrm{I}_i] = c_i$, which are defined by $a = N_{\mathrm{A}}/(V\,N_L)$, $b = N_{\mathrm{B}}/(V\,N_L)$, and $c_i = N_i/(V\,N_L)$ where V is the volume and N_L is Avogadro's number, the number of particles in one mole of substance. The particle numbers N are discrete and non-negative quantities, whereas concentrations are assumed to be continuous because $N_L = 6.023 \times 10^{23}$ is very large.[1]

The equations (A4.2) sustain $(n+1)$ stationary states fulfilling the conditions $\dot{a} = 0, \dot{b} = 0, \dot{c}_i = 0$ for $i = 1, 2, \ldots, n$. Every stationarity conditions for one particular class of replicating molecules $\mathbf{I}_i$

$$\bar{c}_i\,(k_i\bar{a} - (d_i + r)) = 0$$

has two solutions (i) $\bar{c}_i = 0$ and (ii) $\bar{a} = (d_i + r)/k_i$. Since any pair of type (ii) conditions is incompatible,[2] only two types of solutions remain: (i) $\bar{c}_i = 0\,\forall\, i = 1, 2, \ldots, n$, the state of extinction, because no replicating molecule 'survives' and (ii) n states with $\bar{c}_j = (a_0/(d_j + r) - 1/k_j)\, r$ and $\bar{c}_k = 0 \forall k \neq j$. Steady-state analysis through linearisation and diagonalisation of the Jacobian matrix at the stationary points yields the result that only one of the n states is asymptotically stable, in particular the one for the species $\mathbf{I}_m$ that fulfils

$$k_m a_0 - d_m = \max\{a_j k_j - d_j,\ j = 1, 2, \ldots, n\}. \tag{A4.3}$$

Accordingly, species $\mathbf{I}_m$ is selected and we call this state the state of selection. The proof is straightforward and yields simple expressions for the eigenvalues $\lambda_k (k = 0, 1, \ldots, n)$ of the Jacobian matrix when degradation is neglected, $d_j = 0\ (j = 1, 2, \ldots, n)$. For the state of extinction we find

$$\lambda_0 = -r \quad \text{and}\ \lambda_j = k_j\, a_0 - r. \tag{A4.4}$$

[1] An overview of the notation used in this article is found on the last page of this Appendix.

[2] We do not consider degenerate or neutral cases, $d_i = d_j$ and $k_i = k_j$, here (see also section A4.7).

It is asymptotically stable as long as $r > k_m a_0$ is fulfilled. If $r > k_m a_0$ then $r > k_j a_0 \; \forall j \neq m$ is valid by definition because of the selection criterion (A4.3) for $d_j = 0$. For all other n pure states $\{\bar{c}_i = a_0 - r/k_i, \; \bar{c}_j = 0, \; j \neq i\}$ the eigenvalues of the Jacobian are:

$$\lambda_0 = -r,$$
$$\lambda_i = -k_i a_0 + r, \quad \text{and} \tag{A4.5}$$
$$\lambda_j = -\frac{r}{k_i}\left(k_j - k_i\right) \; \forall j \neq i.$$

All pure states except the state at which $\mathbf{I}_m$ is selected (state of selection: $c_m = a_0 - r/k_m, \; c_j = 0, \; j = 1, \ldots, n, j \neq m$) have at least one positive eigenvalue and are unstable. Therefore we observe indeed selection of the molecular species with the largest value of k_j (or $k_j \; a_0 - d_j$, respectively), because only at $\bar{c}_m \neq 0$ are all eigenvalues of the Jacobian matrix negative.

It is worth indicating that the dynamical system (A4.2) has a stable manifold $\bar{y} = \bar{a} + \bar{b} + \sum_{i=1}^{n} \bar{c}_i = a_0$ since $\dot{y} = \dot{a} + \dot{b} + \sum_{i=1}^{n} \dot{c}_i = (a_0 - y)r$. The sum of all concentrations, $y(t)$, follows a simple exponential relaxation towards the steady state $\bar{y} = a_0$:

$$y(t) = a_0 - \big(a_0 - y(0)\big)\exp(-r\,t),$$

with the flow rate r being the relaxation constant.

2. Selection

As shown in the previous section the basis of selection is reproduction in the form of a simple autocatalytic elementary step, $(\mathbf{A}) + \mathbf{I}_i \rightarrow 2\mathbf{I}_i$. We idealise the system by assuming that the material consumed in the reproduction process, $\mathbf{A}$, is present in excess. Therefore, its concentration is constant and can be subsumed as a factor in the rate constant: $f_i = k_i \; [\mathbf{A}]$. In addition we neglect the degradation terms by putting $d_i = 0 \; \forall \; i$. In terms of chemical reaction kinetics selection based on reproduction without recombination and mutation is described by the dynamical system

$$\dot{c}_i = f_i c_i - \frac{c_i}{\sum_{j=1}^{n} c_j(t)}\Phi(t) = c_i\left(f_i - \frac{1}{c(t)}\Phi(t)\right), \; i = 1, 2, \ldots, n. \tag{A4.6}$$

As before the variables $c_i(t)$ are the concentrations of the genotypes I_i, the quantities f_i are reproduction rate parameters corresponding to overall replication rate constants in molecular systems or, in general, the fitness values of the genotypes. A global flux $\Phi(t)$ has been introduced in order to regulate the growth of the system.

Transformation to relative concentrations, $x_i = c_i/c$, and adjusting the global flux $\Phi(t)$ to zero net-growth simplifies further analysis:[3]

$$\dot{x}_i = f_i x_i - x_i \sum_{j=1}^{n} f_j x_j = x_i(f_i - \Phi) \quad \text{with}$$

$$\Phi = \sum_{j=1}^{n} f_j x_j = \bar{f} \quad \text{and } i = 1, 2, \ldots, n. \tag{A4.7}$$

The relative concentrations $x_i(t)$ fulfil $\sum_{j=1}^{n} x_j(t) = 1$ and the flux $\Phi(t)$ is the mean growth rate of the population. Because of this conservation relation only $n - 1$ variables x_i are independent. In the space of n Cartesian variables, $\mathbb{R}^n$, the x variables represent a projection of the positive orthant onto the unit simplex (Figure A4.2)

$$\mathbb{S}_n^{(1)} = \left\{ x_i \geq 0 \forall i = 1, 2, \ldots, n \bigwedge \sum_{i=1}^{n} x_i = 1 \right\}.$$

The simplex $\mathbb{S}_n^{(1)}$ is an invariant manifold of the differential equation (A4.7). This means that every solution curve $\mathrm{x}(t) = (x_1(t), x_2(t), \ldots, x_n(t))$ that starts in one point of the simplex will stay on the simplex forever.

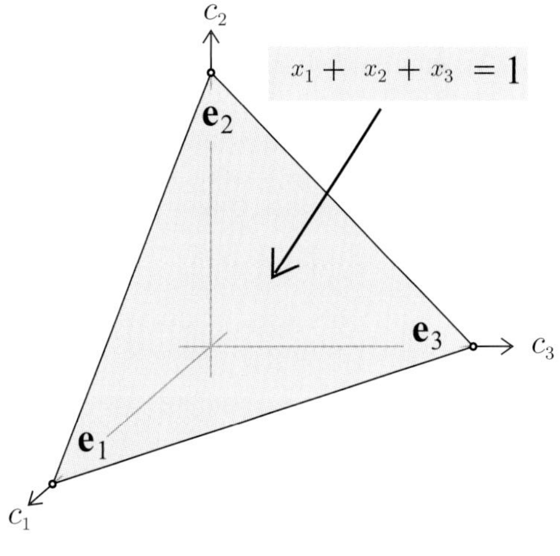

Figure A4.2
The unit simplex.

Shown is the case of three variables (c_1, c_2, c_3) in Cartesian space $\mathbb{R}^{(3)}$ projected onto the simplex $\mathbb{S}_3^{(1)}$. The condition $x_1 + x_2 + x_3 = 1$ defines an equilateral triangle in $\mathbb{R}^{(3)}$ with the three unit vectors, $\mathrm{e}_1 = (1, 0, 0)$, $\mathrm{e}_2 = (0, 1, 0)$ and $\mathrm{e}_3 = (0, 0, 1)$ as corners.

[3] Care is needed for the application of relative co-ordinates, because the total concentration $c(t)$ might vanish and then relative co-ordinates become spurious quantities (see subsection 6.7).

In order to analyse the stability of $\mathbb{S}_n^{(1)}$ we relax the conservation relation $\sum_{i=1}^{n} x_i(t) = c(t)$ and assume that only the conditions

$$\{f_i > 0 \wedge 0 \leq x_i(0) < \infty\} \forall i = 1, 2, \ldots, n,$$

are fulfilled. According to this assumption all rate parameters are strictly positive – a condition that will be replaced by the weaker one $f_i \geq 0 \forall i \neq k \wedge f_k > 0$ – and the concentration variables are non-negative quantities. Stability of the simplex requires that all solution curves converge to the unit simplex from every initial condition, $\lim_{t\to\infty} \left(\sum_{i=1}^{n} x_i(t)\right) = 1$. This conjecture is proved readily: from $\sum_{i=1}^{n} x_i(t) = c(t)$ follows

$$\dot{c} = c(1 - c)\Phi(t) \quad \text{with} \quad \Phi(t) > 0. \tag{A4.8}$$

For $\dot{c} = 0$ we find the two stationary states: a saddle point at $\bar{c} = 0$ and an asymptotically stable state at $\bar{c} = 1$. There are several possibilities to verify its asymptotic stability, we choose to solve the differential equation and find:

$$c(t) = 1 - (1 - c(0)) \exp\left(-\int_0^t \Phi(\tau) d\tau\right).$$

Starting with any positive initial value $c(0)$ the population approaches the unit simplex. When it starts on $\mathbb{S}_n$ it stays there even in the presence of fluctuations.[4] Therefore, we restrict population dynamics to the simplex without loosing generality and characterise the state of a population at time t by the vector $\mathrm{x}(t)$ which fulfils the $\mathbb{L}^{(1)}$ norm $\sum_{i=1}^{n} x_i(t) = 1$.

The necessary and sufficient condition for the stability of the simplex, $\Phi(t) > 0$, enables us to relax the condition for the rate parameters f_i. In order to have a positive flux it is sufficient that one rate parameter is strictly positive provided the corresponding variable is non-zero:

$$\Phi(t) > 0 \implies \exists k \in \{1, 2, \ldots, n\} \text{ such that } f_k > 0 \wedge x_k > 0.$$

For the variable x_k it is sufficient that $x_k(0) > 0$ holds because $x_k(t) \geq x_k(0)$ when all other products $f_j x_j$ were zero at $t = 0$. This relaxed condition for the flux is important for the handling of lethal mutants with $f_j = 0$.

[4] Generalisation to arbitrary but finite population sizes $c \neq 1$ is straightforward: For $\sum_{i=1}^{n} x_i(0) = c_0$ the equation $\dot{x}_i = f_i x_i - (x_i/c_0) \sum_{j=1}^{n} f_j x_j$, $i = 1, 2, \ldots, n$ plays the same role as equation (A4.7) did for $\sum_{i=1}^{n} x_i(0) = 1$.

The time dependence of the mean fitness or flux Φ is given by

$$\frac{d\Phi}{dt} = \sum_{i=1}^{n} f_i \dot{x}_i = \sum_{i=1}^{n} f_i \left(f_i x_i - x_i \sum_{j=1}^{n} f_j x_j \right) =$$

$$= \sum_{i=1}^{n} f_i^2 x_i - \sum_{i=1}^{n} f_i x_i \sum_{j=1}^{n} f_j x_j =$$

$$= \overline{f^2} - \left(\bar{f}\right)^2 = \text{var}\{f\} \geq 0. \tag{A4.9}$$

Since a variance is always non-negative, equation (A4.9) implies that $\Phi(t)$ is a non-decreasing function of time. The value $\text{var}\{f\} = 0$ refers to a homogeneous population of the fittest variant, and then $\Phi(t)$ cannot increase any further. Hence it has been optimised during selection.

It is also possible to derive analytical solutions for equation (A4.7) by a transform called integrating factors ([5], p.322ff.):

$$z_i(t) = x_i(t) \exp\left(\int_0^t \Phi(\tau) d\tau \right). \tag{A4.10}$$

Insertion into (A4.7) yields

$$\dot{z}_i = f_i z_i \;\text{ and }\; z_i(t) = z_i(0) \exp(f_i t),$$

$$x_i(t) = x_i(0) \exp(f_i t) \exp\left(- \int_0^t \Phi(\tau) d\tau \right) \;\text{ with}$$

$$\exp\left(\int_0^t \Phi(\tau) d\tau \right) = \sum_{j=1}^{n} x_j(0) \exp(f_j t),$$

where we have used $z_i(0) = x_i(0)$ and the condition $\sum_{i=1}^{n} x_i = 1$. The solution finally is of the form

$$x_i(t) = \frac{x_i(0) \exp(f_i t)}{\sum_{j=1}^{n} x_j(0) \exp(f_j t)}; \; i = 1, 2, \ldots, n. \tag{A4.11}$$

Under the assumption that the largest fitness parameter is non-degenerate, $\max\{f_i; i = 1, 2, \ldots, n\} = f_m > f_i \forall i \neq m$, every solution curve fulfilling the initial condition $x_i(0) > 0$ approaches a homogenous population: $\lim_{t\to\infty} x_m(t) = \bar{x}_m = 1$ and $\lim_{t\to\infty} x_i(t) = \bar{x}_i = 0 \forall i \neq m$, and the flux approaches the largest fitness parameter monotonously, $\Phi(t) \to f_m$ (examples are shown in Figure A4.3).

Qualitative analysis of stationary points and their stability yields the following results:

(i) The only stationary points of equation (A4.7) are the corners of the simplex, represented by the unit vectors $e_k = \{x_k = 1, x_i = 0 \forall i \neq k\}$,

(ii) only one of these stationary points is asymptotically stable, the corner where the mean fitness Φ adopts its maximal value on the simplex (e_m: $\bar{x}_m = 1$ defined by $\max\{f_i; i = 1, 2, \ldots, n\} = f_m > f_i \forall i \neq m$), one corner is unstable in all directions, a source where the value of Φ is minimal (e_s: $\bar{x}_s = 1$ defined by $\min\{f_i; i = 1, 2, \ldots, n\} = f_s < f_i \forall i \neq s$), and all other $n - 2$ equilibria are saddle points, and

(iii) since $x_i(0) = 0$ implies $x_i(t) = 0 \forall t > 0$, every subsimplex of $\mathbb{S}_n^{(1)}$ is an invariant set, and thus the whole boundary of the simplex consists of invariant sets and subsets down the corners (which represent members of class $\mathbb{S}_1^{(1)}$).

3. Generalised gradient systems

Although $\Phi(t)$ represents a Liapunov function for the dynamical system (A4.7) and its existence is sufficient for the proof of global stability for *selection of the fittest* being the species with the largest f_k value, it is of interest that the differential equation (A4.7) can be interpreted as a generalised gradient system [6, 7, 8] through the introduction of a Riemann-type metric based on a generalised scalar product defined at position x

$$[\mathbf{u}, \mathbf{v}]_{(\mathrm{x})} = \sum_{i=1}^{n} \frac{u_i \, v_i}{x_i}.$$

In a gradient system,

$$\dot{x}_i = \frac{\partial V}{\partial x_i}(\mathrm{x}), \; i = 1, 2, \ldots, n \text{ or } \dot{\mathrm{x}} = \nabla V(\mathrm{x}) \,, \qquad \text{(A4.12)}$$

the potential $V(\mathrm{x})$ increases steadily along the orbits,

$$\frac{dV}{dt} = \sum_{i=1}^{n} \left(\frac{\partial V}{\partial x_i}, \frac{dx_i}{dt} \right) = \sum_{i=1}^{n} \left(\frac{\partial V}{\partial x_i} \right)^2 = \nabla V(\mathrm{x}), \nabla V(\mathrm{x}) \geq 0,$$

and it does so at a maximal rate, since the velocity vector, being equal to the gradient, points at the position of maximum increase of V. In other words, the velocity vector points in the direction of steepest ascent, which is always orthogonal to the constant level sets of V. In gradient systems we observe optimisation of the potential function $V(\mathrm{x})$ along all orbits.

For the purpose of illustration we choose an example, equation (A4.12) with

$$V(\mathrm{x}) = -\frac{1}{2} \sum_{i=1}^{n} f_i x_i^2 \text{ and } \frac{dx_i}{dt} = \frac{\partial V}{\partial x_i}(\mathrm{x}) = -f_i x_i.$$

The time derivative of the potential function is obtained by

$$\frac{dV}{dt} = \sum_{i=1}^{n} \left(\frac{\partial V}{\partial x_i} \right)^2 = \sum_{i=1}^{n} f_i^2 x_i^2 \geq 0.$$

The potential is increasing until it reaches asymptotically the maximal value $V = 0$. Solutions of the differential equation are computed by integration: $x_i(t) = x_i(0) \exp(-f_i t) \forall i = 1, \ldots, n$. The result derived from dV/dt is readily verified, since $\lim_{t \to \infty} x_i(t) = 0 \forall i = 1, \ldots, n$ and hence $\lim_{t \to \infty} V(t) = 0$.

Equation (A4.7), on the other hand, is not an ordinary gradient system: it fulfills the optimisation criterion but the orbits are not orthogonal to the constant level sets of $V(\mathrm{x}) = \Phi(\mathrm{x})$ (see Figure A4.3). In such a situation, it is often possible to achieve the full gradient properties through a generalisation of the scalar product that is tantamount to a redefinition of the angle on the underlying space, $\mathbb{R}^n$ or $\mathbb{S}_n^{(1)}$, respectively. We shall describe here the formalism by means of the selection equation (A4.7) as an example.

The potential function is understood as a map, $V(\mathrm{x}) : \mathbb{R}^n \Rightarrow \mathbb{R}^1$. The derivative of the potential $DV_{(\mathrm{x})}$ is the unique linear map $L : \mathbb{R}^n \Rightarrow \mathbb{R}^1$ that fulfils for all $\mathrm{y} \in \mathbb{R}^n$:

$$V(\mathrm{x} + \mathrm{y}) = L(\mathrm{y}) + o(\mathrm{y}) = DV_{(\mathrm{x})}(\mathrm{y}) + o(\mathrm{y}),$$

where for $o(\mathrm{y})$ holds $o(\mathrm{y}) / \parallel \mathrm{y} \parallel \to 0$ as $\parallel \mathrm{y} \parallel \to 0$. To L corresponds a uniquely defined vector $\mathrm{l} \in \mathbb{R}^n$ such that $\langle l, \mathrm{y} \rangle = L(\mathrm{y})$ where $\langle *, * \rangle$ in the conventional Euclidean inner product is defined by $\langle \mathrm{u}, \mathrm{v} \rangle = \sum_{i=1}^{n} u_i v_i$ for $\mathrm{u}, \mathrm{v} \in \mathbb{R}^n$. This special vector l is the gradient of the potential V, which can be defined therefore by the following mapping of y into $\mathbb{R}^1$:

$$\langle \mathrm{grad} V(\mathrm{x}), \mathrm{y} \rangle = DV_{(\mathrm{x})}(\mathrm{y}) \;\text{ for }\; \mathrm{y} \in \mathbb{R}^n. \tag{A4.13}$$

The conventional Euclidean inner product is associated with the Euclidean metric, $\parallel \mathrm{x} \parallel = \langle \mathrm{x}, \mathrm{x} \rangle^{1/2}$.

It is verified straightforwardly that equation (A4.7) does not fulfill the condition of a conventional gradient (A4.13). The idea is now to replace the Euclidean metric by another more general metric that allows the properties of the gradient system. We introduce a generalised inner product corresponding to a Riemann-type metric where the conventional product terms are weighted by the co-ordinates of the position vector **z**:

$$[\mathbf{u}, \mathbf{v}]_{\mathrm{z}} = \sum_{i=1}^{n} \frac{1}{z_i} u_i v_i. \tag{A4.14}$$

Expression (A4.14), $[*, *]_{\mathrm{z}}$, defines an inner product in the interior of $\mathbb{R}_+^n$, because it is linear in **u** and **v**, and satisfies $[\mathbf{u}, \mathbf{u}]_{\mathrm{z}} \geq 0$ with the equality fulfilled if and only

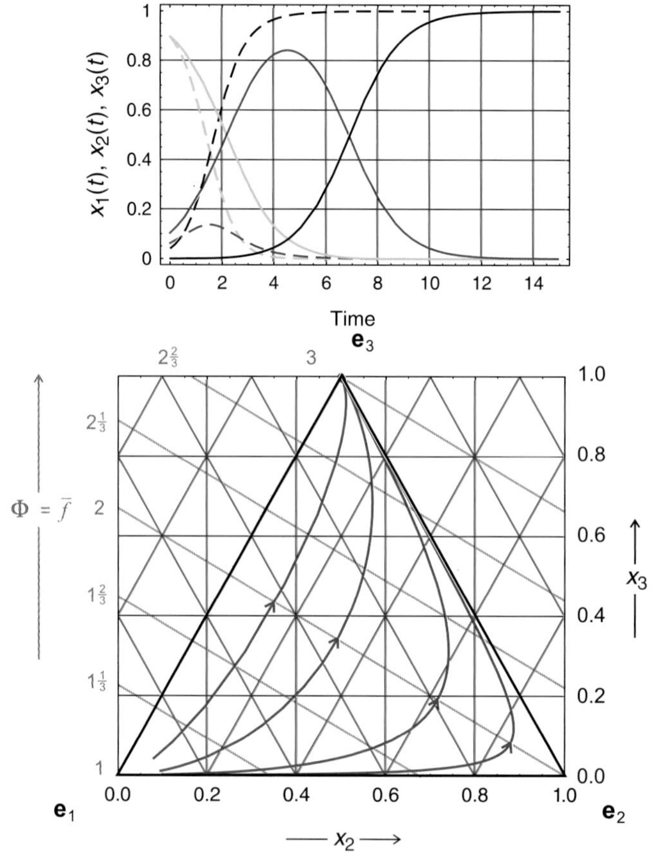

Figure A4.3

Selection on the unit simplex.

In the upper part of the figure we show solution curves $\mathrm{x}(t)$ of equation (A4.7) with $n = 3$. The parameter values are: $f_1 = 1\ [\mathrm{t}^{-1}]$, $f_2 = 2\ [\mathrm{t}^{-1}]$ and $f_3 = 3\ [\mathrm{t}^{-1}]$, where $[\mathrm{t}^{-1}]$ is an arbitrary reciprocal time unit. The two sets of curves differ with respect to the initial conditions:
(i) $\mathrm{x}(0) = (0.90, 0.08, 0.02)$, dotted curves, and (ii) $\mathrm{x}(0) = (0.9000, 0.0999, 0.0001)$, full curves. Color code: $x_1(t)$ green, $x_2(t)$ red and $x_3(t)$ black. The lower part of the figure shows parametric plots $\mathrm{x}(t)$ on the simplex $\mathbb{S}_3^{(1)}$. Constant level sets of Φ are straight lines (grey).

if $\| u\| = 0$. Based on this choice of the inner product we redefine the gradient:

$$\left[\text{Grad}[V(\text{x})], \text{y}\right]_{\text{x}} = DV_{(\text{x})}(\text{y}) \quad \text{for} \quad \text{x}, \text{y} \in \mathbb{R}^n_+. \tag{A4.15}$$

The differential $DV_{(\text{x})}$ is defined completely by $V(\text{x})$ and hence is independent of the choice of an inner product; the gradient, however, is not because it depends on the definition (A4.15). Shahshahani [6] has shown that the relation $d\text{x}/dt = \text{Grad}[\Phi(\text{x})]$ is fulfilled for Fisher's selection equation (A4.19; see section 5) with $\Phi = \sum_{i=1}^{n}\sum_{j=1}^{n} a_{ij}x_i x_j$. As an example for the procedure we consider here the simple selection equation (A4.7) with $\Phi = \sum_{i=1}^{n} f_i x_i$.

The differential equation (A4.7) is conceived as a generalised gradient system and we find:

$$\frac{d\text{x}}{dt} = \begin{pmatrix} x_1(f_1 - \sum_{j=1}^{n} f_j x_j) \\ x_2(f_2 - \sum_{j=1}^{n} f_j x_j) \\ \vdots \\ x_n(f_n - \sum_{j=1}^{n} f_j x_j) \end{pmatrix} = \text{Grad}[V(\text{x})].$$

By application of equation (A4.15) we obtain

$$DV_{(\text{x})}(\text{e}_i) = f_i - \sum_{j=1}^{n} f_j x_j,$$

which can be derived by conventional differentiation from

$$V(\text{x}) = \sum_{i=1}^{n} x_i\,(2f_i - \Phi).$$

By straightforward computation we find the desired result:

$$V(\text{x}) = \sum_{i=1}^{n} x_i(2f_i - \Phi) = 2\Phi - \Phi = \Phi.$$

With the new definition of the scalar product, encapsulated in the definition of 'Grad', the orbits of equation (A4.7) are perpendicular to the constant level sets of $\Phi(\text{x})$.

4. Complementary replication

Often the molecular mechanism of replication proceeds through an intermediate represented by a polynucleotide molecule with a complementary sequence: $(\mathbf{A}) + I_+ \rightarrow I_+ + I_-$ and $(\mathbf{A}) + I_- \rightarrow I_- + I_+$. In analogy to equation (A4.7) and with $f_1 = \tilde{f}_+ [\mathbf{A}]$, $f_2 = \tilde{f}_- [\mathbf{A}]$, $x_1 = [I_+]$, and $x_2 = [I_-]$ we obtain the following differential equation [9–11]:

$$\dot{x}_1 = f_2 x_2 - x_1 \Phi \text{ and}$$
$$\dot{x}_2 = f_1 x_1 - x_2 \Phi \text{ with } \Phi = f_1 x_1 + f_2 x_2. \tag{A4.16}$$

Applying the integrating factor transformation (A4.10) yields the linear equation

$$\dot{z}_1 = f_2 z_2 \text{ and } \dot{z}_2 = f_1 z_1 \text{ or } \dot{\mathrm{z}} = \mathrm{W} \cdot \mathrm{z}; \mathrm{z} = \begin{pmatrix} z_1 \\ z_2 \end{pmatrix}, \mathrm{W} = \begin{pmatrix} 0 & f_2 \\ f_1 & 0 \end{pmatrix}.$$

The eigenvalues and (right-hand) eigenvectors of the matrix W are

$$\lambda_{1,2} = \pm\sqrt{f_1 f_2} = \pm f \text{ with } f = \sqrt{f_1 f_2}, \tag{A4.17}$$
$$\ell_1 = \begin{pmatrix} \sqrt{f_2} \\ \sqrt{f_1} \end{pmatrix} \text{ and } \ell_2 = \begin{pmatrix} -\sqrt{f_2} \\ \sqrt{f_1} \end{pmatrix}.$$

Straightforward calculation yields analytical expressions for the two variables (see paragraph mutation) with the initial concentrations $x_1(0)$ and $x_2(0)$, and $\gamma_1(0) = \sqrt{f_1} x_1(0) + \sqrt{f_2} x_2(0)$ and $\gamma_2(0) = \sqrt{f_1} x_1(0) - \sqrt{f_2} x_2(0)$ as abbreviations:

$$x_1(t) = \frac{\sqrt{f_2}\left(\gamma_1(0)\, e^{ft} + \gamma_2(0)\, e^{-ft}\right)}{(\sqrt{f_1} + \sqrt{f_2})\gamma_1(0)\, e^{ft} - (\sqrt{f_1} - \sqrt{f_2})\gamma_2(0)\, e^{-ft}}$$
$$x_2(t) = \frac{\sqrt{f_1}\left(\gamma_1(0)\, e^{ft} - \gamma_2(0)\, e^{-ft}\right)}{(\sqrt{f_1} + \sqrt{f_2})\gamma_1(0)\, e^{ft} - (\sqrt{f_1} - \sqrt{f_2})\gamma_2(0)\, e^{-ft}}. \tag{A4.18}$$

After a sufficiently long time the negative exponential has vanished and we obtain the simple result

$$x_1(t) \to \sqrt{f_2}/(\sqrt{f_1} + \sqrt{f_2}),\ x_2(t) \to \sqrt{f_1}/(\sqrt{f_1} + \sqrt{f_2}) \text{ as } \exp(-kt) \to 0.$$

After an initial period, the plus-minus pair, $\mathbf{I}_{\pm}$, grows like a single autocatalyst with a fitness value $f = \sqrt{f_1 f_2}$ and a stationary ratio of the concentrations of complementary stands $x_1/x_2 \approx \sqrt{f_2}/\sqrt{f_1}$.

5. Recombination

Recombination of n alleles on a single locus is described by Ronald Fisher's [12] selection equation,

$$\dot{x}_i = \sum_{j=1}^{n} a_{ij} x_i x_j - x_i \sum_{j=1}^{n} \sum_{k=1}^{n} a_{jk} x_j x_k = x_i \left(\sum_{j=1}^{n} a_{ij} x_j - \Phi\right) \tag{A4.19}$$
$$\text{with } \Phi = \sum_{j=1}^{n} \sum_{k=1}^{n} a_{jk} x_j x_k.$$

As in the simple selection case the two conditions $a_{ij} > 0 \forall i, j = 1, 2, \ldots, n$ and $x_i \geq 0 \forall i = 1, 2, \ldots, n$ will guarantee $\Phi(t) \geq 0$. Summation of allele frequencies, $\sum_{i=1}^{n} x_i(t) = c(t)$, yields again equation (A4.8) for $\dot{c}$ and hence for $\sum_{i=1}^{n} x_i(0) = 1$ the population is confined again to the unit simplex.

The rate parameters a_{ij} form a quadratic matrix

$$\mathrm{A} = \begin{pmatrix} a_{11} & a_{12} & \ldots & a_{1n} \\ a_{21} & a_{22} & \ldots & a_{2n} \\ \vdots & \vdots & \ddots & \vdots \\ a_{n1} & a_{n2} & \ldots & a_{nn} \end{pmatrix}.$$

The dynamics of equation (A4.19) for general matrices A may be very complicated [13]. In case of Fisher's selection equation, however, we are dealing with a symmetric matrix for biological reasons,[5] and then the differential equation can be subjected to straightforward qualitative analysis.

The introduction of mean rate parameters $\bar{a}_i = \sum_{j=1}^{n} a_{ij} x_j$ facilitates the forthcoming analysis. The time dependence of Φ is now given by

$$\begin{aligned} \frac{d\Phi}{dt} &= \sum_{i=1}^{n} \sum_{j=1}^{n} a_{ij} \left(\frac{dx_i}{dt} \, x_j + x_i \, \frac{dx_j}{dt} \right) = 2 \sum_{i=1}^{n} \sum_{j=1}^{n} a_{ji} \, x_i \, \frac{dx_j}{dt} \\ &= 2 \sum_{i=1}^{n} \sum_{j=1}^{n} a_{ji} \, x_i \left(\sum_{k=1}^{n} a_{jk} x_j x_k - x_j \sum_{k=1}^{n} \sum_{\ell=1}^{n} a_{k\ell} x_k x_\ell \right) \\ &= 2 \sum_{j=1}^{n} x_j \sum_{i=1}^{n} a_{ji} x_i \sum_{k=1}^{n} a_{jk} x_k - 2 \sum_{j=1}^{n} x_j \sum_{i=1}^{n} a_{ji} x_i \sum_{k=1}^{n} x_k \sum_{\ell=1}^{n} a_{k\ell} x_\ell \\ &= 2 \left(< \bar{a}^2 > - < \bar{a} >^2 \right) = 2 \, \mathrm{var}\{\bar{a}\} \geq 0. \end{aligned} \qquad \text{(A4.20)}$$

Again we see that the flux $\Phi(t)$ is a non-decreasing function of time, and it approaches an optimal value on the simplex. This result is often called Fisher's fundamental theorem of evolution (see, for example, [14]).

Qualitative analysis of equation (A4.19) yields $2^n - 1$ stationary points, which depending on the elements of matrix A may lie in the interior, on the boundary or outside the unit simplex $\mathbb{S}_n^{(1)}$. In particular, we find at most one equilibrium point on the simplex and one on each subsimplex of the boundary. For example, each corner,

[5] The assumption for Fisher's equation is based on insensitivity of phenotypes to the origin of the parental alleles on chromosomes. Phenotypes derived from genotype $\mathbf{a}_i \, \mathbf{a}_j$ are assumed to develop the same properties, no matter which allele, $\mathbf{a}_i$ or $\mathbf{a}_j$, on the chromosomal locus comes from the mother and which comes from the father. New results on genetic diseases have shown, however, that this assumption can be questioned.

represented by the unit vector $e_k = \{\bar{x}_k = 1, x_i = 0 \forall i \neq k\}$, is a stable or unstable stationary point. Where there is an equilibrium in the interior of $\mathbb{S}_n^{(1)}$ it may be stable or unstable depending on the elements of A. In summary, this leads to a rich collection of different dynamical scenarios which share the absence of oscillations or chaotic dynamics. As said above, multiple stationary states do occur and more than one may be stable. This implies that the optimum, which $\Phi(t)$ is approaching, need not be uniquely defined. Instead $\Phi(t)$ may approach one of the local optima and then the outcome of the selection process will depend on initial conditions [14, 15, 16, 17].

Three final remarks are important for a proper understanding of Fisher's fundamental theorem: (i) selection in the one-locus system when it follows equation (A4.19) optimises mean fitness of the population, (ii) the outcome of the process need not be unique since the mean fitness Φ may have several local optima on the unit simplex, and (iii) optimisation behaviour that is susceptible to rigorous proof is restricted to the one locus model since systems with two or more gene loci may show different behaviours of $\Phi(t)$.

6. Mutation

The introduction of mutation into the selection equation (A4.7) based on knowledge from molecular biology is due to Manfred Eigen [9]. It leads to

$$\dot{x}_i = \sum_{j=1}^{n} Q_{ij} f_j x_j - x_i\, \Phi;\; i = 1, 2, \ldots, n \text{ with } \Phi = \sum_{j=1}^{n} f_j x_j = \bar{f}. \tag{A4.21}$$

Mutations and error-free replication are understood as parallel reaction channels, and the corresponding reaction probabilities are contained in the mutation matrix

$$Q = \begin{pmatrix} Q_{11} & Q_{12} & \ldots & Q_{1n} \\ Q_{21} & Q_{22} & \ldots & Q_{2n} \\ \vdots & \vdots & \ddots & \vdots \\ Q_{n1} & Q_{n2} & \ldots & Q_{nn} \end{pmatrix},$$

where Q_{ij} expresses the frequency of a mutation $\mathbf{I}_j \rightarrow \mathbf{I}_i$. Since the elements of Q are defined as reaction probabilities and a replication event yields either a correct copy or a mutant, the columns of Q sum up to one, $\sum_{i=1}^{n} Q_{ij} = 1$, and hence, Q is stochastic matrix. In case one makes the assumption of equal probabilities for $\mathbf{I}_j \rightarrow \mathbf{I}_i$ and $\mathbf{I}_i \rightarrow \mathbf{I}_j$, as it is made for example in the uniform error rate model (see equation (A4.1.29) and [10, 18]), Q is symmetric and hence a bistochastic matrix.[6] The mean fitness or flux Φ is described by the same expression as in the selection-only case

[6] Symmetry in the direction of mutations is commonly not fulfilled in nature. It is introduced as a simplification, which facilitates the construction of computer models for equation (A4.21).

(A4.7). This implies that the system converges to the unit simplex, as it did without mutations. For initial values of the variables chosen on the simplex, $\sum_{i=1}^{n} x_i(0) = 1$, it remains there.

In the replication-mutation system the boundary of the unit simplex, $\mathbb{S}_n^{(1)}$, is not invariant. Although no orbit starting on the simplex will leave it, which is a *conditio sine qua non* for chemical reactions requiring non-negative concentrations, trajectories flow from outside the positive orthant into $\mathbb{S}_n^{(1)}$. In other words, the condition $x_i(0) = 0$ does **not** lead to $x_i(t) = 0 \forall t > 0$. The chemical interpretation is straightforward: If a variant $\mathbf{I}_i$ is not present initially, it can, and depending on Q commonly will, be formed through a mutation event.

6.1 Exact solution

Before discussing the role of the flux Φ in the selection-mutation system, we shall derive exact solutions of equation (A4.21) following a procedure suggested in [19, 20]. At first the variables $x_i(t)$ are transformed as in the selection-only case (A4.10):

$$z_i(t) = x_i(t) \exp\left(\int_0^t \Phi(\tau) d\tau \right).$$

From $\sum_{x=1}^{n} x_i(t) = 1$ follows straightforwardly – again as in the selection-only case – the equation,

$$\exp\left(\int_0^t \Phi(\tau) d\tau \right) = \left(\sum_{i=1}^{n} z_i(t) \right)^{-1}.$$

What remains to be solved is a linear first-order differential equation

$$\dot{z}_i = \sum_{j=1}^{n} Q_{ij} f_j \, z_j \; i = 1, 2, \ldots, n, \tag{A4.22}$$

which is readily done by means of standard linear algebra. We define a matrix $W = \{W_{ij} = Q_{ij} \, f_j\} = Q \cdot F$ where $F = \{F_{ij} = f_i \; \delta_{ij}\}$ is a diagonal matrix, and obtain the differential equation in matrix form, $\dot{z} = W \cdot z$. Provided matrix W is diagonalisable, which will always be the case when the mutation matrix Q is based on real chemical reaction mechanisms, the variables z can be transformed linearly by means of an invertible $n \times n$ matrix $L = \{\ell_{ij}; \; i, j = 1, \ldots, n\}$ with $L^{-1} = \{h_{ij}; \; i, j = 1, \ldots, n\}$ being its inverse,

$$z(t) = L \cdot \zeta(t) \text{ and } \zeta(t) = L^{-1} \cdot z(t),$$

Moreover, the assumption of a symmetric mutation matrix Q is not essential for the analytic derivation of solutions.

such that $\mathrm{L}^{-1} \cdot \mathrm{W} \cdot \mathrm{L} = \Lambda$ is diagonal. The elements of Λ, λ_k, are the eigenvalues of the matrix W. The right-hand eigenvectors of W are given by the columns of L, $\ell_j = (\ell_{i,j};\ i = 1, \ldots, n)$, and the left-hand eigenvectors by the rows of L^{-1}, $\mathrm{h}_k = (h_{k,i};\ i = 1, \ldots, n)$, respectively. These eigenvectors can be addressed as the *normal modes* of selection-mutation kinetics. For strictly positive off-diagonal elements of W, implying the same for Q which says nothing more than every mutation $\mathbf{I}_i \rightarrow \mathbf{I}_j$ is possible, although the probability might be extremely small, the Perron–Frobenius theorem holds (see, for example, [21] and next paragraph) and we are dealing with a non-degenerate largest eigenvalue λ_0,

$$\lambda_0 > |\lambda_1| \geq |\lambda_2| \geq |\lambda_3| \geq \ldots \geq |\lambda_n|, \tag{A4.23}$$

and a corresponding dominant eigenvector ℓ_0 with strictly positive components, $\ell_{i0} > 0 \forall i = 1, \ldots, n$.[7] In terms of components the differential equation in ζ has the solutions

$$\zeta_k(t) = \zeta_k(0) \exp(\lambda_k t). \tag{A4.24}$$

Transformation back into the variables z yields

$$z_i(t) = \sum_{k=0}^{n-1} \ell_{ik}\, c_k(0) \exp(\lambda_k\, t)\,, \tag{A4.25}$$

with the initial conditions encapsulated in the equation

$$c_k(0) = \sum_{i=1}^{n} h_{ki}\, 6f\, z_i(0) = \sum_{i=1}^{n} h_{ki}\, x_i(0). \tag{A4.26}$$

From here we obtain eventually the solutions in the original variables x_i in analogy to equation (A4.11)

$$\begin{aligned} x_i(t) &= \frac{\sum_{k=0}^{n-1} \ell_{ik} c_k(0) \exp(\lambda_k t)}{\sum_{j=1}^{n} \sum_{k=0}^{n-1} \ell_{jk} c_k(0) \exp(\lambda_k t)} \\ &= \frac{\sum_{k=0}^{n-1} \ell_{ik} \sum_{i=1}^{n} h_{ki} x_i(0) \exp(\lambda_k t)}{\sum_{j=1}^{n} \sum_{k=0}^{n-1} \ell_{jk} \sum_{i=1}^{n} h_{ki} x_i(0) \exp(\lambda_k t)}. \end{aligned} \tag{A4.27}$$

6.2 Perron–Frobenius theorem

Perron–Frobenius theorem comes in two versions [21], which we shall now consider and apply to the selection-mutation problem. The stronger version provides

[7] We introduce here an asymmetry in numbering rows and columns in order to point at the special properties of the largest eigenvalue λ_0 and the dominant eigenvector ℓ_0.

a proof for six properties of the largest eigenvector of non-negative primitive matrices[8] T:

(i) The largest eigenvalue is real and positive, $\lambda_0 > 0$,
(ii) a strictly positive right eigenvector ℓ_0 and a strictly positive left eigenvector $\mathbf{h}_0$ are associated with λ_0,
(iii) $\lambda_0 > |\lambda_k|$ holds for all eigenvalues $\lambda_k \neq \lambda_0$,
(iv) the eigenvectors associated with λ_0 are unique up to constant factors,
(v) if $0 \leq \mathrm{B} \leq \mathrm{T}$ is fulfilled and β is an eigenvalue of B, then $|\beta| \leq \lambda_0$, and, moreover, $|\beta| = \lambda_0$ implies $\mathrm{B} = \mathrm{T}$,
(vi) λ_0 is a simple root of the characteristic equation of T.

The weaker version of the theorem holds for irreducible matrices[9] T. All the above given assertions hold except (iii) has to be replaced by the weaker statement

(iii) $\lambda_0 \geq |\lambda_k|$ holds for all eigenvalues $\lambda_k \neq \lambda_0$.

Irreducible cyclic matrices can be used straightforwardly as examples in order to demonstrate the existence of conjugate complex eigenvalues (an example is discussed below). Perron–Frobenius theorem, in its strict or weaker form, holds not only for strictly positive matrices $\mathrm{T} > 0$ but also for large classes of mutation or value matrices ($\mathrm{W} \equiv \mathrm{T}$ being a primitive or irreducible non-negative matrix) with off-diagonal zero entries corresponding to zero mutation rates. The occurrence of a non-zero element $t_{ij}^{(m)}$ in T^m implies the existence of a mutation path $\mathbf{I}_j \rightarrow \mathbf{I}_k \rightarrow \ldots \rightarrow \mathbf{I}_l \rightarrow \mathbf{I}_i$ with non-zero mutation frequencies for every individual step. This condition is almost always fulfilled in real systems.

6.3 Complex eigenvalues

In order to address the existence of complex eigenvalues of the value matrix W we start by considering the straightforward case of a symmetric mutation matrix Q. Replication rate parameters f_i are subsumed in a diagonal matrix: $\mathrm{F} = \{f_i\ \delta_{i,j}; i,j = 1,\ldots,n\}$, the value matrix is obtained as product $\mathrm{W} = \mathrm{Q} \cdot \mathrm{F}$, and, in general, W is

[8] A square non-negative matrix $\mathrm{T} = \{t_{ij}; i,j = 1,\ldots,n; t_{ij} \geq 0\}$ is called *primitive* if there exists a positive integer m such that T^m is strictly positive: $\mathrm{T}^m > 0$, which implies $\mathrm{T}^m = \{t_{ij}^{(m)}; i,j = 1,\ldots,n; t_{ij}^{(m)} > 0\}$.

[9] A square non-negative matrix $\mathrm{T} = \{t_{ij}; i,j = 1,\ldots,n; t_{ij} \geq 0\}$ is called *irreducible* if for every pair (i,j) of its index set there exists a positive integer $m_{ij} \equiv m(i,j)$ such that $t_{ij}^{m_{ij}} > 0$. An irreducible matrix is called *cyclic* with period d if the period of (all) its indices satisfies $d > 1$, and it is said to be acyclic if $d = 1$.

not symmetric. A similarity transformation,

$$F^{\frac{1}{2}} \cdot W \cdot F^{-\frac{1}{2}} = F^{\frac{1}{2}} \cdot Q \cdot F \cdot F^{-\frac{1}{2}} = F^{\frac{1}{2}} \cdot Q \cdot F^{\frac{1}{2}} = W'.$$

yields a symmetric matrix [22], since $F^{\frac{1}{2}} \cdot Q \cdot F^{\frac{1}{2}}$ is symmetric if Q is. Symmetric matrices have real eigenvalues and as a similarity transformation does not change the eigenvalues W has only real eigenvalues if Q is symmetric.

The simplest way to yield complex eigenvalues is by the introduction of cyclic symmetry into the matrix Q in such a way that the symmetry with respect to the main diagonal is destroyed. An example is the matrix

$$Q = \begin{pmatrix} Q_{11} & Q_{12} & Q_{13} & \dots & Q_{1n} \\ Q_{1n} & Q_{11} & Q_{12} & \dots & Q_{1,n-1} \\ Q_{1,n-1} & Q_{1n} & Q_{11} & \dots & Q_{1,n-2} \\ \vdots & \vdots & \vdots & \ddots & \vdots \\ Q_{12} & Q_{13} & Q_{14} & \dots & Q_{11} \end{pmatrix},$$

with different entries Q_{ij}. For equal replication parameters the eigenvalues contain complex n th roots of one, $\gamma_k^n = 1$ or $\gamma_k = \exp(2\pi ik/n), i = 1, \dots, n$, and for $n \geq 3$ most eigenvalues come in complex conjugate pairs. As mentioned earlier symmetry in mutation frequencies is commonly not fulfilled in nature. For point mutations the replacement of one particular base by another one does usually not occur with the same frequency as the inverse replacement, **G** $\rightarrow$ **A** versus **A** $\rightarrow$ **G**, for example. Needless to say, cyclic symmetry in mutation matrices is also highly improbable in real systems. The validity the of Perron–Frobenius theorem, however, is not affected by the occurrence of complex conjugate pairs of eigenvectors. In addition, it is unimportant for most purposes whether a replication-mutation system approaches the stationary state monotonously or through damped oscillations (see next paragraph).

6.4 Mutation and optimisation

In order to consider the optimisation problem in the selection-mutation case, we choose the eigenvectors of W as the basis of a new co-ordinate system (Figure A4.4):

$$x(t) = \sum_{i=1}^{n} x_k(t) e_i = \sum_{k=0}^{n-1} \xi_k(t)\, \ell_k,$$

where the vectors e_i are the unit eigenvectors of the conventional Cartesian co-ordinate system and ℓ_k the eigenvectors of W. The unit eigenvectors represent the corners of $\mathbb{S}_n^{(1)}$ and in complete analogy we denote the space defined by the vectors ℓ_k as $\tilde{\mathbb{S}}_n^{(1)}$. Formally, the transformed differential equation

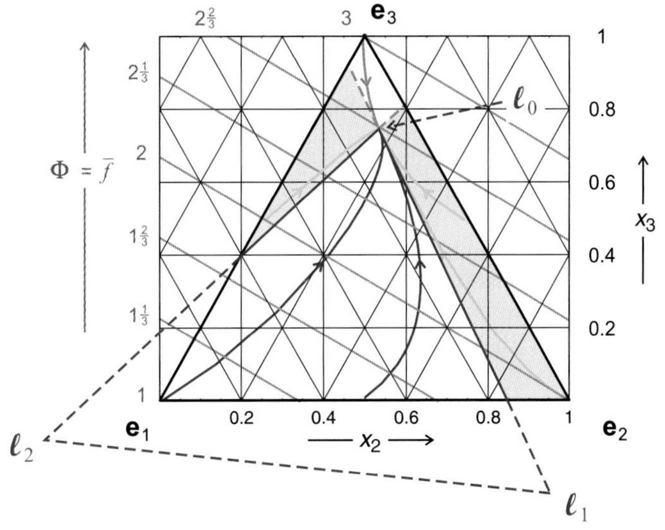

Figure A4.4

The quasi-species on the unit simplex.

Shown is the case of three variables (x_1, x_2, x_3) on $\mathbb{S}_3^{(1)}$. The dominant eigenvector, the quasi-species denoted by ℓ_0, is shown together with the two other eigenvectors, ℓ_1 and ℓ_2. The simplex is partitioned into an *optimization cone* (white, red trajectories) where the mean replication rate $\bar{f}(t)$ is optimised, two other zones where $\bar{f}(t)$ may also decrease (grey) and the *master cone*, which is characterised by non-increasing $\bar{f}(t)$ and which contains the master sequence (white, blue trajectories). Here, $\mathbf{I}_3$ is chosen to be the master sequence. Solution curves are presented as parametric plots x(t). In particular, the parameter values are $f_1 = 1.9[\mathrm{t}^{-1}]$, $f_2 = 2.0[\mathrm{t}^{-1}]$ and $f_3 = 2.1[\mathrm{t}^{-1}]$, and the Q-matrix was assumed to be bistochastic with the elements $Q_{ii} = 0.98$ and $Q_{ij} = 0.01$ for $i, j = \{1, 2, 3\}$. Then the eigenvalues and eigenvectors of W are:

k	λ_k	ℓ_{1k}	ℓ_{2k}	ℓ_{3k}
0	2.065	0.093	0.165	0.742
1	1.958	0.170	1.078	–0.248
2	1.857	1.327	–0.224	–0.103

The mean replication rate $\bar{f}(t)$ is monotonously increasing along red trajectories, monotonously decreasing along the blue trajectory and not necessarily monotonous along green trajectories. Constants level sets of Φ are straight lines (grey).

$$\dot{\xi}_k = \xi_k(\lambda_k - \Phi),\ k = 0, 1, \ldots, n-1 \text{ with } \Phi = \sum_{k=0}^{n-1} \lambda_k \xi_k = \overline{\lambda}$$

is identical to equation (A4.7) and hence the solutions are the same,

$$\xi_k(t) = \xi_k(0) \exp\left(\lambda_k t - \int_0^t \Phi(\tau) d\tau\right),\ k = 0, 1, \ldots, n-1,$$

as well as the maximum principle **on** the simplex

$$\frac{d\Phi}{dt} = \sum_{k=0}^{n-1} \xi_k (\lambda_k - \Phi)^2 = <\lambda^2> - <\lambda>^2 \geq 0. \qquad \text{(A4.9a)}$$

The difference between selection and selection-mutation comes from the fact that the simplex $\tilde{\mathbb{S}}_n$ does not coincide with the physically defined space $\mathbb{S}_n$ (see Figure A4.4 for a low-dimensional example). Indeed only the dominant eigenvector ℓ_0 lies in the interior of $\mathbb{S}_n^{(1)}$: it represent the stable stationary distribution of genotypes or quasi-species [10] towards which the solutions of the differential equation (A4.21) converge. All other $n-1$ eigenvectors, $\ell_1, \ldots, \ell_{n-1}$ lie outside $\mathbb{S}_n^{(1)}$ in the *not physical* range where one or more variables x_i are negative. The quasi-species ℓ_0 is commonly dominated by a single genotype, called the **master sequence** $\mathbf{I}_m$, having the largest stationary relative concentration, $\bar{x}_m \gg \bar{x}_i \forall i \neq m$, reflecting, for not too large mutation rates, the same ranking as the elements of the matrix W, $W_{mm} \gg W_{ii} \forall i \neq m$. As sketched in Figure A4.4 the quasi-species is then situated close to the unit vector e_m in the interior of $\mathbb{S}_n^{(1)}$.

For the discussion of the optimisation behaviour the simplex is partitioned into three zones: (i) the zone of maximization of $\Phi(t)$, the (large) lower white area in figure A4.4 where equation (A4.9a) holds and which we shall denote as *optimisation cone*,[10] (ii) the zone that includes the unit vector of the master sequence, e_m, and the quasi-species, ℓ_0, as corners, and that we shall characterize as *master cone*,[10] and (iii) the remaining part of the simplex $\mathbb{S}_n^{(1)}$ (zones (iii) are coloured grey in Figure A4.4). It is straightforward to prove that increase of $\Phi(t)$ and monotonous convergence towards the quasi-species is restricted to the optimisation cone [23]. From the properties of the selection equation (A4.7) we recall and conclude that the boundaries of the simplex $\tilde{\mathbb{S}}_n^{(1)}$ are invariant sets. This implies that no orbit of the differential equation (A4.21) can cross these boundaries. The boundaries of $\mathbb{S}_n^{(1)}$, on the other hand, are not invariant **but** they can be crossed exclusively in one direction:

[10] The exact geometry of the optimisation cone or the master cone is a polyhedron that can be approximated by a pyramid rather than a cone. Nevertheless we prefer the inexact notion *cone* because it is easier to memorise and to imagine in high-dimensional space.

from outside to inside.[11] Therefore, a solution curve starting in the optimisation cone or in the master cone will stay inside the cone where it started and eventually converge towards the quasi-species, ℓ_0.

In zone (ii), the master cone, all variables ξ_k except ξ_0 are negative and ξ_0 is larger than one in order to fulfill the $L^{(1)}$-norm condition $\sum_{k=0}^{n-1} \xi_k = 1$. In order to analyse the behaviour of $\Phi(t)$ we split the variables into two groups, ξ_0 the frequency of the quasi-species and the rest [23], $\{\xi_k; k = 1, \ldots, n-1\}$ with $\sum_{k=1}^{n-1} \xi_k = 1 - \xi_0$:

$$\frac{d\Phi}{dt} = \lambda_0 \xi_0^2 + \sum_{k=1}^{n-1} \lambda_k^2 \xi_k - \left(\lambda_0 \xi_0 + \sum_{k=1}^{n-1} \lambda_k \xi_k \right)^2 .$$

Next we replace the distribution of λ_k values in the second group by a single λ-value, $\tilde{\lambda}$ and find:

$$\frac{d\Phi}{dt} = \lambda_0^2 \xi_0 + \tilde{\lambda}^2 (1 - \xi_0) - \left(\lambda_0 \xi_0 + \tilde{\lambda}(1 - \xi_0) \right)^2 .$$

After a few simple algebraic operations we find eventually

$$\frac{d\Phi}{dt} = \xi_0 (1 - \xi_0)(\lambda_0 - \tilde{\lambda})^2. \tag{A4.28}$$

For the master cone with $\xi_0 \geq 1$, this implies $d\Phi(t)/dt \leq 0$, as the flux is a non-increasing function of time. Since we are only interested in the sign of $d\Phi/dt$, the result is exact, because we could use the mean value $\tilde{\lambda} = \bar{\lambda} = (\sum_{k=1}^{n-1} \lambda_k \xi_k)/(1 - \xi_0)$, the largest possible value λ_1 or the smallest possible value λ_{n-1} without changing the conclusion. Clearly, the distribution of λ_k-values matters for quantitative results. It is worth mentioning that equation (A4.28) applies also to the quasi-species cone and gives the correct result that $\Phi(t)$ is non-decreasing. Decrease of mean fitness or flux $\Phi(t)$ in the master cone is readily illustrated. Consider, for example, a homogeneous population of the master sequence as the initial condition: $x_m(0) = 1$ and $\Phi(0) = f_m$. The population becomes inhomogeneous because mutants are formed. Since all mutants have lower replication constants by definition, $(f_i < f_m \forall i \neq m)$, Φ becomes smaller. Finally, the distribution approaches the quasi-species ℓ_0 and $\lim_{t \to \infty} \Phi(t) = \lambda_0 < f_m$.

An extension of the analysis from the master cone to zone (iii), where not all ξ_k values with $k \neq 0$ are negative, is not possible. It has been shown by means of numerical examples that $d\Phi(t)/dt$ may show non-monotonous behavior and can go through a maximum or a minimum at finite time [23].

[11] This is shown easily by analysing the differential equation, but follows also from the physical background: no acceptable process can lead to negative particle numbers or concentrations. It can, however, start at zero concentrations and this means the orbit begins at the boundary and goes into the interior of the physical concentration space, here the simplex $\mathbb{S}_n^{(1)}$.

6.5 Mutation rates and error threshold

In order to illustrate the influence of mutation rates on the selection process we apply (i) binary sequences, (ii) the uniform error rate approximation,

$$Q_{ij} = p^{d_{ij}}(1-p)^{\nu - d_{ij}} \tag{A4.29}$$

with d_{ij} being the Hamming distance between the two sequences $\mathbf{I}_i$ and $\mathbf{I}_j$, ν the chain length and p the mutation or error rate per site and replication, and (iii) a simple model for the distribution of fitness values known as *single peak fitness landscape* [24],

$$f_1 = f_m > f_2 = f_3 = \ldots f_n = \bar{f}_{-m} = \frac{\sum_{i=2}^{n} f_i}{1 - x_m},$$

which represents a kind of mean field approximation. The mutants with the master sequence $\mathbf{I}_1$ are ordered in mutant classes: the zero-error class contains only the reference sequence ($\mathbf{I}_1$), the one-error class comprises all single-point mutations, the two-error class all double-point mutations, etc. Since the error rate p is independent of the particular sequence and all molecules belonging to the same mutant class have identical fitness values f_k, it is possible to introduce new variables for entire mutant classes Γ_k:

$$y_k = \sum_{j, \mathbf{I}_j \in \Gamma_k} x_j, \; k = 0, 1, \ldots, \nu, \; \sum_{k=0}^{\nu} y_k = 1. \tag{A4.30}$$

The mutation matrix Q has to be adjusted to transitions between classes [24, 25]. For mutations from class Γ_l into Γ_k we calculate:

$$Q_{kl} = \sum_{i=l+k-\nu}^{\min(k,l)} \binom{k}{i} \binom{\nu - k}{l - i} p^{k+l-2i} (1-p)^{\nu-(k+l-2i)}. \tag{A4.31}$$

The mutation matrix Q for error classes is not symmetric, $Q_{kl} \neq Q_{lk}$ as follows from equation (A4.31).

A typical plot of relative concentrations against error rate is shown in Figure A4.5. At vanishing error rates, $\lim p \to 0$, the master sequence is selected, $\lim_{t\to\infty} y_0(t) = \bar{y}_0 = 1$, and all other error classes vanish in the long time limit. Increasing error rates are reflected by a decrease in the stationary relative concentration of the master sequence and a corresponding increase in the concentration of all mutant classes. Apart from $\bar{y}_0(p)$ all concentrations $\bar{y}_k(p)$ with $k < \nu/2$ go through a maximum and approach pairwise the curves for $\bar{y}_{\nu-k}$ at values of p that increase with p. At $p = 0.5$ the eigenvalue problem can be solved exactly: the largest eigenvalue is strictly positive $\lambda_0 > 0$: it corresponds to an eigenvector ℓ_0, which is the uniform distribution in relative stationary concentrations $\bar{x}_1 = \bar{x}_2 = \ldots = \bar{x}_n = 1/n$, and

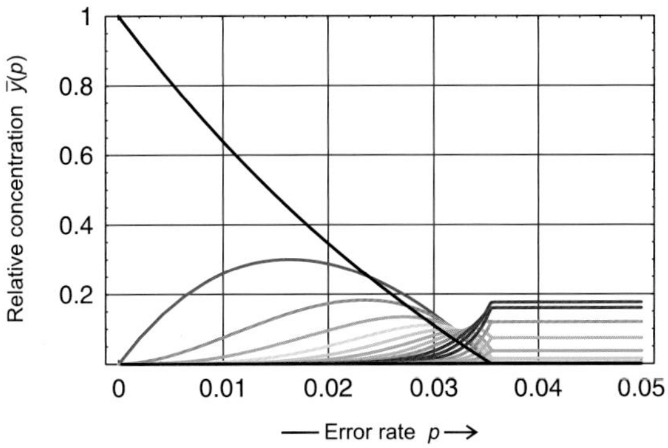

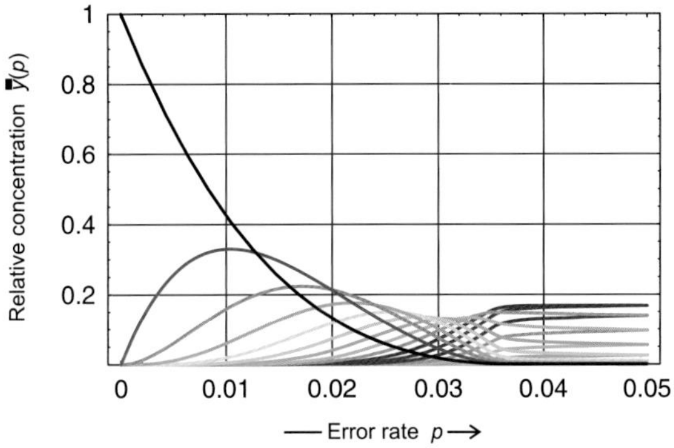

Figure A4.5
Error thresholds in the quasi-species model.

The figures show the stationary distribution of relative concentrations of mutant classes as functions of the error rate, $\bar{y}_k(p)$, for sequences of chain length $\nu = 20$. The population on a *single peak landscape* (upper part, $\sigma = 2$) gives rise to a sharp transition between the ordered regime, where relative concentrations are determined by fitness values f_k and mutation rates Q_{kl} (A4.31), and the domain of the uniform distribution where all error classes are present proportional to the numbers of sequences in them, $|\Gamma_k| = \binom{\nu}{k}$. The colour code is chosen such that the error classes with the same frequency, for example Γ_0 and Γ_ν, Γ_1 and $\Gamma_{\nu-1}$, etc., have identical colours and hence curves with the same colour merge above threshold. The population on a *hyperbolic fitness landscape* (lower part, $\sigma = 1.905$) shows a smoother transition that can be characterised as *weak error threshold*. Careful observation shows that the coalescence of curves with different colours at $p \approx 0.05$ is accidental since they diverge again at higher error rates.

this implies $\bar{y}_k = \binom{\nu}{k}$ for the class variables. The uniform distribution is a result of the fact that at $p = 0.5 = 1 - p$ correct digit replication and errors are equally probable (for binary sequences) and therefore we may characterise this scenario as *random replication*. All other eigenvalues vanish at $p = 0.5$: $\lambda_1 = \lambda_2 = \ldots = \lambda_{n-1} = 0$.

The mutant distribution $\bar{y}(p)$ comes close to the uniform distribution already around $p \approx 0.035$ in Figure A4.5, and stays constant for the rest of the p values $(0.035 < p < 0.5)$. The narrow transition from the *ordered replication* $(0 < p < 0.035)$ to *random replication* $(p > 0.035)$ is called the *error threshold*. An approximation based on neglect of mutational back-flow and using $\ln(1-p) \approx -p$ yields a simple expression for the position of the threshold [9]:

$$p_{\max} \approx \frac{\ln \sigma}{\nu} \quad \text{for small } p. \tag{A4.32}$$

The equation defines a maximal error rate $p_{\max}$ above which no ordered – non-uniform – stationary distributions of sequences exist (see also section 7). In the current example (Figure 5) we calculate $p_{\max} = 0.03466$ in excellent agreement with the value observed in computer simulations. RNA viruses commonly have mutation rates close to the error threshold [26]. Error rates can be increased by pharmaceutical drugs interfering with virus replication and accordingly a new antiviral strategy has been developed, which drives virus replication into extinction either by passing the error threshold [27, 28] or by extinction. Recently, the mechanism of lethal mutagenesis in virus infections has been extensively discussed [29, 30].

Several model landscapes describing fitness by a monotonously decreasing function of the Hamming distance from the master sequence, $f(d)$, are often applied in population genetics, examples are:

$$\text{hyperbolic :} \quad f(d) = f_0 - \frac{(f_0 - 1)(\nu + 1)\, d}{\nu(d+1)},$$

$$\text{linear :} \quad f(d) = f_0 - \frac{(f_0 - 1)\, d}{\nu}, \text{ and}$$

$$\text{quadratic :} \quad f(d) = f_0 - \frac{(f_0 - 1)\, d^2}{\nu^2}.$$

Interestingly, all three model landscapes do not sustain sharp error thresholds as observed with the single peak landscape. On the hyperbolic landscape the transition is less sharp than on the single peak landscape and may be called *weak error threshold*. The linear and quadratic landscapes show rather gradual and smooth transitions from the quasi-species towards the uniform mutant distribution (Figure A4.6). Despite the popularity of smooth landscapes in populations genetics, they are not supported by knowledge derived from biopolymer structures and functions. In contrast, the available data provide strong evidence that the natural landscapes are rugged and properties do not change gradually with Hamming distance.

In order to generalise the results derived from model landscapes to more realistic situations, random variations of rate constants for individual sequences were superimposed upon the fitness values of a single peak landscape – whereby the mean value

$\bar{f}_{-m}$ was kept constant [31, pp. 29–60]. Then, the curves for individual sequences within an error class differ from each other and form a band that increases in width with the amplitude of the random component. Interestingly, the error threshold phenomenon is thus retained and the critical value $p_{\max}$ is shifted to lower error rates. Another very general approach to introduce variation into the value matrix without accounting for the underlying chemical reaction mechanism was taken by Walter Thirring and coworkers [32]. They also found a sharp transition between ordered and disordered domains.

6.6 Population entropies

Population entropies are suitable measures for the width of mutant distributions. For steady states they are readily computed from the largest eigenvector of matrix W:

$$S(p) = -\sum_{i=1}^{2^\nu} \bar{x}_i \ln \bar{x}_i = -\sum_{k=0}^{\nu} \bar{y}_k \left(\ln \bar{y}_k - \ln \binom{\nu}{k} \right), \qquad \text{(A4.33)}$$

where the expression on the right-hand side refers to mutant classes. The pure state at $p = 0$ has zero entropy, $S(0) = 0$. For the uniform distribution the entropy is maximal, and for binary sequences we have

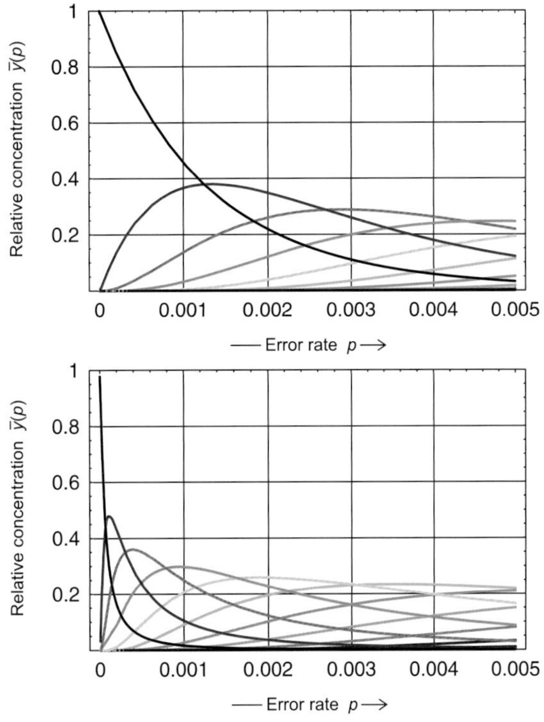

Figure A4.6

Smooth transitions in the quasi-species model.

The two figures show stationary mutant distributions as functions of the error rate, $\bar{y}_k(p)$, for sequences of chain length $\nu = 20$. The upper figure was calculated for a *linear landscape* ($\sigma = 1.333$), the lower figure for a *quadratic landscape* ($\sigma = 1.151$) of fitness values. The transitions are smooth in both cases.

Figure A4.7

Population entropy on different fitness landscapes.

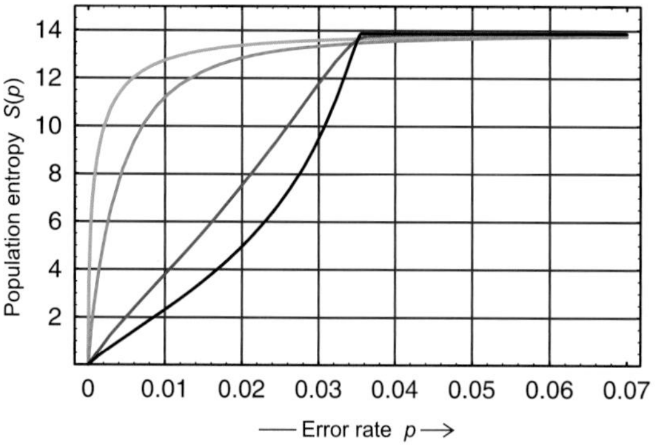

The plot shows the population entropy as functions of the error rate, $S(p)$, for sequences of chain length $\nu = 20$. The results for individual landscapes are colour coded: *single peak landscape* black, *hyperbolic landscape* red, *linear landscape* blue and *quadratic landscape* green. The corresponding values for the superiority of the master sequence are $\sigma = 2,\ 1.905,\ 1.333$ and 1.151, respectively.

$$S_{\max} = S(0.5) = \nu \ln 2.$$

Between these two extremes, $0 \leq p \leq 0.5$, the entropy is a monotonously increasing function of the error rate, p. Figure A4.7 shows the entropy S(p) on the four model landscapes applied in Figures A4.5 and A4.6. The curves reflect the threshold behaviour encountered in the previous paragraphs (Figures A4.5 and A4.6): the entropy on the single peak landscape makes a sharp kink at the position of the error threshold, the curve for the entropy on the hyperbolic landscape has a similar bend at the threshold but the transition is smoother, whereas the entropies for the two other landscapes are curved differently and approach smoothly the maximum value, $S_{\max} = \nu \ln 2$.

6.7 Lethal mutants

It is important to note that a quasi-species can exist also in cases where the Perron–Frobenius theorem is not fulfilled. As an example we consider an extreme case of lethal mutants: only genotype $\mathbf{I}_1$ has a positive fitness value, $f_1 > 0$ and $f_2 = \ldots = f_n = 0$, and hence only the entries $W_{k1} = Q_{k1} f_1$ of matrix W are non-zero:

$$\mathrm{W} = \begin{pmatrix} W_{11} & 0 \ldots 0 \\ W_{21} & 0 \ldots 0 \\ \vdots & \vdots \ddots \vdots \\ W_{n1} & 0 \ldots 0 \end{pmatrix} \quad \text{and} \quad \mathrm{W}^k = W_{11}^k \begin{pmatrix} 1 & 0 \ldots 0 \\ \frac{W_{21}}{W_{11}} & 0 \ldots 0 \\ \vdots & \vdots \ddots \vdots \\ \frac{W_{n1}}{W_{11}} & 0 \ldots 0 \end{pmatrix}.$$

Accordingly, W is not primitive in this example, but under suitable conditions $\bar{\mathrm{x}} = (Q_{11}, Q_{21}, \ldots, Q_{n1})$ is a stable stationary mutant distribution and for $Q_{11} > Q_{j1} \forall j = 2, \ldots, n$ – correct replication occurs more frequently than a particular mutation – genotype $\mathbf{I}_1$ is the master sequence. On the basis of a rather idiosyncratic mutation model consisting of a one-dimensional chain of sequences the claim was raised that no quasi-species can be stable in the presence of lethal mutants and accordingly, no error thresholds could exist [33]. Recent papers [30, 34], however, used a realistic high-dimensional mutation model and presented analytical results as well as numerically computed examples for error thresholds in the presence of lethal mutations.

In order to be able to handle the case of lethal mutants properly we have to go back to absolute concentrations in a realistic physical setup, the flowreactor applied in section 1 and shown in Figure A4.1. We neglect degradation and find for $\mathbf{I}_1$ being the only viable genotype:[12]

$$\begin{aligned} \dot{a} &= -\left(\textstyle\sum_{i=1}^{n} Q_{i1} k_1 c_1\right) a + r(a_0 - a) \\ \dot{c}_i &= Q_{i1} k_1 a c_1 - r c_i,\ i = 1, 2, \ldots, n. \end{aligned} \tag{A4.34}$$

Computation of stationary states is straightforward and yields two solutions, (i) the state of extinction with $\bar{a} = a_0$ and $\bar{c}_i = 0\ \forall\ i = 1, 2, \ldots, n$, and (ii) a state of quasi-species selection consisting of $\mathbf{I}_1$ and its mutant cloud at the concentrations $\bar{a} = r/(Q_{11}k_1)$, $\bar{c}_1 = Q_{11}a_0 - r/k_1$ and $\bar{c}_i = \bar{c}_1(Q_{i1}/Q_{11})$ for $i = 2, \ldots, n$.

As an example we compute a maximum error rate for constant flow, $r = r_0$, again applying the uniform error rate model (A4.29):

$$\begin{aligned} Q_{11} &= (1-p)^{\nu} \text{ and} \\ Q_{i1} &= p^{d_{i1}} (1-p)^{\nu - d_{i1}}, \end{aligned}$$

where d_{i1} again is the Hamming distance between the two sequences $\mathbf{I}_i$ and $\mathbf{I}_1$. Instead of the superiority σ of the master sequence – which diverges since $\bar{f}_{-m} = 0$ because of $f_2 = \ldots = f_n = 0$ – we use the dimensionless carrying capacity η, which can be defined to be

$$\eta = \frac{k_1 a_0}{r_0}$$

for the flowreactor. The value of p, at which the stationary concentration of the master sequence $\bar{c}_1(p)$ and those of all other mutants vanishes, represents the analogue of the error threshold (32), and for the sake of clearness it is called the *extinction threshold*. Using $\ln(1-p) \approx -p$ again we obtain:

$$p_{\max} \approx \frac{\ln \eta}{\nu} \text{ for small } p. \tag{A4.35}$$

[12] We use k_i for the rate constants as in section 1, since $a(t)$ is a variable here.

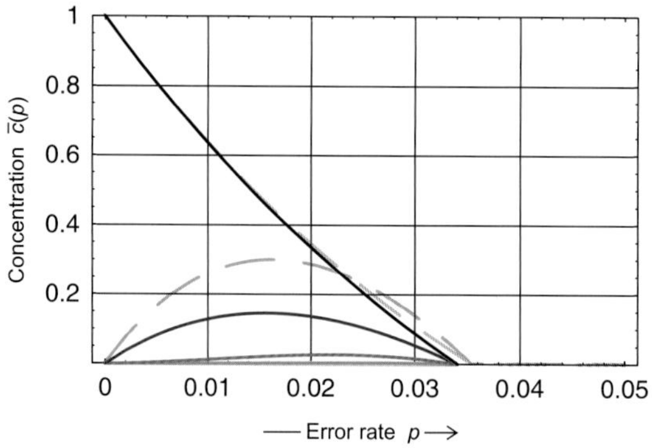

Figure A4.8

Lethal mutants and replication errors. The model for lethal mutants corresponding to a *single peak landscape* with $k_1 = 1$ and $k_2 = \ldots = k_n = 0$ is studied in the flowreactor. The concentrations of the master sequence (black) and the mutant classes (red, dark orange, light orange, etc.; full lines) are shown as functions of the error rate p. For the purpose of comparison the parameters were chosen with $\nu = 20$, $r = 1$, $a_0 = 2$, and $\eta = 2$. The plots are compared to the curves for the master sequence (grey; broken curve) and the one error class (light red; broken curve) in figure A4.5 (*single peak landscape*) with $f_1 = 2, f_2 = \ldots = f_n = 1$, $\nu = 20$, and $\sigma = 2$.

The major difference between the error threshold (A4.32) and the extinction threshold (A4.35) concerns the state of the population at values $p > p_{\max}$: replication with non-zero fitness of mutants leads to the uniform distribution, whereas the population goes extinct in the lethal mutant case. Accordingly, the transformation to relative concentrations fails and equation (A4.7) is not applicable. In Figure A4.8 we show an example for the extinction threshold with $\nu = 20$ and $\eta = 2$. For this case the extinction threshold is calculated from (A4.35) to occur at $p_{\max} = 0.03466$ compared to a value of 0.03406 observed in computer simulations. In the figure we see also a comparison of the curves for the master sequence and the one error class for the single peak landscape and the lethality model. The agreement of the two curves for the master sequences is not surprising since the models were adjusted to coincide in the values $\bar{c}_1(0) = 1$ and $p_{\max} = \ln 2/20$. The curves for the one-error classes show some difference that is due to the lack of mutational backflow in the case of lethal variants.

7. Limitations of the approach

An implicit assumption of the mathematical analysis of Darwinian selection presented here is the applicability of kinetic differential equations to describe selection and mutation in populations. In principle the approach by ordinary differential equations (ODEs) neglects finite size effects and hence is exact in principle for an infinite population size only. Biological populations, however, may be small and low-frequency mutants may be present often in a single copy or very few copies only. The uniform distribution at error rates above the threshold can never be achieved in reality because the numbers of possible polynucleotide sequences – 4^{ν} yielding, for example, 6×10^{45} sequences of tRNA length – are huge compared to typical populations ranging from 10^{6} to 10^{15} individuals in replication experiments with bacteria, viruses or RNA molecules. Typical situations in biology may thus differ drastically from scenarios in chemistry where large populations are distributed upon a few chemical species. Are the results derived from the differential equations then representative for real systems? Two situations can be distinguished: (i) individual mutations are rare events and it is extremely unlikely that the same mutation will occur twice or is precisely reversed after it has occurred, and (ii) mutations are sufficiently frequent and occur in both directions within the time of observation. The second scenario is typical for virus evolution and *in vitro* evolution experiments with molecules. The first case seems to be fulfilled with higher organisms. Bacteria may be in an intermediate situation.

In scenario (i), i.e. at low mutation rates, the exact repetition of a given mutation is of very low probability. Back-mutations, precise inversions of mutations, are also of probability zero for all practical purposes. So-called compensatory mutations are known, but they are not back-mutations, they are rather caused by second mutations that compensate the effect of the first mutation. A phenomenon called Muller's ratchet [35] in population genetics becomes effective in finite populations. Since lost mutants are not replaced, all variants starting with the fittest one will disappear sooner or later, and it is a only matter of time before a situation is reached where all genotypes have been replaced by others no matter what their fitness values were. For a comparison between the error threshold phenomenon and Muller's ratchet see [33].

The frequent mutation scenario (ii) allows for modelling and studying the kinetic equations of reproduction and selection as stochastic processes [25, 36, 37, 38] – examples are multitype branching or birth-and-death processes – as well as for computer simulations [39] (for an overview of stochastic modeling see, for example [40]). In essence, the solutions of stochastic models are time-dependent probability distributions instead of solution curves. The mean or expectation value of the distribution coincides with the deterministic (ODE) solution, since all reactions in the kinetic model are (pseudo) first order. The (relative) width of the distribution increases with growing mutation rate and deceasing population size, and the error threshold phenomenon is reproduced as a superposition of error propagation and finite size effects. The expression for the error threshold can be readily extended to finite populations [25]. Formation of stable quasi-species requires a replication fidelity that is higher the smaller the population size.

At error rates above threshold the kinetic ODEs predict the uniform distribution of sequences as stationary solution of equation (A4.21).[13] Differences in fitness values do not matter under these conditions and there is no preferred master sequence. Realistic populations are far too small to form uniform distributions of sequences and hence the deterministic model fails. Below threshold the quasi-species can be visualised as a localisation of the population in some preferred region of sequence space with high fitness values (or at least one particularly high fitness value) [18, 41]. Above threshold the population is no more localised and drifts randomly in sequence space.[14] At the same time, populations are also too small to occupy a coherent region in sequence space and break up into smaller clones, which migrate in different directions as described for the neutral evolution case [3, 44].

How relevant is the error threshold in realistic situations? According to the results presented in section 6.5 the question boils down to an exploration of natural fitness landscapes: are biopolymer landscapes rugged or smooth? All evidence obtained so far points towards a rather bizarre structure of these landscapes. Single nucleotide exchanges may lead to large effects, small effects or no consequences at all, as in the case of neutral mutations. Since biomolecules are usually optimised with respect to their functions within an organism, most mutations have deleterious effects or no effect. Biopolymer landscapes have three characteristic features, which are hard to visualize: (i) high dimensionality (ii) ruggedness and (iii) neutrality. Where equally fit genotypes are nearest or next nearest neighbors in sequence space they form joint quasi-species as described in [23]. When they are not closely related, however, neutral evolution in the sense of Motoo Kimura is observed [45]. In the case of neutrality in genotype space a selection model can still be formulated in phenotype space [46, 47]. The variables are concentrations of phenotypes that are obtained by lumping together all concentrations of genotypes, which form the same phenotype. Then, an analysis similar to the one presented here can be carried out. The genotypic error threshold is relaxed and the system gives rise to a phenotypic error threshold below which the fittest or master phenotype is conserved in the population. The ODE model is readily supplemented by a theory of phenotype evolution based on new concept of evolutionary nearness of phenotypes in sequence space [4, 48, 49], which is confirmed by computer simulations of RNA structure optimization in a flowreactor of the type shown in Figure A4.1 [4, 48, 50]. Random drift of populations on neutral subspaces of sequence space has also been dealt with [50]. A series of snapshots shows the spreading of a population that breaks up into individual clones

[13] As mentioned before the uniform distribution is the exact stationary solution of equation (A4.21) for equal probabilities of correct and incorrect incorporation of a nucleotide, which is the case at an error rate $p = 1 - p = 0.5$ for binary sequences.

[14] The mutation rate can be seen as an analogue to temperature in spin systems and the error threshold corresponds to a phase transition. The relation between the selection-mutation equation and spin systems was studied first by Ira Leuthäusser [42, 43].

in full agreement with earlier models [44, 51]. Computer simulations were also successful in providing evidence for the occurrence of error thresholds in stochastic replication-mutation systems [52].

References

1. Watts, A. and Schwarz, G. (eds.) (1997). "Evolutionary Biotechnology – From Theory to Experiment". *Biophys. Chem.* 66/2–3: 67–284.
2. Fontana, W. and Schuster, P. (1987). "A Computer Model of Evolutionary Optimization". *Biophys. Chem.* 26: 123–147.
3. Huynen, M. A., Stadler, P. F. and Fontana, W. (1996). "Smoothness Within Ruggedness. The Role of Neutrality in Adaptation". *Proc. Natl. Acad. Sci. USA* 93: 397–401.
4. Fontana, W. and Schuster, P. (1998). "Continuity in Evolution. On the Nature of Transitions". *Science* 280: 1451–1455.
5. Zwillinger, D. (1998). "Handbook of Differential Equations", 3rd edn. Academic Press, San Diego.
6. Shahshahani, S. (1979). "A New Mathematical Framework for the Study of Linkage and Selection". *Mem. Am. Math. Soc.* 211.
7. Sigmund, K. "The Maximum Principle for Replicator Equations". In: "Lotka-Volterra Approach to Cooperation and Competition in Dynamical Systems", Ebeling, W. and Peschel, M. (eds). Akademie-Verlag, Berlin, pp. 63–71.
8. Schuster, P. and Sigmund, K. (1985). "Dynamics of Evolutionary Optimization". *Ber. Bunsenges. Phys. Chem.* 89: 668–682.
9. Eigen, M. (1971). "Selforganization of Matter and the Evolution of Biological Macromolecules". *Naturwiss.* 58: 465–523.
10. Eigen, M. and Schuster, P. (1977). "The Hypercycle. A Principle of Natural Self-Organization. Part A: Emergence of the Hypercycle". *Naturwiss.* 64: 541–565.
11. Eigen, M. and Schuster, P. (1978). "The Hypercycle. A Principle of Natural Self-Organization. Part B: The Abstract Hypercycle". *Naturwiss.* 65: 7–41.
12. Fisher, R. A. (1930). "The Genetical Theory of Natural Selection". Oxford University Press, Oxford.
13. Schuster, P. and Sigmund, K. (1983). "Replicator Dynamics". *J. Theor. Biol.* 100: 533–538.
14. Ewens, W. J. (1979). "Mathematical Population Genetics", volume 9, "Biomathematics Texts". Springer-Verlag, Berlin.
15. Akin, E. (1979). "The Geometry of Population Genetics", volume 31, "Lecture Notes in Biomathematics". Springer-Verlag, Berlin.

16. Hofbauer, J. and Sigmund, K. (1988). "Dynamical Systems and the Theory of Evolution". Cambridge University Press, Cambridge.
17. Schuster, P. (1988). "Potential Functions and Molecular Evolution". In: "From Chemical to Biological Organization". Springer Series in Synergetics, volume 39, Markus, M., Müller, S. C. and Nicolis, G. (eds). Springer-Verlag, Berlin, pp. 149–165.
18. Eigen, M., McCaskill, J. and Schuster, P. (1989). "The Molecular Quasispecies". *Adv. Chem. Phys.* 75: 149–263.
19. Thompson, C. J. and McBride, J. L. (1974). "On Eigen's Theory of the Self-organization of Matter and the Evolution of Biological Macromolecules". *Math. Biosci.* 21: 127–142.
20. Jones, B. L., Enns, R. H. and Rangnekar, S. S. (1976). On the Theory of Selection of Coupled Macromolecular Systems. *Bull. Math. Biol.* 38: 15–28.
21. Seneta, E. (1981). "Non-negative Matrices and Markov Chains", 2nd edn. Springer-Verlag, New York.
22. Rumschitzki, D. S. (1987). "Spectral Properties of Eigen Evolution Matrices". *J. Math. Biol.* 24: 667–680.
23. Schuster, P. and Swetina, J. (1988). "Stationary Mutant Distribution and Evolutionary Optimization". *Bull. Math. Biol.* 50: 635–660.
24. Swetina, J. and Schuster, P. (1982). "Self-Replication with Errors – A Model for Polynucleotide Replication". *Biophys. Chem.* 16: 329–345.
25. Nowak, M. and Schuster, P. (1989). "Error Thresholds of Replication in Finite Populations. Mutation Frequencies and the Onset of Muller's Ratchet". *J. Theor. Biol.* 137: 375–395.
26. Drake, J. W. (1993). "Rates of Spontaneous Mutation Among RNA Viruses". *Proc. Natl. Acad. Sci. USA* 90: 4171–4175.
27. Eigen, M. (2002). "Error Catastrophe and Antiviral Strategy". *Proc. Natl. Acad. Sci. USA* 99: 13374–13376.
28. Domingo, E. (ed.) (2005). "Virus Entry Into Error Catastrophe as a New Antiviral Strategy". *Virus Res.* 107(2): 115–228.
29. Bull, J. J., Sanjuan, R. and Wilke, C. O. (2007). "Theory for Lethal Mutagenesis for Viruses". *J. Virol.* 81: 2930–2939.
30. Tejero, H., Marín, A. and Montero, F. (2010). "Effect of Lethality on the Extinction and on the Error Threshold of Quasispecies". *J. Theor. Biol.* 262: 733–741.
31. Phillipson, P. E. and Schuster, P. (2009). "Modeling by Nonlinear Differential Equations. Dissipative and Conservative Processes", volume 69, World Scientific Series on Nonlinear Science A. World Scientific, Singapore.
32. Marx, C., Posch, H. A. and Thirring, W. (2007). "Emergence of Order in Selection-Mutation Dynamics". *Phys. Rev. E*, 75: 061109.
33. Wagner, G. P. and Krall, P. (1993). "What is the Difference Between Models of Error Thresholds and Muller's Ratchet?" *J. Math. Biol.* 32: 33–44.

34. Takeuchi, N. and Hogeweg, P. (2007). "Error-Thresholds Exist in Fitness Landscapes with Lethal Mutants". *BMC Evol. Biol.* 7:e15: 1–11.
35. Muller, H. J. (1932). "Some Genetic Aspects of Sex". *Am. Nat.* 66: 118–138.
36. Schuster, P. and Sigmund, K. (1984). "Random Selection – A Simple Model Based on Linear Birth and Death Processes". *Bull. Math. Biol.* 46: 11–17.
37. McCaskill, J. S. (1984). "A Stochastic Theory of Macromolecular Evolution". *Biol. Cybern.* 50: 63–73.
38. Demetrius, L., Schuster, P. and Sigmund, K. (1985). "Polynucleotide Evolution and Branching Processes". *Bull. Math. Biol.* 47: 239–262.
39. Gillespie, D. T. (2007). "Stochastic Simulation of Chemical Kinetics". *Annu. Rev. Phys. Chem.* 58: 35–55.
40. Blythe, R. A. and McKane, A. (2007). "Stochastic Models of Evolution in Genetics, Ecology and Linguistics". *J. Stat. Mech.: Theor. Exp.*, P07018.
41. McCaskill, J. S. (1984). "A Localization Threshold for Macromolecular Quasispecies from Continuously Distributed Replication Rates". *J. Chem. Phys.* 80: 5194–5202.
42. Leuthäusser, I. (1986). "An Exact Correspondence Between Eigen's Evolution Model and a Two-Dimensional Ising System". *J. Chem. Phys.* 84: 1884–1885.
43. Leuthäusser, I. (1987). "Statistical Mechanics of Eigen's Evolution Model". *J. Stat. Phys.* 48: 343–360.
44. Derrida, B. and Peliti, L. (1991). "Evolution in a Flat Fitness Landscape". *Bull. Math. Biol.* 53: 355–382.
45. Kimura, M. (1983). "The Neutral Theory of Molecular Evolution". Cambridge University Press, Cambridge.
46. Reidys, C., Forst, C. and Schuster, P. (2001). "Replication and Mutation on Neutral Networks". *Bull. Math. Biol.* 63: 57–94.
47. Takeuchi, N., Poorthuis, P. H. and Hogeweg, P. (2005). "Phenotypic Error Threshold – Additivity and Epistasis in RNA Evolution". *BMC Evol. Biol.* 5:e9:1–9.
48. Fontana, W. and Schuster, P. (1998). "Shaping Space. The Possible and the Attainable in RNA Genotype-Phenotype Mapping". *J. Theor. Biol.* 194: 491–515.
49. Stadler, B. R. M., Stadler, P. F., Wagner, G. P. and Fontana, W. (2001). "The Topology of the Possible: Formal Spaces Underlying Patterns of Evolutionary Change". *J. Theor. Biol.* 213: 241–274.
50. Schuster, P. (2003). "Molecular Insight into the Evolution of Phenotypes". In: "Evolutionary Dynamics – Exploring the Interplay of Accident, Selection, Neutrality, and Function", Crutchfield, J. P. and Schuster, P. (eds). Oxford University Press, New York, pp. 163–215.
51. Higgs, P. G. and Derrida, B. (1991). "Stochastic Models for Species Formation in Evolving Populations". *J. Phys. A: Math. Gen.* 24: L985–L991.
52. Kupczok, A. and Dittrich, P. (2006). "Determinants of Simulated RNA Evolution". *J. Theor. Biol.* 238: 726–735.

Notation

building blocks and degradation products	$\mathbf{A}, \mathbf{B}, \ldots$
numbers of particles of $\mathbf{A}, \mathbf{B}, \ldots$	$N_{\mathbf{A}}, N_{\mathbf{B}}, \ldots$
concentrations of $\mathbf{A}, \mathbf{B}, \ldots$	$[\mathbf{A}] = a, [\mathbf{B}] = b, \ldots$
replicating molecular species	$I_1, I_2, \ldots$
numbers of particles of $I_1, I_2, \ldots$	$N_1, N_2, \ldots$
concentrations of $I_1, I_2, \ldots$	$[I_1] = c_1, [I_2] = c_2, \ldots$
relative concentrations of $I_1, I_2, \ldots$	$[I_1] = x_1, [I_2] = x_2, \ldots$
partial sums of relative concentrations	$y_k = \sum_i x_i$
flow rate in the CSTR	r
influx concentration into the CSTR	a_0
rate parameters	$d_i, k_i, f_i, \ldots \quad i = 1, 2, \ldots$
global regulation flux	$\Phi(t)$
chain length of polynucleotides	ν
superiority of the master sequence I_m	$\sigma_m = \frac{f_m(1-x_m)}{\sum_{i \neq m} f_i}$
population entropy	$S = \sum_i x_i \ln x_i$

Appendix A5

Kinetics of Multistep Replication

This incursion into the theory of the multistep cycles of replication is concerned with a very important question. Selection is required for evolutionary optimisation and we learned in Section 4.7 that selection is a physical consequence of inherent autocatalytic reproduction, which we described by a relatively simple linear autocatalytic ansatz. Moreover, we have seen that natural selection in the Darwinian sense is quite narrowly confined to such a description. How can such a complicated mechanism, as is to be expected for a template-instructed and enzyme-catalysed process of macromolecular polymerisation, be described by an ansatz as simple as $dx_i/dt = k_i x_i$, where x_i is the relative population variable (with the sum of all x_k being equal to 1), t is time and k_i is a quasi-first-order rate constant of the autocatalyst "i"? The validity of this ansatz becomes phenomenologically manifest in the observation of a precisely exponential initial growth of clones under suitable experimental conditions. So let us look in more detail at the multistep mechanism of such a reaction cycle, which was sketched in Figure 5.3.1 and is explained in more detail in Figure A5.1 and Table A5.1. I have chosen this picture for the text not in order to deter the reader, but rather to convey some feeling for the true complexity that we often meet in biochemical reactions. If the sequence is that of a single gene, the replication mechanism includes thousands of individual steps. If a genome of a complete organism, such as the bacterium *E. coli*, is involved, then the total number of steps might include as many as millions of single reaction steps. In Table A5.1 the different rate constants involved are identified by the function to which they refer.

The rates for quite a number of these systems have been determined in our laboratory by the late Christoph Biebricher and his co-workers. For viruses, and in particular for the plus-strand virus Q_β, such data are discussed in the papers already cited in the text, where references to further publications by Biebricher and his school can be found. Generally, the reaction rates have an initial exponential phase due to autocatalytic production of a new template. However when templates reach the concentration range of enzymes there is transition to linear growth of the template. As in the Michaelis–Menten mechanism, the active enzyme–template

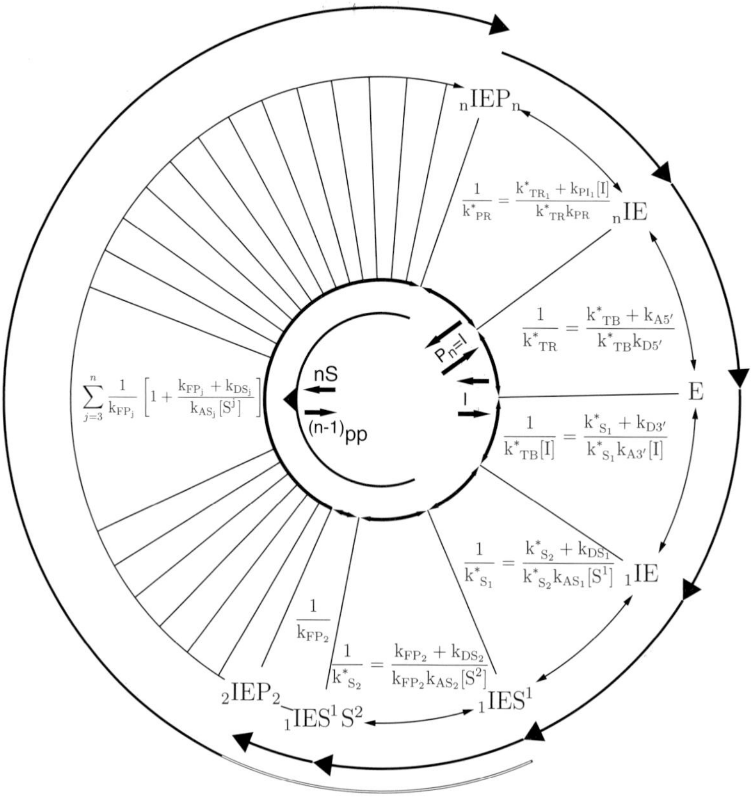

Figure A5.1

The individual steps of a replication cycle

The picture refers to Figure 5.3.1, where for palindromic sequences the two cycles (for the plus and minus strand, respectively) coalesce into a single cycle. The various rate parameters (explained in Table A5.1) are written in explicit form.

complex reaches a constant stationary level and the template is merely a linearly growing product of the reaction; in this range, the same steady-state assumptions as made in classical enzyme kinetics apply. The overall rate constants then reflect the steady-state behaviour of saturation, as do corresponding terms in the classical Michaelis–Menten mechanism, but they are of much more complex composition.

Table A5.2 gives a summary of the expressions obtained. They refer to the palindromic reaction cycle of replication depicted in Figure A5.1. It is a simplification of the tandem cycle shown in the text (Figure 5.3.1) insofar as it assumes the plus and minus strands to have palindromic sequences, which allows the tandem double cycle to coalesce to a single palindromic cycle. Otherwise the corresponding steps in both cycles might have different rate constants, which at steady state yield an overall rate constant that is the geometric mean of the overall rate parameters of the plus

Table A5.1 Kinetics of RNA replication

Mathematical symbols	
n	Length of nucleotide chain
j	Index denoting stage of elongation
λ	Eigenvalues
K	Matrix of rate constants
E	Unit matrix
Species symbols	
I	Free template (information carrier)
P_n	Replica = new template
E	Free enzyme
$_1IE$	Active enzyme – template complex
S^j	Substrate at position j of replica
$_1IES^1$	First replication complex
$_1IES^1S^2$	Geminal association complex
$_jIEP_j$	Chain elongation complex (also I_j)
$_jIEP_jS^{j+1}$	Substrate recognition complex ($2 \leq j \leq n-1$)
$_nIE$	Inactive enzyme–template complex
Concentrations	
$[I_0]$	Total template concentration
$[E_0]$	Total enzyme concentration
$[E_0^+]$	Total enzyme concentration associated with plus strand
$[E_0^-]$	Total enzyme concentration associated with minus strand
$[S_0]$	Effective substrate concentration
Elementary reaction rate constants[a]	
k_{ASj}	Association of substrate S^j
k_{AS}	Average association of substrate (1×10^8)
k_{DSj}	Dissociation of substrate S^j (1×10^5 for j = 1)
k_{DS}	Average dissociation of substrate S (2×10^4 for $j \neq 1$)
k_{FPj}	Formation of phosphodiester linkage by incorporation of substrate S^j into replica (I for j = 2)
k_{FP}	Average phosphodiester linkage formation (5 for j >2)
k_{DPj}	Pyrophosphorolysis at position j
k_{DP}	Average pyrophosphorolysis (2×10^3)
k_{PR}	Product release (0.5)
k_{PI}	Product inhibition (2.5×10^5)
$k_{5'3'}$	Direct reactivation of enzyme–template complex (0)
$k_{3'5'}$	Reverse direct reactivation (0)
$k_{D5'}$	Dissociation of enzyme from inactive position on template (1×10^{-2})
$k_{D3'}$	Dissociation of enzyme from active position (1×10^{-5})
$k_{A5'}$	Association of enzyme at inactive position of template (1×10^7, 1×10^5)
$k_{A3'}$	Association of enzyme at active position of template (1×10^7)

(Cont.)

Table A5.1 (Cont.)

Composite rate parameters	
k^*_{12}	Initiation rate parameter
k^*_{PR}	Product release
k^*_{TR}	Template release and enzyme reactivation
k^*_{TB}	Template binding
k^*_{T1}	Substrate turnover and product release
k^*_{T2}	5′3′ reactivation
$k^*_{A5'}$	$k_{A5'}\ [E_0]$
k^*_{S1}	Binding of first substrate
k^*_{S2}	Binding of second substrate
k^*_{SB}	Successive substrate binding and incorporation
κ	Growth rate
κ_1	Overall rate of replica synthesis
κ_j	Substrate incorporation rate constant
κ_n	Net reactivation rate constant
Other parameters	
v	Steady-state RNA production rate
v_{max}	Saturation steady-state RNA production rate
k_T	Turnover number
k_c	Competition rate constant
K_M	Michaelis constant
K_S	Average substrate association constant
K'_S	Average substrate association constant, including substrate-mediated reactivation
$K_{3'5'}$	Coupling to reactivation
K_H	Inhibition constant
$K_{\Sigma IE}$	Overall template–enzyme interaction constant

[a] Standard values for the simulations are in parentheses. Units are moles per liter and seconds

and minus cycles. For non-steady states the tandem cycle represents a cross-catalytic rather than a true autocatalytic system.

In order to create some way of imagining the confusing-looking expressions in Figure A5.1 (which in fact describe a quite lucid mechanism) let us look at some limiting cases of the overall rate equation:

$$\text{reaction velocity: } v = \frac{k_T\ [E_0]\ [S_0]}{K_M + [S_0]\ \{1 + K_H\ [I]\}}$$

Table A5.2 Summary of the Rate Expressions obtained

$$v = \frac{k_T[E_0][S_0]}{K_M + [S_0]\{1 + K_H[I]\}}$$

$$\frac{1}{k_T} = \underbrace{\sum_{j=2}^{n} \frac{1}{k_{FP_j}} + \frac{1}{k_{PR}} + \frac{1}{k_{D5'}}}_{\substack{k^*_{T1} \\ \text{substrate turnover and} \\ \text{product release}}} \underbrace{\left(1 + \frac{k_{A5'}}{k_{A3'}}\right) + \frac{1}{k_{A3'}[I]}}_{\substack{k^*_{T2} \\ 5'3' \text{ reactivation}}}$$

$$K_M = k_T \left[\frac{1 + K_{3'5'}}{k^*_{12}} + \frac{1}{k^*_{SB}}\right]$$

$$\frac{1}{k^*_{12}} = \frac{[S_0]}{k_{AS_1}[S^1]} \left\{1 + \frac{(k_{FP_2} + k_{DS_2})k_{DS_1}}{k_{FP_2}k_{AS_2}[S^2]}\right\}$$

$$K_{3'5'} = \frac{k_{D3'}k_{A5'}}{k_{A3'}k_{D5'}} + \frac{k_{D3'}}{k_{A3'}[I]}$$

$$\frac{1}{k^*_{SB}} = \sum_{j=2}^{n} \frac{k_{FP_j} + k_{DS_j}}{k_{FP_j}} \frac{[S_0]}{k_{AS_j}[S^j]}$$

$$K_H = \frac{k_{PI}}{k_{PR}} \frac{k_T}{k_{D5'}} \left\{1 + \frac{k_{A5'}}{k_{A3'}} \left(1 + \frac{k_{D3'}}{k^*_{12}[S_0]}\right)\right\}$$

$$[I] = [I_0] - [E_0] - [{}_nIEP_n]$$

Case 1: Assume that the chain length n is large, so that elongation is rate-limiting. The sum terms then dominate all the others. Let k_{FP} represent an average k_{FPJ}. Then the expressions in Table A5.2 simplify to:

$$k_T = k_{FP}/n$$

$$K_M = K_S = k_{DS}/k_{AS}$$

$$v = \frac{k_{FP}[E_0][S_0]/n}{[S_0] + K_S}$$

Case 2: Assume that the substrate concentration is lowered until initiation becomes rate-limiting, i.e. until $k^*_{12} \ll k^*_{SB}$, and that reactivation is irreversible. The rate then becomes:

$$v = \frac{k_{FP}[E_0][S_0]^2}{K_S^2}$$

Case 3: Assume that reactivation, i.e. dissociation of 5′-bound template and association of 3′-bound template, is rate-limiting. Two kinetically distinguishable mechanisms are possible, depending on whether substrate (e.g. GTP) binding is involved in reactivation. If it is not, then the rate is

$$v = \frac{k_{D5'}\,[E_0]\,[S_0]}{[S_0] + nK_S k_{D5'}/k_{FP}}$$

where reassociation has been neglected, such that $k_T = k_{D5'}$; $1/k^*_{12}$ has been neglected compared with $1/k^*_{SB}$, so that $K_M = k_T/k^*_{SB}$ can be simplified to $nK_S k_{D5'}/k_{FP}$.

Case 4: If reactivation is rate-limiting and substrate binding is required, the rate is

$$v = \frac{k_{D5'}\,[E_0]\,[S_0]}{[S_0] + K'_S}$$

in which a new Michaelis constant, K'_S, appears, given by the fact that the expression $1/k^*_{SB}$ changes to $1/k^*_{SB} = K_S(1/k_{D5'} + n/k_{FP}) \cong K_S/k_{D5'}$, while k_T remains unchanged.

It can be seen that cases 3 and 4 show a common turnover number but different Michaelis constants, while cases 1 and 4 have a common Michaelis constant but different turnover numbers.

A further, more detailed, discussion with experimental tests and computer simulations can be found in the series of papers quoted in the text.

I shall now proceed to the initial phase of exponential amplification of templates. It demonstrates the fundamental rôle of replication as being the basis of natural selection. A rigorous theoretical treatment will show that the kinetic equations used for a theory of Darwinian selection can indeed be used in the straightforward exponential form, as was done in Chapter 4. In addition, it shows in detail how exponential growth builds up deterministically in early phases.

The replication cycle of Figure A5.1 can be reformulated as in Figure A5.2 and described by a system of coupled linear differential equations:

$$\frac{d}{dt}\mathbf{X}(t) = \mathbf{K}\mathbf{X}(t) \tag{A5.1}$$

where **X**(t) is a column vector whose components x_j are the concentrations of the RNA species I_j and **K** is the matrix of rate coefficients explained in Table A5.1. The solution to equation (A5.1) is

$$x_j(t) = a_{j1}e^{\lambda_1 t} + a_{j2}e^{\lambda_2 t} + \ldots + a_{jn}e^{\lambda_n t} \tag{A5.2}$$

where the eigenvalues λ_m (m = 1, 2, ..., n) are the solutions of the characteristic equation (Det = determinant, E = unity matrix)

$$\mathrm{Det}(\mathbf{K} - \lambda\mathbf{E}) = 0 \tag{A5.3}$$

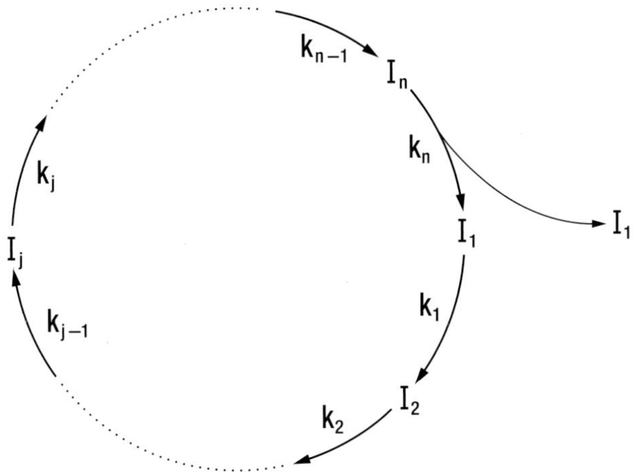

$$
\begin{aligned}
\frac{d[I_1]}{dt} &= 2k_n[I_n] - k_1[I_1] \\
\frac{d[I_2]}{dt} &= k_1[I_1] - k_2[I_2] \\
&\;\;\vdots \\
\frac{d[I_n]}{dt} &= k_{n-1}[I_{n-1}] - k_n[I_n]
\end{aligned}
\qquad
\mathbf{K} = \begin{pmatrix}
-k_1 & 0 & \cdot & \cdot & \cdot & \cdot & \cdot & 0 & 2k_n \\
k_1 & -k_2 & 0 & \cdot & \cdot & \cdot & \cdot & \cdot & 0 \\
0 & k_2 & -k_3 & \cdot & \cdot & \cdot & \cdot & \cdot & \cdot \\
\cdot & \cdot & \cdot & \cdot & \cdot & \cdot & \cdot & \cdot & \cdot \\
\cdot & \cdot & \cdot & \cdot & \cdot & \cdot & \cdot & \cdot & \cdot \\
\cdot & \cdot & \cdot & \cdot & \cdot & \cdot & \cdot & \cdot & \cdot \\
\cdot & \cdot & \cdot & \cdot & \cdot & \cdot & \cdot & \cdot & \cdot \\
\cdot & \cdot & \cdot & \cdot & \cdot & \cdot & k_{n-2} & -k_{n-1} & 0 \\
0 & \cdot & \cdot & \cdot & \cdot & \cdot & 0 & k_{n-1} & -k_n
\end{pmatrix}
$$

Figure A5.2

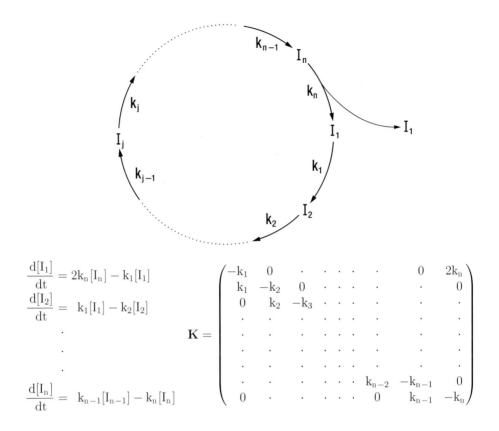

The initial phase of replication

The rate parameters are written in simplified form. The coupled rate equations are written in matrix form. They yield exponential growth for the whole system.

while the a_{jm} are the coefficients of the diagonalised equations. For all rate constants k_j equal to a common k the eigenvalues are

$$\lambda_m = k(2^{1/n}e^{2\pi im/n} - 1) \tag{A5.4}$$

The factor $e^{2\pi im/n}$ is an nth root of unity, $2^{1/n}$ is the absolute value of the nth root of 2 and i is the imaginary number $\sqrt{-1}$. The eigenvalues λ_m belong to a column vector of normal modes $\mathbf{Y}(t)$ whose n components $y_m(t)$ satisfy the eigenvalue equations:

$$\frac{d}{dt}y_m(t) = \lambda_m y_m(t) \tag{A5.5}$$

such that

$$y_m(t) = y_m(0)e^{\lambda_m t} \tag{A5.6}$$

(Eigen and de Maeyer 1974). The transformations from concentrations to normal modes and vice versa:

$$\mathbf{Y} = \mathbf{MX} \text{ and } \mathbf{X} = \mathbf{M}^{-1}\mathbf{Y} \tag{A5.7}$$

with

$$\mathbf{M} \cdot \mathbf{M}^{-1} = \mathbf{E} \tag{A5.8}$$

introduce the transformation matrices representing the linear combinations

$$y_m(t) = m_{m1}x_1 + m_{m2}x_2 + \ldots + m_{mn}x_n \tag{A5.9}$$

and

$$x_j(t) = m'_{j1}y_1 + m'_{j2}y_2 + \ldots m'_{jm}y_m + \ldots + m'_{jn}y_n \tag{A5.10}$$

where $m_{mj} = q_m^{1-j}$ and $m'_{jm} = (1/n)q_m^{j-1}$ are the matrix elements of M and M^{-1}, respectively, and $q_m = 2^{-1/n}e^{-2\pi i m/n}$. (The first subscript refers to the row, the second to the column of each matrix.) The elements m_{mj} are related to the eigenvalues through:

$$m_{mj} = (\lambda_m/\kappa + 1)^{j-1} \tag{A5.11}$$

The full matrices are thus

$$\mathbf{M} = \begin{pmatrix} 1 & 2^{\frac{1}{n}}e^{2\pi i\frac{1}{n}} & 2^{\frac{2}{n}}e^{2\pi i\frac{2}{n}} & \cdots & 2^{\frac{n-1}{n}}e^{2\pi i\frac{n-1}{n}} \\ 1 & 2^{\frac{1}{n}}e^{2\pi i\frac{2}{n}} & 2^{\frac{2}{n}}e^{2\pi i\frac{4}{n}} & \cdots & 2^{\frac{n-1}{n}}e^{2\pi i\frac{2(n-1)}{n}} \\ 1 & 2^{\frac{1}{n}}e^{2\pi i\frac{3}{n}} & 2^{\frac{2}{n}}e^{2\pi i\frac{6}{n}} & \cdots & 2^{\frac{n-1}{n}}e^{2\pi i\frac{3(n-1)}{n}} \\ \vdots & \vdots & \vdots & \vdots & \vdots \\ 1 & 2^{\frac{1}{n}} & 2^{\frac{2}{n}} & \cdots & 2^{\frac{n-1}{n}} \end{pmatrix} \tag{A5.12}$$

$$\mathbf{M}^{-1} = \frac{1}{n}\begin{pmatrix} 1 & 1 & 1 & \cdots & 1 \\ 2^{-\frac{1}{n}}e^{-2\pi i\frac{1}{n}} & 2^{-\frac{1}{n}}e^{-2\pi i\frac{2}{n}} & 2^{-\frac{1}{n}}e^{-2\pi i\frac{3}{n}} & \cdots & 2^{-\frac{1}{n}} \\ 2^{-\frac{2}{n}}e^{-2\pi i\frac{2}{n}} & 2^{-\frac{2}{n}}e^{-2\pi i\frac{4}{n}} & 2^{-\frac{2}{n}}e^{-2\pi i\frac{6}{n}} & \cdots & 2^{-\frac{2}{n}} \\ \vdots & \vdots & \vdots & \vdots & \vdots \\ 2^{-\frac{n-1}{n}}e^{-2\pi i\frac{n-1}{n}} & 2^{-\frac{n-1}{n}}e^{-2\pi i\frac{2(n-1)}{n}} & 2^{-\frac{n-1}{n}}e^{-2\pi i\frac{3(n-1)}{n}} & \cdots & 2^{-\frac{n-1}{n}} \end{pmatrix} \tag{A5.13}$$

M in the expression given above has the form of an $n \times n$ matrix, the kth row of which appears to be a sequence of powers: $1, x_k, x_k^2, x_k^3 \ldots x_k^{n-1}$ where $x_k = 2^{1/n}$ exp $(2\pi i\, k/n)$. This is formally identical with the matrix of a Vandermonde determinant.

The coefficients a_{jm} of equation (A5.2) can be calculated from M and M^{-1} by

$$a_{jm} = m'_{jm} \sum_{i=1}^{n} m_{ji} x_i(0) \qquad (A5.14)$$

We can now consider the time dependence of the concentrations x_j. The eigenvalue spectrum (equation (A5.4)) can be converted to the trigonometric form

$$\lambda_m = k\{2^{1/n}[\cos(2\pi m/n) + i \sin(2\pi m/n)] - 1\} \qquad (A5.15)$$

The representation in Figure A5.3 shows the physical significance of the different real eigenvalues. Real eigenvalues arise when $\sin(2\pi m/n)$ is 0, which happens only for m = n and (for even n only) for m = n/2. (The value m = 0 does not occur, since we enumerate the eigenvalues from 1 to n.) The real positive eigenvalue for m = n is

$$\lambda_n = k\{2^{1/n} - 1\} \xrightarrow{\text{for large n}} (k/n)\ln 2 \qquad (A5.16)$$

while the real negative one for even n is

$$\lambda_{n/2} = -k\{2^{1/n} + 1\} \xrightarrow{\text{for large n}} -2k \qquad (A5.17)$$

The other eigenvalues are complex and occur in unique conjugate pairs. The non-degeneracy of the eigenvalues corresponds to the non-zero determinant of **M**: Det **M** = $(2n)^{n/2}/2^{1/2}$

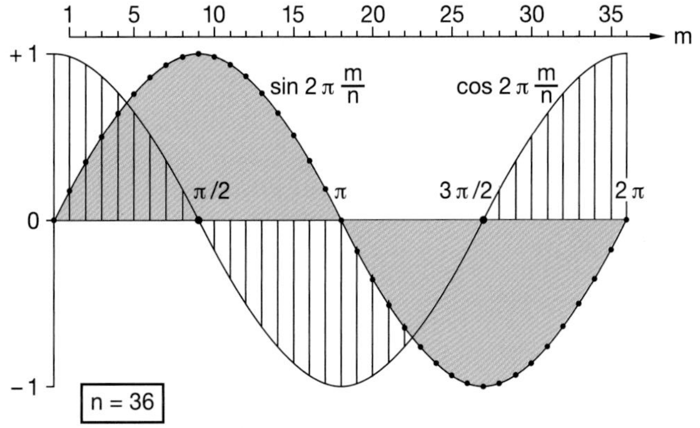

Figure A5.3

Roots of unity for n = 36. Real arts are generated by the cosine function and imaginary parts by the sine function. For even n the sine function is zero for m = n/2 and m = n, giving one positive (m = n, $\cos(2\pi m/n) = +1$) and one negative (m = n/2, $\cos(2\pi m/2n) = -1$) real eigenvalue. For odd n only the positive real eigenvalue occurs. Larger values of n imply finer graduations, i.e., more eigenvalues.

Complex eigenvalues are equivalent to oscillatory terms. We now consider how they superimpose in order to yield the temporal evolution of the concentration.

If we evaluate equation (A5.15) we find that for $n < 29$ only negative real parts of λ_m result for all integers m (Figure A5.4). However, with $n = 29$ the real part of λ_m becomes positive for $m = 1$ and $m = n - 1$, and more such positive increments of the real part appear at larger values of n. This means that the amplitudes of some oscillatory normal modes increase with time. This growth is always slower, however, than that due to the positive real eigenvalue λ_n. We call the normal mode associated with λ_n the growth term; all oscillatory modes die out with respect to it, although some of them persist for quite some while. To demonstrate this we proceed in the following way.

1) Combining equations (A5.6) and (A5.10) yields

$$x_j(t) = \sum_{m=1}^{n} m'_{jm} y_m(t) = \sum_{m=1}^{n} m'_{jm} y_m(0) e^{\lambda_m t} \tag{A5.18}$$

2) Assume that the reaction is started by adding an initial amount c_0 of free template (I_1) to the incubation mixture:

$$x_1(0) = c_0$$
$$x_j(0) = 0 \qquad j = 2, 3, \ldots, n \tag{A5.19}$$

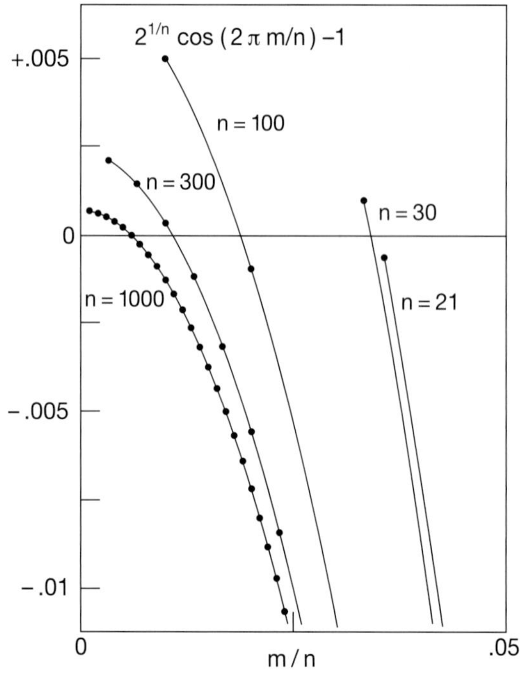

Figure A5.4

Real parts of eigenvalues. The function $2^{1/n} \cos(2\pi m/n) - 1$ changes relatively little for different large values of n (left), but its fine structure shows qualitative differences near $m = 0$ and $m = n$. For $n < 29$ all values of this function at integral values $m = 1, 2 \ldots (n - 1)$ remain negative. For $n \geq 29$, however, positive values appear near $m = 1$ and $m = n - 1$, the numbers of which increase with increasing n (as shown for $n = 21$, 30, 100, 300 and 1000).

yielding, according to equation (A5.9),

$$y_m(0) = c_0 \quad \text{for} \quad m = 1, 2, 3, \ldots, n \tag{A5.20}$$

since $m_{m1} = 1$ for all m (equation (A5.12)).

3) The concentration variables best accessible to experiments are the sum of all template (free or complexed) concentrations $\sum_{j=1}^{n} x_j(t)$ and the sum over all nucleotides present in some polymeric form $\frac{1}{n}\sum_{j=1}^{n}(n+j)x_j(t)$, in which we have normalised to the completed template length by including the factor (1/n). In the first case the new replica is not counted until it is complete and dissociated from the template, while in the second case everything linked in polymeric chains is counted. (There are methods for determining either of these sums. Radioactive labelling methods are associated with the second one (cf. Biebricher *et al.* 1981b).)
4) The first sum is found by combining equations (A5.18) and (A5.20):

$$\sum_{j=1}^{n} x_j(t) = \sum_{j=1}^{n}\sum_{m=1}^{n} m'_{jm} \; c_0 e^{\lambda_m \cdot t} \tag{A5.21}$$

which can be simplified by permuting the order of summation and factoring to

$$\sum_{j=1}^{n} x_j(t) = c_0 \sum_{m=1}^{n} e^{\lambda_m \cdot t} \sum_{j=1}^{n} m'_{jm} \tag{A5.22}$$

Likewise for the second sum

$$\frac{1}{n}\sum_{j=1}^{n}(n+j)x_j(t) = \frac{c_0}{n}\sum_{m=1}^{n} e^{\lambda_m \cdot t}\sum_{j=1}^{n}(n+j)m'_{jm} \tag{A5.23}$$

5) The two concentration sums are seen to involve summation down columns of M^{-1}. The first is the geometric series

$$\sum_{j=1}^{n} m'_{jm} = \frac{1}{n}\sum_{j=1}^{n} q_m^{j-1} = \frac{1}{2n}\frac{1}{(1-q_m)} \tag{A5.24}$$

The second is the arithmetic–geometric series

$$\sum_{j=1}^{n}(n+j)m'_{jm} = \frac{1}{2n}\frac{1}{(1-q_m)^2} \tag{A5.25}$$

Equations (A5.24) and (A5.25) are independent of n because $q_m^n = 1/2$.

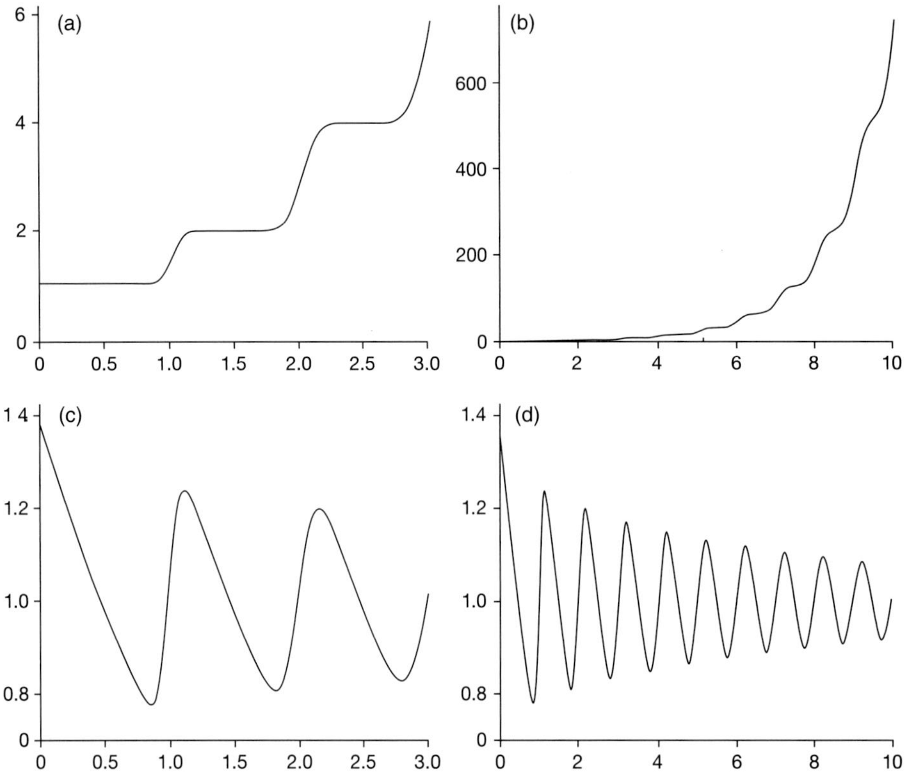

Figure A5.5

(a) and (b): Time-dependence of the total concentration of free and complexed complete templates for $n = 200$.
(c) and (d): Curves (a) and (b) normalised by division by the growth term. Ordinates are in units of the initial concentration $x_1(0) = C_0$, abscissa in units of n/k. The modulation factors decay to 1 at long times after coherent exponential growth is established. They differ from 1 at $t = 0$ because the initial induction phase, where the modulations are at first unsymmetric and then sinusoidal, delays onset of exponential growth.

6) The terms for m and (n – m) are conjugate and can be combined to yield purely real functions. I reported the quite lengthy sum expressions in explicit form as supplementary material for one of the papers quoted in the text, so I can dispense with reproducing such detail in this book. Instead I prefer to show some graphical representations of the results for n = 200 in Figures A5.5 and A5.6, yielding a more lucid insight into the physical content of those expressions.

The total concentration of complete RNA chains $\sum_{j=1}^{n} x_j(t)$ is seen to start at the initial concentration c_0 and increase stepwise in a geometric progression, while the total amount of macromolecular RNA (normalised to the length of a complete

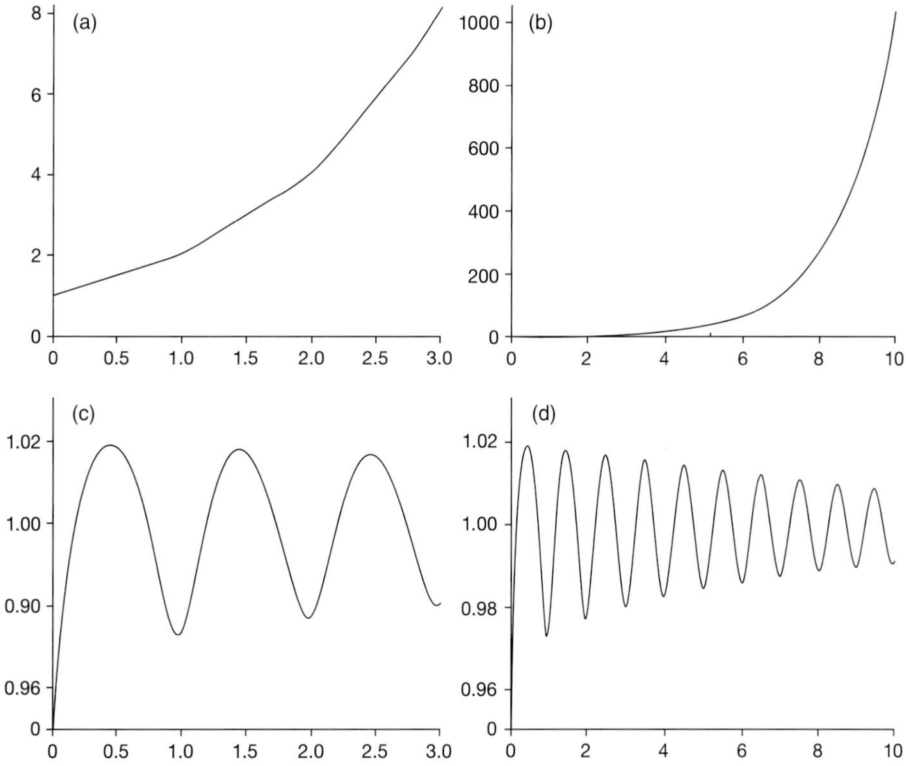

Figure A5.6

(a) and (b): Time-dependence of the total amount of polymer, scaled to complete chain length n = 100.
(c) and (d): Co-ordinates as in Fig. A5.5. The step-like behaviour of the complete template profiles (Figures A5.5a and b) is replaced by smoother, total polymer profiles, showing more closely exponential growth and smaller modulation factors.

chain) $\frac{1}{n}\sum_{j=1}^{n}(n+j)x_j(t)$ starts linearly and increases its slope stepwise in a geometric progression. The steplike changes are thus superimposed on the underlying exponential growth, which they finally modulate by a sinusoidal function with decreasing amplitude. An expansion of the expressions for large values of n shows that the oscillatory mode with $m = 1$ persists much longer than any of the modes with $m > 1$, its decay constant being proportional to $(m/n)^2$. It is seen that the oscillatory modes superimposed on the growth asymptotically reduce to a sinusoidal form represented by the longest persisting oscillatory mode.

The particular oscillatory behaviour found in this model is a consequence of the assumed degeneracy of the k values. Such degeneracy may be formally applied to multiplying systems with internal maturation, yielding the experimentally well-studied curves of synchronised growth. With regard to the mechanisms studied in this appendix they only refer to a limiting case. One (or several) rate-limiting steps in the cycle would result in stronger attenuation of the oscillations. In such a case the reaction would quickly proceed to the barrier and accumulate material there until its overflow starts coherent growth.

Summary and assessment

In order to grasp the content of this appendix one does not have to go into all the details of the mathematical deliberations shown above. The essential lesson is that, despite the complexity of the multistep mechanism of replication, the initial growth phase in which selection occurs can be formally represented by an exponential growth term e^{kt} where k is some overall rate constant of the multiplication process. Transition to linear growth is then simply a consequence of enzyme saturation as we have already experienced with the Malthusian limitations of growth, expressed by the logistic equation originally derived by the Belgian sociologist Pierre-Francois Verhulst. There it was meant to describe the population growth of self-reproducing (living) individuals, which in all its chemical details is a process even more complex than the one described above. One may object, of course, that reproduction of even multicellular organisms is a synchronised process, having a defined time constant. Yes, that's true for many of those processes, but our multistep system, in the absence of any particular superimposed synchronisation mechanism, has a constant doubling time too. In other words, it is not the complexity of the mechanism, but the (apparent) reaction-kinetic order of the rate-limiting individual chemical steps involved that determine the kinetic order of the overall process.

Over and above this, some of the details are of special interest. Particularly worth mentioning is the discrete nature of the early steps of reproduction, and also the effect of these steps in causing a decaying periodicity of the initial growth curve. This depends, of course, on the detailed mechanism of the multistep process, but it could mean that an ultimately exponential growth process cannot simply be extrapolated back to its initial phase. Something similar may also be observed in a decay process if it starts in a stepwise manner. Here, observations must be made within times of the order of magnitude of the time constants of the first individual steps.

As an example, take the decay of the proton, which is supposed to occur with a half-time of the order of magnitude of 10^{32} years. If decay truly occurs in one step, observation of a set of 10^{32} protons (say a tank of 10^{32} water molecules, corresponding to almost 10^4 tons of water) – apart from the expected stochasticity – may yield one single proton decay in a year, as would be expected from an extrapolation of the exponential function to its linear start. In order to be sure of this, one has to know precisely the mechanism of the decay process. In our universe, no proton can be older than about 10^{11} years, the time elapsed since the Big Bang. The first process of formation of quark structures and so on in their mechanisms are totally unknown. What kind of phase transitions, for instance, took place during the first 10^{-40} to 10^{-30} seconds after the Big Bang? The lifetime of the proton has been estimated on the basis of very general presuppositions, and I am reminded of Wheeler's words: "Symmetries hide the mechanisms involved." A negative result of the present experimental observations, currently being carried out at various locations, could mean that we still do not know enough about the mechanisms involved. However, this is only the view of a complete outsider and layman in particle physics. It was prompted by a more complete mathematical analysis of linear extrapolation for exponential growth and decay processes of the kind considered above. In this connection I should emphasise again that all the above deliberations are of a deterministic nature. Nevertheless, for exponential processes stochastic theory generally confirms the first moments, i.e. the average behaviour.

Author Index

Aarts, E. 351
Abraham, Max 20
Adair, Gilbert Smithson 150
Adam, Gerold 402
Adleman, Leonard 261
Albee, Edward 139
Altman, Sidney 415, 489
Ampère, André Marie 21
Anderson, Carl 58, 61
Anderson, Phil 574
Avery, Oswald Theodore 476
Avogadro, Amedeo 389, 626

Babbage, Charles 577
Bak, Per 574
Balmer, Johann Jakob 34
Barnes, V.E. 67
Barnett, Lincoln (*New York Times* editor) 12
Barrow, John D. 125
Basov, Nikolai G. 302
Bauer, Simon 648
Bayes, Thomas 167
Becker, Richard 39, 176
Becquerel, Antoine Henri 18, 55
Bekenstein, Jacob 210, 211, 213, 595, 596
Bell, John Stuart 47
Bennett, Charles 48, 229, 279, 280, 283
Bentley, W.A. 109
Benzer, Seymour 479
Berg, Alban 430
Berkeley, Edmund C. 579
Bernoulli, Jacob 166, 652
Berry, R. Stephen 203, 205, 643
Bethe, Hans 2, 70, 75, 578
Bianchi, Luigi 92
Biebricher, Christof 423, 453–4, 455, 456, 509, 511, 701, 711
Billroth, Theodor 579
Birkhoff, George David 281
Bjerrum, Niels 330
Bjorken, James 60
Black, Max 170
Bogomol'nyi, Eugéne 212
Bohr, Aage N. 57
Bohr, Niels 4, 34, 37, 273
Boltzmann, Ludwig 21, 26, 142, 147, 162, 164, 173, 174, 208, 214, 215, 220, 221, 229, 354, 587, 639
Bolyai, Wolfgang 518
Bondi, Hermann 101
Bonhoeffer, Karl Friedrich 426
Born, Max 27, 37, 273, 275, 329
Borštnik, Branko 294
Bose, Satyendra Nath 64, 169, 205
Boyle, Robert 154, 389, 643
Brahe, Tycho 73
Brahms, Johannes 579
Breit, Gregory 40
Bremermann, Hans 291
Brenner, Sydney 458, 459
Briemann, Leo 239
Brillouin, Léon 231, 232, 447
Brown, Robert 388, 394
Bruckner, Anton 579
Bruegel, Pieter (the Elder) 236
Buckingham, R.A. 332
Burks, Arthur 578
Buss, Leo 564
Butenandt, Adolf 322
Butler, Samuel 578, 579
Byron, Lord 577
Byron Lovelace, Ada 577

Calder, Nigel 619
Campbell, Jeremy 233
Cantor, Georg 21, 285, 392, 662, 663
Carathéodory, Constantin 7, 143
Carpenter, Gail 587
Cartan, Elie Joseph 15, 91, 92, 110, 358
Carter, Brandon 125
Casimir, Jan Hendryk 332
Casti, John 279, 429
Cech, Thomas 415, 489
Chaitin, Gregory 278, 280, 282, 285
Chamberlin, Owen 61
Chandrasekhar, Subrahmanyan 76
Changeux, Jean-Pierre 149, 153, 286
Cheeseman, Peter 170

Church, Alonzo 284, 565
Churchland, Patricia 582, 587
Clapeyron, Emile 576
Clausius, Rudolf 4, 6, 139, 147, 208, 211, 215
Compton, Arthur Holly 32, 33
Connes, Alain 286, 287
Conway, John Horton 377, 578
Copernicus, Nicolaus 73, 548
Coulomb, Charles-Augustin de 627
Courant, Richard 273, 276, 286, 651
Cowan, Clyde L. 5, 58
Cramer, Friedrich 292
Crick, Francis 292, 457, 458, 459, 479, 480
Cronin, James Watson 113
Crothers, Donald 645
Crow, James 478
Crow, J.E. 528
Cullman, G. 298
Curie, Marie 19, 203, 620
Curie, Pierre 19, 203, 620, 644
Czerny, V. 349

Dalgarno, Lynn 515, 516
Darwin, Charles 300, 301, 409, 422, 480, 497, 503, 524, 573, 581, 588, 597, 639
Darwin, Erasmus 579
Davidson, Kean 28
Davies, Paul 437
Davisson, Clinton Joseph 40
Dawkins, Richard 343, 437, 545, 547, 551
Dayhoff, Margaret O. 294, 295
de Broglie, Louis 32, 36
Debye, Peter 36, 70, 201, 273, 330, 398, 476, 646, 648
de Chéseaux, Jean-Philippe Loys 86
de Duve, Christian 369, 489
de Fermat *see* Fermat, Pierre de
Dehmelt, Hans 273
de Laplace *see* Laplace, Pierre Simon de
Delbrück, Max 402, 573
de Maeyer, Leo 399
Dembski, William 437
de Moivre *see* Moivre, Abraham de
Demokritos 21
Dennett, Daniel 497, 501, 547, 548
de Vries, Hugo 503
Dicke, Robert 125
Dickerson, R.E. 326
Dirac, Paul Adrien Maurice 35, 37, 58, 61, 91, 110, 169, 273, 634
Dirichlet, Peter Gustav Lejeune 90, 273
Dobzhansky, Theodosius 302, 489
Doppler, Christian Johann 82
Dress, Andreas 364, 423, 490
Dyson, Freeman 62, 494, 573, 574, 575
Dyson, George B. 578, 579, 589

Ebeling, Werner 220, 549
Edelman, Gerald 588
Ehrenfest, Paul 163, 179, 180, 640, 647
Ehrenfest, Tatjana 163, 179, 180
Eibl, Hansjörg 485
Eigen, Gerald 45
Eigen, Manfred 680
Einstein, Albert 2, 8, 36, 37, 73, 99, 121, 145, 169, 205, 228, 388, 389, 390, 392, 397, 632
Eisenschitz, Robert 143
Engel, Ernst 319
Enns, Richard H. 411
Epimenides 276
Eschenmoser, Albert 250, 340, 341, 495
Escher, Maurits Cornelius 360
Eucken, Arnold 153, 426
Euler, Leonhard 73, 265, 379, 379, 652
Everett, Hugh 47

Falk, Gottfried 143
Fano, Robert 233, 253, 295
Faraday, Michael 21
Feigenbaum, Mitchell 430
Fermat, Pierre de 166, 262, 263, 264, 280, 283
Fermi, Enrico 5, 60, 169
Feynman, Arlene 564
Feynman, Richard 3, 47, 62, 63, 207, 209, 533, 563
Fibonacci, Leonardo 512
Fick, Adolf 387, 388
Fisher, Michael 204, 648
Fisher, Ronald A. 503, 505, 523, 528, 677, 678
Fitch, Val Logsdon 113
Fock, Vladimir Aleksandrovich 329
Fokker, Adriaan Daniël 528, 529, 619
Fontana, Walter 564
Fourier, Joseph 633, 634
Fowler, William A. 88
Fox, Sidney 485, 565, 568
Franck, James 27, 273
Franklin, Benjamin 273
Freud, Sigmund 480
Friedman, Jerome Isaac 58
Friedmann, Alexandr Alexandrovich 100, 101
Frobenius, Ferdinand Georg 412, 618, 682
Fuhr, Ellen 27

Gábor, Dennis 230, 231
Galileo Galilei 73
Galois, Évaríste 404
Gamow, George (Georgi Antonovich) 101, 229, 458
Gardiner, William C. 511
Garey, Michael 345
Gatlin, Lila 291
Gauss, Carl Friedrich 90, 249, 262, 273, 390, 518, 591, 592, 652, 661
Gebinoga, Michael 511, 514, 515
Gehring, Walter 213
Geiger, Hans 57
Gelatt jr., C.D. 349

Gell-Mann, Murray 47, 58, 65, 67, 204, 279, 280, 574, 648
Genzel, Reinhard 77
Genz, Henning 121
George II (King of England and Elector of Hanover) 273
Germer, Lester Halbert 40
Ghadiri, Reza 573
Ghahramani, Zoubin 168
Gibbs, Josiah Willard 21, 143, 153, 159, 173, 174, 215
Gilbert, Walter 479
Giovanni, Don 377
Glaser, Daniel 58
Glashow, Sheldon 60, 67
Gödel, Kurt 21, 277, 282, 285, 287
Goeppert-Mayer, Joe 53
Goeppert-Mayer, Maria 53
Goldbach, Christian 280
Goldenfeld, Nigel 204, 619, 640, 646, 647
Gold, Thomas 101
Goodwin, Brian 443
Gould, Stephen Jay 477, 478, 526, 553
Green, Bryan 118
Green, Michael (of Queen Mary's College, London) 117
Griesinger, Christian 319
Griffith, John 458
Grimm brothers
  Grimm, Jacob 275
  Grimm, Wilhelm 275
Grossberg, Stephen 587
Gross, David 117
Grossmann, Marcel 89
Grossmann, Siegfried 431
Gutenberg, Johann Gensfleisch 575
Guth, Alan 107

Haeckel, Ernst 107, 108
Haensch, Theodore 305
Hahn, Otto 275
Haken, Hermann 302, 304, 308
Haldane, John B.S. 503, 523
Hamilton, William Rowan 38, 110, 173, 635
Hamming, Richard 255, 257, 259, 356, 360, 372, 374, 375, 376, 379, 382, 384, 405, 549, 592
Hanslick, Eduard 579
Harrison, E.R. 87
Harris, Sidney 12
Harriot, Thomas 591
Hartl, James B. 17, 47, 209
Hartree, Douglas R. 329
Harvey, Jeffrey 117
Hausdorff, Felix 392, 394
Hawking, Stephen 17, 207, 208, 209, 210, 211, 213, 214, 477, 595, 596
Haxel, Otto 54
Hebb, Donald 587
Hegel, Georg Wilhelm Friedrich 476
Heine, Heinrich 272
Heisenberg, Werner 37, 70, 273, 386, 495, 496, 633, 638
Heitler, Walter H. 329
Hermite, Charles 379
Hertwig, Oscar 476, 477
Hertz, Gustav 27
Hertzsprung, Ejnar 82
Higgs, Peter 72, 107
Hilbert, David 21, 37, 116, 273, 275, 277, 282, 354, 633
Hill, Terrel L. 332
Hindemith, Paul 564
Hindemith, Rudolf 564
Hitler, Adolf 26, 276, 498
Hitlin, David 114
Hofstadter, Robert 57, 58, 96, 276
Hofacker, Ludwig 294
Hoffmann-Berling, H. 514
Hoffmann, Roald 329
Hookman, John 229
Hopfield, John 556, 587, 588
Hoyle, Sir Fred 88, 101
Hubble, Edwin Powell 83, 100, 101
Hückel, Erich 330
Huffman, David A. 295
Humphreys, W.J. 109
Hund, Friedrich 37, 49
Huygens, Christian 166

Icke, Vincent 107
Ilgenfritz, Georg 151
Ise, Norio 645
Ising, Ernst 201, 588, 645, 646

Jacobi, Carl 669
Jaffe, Rona 423
Jantsch, Erich 488
Jaynes, Edwin Thompson 144, 147, 153, 158, 169, 211
Jeans, James H. 24
Jensen, Hans Daniel 54
Johnson, David 345
Jones, Billy L. 411
Jordan, Pascual 37, 62
Josse, J. 291
Jost, Wilhelm 481
Joule, James Prescott 6, 626
Joyce, Gerald 489, 520, 522, 523
Joyce, James 58
Judson, Horace 460–461
Jukes, Thomas H. 291, 298, 524, 526
Jung, Herbert 143

Kac, Marc 408, 411, 430, 564
Kadanoff, Leo P. 648
Kaiser, A.D. 291
Kaissling, Karl Ernst 324
Kaku, Michio 117, 118
Kaluza, Theodor 116
Kanada, Yasumasa 378

Kant, Immanuel 86, 116
Kapur, J.N. 169
Kästner, Erich 498
Katchalsky, Ephraim 402
Katzir, Ephraim 481
Kauffman, Stuart 553, 556, 566, 574
Kaufmann, Walter 18
Kendall, Henry Way 58
Kendrew, John 642
Kepler, Johannes 73
Kesavan, H.K. 169
Kettling, Ulrich 586
Khinchin, Alexander 239
Kimura, Motoo 195, 294, 411, 523, 524, 526, 528, 529, 530, 532, 574, 619, 696
King, Jack L. 524, 526
Kippenhahn, Rudolf 85
Kirkpatrick, S. 349
Kirschner, Kasper 151
Klein, Felix 273, 663
Klein, Oskar 116
Kolmogorov, Andrey N. 168, 278, 426, 435, 529, 619
Koltermann, Andre 586
Kopfermann, Hans 273
Kornberg, Arthur 291
Korst, J. 351
Koshland, Daniel E. 149
Kosko, Bart 170
Kovalevskaya, Sofya 434
Kuhn, Thomas S. 168
Kullback, Solomon 281
Küppers, Bernd-Olaf 437

Labouygues, J.M. 298
Lagrange, Joseph Louis 110, 169
Lamarck, Jean-Baptiste 497, 573, 579
Lambert, Johann Heinrich 378
Lam, Bianca 523
Lamb, Jr., Willis Eugene 63, 64, 65, 329
Landau, Edmund 273, 276
Landauer, Rolf 229
Landau, Lev Davidovich 203, 646
Landes-Hindemith, Maria 564
Langevin, Paul 398
Langton, Christopher 580
Laplace, Pierre Simon de 166, 168, 169
Leach, Andrew R. 329
Lederman, Leon 58, 60, 72, 107
Lee, Tsung Dao 67, 112
Leibler, Richard 281
Leibniz, Gottfried Wilhelm von 60, 107, 590
Leighton, R.B. 62
Lemaître, Georges Edouard 101
Lenard, Philipp 31
Lennard-Jones, John E. 332
Lerner, Richard 341
Lessing, Gotthold Ephraim 575
Leukippos 21
Levine, R.D. 294
Lévy, Pierre 421
Liapunov, Aleksandr 674
Lie, Marius Sophus 66
Lifson, Shneior 573, 575, 645
Ligeti, György 433
Lincoln, Tracey A. 520, 522
Lindemann, Carl Louis Ferdinand von 378, 379
London, Fritz W. 329, 332, 643
Long, Frank 648
Lord Kelvin *see* Thomson, William
Lorentz, Hendrik Antoon 10, 17, 20, 632
Lorenz, Edward 427
Loschmidt, Joseph 164, 389, 626
Low, Francis 204
Lüders, Günter 113
Lumry, Rufus 645
Lwoff, André 207
Lyapunov, Aleksandr Mikhailovich 434
Lynen, Feodor 151

Mach, Ernst 89, 389
MacLaurin, Colin 117
Maier-Leibnitz, Heinz 95
Malécov, Gustave 524, 528
Malthus, Thomas Robert 428, 501, 597
Mandelbrot, Benoit 238, 392, 419, 420, 421, 422, 432, 435, 505
Mann, Thomas 109, 475, 620
Marchetti, Cesare 429
Margenau, Henry 354
Margulis, Lynn 481, 490
Mariotte, Edme 154, 643
Markov, Andrei Andreyevich 237, 238, 291, 529
Marois, Maurice 354
Martinec, Emil 117
Mather, J.C. 28
Matthaei, Heinrich 459
Maxwell, James Clerk 20, 21, 176, 227, 228
Mayer, Julius Robert von 4
Maynard-Smith, John 356, 503
Mayr, Ernst 498, 526
May, Robert (Lord) 428, 554, 556
McBride, John L. 411
McCaskill, John 417, 423, 526, 588
McCulloch, Warren 587
McKay, Donald 213, 437
McMillan, Brockway 239
Meisenheimer, Klaus 46
Meixner, Joseph 147, 505
Mendeleyev, Dimitri Ivanovich 49
Mendel, Gregor 503, 524, 547
Mendelssohn-Bartholdy, Felix 273
Mendelssohn-Bartholdy, Rebecca 273
Menten, Maud 509, 702
Metropolis, N. 349
Mewes, Hans-Werner 619
Meyer, Julius Lothar 48
Michaelis, Leonor 508, 509, 702
Michelson, Albert Abraham 9
Midgley, Thomas 481

Miescher, Friedrich 476
Miller, Stanley 485
Minkowsky, Hermann 12, 209, 273, 631
Miro, Joan 394
Misner, Charles W. 93, 209
Moivre, Abraham de 166
Monod, Jacques 149, 153, 545
Morgenstern, Oskar 578
Morley, Edward Williams 9
Morowitz, Harold 485
Morse, Samuel Finley Breese 249, 252, 591, 592
Moses, R.E. 514
Mössbauer, Rudolf Ludwig 95
Mottelson, Ben R. 57
Mozart, Wolfgang Amadeus 429
Muller, Hermann Joseph 695
Mulliken, Robert Saunders 49
Münchhausen, Hieronymus Carl Friedrich von ("lying baron") 276

Neddermeyer, S. 58
Néel, Louis Eugène Félix 645
Ne'eman, Yuval 65
Nernst, Walter 275
Newton, Isaac 71, 73, 631
Nicholson, A.J. 428
Nicolaou, Kyriacos C. 327
Nicolis, Gregoire 435
Nieselt-Struwe, Katja 368, 588
Nirenberg, Marshall 459
Noether, Emmy 4, 273, 276

Ochoa, Severo 291
Ohno, Susumu 619
Ohta, Tomoko 524, 530
Okazaki, Reiji 561
Okazaki, Tsuneko 561
Olbers, Heinrich Wilhelm Matthias 86
Onsager, Lars 201, 398, 505, 646
Oppenheimer, Robert 329
Orgel, Leslie 457, 458, 479, 489, 520
Osborn, Mary 483
Ostwald, Friedrich Wilhelm 389
Otha, Tomoko 619
Ouellette, Robert J. 323

Pais, Abraham 44, 389, 397
Paley, William 545
Parmenides 7
Pascal, Blaise 166
Pauling, Linus Carl 49
Paul, Wolfgang 273
Pauli, Wolfgang 4, 37, 49, 54, 55, 70, 76, 143, 273, 328, 386
Peitgen, Heinz Otto 433, 553
Penrose, Roger 209, 211, 285, 577, 598
Penzias, Arno 26, 101, 103
Perl, Martin L. 58
Perrin, Jean-Baptiste 389
Perron, Oskar 412, 618, 682
Perutz, Max 642
Pitts, Walter 587
Planck, Max 24, 27, 30, 36, 68, 116, 220, 275, 528, 619, 623
Plato 287
Podolski, Boris 44, 121
Poincaré, Jules Henri 9, 162, 164, 188, 427, 435
Poisson, Siméon Denis 548, 635
Polya, George 395, 402
Popper, Karl 99
Post, Emil 284, 285, 565
Pound, Robert Vivian 96, 99
Prasad, Manoj 212
Prelog, Vladimir 341
Price, Huw 210
Prigogine, Ilya 164, 208, 220, 435, 505
Prokhovov, Aleksander M. 302
Ptolemy 549
Pythagoras 379, 630

Quinkert, Gerhard 319

Rabi, Isidor Isaac 64
Rainwater, Leo James 57
Ramanujan, Srinivasa 118
Ramsay, William 49
Rangnekar, Sada S. 411
Rayleigh, Lord 24, 49
Ray, Thomas 580
Rebka Jr., Glen 96, 99
Rechenberg, Ingo 356
Rees, Martin (*Just Six Numbers*) 119
Reid, Alexander 40
Reid, Constance 276
Reines, Frederick 5, 58
Rellich, Franz 276
Rice, Stuart A. 203, 643
Rich, Alex 458
Richardson, C.C. 514
Richter, Burton 60
Richter, Peter 86, 401, 433, 553
Riemann, Georg Friedrich Bernhard 89, 273, 505
Rivest, Ronald L. 261
Robbins, Herbert 651
Rohlf, J.W. 26
Rosenblatt, Frank 587
Rosenbluth, A. 349
Rosenbluth, M. 349
Rosen, Nathan 44, 121
Rössler, Otto 427
Rohm, Ryan 117
Ross, John 203, 556, 588, 643
Rubbia, Carlo 67
Ruch, Ernst 197
Rudolf August, Duke of Wolfenbüttel 591
Ruelle, David 427
Rumschitzki, David 412, 423
Runge, Kai 46
Russell, Bertrand A.W. 20, 275
Russell, Henry Norris 82

Rutherford, Ernest 34, 55, 56, 57
Ružička, Leopold 495, 496

Sadi Carnot, Nicolas Léonard 5, 155, 576
Sagan, Carl 481
Sagan, Dorion 481
Saint-Hilaire, Etienne Geoffroy 579
Salam, Abdus 67
Sands, M. 62
Scatchard, George 395
Schäfer, Fritz Peter 305
Scheherazade 378
Scheraga, Harold 648
Scherrer, Klaus 562
Schmidt, Arnold 276
Schmitt, Francis Otto 437
Schoenberg, Arnold 430
Scholem, Gershom 275
Schoute, Pieter Hendrik 188, 424
Schrödinger, Erwin 36, 44, 196, 207, 475, 480, 504, 573, 633, 638
Schroeder, Manfred 394
Schulz, Peter 341
Schuster, Peter 356, 422, 424, 505, 511, 526, 533, 534, 536, 549, 580, 588, 619
Schwartz, Jacob 564
Schwartz, Melvin 58
Schwarz, Gerhard 645
Schwarz, John 117
Schwarzschild, Karl 74, 77, 89, 99
Schweber, Silvan S. 62
Schwinger, Julian Seymons 62
Segré, Emilio 61
Sejnowski, Terrence 354, 582, 587
Shahshahani, Siavash 435, 436, 505, 674, 677
Shakespeare, William 437
  Hamlet 437, 446
  Polonius 437
Shamir, Adi 261
Shannon, Claude 174, 221, 232, 233, 238, 239, 253, 281, 287, 295, 300, 301, 592
Shapley, Harlow 73
Shine, John 515, 516
Siegel, Carl Ludwig 273, 275
Simon, Sir Franz (Eugen) 207
Sinai, Yakov 426, 435
Sloane, Neil James Alexander 377
Smolin, Lee 124, 597
Smoluchowski, Marian Ritter von Smolan- 396, 398, 401, 402
Smoot, George 26, 101
Snow, C.P. 448
Solomonoff, Ray 278
Sommerfeld, Arnold 39, 70, 176
Sommerfield, Charles 212
Sparnaay, Marcus J. 332
Spiegelman, Sol 442, 443, 446, 453, 480, 511
Stefan, Josef 26, 147, 220
Steinberger, Jack 58
Stewart, Ian 263, 344, 651
Stewart, Jan 261, 262, 285
Stirling, James 177, 348
Stokes, George Gabriel 389
Strominger, Andrew 211, 212, 213, 595
Strunk, Günter 453
Subak-Sharpe, J.H. 292
Suess, Hans 54
Susskind, Leonard 597, 598
Swartz, M.N. 291
Swetina, Jörg 533, 534, 619
Szathmáry, Eörs 557
Szilárd, Leó 228, 231
Szostak, Jack 489, 520

Takens, Floris 427
Tarazona, Pedro 535, 536
Taylor, Brook 117
Taylor, Richard 58
Tchaikovsky, Peter 579
Teller, Augusta 349
Teller, Edward 349
Thiele, Thorvald Nicolai 388
Thompson, Colin J. 411
Thomson, George Paget 40
Thomson, Joseph John 18
Thomson, William (Lord Kelvin) 6
Thorne, Kip S. 93, 209
Ting, Samuel 60
Tinguely, Jean 589
Tipler, Frank J. 125
Toennies, Jan Peter 47
Tolstoy, Leo 307
Tomonaga, Sin-itivo 62
Townes, Charles H. 302
Träuble, Hermann 485
Trautner, T.A. 292
Tümmler, Burkhardt 482, 484
Turing, Alan Mathison 283, 284, 285, 346, 565, 576, 577, 578
Tyndall, John 4

Ulam, Stanislaw 578
Usher, David 554

Vafa, Cumrun 211, 212, 213, 595
Van der Meer, Simon 67
Vandermonde, Alexandre-Théophile 511, 512
van der Waals, Johannes Diderik 53, 160, 202, 643
van Duuren, Hendrik 253
Vecchi, M.P. 349
Verhulst, Pierre-Francois 428, 714
Volkenstein, Mikhail 220
von Fraunhofer, Joseph 83
von Goethe, Johann Wolfgang *480*, 579, 623
von Kiedrowski, Günter 557
von Laue, Max 275
von Lindemann, Carl Louis Ferdinand *see* Lindemann, Carl Louis Ferdinand von
von Mises, Richard 168

von Neumann, John 174, 232, 283, 355, 505, 507, 576, 577, 578
von Seeliger, Hugo 89
Vosberg, H.P. 514

Wagner, Richard 579
Wallace, Alfred Russel 500
Wallich, George C. 479, 497
Wang, Katja 588
Watson, James D. 458, 459, 480, 526
Watt, James 6, 430
Weaver, Warren 232
Webern, Anton 430
Weber, Wilhelm 249, 591, 592
Weierstrass, Karl Theodor 394
Weinberg, Steven 40, 47, 105, 477
Weiner, Norbert 394, 395, 402, 405
Weismann, August Friedrich Leopold 497, 502, 547
Weissmann, Charles 513, 572
Weiss, Pierre-Ernest 645
Weizsäcker, Carl Friedrich von 2, 53, 75, 208, 436
Weizsäcker, Christine v. 281
Weizsäcker, Ernst Ulrich 281
Weyl, Hermann 273, 276
Wheeler, John Archibald 15, 48, 91, 93, 118, 123, 124, 209, 210, 358, 590, 592, 595, 622
Whitehead, Alfred N. 275
Widom, Ben 204, 648
Wiegand, Clyde 61
Wiener, Norbert 233
Wigner, Eugene 40, 54, 211, 301, 407, 578
Wiles, Andrew 262
Williams, Garnett P. 435
Wilson, Charles Thomson Rees 58
Wilson, Edward O. 481, 553
Wilson, Kenneth G. 204, 620, 648
Wilson, Robert W. 26, 101, 103
Winkler-Oswatitsch, Ruthild 180, 191, 193, 201, 364, 439, 490, 495, 518, 526
Witten, Edward 117
Wittgenstein, Ludwig 20, 122, 168, 170, 275, 422
Wolfskehl, Paul 37
Wollaston, William Hyde 83
Woodward, Robert Burns 329
Woolf, Virginia 139
Wright, Sewall 503, 523, 528
Wu, Chien-Shung 112
Wyman, Jeffries 149, 153

Yang, Chen Ning 67, 112
Yockey, Hubert 239, 292, 293, 299
Young, Albert 198
Ypsilantis, Thomas 61

Zadeh, Lofti 170
Zeeman, Pieter 64
Zeh, Dieter 48
Zeilinger, Anton 47
Zermelo, Ernst 164
Zimm, Bruno 645
Zurek, Wojtek 48
Zweig, George 65

Subject Index

Abelian group 110
Abelian group U(1) 65
active transport 485
adaptability 575
affinities of chemical forces 504
affinity 504
AIDS virus (HIV) 366
algebraic topology 357
allosteric binding curve 148
allosteric enzymes 148, 642
  regulatory mechanism 641
  responding 148
  "supply and demand" 148
allosteric protein 642
allozymes 526
"Alpher–Bethe–Gamow" paper 101
Alzheimer's disease 642
antibodies 498
antiferromagnet 203, 644
antigravity 119
anti-knocking agents 481
antimatter 61
antineutrino 113
antineutrino νe 4
antiparticles 61
antipartner 113
antiproton 61
aperiodic crystal 480
A Principle of Natural Self-Organisation (by Manfred Eigen and Peter Schuster) 511
artificial life 580
associative networks 587
asymptotic stability 672, 674
autocatalytic reproduction 701
Avogadro's (or Loschmidt's) constant 389, 626
axiomatic theory of thermodynamics 143

BABAR collaboration, Stanford Linear Accelerator 114
"barns" 397
baryon number 115
baryons 65
β-decay 4, 112
β^+-decay 5
Beilstein, The 319
Bekenstein–Hawking entropy 211, 213
Bekenstein–Hawking formula 596
Bell Telephone Laboratories 232
Bernoulli numbers 652
Bethe–von Weizsäcker cycle 75
bifurcations 435, 563
  diagram 433
  "pitchfork" 435
  rate 431
"big bang", singularity of 209
"big crunch", singularity of 209
binary logic 592
binomial distribution 533
biochemical reaction 701
biological membranes 206
black body radiation 147
black dwarf 76
black hole 76, 206, 628
  in our galaxy 78
block code 591
B mesons 113
Bohr festivals 37
Bohr's model 34
Boltzmann factor 24
Boltzmann machine 354, 587
Boltzmann's constant, k24 142
Bombyx mori (female silkworm moth) 322
Boolean networks 574
Borštnik–Hofacker distance 294
bosons 60, 64, 67
  and fermions, symmetry between 386
Bose–Einstein condensation 205
Bose–Einstein statistics 64, 169
boson 115
bottom quarks (b) 113
bottom-up logical reflection 60
B-physics 114
BPS theory 212
"branes" 116, 212
broken symmetries 204
Brownian motion 388
  physics of 394

calorie (amount of energy necessary to heat 1 gram of water from 15 to 16 [°C]) 626
calorimeter 142
canonical transformations 173
Cantor's diagonal argument 285
capsid proteins 516
carbon cycle 75
cardinality 663
Carnot cycle 155
Cartan's Principle 92
Cartesian vectors 360
catalytic antibodies 551
catalytic reaction cycles 488
cathode rays 31
cellstat, the 502, 555
CFR *see* constant flow reactor (CFR)
"chance and necessity" 545
Chandrasekhar limit 76
channel capacity 245
chaos
  deterministic 427, 430, 558, 563
  edge of 555, 556
    criticality at the edge of 574
  truly "random" 563
chaotic solutions 429
charge conjugation 112
charm quark 115
chemical complexity 319
  potential 355
chemical interactions 627
chemical relaxation spectrometry 147
Clarkia pulchella (primrose) 388
cobalt-60 nuclei 112
COBE (Cosmic Background Explorer) 26
Cobe satellite 101
codons, comma-free reading of 459
combination 168
  modular 369, 370
commutation relations 637
commutator 637
compartmentation 513
complexity 701
  combinatorial 372
  dynamical 427
  exponential 344, 369, 446
computational networks 587
concatenations 239
conjugate 634
conservation principle (the first law) 142
constant flow reactor (CFR) 504
co-operativity 201
Copenhagen interpretation 44
Copernican revolution 548
correspondence 91
Cosmic Background Explorer *see* COBE
cosmological constant 119
Coulomb's law 70, 627
covariance 91
CPT invariance 113
CP transformation 113
CPT reversal 113
CPT theorem 113
  invariance under CPT reversal 113
CP violation 113
Creutzfeldt–Jakob disease 642
Crick's central dogma 292
critical exponent 204
critical opalescence 203, 644
critical phenomena 640
  universality of 203
critical point 204
"cross-entropy" 281
Curie point 644
Curie temperature 203
crypticity 279, 283
cryptography 257
crystallisation of sodium thiosulphate 205

Darwin among the Machines (by George B. Dyson) 578
Darwinian machines 588
Darwinian mechanism 409
Darwinian model 597
Darwinism 573
  neo-Darwinism 503, 524
  neural/evolutionary 581, 588
de Broglie material waves 36
decay 504
degrees of freedom 142
"denumerable" 663
depth 279, 283
deterministic chaotic behaviour 558
diagrams 62
Dirac's theory of the electron 110
DNA chain of the *E. coli* genome 333
"DNA computing" 578
DNA–protein world 521
diffusion
  dimensionality 395
  fractal dimension of 388
  one-dimensional 401
  recurrence 395
  self-similar nature of 388
  "space-filling" nature of 388, 395
dispersion forces 332
distance squared / time relation 389
Doppler effect 82
  optical Doppler effect 82
  relativistic Doppler shift of light 84
  shifts 83, 101

Ehrenfest's model 647
Ehrenfest's urn game 180
Eigen's selection equation 680, 682
eigenvalues 37, 147, 669, 678, 682, 683, 688
  complex 683
  problem 635
  spectrum of 635
eigenvector 678, 682, 683, 684, 691
Einstein covariance 632

Einstein–de Sitter universe 100
Einstein–Podolsky–Rosen experiment 121
Einstein's "reinvention" of Riemannian
geometry 37
Einstein's theory 94
general-relativistic equation of gravitation 90
electromagnetic radiation 627
"electron hole" 61
electron volt 626
electrophoresis 526
electroporation 485
elongation factors
EF-Ts 513
EF-Tu 513
ribosomal protein S1 513
encapsulation 482
"cells first" slogan 482, 484
energy-momentum tensor $T_{\mu\nu}$ 91
English is "hotter" than German 308
Enigma cipher code 577
"entangled" 44
entropy 139
-driven assembly of lipid molecules 206
maximum 172
mixing 156
mutual 247
production 555
equation, Fokker–Planck 528, 529
equivalence 91
ergodicity 163
error catastrophe 511
error rates, uniform 680, 688
error threshold 532, 556, 688, 690, 693, 695, 696
no deleterious error threshold 293
errors, accumulation of 518
Euclidean space 355, 373
Euler numbers 652
event horizon 77
evolution
biological 506
by natural selection 213, 407
machines 504
molecular evolution of life 598
evolutionary biotechnology 580
evolutionary optimisation 639
evolutionary potential 501
existence 446
expectation value 636
"exponential complexity" 344
exponentially complex structure of sequence 370
exponential population growth 518
exponential space 372
"exponential time" 344
"extensive" complement 142
extensive function 143
extensive variables 143
extinction threshold 693
extremum principle for entropy (the second
law) 142
exon 563

fairy tale 272
Fano bound 233
feedback 593
Fermat's last theorem 37, 280, 283
Fermat's "little" theorem 264
Fermi–Dirac statistics 64, 169
Fermi laboratory 60
fermion partner 115
fermions 60, 64, 65
and bosons, symmetry between 386
ferromagnet 203
Feynman diagrams 62, 63
Fibonacci series 512
Fick's first equation 387
Fick's second equation 388
fidelity of reproduction 533
Finnegan's Wake (by James Joyce) 59
Fisher's fundamental theorem 679
Fisher's selection equation 677, 678
fitness 575
mean 671, 680
fitness landscape 588
model 690
mountain ridges of 538
natural 696
single peak 688
flowreactor 667, 693
fluctuations 647
large-scale 648
flux vector 388
Fokker–Planck equation 528, 529
Font's point, Anza Borrego desert 553
formalist 287
Fourier transformation 633, 634
fractal dimension 392
fractal structure 432
freedom, degrees of 142
Friedmann's equation 100, 101
Friedmann's solution 100
"frustrated" states 587
fundamental growth laws 498, 499, 500, 501
fundamental interactions, unification of the
four 596
fusion 2
fusion reaction 75
fuzzy logic 170

Galilean transformations 9
Galois field 404
gauge invariance 63
gauge postulates 115
gauge symmetry 115
gauge theory 62, 63, 648
most accurate theory 63
"gelation" 642
generalised gradient system 674
general relativity 77
genetic code 592
age of 490
genetic engineering 526

genetic information
complementary letters A, U and G, C 452
"goal-directed activity" 445
semantic nature of 445
genetic recombination 503
genetic variability 526
genomic "book of law" 513
genotype 521, 670, 679, 686, 692, 695
genotype–phenotype dichotomy 511
genotype space 696
genotypic information 513
geometric progression 428
geometric ratio of increase 501
geometrodynamics 358
geometry, as an emerging "quality" of physics 16
"German" physics 31
global flux *see* fitness, mean
gluons 61
glyceraldehyde-3-phosphate dehydrogenase 151
Gödel's incompleteness theorem 282, 285
Goldbach conjecture 280
grammar 238
Gram-negative aerobic bacterium *Pseudomonas putida* 484
"gravitational crunch" 74
gravitation disorder 206
gravitational energy 628
gravitational law 73
gravitational pull between proton and electronic in a hydrogen atom 74
growth 504
G-U wobble pairs 523

hadronic particles 65
hadrons 65
haemoglobin molecule 148, 641
Hamiltonian 635
invariance 110
mechanics 173
Hamming
distance 360, 372, 374, 549
"kissing point" 375, 377
space 405
spheres 257, 259, 375
Hamming's concept 257
Hamming's paradox 379
"handedness" 111
hapten 498
hard sphere 643
Hausdorff dimension 394
heat 142
capacity 204
reversible supply of 141
Hebbian synapses 587
Heisenberg's quantum mechanics 633
Heisenberg's "reinvention" of matrix calculus 37, 62
Heisenberg's theory 638
Hertzsprung–Russell diagram, the 81, 82
heterotic superstring theory 117, 118
hierarchical models 372
Higgs boson 8, 68, 72
"God particle" 72
Higgs field 72
Higgs mechanism 107
Hilbert space 354, 633
homeostasis 410, 507, 554
Hopf bifurcation, in Prigogine's "Brusselator" 435
Hopfield's model 587, 588
Hopfield's neural network 556
models of 556
H theorem 164
Hubble's constant 85
Hubble's law 84, 85
relativistic nature of 85
Huffman code 295
human genome 333
hydrophobic interaction 206
hyperbolic growth 515
hyperbolic increase 517, 518
hypercube 15, 374, 658
ν-dimensional 357
hypercycles 414, 513, 514, 518
catalytic 511
hypercyclic mechanism 516
hypercyclic organisation 513
hyperface 373, 658
hyperplanes 357, 373, 658
hyperspace 372, 387

ideal-gas law of Boyle and Mariotte 154, 389
immense connectivity 369
influenza A virus 367
influenza virus 366
information 386, 593
concatenations 239
discrete non-Euclidean 385
kth extension of the code 239
temperature of 306
theory 239
origin of 507, 598
information space 372
concept of 368
"getting from here to there" 423
"instantaneous action" 74
Institut de la Vie 354
intelligent design 552
intensive variables 143
internal homeostatic regulation 518
introns 563
Ising model
one-dimensional 201
two-dimensional 588, 646
Isospin symmetry 115

Jaynes' enunciation 169
Joule 626
Jukes' evolutionary model 298
Just Six Numbers (by Martin Rees) 119

kaon 210
kinetic collision approach (German: *Stossansatz*) 164
"kinetics of phage infection" 504
Klosters, Winter Seminar 590
K-meson 113
"Knabenphysik" (boy's physics) 37
$K^0 = ds$ 115
Kolmogorov's backward equation 529
Kolmogorov's forward equation 529
Kolmogorov's fundamental contribution 529
Kolmogorov–Sinai entropy 424, 426, 435
Kullback–Leibler measure of "cross-entropy" 281

lagging strand 562
Lagrangian invariance 110
Lagrangian multipliers 169
Lamarckism 573
λ-point 204
Lamb shift 63, 329
Landau–Oppenheimer–Volkov limit 76
Landau's theory 646
Large Magellanic Cloud 76
laser 303
leading-strand 562
Lederman's question 107
lethal mutants 672, 690, 692, 693, 694
Liapunov function 674
Lie groups 110
"life games" 578
linear ansatz 409
linear autocatalysis 501
living cells 488
logistic algorithm 429
logistic equation 428, 714
logistic form: $x_{t+1} = x_t + kx_t - kx_t^2$ 428
logistic parabola (or "quadratic map") 428
London forces 332
long half-axis of the prolate ellipsoid 402
loop 116
looped shape 596
Lorentz transformation 17, 632
Loschmidt's constant *see* Avogadro's (or Loschmidt's) constant
loss of memory 240
Luddites 589
luminosity 82
Lyman α line 86

MacLaurin's theorem 117
"magic numbers" 53, 54
magnetic moment 113
  large 587
Malthus, doctrine of 501
Malthusian point of view 597
Mandelbrot's raindrop problem 422, 435
MANIAC-I 577
"many worlds" interpretation 47
Markov
  entropy 237, 291
  process 238
  systems 529
mass action 527
mass energies 628
mathematical induction 651
matrix
  bistochastic 680
  cyclic 683
  fitness 681
  irreducible 683
  Jacobian 669
  mutation 680, 683
  primitive 683
  sequence 686
  stochastic 680
  value 681
matter
  animate 575
  excess of 113
  in great excess over antimatter 113
matter–antimatter transformation 112
maximum entropy 172
Maxwell's
  Demon 227
  distribution law 228
  velocity distribution 176
McCulloch–Pitts neuron 587
mean field approximation 688
meaning 445
  generation of 205
mechanism of RNA replication 508
meiosis, chromosome number in 476
Mendelian laws 524
mental consciousness 546
mesons 113
  B 113
  bottom 113
  charm 113
  down 113
  K 113
  up 113
messenger RNA (mRNA) 489
metric tensor $G_{\mu\nu}$ of curvature 91
Michaelis–Menten expression 509
Michaelis–Menten mechanism 702
Michelson–Morley experiment 9
microcanonical ensemble 173
microscopically reversible 504
microstate on a probability basis 164
Minkowski space-time 631
mixing 168
modular combination 369, 370
modulo arithmetic 265
mole 626
momentum representation 633
monochromatic coherence 303
Morse code 252
Mössbauer effect 96
Mozart 429
  creative output of 429

M-theory 596
Muller's ratchet 695
multistep mechanism 701
multistep replication 701
muon 64, 113
muon-associated neutrinos 113
mutagenic reproduction 503
myoglobin 642

natural numbers, adding only one in each step 547
natural selection 406, 407, 414
as a Physical Law 406
evolution by 213
existence 446
first-order phase transition in 414
n-dimensional simplex 385
Néel temperature 203
neo-Darwinian school 403
neo-Darwinian theory 524
nerve cells
firing pattern of 446
selectively enforcing synaptic contacts 446
neural Darwinism 588
neutral allele replacement 526
neutral evolution 696
neutral mutants 523
neutral networks 549, 588
neutrino 4
Newtonian mechanics 631
Newton's law 71, 73, 74
"no boundary boundary conditions" 209
"non-Abelian" group 110
non-Abelian group SU(3) 65
non-Abelian symmetries 110
"non-deterministic polynomial" 344
"hard" problems 346
problems 346
non-differentiable curves 394
"non-hostile" competition 597
non-linear "nucleation" phase 558
non-linear optics 53
non-overlapping codes 459
normal modes of reaction 147
nuclear binding energy 55
nucleation 488
nucleic acids (German: *Kernsäuren*) 476
nucleus 476

observables 634
observer participancy 593
octane numbers of petrol 481
Okazaki fragments 561
$\Omega^- = 955$ has a strangeness of -3 115
"once for all time" 557
On the Nature of Mathematical Reality (by Alain Connes) 286
organisation 575
origin 593
origin of life 290, 489, 554
oscillation 684

palaeontology 477
palindromic loop 454
parallax method, the 81
parity 111
violation of in weak interactions 112
particle-scattering, "cross-section" 397
particle symmetries, pair nature of the property to be conserved 115
parsec (pc) 80
partially ordered 429
participatory universe 593
partition 168
past chemistries 481
Pauli principle 49, 55, 328
Pauli's exclusion principle 49
period doubling 431
periodic table of the elements 48, 49
Perron–Frobenius theorem 682
personal freedom of belief 553
perturbation theory 204
phage Qβ 513
phase space 163, 173
elementary cell of 163
phase transition 535, 536
between liquids and gases 202
critical 203
first-order 535, 536
physical 639
phenotype 521, 679
phenotype space 696
phenotypic function 513
pheromone 322
pheromone problem 402
physics of condensed matter 53
Pied Piper of Hamlin 39
pion 113
Planck's
constant 24
dimensions 68, 116
mass 68
radiation 220
"revolutionary" idea 24
planetary nebula 76
Platonists 287
Poincaré recurrence time 162
point-like protons 57
Poisson brackets 635
Poissonian (i.e. random) distribution 548
polio-1 virus 366
polymerases 452
polynomial time 344, 446
population entropy 691
population inversion 303
poration of the compartment 485
positional representation 633
positron 61
potency (German: *Mächtigkeit*) 663

potential alternative histories 47
potential function 674
potential gradients 423
Pound–Rebka experiment 98, 99
power law 203, 393
pp cycle 75
precise formula 57
primordial genes 370
Princeton string quartet 117
principle of least action 91
prion proteins 642
probability amplitudes 633, 634
"proteinoids" 565
"proteins first" hypothesis 565
"proteins first" slogan 485
Ptolemaic system 549
purines 360, 456
pyrimidines 360, 456
Pythagoras 630
Pythagoras' theorem 379

Qβ RNA, mini-variant of 455
QED and the Men Who Made It (by Silvan S. Schweber) 62
quantum computer 577
  MANIAC-I 577
quantum electrodynamics (QED) 61–2
quantum field theory 648
quantum-mechanical tunnelling 55
quantum mechanics 36, 633
quarks 57, 58
  bottom (b) 113
  charm 115
  strange 115
quasi-linear autocatalysis 507, 511
quasi-linear processes 506
quasispecies 372, 686, 690, 692, 693, 696
  model 525, 588
  molecular, build up of 535
  selection, computer simulation of 534
  theory 416

radiolarian 108
  see also Sagenoscena stellata
radon 49
random replication 689
random walk 388
  mechanism 527
rational numbers 664
Rayleigh–Jeans formula 24
reaction–diffusion coupling 564
reaction, extent of 504
recognition pattern 452
recoiling 4
recombination 678
recombination shuffling 503
red giant 75
red shift 85
refractive index 203
relativistic quantum field theory 204
renormalisation groups 204, 640, 648
replicases 452
reproduction, higher-order terms in 409
retrovirus 498
"reversible supply of heat" 141
ribonuclease A 453
(ribo-)nucleases 452
ribonuclease T_1 453
ribosomal translation system 523
ribozyme 488
Riemannian metric 674
RNA world 488, 489, 511, 521
Rosenblatt perceptron 587
rotational ellipsoid 401
RSA system 261, 267
Rubik's cube 110

Sagenoscena stellate 108
scalar product 630
scalars 629
  3-scalars 631
scaling law 143, 204, 431
Schrödinger's "aperiodic crystal" 480
Schrödinger's cat 44
Schrödinger's quantum mechanics 633
Schrödinger's theory 638
sea-urchin egg 476
selection 168, 536, 649
selection equation
  Eigen's 680, 682
  Fisher's 677, 678
selection of the fittest 532
selective advantage 549
selective behaviour 498, 499, 500
selective power 501
self-reproducing automaton 507, 578
self-similarity 432
sequence analysis 535
sequence space 696
serial transfer 442
  Spiegelman's work in 446
set theory 663
Shakespeare's *Hamlet* 437, 446
Shakespeare's Polonius 437
Shannon entropy 281
Shannon–Fano coding 233, 295
Shannon's theory 287
Shine–Dalgarno box 515
Shine–Dalgarno sequence of the replicase cistron 516
sickle-cell anaemia 642
simplex 424
"site-specific mutagenesis" 526
Snow Crystals (by W.A. Bentley and W.J. Humphreys) 109
source of inheritance 476
space-time
  curvature of 94
  encapsulation of 482
  four-vector representation of 16

space
  cellular 373
  Euclidean 355, 373
  exponential 372
  Hamming 405
  Hilbert 354
  hyperspace 372, 387
  information 372, 588
  metric 373
  n-dimensional simplex 385
  phase 163
    elementary cell of 163
  physical 404
  point 373
  specified 446
  survival in 387
special relativity 89
spectrum of relaxation times 117
spheroidal co-ordinates 401
spin-glass 587
spin-glass equations 587
Stanford Linear Accelerator *see* BABAR collaboration, Stanford Linear Accelerator
state 633
state of, selection 670
states, multiple stationary 680
states, stationary 669
statistically self-similar behaviour 392
Stefan–Boltzmann law 87
Stirling's approximation 177
stochastic process 695
Stossansatz (kinetic collision approach) 164
"strangeness" 115
strange quark 115
string 116
string-like shape 596
string theory 595
superconductivity 205
superfluidity phenomena 203
supernova 76
superstring theory 116, 212, 596
supersymmetric nature 115
supersymmetry 115, 386
S values of +1 115
"swelling" of the universe 101
symmetry group 115
symplectic structure 634
syntax 238
synthesis, modern 524
"synthetic theory" 503
system, closed 555
system, dynamical 670, 674

tau 64
Taylor's theorem 117
telecommunication 232
temperature 141
temperature dependence of heat capacity 204
tensorial representation 91
tensors 630
  4-tensors 632
  3-tensors 631
The Magic Mountain (by Thomas Mann) 109
Theory of the Heavens (by Immanuel Kant) 86
thermal energy 626
"thermal equilibrium" 145
thermodynamic limit 536, 642, 647
thermodynamics 143, 144
three-body interaction 427
time arrow 213
"time complexity function" 345
time development 637
titration end point 517
top-down input 60
Tower of Babel 237
Tower of Hanoi game 352
T-physics 114
trajectory of the phase point 163
transfer RNAs (tRNA) 521
transformation, integrating factor 673, 678, 681
transitions 360
transversions 360
triangulation method 79
Turing–Gödel conclusion 285
Turing machine 577, 578
  non-determinism of 346
Turing–Post program 284, 285
  "halting problem" 284
Turing test 577
"two cultures" (C. P. Snow) 448

ultraviolet catastrophe 24
"ultimately elementary" 57
"ultimate machine" 579, 588
uncertainty principle 637
uncertainty relation 38
uniform chiralities of biological macromolecules 506
uniform distribution 688
unit simplex 671, 681
University of Bielefeld, mathematics department 364

vagueness, philosophy of 170
Vandermonde determinant 511, 512
van der Waals
  equation 160, 202, 643
  forces 53, 643
van Duuren code 253
variable-length code 591
variables 143
vectorial representation 91
vectors 629
  Cartesian 630
  4-vectors 632
  3-vectors 631
Verhulst process, bifurcation diagram of 433
viable matter 590

viruses
  AIDs (HIV) 366
  HIV-1 367
    variability of 368
  HIV-2 367
    variability of 368
  influenza A 367
  influenza 366
  polio-1 366
  retroviruses 498
von Neumann machine 578

War and Peace (by Leo Tolstoy) 307
Watson–Crick model 458
wave function 633
wave-particle dualism 37
weak forces 61
weak interactions, violations of parity in 112
Webster's dictionary 7
Weiner processes 402, 405
white dwarf 76
wild type 454
  plus and minus sequences of 454
"Wittgenstein ladder" 122
Woodward–Hoffmann rule 329

Yang–Mills field 116

Z bosons 61
Zeeman splitting 64
  most convincing proof 64
Zoonomia 579